Lecture Notes in Mechanical Engineering

Lecture Notes in Mechanical Engineering (LNME) publishes the latest developments in Mechanical Engineering—quickly, informally and with high quality. Original research reported in proceedings and post-proceedings represents the core of LNME. Volumes published in LNME embrace all aspects, subfields and new challenges of mechanical engineering. Topics in the series include:

- Engineering Design
- Machinery and Machine Elements
- Mechanical Structures and Stress Analysis
- Automotive Engineering
- Engine Technology
- Aerospace Technology and Astronautics
- Nanotechnology and Microengineering
- Control, Robotics, Mechatronics
- MEMS
- Theoretical and Applied Mechanics
- Dynamical Systems, Control
- Fluid Mechanics
- Engineering Thermodynamics, Heat and Mass Transfer
- Manufacturing
- Precision Engineering, Instrumentation, Measurement
- Materials Engineering
- Tribology and Surface Technology

To submit a proposal or request further information, please contact the Springer Editor of your location:

China: Ms. Ella Zhang at ella.zhang@springer.com
India: Priya Vyas at priya.vyas@springer.com
Rest of Asia, Australia, New Zealand: Swati Meherishi at swati.meherishi@springer.com
All other countries: Dr. Leontina Di Cecco at Leontina.dicecco@springer.com

To submit a proposal for a monograph, please check our Springer Tracts in Mechanical Engineering at http://www.springer.com/series/11693 or contact Leontina.dicecco@springer.com

Indexed by SCOPUS. All books published in the series are submitted for consideration in Web of Science.

More information about this series at http://www.springer.com/series/11236

Pinakeswar Mahanta · Pankaj Kalita · Anup Paul ·
Abhik Banerjee
Editors

Advances in Thermofluids and Renewable Energy

Select Proceedings of TFRE 2020

Springer

Editors
Pinakeswar Mahanta
Department of Mechanical Engineering
Indian Institute of Technology Guwahati
Guwahati, Assam, India

Pankaj Kalita
Department of Mechanical Engineering
Indian Institute of Technology Guwahati
Guwahati, Assam, India

Anup Paul
Department of Mechanical Engineering
National Institute of Technology Arunachal Pradesh
Yupia, Arunachal Pradesh, India

Abhik Banerjee
Department of Electrical Engineering
National Institute of Technology Arunachal Pradesh
Yupia, Arunachal Pradesh, India

ISSN 2195-4356 ISSN 2195-4364 (electronic)
Lecture Notes in Mechanical Engineering
ISBN 978-981-16-3496-3 ISBN 978-981-16-3497-0 (eBook)
https://doi.org/10.1007/978-981-16-3497-0

This Springer imprint is published by the registered company Springer Nature Singapore Pte Ltd.
The registered company address is: 152 Beach Road, #21-01/04 Gateway East, Singapore 189721, Singapore

Preface

The Mechanical Engineering Department and Electrical Engineering Department of NIT Arunachal Pradesh have jointly organized an International Conference on *Recent Trends in Developments of Thermo-fluids and Renewable Energy (TFRE 2020)* during 26–28 November 2020. Due to the present pandemic situation of COVID-19, the conference was held online. A total of 59 out of 145 papers received across the globe were considered for the proceedings. All the papers were peer-reviewed. Altogether, 11 keynote lectures were delivered by speakers of international repute in various sessions. A total of 3 keynote lectures were included in the proceedings. The participants had a fruitful scientific and technical discussion on all aspects of thermofluids, generation and storage of renewable energy, transmission systems, etc. The proceeding content covers the following themes:

1. Bio-Heat Transfer, Computational Fluid Dynamics, Refrigeration and Air Conditioning
2. Bio-Fuel, Combustion, Aerodynamics
3. Solar and Renewable Energy and Energy Storage
4. Control Mechanism for Constant Power Generation, Distributed Generation, Electric Vehicle, Hybrid Power System, Hydropower System

The contents of this proceedings reveal the current scientific research activities in thermofluids and renewable energy. We hope they form a useful starting point for beginners as well as practitioners working in the field of thermofluid and renewable energy.

Yupia, India

Pinakeswar Mahanta
Pankaj Kalita
Anup Paul
Abhik Banerjee

Contents

Bio Heat Transfer, Computational Fluid Dynamics, Refrigeration and Air Conditioning

Perturbation and MAPLE Quadrature Computation of Thermosolutal Dissipative Reactive Convective Flow in a Geothermal Duct with Robin Boundary Conditions 3
J. C. Umavathi, O. Anwar Bég, B. Vasu, and Rama S. R. Gorla

Fluid–Structure Interaction Analysis in an Atherosclerosis Carotid Artery 23
Gineesh M. Pillai, J. S. Jayakumar, and R. Ajith Kumar

Effect of Substrate Inclination on Post-impact Dynamics of Droplets . . . 37
N. Sahoo, D. Samanta, and P. Dhar

Modelling and Analysis of Dusty Fluid Flow Past a Vertical Surface with Exothermic and Endothermic Kind of Chemical Reactions 45
Debasish Dey and Barbie Chutia

Entropy Generation Analysis and Slip Effects on Nano Fluid Flow with Heat and Mass Transfer 59
Debasish Dey and Madhurya Hazarika

Rheology of Power-Law Nano-Fluid Flow Past a Stretching Sheet 73
Debasish Dey and Bhagyashree Mahanta

Nanoparticle-Assisted Multilayered Photothermal Therapy Concerning Countercurrent Blood Flow: A Numerical Study 85
Abhijit Paul and Anup Paul

Energetic and Exergetic Analysis of an Ejector-Based Green Refrigeration System Employing Evaporative Cooling in the Gas Cooler 99
Biswajit Gorai and Satyabrata Sahoo

Influence of Pulsating Flow on Thermal Characteristics in a Triangular Sharp-Edged Wavy Channel 113
Rakesh Kumar, Kapil Dev, Ranjan Kumar, Pulak Sen, and Dipak Sen

Numerical Modelling of Porous Square Cavity Heated on Vertical Walls in Presence of Magnetic Field 127
L. Jino, A. Vanav Kumar, Swapnali Doley, M. Berlin, and P. K. Mohanty

Thermal Effects of CNTs Nanoliquid Film Flow During Spin Coating 139
Swatilekha Nag and Susanta Maity

Performance Improvement Techniques in Shell-and-Tube Type of LHS Unit 153
Berihu Gebreyohannes Abreha, Pinakeswar Mahanta, and Gaurav Trivedi

Thermal Response of Nanoparticle Infused Tissue Phantoms During Microwave Heating—A Comparative Study 165
Abishek Sarkar, Mukesh Mehta, Shubham Gupta, and Anup Paul

Study of Bubble Dynamics in Aerated Liquid Storage Tanks Opened to Atmosphere 175
Raj Sarath and J. S. Jayakumar

Unsteady Casson Fluid Flow Past a Stretching Sheet Subject to Non Linear (Quadratic) Free Convection Along with Suction 191
R. Balamurugan and A. Vanav Kumar

Bio Fuel, Combustion, Aerodynamics

Wind Tunnel Experiments to Estimate the Performance of a Novel Arc-Elliptical-Bladed Savonius Wind Rotor 211
Parag K. Talukdar, Nur Alom, Umang H. Rathod, Vinayak Kulkarni, Palash Saikia, and Deva Kanta Rabha

Design and Performance Analysis of Hydrokinetic Turbine with Aerodynamic Stall Model 221
Mahendra Kumar Gupta and P. M. V. Subbarao

Energy and Exergy Analyses of a Gas Turbine and Reheat-Regenerative Steam Turbine Integrated Combined Cycle Power Plant 233
J. Nondy and T. K. Gogoi

Comparative Analysis of Savonius Type Ultra-Micro Hydrokinetic Turbine of Experimental and Computational Investigation 249
Thochi Seb Rengma and P. M. V. Subbarao

Effect of Biomass Blending Ratio and Catalysts in the Kinetic Modelling of Coal and Duckweed Co-Pyrolysis 261
Munmi Bhattacharyya, Pinakeswar Mahanta, and Kaustubha Mohanty

Integral Sliding Mode Controller Design for Francis Turbine Electrohydraulic IGV System 277
J. Vinod and Bikash Kr. Sarkar

Bio-Waste Fired Gas Turbine and Transcritical Co_2 Cycle Based Combined Power Plant: Thermodynamic, Economic and Environmental Performance Assessment 287
P. Mondal, S. Samanta, S. Ghosh, and S. N. Barman

Production of Hydrogen Gas from Al-Water Reaction in the Presence of Aqueous NaOH—An Experimental Approach 303
Biswajyoti Das, Pinakeswar Mahanta, and P. S. Robi

Thermal Cracking of High-Density Polyethylene Wastes and Production of Liquid Hydrocarbon 311
Awinash Kumar, Vikeshachand Prasad Singh, and Pradip Lingfa

A Review on Indigenous Freshwater Microalgae Isolated from Natural Habitats of Arunachal Pradesh, India, as a Biodiesel Source 321
Nabam Hina Papu and Pradip Lingfa

Experimental Investigation of Feasibility of Using Biodiesel Produced from Nahar, Castor, and Rice Bran Seeds in CI Engine 339
Animesh Das, Nibedita Das, and Manjula Das Ghatak

Effect of EGR on Performance and Emission Characteristics of a Diesel Engine Fueled with Yellow Oleander Seed Oil Biodiesel 351
Mridul Deka, Pradeep Kumar Mahanta, and Nabajit Dev Choudhury

Synthesis of a Low Cost, Natural Base Solution for CO_2 Scrubbing from Biogas and its Comparison with KOH 365
Deep Bora, Lepakshi Barbora, and Pinakeswar Mahanta

Vegetable Oil as Fuel in Ci Engine with and Without Exhaust Gas recirculation—A Review 377
S. S. Mohapatra, S. Choudhury, and Binayak Pattanayak

Extraction of Silica (Sio_2) from Coal Fly Ash by Leaching and Sintering Technology 389
Rajaram Mohanty, Shantanu Kumar Mishra, Siba Shankar Mohapatra, Sasanka Choudhury, and Binayak Pattanayak

Extraction of SIO_2 from Coal Fly Ash by Using Sodium Meta Silicate Leaching and Sintering Technology 397
Suryaranjan Samantaray, Shantanu Kumar Mishra, Siba Shankar Mohapatra, and Binayak Pattanayak

Effect of Obstruction Blockage Ratio in a Suddenly Expanded Micro-Combustor 407
Arees Qamareen, Shah Shahood Alam, and Mubashshir Ahmad Ansari

Thermodynamic Study for Performance Enhancement of Supercritical Carbon Dioxide Cycle 417
Amit Kanaujiya and Onkar Singh

The Influence of Pressure and Bed Inventory in Hydrodynamics of Fluidized Bed Unit 435
A. Kumar, N. Mahanta, and A. Mahapatro

Study of Agricultural Product Drying in a Rotating Fluidized Bed in Static Geometry Dryer With and Without Slits 443
Pavitra Singh, Pankaj Kalita, and Pinakeswar Mahanta

Recent Trends in Biomass Gasification in Germany 453
A. Gradel and Tobias Plessing

Solar and Renewable Energy, Energy Storage

Mathematical Modeling and Simulation of the Drying Process of Tea Leaves in a Batch-Type Tray Dryer 469
M. Das and Deva Kanta Rabha

Design and Development of a Beach Sand Solar Collector 479
Animesh Borgohain and Kaustav Bharadwaz

Spectrum-Based 3D Thermal Modeling of the PV Module: Simulation and Experimental Study 493
H. Brahma and N. Sarmah

Duckweed Assisted Sediment Microbial Fuel Cell for Powering Small Scale Devices 503
Arup Dutta, Lepakshi Barbora, Aparupa Thakuria, Pranab Goswami, and Davard Stom

Investigation and Modeling for Energy Consumption During Conventional Machining: A Case Study 511
Shubhajit Das, Sangeeta Das, and Kakoli Roy

Investigation of Thermo-Hydraulic Performance for Different Arrangements of Ribs in Rectangular Solar Air Channel 521
J. Singh, A. Sharma, and R. Chauhan

Repowering of Wind Farm and Its Economic Feasibility: A Case Study 535
Suriti Kumari, K. Boopathi, and Nabin Sarmah

Computational Performance Analysis of the Perforated and Flat Plates Double Pass Solar Air Heaters 549
P. P. Dutta, P. Goswami, A. Sharma, Polash P. Dutta, and M. G. Baruah

Numerical Analysis and Performance Evaluation of Cast Steel Based Sensible Heat Energy Storage System 563
Sanjeev K. Vishwakarma, Sukumar Pati, and Siddhartha Kosti

Comparison of System Performance of Different Types of Ice-Battery for Cooling Applications in Industries 577
Tushar Sharma, Robert Honke, and Tobias Plessing

An Overview of the Methods of Enhancement of the Reaction Kinetics of Al-Water Reaction in an Aqueous Medium and the Prospect of the Economic Viability 591
Biswajyoti Das, Pinakeswar Mahanta, and P. S. Robi

Control Mechanism for Constant Power Generation, Distributed Generation, Electric Vehicle, Hybrid Power System, Hydro Power System

A Novel Architecture for Cluster Based False Data Injection Attack Detection and Location Identification in Smart Grid 599
S. Mallikarjunaswamy, N. Sharmila, G. K. Siddesh, K. R. Nataraj, and M. Komala

Speed Control of PMBLDC Motor Using Rotor Position Rotor Speed PWM Technique 613
Rajen Pudur and Brajagopal Datta

Analysis of Savonius Rotor with Multiple Blades for Hydrokinetic Application 623
Rajen Pudur, Mrinal Kanti Rajak, and Shadab Zafar

A Sensor-Based Improved MPPT Technique Under Partial Shading Condition 635
Boni Satya Varun Sai, Sarang A. Khadtare, and Debashis Chatterjee

A Modified ERA Algorithm for Mode Estimation in Power System 645
Angelica Wartina and Shekha Rai

A Wind-PV Hybrid Generation Scheme for Grid-Isolated Remote Application 657
Sarang A. Khadtare, Boni Satya Varun Sai, and Debashis Chatterjee

Comparison of Multi-carrier MLI, CHB-MLI, SHEPWM and SPWM Inverters for PV-Grid Integration 669
Swati Suman, Rupali Mohanty, and Debashis Chatterjee

Comparison Analysis of THD Optimization Using PSO and Jaya Algorithm in Reduced H-bridge Multilevel Inverter 683
Rupali Mohanty, Swati Suman, and Debashis Chatterjee

Stability Improvement of Renewable Source Integrated Hybrid System Using SVC Controller and ALO Technique 697
Gomar Bam, Pabitra Kumar Guchhait, and Abhik Banerjee

About the Editors

Prof. Pinakeswar Mahanta is a Professor in the Mechanical Engineering Department, Indian Institute of Technology Guwahati. He has graduated in Mechanical Engineering from Regional Engineering College Rourkela. He has completed his master degree (M.Tech.) in Mechanical Engineering from Indian Institute of Technology Kharagpur and Ph.D. in Mechanical Engineering from Indian Institute of Technology Guwahati. He has more than 20 years of teaching experience in the field of conjugate radiation, renewable energy (biodiesel, biomass gasification and biogas technology), circulating fluidized bed technology, energy conservation, combustion and gasification of solid fuel, refrigeration and air conditioning besides 12 years of industry experience. He has published more than 130 research papers in various reputed journals and conferences, and 10 book chapters. He is a recipient of India Distinguished Visiting Fellow awarded by the University of Nottingham, UK, in the year 2010. He is also awarded JSPS Fellowship by invitation by Gifu University, Japan, in December 2017. He has also visited several times to Hof University of Applied Sciences, Germany, as resource person. He has guided twenty Ph.D. students and number of master's students. Currently, he is a Director of National Institute of Technology, Arunachal Pradesh.

Dr. Pankaj Kalita is Assistant Professor in the Centre for Energy, Indian Institute of Technology Guwahati. He has graduated from Jorhat Engineering College (Under Dibrugarh University, Assam) in Mechanical Engineering. He has completed his master degree (M.Tech.) in Mechanical Engineering (specialization: Fluids and Thermal Engineering) and Ph.D. in Energy from IIT Guwahati. He worked at Tezpur University as Assistant Professor in the Department of Energy before joining at IIT Guwahati. He has more than 6 years of teaching and 16 years of research experience in the field of renewable energy (solar energy, thermochemical and biochemical conversion of biomass and energy storage). He has published more than 30 peer-reviewed research papers in various reputed journals, 8 book chapters and 15 conference papers. He is a recipient of India Distinguished Visiting Fellow awarded by the University of Nottingham, UK, in the year 2010. He was also selected for attending prestigious India-UK young scientist networking conference on low carbon technologies held during 2–3 December, 2009, Kolkata, organized by British Council

Kolkata. He has also represented IIT Guwahati in the 2nd Winter School held at Gifu University, Japan, in December 2016. Currently he is guiding ten Ph.D. students and three master's students. He has successfully implemented seven research projects and twelve consultancy projects funded by different government and private agencies in various fields of renewable energy. At present two research projects are in progress in the areas of energy storage, and renewable energy integration for remote electrification. He has organized three short-term courses and a national conference sponsored by TEQIP in the areas of clean energy and waste to energy successfully. He has also organized a GIAN course on Advances in Combustion and Gasification Technology successfully. His current areas of research include solar energy conversion, thermochemical and biochemical conversion, energy management, energy storage (li-ion and thermal energy) and integration of renewable energy for remote electrification.

Dr. Anup Paul is Assistant Professor in the Department of Mechanical Engineering, National Institute of Technology Arunachal Pradesh. He obtained his B.Tech. (Mechanical Engineering) from North Easter Hill University, Shillong; M.Tech. (Eng.) from Biju Patnaik University of Technology, Odisha, and Ph.D. from the Indian Institute of Technology Madras, Chennai. He has worked as Assistant Professor in the Department of Mechanical Engineering at Tezpur University before joining NIT Arunachal Pradesh in 2015. He is working in the field of biological heat transfer and fluid flow, laser mediated hyperthermia treatment in presence of nanoparticles, thermal management etc. He has previously worked on thermal ablation of bio-tissues embedded with large blood vessels during NIR Laser assisted plasmonic photo-thermal therapy. Currently, he is working on laser assisted photo-thermal heating of skin embedded with vascular trees, microwave tumor ablation, heat and flow maldistribution in microchannel heat sink, thermo-hydro dynamics of droplet impacts. He is guiding two Ph.D. students and has published 12 papers in both journals and conferences. He has received Early Career Research Award by Science and Engineering Research Board.

Abhik Banerjee received his Ph.D. degree from Indian Institute of Technology (Indian School of Mines), IIT (ISM) Dhanbad, Jharkhand, India in 2015. Previously, he was working in the capacity of assistant professor in the department of Electrical Engineering, Asansol Engineering College, Asansol, West Bengal, India. He joined NIT Arunachal Pradesh as Assistant Professor in EE Department in 2015. Currently, he is working as assistant professor (in Grade-I) and works as Head of the Department in Electrical Engineering Department in National Institute of Technology, Arunachal Pradesh, Yupia, India. Dr. Banerjee has published many articles in journals published by Elsevier and Springer. He has also presented 15 papers at national and international conferences and 7 book chapters in Springer book series. Dr. Banerjee has conducted 04 (four) workshop as a Convener and several other workshops as Co-convenor in EE Department. Dr. Banerjee is a reviewer of international journals like *International Journal of Electric Power and Energy System* (IJEPES), *ISA Transactions, Expert Systems* (Wiley) and many more. He got the outstanding reviewer award from the IJEPES journal under ELSEVIER. Dr. Banerjee has also worked as the lead guest editor of renowned *Hindawi* journal. He is guiding 6 Ph.D. research scholars in various field of power system and renewable energy. His research interest includes reactive power control, load tracking, distributed generation, renewable energy, hybrid generation etc. Dr. Banerjee is also associated with the Institution of Engineers (India) from 2015.

Bio Heat Transfer, Computational Fluid Dynamics, Refrigeration and Air Conditioning

Perturbation and MAPLE Quadrature Computation of Thermosolutal Dissipative Reactive Convective Flow in a Geothermal Duct with Robin Boundary Conditions

J. C. Umavathi, O. Anwar Bég, B. Vasu, and Rama S. R. Gorla

Nomenclature

A	Constant
Bi_1, Bi_2	Biot numbers at the left and right channel walls
Br	Brinkman number
C_F	Forchheimer drag term
c_p	Specific heat at constant pressure
D	Hydraulic diameter, $2L$
g	Acceleration due to gravity
h_1, h_2	External heat transfer coefficients
I	Inertial parameter
K	Thermal conductivity of the fluid
K_m	Dimensional chemical reaction parameter
L	Channel width
Nu_1, Nu_2	Nusselt numbers at the left and right channel walls
P	Pressure

J. C. Umavathi
Dipartimento di Ingegneria dell, Università degli Studi della Campania "Luigi Vanvitelli", Via Roma 29, Caserta, Italy

O. A. Bég
Department of Mechanical Engineering, SEE, University of Salford, Engineering Science, ManchesterSalford M5 4WT, UK

B. Vasu (✉)
Department of Mathematics, Motilal Nehru National Institute of Technology Allahabad, Uttar Pradesh, Prayagraj, India
e-mail: bvasu@mnnit.ac.in

R. S. R. Gorla
Department of Aeronautics and Astronautics, Air Force Institute of Technology, Wright Patterson Air Force Base, Dayton, OH 45433, USA

P. Mahanta et al. (eds.), *Advances in Thermofluids and Renewable Energy*, Lecture Notes in Mechanical Engineering, https://doi.org/10.1007/978-981-16-3497-0_1

$P = p + \rho_0 g X$	Difference between the pressure and the hydrostatic pressure
Re	Reynolds number
R_T	Temperature difference ratio
S	Dimensionless parameter
T	Temperature
T_1 , T_2	Reference temperatures of the external fluid
T_0	Reference temperature
U	Velocity component in the X-direction
U_0	Reference velocity
u	Dimensionless velocity in the X-direction
X	Stream-wise coordinate
Y	Transverse coordinate
y	Dimensionless transverse coordinate

Greek Symbols

α	Chemical reaction parameter
β_T	Thermal expansion coefficient
β_C	Concentration expansion coefficient
ΔT	Reference temperature difference
ε	Dimensionless parameter
λ_1	Thermal Grashof number
λ_2	Solutal (species) Grashof number
θ	Dimensionless temperature
ϕ	Dimensionless concentration
ρ	Density of the fluid
ρ_0	Value of the mass density when $T = T_0$
μ	Dynamic viscosity
υ	Kinematic viscosity
σ	Porous media permeability parameter.

1 Introduction

Thermosolutal buoyancy-driven flows are mobilized by variations in density variation induced by species concentration and temperature. Such flows which feature coupled heat and mass transfer often arise in porous media. Applications of these "double-diffusive" flows include geothermal energy systems (Fig. 1), contamination fate in soils, petrochemical reservoir dynamics and chromatography. Double-diffusive natural convection in confined porous media is also of relevance in fuel combustion

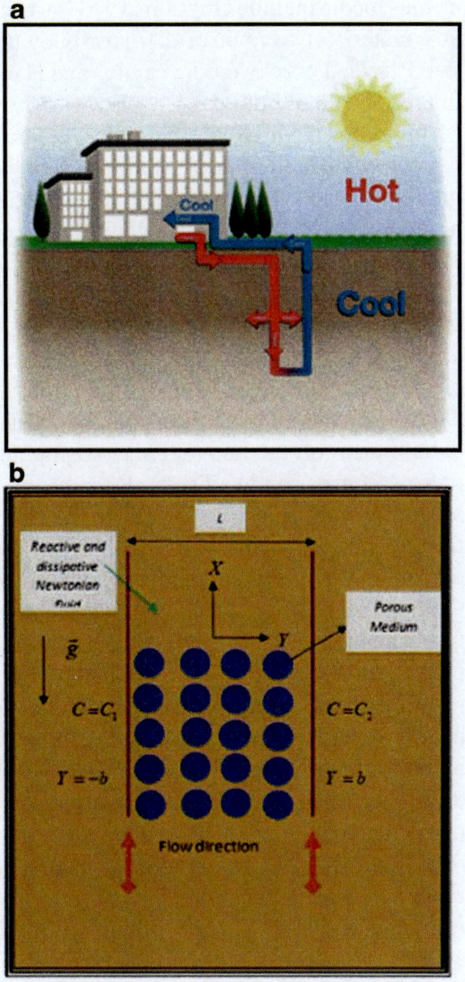

Fig. 1 **a** Schematic of a geothermal duct with porous media and reactive flow. **b** Schematic diagram of thermosolutal reactive convection in vertical porous medium geothermal duct

(e.g., forest fires), thermal insulation, materials processing, room ventilation, packed-bed chemical reactors, etc.

Thermosolutal transport in porous media has been reviewed extensively by Kaviany [1], Nield and Bejan [2] and Ingham and Pop [3], who have provided an excellent range of modeling approaches and identified many new emerging areas of application of both internal flows (cavities, channels and heat pipes) and external boundary layer flows (polymer fabrication, spray deposition, filtration, etc.). Many further geological applications of heat and mass transfer in permeable media have been reviewed by Phillips [4]. Interesting studies of double-diffusive convection in porous media include Gupta et al. [5] (on transient micropolar flow in isotropic permeable materials), Mamou et al. [6] (on buoyancy-driven heat and mass transfer in a saturated vertical porous media cavity) and Narayana and Murthy [7] (on cross-diffusion in thermosolutal boundary layer flow in homogenous porous media).

In porous media hydrodynamics, the traditional approach is to use the Darcy law which provides the linear relationship between the pressure drop across the porous medium and the Darcian velocity. This model is, however, only valid for low Reynolds numbers, i.e., viscous-dominated "creeping" flows. It has been deployed extensively in recent years in a multiplicity of applications. Relevant studies include Prasad and Kulacki [8], Vasseur et al. [9] and Umavathi et al. [10]. Two popular methodologies for non-Darcy models have emerged, i.e., *Brinkman-extended Darcy models* [11] and *Forchheimer-extended Darcy models* [12]. Bég et al. [13] conducted network electrothermal computational simulations of Sakiadis flow in a thermally stratified Darcy–Forchheimer medium with the PSPICE code. Bég et al. [14] applied the differential transform method (DTM) to compute buoyancy effects in dusty physiological thermo-convection in a Darcy–Forchheimer vertical porous medium channel. These studies all confirmed the significant modification in transport characteristics both in the bulk flow and at the walls induced with Forchheimer quadratic drag. Experimental corroboration of this model has also been presented by Sener et al. [15].

In many geological and engineering systems featuring double-diffusive convection, chemical reaction effects arise. Mathematical models of reactive flows usually feature either homogeneous or heterogeneous chemical reactions. Homogeneous reactions occur in one phase only, whereas heterogeneous reactions occur in two or more phases. Anjali Devi and Kandasamy [16] investigated natural convection boundary layer flow with higher-order chemical reaction. Postelnicu [17] used a finite difference technique to study the influence of order of homogenous chemical reaction, chemical reaction rate parameter and sustentation parameter on thermosolutal reactive flow in a Darcy medium with cross-diffusion effects. Nguyen et al. [18] used Fourier spectral element and hybrid Adams–Bashforth and backward Euler numerical schemes to investigate the thermal convection in a fluid-saturated non-Darcy anisotropic porous medium generated by a surface nth-order irreversible reaction. They showed that Forchheimer drag and chemical reaction modify the momentum, heat and species diffusion characteristics substantially.

Convective heat exchange at a bounding surface is also an important consideration in thermosolutal convection flows. It arises in transpiration cooling processes, cooling or heating of geological strata, material drying, heat exchangers, geothermal bore wells, etc. Very few studies have incorporated the *Robin boundary condition,* i.e., mixed boundary condition. Also known as the boundary condition of the third kind, inclusion of this type of boundary condition has been shown to produce results which deviate non-trivially from those computed with classical isothermal or isosolutal boundary conditions. Umavathi et al. [19] further extended the Zachini model to consider non-Darcy effects.

In the present investigation, a mathematical model is developed to study the collective effects of first-order homogeneous chemical reaction on double-diffusive convection in a viscous fluid flowing in a vertical duct containing an isotropic, homogenous porous medium. The Brinkman–Forchheimer-extended Darcy model is employed, and Robin boundary conditions are imposed. This work therefore further generalizes the study in [20] to consider chemical reaction and viscous heating effects. Viscous heating is known to arise in a number of geological applications including mantle convection and enhanced oil recovery. Both perturbation and numerical solutions are presented, the former for the case of small Brinkman number (dissipation parameter) in the absence of Forchheimer inertial drag. Extensive visualization of the influence of non-Darcy, buoyancy, reaction, dissipation and boundary conditions on momentum, species and heat transfer characteristics is provided.

2 Mathematical Geothermal Reactive Duct Flow Model

Steady-state, incompressible and fully developed flow driven by buoyancy due to temperature and concentration gradients in a vertical channel containing a saturated non-Darcy porous medium is considered. Viscous dissipation and first-order homogenous chemical reaction of the solute are assumed. An (X, Y) coordinate system is employed, and the origin is located at the mid-plane of the channel. The distance between the plates is L as shown in the geothermal duct model, i.e., Figure 1. The fluid properties are assumed to be constant except for density variations in the buoyancy force terms. Thermal dispersion, cross-diffusion and stratification effects are neglected. In addition, the concentration of the solute constituent in the solution that saturates the porous medium at the left wall is C_1 and at the right wall is C_2 in such way that $C_2 \geq C_1$. The velocity is taken as zero on the walls of the duct. The duct walls are infinite in the X-direction. Therefore, the flow becomes one-dimensional along the X-axis, and hence, velocity is a function of Y only. The governing equations for momentum, energy and species conservation with the Darcy–Forchheimer model after adopting the Boussinesq approximations may be shown to take the form:

$$\rho_0 g \beta_T (T - T_0) + \rho_0 g \beta_C (C - C_0) - \frac{\partial p}{\partial X} + \mu \frac{d^2 U}{dY^2} - \frac{\mu}{\kappa} U - \frac{\rho C_F}{\sqrt{\kappa}} U^2 = 0 \quad (1)$$

$$K \frac{\mathrm{d}^2 T}{\mathrm{d}Y^2} + \mu \left(\frac{\mathrm{d}U}{\mathrm{d}Y} \right)^2 + \frac{\mu}{\kappa} U^2 = 0 \quad (2)$$

$$D_m \frac{\mathrm{d}^2 C}{\mathrm{d}Y^2} - K_m C = 0 \quad (3)$$

The thickness of the channel walls is neglected, and the walls are assumed to exchange convective heat with the external fluid. The convective heat coefficient is taken as h_1 at the left wall and h_2 at the right wall of the duct (channel). The reference temperature is T_1 at $Y = -L/2$ and T_2 at $Y = L/2$ such that $T_2 \geq T_1$. The pressure gradient is treated as constant $\left(\frac{\partial P}{\partial X} = A\right)$. With these assumptions, the conditions on the boundary for the velocity, temperature and concentration now can be written as

$$U\left(-\frac{L}{2}\right) = U\left(\frac{L}{2}\right) = 0 \tag{4}$$

$$-K\left(\frac{dT}{dY}\right)_{Y=-\frac{L}{2}} = h_1\left[T_1 - T\left(X, -\frac{L}{2}\right)\right],$$
$$-K\left(\frac{dT}{dY}\right)_{Y=\frac{L}{2}} = h_2\left[T\left(X, \frac{L}{2}\right) - T_2\right] \tag{5}$$

$$C\left(-\frac{L}{2}\right) = C_1, \quad C\left(\frac{L}{2}\right) = C_2 \tag{6}$$

It is pertinent to invoke the following dimensionless parameters:

$$u = \frac{U}{U_0}, \theta = \frac{T - T_0}{\Delta T}, y = \frac{Y}{D}\mathrm{Re} = \frac{U_0 D}{\upsilon}\mathrm{GR_T} = \frac{g\beta_T \Delta T D^3}{\upsilon^2},$$
$$\mathrm{GR_C} = \frac{g\beta_C \Delta T D^3}{\upsilon^2}, \mathrm{Br} = \frac{\mu U_0^2}{K \Delta T}, R_T = \frac{T - T_0}{\Delta T}, \mathrm{Bi}_1 = \frac{h_1 D}{K},$$
$$\sigma = \frac{D}{\sqrt{\kappa}}, \mathrm{Bi}_2 = \frac{h_2 D}{K}, U_0 = \frac{-AD^2}{48\mu}, D = 2L\alpha = \frac{K_m D^2}{D_m},$$
$$S = \frac{\mathrm{Bi}_1\mathrm{Bi}_2}{\mathrm{Bi}_1\mathrm{Bi}_2 + 2\mathrm{Bi}_1 + 2\mathrm{Bi}_2}, \phi = \frac{C - C_0}{\Delta C}, I = \frac{\rho C_F D^2 U_0}{\mu\sqrt{\kappa}},$$
$$\lambda_1 = \frac{GR_T}{\mathrm{Re}} T_0 = \frac{T_1 + T_2}{2} + S\left(\frac{1}{\mathrm{Bi}_1} - \frac{1}{\mathrm{Bi}_2}\right)(T_2 - T_1), \lambda_2 = \frac{\mathrm{GR_C}}{\mathrm{Re}},$$
$$\Delta T = T_2 - T_1, \Delta C = C_2 - C_1, C_0 = \frac{C_1 + C_2}{2} \tag{7}$$

Implementing Eq. (7) in Eqs. (1)–(6) yields the following system of coupled ordinary differential equations:

$$\frac{d^2u}{dy^2} + \lambda_1\theta + \lambda_2\phi + 48 - \sigma^2 u - I u^2 = 0 \tag{8}$$

$$\frac{d^2\theta}{dy^2} + \mathrm{Br}\left(\frac{du}{dy}\right)^2 + \sigma^2 \mathrm{Br}\, u^2 = 0 \tag{9}$$

$$\frac{d^2\phi}{dy^2} - \alpha^2\phi = 0 \tag{10}$$

The associated boundary conditions are as follows:

$$u\left(-\frac{1}{4}\right)=u\left(\frac{1}{4}\right)=0 \tag{11}$$

$$\left(\frac{\mathrm{d}\theta}{\mathrm{d}y}\right)_{y=-\frac{1}{4}}=\mathrm{Bi}_1\left(\theta\left(-\frac{1}{4}\right)+\frac{R_{\mathrm{T}}S}{2}\left(1+\frac{4}{Bi_1}\right)\right),$$

$$\left(\frac{\mathrm{d}\theta}{\mathrm{d}y}\right)_{y=-\frac{1}{4}}=\mathrm{Bi}_2\left(-\theta\left(\frac{1}{4}\right)+\frac{R_{\mathrm{T}}S}{2}\left(1+\frac{4}{\mathrm{Bi}_2}\right)\right) \tag{12}$$

$$\phi\left(-\frac{1}{4}\right)=-0.5,\quad \phi\left(\frac{1}{4}\right)=0.5 \tag{13}$$

Here, all parameters are defined in the notation.

3 Perturbation Solution Method

The closed-form solution of the linear species diffusion Eq. (10) is readily obtained as follows:

$$\phi=\frac{\mathrm{Sinh}(\alpha\, y)}{2\,\mathrm{Sinh}(\alpha/4)} \tag{14}$$

However, the problem represented by Eqs. (8) and (9), i.e., the transformed momentum and energy equations, is nonlinear and has no closed-form solutions. Therefore, approximate solutions are determined using a regular perturbation method (RPM). These solutions are valid (in the absence of inertial effects) only for small values of the dissipation parameter, i.e., Brinkman number (*Br*) which is taken as the perturbation parameter. The solutions of Eqs. (8) and (9) can be expressed as

$$u=u_0+\mathrm{Br}\,u_1+\mathrm{Br}^2u_2+\mathrm{Br}^3u_3+\cdots \tag{15}$$

$$\theta=\theta_0+\mathrm{Br}\,\theta_1+\mathrm{Br}^2\theta_2+\mathrm{Br}^3\theta_3+\cdots \tag{16}$$

Substituting the above expansions (15) and (16) in Eqs. (8), (9), (11) and (12) and equating the like powers of Br, one obtains the following equations:

Zeroth-order equations ($\mathrm{Br}=0$)

$$\frac{d^2u_0}{\mathrm{d}y^2}-\sigma^2u_0+\lambda_1\theta_0+\lambda_2\phi+48=0 \tag{17}$$

$$\frac{d^2\theta_0}{\mathrm{d}y^2}=0 \tag{18}$$

$$u_0\left(-\frac{1}{4}\right) = u_0\left(\frac{1}{4}\right) = 0 \tag{19}$$

$$\left(\frac{\mathrm{d}\theta_0}{\mathrm{d}y}\right)_{y=-\frac{1}{4}} = \mathrm{Bi}_1\left(\theta_0\left(-\frac{1}{4}\right) + \frac{R_\mathrm{T}S}{2}\left(1 + \frac{4}{\mathrm{Bi}_1}\right)\right),$$
$$\left(\frac{\mathrm{d}\theta_0}{\mathrm{d}y}\right)_{y=-\frac{1}{4}} = \mathrm{Bi}_2\left(-\theta_0\left(\frac{1}{4}\right) + \frac{R_\mathrm{T}S}{2}\left(1 + \frac{4}{\mathrm{Bi}_2}\right)\right) \tag{20}$$

The closed-form solutions of Eqs. (17) and (18) using boundary conditions (19) and (20) emerge as

$$\theta_0 = c_3 + c_4 y \tag{21}$$

$$\begin{aligned} u_0 &= c_5 \mathrm{Cosh}(\sigma y) + c_6 \mathrm{Sinh}(\sigma y) \\ &\quad + d_1 \mathrm{Sinh}(\alpha y) + d_2 y + d_3 \end{aligned} \tag{22}$$

First-order equations (Br = 1)

$$\frac{d^2 u_1}{\mathrm{d}y^2} - \sigma^2 u_1 + \lambda_1 \theta_1 = 0 \tag{23}$$

$$\frac{d^2 \theta_1}{\mathrm{d}y^2} + \left(\frac{\mathrm{d}u_0}{\mathrm{d}y}\right)^2 + \sigma^2 u_0^2 = 0 \tag{24}$$

$$u_1\left(-\frac{1}{4}\right) = u_1\left(\frac{1}{4}\right) = 0 \tag{25}$$

$$\left(\frac{\mathrm{d}\theta_1}{\mathrm{d}y}\right)_{y=-\frac{1}{4}} = \mathrm{Bi}_1(\theta_1)_{y=-\frac{1}{4}},$$
$$\left(\frac{\mathrm{d}\theta_1}{\mathrm{d}y}\right)_{y=\frac{1}{4}} = -\mathrm{Bi}_2(\theta_1)_{y=\frac{1}{4}} \tag{26}$$

The closed-form solutions of Eqs. (23) and (24) using boundary conditions (25) and (26) are as follows:

$$\begin{aligned} \theta_1 &= g_1 \mathrm{Cosh}(2\sigma y) + g_2 \mathrm{Sinh}(2\sigma y) \\ &\quad + g_3 \mathrm{Cosh}(2\alpha y) + g_4 \mathrm{Cosh}((\sigma + \alpha)y) \\ &\quad + g_5 \mathrm{Cosh}((\sigma - \alpha)y) + g_6 y\, \mathrm{Sinh}(\alpha y) \\ &\quad + g_7 \mathrm{Cosh}(\alpha y) + g_8 \mathrm{Sinh}(\sigma y) \\ &\quad + g_9 \mathrm{Cosh}(\sigma y) + g_{10} y^4 + g_{11} y^3 + g_{12} y^2 \\ &\quad + c_7 y + c_8 \end{aligned} \tag{27}$$

$$
\begin{aligned}
u_1 = {} & c_9\mathrm{Cosh}(\sigma y) + c_{10}\mathrm{Sinh}(\sigma y) \\
& + f_1\mathrm{Cosh}(2\sigma y) + f_2\mathrm{Sinh}(2\sigma y) \\
& + f_3\mathrm{Cosh}(2\alpha y) + f_4\mathrm{Cosh}((\sigma+\alpha)y) \\
& + f_5\mathrm{Cosh}((\sigma-\alpha)y) + f_6 y\,\mathrm{Sinh}(\alpha y) \\
& + f_7 y\,\mathrm{Cosh}(\sigma y) + f_8 y\,\mathrm{Sinh}(\sigma y) \\
& + f_9\mathrm{Cosh}(\alpha y) + f_{10}y^4 + f_{11}y^3 \\
& + f_{12}y^2 + f_{13}y + f_{14}
\end{aligned} \tag{28}
$$

4 Maple Quadrature Numerical Solutions

The approximate analytical solutions obtained are valid only for small values of Brinkman number and in the absence of inertia forces. To relax this condition on Brinkman number and to evaluate the effects of porous medium inertial drag (Forchheimer second-order impedance) on the flow, the momentum and energy Eqns. (8) and (9) are solved numerically using an efficient Runge–Kutta procedure with an appropriate shooting method. This may be executed in any number of symbolic software including Maple, Mathematica and MATLAB. Details of this technique are given in [21–23]. **MAPLE** has been used in the geothermal duct simulation. The stepping formulae although designed for nonlinear problems are even more efficient for any order of linear differential equation and are summarized below:

$$k_0 = f(x_i, y_i), \tag{29}$$

$$k_1 = f\left(x_i + \frac{1}{4}h,\ y_i + \frac{1}{4}hk_0\right), \tag{30}$$

$$k_2 = f\left(x_i + \frac{3}{8}h,\ y_i + \left(\frac{3}{32}k_0 + \frac{9}{32}k_1\right)h\right), \tag{31}$$

$$k_3 = f\left(x_i + \frac{12}{13}h,\ y_i + \left(\frac{1932}{2197}k_0 - \frac{7200}{2197}k_1 + \frac{7296}{2197}k_2\right)h\right), \tag{32}$$

$$k_4 = f\left(x_i + h,\ y_i + \left(\frac{439}{216}k_0 - 8k_1 + \frac{3860}{513}k_2 - \frac{845}{4104}\right)h\right), \tag{33}$$

$$k_5 = f\left(x_i + \frac{1}{2}h,\ y_i + \left(\begin{array}{c} -\frac{8}{27}k_0 + 2k_1 - \\ \frac{3544}{2565}k_2 + \frac{1859}{4101}k_3 - \frac{11}{40}k_4 \end{array}\right)h\right) \tag{34}$$

$$y_{i+1} = y_i + \left(\frac{25}{216}k_0 + \frac{1408}{2565}k_2 + \frac{2197}{4101}k_3 - \frac{1}{5}k_4\right)h, \tag{35}$$

$$z_{i+1} = z_i + \left(\begin{array}{c} \frac{16}{135}k_0 + \frac{6656}{12825}k_2 + \frac{28561}{56430}k_3 \\ -\frac{9}{50}k_4 + \frac{2}{55}k_5 \end{array} \right) h \tag{36}$$

Here, y denotes fourth-order Runge–Kutta phase, and z is the fifth-order Runge–Kutta phase. An estimate of the error is achieved by subtracting the two values obtained. If the error exceeds a specified threshold, the results can be re-calculated using a smaller step size. The approach to estimating the new step size is shown below:

$$h_{\text{new}} = h_{\text{old}} \left(\frac{\varepsilon h_{\text{old}}}{2|z_{i+1} - y_{i+1}|} \right)^{1/4}. \tag{37}$$

The solutions obtained by the numerical scheme (RKM) are used to validate the perturbation solutions (RPM), for small values of Brinkman number. Excellent correlation is obtained as described in due course (Tables 3 and 4).

5 Energy Design Quantities—Skin Friction and Nusselt Number

The dimensionless skin friction (surface shear stress, i.e., velocity gradient) and Nusselt number (surface heat transfer rate, i.e., temperature gradient) are given by

$$\tau_1 = \left(\frac{\mathrm{d}u}{\mathrm{d}y} \right)_{y=-\frac{1}{4}}, \quad \tau_2 = \left(\frac{\mathrm{d}u}{\mathrm{d}y} \right)_{y=\frac{1}{4}} \tag{38}$$

$$\mathrm{Nu}_1 = \left(\frac{\mathrm{d}\theta}{\mathrm{d}y} \right)_{y=-\frac{1}{4}}, \quad \mathrm{Nu}_2 = \left(\frac{\mathrm{d}\theta}{\mathrm{d}y} \right)_{y=\frac{1}{4}} \tag{39}$$

6 Results and Discussion

Figures 2 and 3 display the profiles for the velocity, temperature and solute concentration obtained with the numerical method (RKM). These results illustrate the effects of Brinkman number Br, thermal Grashof number λ_1, solutal (concentration) Grashof number λ_2, chemical reaction parameter α, Forchheimer inertial parameter I for different values of porous parameter σ, Biot numbers and temperature difference ratio R_T. Equations (8) to (10) along with boundary conditions (11) to (13) are solved analytically using the regular perturbation method (RPM) which is valid when Brinkman number is less than one and numerically with the Runge–Kutta and shooting method (RKM) which can be employed for any values of Brinkman number.

Fig. 2 Plots of u for different values of λ_1

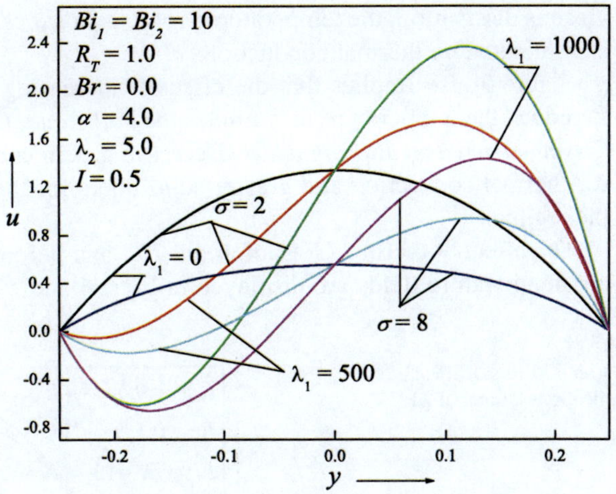

Fig. 3 Plots of θ for different values of Br

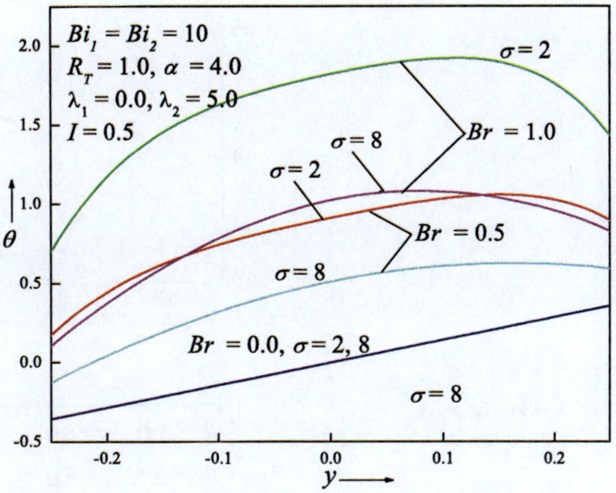

The effects of thermal Grashof number λ_1 and porous medium parameter σ on the velocity and temperature characteristics are shown in Figs. 2 and 3. In the absence of viscous dissipation (Br $= 0$), there is a flow reversal near the cold wall for strong thermal buoyancy cases with values of $\lambda_1 = 500, 1000$. Figure 2 also indicates that increasing porous medium parameter σ decelerates the flow for all values of thermal Grashof number, λ_1. $\sigma = \frac{D}{\sqrt{\kappa}}$, and evidently, this parameter is inversely proportional to the medium permeability, κ. It arises in the Darcian linear impedance term in the normalized momentum Eq. (8), i.e., $\sigma^2 u$. Clearly, as σ *increases, the medium permeability is reduced, and this increases the resistance of porous medium fibers (fabric) to the flow. Velocity is therefore depleted.* The temperature profiles as seen in Fig. 3 are drawn for different values of Brinkman number Br in the absence of thermal Grashof number λ_1 for equal and unequal Biot numbers. In the absence of viscous dissipation, the temperature profile is linear as the heat transfer contribution is dominated by thermal conduction.

Figure 3 also implies that the effect of increasing permeability parameter σ is to reduce the temperature magnitudes *for both equal (symmetric case) and unequal (asymmetric case) Biot numbers*. Decreasing permeability offers less fluid volume for thermal convection and greater solid fibers for thermal conduction. This cools the regime.

The effect of thermal Grashof number λ_1 and porous parameter σ on the velocity and temperature fields are displayed in Figs. 4 and 5 for equal Biot numbers. The

Fig. 4 Plots of u for different values of λ_1

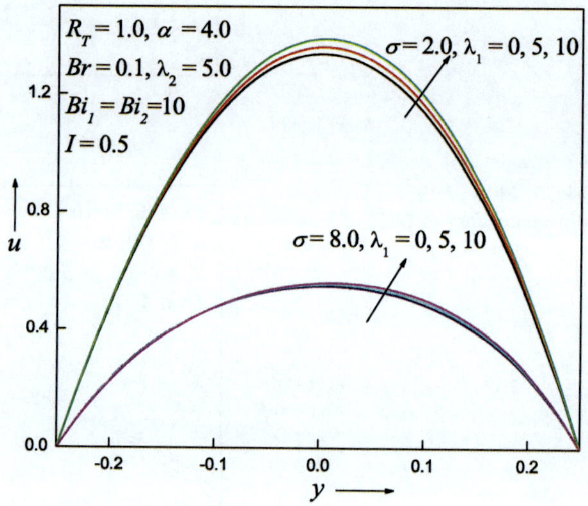

Fig. 5 Plots of θ for different values of λ_1

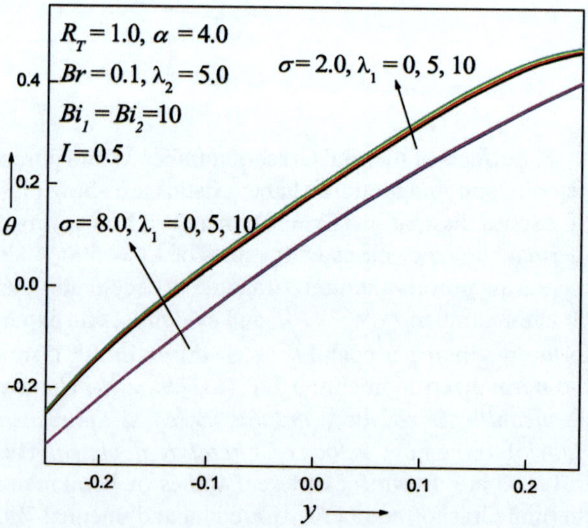

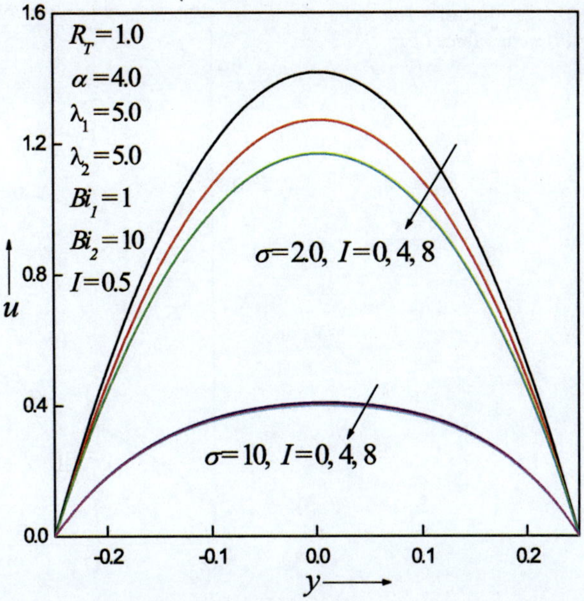

Fig. 6 Plots of *u* for different values of *I*

effect of thermal Grashof number λ_1 is to enhance the velocity and temperature fields for both equal. This is due to the fact that with higher values of thermal Grashof number there is an enhancement in thermal buoyancy force in the body force term + $\lambda_1\theta$ in Eq. (8) which serves to accelerate the flow, as noted in many studies including Gebhart et al. [24].

Figures 6 and 7 display the effect of Forchheimer inertial parameter I and porous medium parameter σ on the velocity and temperature distributions for unequal Biot numbers. The effect of inertial parameter is to decelerate the flow for unequal Biot numbers. The impact of inertial parameter I is more prominent at smaller values of porous parameter σ as noted by Lai and Kulacki [25]. Clearly, the neglection of Forchheimer second-order drag leads to overpredictions in the velocity magnitudes. Temperatures are also strongly suppressed with increasing inertial parameter, *I,* and the distributions for the both equal (symmetric) and unequal (asymmetric) Biot number cases are distinctly *monotonic* in nature at high permeability parameters (σ), whereas they are morphed into *parabolic profiles* at lower values of σ.

The values of skin friction and Nusselt numbers are shown in Tables 1 and 2, respectively, for all the emerging thermophysical parameters for both equal and unequal Biot numbers. For equal Biot numbers, as the Brinkman number Br and chemical reaction parameter α are increased, skin friction increases at the left wall, whereas it decreases at the right wall with greater values of thermal Grashof number λ_1, solutal (concentration) Grashof number λ_2, porous parameter σ and inertial parameter I. The skin friction at the right wall increases in magnitude with an elevation in values of thermal Grashof number, solutal Grashof number and Brinkman number, whereas the converse behavior is observed with increasing the values of

Fig. 7 Plots of θ for different values of I

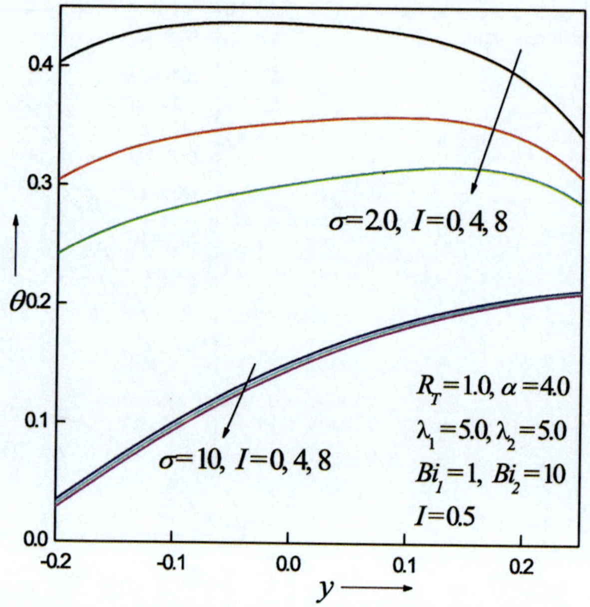

chemical reaction parameter, porous parameter and inertial parameter. A similar trend is computed for the asymmetric case of unequal Biot numbers.

The rate of heat transfer, i.e., Nusselt number Nu_1 at the left wall increases with greater values of thermal Grashof number, Brinkman number and chemical reaction parameter, whereas it decreases with concentration Grashof number, porous medium parameter and Forchheimer inertial parameter. The Nusselt number Nu_2 at the right wall, however, decreases with increasing the values of thermal Grashof number, concentration Grashof number and Brinkman number, whereas it increases with chemical reaction parameter, porous medium parameter and Forchheimer inertial parameter. For unequal Biot numbers, Nu_1 increases with thermal Grashof number, Brinkman number and chemical reaction parameter, whereas it is reduced with increasing concentration Grashof number, porous medium parameter and Forchheimer inertial parameter. At the right wall, Nu_2 increases with thermal Grashof number, concentration Grashof number and Brinkman number, whereas it decreases with chemical reaction parameter, porous medium parameter and Forchheimer inertial parameter.

The results illustrated in Figs. 2 and 3 and Tables 1 and 2 are evaluated by solving Eqs. (8)–(10) along with the boundary conditions (10)–(13) using the Runge–Kutta–shooting method (RKM). To verify the solutions obtained by Runge–Kutta–shooting method, a comparison is made with the analytical solutions obtained using regular perturbation method (RPM) in the absence of inertial effects.

Table 1 Values of skin friction for $R_T = 1.0$, $\mathrm{Br} = 0.1$, $\lambda_1 = 5.0$, $\lambda_2 = 5.0$, $\alpha = 4.0$, $\sigma = 4$, $I = 2.0$

	$\mathrm{Bi}_1 = 10$, $\mathrm{Bi}_2 = 10$		$\mathrm{Bi}_1 = 1$, $\mathrm{Bi}_2 = 10$	
	τ_1	τ_2	τ_1	τ_2
λ_1				
0	8.20961740	−9.66979819	8.20961741	−9.66979817
1	8.02258775	−9.90405984	8.07487173	−9.90808428
10	8.17853069	−10.20209829	8.76825020	−10.28391936
15	8.17471594	−10.48219512	9.12954560	−10.64964620
λ_2				
0	8.92198237	−9.19971245	9.20543493	−9.23294741
10	8.55619963	−9.56564540	8.83568246	−9.59665599
15	8.37333986	−9.74864341	8.65096071	−9.77860981
20	8.19050118	−9.93166244	8.46634234	−9.96062885
Br				
0	8.61880003	−9.26093744	8.69674926	−9.18302219
0.1	8.73908047	−9.38266842	9.02050713	−9.41476862
0.25	8.93370857	−9.57956463	9.60097625	−9.82858396
0.5	9.30448798	−9.95440573	11.00456636	−10.8217036
α				
2	8.73066964	−9.39108481	9.01196010	−9.42310853
4	8.73908047	−9.38266842	9.02050713	−9.41476862
6	8.75094236	−9.37079924	9.03255979	−9.40300648
8	8.76422503	−9.35750902	9.04605371	−9.38983481
σ				
2	10.51822810	−11.18994069	10.89017888	−11.2780889
4	8.73908047	−9.38266842	9.02050713	−9.41476862
6	6.93569686	−7.53849292	7.13590776	−7.52731639
8	5.52045106	−6.07667328	5.66585566	−6.04313955
I				
0	8.95016417	−9.59937974	9.25515002	−9.64623862
4	8.55273965	−9.19117075	8.81516225	−9.21177612
8	8.23576153	−8.86498898	8.46924677	−8.86890439
12	7.97319433	−8.59435797	8.18553630	−8.58680505

Table 2 Values of Nusselt number for $R_T = 1.0$, $\text{Br} = 0.1$, $\lambda_1 = 5.0$, $\lambda_2 = 5.0$, $\alpha = 4.0$, $\sigma = 4$, $I = 2.0$

	$\text{Bi}_1 = 10, \text{Bi}_2 = 10$		$\text{Bi}_1 = 1, \text{Bi}_2 = 10$	
	Nu_1	Nu_2	Nu_1	Nu_2
λ_1				
0	2.20590246	0.58641333	0.96508232	−0.65440681
1	2.20274019	0.57056862	0.96624557	−0.67561102
10	2.24398751	0.51993840	1.01314568	−0.83307637
15	2.26612337	0.48278372	1.04479064	−0.94909266
λ_2				
0	2.25330114	0.59097032	1.00103041	−0.71926289
10	2.23799593	0.57332905	0.99374506	−0.72683200
15	2.23083496	0.56401640	0.99032844	−0.73141408
20	2.22400171	0.55437568	0.98706252	−0.73652758
Br				
0	1.42857143	1.42857143	0.62500000	0.62500000
0.1	2.24548465	0.58231369	0.99731231	−0.72278159
0.25	3.56829007	−0.78642688	1.66717021	−3.12403029
0.5	6.09162205	−3.39234347	3.29821141	−8.86814262
α				
2	2.24495965	0.58170685	0.99706236	−0.72304286
4	2.24548465	0.58231369	0.99731231	−0.72278159
6	2.24622029	0.58315811	0.99766245	−0.72242132
8	2.24703481	0.58408524	0.99805013	−0.72203008
σ				
2	2.43406463	0.38852539	1.08684360	−1.04457598
4	2.24548465	0.58231369	0.99731231	−0.72278159
6	2.04557087	0.78782259	0.90382367	−0.38627518
8	1.88601903	0.95207874	0.83031980	−0.12121522
I				
0	2.30008798	0.52648241	1.02363517	−0.81726044
4	2.19887822	0.62999707	0.97505296	−0.64284628
8	2.12303228	0.70765867	0.93919988	−0.51400837
12	2.06347537	0.76870215	0.91133820	−0.41380703

7 Conclusions

Thermosolutal dissipative chemically reacting flow in a vertical geothermal duct channel containing a non-Darcy porous medium with Robin boundary conditions has been studied theoretically. The transformed, non-dimensional equations for momentum, energy and species conservation have been solved both with a perturbation method (for low Brinkman number in the absence of Forchheimer drag) and with Runge–Kutta quadrature and a shooting method in the presence of inertial effects. Good correlation of both solutions has been demonstrated in the absence of inertial effects. The simulations have shown that for the small values of Brinkman number, the analytical and numerical solutions agree, and the error increases as Brinkman number increases. For the non-Darcy case, an increase in Brinkman number and thermal Grashof number induces significant flow acceleration. With increasing solutal (concentration) Grashof number, the velocity distribution is less significantly affected, and the flow decreases at the cold wall and increases at the hot wall for both equal and unequal Biot numbers. An increase in homogenous first-order chemical reaction parameter decreases the velocity magnitudes. Increasing porous medium (Darcy inverse permeability) parameter and Forchheimer second-order inertial parameter both inhibits flow and manifests in strong retardation. The skin friction at the left wall is elevated with greater viscous dissipation (i.e., Brinkman number), chemical reaction parameter, whereas it is reduced with increasing thermal Grashof number, concentration Grashof number, porous medium and Forchheimer inertial parameters for both equal and unequal Biot numbers. At the right wall, the skin friction is enhanced with greater thermal Grashof number, concentration Grashof number and Brinkman number, and the contrary effect is induced with higher values of chemical reaction parameter, porous medium and Forchheimer inertial parameter. The Nusselt number at the left wall is boosted with greater thermal Grashof number, Brinkman number, chemical reaction parameter, whereas it is suppressed with greater concentration Grashof number, porous medium and Forchheimer inertial parameter for equal Biot numbers. At the hot wall, the rate of heat transfer (Nusselt number) decreases with an increase in the values of thermal Grashof number, (solutal) concentration Grashof number and Brinkman number, and the opposite behavior is generated with an increase in the chemical reaction parameter, porous medium and Forchheimer inertial parameters. Overall the inclusion of viscous dissipation, chemical reaction and Robin (mixed) boundary conditions modifies the transport phenomena characteristics significantly and provides for more realistic simulations of geological (geothermic) double-diffusive convection flows. The current study has been confined to Newtonian fluids. Future studies will consider non-Newtonian models and more complex porous media models of relevance to geothermal energy systems and will be reported soon.

References

1. Kaviany, M. (1995). *Principles of heat transfer in porous media* (2nd ed.). New York: Springer.
2. Nield, D. A., & Bejan, A. (2013). *Convection in porous media* (4th ed.). New York: Springer.
3. Ingham, D. B., & Pop, I. (2002). *Transport phenomena in porous media* (2nd ed.). Oxford: Pergamon.
4. Phillips, O. M. (2009). *Geological fluid dynamics: Sub-surface flow and reactions*. UK: Cambridge University Press.
5. Gupta, D., Kumar, L., Anwar Bég, O., & Singh, B. (2014). Finite element analysis of transient heat and mass transfer in microstructural boundary layer flow from porous stretching sheet. *Computational Thermal Sciences, 6*, 155–169.
6. Mamou, M., Vasseur, P., & Bilgen, E. (1995). Multiple solutions for double-diffusive convection in a vertical porous enclosure. *International Journal of Heat and Mass Transfer, 38*, 1787–1798.
7. Lakshmi Narayana, P. A., & Murthy, P. V. (2008). Soret and Dufour effects on free convection heat and mass transfer from a horizontal flat plate in a Darcy porous medium. *ASME Journal of Heat Transfer, 130*, 104504
8. Prasad, V., & Kulacki, F. A. (1985). Natural convection in porous media bounded by short concentric vertical cylinders. *ASME Journal of Heat Transfer, 107*, 147–154.
9. Vasseur, P., Wang, C. H., & Sen, M. (1990). Natural convection in an inclined rectangular porous slot: Brinkman extended Darcy model. *ASME Journal of Heat Transfer, 112*, 507–511.
10. Umavathi, J. C., Kumar, J. P., Chamkha, A. J., & Pop, I. (2005). Mixed convection in a vertical porous channel. *Transport in Porous Media, 61*, 315–335.
11. Brinkman, H. C. (1947). On the permeability of media consisting of closely packed porous particles. *Applied Science Research, 1*, 81–86.
12. Forchheimer, P. (1901). Wasserbewegung durch Boden. *Z. Ver. Deut. Ing., 45*, 1736–1741.
13. Anwar Bég, O., Zueco, J., & Takhar, H. S. (2008). Laminar free convection from a continuously moving vertical surface in a thermally stratified, non-Darcian high-porosity medium: Network numerical study. *nternational Communications in Heat and Mass Transfer, 35*(7), 810–816.
14. Bég, T. A., Rashidi, M. M., Anwar Bég, O., & Rahimzadeh, N. (2013). DTM semi-numerical simulation of biofluid-particle suspension flow and heat transfer in non-Darcian porous media. *Computer Methods Biomechanics Biomedical Engineering, 16*, 896–907.
15. Sener, M., et al. (2016). Forchheimer forced convection in a rectangular channel partially filled with aluminum foam. *Experimental Thermal and Fluid Science, 75*, 162–172.
16. Anjali Devi, S. P., & Kandasamy, R. (1999). Effects of chemical reaction, heat and mass transfer on laminar flow along a semi-infinite horizontal plate. *Heat Mass Transfer, 35*, 465–467.
17. Postelnicu, A. (2007). Influence of chemical reaction on heat and mass transfer by natural convection from vertical surfaces in porous media considering Soret and Dufour effects. *Heat Mass Transfer, 43*, 595–602.
18. Nguyen, H. D., et al. (1996). Unsteady non-Darcy reaction-driven flow from an anisotropic cylinder in porous media. *Chemical Engineering Science, 51*, 4963–4977.
19. Umavathi, J. C., Prathap Kumar, J., & Sultana, J. (2012). Mixed convection flow in a vertical channel with boundary conditions of the third kind in the presence of heat source/sink. *Applied Mathematics and Mechanics, 33*, 1015–1034.
20. Umavathi, J. C., & Veershetty, S. (2012). Non-Darcy mixed convection in a vertical porous channel with boundary conditions of third kind. *Transport Porous Media, 95*, 111–131.
21. Umavathi, J. C., & Sheremet, M. A. (2016). Mixed convection flow of an electrically conducting fluid in a vertical channel using Robin boundary conditions with heat source or sink. *European Journal of Mechanics-B/Fluids, 55*, 132–145.
22. Anwar Bég, O., Basir, M. F. M., Uddin, M. J., & Ismail, A. M. (2017). Numerical study of slip effects on asymmetric bioconvective nanofluid flow in a porous microchannel with an expanding/contracting upper wall using Buongiorno's model. *J. Mechanics in Medicine and Biology, 17*, 1750059.1–1750059.28.

23. Bég, O. A., Bég, T. A., Karim, I., Khan, M. S., Alam, M. M., Ferdows, M., & Shamshuddin, M. D. (2019). Numerical study of magneto-convective heat and mass transfer from inclined surface with Soret diffusion and heat generation effects: a model for ocean magnetic energy generator fluid dynamics. *Chinese Journal of Physics, 60*, 167–179. https://doi.org/10.1016/j.cjph.2019.05.002.
24. Gebhart, B., et al. (1988). *Buoyancy-induced flows and transport*. Washington, USA: Hemisphere.
25. Lai, F. C., & Kulacki, F. A. (1987). Non-Darcy convection from horizontal impermeable surfaces in saturated porous media. *International Journal of Heat and Mass Transfer, 30*, 2189–2192.

Fluid–Structure Interaction Analysis in an Atherosclerosis Carotid Artery

Gineesh M. Pillai, J. S. Jayakumar, and R. Ajith Kumar

Nomenclature

C	Damping factor (Kns/m)
d	Incompressibility (dimensionless)
f	External body force (N/m^3)
$\bar{I}_1$	First-order strain rate (dimensionless)
$\bar{I}_2$	Second-order strain rate (dimensionless
J	Elastic deformation gradient (Pa)
K	Structural stiffness (N/m)
m	Structural mass (Kg)
P	Force vector (N)
$\dot{U}$	Velocity vector (m/s)
U	Displacement vector (m)
v	Velocity vector (m/s)
W	Strain energy potential (J)

Greek Symbols

η	Viscosity of blood (Ns/m^2)
ρ	Density of blood (Kg/m^3)
σ	Stress tensor (N/m^2)

G. M. Pillai · J. S. Jayakumar (✉) · R. A. Kumar
Department of Mechanical Engineering, Amrita Vishwa Vidyapeetham, Amritapuri, India

P. Mahanta et al. (eds.), *Advanços in Thermofluids and Renewable Energy*, Lecture Notes in Mechanical Engineering, https://doi.org/10.1007/978-981-16-3497-0_2

1 Introduction

Atherosclerosis, or atherosclerotic cardio vascular disease, is a life-threatening condition that can start its early development during childhood. It is a process by which minute deposits of fatty materials known as "plaque" build up inside the walls of the arteries and reduces or completely blocks the oxygen-rich blood flow overtime. In extreme conditions, such cases can lead to a life-threatening heart attack or a sudden stroke. Atherosclerosis comes with a variety of risk factors that include high blood pressure, obesity, high cholesterol, smoking, physical inactivity, and diabetes. The hemodynamics in carotid artery results in limited blood supply to the area surrounding the partially blocked artery, which can end up degrading and potentially killing the neighboring tissues. Significant damage to brain tissues can result in ischemia, along with high wall shear stress and high blood pressure. In atherosclerotic plaque carotid, artery causes the plaque rupture and it leads to complete block in blood vessel.

There are many studies reported on carotid artery disease in the literature. Some of the studies are based on the experimental methods, and others are based on numerical investigations. In the experimental studies reported in the past, the focus was to examine the effects of mechanical forces using particle image velocimetry (PIV) method. Perktold et al. [1] conducted the numerical study of pulsatile, non-Newtonian blood flow in three dimensions on common carotid artery (CCA) under different bifurcation angles using finite element analysis (FEA). This study focuses on the flow and stress properties of different sinus angle variations. They concluded that the stress is affected on the sinus, due to the complex flow at the reattachment zone, flow separation and high wall shear stress on the blood vessels. All of these are important in atherogenesis. Perktold with Rappitsch [2] investigated numerical simulations of local blood flow and vessel mechanics in a carotid artery bifurcation to determine the mechanical stresses in the CCA wall. Urquiza et al. [3] investigated CCA blood flow using multidimensional modeling. This work involves a comparison of the compliant 1D and 3D models of CCA bifurcation using finite element method. They concluded that the blood flow patterns contribute to the development of vascular disease. Zhao et al. [4] conducted a numerical study based on both blood effects and vessel effects employing a realistic physiological model of human CCA bifurcation. Their model was generated from an MRI data and in order to get the results, they implemented combined commercial code of CFX4 and ABAQUS. They revealed that the wall shear stress reduces with respect to pulsatile blood flow and wall movement. A CFD study carried out by Gharahi et al. [5] compares various viscosity models for hemodynamic study on different geometries of human carotid artery that extracted from MRI data (patient-specific study). Sousa et al. [6] conducted fluid–structure interaction (FSI) study of bloodstream in a non-stenosed basic carotid artery bifurcation. This investigation was completed by utilizing vascular geometries that made from Doppler ultrasound pictures. In this study, they explored the impact of hemodynamics of blood, pressure dissemination on the bifurcation region, and lowermost extreme wall shear stress (WSS) at the bifurcation region.

From the literature survey, it is found that the following aspects of atherosclerosis CCA were not studied, viz. plaque rupture in the common carotid artery, hyper-elasticity nature of blood vessel, three layers of the carotid artery wall, and the nature of plaque. Figure 1c shows a typical picture of atherosclerosis CCA (complex atheroma). These aspects being important from medical point of view are taken up in the present study for a numerical investigation. This current study is on the CCA, which is situated in the neck region of the human body (Fig. 1a). CCA supplies blood, oxygen, and nutrients for the brain. A blockage in a CCA artery can cause brain ischemia, brain hypoxia, or brain infarction/ ischemic stroke. This study also aims to provide a dynamic model for the common carotid artery to analyze the flow effects and predict the plaque initiation. This model could predict the chance of initiation of another plaque formation inside the blood vessel.

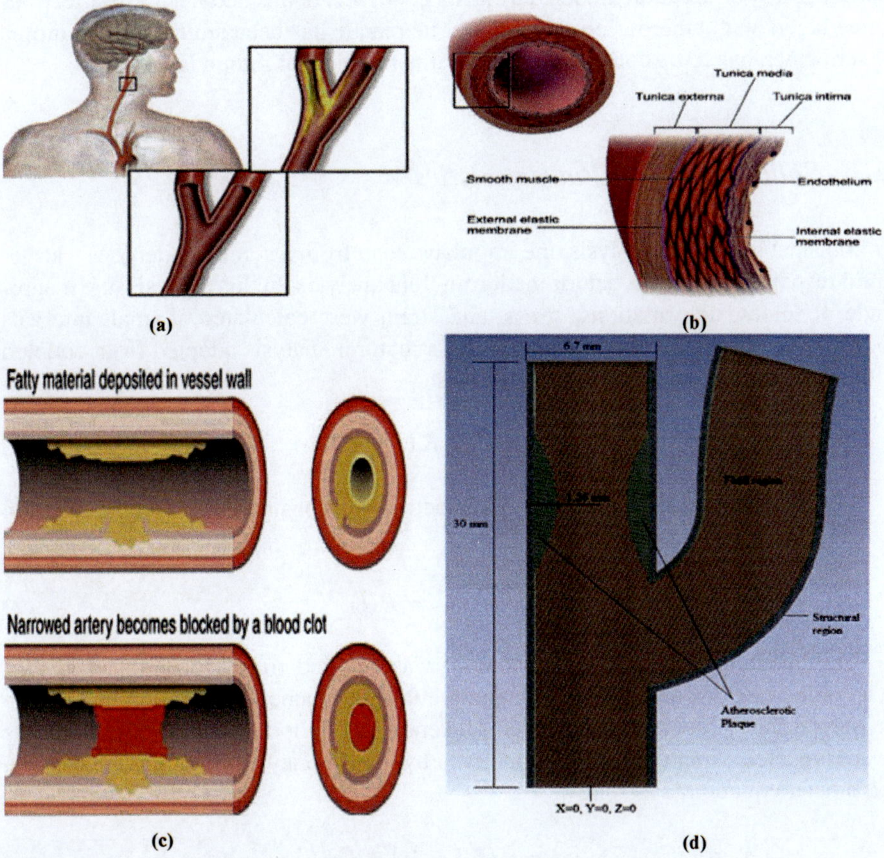

Fig. 1 **a** Atherosclerosis in the carotid artery [15]. **b** Three layers of artery wall. **c** Atherosclerotic CCA cross section. **d** Artery narrowed by plaque [15]

2 Methodology

2.1 Geometry

The artery is modeled as 3D, three layered, (tunica intima or intima, tunica media or media, and tunica externa or adventitia) shown in Fig. 1b, and the plaque is modeled as three-dimensional circular shape, the inner diameter of the stenosis region is 5.72 mm, and the length of the plaque is 5 mm (plaque shoulder). The maximum height of the plaque is 1.26 mm and normal diameter of blood vessel is 6.7 mm; each layer has different hyperelastic property and thickness such as: 0.23, 0.31, and 0.34 mm [7] for intima, media, and adventitia, respectively. The total length of the model is 30 mm, shown in Fig. 1d. Figure 1d shows that the cross-sectional view of atherosclerosis carotid artery bifurcation modeled by ANSYS V19.1. In this geometry, the artery has three layers with atherosclerotic plaque. The plaque has been situated in the intima layer, which has to be considered as the same property of intima layer.

2.2 Solid Structure Characteristics

In transient structural analysis, the impulsive load from a source undergoes a structural response which was determined from fluid analysis. In this analysis, the magnitude of forces, deformations, stress, and strain were calculated at small intervals of time. The basic governing equation for structural analysis adopted from coupled fluid–structure interaction [8] is as follows:

$$m\ddot{U} + C\dot{U} + KU = P(t). \tag{1}$$

In this analysis, the structural damping factor is not considered, so that the equation changes to

$$m\ddot{U} + kU = P(t) \tag{2}$$

This modeling of three-layer material was adopted from Holzapfel et al. [9]. The properties displayed by the plaque are taken as same as the intima layer. The hyperelasticity property of each layer is taken from Gholipour et al. [7] The materials were modeled using the formulation given by the Mooney-Rivlin 5 parameter. The parameter equation is as follows;

$$\begin{aligned} W = &-C_{01}(3 - \bar{I}_2) - C_{10}(3 - \bar{I}_1) + C_{11}(3 - \bar{I}_1)(3 - \bar{I}_2) \\ &+ \frac{1}{d}(1 - J)^2 + C_{20}(3 - \bar{I}_1)^2 + C_{02}(3 - \bar{I}_2)^2 \end{aligned} \tag{3}$$

Table 1 Mooney–Rivlin 5-parameter model material constants [8]

Material	C_{10}	C_{01}	C_{20}	C_{11}	C_{02}
Intima	−2.04E + 05	2.03E + 05	1.37E + 06	−3.71E + 06	2.67E + 06
Media	−1.17E + 05	1.28E + 05	2.24E + 05	−6.72E + 05	5.69E + 05
Adventitia	−1.89E + 05	2.02E + 05	4.59E + 05	−1.38E + 06	1.34E + 06

In this equation, there are five material constants. These were derived from the curve fitting data as shown in Table 1. The density of each layer is considered as 1000 kg/m^3 Chan et al. [10].

2.3 *Fluid Characteristics*

Fluid flow domain was mathematically modeled by the Navier–stokes equations. The conservation of mass, momentum, for an incompressible fluid in 3D, can be expressed as:

$$\nabla.v = 0 \tag{4}$$

$$\frac{\rho\partial u}{\partial t} + \rho v \nabla v = \nabla.\sigma + f \tag{5}$$

The external body force is considered as zero. The stress tensor can be expressed as:

$$\sigma = -PI + \tau \tag{6}$$

The blood has been modeled as transient pulsatile motion, with a non-Newtonian behavior. In order to modeling of its non-Newtonian nature, power law is enabled, which is taken from Mandal et al. [11].

$$\eta = \eta_0 \gamma^{n-1} \tag{7}$$

where, $\eta_0 = 0.035, n = 0.6, T = 310$K Density of blood is taken as $\rho_{\text{Blood}} = 1060$ kg/m^3, η denotes viscosity of blood from cilia et al. [12]. The blood flow is treated as transient pulsatile flow profile exceeds up to 0 to 0.8 s, i.e., for one heartbeat. The inlet flow velocity range of common carotid artery is 0.5 to 0.8 ± 0.3 m/s, and the inlet velocity profile has shown in Fig. 4 Perktold et al. [1].

3 Solution Method

The important portion of this study is fluid–structure interaction. The common carotid artery is modeled within the ANSYS using the two-way FSI system coupling method. The general condition is transient. Structural analysis is done by transient structural analysis, and fluid analysis is done in the Fluent analysis. These systems are coupled with "system coupling." The data is transferred from fluid to structural and structural to fluid. The boundary is not rigid, therefore "Dynamic mesh" method was used for this simulation. In dynamic mesh parameter, it should be "smoothing and remeshing," for dynamic mesh parameter, diffusion method was used, and the value was taken as 0.45. Because if diffusion parameter is zero, mesh will equally be displaced across the domain. The maximum value of diffusion parameter is 2. When maximum displacement will be felt in the far field boundary, a value between 0.4 and 0.45 is good enough. For remeshing, the local cell and local face providing mesh scale parameters from mesh scale info were used. First mesh would distort itself and try to smooth the mesh first as possible as it could. However, once the quality of mesh on the faces across the limit of 0.7 or in case of volume reached is beyond the 0.9, remeshing will be done. The smoothing algorithm will work and beyond these limits of the mesh quality, remeshing will take place. The solution methods used as follows: For the spatial discretization, coupled scheme is used, and least square cell-based gradients are used for this solution. Pressure is discretized in second-order upwind scheme, and for transient formulation, first-order implicit method is used. The mesh is tetrahedral type elements and time step is set to 0.001 s. Fluid solid interface is set to system coupling region, the blood is enter from the bottommost portion of artery. The coupling system transfers converged values of the fluid model solver to the structural model solver. The interpolation of pressure (obtained from fluid solver) is the input data of structural model solver. The converged value of structural analysis is transfer to the fluid model solver. The interpolation of displacement (obtained from transient structural analysis) will be the input data of fluid model solver. This process will continue till the end of the time step.

3.1 Numerical Validation

Gholipour et al. [7] had developed a simple model of coronary artery. Their study has demonstrated the effects of three layers of artery and its hyperelasticity where blood displayed a non-Newtonian behavior. They discovered that the maximum wall shear stress acts at the tip of the plaque and its value is 145 Pa. We recreated the same model and corresponding factors to get the results for the present study. The wall shear stress distribution with respect to time for Gholipour et al. and present work is plotted in Fig. 2b. Figure 2b shows that current result (wall shear stress distribution) is in good agreement with that of Gholipour et al. and hence, our numerical model is validated. For CCA geometry, the mesh element size is the same in both fluid

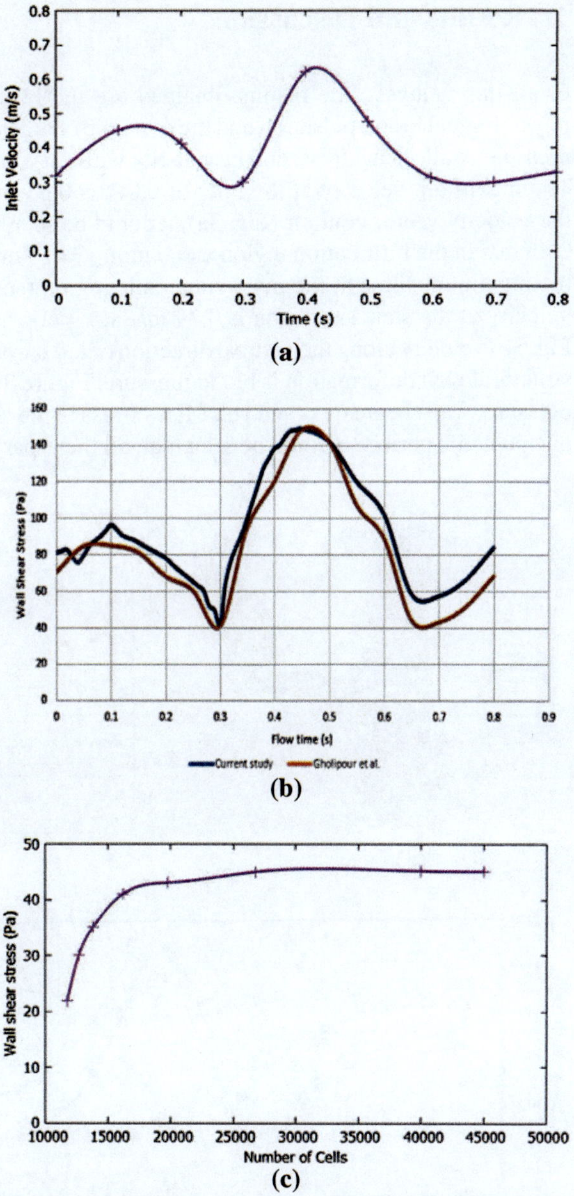

Fig. 2 **a** Inlet velocity profile perktold et al. [7]. **b** Numerical validation of WSS with Gholipour et al. [8]. **c** Grid independency study

and structural analysis with the tetrahedral mesh-type method. The systematic grid independency study is done by changing the grid count. Figure 2c shows that for a representative accuracy of WSS is attained with the cell numbers 28915. This study reveals that the precision will not be increased even if the cell number increased beyond the above-mentioned cell number.

4 Results and Discussions

From this analysis, the results obtained are the blood vessel deformation due to physiological blood pulsation and the dynamic characteristics of blood such as blood pressure, wall shear stress, and its velocity with varying time for one pulsating period The maximum velocity at the inlet of CCA is 0.65 m/s as shown in Fig. 2a. From the velocity vector contour (Fig. 3a), it could be seen that the velocity is reduced to 0.45 m/s in the bifurcation region $x = 0$ mm, $y = 15$ mm, $z = 0$ mm. Velocity attains the maximum value at the most vulnerable region (stenosis region). The maximum velocity at the stenosis region is 0.79 m/s at $x = 0$, $y = 20$ mm, $z = 0$ as shown in Fig. 3b (velocity along the vertical direction of left carotid artery). The main source of structural wall deformation is blood pressure. Figure 3c shows the variation of blood pressure, wall shear stress and von Mises stress in the CCA bifurcation region during one pulsatile period. It could be seen that, all these parameters attain their respective

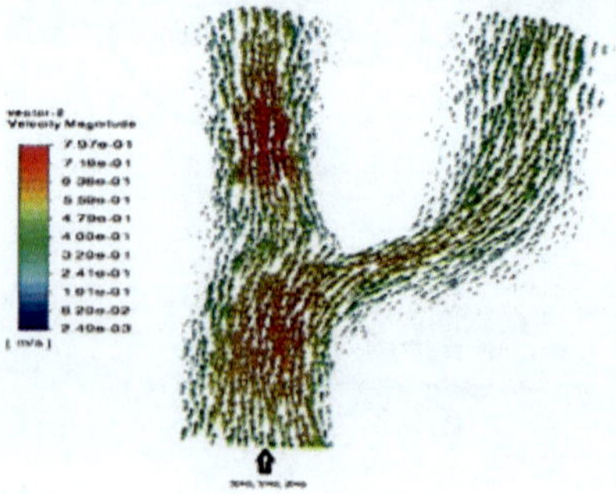

(a)

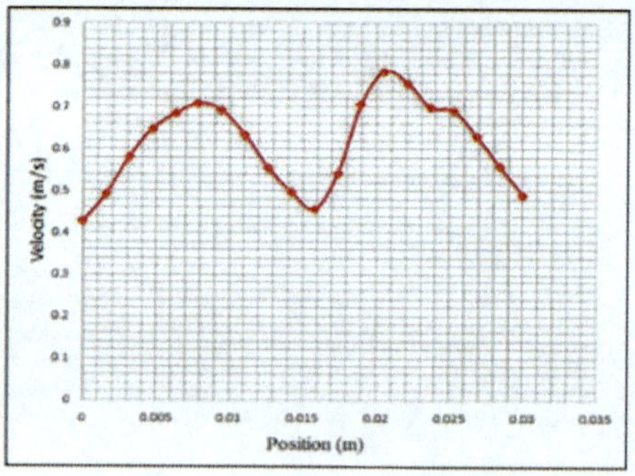

(b)

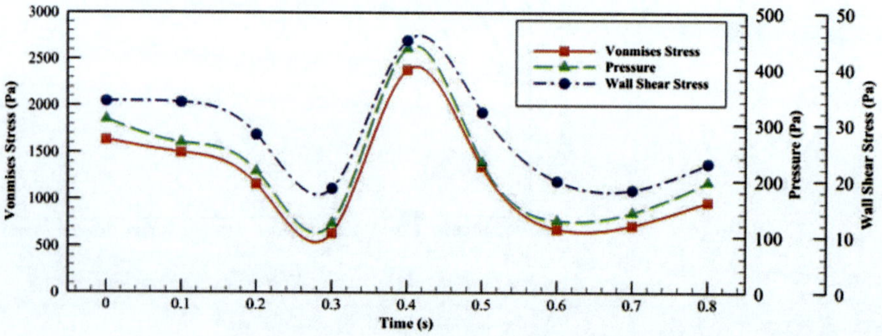

(c)

Fig. 3 **a** Velocity vector contour ($t = 0.4$ s). **b** Velocity position (vertical graph $x = 0$, $y = 15$ mm, $z = 0$) **c** Numerical results of wall shear stress ($t = 0.4$ s, $x = 20$ mm, $y = 20$ mm and $z = 0$), effective von Mises stress ($t = 0.4$ s, $x = 2.715$ mm. $y = 15$ mm, $z = 0$), pressure ($t = 0.4$ s, $x = 2.715$, $y = 17$ mm, $z = 0$) with respect to time maximum values at $t = 0.4$ s. The presence of plaque reduces the blood pressure. The pressure contour illustrates that the minimum pressure region is at the maximum stenosis region or the middle portion of atherosclerotic plaque. The minimum blood pressure (9.57 pa) is obtained at $x = 2.175$ mm, $y = 20$ mm, $z = 0$ from Fig. 4a. The location of maximum pressure (403 Pa) is between the bifurcation wall and the point of initiation of stenosis region ($x = 2.175$ mm, $y = 15$ mm, $z = 0$ at $t = 0.4$ s) as marked in Fig. 4a. As mentioned before, temporal variation of wall shear stress is shown in Fig. 4b with the maximum of 45 Pa at $t = 0.4$ s, and $x = 1.675$ mm, $y = 20$ mm, $z = 0$. The highest wall shear stress occurs at the neck region (stenosis region) or the tip of the plaque which could cause the fibrous cap to rupture. The von Mises stress is used to predict yielding of materials under complex loading [13]. In this study, the source of the load is high blood pressure. The pulsating blood flow exerts some forces on the blood vessels layer by layer. Due to this, the blood vessel undergoes a vascular dilation and subsequent contraction because of its hyperelastic nature. There are three layers such as intima, media, and adventitia wherein each one has different elastic properties. In these layers, adventitia is more rigid than the other two. Media layer more flexible than other two layers and the innermost layer is intima. The atherosclerotic plaque is situated in the intima layer. Von Mises stress contour in the intima layer is shown in Fig. 5a. The von Mises stress contour represents the maximum stress values at the maximum elongation and maximum contraction stages of blood vessel. Variation of average value of von Mises stress (effective equivalent stress) with respect to time is shown in Fig. 3c which shows that the maximum averaged value is 2480 pascals at $t = 0.4$ s and $x = 2.715$ mm, $y = 17$ mm, $z = 0$. From these results, it could be inferred that the blood flow is greatly affected by the presence of atherosclerotic plaque. Because of the plaque, the value of von Mises stress is minimum in posterior regions (media, adventitia). In general, the vascular dilation and contraction of blood vessel regulate the blood flow. It could be seen that the von Mises stress value attains a minimum for all the layers as shown in Fig. 5b, c in their respective plaque regions. The self-contraction of the media layer is responsible for the blood flow from CCA to capillaries. The presence of atherosclerosis causes a deficiency in active contraction or elongation of blood vessel layers.

Due to this problem, the blood does not reach the cells and organs in the human body and causes ischemic stroke. Falk shaw et al. [14] conducted an experimental study which reveals that plaque rupture condition range is about 45–50% of vessel stenosis level. In the current study, 45% of stenosis level condition of blood vessel is considered which denotes a possible rupture condition of plaque. Therefore, the present results are expected to provide valuable insight in to the diagnostics of atherosclerosis, by analyzing various stenosis conditions, which helps to diagnosis in atherogenesis.

Fig. 4 **a** Pressure distribution (contour) in of CCA bifurcation at $t = 0.4$ s. **b** Wall shear stress distribution (contour) in CCA bifurcation at $t = 0.4$ s

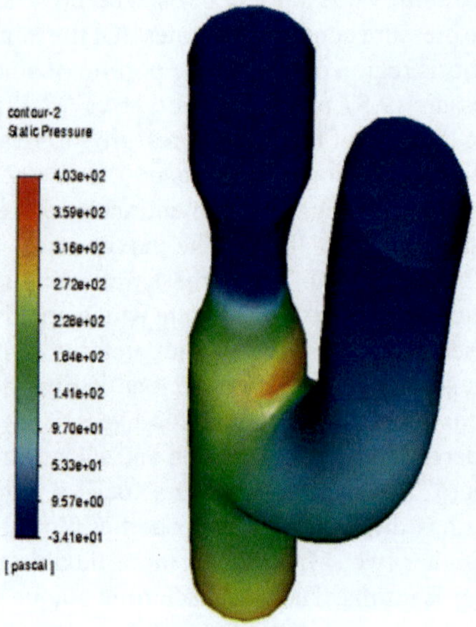

(a)

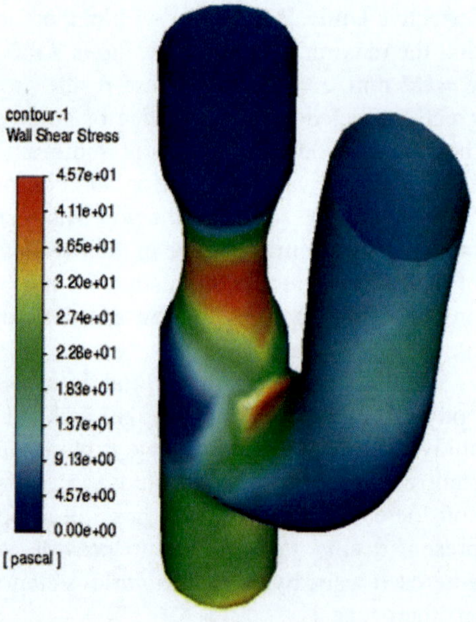

(b)

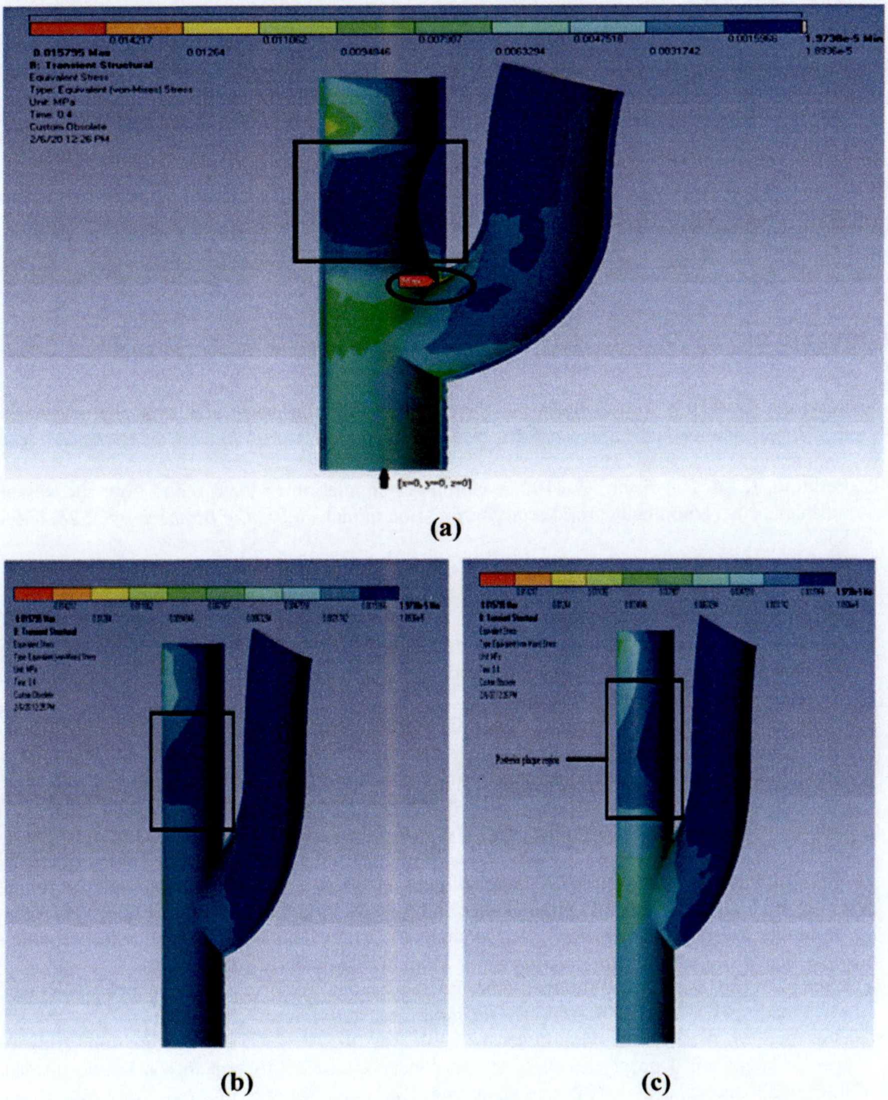

Fig. 5 **a** Von Mises stress in intima layer at $t = 0.4$ s, **b** von Mises stress in media layer $t = 0.4$ s, and **c** von Mises stress at adventitia at $t = 0.4$ s

5 Conclusion

This work succeeded to develop a numerical model of an atherosclerotic carotid artery. The model developed is expected to help in the examination of the effect of different factors on a diseased carotid artery. More specifically, the model for the artery and plaque could help to predict the initiation of stroke, or brain ischemia accurately. The blood pulsation, non-Newtonian flow of blood, and three layers of an artery are essential for predicting the consequences of atherosclerosis artery. These parameters that exert their influences on the atherosclerotic carotid artery can be considered as a realistic representation of plaque behavior. Results indicate that stenosis condition of blood flow causes a reduction in blood pressure by 54%. In the stenosis region, von Mises stress is minimum. Results further indicate that the maximum wall shear stress acts on the fibrous cap (plaque tip).

References

1. Perktold, K., Resch, M., & Florian, H. (1991). Pulsatile non-Newtonian flow characteristics in a three-dimensional human carotid bifurcation model. *ASME Journal of Biomechanical Engineering, 113*, 464–475.
2. Perktold, K., & Rappitsch, G. (1995). Computer simulation of local blood flow and vessel mechanics in a compliant carotid artery bifurcation model. *Journal of Biomechanics, 28*, 845–856.
3. Urquiza, S. A., Blanco, P. J., & Venere, M. J. (2005). Multidimensional modelling for the carotid artery blood flow. *Computer Methods in Applied Mechanics and Engineering, 195*(2006), 4002–4017.
4. Zhao, S. Z., Xu, X. Y., Hughes, A. D., & Thom, S. A. (2000). Blood flow and vessel mechanics in a physiologically realistic model of a human carotid arterial bifurcation. *Journal of Biomechanics, 33*(2000), 975–984.
5. Gharahi, H., Zambrano, B. A., Zhu, D. C., DeMarco, J. K., & Baek, S. (2016). Computational fluid dynamic simulation of human carotid artery bifurcation based on anatomy and volumetric blood flow rate measured with magnetic resonance imaging. *Int J Adv Eng Sci Appl Math, 8*(1), 46–60. https://doi.org/10.1007/s12572-016-0161-6.
6. Sousa, L. C., Castro, C. F., António, C. C., & Azevedo, E. (2009). Fluid-structure interaction modeling of blood flow in a non-stenosed common carotid artery bifurcation. In *Proceedings of the 7th International Conference on Mechanics and Materials in Design*. Albufeira/Portugal 11–15 June 2017. Editors J.F. Silva Gomes and S.A. Meguid. Publ. INEGI/FEUP (2017)
7. Gholipour, M. H., Ghayesh, A. Z., & Mahajan, R. (2018). Three-Dimensional Biomechanics of coronary artery. *International Journal of Engineering Science, 130*, 93–114.
8. Raja, R. S. (2012). Coupled fluid structure interaction analysis on a cylinder exposed to ocean wave loading. CHALMERS, *Applied Mechanics*, Master's Thesis 2012, 55
9. Holzapfel, G. A., Sommer, G., Gasser, C. T., & Regitnig, P. (2005). Determination of layer-specific mechanical properties of human coronary arteries with nonatherosclerotic intimal thickening and related constitutive modeling. *American Journal of Physiology-Heart and Circulatory Physiology, 289*, H2048–H2058.
10. Chan, W., Ding, Y., & Tu, J. (2007). Modeling of non-Newtonian blood flow through a stenosed artery incorporating fluid-structure interaction. *Anziam Journal, 47*, 507–523.
11. Mandal, M. S., Mukhopadhyay, S., & Layek, G. (2012). Pulsatile flow of an incompressible, inhomogeneous fluid in a smoothly expanded vascular tube. *International Journal of Engineering Science, 57*, 1–10.
12. Cilla, M., Borrás, I., Pena, E., Martínez, M., & Malve, M. (2015). A parametric model for analysing atherosclerotic arteries: On the FSI coupling. *International Communications in Heat and Mass Transfer, 67*, 29–38.
13. Lian, H., Christiansen, A. N., Tortorelli, D. A., Sigmund, O., & Aage, N. (2017). Combined shape and topology optimization for minimization of maximal von Mises stress. *Structural and Multidisciplinary Optimization, 55*, 1541–1557. https://doi.org/10.1007/s00158-017-1656-x
14. Falk, E., Shah, P. K., & Fuster, V. (1995). Coronary plaque disruption. *Circulation, 92*, 657–671.

Online document

15. https://www.cookmedical.com/peripheral-intervention/health-and-wellness-ways-to-prevent-or-ease-the-symptoms-of-pad/

Effect of Substrate Inclination on Post-impact Dynamics of Droplets

N. Sahoo, D. Samanta, and P. Dhar

Nomenclature

- v Velocity of droplet (m/s)
- d Diameter of droplet (m)
- t Time

Greek Symbols

- ρ Density (dimensionless)
- β Elongation factor (dimensionless)
- ϕ Tilted angle

Subscripts

- i Pre-impact
- l Liquid

N. Sahoo (✉) · D. Samanta
Department of Mechanical Engineering, Indian Institute of Technology Ropar, Ropar, Punjab 140001, India

P. Dhar
Department of Mechanical Engineering, Indian Institute of Technology Kharagpur, Kharagpur, West Bengal 721302, India

P. Mahanta et al. (eds.), *Advances in Thermofluids and Renewable Energy*, Lecture Notes in Mechanical Engineering, https://doi.org/10.1007/978-981-16-3497-0_3

1 Introduction

Drop impact dynamics on surfaces is a complex phenomenon due to interplay of fluid mechanics, surface physics, and interfacial chemistry. This impact process exhibits a diverse set of intriguing behavior depending upon the wettability of the solid surface with respect to the fluid [1, 2] and the fluid properties [3–6]. Oblique impact studies may be useful in applications like spraying of pesticides at random angles and impact of supercooled water drops on airplane leading to frost formation [7–9]. In IC engines too, fuel injection system sprays fuel obliquely toward the combustion zone [10]. In addition, typical applications such as spray coating, painting, cooling, or inkjet printing are more likely to occur at oblique angles instead of normal impact. Sikalo et al. [11] reported asymmetry in the spreading factor, i.e., the differences in the deformation at the front and back of the droplet after impact. Chiarot et al. [12] and Zheng et al. [13] studied grazing impact of high velocity continuous drop streams on inclined SH surfaces and observed that the shape and structure of the rebounding stream are influenced by the frequency of the drop ejection and velocity. Yeong et al. [14] investigated the dependence of drop dynamics on Weber number. Antonini et al. [15] distinguished six distinct impact regimes at wide range of We for drop impact on tilted hydrophobic and superhydrophobic (SH) substrates. LeClear et al. [16] observed the transition from the superhydrophobic Cassie–Baxter regime to the fully wetted Wenzel regime while studying the impact of water drops on inclined textured SH surfaces.

While a lot of attention has been paid to the spreading dynamics on SH surfaces, the dynamics of lift off of the drops upon inclined surfaces are rarely studied. Therefore, the present study highlights the role of non-dimensional numbers like Reynolds number (Re) and Weber number (We) and surface inclination upon impact of droplets. Experiments have been carried out with different fluids to observe the post-impact elongation characteristics of droplets and the dynamics vis-à-vis physical properties.

2 Methodology

2.1 Experimental Details

The experimental setup consists of a drop dispenser controller (Holmarc Opto-Mechantronics Pvt. Ltd., India) that maintains constant volume of droplet discharged from a syringe pump as shown in Fig. 1. The drop impact images were recorded using a high speed camera (Photron FASTCAM SA4) mounted with a G-type AF-S macrolens of focal length 105 mm (Nikkor, Nikon). The images were taken at 1024 × 1024 pixels resolution at 3600 frame per second. Experiments were performed at ambient conditions (25 °C) on SH surfaces. For impact study, sterile glass slides were thoroughly cleaned with acetone and DI water and then dried in hot air oven. The SH surfaces were created on similar glass substrates using superhydrophobic spray

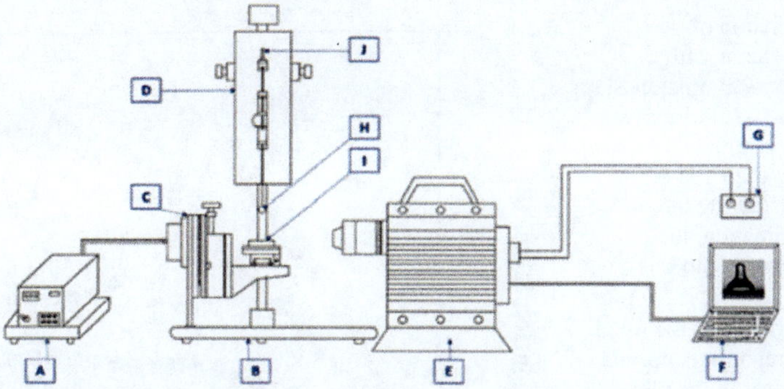

Fig. 1 Schematic of the experimental setup [17] (A) drop dispenser controller (B) base (C) substrate inclination apparatus with backlight arrangement (D) syringe pump (E) high-speed camera (F) laptop (G) power source (H) drop (I) target surface (J) syringe

coating (Ultra Tech International Inc., USA). In the present study, three fluids such as DI water, SiO_2 water nanocolloids (2.5 and 5% wt.), and different concentrations of sodium dodecyl sulfate (SDS, Merck, India) dissolved in DI water (0.25 and 0.50% of the critical micelle concentration (CMC)) were used as test fluids.

3 Results and Discussions

The present study focuses on experiments with different test fluids, highlighting the role of surface inclination and Weber number (We) and Reynolds number (Re). The Weber number has been defined based on the initial velocity (v_i), density (ρ_l), diameter (d_i), and fluid surface tension (σ_l) (Subscripts *i*, l stand for initial condition before impact and liquid, respectively). The Weber number (We) was varied between 10 and 127 (We = 10–89 for water, We = 12–127 for sodium dodecyl sulfate (SDS) surfactant solutions (0.25 and 0.5 wt % of critical micelle concentration (CMC), and We = 10–89 for aqueous solutions of silica particles (2.5 and 5 wt %)). Similarly, the Reynolds number based on initial velocity is defined as Re = . Re was varied from 28 to 4873, (Re = 1624–4873 for water, Re = 28–1105 for silica colloids and Re = 1411–4587 for surfactant solutions). The initial diameter of droplets of DI water, 0.25 and 0.5 wt. % CMC are considered as 2.9, 2.71, and 2.52 mm, respectively. The initial diameters of 2.5 and 5 wt. % of silica colloidal solutions are almost similar to the initial diameters of water droplets. The measurement uncertainty of initial diameter of droplets for test fluids is within 5% error.

Figure 2 elucidates the elongation dynamics of post-impact drops on different inclination angles on SH surfaces. The elongation factor (β) is expressed as the ratio of height of post-impact drop bouncing off from the target to initial drop diameter. The non-dimensional time (τ) is expressed as the ratio of the product of time (t) at which post-impact image is considered and initial velocity (v_i) to the initial droplet diameter

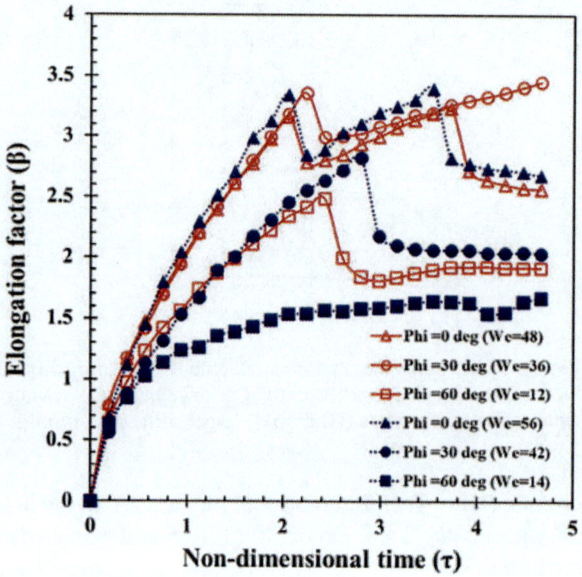

Fig. 2 Variation of elongation factor with non-dimensional time on SH surface. The test fluids are surfactant solutions considering 0.25 and 0.5 CMC. Considering the surface inclination, the normal Weber number is defined as $We_n = \rho_l v_i^2 d_i$ (cos $\Phi)^2/\sigma_l$, where ϕ is the tilted angle. We represents normal Weber number in the plot

(d_i). It is observed that the elongation factor for droplets of 0.25 and 0.5 wt. % CMC solution increases up to certain time interval followed by a sharp decline. The sharp decline occurs due to emission of secondary droplets during vertical acceleration of the droplets after impact. The vertical momentum and reduction in interfacial energy promotes pinching off event (Fig. 3). At $\phi = 0°$, for 0.25 CMC surfactant solution, formation of secondary droplet twice at time $t = 13.77$ and 18.63 ms corresponds to twice decrease in β (Fig. 2).

In case of water at initial impact velocity $v_i = 1$ m/s, elongation factor follows the same trend as droplets of surfactant solutions (Fig. 4). The emission of secondary droplets occurs at $\phi = 0°$, but this behavior is diminished with increasing surface inclinations. The reduction in normal momentum to the oblique plane inhibits the formation of secondary droplet formation while lifting off from the surface.

Figure 5 investigates the elongation dynamics of three test fluids such as water, silica colloid solutions and surfactant solutions at We = 40 and $\phi = 0°$ for SH surfaces. It is found that the secondary droplet is formed earlier in case of surfactant solutions compared to water and SiO_2 colloidal solution. This is obvious due to the fact that surfactants reduce the interfacial energy, thereby promoting the formation of secondary droplets.

4 Conclusion

The present study investigates the temporal variation of elongation factor using different test fluids on tilted SH surfaces. It is observed that the elongation factor

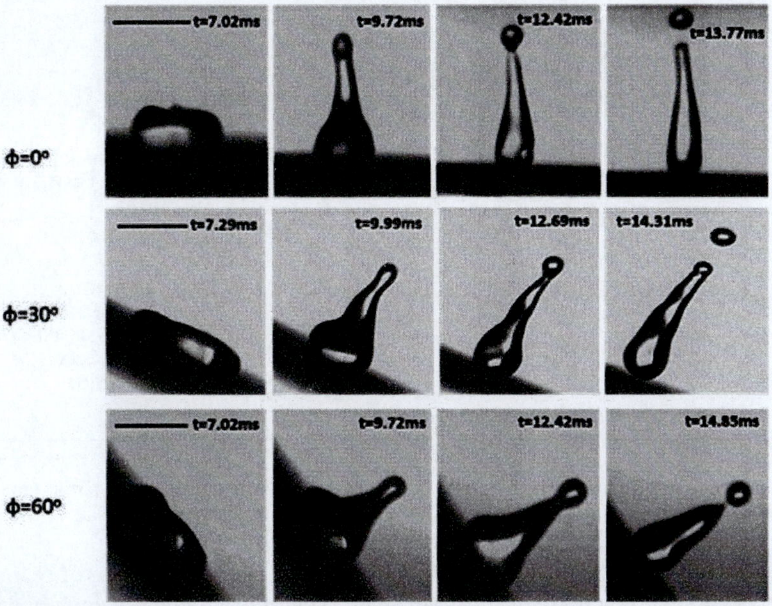

Fig. 3 Post-impact droplet images of surfactant solution (0.25 CMC) at We = 40 on SH surfaces. The magnitude of scale bar is equal to 2.71 mm for each figure

Fig. 4 Variation of elongation factor with non-dimensional time on SH surface. The test fluid is water and We represents the normal Weber number. Inset (a–c) elongation dynamics of water drops post-impact on horizontal SH surface at $t =$ 16.2 ms

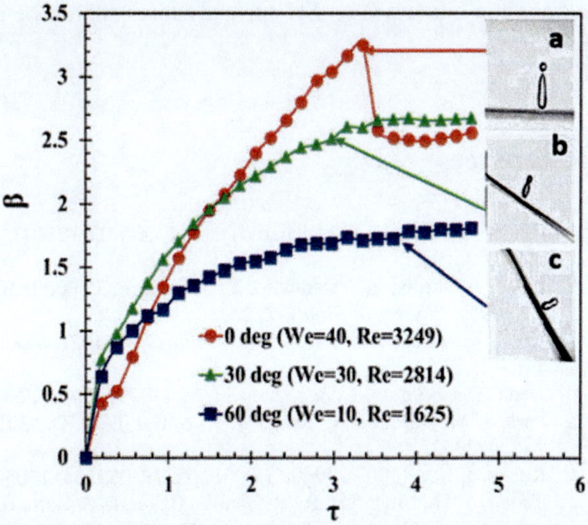

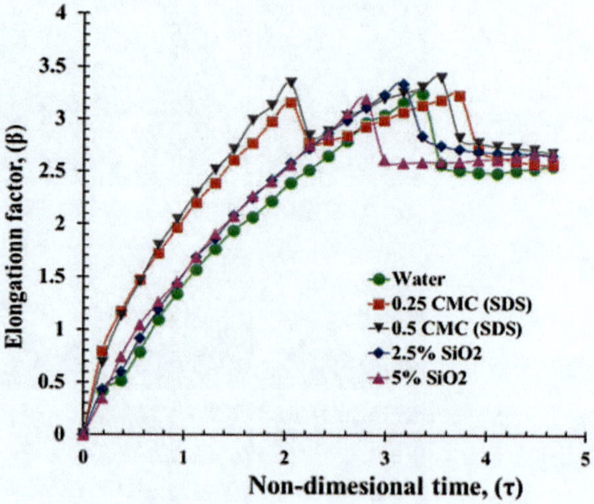

Fig. 5 Elongation dynamics of different test fluids at $\phi = 0°$ and We = 40 fo SH surface

decreases with increase in surface inclination. Again, the suppression of secondary droplet occurs with increase in substrate angle. The time at which secondary droplet departs from the original droplet depends upon the fluid properties. The reduction in normal momentum and interfacial energy are the two key parameters for the above events. This study enables to find the critical parameters for comprehensive study of elongation outcomes on SH surfaces.ables should be pasted within the text column as follows.

References

1. Clanet, C., Béguin, C., Richard, D., & Quéré, D. (2004). *Journal of Fluid Mechanics, 517*, 199–208.
2. Fukai, J., Shiiba, Y., Yamamoto, T., Miyatake, O., Poulikakos, D., Megaridis, C. M., & Zhao, Z. (1995). *Physics of Fluids, 7*, 236.
3. Koch, K., & Barthlott, W. (2009). *Philosophical Transactions of the Royal Society A, 367*, 1487–1509.
4. Ueda, T., Enomoto, T., & Kanetsuki, M. (1979). *Bullet JSME, 22*, 724–732.
5. Park, J. Y., Gardner, A., King, W. P., & Cahill, D. G. (2014). *The Journal of Heat Transfer, 136*, 092902.
6. Roisman, I. V. (2009). *Physics of Fluids, 21*, 052104–1–052104–11.
7. Yeong, Y. H., Mudafort, R., Steele, A., Bayer, I., & Loth, E. (2012). In *4th AIAA Atmospheric and Space Environments Conference*, New Orleans, Louisiana, 2012.
8. Politovich, M. K. (1989). *The Journal of Applied Meteorology, 28*, 856–868.
9. Massinon, M., & Lebeau, F. (2012). *Biosystems Engineering, 112*, 56–64.
10. Chen, R., Chiu, S. L., & Lin, T. H. (2007). *Experimental Thermal and Fluid Science, 32*, 587–595.
11. Šikalo, Š, Tropea, C., & Ganic, E. N. (2005). *Journal of Colloid and Interface Science, 286*, 661–669.
12. Chiarot, P. R., & Jones, T. (2010). *Experiments in Fluids, 49*, 1109–1119.

13. Zheng, L., Li, Z., Bourdo, S., Khedir, K. R., Asar, M. P., Ryerson, C. C., & Biris, A. S. (2011). *Langm, 27*, 9936–9943.
14. Yeong, Y. H., Burton, J., & Loth, E. (2014). *Langm, 30*, 12027–12038.
15. Antonini, C., Villa, F., & Marengo, M. (2014). *Experiments in Fluids, 55*, 1713–1–9.
16. LeClear, S. LeClear, J., Abhijeet, Park, K. C., & Choi, W. (2016). *Journal of Colloid and Interface Science, 461*, 114–121.
17. Sahoo, N., Khurana, G., Harikrishnan, A. R., Samanta, D., & Dhar, P. (2021). Post impact droplet hydrodynamics on inclined planes of variant wettabilities. *European Journal of Mechanics/B Fluids, 79*, 27–37.

Modelling and Analysis of Dusty Fluid Flow Past a Vertical Surface with Exothermic and Endothermic Kind of Chemical Reactions

Debasish Dey and Barbie Chutia

Nomenclature

B	Pre-exponential factor simply prefactor (constant)
B_0	Magnetic field
C_f	Friction factor
C_p	Dust concentration
c_p, c_m	Specific heat of fluid and dust particles, respectively
D_m	Coefficient of mass diffusivity
d	The radius of the dust particles
E	Activation energy (dimensionless)
E_a	Activation energy
$Ec = \frac{u_0^2}{(T_w - T_\infty)C_p}$	Eckert number
$G_0\left(= \frac{g\beta_1\delta^2 T_\infty}{U_0\nu}\right)$	Grashof number
$G_m\left(= \frac{g\beta_2\delta^2 \Delta C}{U_0\nu}\right)$	Modified Grashof number
g	Acceleration due to gravity
$H\left(= \frac{\delta\delta'}{\nu}\right)$	Constant
K	Stokes' resistance coefficient
k_r	The chemical reaction rate constant
k^*	Mean absorption coefficient
L	Length scale for a steady flow
$M\left(= \frac{\sigma B_0^2\delta^2}{\mu}\right)$	Magnetic interaction parameter
M_w	Mass flux
m	Mass of the dust particle
$N \rightarrow$	Radiation parameter (dimensionless)
N_0	The number density of dust particles

D. Dey (✉) · B. Chutia
Department of Mathematics, Dibrugarh University, Dibrugarh, Assam 786004, India

P. Mahanta et al. (eds.), *Advances in Thermofluids and Renewable Energy*, Lecture Notes in Mechanical Engineering, https://doi.org/10.1007/978-981-16-3497-0_4

$N_1 = (\Delta C/(\rho c_p \Delta T))$	The ratio of concentration/temperature
N_u	Nusselt number
n	Unit less constant exponent fixed rate constant
$\Pr = \frac{\rho \nu C_p}{\kappa}$	Prandtl number
q_r	Radiative heat flux
q_w	Heat flux
$R\left(= \frac{N_0 m}{\rho}\right)$	Particle concentration parameter
Re	Reynolds number
$S = \frac{KN_0\delta^2}{\mu}$	Permeability parameter
$S_c\left(= \frac{\nu}{D_m}\right)$	Schmidt number
S_h	Sherwood number
T, T_p	The temperature of fluid and dust particles, respectively
$T_w,\ T_\infty$	The temperature at the wall and a large distance, respectively
t	Time
U_0	Uniform velocity
u, v	Velocity components of fluid phase
u_p, v_p	Velocity components of dust phase
v_0	Suction

Greek Symbols

$\alpha\left(= \frac{\delta^2}{\upsilon \tau_T}\right)$	Temperature relaxation time parameter
$\alpha_1\left(= \frac{\delta^2}{\upsilon \tau_c}\right)$	Concentration relaxation time parameter
$\alpha_2\left(= \frac{\delta^2}{\upsilon \tau_v}\right)$	Velocity relaxation time parameter of dust particles
$\beta(\pm 1)$	Exothermic/Endothermic parameter
$\beta_1,\ \beta_2$	The coefficients of volume expansions for temperature and concentration, respectively
$\gamma = (T_w - T_\infty)/T_\infty$	Temperature relative parameter
$\delta(t)\left(= \sqrt{2H\nu t + L^2}\right)$	Length scale for unsteady flow
$\varepsilon = \frac{1}{\phi_1}$	Volume fraction parameter for dust phase (dimensionless)
$\bar{\varepsilon} = \frac{1}{1-\phi_1}$	Volume fraction parameter for fluid phase (dimensionless)
η	Similarity variable
θ	The temperature of the fluid (dimensionless)
θ_p	Dust temperature (dimensionless)
κ	Thermal conductivity
λ^2	Chemical reaction rate constant (dimensionless)
μ	Dynamic viscosity of the fluid
ν	Kinematic viscosity of the fluid

ρ	Density of the fluid
ρ_{p}	Density of dust phase,
$\rho_{\mathrm{r}}\left(=\frac{\rho_{\mathrm{p}}}{\rho}\right)$	Relative density
σ	Electrical conductivity of the fluid
σ^*	Stefan Boltzmann constant
τ	Shear Stress
τ_{c}	Concentration equilibrium time
τ_{T}	Thermal equilibrium time
τ_{v}	Relaxation time of the dust particles
ϕ	Concentration of the fluid (Dimensionless)
ϕ_1	Volume fraction of the dust particles
ϕ_p	Dust concentration (Dimensionless

Superscripts

n	Nth power

Subscripts

p	Dust phase

1 Introduction

The study of dust particles on viscous fluid flows has many applications in various fields such as petroleum industry, crude oil purification, soil erosion process, blood flows through arteries and paint spraying. The equations for dusty fluid flow have been formulated by Saffman [1]. Dusty fluid flow past a semi-infinite plate has been studied by Dutta and Mishra [2]. Study of heat transfer of memory two-phase fluid flow past a horizontal surface has been carried out by Dey and Baruah [12]. Analysis of the MHD flow of two-phase fluid with dust particles in the presence of suction has been discussed by Vajravelu and Nayfeh [3]. Singh and Singh [4] have analysed MHD flow of free convective dusty fluid flow under the influence of volume fraction past through parallel vertical plates. Dey [5, 6] discussed the presence of volume fraction in a hydromagnetic dusty fluid flow. Many researchers have illustrated dusty fluid flow under the influence of chemical reactions which has a wide range of importance in different areas like cooling effects of air conditioners, environmental pollutants,

MHD generators, pumps, and many more. Numerical study of heat and mass transfer of study dusty fluid flow undergoing chemical reactions has been investigated by Izani and Ali [7]. Sagar and Jha [8] have formulated the problems of two-phase fluid flow with the effects of chemical reactions.

Many systems undergoing chemical reactions with finite activation energy are useful in geothermal and oil reservoir industries. Nowadays, several scientists have been carried the problems of fluid flow under the influence of chemical reactions along with Arrhenius activation energy. The term activation energy was first given by Svante Arrhenius [9] in 1889. The modified Arrhenius law, $k_r^2 = B\left(\frac{T}{T_\infty}\right)^n \exp\left(-\frac{E_a}{kT}\right)$. It is the minimum energy required to start a chemical reaction and essential to proceed with a reaction at a reasonable rate. The problems of mass transfer of fluid flow with the involvement of different types of chemical reactions are tough to solve. But if it is limited to binary type, then a lot of problems can be solved. Chemical reactions can be categorized as exothermic and endothermic kinds of reactions based on net heat transfer. An exothermic reaction is a spontaneous process in which heat releases from the body, while in endothermic reactions, heat is absorbed and energy must be added for the reaction to happen. Bestman [10] has formulated the problem of binary chemical reactions in boundary layer flow with the effects of activation energy. Later, the impact of chemical reactions on unsteady flow with nth order Arrhenius energy past a vertical surface has been studied by Makinde et al. [11]. Recently, Maleque [13] has discussed the heat and mass transfer of unsteady fluid flow past a porous vertical surface with the effects of activation energy, binary reactions and thermal radiation. The present study of our paper is the extension work of Maleque to the dusty fluid flow. There has not been any work done on dusty fluid flow with the effects exothermic/endothermic to our knowledge. We have discussed the influence of exothermic/endothermic chemical reactions along with activation energy on dusty fluid flow with volume fraction, thermal radiation, suction and magnetic fields. The problem has been solved numerically using MATLAB built-in Bvp4c solver scheme. The results of the flow patterns are shown graphically and coefficient of skin friction, heat transfer rate and mass transfer rate are presented through tables.

2 Mathematical Formulation of the Problem

Let us consider an unsteady hydromagnetic two-phase fluid flow with dust particles past through a vertical porous surface. The boundary wall is assumed to be extended infinitely such that all the variables are independent of x. The flat surface/plate is taken along x-axis, and y-axis is taken perpendicularly to the plate which is shown in Fig. 1. The dusty fluid flow with volume fraction under the influence of exothermic/endothermic kinds of chemical reactions along with activation energy, thermal radiation and uniform magnetic field (B_0) is forced perpendicular to the plate which is moving in uniform velocity (U_0).

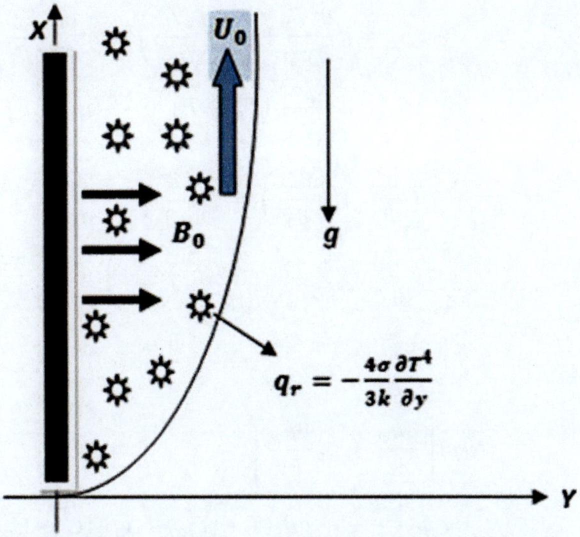

Fig. 1 Flow configuration of the problem

Some of the assumptions are taken into consideration for the dust particles such as dust particles are assumed to be undeformable, non-conducting, spherically shaped, uniform in size and density is taken as constant during the flow. Induced magnetic field is negligible compared to applied magnetic field as we have considered the fluid to be conducting.

3 Equations

Considering the above assumptions, the governing equations of the following problem are as follows:

$$\frac{\partial v}{\partial y} = 0 \tag{1}$$

$$(1-\phi_1)\left[\frac{\partial u}{\partial t} + v\frac{\partial u}{\partial y}\right] = (1-\phi_1)\left[v\frac{\partial^2 u}{\partial y^2} + g\beta_1(T-T_\infty) + g\beta_2(C-C_\infty) - \frac{\sigma B_0^2 u}{\rho}\right] + \frac{KN_0}{\rho}(u_p - u) \tag{2}$$

$$\rho c_p\left(\frac{\partial T}{\partial t} + v\frac{\partial T}{\partial y}\right)$$

$$= \kappa \frac{\partial^2 T}{\partial y^2} + \rho k_r^2 \left(\frac{T}{T_\infty}\right)^n \exp\left[\frac{-E_a}{k' T}\right](C - C_\infty)$$
$$+ \frac{\rho_p c_m}{\tau_T}(T_p - T) + \frac{\rho_p}{\tau_v}(u_p - u)^2 - \frac{\partial q_r}{\partial y} \tag{3}$$

$$\left(\frac{\partial C}{\partial t} + v \frac{\partial C}{\partial y}\right) = D_m \frac{\partial^2 C}{\partial y^2} + \frac{\rho_p}{\rho \tau_c}(C_p - C)$$
$$- k_r^2 \left(\frac{T}{T_\infty}\right)^n \exp\left[\frac{-E_a}{k' T}\right](C - C_\infty) \tag{4}$$

$$\frac{\partial v_p}{\partial y} = 0 \tag{5}$$

$$N_0 m\left[\frac{\partial u_p}{\partial t} + v_p \frac{\partial u_p}{\partial y}\right]$$
$$= \phi_1\left[\frac{\partial^2 u}{\partial y^2} + g\beta_1(T - T_\infty) + g\beta_2(C - C_\infty)\right] + K N_0 (u_p - u) \tag{6}$$

$$\rho_p c_m \left(\frac{\partial T_p}{\partial t} + v_p \frac{\partial T_p}{\partial y}\right)$$
$$= -\frac{\rho_p c_m}{\tau_T}(T_p - T) \tag{7}$$

$$\frac{\partial C_p}{\partial t} + v_p \frac{\partial C_p}{\partial y} = \frac{1}{\tau_c}(C - C_p) \tag{8}$$

The boundary conditions of the following problem are given as:

$$\left.\begin{array}{l} u = u_p,\ v = v(t),\ T = T_w,\ C = C_w; \\ y = 0,\ t > 0 \\ u \to 0,\ T \to T_\infty,\ C \to C_\infty,\ u_p \to 0,\ T_p \to T_\infty, \\ C_p \to C_\infty;\ y \to \infty, t > 0 \end{array}\right\} \tag{9}$$

4 Method of Solution

We introduce the following similarity transformations for converting the above governing equations to corresponding ordinary differential equations:

$$\left.\begin{array}{l} \eta = \frac{y}{\delta(t)},\ \frac{u}{u_0} = f(\eta),\ \frac{T - T_\infty}{T_w - T_\infty} = \theta(\eta), \\ \frac{C - C_\infty}{C_w - C_\infty} = \phi(\eta),\ \frac{u_p}{u_0} = F(\eta), \\ \frac{T_p - T_\infty}{T_w - T_\infty} = \theta_p(\eta),\ \frac{C_p - C_\infty}{C_w - C_\infty} = \phi_p(\eta), \end{array}\right\} \tag{10}$$

From Eqs. (1) and (5), we get $v(t) = -\frac{v_0 \nu}{\delta(t)}$ and $v_p(t) = -\frac{v_{p0}\nu}{\delta(t)}$ respectively, where $v_0 = v_{p_0}$.

Under Rosseland approximation, the value of the radiative heat flux vector is given by

$$q_r = \frac{4\sigma^*}{3k^*}\frac{\partial T^4}{\partial y} \tag{11}$$

Let us assume small temperature differences between fluid temperature T and free stream temperature T_∞, and expanding T^4 in Taylors series about T_∞ to get the linear function, we have

$$T^n = (1-n)T_\infty^n + nT_\infty^{n-1}T + \cdots \tag{12}$$

By substituting Eqs. (11), (12) in Eq. (3) and equations represented by (10) in Eqs. (2), (3), (4), (6), (7) and (8), we get the following non-linear ordinary differential equations:

$$f'' + (\eta H + v_0)f' - M\varepsilon f + S\varepsilon(F - f) + G_0\gamma_T\theta + G_m\gamma_c\phi = 0 \tag{13}$$

$$\theta'' + \frac{\Pr}{1+N}\left((\eta H + v_0)\theta' + \beta\lambda^2 N_1(1 + n\gamma_T\theta)\exp\left(-\frac{E_a}{1+\gamma_T\theta}\right)\phi \right.$$
$$\left. + \frac{c_m}{c_p}\rho_r\alpha\big(\theta_p - \theta\big) + E_c\rho_r\alpha(F - f)^2\right) = 0 \tag{14}$$

$$\phi'' + Sc(\eta H + v_0)\phi' - Sc\lambda^2(1 + n\gamma_T\theta)\exp\left(-\frac{E_a}{1+\gamma_T\theta}\right)\phi$$
$$+ Sc\rho_r\alpha_1\big(\phi_p - \phi\big) = 0 \tag{15}$$

$$f'' + R\bar{\varepsilon}\big(\eta H + v_{p_0}\big)F' + G_0\gamma_T\phi + G_m\gamma_c\phi + S\bar{\varepsilon}(f - F) = 0 \tag{16}$$

$$\big(\eta H + v_{p_0}\big)\theta_p' = \alpha\big(\theta_p - \theta\big) \tag{17}$$

$$\big(\eta H + v_{p_0}\big)\phi_p' = \alpha_1\big(\phi - \phi_p\big) \tag{18}$$

where a prime denotes differentiation with respect to η.

From boundary condition given in (9), we have the transformed non-dimensional boundary conditions as follows:

$$\left.\begin{aligned} &f(0) = 1,\ \theta(0) = 1,\ \phi(0) = 1,\\ &f(\infty) = 0,\ \theta(\infty) = 0,\ \phi(\infty) = 0,\\ &F(\infty) = 0,\ \theta_p(\infty) = 0,\ \phi_p(\infty) = 0. \end{aligned}\right\} \tag{19}$$

The quantities such as coefficient of skin friction, Nusselt number and Schmidt number have importance in these types of non-linear and engineering problems which are defined as:

$$\frac{1}{2}C_f = \frac{\tau}{\rho U_0^2} => \frac{1}{2}R_e C_f = -f'(0),$$
$$N_u = \frac{q_w \delta}{\kappa \Delta T} = -\theta'(0),$$
$$S_h = \frac{M_w \delta}{D_M \Delta C} = -\phi'(0)$$

where

$$\text{Wall skin friction}(\tau) = -\mu\left(\frac{\partial u}{\partial y}\right)_{y=0} = -\frac{\mu U_0}{\delta}f'(0),$$
$$\text{Heat flux at the wall}(q_w) = -\kappa\left(\frac{\partial T}{\partial y}\right)_{y=0} = -\frac{\kappa \Delta T}{\delta}\theta'(0),$$
$$\text{Mass flux at the wall}(M_w) = -D_M\left(\frac{\partial C}{\partial y}\right)_{y=0} = -\frac{D_M \Delta C}{\delta}\phi'(0).$$

5 Results and Discussion

In this section, we have discussed the numerical solutions of unsteady dusty fluid flow patterns with the effects of exothermic/endothermic parameter (β), activation energy parameter (E_a) and chemical reaction rate parameter (λ) through graphs and tables. The other parameters are taken as Gr = 10, Gm = 1, $N = 0.5$, Pr = 0.71, $M = 0.5$, Sc = 0.6, $\lambda = 5$, $n = 0.5$, $\nu_0 = 3$, $\gamma = 0.5$, $E_a = 1$, $\bar{\varepsilon} = 1.2$, $\varepsilon = 5$, Ec = 1, $R = 0.5$. Some of the parameters taken in the discussion part have physical interpretations. Positive values of Gr and Gm (Grashof numbers for heat and mass transfers) represent that temperature and concentration at the surface is greater than temperature and concentration at the free stream. Pr value '0.71' resembles the case of air, and Sc = 0.6 corresponds to water vapour.

Figure 2 describes the velocity profile of the fluid phase in the presence of chemical reaction rate for both the cases of exothermic ($\beta = 1$) and endothermic ($\beta = -1$)kinds of chemical reactions. From Fig. 2, it is seen that the reaction rate (in exothermic reactions) helps to accelerate the fluid motion. While for endothermic process, fluid motion slows down with a push up in the values of reaction rate.

Figure 3 depicts the velocity profile of the dust phase with the effects of chemical reaction rate for $\beta = 1$ and $\beta = -1$. We have observed that during the exothermic case, velocity of dust phase enhances along with fluid phase, but an opposite behaviour is seen during endothermic cases.

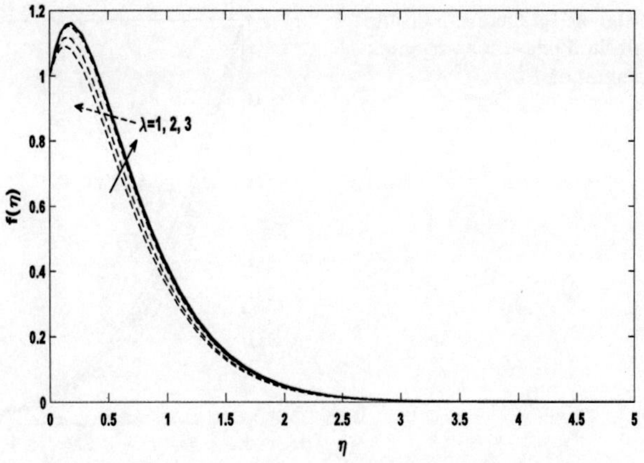

Fig. 2 Velocity profile of fluid phase for varying values of λ

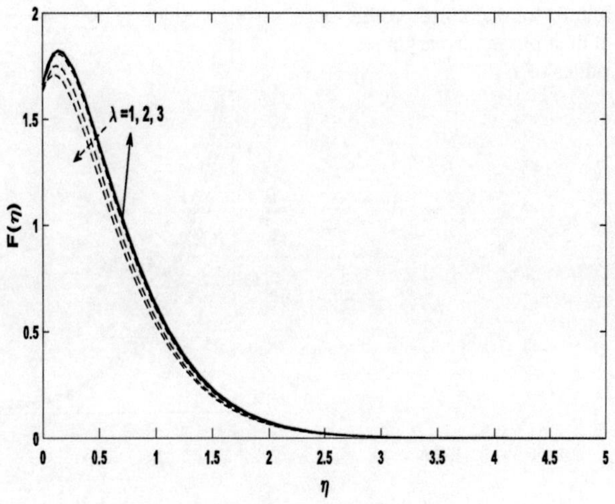

Fig. 3 Velocity profile of dust phase for varying values of λ

The relationship between temperature profiles of fluid and dust phases with activation energy parameter is shown for both kinds of reactions in Figs. 4 and 5, respectively. Activation energy parameter corresponds to the requirement of minimum energy to process a chemical reaction. During exothermic chemical reaction (release of energy in the form of heat and light), temperature of the fluid experiences retarding trend (Fig. 4), but an opposite phenomenon is experienced during endothermic chemical reaction (it absorbs heat from the surroundings). These physical phenomena are seen in both the figures. So, the results obtained from our work are in perfect agreement with the theoretical works. Further, these figures reflect that with the enhancement in the values of activation energy, the temperature of both the phases diminishes for $\beta = 1$. But the opposite trend is seen for $\beta = -1$.

Figure 6 shows the effects of E_a on concentration profiles of fluid phase for the cases $\beta = 1$ and $\beta = -1$ respectively. We have seen that there is deposition of fluid

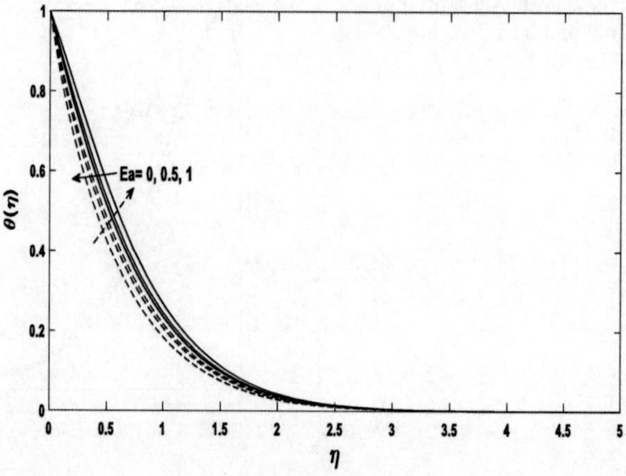

Fig. 4 Temperature profile of fluid phase for varying values of E_a

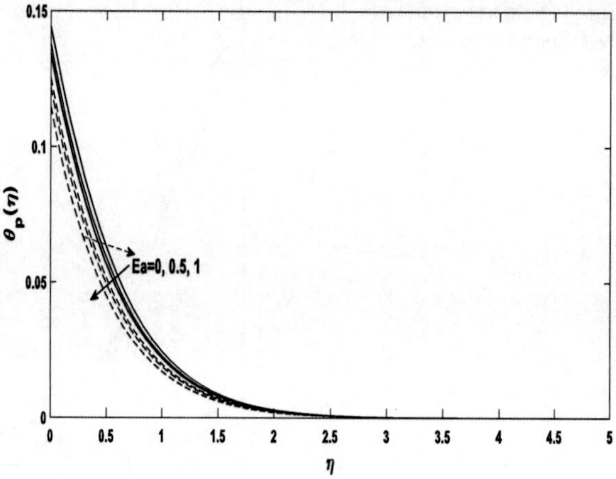

Fig. 5 Temperature profile of dust phase for varying values of E_a

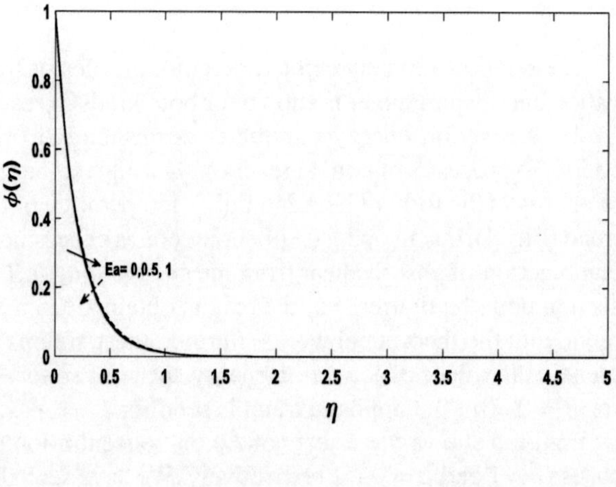

Fig. 6 Concentration profile of fluid phase for varying values of E_a

particles with the enhancement of activation energy parameter for exothermic cases ($\beta = 1$), but an opposite scenario is observed during endothermic cases ($\beta = -1$. A comparison of our work with Maleque [13] has been shown by Fig. 7, and it displays a good agreement with the published work.

Table 1 has shown the values of skin friction coefficient, rates of heat and mass transfers for various values of E_a (activation energy parameter) and λ (chemical reaction rate parameter). We have seen from the table that the coefficient of skin friction $f'(0)$ or viscous drag surface reduces with exothermic cases and a reverse picture is seen during endothermic cases. Also, the chemical reaction rate parameter

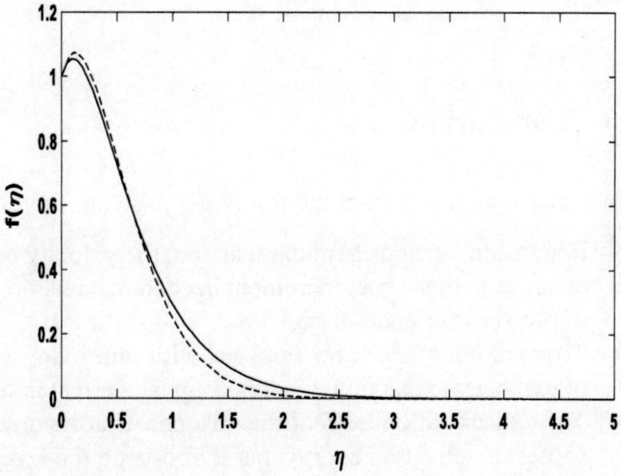

Fig. 7 Comparison of velocity profiles of the present study and Maleque for $\lambda = 5$

Table 1 Effects on the coefficient of skin friction, rate of heat transfer and rate of mass transfer for different values of E_a and λ for exothermic and endothermic reactions

β	$E = E_a$	$-f'(0)$	$-\theta'(0)$	$-\phi'(0)$
	0	2.2144	0.8010	4.0160
1	0.5	2.0405	0.9538	4.0079
	1	1.9214	1.0608	4.0023
	0	1.1220	1.8079	3.9631
−1	0.5	1.3008	1.6533	3.9716
	1	1.4199	1.5471	3.9773
	λ	$-f'(0)$	$-\theta'(0)$	$-\phi'(0)$
	1	1.7965	1.2891	2.3976
1	2	1.8241	1.2508	2.6609
	5	1.9214	1.0608	4.0023
	1	1.7566	1.3175	2.3974
−1	2	1.6831	1.3566	2.6589
	5	1.4199	1.5471	3.9774

helps to increase the shear stress at the surface for the exothermic case. The rate of heat transfer $\left(-\theta'(0)\right)$ or Nusselt number enhances with improvement in the values of activation energy during exothermic kind of chemical reactions while the opposite effect goes for the endothermic process.

But the rate of mass transfer $\left(-\phi'(0)\right)$ or Sherwood number slows down with a raise in the values of E_a for $\beta = 1$and boosts up for $\beta = -1$. The reaction rate parameter reduces the Nusselt number (Nu) for $\beta = 1$, whereas an increase in values of Nu can be seen for $\beta = -1$ Sherwood number boosts up for both exothermic and endothermic cases with the increment in reaction rate parameter. Physically this is possible because the chemical reaction contributes to reducing the mass accumulation rate.

6 Conclusions

We have concluded from the following problem:

- For an exothermic chemical reaction, the velocity profile of fluid and dust phases enhance as there is an increment in chemical reaction rate parameter. While it gets diminishes for endothermic type.
- Thermal boundary layer thickens with improving values of reaction rate in case of exothermic reaction, whereas opposite effect is seen for endothermic reaction.
- The concentration level of the dust phase slows down as there is a push up in the values of activation energy, but it boosts up the level of concentration for fluid in case of exothermic kind.
- Rate of heat transfer enhances with increasing values of activation energy for the case of exothermic.
- Rate of mass transfer falls off as the reaction rate parameter increases and boosts up for decreasing values of activation energy parameter in case of endothermic reaction.

References

1. Saffman, P. G. (1962). On the stabilities of laminar flow of a dusty gas. *Journal of Fluid Mechanics, 13*, 120–128.
2. Dutta, N., & Mishra, S. K. (1982). Boundary layer flow of a dusty fluid over a semi-infinite flat plate. *Acta Mechanica, 42*, 71–83.
3. Vajravelu, K., & Nayfeh, J. (1992). Hydromagnetic flow of a dusty fluid over a stretching sheet, *International Journal of Non-Linear Mechanics, 27*, 937–945.
4. Singh, N. P., & Singh, A.K. (2001). MHD effects on heat and mass transfer in flow of a dusty viscous fluid with volume fraction. *Indian Journal of Pure and Applied Physics, 39*, 496–509.
5. Dey, D. (2016). Dusty hydromagnetic oldryod fluid flow in a horizontal channel with volume fraction and energy dissipation. *International Journal Heat and Technology, 34*, 415–422.

6. Dey, D. (2016). Non-Newtonian effects on hydromagnetic dusty stratified fluid flow through a porous medium with volume fraction. *Proceedings of the National Academy of Sciences India Section A: Physical Sciences, 86*, 47–56.
7. Izani, S. N. H., & Ali, A. (2017). Heat and mass transfer of steady magnetohydrodynamics mixed convection of dusty fluid flow with chemical reaction past an exponentially stretching sheet. *Malaysian Journal of Fundamental and Applied Sciences, 13*, 60–67.
8. Sagar, Y. K., & Jha, R. (2016). Unsteady MHD flow of dusty viscous fluid through porous medium with chemical reaction. *International Journal of Engineering and Mathematical Sciences, 10*(1), 45–54.
9. Moss, J. S., & Zapach, T., Tencer, M. (2004). Arrhenius average temperature: The effective temperature for non-fatigue weatout and long term reliability in variable thermal conditions and climates. *IEEE Transactions on Components and Packaging Technologies, 27*, 602–607.
10. Bestman, A. R. (1991). Radiaactive heat transfer to flow of a combustible mixture in a vertical pipe. *International Journal of Energy Research, 15*, 179–184.
11. Olanrewaju, P. O., & Charles, W. M., Makinde, O. D. (2011). Unsteady convection with chemical reaction and radiative heat transfer past a flat porous plate moving through a binary mixture. *Africka Matematika, 22*, 65–78.

DOI Documents

12. Dey, A. J., & Barsuah, D. (2019). Dusty memory fluid flow through a horizontal channel with energy transfer. *AIP Conference Proceedings, 2061.*

Conference Proceedings

13. Maleque, K. A. (2013) Effects of exothermic/endothermic chemical reactions with Arrhenius activation energy on MHD free convection and mass transfer flow in presence of thermal radiation. *Journal of Thermodynamics*, 11. https://doi.org/10.1155/2013/692516.

Entropy Generation Analysis and Slip Effects on Nano Fluid Flow with Heat and Mass Transfer

Debasish Dey and Madhurya Hazarika

Nomenclature

Br	Brinkman number
C	Species concentration
C_p	Specific heat at constant pressure
g	Acceleration due to gravity
D_B	Brownian diffusion coefficient
D_T	Thermophoretic diffusion coefficient.
e	Velocity slip
m	Stretching index
Nb	Brownian motion parameter
Nt	Thermopherosis parameter
Bi	Biot number
Gr	Local thermal Grashof number
Gm	Local modified Grashof number
M	Magnetic field parameter
Re	Reynolds Number
λ_1	Thermal bouncy parameter
λ_1^*	Nanoparticle bouncy parameter
Pr	Prandtl number
S_{G_0}	Characteristic entropy generation rate
T	Temperature component
u	Velocity component along x-axis
v	Velocity component along y-axis
α	Volume fraction (dimensionless

D. Dey · M. Hazarika (✉)
Department of Mathematics, Dibrugarh University, Dibrugarh, Assam 786004, India

P. Mahanta et al. (eds.), *Advances in Thermofluids and Renewable Energy*, Lecture Notes in Mechanical Engineering, https://doi.org/10.1007/978-981-16-3497-0_5

Greek Symbols

η	Efficiency (dimensionless)
μ	Viscosity
k	Thermal conductivity
ν	Kinematic viscosity
σ	Electric conductivity
σ^*	Stephen-Boltzmann constant
Ω	Dimensionless reference temperature
λ	Heat generation/absorption parameter
γ	Spin gradient viscosity
η	Dimensionless coordinate

Subscripts

nf	Nano fluid
f	Base fluid

1 Introduction

In recent years, the study of nano-sized particles in fluid flows has attracted many researchers due to their numerous applications in industrial fields such as pumps, flow meters and heat exchangers. Addition of nano-sized particles into classical fluid results in enhancement of the thermal conductivity of heat transfer process. Choi [1] has introduced the term nano fluid in which nano-sized particles are mixed with the base fluid. The broad range of current and future applications involving nanofluid motivated many researchers to derive and analyze the enhancement of thermal conductivity. Buongiorno [2] and Shukla [3] have discussed on nano fluid flows, and their numerical approach to find out the enhancement of the thermal performance. The movement of nano fluid over a stretching surface with convective boundary conditions has been studied by Rana and Bhargava [4] and Rahman and Eltayeb [5]. Dey and Baruah [6], Dey [7] have studied the viscoelastic effects of nano fluid flow in rotating systems.

Flow and heat transfer with slip has several applications in many engineering and industrial processes. Navier [8] has introduced the concept of no-slip condition to incorporate the possibility of fluid at the solid boundary. The slip effect of second-grade fluid flow has been investigated by Chauhan [9] over shrinking and stretching sheets. Das et al. [10] have investigated the slip effect on MHD flow over a flat plate with variable fluid properties. Zhu et al. [11] have discussed the analytical study of nano fluid flow with wall heat flux and nonlinear velocity slip.

Entropy generation involves the second law of thermodynamics. Physically, entropy generation is associated with the thermodynamics irreversibilities. Increasing the entropy generation rate results in destroying the useful work of thermodynamic processes. So we have to minimize the entropy generation rate to obtain the maximum available energy. Bejan [12] has reported the entropy generation minimization method to measure the available energy. Ellahi et al. [13] have investigated the effects of nano-sized particles on entropy generation rate with mixed convection. Govindaraju et al. [14] have analyzed the second law of thermodynamics on slip slow of nano fluid over a stretching sheet. The entropy generation analysis of two-phase nano fluid flow has been studied by Noghrehabadi [15]. Analytical modeling of entropy generation induced by a stretching surface has been investigated by Seth et al. [16], Govindaraju*et al.* [17] and Afridi and Qasim [18]. Anand and Chrestov [19], Dey and Hazarika [20] have studied the analysis of entropy generation and the enhancement of heat transfer in the presence of pressure-driven non-Newtonian fluid flow.

2 Mathematical Formulation

Let us consider a steady two-dimensional convective boundary layer flow of a nano fluid over a flat horizontal stretching sheet. The sheet is stretched with the velocity of the form $u_w = ax^m$. A transverse magnetic field $B(x) = B_0 x^{m-1}$ is applied normally to the stretching sheet. The magnetic Reynolds number is assumed to be very small. The strength of the electric field is expected to be zero. On the surface of the sheet, there is the presence of velocity slip, and the sheet is permeable in nature. Under these assumptions and applying Boussinesq approximation, the governing equations are of the form [21] (Fig. 1).

$$\frac{\partial u}{\partial x} + \frac{\partial v}{\partial y} = 0, \tag{1}$$

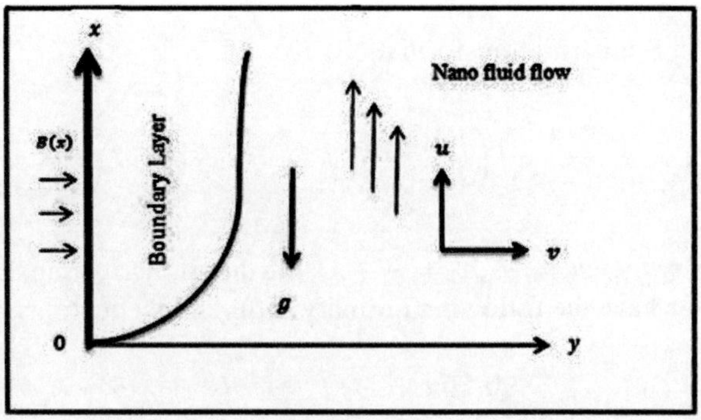

Fig. 1 Schematic diagram of the problem

$$\rho_{\rm f}\left(u\frac{\partial u}{\partial x}+v\frac{\partial u}{\partial y}\right)=\mu_{\rm nf}\frac{\partial^2 u}{\partial y^2}+\left[(1-C_\infty)\rho_{f\infty}g\beta_{\rm T}(T-T_\infty)\right.$$
$$\left.-(\rho-\rho_{f\infty})g\beta_{\rm c}(C-C_\infty)\right]-\sigma B^2(x)u-\frac{\mu_{\rm nf}u}{k_0}, \tag{2}$$

$$\left(u\frac{\partial T}{\partial x}+v\frac{\partial T}{\partial y}\right)=\alpha_1\frac{\partial^2 T}{\partial y^2}+\tau\left[D_{\rm B}\frac{\partial C}{\partial y}\frac{\partial T}{\partial y}-\frac{D_{\rm T}}{T_\infty}\left(\frac{\partial T}{\partial y}\right)^2\right]$$
$$-\frac{1}{\rho_{\rm f}c_{\rm f}}\frac{\partial q_{\rm r}}{\partial y}+\frac{\nu_{\rm f}}{k_{\rm f}C_{\rm f}}u^2+\frac{\sigma B^2(x)}{\rho_{\rm f}c_{\rm f}}u^2, \tag{3}$$

$$u\frac{\partial C}{\partial x}+v\frac{\partial C}{\partial y}=D_B\frac{\partial^2 C}{\partial y^2}+\frac{D_T}{T_\infty}\left(\frac{\partial^2 T}{\partial y^2}\right)-k_1(x)(C-C_\infty) \tag{4}$$

Boundary conditions:

$$y=0: u=u_{\rm w}+u_{\rm s},\ v=v_{\rm w},\ k_{\rm f}\frac{\partial T}{\partial y}=h_{\rm f}(T_{\rm w}-T_\infty),\ C=C_{\rm w}$$
$$y\to\infty: u=0,\ T=T_\infty,\ D_{\rm B}\frac{\partial C}{\partial y}+\frac{D_{\rm T}}{T_\infty}\frac{\partial T}{\partial y}=0 \tag{5}$$

Here, u, v are the components of velocity along x and y direction, T and C are the components of temperature and concentration. $\rho_{\rm nf}$, $\mu_{\rm nf}$ are the density and viscosity of the nano fluid. $C_{\rm w}$ and $T_{\rm w}$ are the species concentration and the temperature value at the surface of the sheet. α_1 is the thermal diffusivity, $q_{\rm r}$ is the radiative heat flux, $(\rho C)_{\rm f}$ and $(\rho C)_{\rm nf}$ are the heat capacity of the base fluid and nano fluid. $\tau=(\rho C)_{\rm f}/(\rho C)_{\rm nf}$ is termed as the ratio between the heat capacity of base fluid and the nano fluid. $D_{\rm B}$ and $D_{\rm T}$ are the Brownian diffusion coefficient and the thermophoretic diffusion coefficient.$k_1(x)$ is the chemical reaction parameter. $k_{\rm nf}$ is the thermal conductivity of the nano fluid.

On using the Rosseland diffusion approximation, the radiative diffusion heat flux is given by the following:

$$q_{\rm r}=-\frac{4\sigma^*}{3k_1}\frac{\partial T^4}{\partial y} \tag{6}$$

Similarity transformation

$$\eta=y\sqrt{\left(\frac{b}{\nu}x^{m-1}\right)},\ \psi=\sqrt{b\nu}f(\eta),\ \theta=\frac{T-T_\infty}{T_{\rm w}-T_\infty},\ \phi=\frac{C-C_\infty}{C_{\rm w}-C_\infty} \tag{7}$$

where$u=\frac{\partial\psi}{\partial y}$, $v=-\frac{\partial\psi}{\partial x}$.Using the similarity transformations (7) on Eqs. (1)–(4), we have the following ordinary differential equations.

$$f'''+0.5(m+1)ff''-mf'^2+\lambda_1\theta-\lambda_1^*\phi-\left(M^2+K\right)f'=0 \tag{8}$$

$$\theta'' + \Pr \mathrm{Nb}\theta'\phi' + 0.5(m+1)\Pr f\theta' + \Pr \mathrm{Nt}\theta'^2 + \left(\frac{4}{3R}\right)\left[\{1 + (T_\mathrm{r} - 1)\theta\}^2\theta'\right]' = 0, \tag{9}$$

$$\phi'' + 0.5(m+1)\mathrm{Lef}\phi' + \left(\frac{\mathrm{Nt}}{\mathrm{Nb}}\right)\theta'' - K_1\phi = 0, \tag{10}$$

Relevant boundary conditions are

$$\begin{aligned} &\eta = 0 : f = f_\mathrm{w}, f' = 1 + ef'', \theta' = -\mathrm{Bi}(1-\theta), \phi = 1 \\ &\eta \to \infty : f \to 0, \theta = 0, \mathrm{Nb}\theta' + \mathrm{Nt}\phi' = 0 \end{aligned} \tag{11}$$

Here, $\lambda_1 = \frac{\rho f_\infty(1-C_\infty)\mathrm{Gr}}{\rho_\mathrm{f}\mathrm{Re}^2}$ and $\lambda_1^* = \frac{\rho_\mathrm{p}-\rho_{f\infty}}{\rho_\mathrm{f}}\frac{\mathrm{Gm}}{\mathrm{Re}^2}$ are the thermal bouncy parameter and the nanoparticle bouncy parameter. $\mathrm{Nb}\left(= \frac{\tau D_\mathrm{B}(C_\mathrm{W}-C_\infty)}{\nu}\right)$ and $\mathrm{Nt} = \left(\frac{\tau D_\mathrm{B}(T_\mathrm{W}-T_\infty)}{T_\infty \nu}\right)$ are the Brownian motion parameter and the thermopherosis parameter. $\mathrm{Le} = \frac{\nu}{D_\mathrm{B}}$ is the Lewis number, $\Pr = \frac{\nu}{\alpha 1}$ is the Prandtl number, $M = B_0\sqrt{\frac{\sigma}{a\rho_\mathrm{f}}}$ is the magnetic parameter, R is the conduction radiation parameter, $K_1 = \frac{k_1\nu}{aD_\mathrm{B}}$ is the chemical reaction parameter and K is the permeability parameter.$\mathrm{Gr}\left(= g\beta_\mathrm{T}(T_\mathrm{w} - T_\infty)x^3/\nu^2\right)$ and $\mathrm{Gm} = g\beta_\mathrm{T}(C_\mathrm{w} - C_\infty)x^3/\nu^2)$ are the local thermal Grashof number and the local modified Grashof number. f_w and e are the suction/ injection parameter and the velocity slip parameter. The local skin friction coefficient (C_f), Nusselt number (Nu_x) and Sherwood number can be determined as

$$\begin{aligned} C_\mathrm{fr} &= C_\mathrm{f}\frac{1}{\sqrt{\mathrm{Re}}} = f''(0), \\ \mathrm{Nu}_{rx} &= \mathrm{Nu}_x\frac{1}{\sqrt{\mathrm{Re}}} \\ &= -\left[1 + \frac{4}{3R}\{(1 + (T_r - 1))\theta(0)\}^3\right]\theta'(0) \\ \mathrm{Sh}_{rx} &= \mathrm{Sh}_x\frac{1}{\sqrt{\mathrm{Re}}} = -\phi'(0) \end{aligned} \tag{12}$$

3 Entropy Generation

The volumetric rate of entropy generation in the presence of magnetic field with heat and mass transfer is given by

$$S_{\mathrm{G}} = \frac{k_{\mathrm{nf}}}{T_\infty}\left(1 + \frac{16\sigma_1^* T_\infty^2}{3k_{\mathrm{nf}}^* k_{\mathrm{nf}}}\right)\left(\frac{\partial T}{\partial y}\right)^2 + \frac{\mu_{\mathrm{nf}}}{T_\infty}\left(\frac{\partial u}{\partial y}\right)^2 + \frac{\sigma B_0^2}{T_\infty}u^2$$
$$+ \frac{RD}{C_\infty}\left(\frac{\partial C}{\partial y}\right)^2 + \frac{RD}{T_\infty}\left(\frac{\partial C}{\partial x}\frac{\partial T}{\partial x} + \frac{\partial C}{\partial y}\frac{\partial T}{\partial y}\right) \tag{13}$$

Here, the first term of Eq. (13) specifies the entropy generation due to heat transfer, second term is due to momentum diffusion, the third term is due to magnetic field and the last two terms denote the local entropy generation due to mass transfer and the combined effect of heat and mass transfer. Using the similarity transformations, the expression for the dimensionless entropy generation number is given by:

$$\mathrm{Ns} = \frac{S_G}{S_{G_0}} = \left(1 + \frac{4}{3R}\right)\theta'^2 + \frac{K\,\mathrm{Br}}{\Omega}f''^2 + M\,\mathrm{Br}\frac{f'^2}{\Omega}$$
$$+ \lambda_1'\left(\frac{\chi}{\Omega}\right)\left(\frac{\chi}{\Omega}\phi'^2 - \phi'\theta'\right) \tag{14}$$

where $\lambda_1' = RD\frac{C_\infty}{k_{\mathrm{nf}}}$, $S_{G_0} = \frac{k_{\mathrm{nf}}(T_{\mathrm{w}} - T_\infty)^2}{l^2 T_\infty^2}$)

4 Numerical Solution

The system of ordinary differential Eqs. (8)–(10) subjected to the relevant boundary conditions can be solved numerically using MATLAB built-in bvp4c solver technique. The set of nonlinear differential equations are reduced to a set of first-order differential equations.

5 Results and Discussions

The nonlinear differential equations with the boundary conditions are solved using MATLAB built-in bvp4c solver technique. Table 1 depicts the numerical values of Nusselt number $\left(-\theta'(0)\right)$ and Sherwood number $\left(-\phi'(0)\right)$. To validate the results, we have to redefine the boundary conditions $D_{\mathrm{B}}\frac{\partial C}{\partial y} + \frac{D_{\mathrm{T}}}{T_\infty}\frac{\partial T}{\partial y} = 0$ as $\theta \to 0, \phi \to 0$ at $y \to \infty$. Numerical results show a good agreement with the previous work done by Rahman and Eltayeb [5].

The research in fluid dynamics is mainly focused on checking the parameters responsible for enhancing shear stress at the surface. The modeling should be done based on the above parameters and special check must be given on R, Bi and Nt because these lead to rise of the shear stress. Higher values of shear stress can cause damage to the surface. Table 2 depicts the numerical values of skin friction coefficient, Nusselt number and Sherwood number with the effects of suction/injection parameter f_{w}, velocity slip parameter e, conduction radiation parameter R, Biot number Bi

Table 1 Comparison the values of Nusselt number (Nu_{rx}) and Sherwood number (Sh_{rx}) with the previous results of Rahman and Eltayeb [5] and Das et al. [10] by considering $e = 0, m = 1, f_w = 0, K_r = 0, \lambda = \lambda^* = 0$, R and $Bi- > \infty$

	Nu_{xx}		Sh_{rx}	
Nt	Rahman and Eltayeb([5])	Present results	Rahman and Eltayeb ([5])	Present results
0.1	0.952376	0.9524	2.129393	2.1293
0.2	0.693174	0.6932	2.274020	2.2738
0.3	0.520079	0.5201	2.528636	2.5284
0.4	0.402581	0.4028	2.795167	2.7949
0.5	0.321054	0.3211	3.035139	3.0347

Table 2 Effects of various parameters on skin friction coefficient (C_{fr}), Nusselt number (Nu_{rx}) and Sherwood number (Sh_{rx})

f_w	e	R	Bi	Nt	C_f	Nu_x	Sh_{rx}
−1	0.3	2	5	0.2	1.2922	0.0205	−0.0308
−0.5	–	–	–	–	1.1556	0.1258	−0.1888
0	–	–	–	–	0.9160	0.2477	−0.3716
0.5	–	–	–	–	0.6520	0.3840	−0.5760
1	–	–	–	–	0.3734	0.5330	−0.7995
0.2	0.0	2	5	0.2	1.4601	0.3043	−0.4564
–	0.1	–	–		1.1546	0.3028	−0.4542
–	0.2	–	–		0.9540	0.3015	−0.4509
–	0.3	–	–		0.8125	0.3006	−0.4497
0.2	0.3	2	5	0.2	0.8125	0.3006	−0.4509
–	–	3			0.8438	0.3043	−0.4563
–	–	4			0.8803	0.2913	−0.4569
–	–	5			0.9092	0.2790	−0.4184
0.2	0.3	2	2	0.2	0.6565	0.2694	−0.4040
–	–	–	3	–	0.7390	0.2859	−0.4288
–	–	–	4	–	0.7841	0.2949	−0.4423
–	–	–	5	–	0.8125	0.3006	−0.4509
0.2	0.3	2	5	0.1	0.7475	0.2976	−0.1488
–	–	–	–	0.2	0.7800	0.2991	−0.2991
–	–	–	–	0.3	0.8774	0.3006	−0.4509
–	–	–	–	0.5	0.9442	0.3036	−0.7590

and thermophoresis parameter Nt. In our investigations, the default values of the parameters are taken as $e = 0.3, \lambda = 10, \lambda^* = 5, R = 2, Nb = 0.3$, $\mathrm{Nt} = 0.2, \mathrm{Le} = 5, \mathrm{Pr} = 6.21, T_r = 2, M = 2$.

From Table 2, it is noticed that the skin friction coefficient (C_{fr}) and the Nusselt number ($\mathrm{Nu_x}$) and Sherwood number (Sh_{rx}) reduce as the slip parameter changes from 0 to 0.3. It signifies that the increasing values of velocity slip lead to diminish the wall shearing stress. It can be witnessed from Table 2 that the wall shearing stress rises with the increase in conduction radiation parameter (R), Biot number (Bi) and thermopherosis parameter (Nt), but an opposite trend is observed in the case of suction/injection parameter. In the case of heat transfer rate, it is inferred from the table that the slip parameter, suction injection parameter, radiation parameter, Biot number and the thermophoresis parameters were able to increase the rate of heat transfer in this problem. Mass transfer rate enhances with increasing values of f_w and e but it shows reverse effect in case of R, Bi and Nt.

Figure 2 shows the influence of e on the dimensionless velocity component $f'(\eta)$. It is seen that the fluid is accelerated in the boundary layer region as velocity slip increases. Physically, it is obvious that as increasing values of slip parameter give a positive impact on the slip at the surface, it results the reduction of the velocity of the fluid flow. Figures 3 and 4 reveal that the velocity profile enhances near the boundary layer as Biot number (Bi) and thermopherosis parameter (Nt) increase. From Fig. 3, it is seen that with the rise in Bi the thickness boundary layer increases beyond this layer is behaving like uniform stream. Figure 5 indicates that an increase in Biot number leads to enhance the distribution of the temperature profile.

Figures 6, 7, 8 and 9 present the distribution of entropy generation for various values of suction /injection parameter (f_w), Biot number (Bi), velocity slip parameter (e) and thermopherosis parameter (Nt). It is observed from Fig. 6 that the entropy generation rate increases with the increasing values of Bi. Entropy generation profile

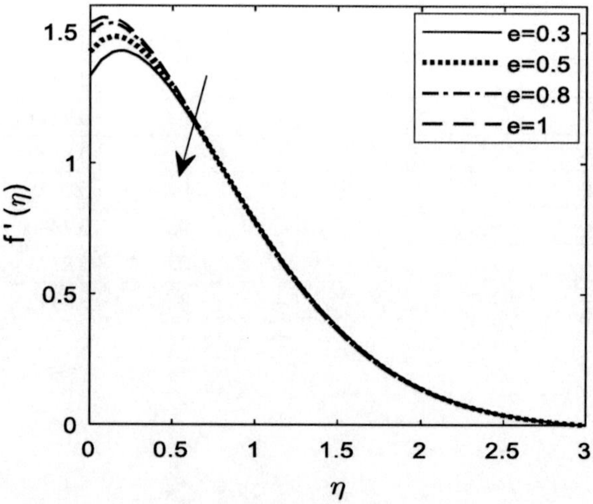

Fig. 2 Distribution of velocity profile against η for different values of e

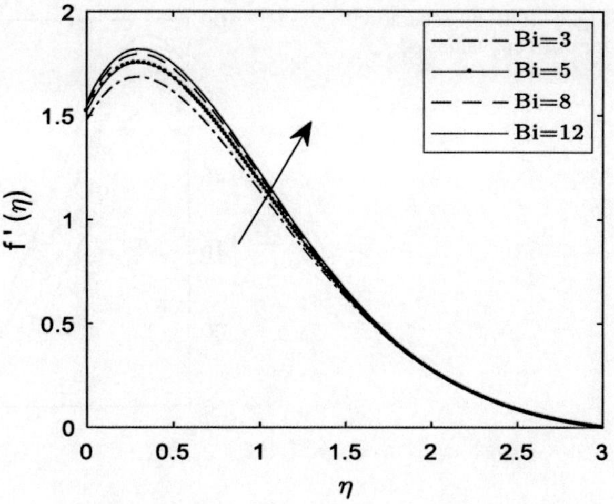

Fig. 3 Distribution of velocity profile against η for different values of Bi

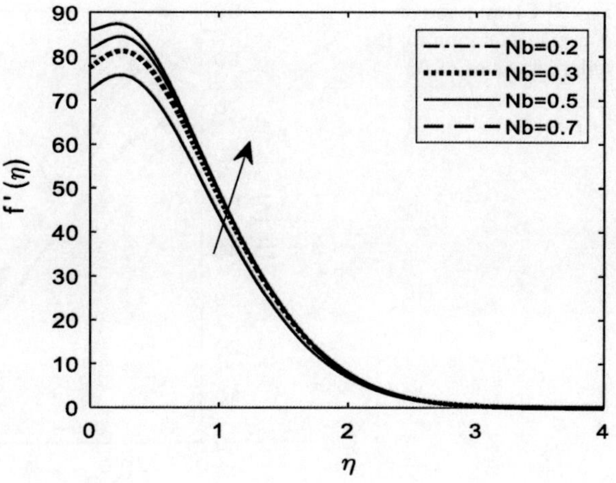

Fig. 4 Velocity profile against η for different values of Nb

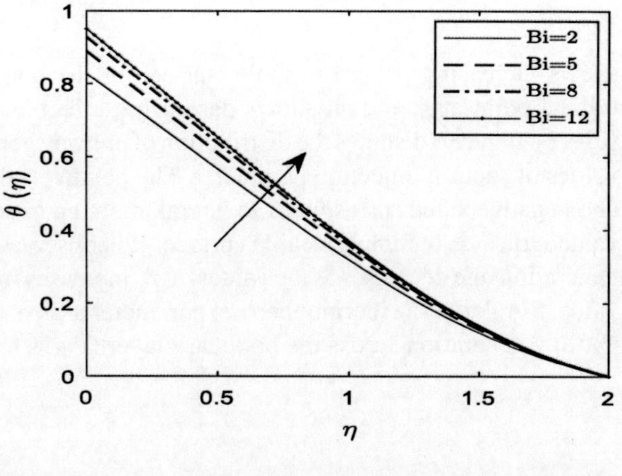

Fig. 5 Distribution of temperature profile for different values of Bi

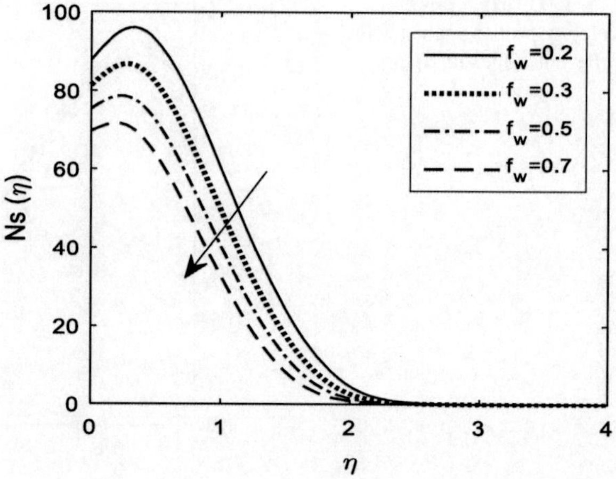

Fig. 6 Entropy generation for different values of f_w

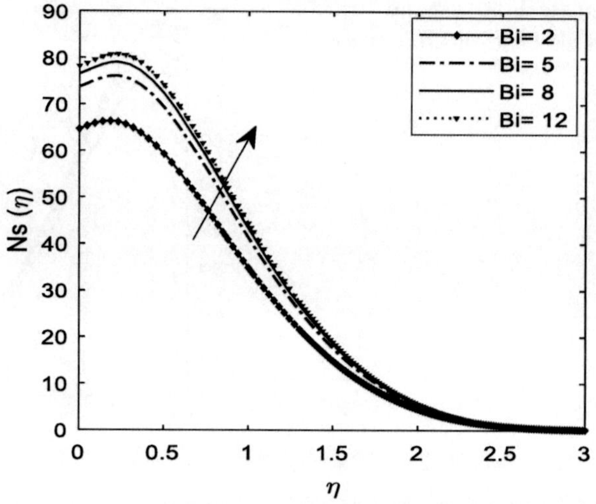

Fig. 7 Distribution of entropy generation for the different values of Bi

shows increasing pattern near the surface of the boundary layer as it reaches the critical point at $\eta = 1$, it shows decreasing effect as velocity slip parameter rises (Fig. 8). Figure 6 displays the distribution of entropy generation against η for different values of suction /injection parameter. The positive values of f_w implies the suction and negative value corresponds to lateral injection over permeable surface, and zero value indicates the impermeable surface. It is observed from Fig. 6 that the entropy generation rate decreases as the values of f_w increases from negative values to positive value. Similarly, the thermopherosis parameter assists to decrease the distribution of entropy generation across the boundary layer (Fig. 9).

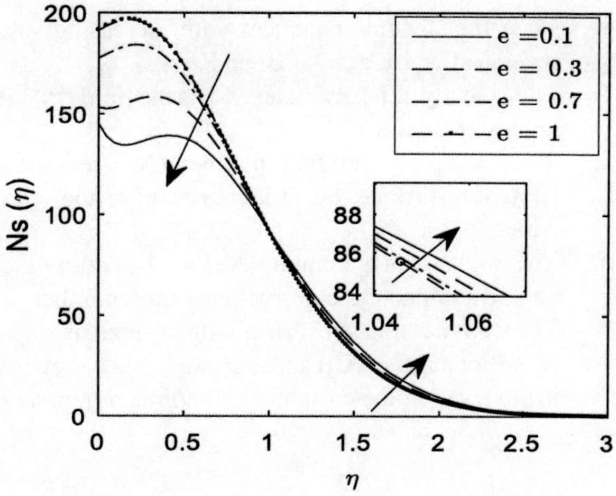

Fig. 8 Distribution of entropy generation against the displacement variable η

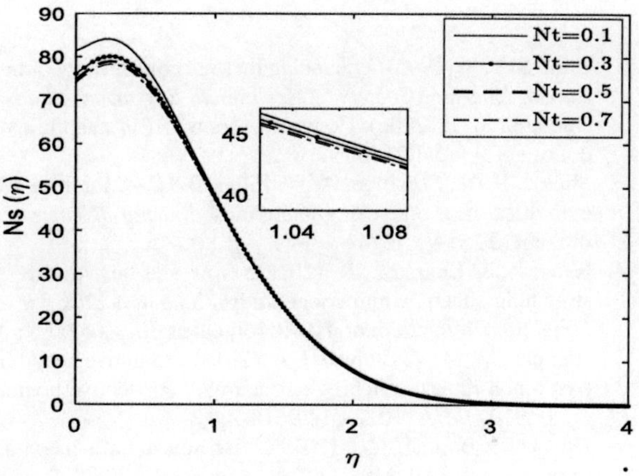

Fig. 9 Distribution of entropy generation against η for various values of Nt

6 Conclusions

The presence of nanoparticles in base fluid helps to enhance heat transfer characteristics, and by doing this, it might change the entropy generation rate. So, with this study, we can predict the changes in entropy generation due to various flow parameters. This also helps us to choose flow parameters characteristics in such a way that the entropy generation should not be too high. The following conclusions are drawn from these observations are listed below.

1. The increasing values of velocity slip leads to diminish the wall shearing stress.
2. It is inferred from the table that the slip parameter, suction injection parameter, Biot number and thermophoresis parameters were able to increase the rate of heat transfer in this problem.

3. Mass transfer rate enhances with increasing values of f_w and e, but it shows reverse effect in case of Bi and Nt.
4. The velocity slip parameter resists the fluid flow. It increases with the increasing values of Bi and Nb.
5. The entropy generation number (Ns) reduces in the neighborhood of the stretching surface, but it increases after the critical point obtain at the region where $\eta > 1$.
6. Themopherosis parameter(Nt) and suction/injection parameter influence a negative impact on entropy generation number.
7. The wall shearing stress rises with the increase in conduction radiation parameter (N), Biot number (Bi) and thermopherosis parameter (Nt), but an opposite trend is observed in the case of suction/injection parameter.

References

1. Choi, S. U. S. (1995). Enhancing thermal conductivity of fluids with nanoparticles. *The overall rank of American Society of Mechanical Engineers, Fluids Engineering Division FED.*
2. Buongiorno, J. (2006). Convective transport in nanofluids. *Journal of Heat Transfer*. https://doi.org/10.1115/1.2150834
3. Shukla, K. N., Solomon, A. B., Pillai, B. C., & Ibrahim, M. (2010). Thermal performance of cylindrical heat pipe using nanofluids. *Journal of Thermophysics and Heat Transfer*. https://doi.org/10.2514/1.48749
4. Rana, P., & Bhargava, R. (2012). Flow and heat transfer of a nanofluid over a nonlinearly stretching sheet: A numerical study. *Communications in Nonlinear Science and Numerical Simulation*. https://doi.org/10.1016/j.cnsns.2011.05.009
5. Rahman, M. M., & Eltayeb, I. A. (2013). Radiative heat transfer in a hydromagnetic nanofluid past a non-linear stretching surface with convective boundary condition. *Meccanica*. https://doi.org/10.1007/s11012-012-9618-2
6. Dey, D., & Baruah, A. J. (2019). Visco-elastic effects on nano-fluid flow in a rotating system in presence of hall current effect. *Advances in Intelligent Systems and Computing*. https://doi.org/10.1007/978-981-13-1951-8_52
7. Dey, D. (2018). Viscoelastic fluid flow through an annulus with relaxation, retardation effects and external heat source/sink. *Alexandria Engineering Journal, 57*, 995–1001. https://doi.org/10.1016/j.aej.2017.01.039
8. Navier, C. L. (1823). Mémoire sur les lois du mouvement des fluides, Mémoires l'Académie Des Sci. l'Institut Fr.
9. Chauhan, D. S., & Olkha, A. (2011). Slip flow and heat transfer of a second-grade fluid in a porous medium over a stretching sheet with power-law surface temperature or heat flux. *Chemical Engineering Communications*. https://doi.org/10.1080/00986445.2011.552034
10. Das, K., Duari, P. R., & Kundu, P. K. (2015). Numerical simulation of nanofluid flow with convective boundary condition. *Journal of the Egyptian Mathematical Society*. https://doi.org/10.1016/j.joems.2014.05.009
11. Zhu, J., Chu, P., & Sui, J. (2018). Exact analytical nanofluid flow and heat transfer involving asymmetric wall heat fluxes with nonlinear velocity slip. *Mathematical Problems in Engineering*. https://doi.org/10.1155/2018/3094980
12. Bejan, A. (1982). Second-law analysis in heat transfer and thermal design. *Advances in Heat Transfer*. https://doi.org/10.1016/S0065-2717(08)70172-2

13. Ellahi, R., Hassan, M., & Zeeshan, A. (2015). Shape effects of nanosize particles in Cu - H 2O nanofluid on entropy generation. *International Journal of Heat and Mass Transfer*. https://doi.org/10.1016/j.ijheatmasstransfer.2014.10.041
14. Govindaraju, M., Ganga, B., & Abdul Hakeem, A. K. (2017). Second law analysis on radiative slip flow of nanofluid over a stretching sheet in the presence of Lorentz force and heat generation/absorption. *Frontiers in Heat and Mass Transfer*. https://doi.org/10.5098/hmt.8.10.
15. Noghrehabadi, A., Saffarian, M. R., Pourrajab, R., & Ghalambaz, M. (2013). Entropy analysis for nanofluid flow over a stretching sheet in the presence of heat generation/absorption and partial slip. *Journal of Mechanical Science and Technology*. https://doi.org/10.1007/s12206-013-0104-0
16. Seth, G. S., Bhattacharyya, A., Kumar, R., & Chamkha, A. J. (2018). Entropy generation in hydromagnetic nanofluid flow over a non-linear stretching sheet with Navier's velocity slip and convective heat transfer. *Physics of Fluids*. https://doi.org/10.1063/1.5054099
17. Govindaraju, M., Akilesh, M., Abdul Hakeem, A. K. (2018). Analysis of entropy generation for nanofluid flow over a stretching sheet with an inclined magnetic field and uniform heat source/sink. *Journal of Physics: Conference Series*. https://doi.org/10.1088/1742-6596/1139/1/012010.
18. Afridi, M. I., & Qasim, M. (2018). Comparative study and entropy generation analysis of Cu–H2O and Ag–H2O nanofluids flow over a slendering stretching surface. *Journal of Nanofluids*. https://doi.org/10.1166/jon.2018.1488
19. Anand, V., & Christov, I. C. (2019). On the enhancement of heat transfer and reduction of entropy generation by asymmetric slip in pressure-driven non-Newtonian microflows. *Journal of Heat Transfer*. https://doi.org/10.1115/1.4042157
20. Dey, D., & Hazarika, M. (2020). Entropy generation of hydro-magnetic stagnation point flow of micropolar fluid with energy transfer under the effect of uniform suction/injection, *Latin American Applied Research, 50*(3), 209–214
21. Das, K., Chakraborty, T., Kumar Kundu, P. (2016). Slip effects on nanofluid flow over a nonlinear permeable stretching surface with chemical reaction. *Proceedings of the Institution of Mechanical Engineers, Part C: Journal of Mechanical Engineering Science*. https://doi.org/10.1177/0954406215595654.

Rheology of Power-Law Nano-Fluid Flow Past a Stretching Sheet

Debasish Dey and Bhagyashree Mahanta

Nomenclature

A	Energy dissipation parameter
a, b	Constants
B	Variable magnetic field
C	Concentration of the nanoparticles
C_p	Specific heat capacity
C_w	Wall concentration
D_B	Brownian coefficients
D_T	Thermophoresis diffusion coefficients
E	Power law parameter
K	Flow consistency index
K^*	Mean absorption coefficient
K	Thermal conductivity
K_0	Product of viscosity and specific heat
L	Length of the sheet
L_e	Lewis number
M	Magnetic parameter
n	Flow behaviour index
N_b	Brownian motion parameter
N_t	Thermophoresis parameter
q_r	Radiative heat flux
R_d	Thermal radiation parameter
T	Temperature of the fluid
T_∞	Ambient Fluid temperature
T_w	Wall temperature
U_∞	Free stream velocity

D. Dey (✉) · B. Mahanta
Department of Mathematics, Dibrugarh University, Dibrugarh, Assam 786004, India

P. Mahanta et al. (eds.), *Advances in Thermofluids and Renewable Energy*, Lecture Notes in Mechanical Engineering, https://doi.org/10.1007/978-981-16-3497-0_6

U_w Wall velocity
uv Velocity components

Greek Symbols

ρ Density of the fluid
σ Conductivity of the fluid
τ Ratio of heat capacities of nano-fluid to base fluid
μ Viscosity coefficient
$\upsilon = \frac{\mu}{\rho}$ Kinematic viscosity
σ^* Stefan-Boltzman constant

1 Introduction

Following the enrichment of the use of nanotechnology, an innovative kind of fluid having heat transfer called nano-fluids having particle size 1–100 nm that are evenly and firmly suspended in a fluid has been a field of scientific attentiveness and pioneering applications for many researchers in the industries. Due to the properties of suspended particles, they boost the thermal functioning more than conventional fluids. Few appliances such as heat exchangers and tools cooling have been widely studied lately. Over the past few years, many researchers and investigators studied the behaviour of heat and mass transfer of nano-fluids magnificently. Besides, they made a remarkable note that by using various types of nano-fluids temperature distribution changes lead to many heating and cooling effects. Also, to study a wide and important range of shear rates, use of power-law fluid have also been done by many, making it a very helpful model numerically as well as analytically.

Das [1] has evaluated the transfer of heat of a nano-fluid past a permeable stretching sheet along with partial slip flow. Under the stimulus effect of Bio-convection, Dey [2] has analysed numerically the MHD flow of nano-fluid past a vertical cylinder by using a very effective MATLAB built-in bvp4c methodology. In presence of Hall current effects, a study on nano-fluid flow and the effects of viscoelastic fluid past a porous medium has been completed by Dey and Baruah [3] past a porous medium by using perturbation method. Followed by power-law fluid, Dey and Deb [4] have beautifully studied the MHD flow with dissipation of energy past a porous medium using MATLAB built-in bvp4c method. Elgazery [5] has obtained a pioneered numerical result on the nanoparticles and its effects of heat transfer past an unsteady stretching surface together with magnetic field. By considering shear thinning fluid as the base fluid Ellahi et al. [6] took the copper nanoparticles to investigate the effects of power law fluid on nano-fluid flow. Grace et al. [7] while exploring the nano-fluid flow past a stretching sheet undergoing thermal radiation found that there is a boost in thermal conductivity and a downfall in heat flux. Haya et al. [8] have studied well on the nanoparticles and its effect in power law fluid accompanied by stretching sheet placed vertically. Irfan et al [9] have presented the variable viscosity effects and thermal conductivity in the nano-fluid flow past a stretching sheet. Khan and Gorta [10] have numerically presented the study of mass and heat transfer in nano-fluids flow past a stretching wall by using a finite difference method. Mishra et al. [11] studied the phenomenon of power law fluid past a stretching sheet in presence of heat source numerically.

Persuaded by the literature and hard works of the above authors, our present paper revolves around the study of power law fluid model past a stretching sheet of nano-fluid flow, where the effects of the Brownian motion and thermophoresis are assimilated with it. With defined boundary conditions and similarity transformation, concerned partial differential equations are converted to ordinary differential equations and solved numerically by using MATLAB built-in bvp4c methodology. The major significance of the power law fluid model as it helps in describing the behaviour of the fluid and correlates various investigational data extensively across the diverse range of shear stress. The originality of our paper is to study the rheology of power law nano-fluid flow past a stretching sheet which is different from Janes et al. The flow analysis of the problem is described in the next section. Followed by Sect. 3, the methodology used to solve the current problem is explained. Results in the form of graphs and tables obtained from the above work is depicted and expressed in Sects. 4 and 5 respectively. In Sect. 6, conclusions of the paper so far is written in a short and concise way for the better understanding, followed by the final section which contains the references that helped us for this paper.

2 Problem Formulation

We consider a nano-fluid flow past a stretching sheet with thermal radiation. The flow is taken to be laminar. The constitutive equation is administrated by Power law fluid model of nano-fluid flow. The velocity is taken as u and T_∞ as the surrounding temperature.

The governing equations that are established from the fluid motion are:

$$\frac{\partial u}{\partial x} + \frac{\partial v}{\partial y} = 0 \tag{1}$$

$$u\frac{\partial u}{\partial x} + v\frac{\partial u}{\partial y} = U_\infty \frac{dU_\infty}{dx} + \frac{K}{\rho}\frac{\partial}{\partial y}\left(\frac{\partial u}{\partial y}\right)^n - \frac{\sigma B^2}{\rho}(u - U_\infty) \tag{2}$$

$$u\frac{\partial T}{\partial x}+v\frac{\partial T}{\partial y}=\frac{1}{\rho C_p}\frac{\partial}{\partial y}\left(\frac{k\partial T}{\partial y}\right)+\tau\left(D_B\frac{\partial T}{\partial y}\frac{\partial C}{\partial y}+\frac{D_T}{T_\infty}\left(\frac{\partial T}{\partial y}\right)^2\right)$$
$$-\frac{1}{\rho C_p}\frac{\partial q_r}{\partial y}+\frac{\mu}{\rho C_p}\left(\frac{\partial u}{\partial y}\right)^{n+1} \tag{3}$$

$$u\frac{\partial C}{\partial v}+v\frac{\partial C}{\partial y}=D_B\frac{\partial^2 C}{\partial y^2}+\frac{D_T}{T_\infty}\frac{\partial^2 T}{\partial y^2}. \tag{4}$$

The appropriate boundary conditions that satisfy the above equations are:

$$\left.\begin{array}{l} u=U_w(x)=ae^{\frac{x}{L}},\, v=0,\, T=T_w,\, C=C_w \text{ at } y=0 \\ u\to U_\infty=be^{\frac{x}{L}},\, T\to T_\infty,\, C\to C_\infty \text{ at } y\to\infty \end{array}\right\} \tag{5}$$

The introduced similarity transformations to be used in the above equations are given as:

$$\left.\begin{array}{l} \eta=\sqrt{\frac{a}{2\upsilon L}}e^{\frac{x}{2L}}y,\, \psi=\sqrt{2a\upsilon L}e^{\frac{x}{2L}}f(\eta),\, \theta=\frac{T-T_\infty}{T_w-T_\infty}, \\ \phi=\frac{C-C_\infty}{C_w-C_\infty},\, u=ae^{\frac{x}{L}}f'(\eta),\, v=-\sqrt{\frac{\upsilon a}{2L}}e^{\frac{x}{2L}}(f(n)+nf'(\eta)),\, q_r=-\frac{4\sigma^*}{3K^*}\frac{\partial T^4}{\partial y} \end{array}\right\} \tag{6}$$

We consider that the temperature deviation of the flow is very small so we expand T^4 using Taylor's series about T_∞, i.e.$T^4\cong 4T_\infty^3-3T_\infty^4$.
Using the above data and also Eq. (6) in Eqs. (1) to (4) the following results yields:

$$2f'^2-ff''-2\lambda^2+Ef'''f'''^{n-1}+M(f'-\lambda)=0 \tag{7}$$

$$f\theta'+\left(1+\frac{4}{3}R_d\right)\left(\frac{k}{k_0}\theta''\right)+N_b\theta'\phi'+N_t\theta'^2$$
$$+Af''^{n+1}=0 \tag{8}$$

$$\phi''+\frac{N_t}{N_b}\theta'+L_e(f\phi'-f'\phi)=0 \tag{9}$$

where $\lambda=\frac{b}{a}$, $E=\frac{K}{\rho}\frac{a^{\frac{3n}{2}-2}e^{\frac{3nx-4x}{2}}}{2^{\frac{n-2}{2}}\upsilon^{\frac{n}{2}}L^{\frac{n}{2}-1}}n$, $M=\frac{2\sigma B_0^2 L}{\rho a}$, $R_d=\frac{4\sigma^* T_\infty^3}{k_0 k^*}$,

$$k_0=\rho C_p\upsilon,\, N_b=\frac{\tau D_B(C_w-C_\infty)}{\upsilon},\, N_t=\frac{\tau D_T(T_w-T_\infty)}{T_\infty\upsilon},$$

$$A=\frac{\mu}{\rho C_p}\frac{a^{\frac{3n+1}{2}}e^{\frac{x}{2L}(3n+1)}}{(T_w-T_\infty)2^{\frac{n-1}{2}}\upsilon^{\frac{n+1}{2}}L^{\frac{n-1}{2}}},\, L_e=\frac{\upsilon}{D_B},$$

The suitable boundary conditions are:

$$\begin{aligned} & f(0) = 0, f'(0) = 1, \theta(0) = 1, \\ & f'(\infty) = \lambda, \theta(\infty) = 0, \\ & \varphi(0) = 1, \varphi(\infty) = 0 \end{aligned} \tag{10}$$

The parameters E and A are dependent on the values of 'n'. Both E and A increase with n. The set of values of E and A for shear thinning fluid ($n < 1$) are taken different from shear-thickening fluid ($n > 1$). We have taken set of values of E and A for shear thickening fluid is greater than shear-thinning fluids.

3 Method of Solution

In this section, we adopt MATLAB built-in bvp4c method of the form $y' = f(x, y, c)$ with the relevant boundary conditions $\theta(y(a), y(b), c)$ and $\phi(y(a), y(b), c)$ where $a \leq x \leq b$. Consequently, we have the first-order differential equations as shown below:

$$\left.\begin{aligned} & \text{Let } f = y_1, f' = y_2, f'' = y_3 \\ & y_1' = f' = y_2 \\ & y_2' = f'' = y_3 \\ & y_3' = f''' = \frac{-2y_2^2 + y_1 y_3 + 2\lambda^2 - M(y_2 - \lambda)}{E y_3^{n-1}} \end{aligned}\right\} \tag{11}$$

This is obtained by extracting f''' from Eq. (7).
Similarly, we again suppose

$$\left.\begin{aligned} & \theta = y_4, \theta' = y_5 \\ & y_4' = \theta' = y_5 \\ & y_5' = \theta'' = \frac{(-y_1 y_5 - N_b y_5 y_7 - N_t y_5^2 - A y_3^{n+1}) k_0}{(1 + \frac{4}{3} R_d) k} \end{aligned}\right\} \tag{12}$$

This is obtained by separating θ'' from Eq. (8).
Again, let

$$\left.\begin{aligned} & \phi\prime = y_6, \phi\prime = y_7 \\ & y_6' = \phi\prime = y_7 \\ & y_7' = \phi\prime\prime = -\frac{N_t}{N_b}\theta\prime\prime + L_e(y_1 y_7 - y_2 y_6) \end{aligned}\right\} \tag{13}$$

This is obtained by separating $\phi\prime\prime$ from Eq. (9).
The formulated boundary conditions for the above equations are:

$$y_1(0) = 0,\ y_2(0) = 1,\ y_4(0) = 1,\ y_6(0) = 1$$
$$y_2(1) = \lambda,\ y_4(1) = 0,\ y_6(1) = 0$$

4 Results and Discussion

To get a strong visual of the physical problem, a numerical survey has been done graphically to explore the study on rheology of MHD nano-fluid flow along with the effects of shear thinning and shear thickening past a stretched sheet. Influences of various parameters on velocity, temperature and concentration distribution are shown graphically.

In this work, we have taken the value of λ fixed as 0.1. This interprets that stretching is considered superior over free-stream parameter. Here, we have considered $\lambda = 0.1,\ k_0 = 0.1,\ k = 0.1,\ R_d = 0.1,\ L_e = 1$.

Figure 1a explains the change in the velocity distribution due to the effect of the magnetic parameter (M) as the other parameters are changed. Due to the increase in the magnetic field, a downfall in the velocity is observed since there is an increase in the thickness of boundary layer of velocity in the case of shear-thinning fluid, i.e. $n < 1$. Physically, as magnetic parameter is applied a drag-like force comes into action which in turn reduces the flow velocity within the boundary layer. An opposite trend can be seen in Fig. 1b in the case of shear thickening fluid (i.e. $n > 1$), i.e. a decrease in velocity boundary layer is noticed, and as a result, acceleration in fluid velocity is observed.

Figure 2a and b depicts the effect of Brownian motion parameter against temperature profile. The reason behind the increment of the temperature as the parameter increases is that the random collision among the nanoparticles in the fluids, resulting in the generation of extra energy which in turn boosts the rate of heat transfer. Also, an increase in thermal boundary layer is noted as there is an upsurge in the values of Brownian motion parameter (N_b) in presence of shear thinning ($n < 1$) and shear thickening ($n > 1$) as shown in Fig. 2a and b.

The behaviour of increasing thermophoresis parameter N_t on concentration profile of the nano-fluid is depicted graphically. This parameter tends to decrease in the viscosity of the fluid and thus acceleration in velocity is seen. As a result of this, increase in boundary layer of concentration is monitored. Thus, we come to an analysis that concentration of the fluid decreases as N_t rises. This nature can be seen in both the cases of shear thinning and shear thickening fluid shown in Fig. 3a and b.

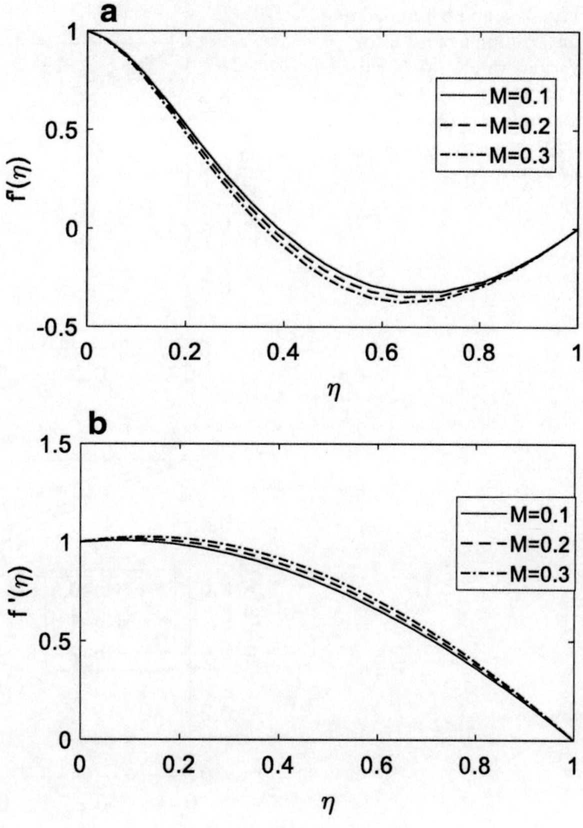

Fig. 1 **a** and **b** Velocity distribution against η for various values of M when $n = 0.5$ ($E = 0.1, A = 0.1$) and $n = 1.5$ ($E = 0.5, A = 0.5$), respectively (Taking $N_b = 0.1, N_t = 0.1$)

5 Tables

To ensure the correctness and validity of our results, a comparison of our data with the data given in Mishra et al. [11] when $Rd = 0.1$, $E = 0.1$, $N_b = 1$, $N_t = 1$, $A = 0.1$ for various values of magnetic parameter M when $n = 0.5$ is tabulated in Table 1 i.e. numerical values for velocity gradient across a stretched sheet dealing with nano-fluid particles have been interpreted. These values illuminate a good deal with the compared data, and the results authenticate our calculated numerical values for velocity gradient. Also, various values for M with different values of E and A in case of $n = 1.5$ is also shown in the same table.

The study of variation in flow of velocity between adjacent layers which is called velocity gradient is very important in the study of fluid dynamics. It also plays a characteristic role in drag formation. When velocity gradient is zero, the flow is undisturbed, and as the value increases, variation in flow displacement variables is seen. The physical meaning of velocity gradient is the deformation rate per unit area.

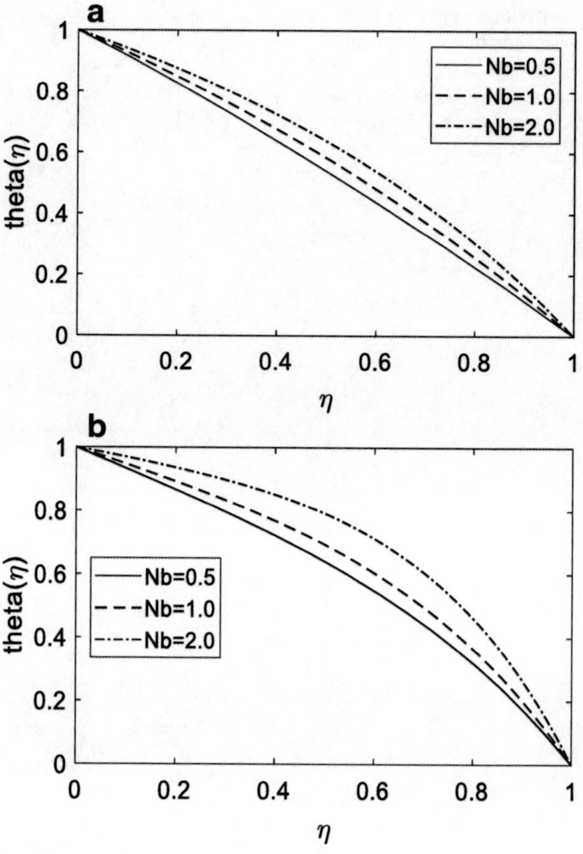

Fig. 2 **a** and **b** Temperature distribution against η for various values of N_b when $n = 0.5$ ($E = 0.1, A = 0.1$) and $n = 1.5$ ($E = 0.5, A = 0.5$), respectively (Taking $M = 0.1, N_t = 0.1$).

From Table 1, we can see that as magnetic parameter M increases, velocity gradient is enhanced when $n = 0.5$, while an opposite nature is seen when $n = 1.5$.

Knowledge in importance of shear stress is worth during the study of non-Newtonian viscous fluid in the research field. It is due to the fact that shear stress is directly proportional to the viscosity of the fluid, the shear stress is responsible for the damage at the surface due to more friction. The main aim of our study is to lower the shear stress and to study about the factors and parameters which can lower the shear stress. Subsequently, a rise in shear stress is observed as M increases in case of shear-thinning fluid, and a fall is perceived when M increases in case of shear-thickening fluid as shown in Table 2.

Fig. 3 a and b: Concentration distribution against η for various values of N_t when $n = 0.5$ ($E = 0.1, A = 0.1$) and $n = 1.5$ ($E = 0.5, A = 0.5$), respectively (Taking $M = 0.1$, $N_b = 0.1$)

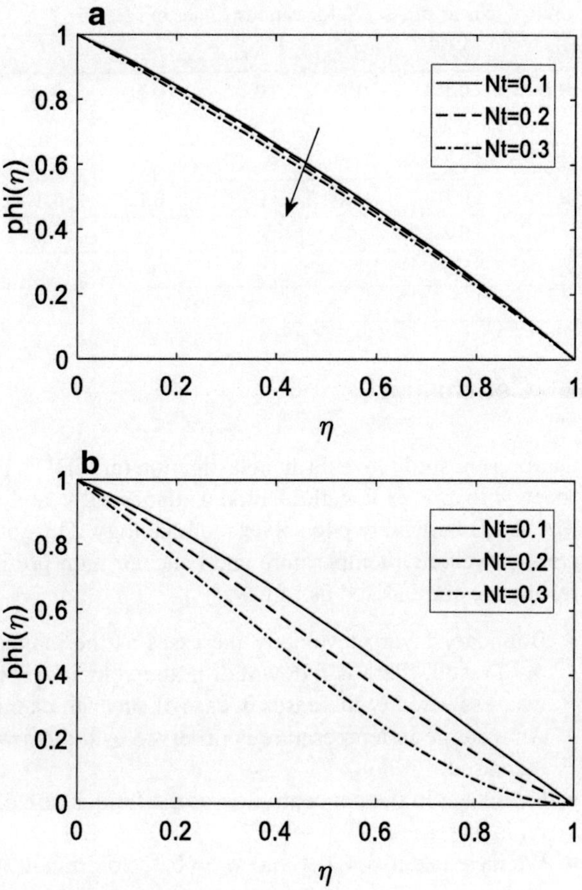

Table 1 Velocity gradient f'' for various values of M

Rd	*M*	*E*	N_b	N_t	*A*	$n = 0.5$		$n = 1.5$
						Mishra et al. [11]	Present Paper	
0.1	0	0.1	0.1	0.1	0.1	1.1671	1.1551	
	1					1.9744	1.9492	
	2					2.6355	2.6655	
0.1	1	0.5	0.1	0.1	0.8			1.1822
	2							1.0970
	3							1.0237

Table 2 Shear stress f'''^{n} for various values of M

Rd	M	E	N_b	N_t	A	$n = 0.5$	$n = 1.5$
0.1	0.1	0.1	0.1	0.1	0.1	0.0509	
	0.2					0.0647	
	0.3					0.1985	
0.1	0.1	0.5	0.1	0.1	0.8		
	0.2						1.2763
	0.3						1.2188
							1.1711

6 Conclusion

Our present study reveals an investigation on MHD nano-fluid flow past a stretching sheet with power-law fluid model numerically and shown graphically by using MATLAB built-in bvp4c solver methodology. The consequences of various parameters on velocity, temperature and concentration profiles are scrutinised. The main results are emphasised as follows:

- Boundary layer of velocity increases as the magnetic parameter (M) upsurges. As a result, there is a downfall in the velocity profile in case of shear thinning, whereas velocity increases in case of shear-thickening fluid.
- An upsurge in temperature is observed as the Brownian motion parameter N_b is increased.
- Shrinkage in the concentration of the fluid is noted as thermophoresis parameter N_t is increased.
- We have taken $k = 0.1$ and $k_0 = 0.1$. So, Prandtl number (Pr) $= \frac{k}{k_0} = 1$, which means rate of momentum and dissipation of heat through the fluid is same.

References

1. Das, K. (2014). Nano- fluid over a non- linear permeable stretching sheet with partial slip. *Journal of the Egyptian Mathematical Society, 23*(2), 451–456. https://doi.org/10.1016/j.joems.2014.06.014
2. Dey, D. (2018). Modelling and analysis of Bio- convective nano- fluid flow past a continuous moving vertical cylinder. *Emerging Technologies in Data Mining and Information Security: Advances in intelligent systems and computing, 755*, 331–340. https://doi.org/10.1007/978-981-13-1951-8_30
3. Dey, D., & Baruah, A. I. (2019). Visco- elastic effects on nano- fluid flow in a rotating system in presence of hall- current effects. *Proceeding of IEMIS 2018, 1*, 575–585. https://doi.org/10.1007/978-981-13-1951-8_52
4. Dey, D., & Deb, H. R. (2018). Hydromagnetic flow of power law fluid in a porous medium with energy dissipation: A numerical approach. *Italian Journal of Engineering Science: Technicaltaliana, 61+1*, 130–134. https://doi.org/10.18280/ijes.620211

5. Elgazery, N. S. (2019). Nano- fluids flow over a permeable unsteady stretching surface with non-uniform heat source/ sink in the presence of inclined magnetic field. *Journal of the Egyptian Mathematical Society, 27*(9). https://doi.org/10.1186/s42787-019-0002-4
6. Ellahi, R., Hassan, M., & Zeeshan, A. (2015). A study of heat transfer in power law nanofluid. *Thermal Science, 20*(6), 2015–2026. https://doi.org/10.2298/TSCI150524129E
7. Grace, M., Mureithi, E. W., & Janes, M. N. (2019). Unsteady MHD flow of nanofluid with variable properties over a stretching sheet in the presence of thermal radiation and chemical reaction. Hindawi. https://doi.org/10.1155/2019/7392459
8. HayatT, H. M., & Alsaedi, A. (2016). Flow of a power law nanofluid past a vertical stretching sheet with a convection boundary condition. *Journal of Applied Mechanics and Technical Physics, 57*(1), 173–179. https://doi.org/10.1134/s0021894416010193
9. Irfan, M., Farooq, M. A., & Iqra, T. (2019). Magnetohydrodynamic free stream and heat transfer of nanofluid flow over an exponentially radiating stretching sheet with variable fluid properties. *Frontiers in Physics, 7*, 186. https://doi.org/10.3389/fphy.2019.00186
10. Khan, W. A., & Gorta, R. S. R. (2012). Heat and mass transfer in power law nanofluids over a non-isothermal stretching wall with convective boundary condition. *Journal of Heat Transfer, 134*(11). https://doi.org/10.1115/1.4007138
11. Mishra, S. R., Baag, S., Dash, G. C., & Acharya, M. R. (2019). Numerical approach to MHD flow ofPower law fluid on a stretching sheet with non- uniform heat source. *Nonlinear Engineering, 9*(1). https://doi.org/10.1515/nleng-201-0026

Nanoparticle-Assisted Multilayered Photothermal Therapy Concerning Countercurrent Blood Flow: A Numerical Study

Abhijit Paul and Anup Paul

Nomenclature

t	Laser exposure time (s)
R	Beam radius (mm)
D	Diameter of rod (m)
s	Number of nanoparticles per unit volume, (m^{-3})
T	Temperature (°C)
Q	Vomumetric heat generation rate (W/m^3)
U	Universal gas constant (J/mol K)
C_d	Damaged concentration of protein
C_0	Undamaged concentration of protein
∇	Gradient
v	Blood flow velocity
C	Specific heat (J/kg K)
k	Thermal conductivity (W/m K)
x, y, z	Coordinates (mm)
I_0	Irradiated laser intensity (W/m^2)
ω_b	Blood perfusion rate (s^{-1})
Ω	Damage integral
h	Ambient convective heat transfer coefficient, (W/m^2 K)
G	Activation energy (J/mol K)
A	Pre-exponential factor (s^{-1})
P	Pressure (Pa)
S	Thermophysical and optical properties
p, n	Consistency and power law index
q	Heat flux
f	Volume force

A. Paul · A. Paul (✉)
Department of Mechanical Engineering, NIT Arunachal Pradesh, Yupia 791112, India

P. Mahanta et al. (eds.), *Advances in Thermofluids and Renewable Energy*, Lecture Notes in Mechanical Engineering, https://doi.org/10.1007/978-981-16-3497-0_7

Greek Symbols

η	Volumetric concentration (dimensionless)
ρ	Density (kg/m^3)
ψ	Shear stress in blood medium
τ_q	Heat flux phase lag
τ_T	Temperature gradient phase lag
α	Absorption coefficient (m^{-1})
β	Scattering coefficient (m^{-1})

Subscripts

amb	Ambient
np	Nanoparticle
mix	Mixture
met	Metabolic
perf	Perfusion
b	Blood
t	Tissue

Abbreviations

IT	Intratumoral
IV	Intravenous
IR	Infrared
AuNs	Gold nanoshells

1 Introduction

Day by day cancer has turned out to be a worldwide threat for humankind. Nowadays, skin cancer exhibits a common cause of cancer deaths. Thus, the study of various cancer treatment methods is the prime focus of most researchers. Light sensitive agent-assisted photothermal therapy has arrived as an effective treatment procedure, applying laser technology.

The laser is a collimated, coherent and monochromatic source of light. Therefore, laser is one of the best tools [1–4] for treating cancer precisely either directly or via optical catheter. During photothermal therapy in clinic, surgeons often face difficulties in proper assessment of the extent of thermal ablation. Thus, the computational justification of a certain prototype before real clinical application often enhances the clinical safety.

In spite of many advantages, the complete acceptance of laser technology in noninvasive cancer treatments is still a challenge [1] due to the limitations of unfocused energy diffusion. Various works have been executed on precise thermal damage of tumor, preserving the surrounding healthy tissues. Recently laser-based photothermal therapy incorporating nanotechnology has drawn a wider focus. During the infusion of nanoparticles through veins, most of the particles get entrapped [2] in other organs despite the targeted zone. Hence, it is a challenge to deliver the nanoparticles properly to the targeted site during intravenous infusion scheme. Literature [1, 2] shows that the intratumoral infusion scheme of nanoparticles dominates over intravenous infusion scheme during laser therapy.

Different heat transmission models are applied by various authors [5, 6] to predict thermal behavior of bio-tissues under external agitation. Pennes bioheat model is the most popular model considering heat transmission at an infinite speed, although this model is a deficit of accurate prediction within inhomogenous tissue medium. The biological tissue medium exhibits a non-Fourier thermal behavior with heat propagation at a finite speed. Many authors have witnessed a fluctuating thermal history [7, 8] in nonhomogeneous tissue medium applying non-Fourier DPL model. In fact, several researchers have revealed accurate thermal prediction in the case of DPL model compared to classical heat transfer model. For accurate prediction of thermal history in biological tissue, it is also very important to incorporate blood vessels in the heat transfer model. Additionally, the metabolic effect in human body exhibits a heat source during photothermal treatment. In Pennes bioheat model, the effect of all micro capillaries has been considered as a scalar blood perfusion term.

The present literature explains about the nanoparticle-based photothermal behavior of tissues during laser-assisted noninvasive cancer treatment. However, studies so far have failed to predict the thermal effect of biological tissues under the intratumoral infusion scheme of nanoparticles considering the countercurrent blood flow within multilayer tissue model. Therefore, in present work, an attempt has been executed to analyze the thermal response in triple-layered tissue model infusing gold nanosphere embedded in countercurrent blood vessel. The prime aim is to perform a comparative study between intratumoral and intravenous infusion schemes of nanoparticles in terms of localized thermal heating of targeted region, sparing the surrounding healthy tissue. Both Pennes and dual-phase lag bioheat model are applied to analyze the heat transfer process within the respective tissue model. Besides, the non-Newtonian power law has been utilized for analyzing the heat transfer of blood flow.

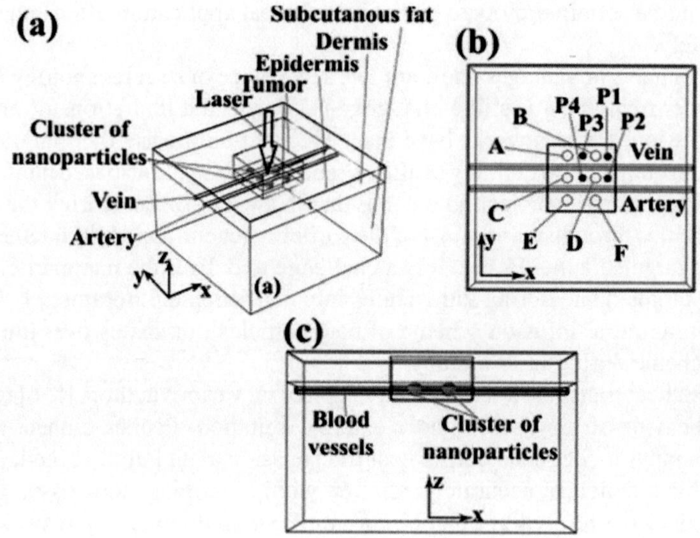

Fig. 1 Demonstration of the computational domain **a** Isometric view **b** top view, **c** side view, where A, B, C, D, E and F denote the cluster of nanoparticles

2 Mathematical Model

2.1 Physical Domain

A three-dimensional triple-layered tissue model has been selected for the present numerical study and is shown in Fig. 1. The dimensions and various optical and thermal properties of tissue model are given in Table 1 which are collected from various references [10–13]. The blood vessels bear a radius of 0.24 mm and are located at a depth of 2.4 mm from tissue top surface. The radius of each AuNp clusters are chosen to be 0.5 mm [10–12] and respective locations are given as A, B, C, D, E and F in the model and is expressed as (x, y, z): (−1.25, 2, −2.3); (1.25, 2, −2.3); (−1.25, 0, -2.3); (1.25, 0, −2.3); (−1.25, −2, −2.3); (1.25, −2, −2.3), respectively, in mm. P1, P2, P3 and P4 are indicated in Fig. 1b as (x, y, z): (2, 2, 0), (2, 0, 0), (0, 2, 0), (0, 0, 0) for the following discussion.

2.2 Physical Description

To analyze the thermal response of tissue under laser heating, a transient heat transfer momentum equation is solved simultaneously considering various initial and boundary conditions. To simplify the problem, the following assumptions are considered.

Table 1 Dimensions and physical properties of tissue medium [10–12]

	Subcutaneous fat	Dermis	Epidermis	Tumor	AuNs
Width and breadth (mm)	20	20	20	6	–
Thickness (mm)	10	2	0.08	3	–
Density, ρ (kg/m^3)	1210	1090	1200	1030	19,320
Thermal conductivity, k (W/m K)	0.194	0.42	0.24	0.558	317
Specific heat, C (J/kg K)	2240	3350	3950	3582	128
Blood perfusion, ω_b (1/s)	0.0031	0.0031	0	0.0063	–
Metabolic heat generation rate, Q_{met} (W/m^3)	368	368	368	3680	–
Scattering coefficient, b (1/m), 549 nm	5	5	5	500	45.94×10^6
Absorption coefficient, a (1/m), 549 nm	0.24	0.24	0.9	80	67.88×10^6

1. Constant thermal and optical properties of the tissue medium.
2. No phase change and chemical reaction within the tissue domain.
3. Smooth condition at the interface of each tissue layer.
4. Isotropic tissue medium.
5. Laminar and incompressible blood flow.

The classical bioheat equation of Pennes is given as,

$$\rho C \frac{\partial T}{\partial t} = -\nabla \cdot q(r, z, t) + \rho_b C_b \omega_b (T_b - T) + Q_{met} + Q_{laser}(r, z) \quad (1)$$

$$q(r, z, t) = -k \nabla T(r, z, t) \quad (2)$$

where ρ is the tissue density (kg/m^3), C is the tissue capacity of heat at constant pressure (J/kg K), k is the thermal conductivity (W/m K) of tissue, T is the temperature (°C) of tissue domain, ω_b is the blood perfusion rate (1/s), the subscript b defines blood domain, Q_{met} is the heat generation rate of metabolism, Q_{laser} is the term of external heat source (W/m^3).

The deposition of external heat ($Q_{laser}(r, z)$) is formulated by the modified Beer–Lambert law, given as,

$$Q_{laser}(r, z) = \alpha I_0 \exp\left\{ \frac{-r^2}{2(R(0))^2 \exp(\beta z)} \right\} \exp\{-(\alpha + \beta) z\} \quad (3)$$

$$r = \sqrt{x^2 + y^2} \quad (4)$$

where I_0 is the laser energy intensity at upper surface of the tissue in W/m^2, $R(0)$ is the beam radius at tissue upper surface ($r, z = 0$), α, β are the coefficient of absorption and scattering, respectively. In present study, I_0 is taken as 32,000 W/m^2 whereas $R(0)$ is taken as 2.5 mm [10].

The DPL model in Cartesian coordinate form can be expressed as,

$$q(x, y, z, t + \tau_q) = -k\nabla T(x, y, z, t + \tau_T) \tag{5}$$

Applying divergence of Eq. (5) and placing $\nabla \cdot q$ in Eq. (1), the DPL model can be expressed as

$$\begin{gathered}\tau_q \rho C \frac{\partial^2 T}{\partial t^2} + \left(\tau_q \omega_b \rho_b C_b + \rho C\right)\frac{\partial T}{\partial t} = k\nabla^2 T + \tau_T k \nabla^2 \frac{\partial T}{\partial t} + \\ \omega_b \rho_b C_b (T_b - T) + Q_{\text{laser}} + Q_{\text{met}} + \tau_q \frac{\partial Q_{\text{met}}}{\partial t} + \tau_q \frac{\partial Q_{\text{laser}}}{\partial t}\end{gathered} \tag{6}$$

where τ_q and τ_T define the phase lags due to heat flux and temperature gradient, respectively. For both healthy tissue and tumor section, τ_q and τ_T are considered as 16 s and 0.043 s, respectively, and the same for AuNs is considered as 0.744 ps and 89.28 ps, respectively [10]. A constant value of τ_q and τ_T is considered for all the layers of tissue.

The changed thermophysical and optical properties of tissue due to the doping of gold nanospheres (AuNp) (r = 40 nm) are addressed applying Eqs. (7) and (8) for the case of IV only [14].

$$S_{\text{mix}} = \eta S_{\text{np}} + (1 - \eta) S_t \tag{7}$$

$$\frac{1}{k_{\text{mix}}} = \frac{\eta}{k_{\text{np}}} + \frac{1 - \eta}{k_t} \tag{8}$$

where η is the volume concentration of nanoparticles and can be expressed by the below relation mentioning the total number of AuNp per unit volume (s) and radius of AuNp (r_{np}), respectively.

$$\eta = s\left(\frac{4\pi r_{\text{np}}^3}{3}\right) \tag{9}$$

In the present work, η for the intravenous case is considered as 3.6×10^{-5} [11]. On the other hand, for the IT scheme, the AuNp clusters are taken as subdomains into the tissue domain and the bioheat equation is computed individually.

The initial condition of temperature within the tissue domain is

$$T_i(x, y, z, 0) = 37\ ^\circ\text{C} \tag{10}$$

The upper plane of the tissue (z = 0) is given to be Fourier convective boundary condition for the Fourier model (Eq. 1) and the non-Fourier convective boundary condition for the DPL model (Eq. 16), respectively

$$-n.(-k\nabla T) = h_{\text{amb}}(T_{\text{amb}} - T) \tag{11}$$

$$-n.\left(-k\nabla T - \tau_T k\nabla\frac{\partial T}{\partial t}\right) = h_{\text{amb}}\left(T_{\text{amb}} - T - \tau_q\frac{\partial T}{\partial t}\right) \tag{12}$$

The remaining planes are kept to be adiabatic

$$-n.(-k\nabla T) = 0 \tag{13}$$

The velocity and temperature of blood flow have been computed from the following equations,

$$\nabla \cdot v = 0 \tag{14}$$

$$\rho\left(\frac{\partial v}{\partial t} + v \cdot \nabla v\right) = -\nabla p + \nabla \cdot \Gamma + f \tag{15}$$

$$\rho C_b\frac{\partial T}{\partial t} + \rho C_b v \cdot \nabla T = \nabla \cdot (k\nabla T) + Q_{\text{laser}}(r, z) \tag{16}$$

where v, ρ, Γ, p, f are velocity, density, stress tensor, pressure and body force term, respectively, whereas $Q_{\text{laser}}(r, z)$ is the applied external heat source (Eq. 3).

The non-Newtonian power law for the blood flow is expressed by the following [11] shear stress (ψ), given as

$$\psi_{xy} = p\left(\frac{\mathrm{d}v}{\mathrm{d}y}\right)^n = p\left|\frac{\mathrm{d}v}{\mathrm{d}y}\right|^{n-1}\frac{\mathrm{d}v}{\mathrm{d}y} \tag{17}$$

where n and p are power law index and consistency, respectively.

In the present problem, the blood velocity at the inlet of artery and vein is kept as 0.03 m/s [11]. Whereas the temperature at the inlet section of artery and vein is given as 37 °C. The outlet pressure of the blood vessels is kept as 80 mm of Hg.

The damage parameter (Ω) [15] has been expressed as,

$$\Omega = \ln\left(\frac{C_0}{C_0 - C_d}\right) \tag{18}$$

where C_0 is the concentration of protein in healthy tissue,C_d is the damaged concentration of protein after therapy. The denaturation of protein happens at $\Omega \geq 0.53$[11], whereas the permanent cell death happens at $\Omega \geq 1$. The results of the DPL bioheat model are applied to compute the damage integral applying the Arrhenius equation. The thermal damage rate can be defined as

$$\frac{d\Omega}{dt} = Ae^{(-G/UT(x,y,z,t))} \tag{19}$$

where G is the energy of activation, considered as 6.28×10^5 J/mol, U is the universal gas constant in J/mol K, T is the tissue temperature in K, A is the pre-exponential factor which is taken as 3.1×10^{98} s^{-1}.

2.3 Numerical Scheme and Validation

In the present study, the bioheat equations were computed by applying finite element method using COMSOLTM Multiphysics software. The present computational results were validated with the previous Paul et al. [11] experimental results and are shown in Fig. 2.

3 Results and Discussion

To analyze the thermal characteristics of tissue infusing nanoparticles (AuNp) under laser therapy, two separate schemes, viz. intravenous (IV) and intratumoral (IT), have been implemented in the present work. The simulated domain becomes more realistic with triple-layered tissue structure embedding countercurrent blood flow. The consequence of two different nanoparticle infusion techniques in terms of localized thermal damage of tumor preserving the surrounding healthy tissue, considering the role of blood flow through artery and vein under external laser irradiation has been analyzed numerically. The thermal phase lag of biological tissue to respond to external heating has been simulated in the current bioheat model by applying dual-phase lag (DPL) model.

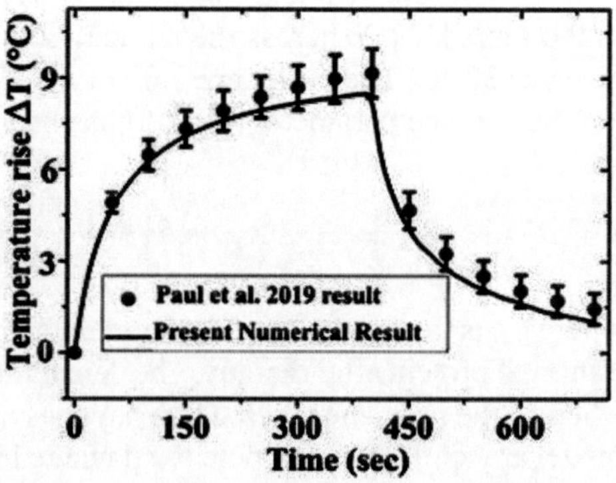

Fig. 2 Variation of predicted (present study) and measured [6] temperature difference with respect to time in the tissue medium

3.1 Thermal Response of Tissue Under Localized Nanoparticle Injection

The spatial temperature history within the malignant and normal tissue has been computed to predict the span of identical necrosis into the tumor site under external heating. The 2-D temperature plot on *x*–*y* plane at various positions (e.g., $z = 0$, −1 mm, −2 mm, −3 mm) with respect to tissue upper plane for the schemes of IV and IT at $t = 300$ s is depicted in Fig. 3. The utmost temperature of *x*–*y* plane at $z = 0$, −1 mm, −3 mm and −4 mm are revealed to be 71.4 °C, 50 °C, 44 °C and 42°C for IV and 56 °C, 58 °C, 71.4 °C and 63 °C for IT, respectively. The plasmonic effect at the surface of the gold nanoparticles under external laser heating implies a fast conversion of exterior light energy to heat energy subsequently dissipating the same to the neighboring normal tissue. The extent of threshold temperature profile ($T >$

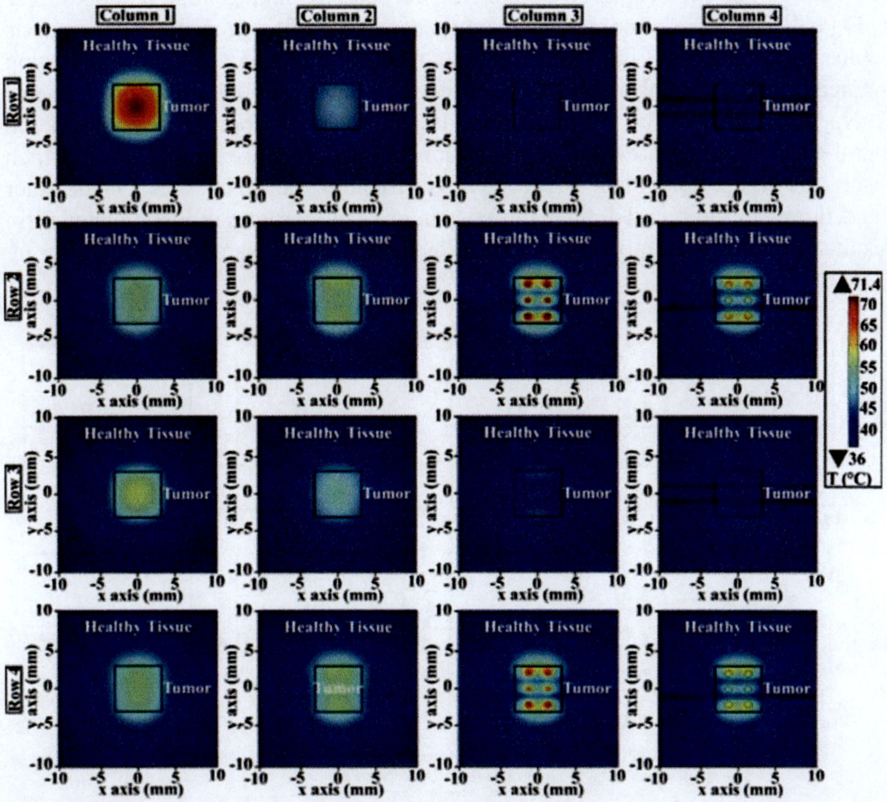

Fig. 3 Tissue temperature history on *x*–*y* plane at different depths of $z = 0$ mm [Column 1], $z = -1$ mm [Column 2], $z = -2$ mm [Column 3], $z = -3$ mm [Column 4], for different schemes of IV [Row 1 and 3] and IT [Row 2, 4] applying Pennes [Row 1 and 2] and DPL [Row 2 and 4] model at $t = 300$ s

50 °C) [15] has been revealed to be gradually diminished with the increasing depth for the scheme of IV (Fig. 3 [Row 1 and 3]) ensuing a nonuniform profile of necrosis along the depth of tumor. In case of IV, the absorption of applied laser energy within the tissue domain progressively decreases its power along its path. A uniform profile of necrotic temperature has been revealed in the whole tumor in case of IT (Fig. 3 [Row 2 and 4]). The tumor confined gathering of nanoparticles provides an option in the case of IT to accumulate the photon energy up the tumor's lower surface unlike decreasing its power due to diffusion. Moreover, the countercurrent blood flow with the interior body temperature of 37 °C affects a significant heat loss within the tissue structure for the cases of IV in comparison to IT as shown in Fig. 3. During the intratumoral doping scheme, the gathered AuNp attains more photon energy into itself causing a targeted temperature raise at the cluster regime thereby diminishing the heat sink effect of blood vessels.

The resulting extent of necrosis obtained within the tissue model for the schemes of IV and IT has been explored by showing the necrotic damage integral contour ($\Omega \geq 1$) [16]. The corresponding 2-D necrotic damage profile on x–y plane at various depths are depicted in Fig. 4 for $t = 300$ s. A uniform damage of tumor preserving the neighboring normal tissue has been revealed during the IT infusion scheme of AuNp in contrast to the scheme of IV. For intravenous doping, it is considered that the nanoparticles are homogenously spread into the entire tumor section which results the dissipation of heat energy to the surrounding normal tissues. On the other hand, during intratumoral doping, the existence of nanoparticles in accumulated form results in binding the heat energy and thereby dissipating the same to the neighboring

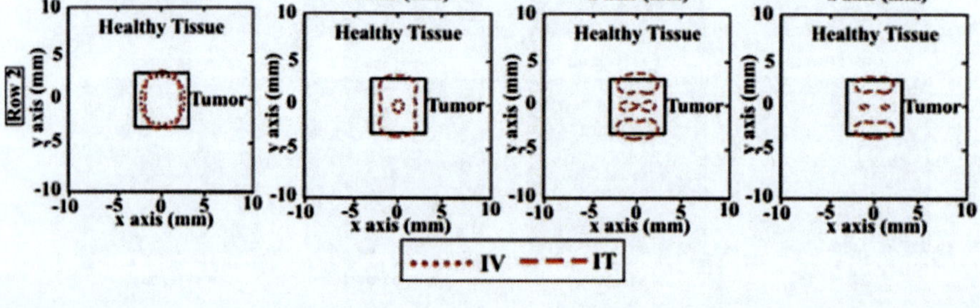

Fig. 4 Necrotic damage contour plots of tissue on x–y plane at different depths of $z = 0$ mm [Column 1], $z = -1$ mm [Column 2], $z = -2$ mm [Column 3], $z = -3$ mm [Column 4] applying Pennes [Row 1] and DPL [Row 2] model under different infusion schemes of nanoparticles, viz. IV and IT at $t = 300$ s

tumor regime. Thus, the targeted injection technique of AuNp in the scheme of IT affects a more uniform and concise ablation of tumor in contrast to the IV scheme.

3.2 *Effect of Non-fourier Bioheat Model Over the Fourier Model*

To explore the inhomogeneous nature of tissue model in the presence of large blood vessel incorporating clusters of nanoparticles under laser illumination, the variation of the temperature history with the exposure time nearby the cluster position is depicted in Fig. 5. The DPL heat transfer model has been implemented in comparison to the Pennes model to compute the thermal behavior of tissues during the laser heating. For DPL model, in both the schemes of IV and IT, a lag in stiff temperature variation has been revealed at the instant of laser on and off as revealed in Fig. 5 (Row 2), due to the limited speed of thermal wavefront. On the other hand, this lag of the stiff temperature variation has not been witnessed in the case of Fourier-based model. Considering the IV scheme, the temperature magnitude at the upper surface of tissue touches the ablation limit (≥ 51°C) in lesser time with respect to the IT scheme as shown in Fig. 5. The homogenous spreading of nanoparticles during IV scheme

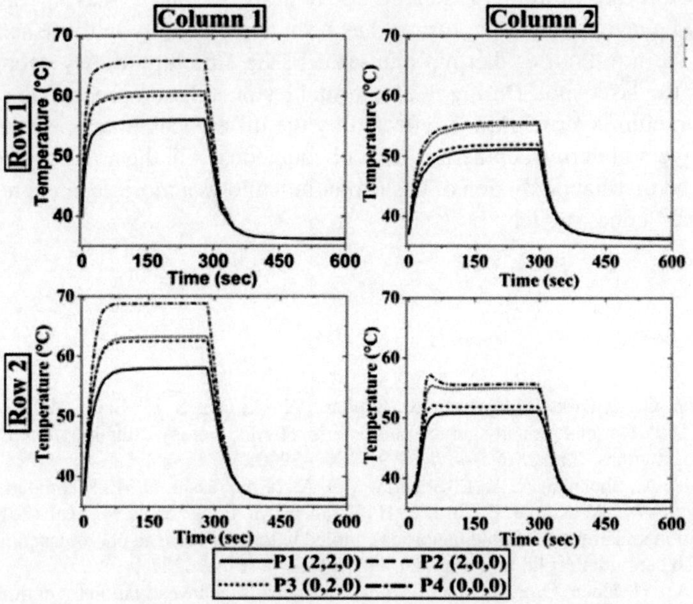

Fig. 5 Temporal plots of temperature at different locations of tissue top surface, viz. P1, P2, P3 and P4, considering the schemes of IV [Column 1] and IT [Column 2] applying the Pennes [Row 1] and DPL [Row 2] model

affects a higher photonic energy immersion at the tissue top surface, in contrast to the IT scheme. For the DPL model, a fluctuation in thermal plot with respect to time has been revealed in the scheme of IT (shown in Fig. 5 [Column 2, Row 2]) prior to achieving the equilibrium condition. The main reason behind this incident is because of the nonhomogeneous tissue structure embedded in clusters of AuNp and also due to the consideration of very small magnitude [17] of 0.043 s as well as a very high magnitude of 16 s. Moreover, no fluctuation in thermal plot has been revealed in the scheme of IV considering DPL model, as shown in Fig. 5 (Column 1, Row 2).

4 Conclusion

A computational study of the thermal behavior of triple-layered tissue structure during laser-based tumor necrosis process has been done signifying the heat sink effect of large blood vessel. To attain a target-specific ablation with minimum healthy tissue damage, considering the countercurrent blood flow, cases like intravenous (IV) and intratumoral (IT) doping of nanoparticles have been applied in the present study. For the fixed protocol parameters, the IT scheme witnessed a more uniform necrotic temperature and damage history to the entire tumor regime preserving the surrounding healthy tissue, in contrast to the IV scheme. A lag in thermal responses has been witnessed in the DPL scheme for both the IV and IT cases, whereas the fluctuating behavior of thermal history has been depicted only in the scheme of IT. However, the non-Fourier thermal nature of tissue structure slowly decreases far away from the laser spot. During photothermal lesion ablation process, the effect of stratum corneum layer of skin is reflected by the differences in temperature plot at different layers of dermis, epidermis and subcutaneous. All, the non-Fourier bioheat model-based thermal prediction of tissue medium allows a more accurate result than the classical Pennes model.

References

1. Maltzahn, G. V., Park, J., Agarwal, A., Bandaru, N. K., Das, S. K., Sailor, M. J., & Bhatia, S. N. (2009). Computationally guided photothermal tumor therapy using long-circulating gold nanorod antennas. *Cancer Research, 69*(9), 3892–3900.
2. Sayed, M. A., Shabaka, A. A., El-Shabrawy, O. A., N.A. Yassin, N. A., Mahmoud, S. S., El-Shenawy, S. M., Al-Ashqar, E., Eisa, W. H., Farag, N. M., El-Shaer, M. A., et al. (2013). Tissue distribution and efficacy of gold nanorods coupled with laser induced photoplasmonic therapy in ehrlich carcinoma solid tumor model. *PLOS One, 8*(10), e76207 1–9.
3. Ren, Y., Qi, H., Chen, Q., & Ruan, L. (2017). Thermal dosage investigation for optimal temperature distribution in gold nanoparticle enhanced photothermal therapy. *International Journal of Heat and Mass Transfer, 106*, 212–221.
4. Dua, R., & Chakraborty, S. (2005). A novel modeling and simulation technique of photo–thermal interactions between lasers and living biological tissues undergoing multiple changes in phase. *Computers in Biology and Medicine, 35*, 447–462.

5. Jaunich, M., Raje, S., Kim, K., Mitra, K., & Guo, Z. (2008). Bio-heat transfer analysis during short pulse laser irradiation of tissues. *International Journal of Heat and Mass Transfer, 51*(23–24), 5511–5521.
6. Paul, A., Narasimahan, A., Kahlen, A., & Das, S. K. (2014). Temperature evolution in the tissues embedded with large blood vessels during photo-thermal heating. *Journal of Thermal Biology, 41*, 77–87.
7. Kumar, S., & Srivastava, A. (2015). Thermal analysis of laser-irradiated tissue phantoms using dual phase lag model coupled with transient radiative transfer equation. *International Journal of Heat and Mass Transfer, 90*, 466–479.
8. Kumar, S., & Srivastava, A. (2017). Finite integral transform-based analytical solutions of dual phase lag bio-heat transfer equation. *Applied Mathematical Modelling, 52*, 378–403.
9. Wongchadakul, P., Rattanadecho, P., & Wessapan, T. (2018). Implementation of a thermo-mechanical model to simulate laser heating in shrinkage tissue (effects of wavelength, laser irradiation intensity, and irradiation beam area). *International Journal of Thermal Sciences, 134*, 321–326.
10. Paul, A., & Paul, A. (2018). Computational study of photo-thermal ablation of large blood vessel embedded tumor using localized injection of gold nanoshells. *Journal of Thermal Biology, 78*, 329–342.
11. Paul, A., & Paul, A. (2019). Thermomechanical analysis of a triple layered skin structure in presence of nanoparticle embedding multi-level blood vessels. *International Journal of Heat and Mass Transfer, 148*, 119076. https://doi.org/10.1016/j.ijheatmasstransfer
12. Paul, A., & Paul, A. (2020). In-vitro thermal assessment of vascularized tissue phantom in presence of gold nanorods during photo-thermal therapy. *Journal of Heat Transfer ASME*. https://doi.org/10.1115/1.4047371
13. Cetingul, M. P., & Herman, C. (2011). Quantification of the thermal signature of melanoma lesion. *International Journal of Thermal Science, 50*, 421–431.
14. Wang, X., Gao, X., & Liu, J. (2010). Monte-Carlo simulation on gold nanoshells enhanced laser interstitial thermal therapy on target tumor. *Journal of Computational and Theoratical Nanoscience, 7*, 1025–1031.
15. Saccomandi, P., Schena, E., Caponero, M. A., Di Matteo, F. M., Martino, M., Pandolfi, M., et al. (2012). Theoretical analysis and experimental evaluation of laser-induced interstitial thermotherapy in ex vivo porcine pancreas. *IEEE Transactions on Biomedical Engineering, 59*(9), 2958–2964.
16. Welch, A. J. (1984). The thermal response of laser irradiation tissue. *IEEE Journal of Quantum Electronics, 20*(12), 1471–1481.
17. Patidar, S., Kumar, S., Srivastava, A., & Sing, S. (2016). Dual phase lag model-based thermal analysis of tissue phantoms using lattice Boltzmann method. *International Journal of Thermal Science, 103*, 41–56.

Energetic and Exergetic Analysis of an Ejector-Based Green Refrigeration System Employing Evaporative Cooling in the Gas Cooler

Biswajit Gorai and Satyabrata Sahoo

Nomenclature

c_p	Sp. heat at cont. pressure (kJ/kg-K)
DBT	Dry bulb temperature (°C)
GC	Gas Cooler
h	Enthalpy (kJ/kg)
$\dot{I}$	Exergy destruction (kW)
$\dot{m}$	Mass flow rate (kg/s)
P	Pressure (kPa)
$\dot{Q}$	Heat transfer rate (kW)
RH	Relative humidity (dimensionless)
s	Entropy (kJ/kg-K)
T	Temperature (°C)
u	Velocity (m/s)
$\dot{W}$	Power Input (kW)
WBT	Wet bulb temperature (°C)

Greek Symbols

$\dot{\psi}$	Exergy (kW)
η	Efficiency (dimensionless)
μ	Entrainment ratio (dimensionless)
Δ	Difference
v	Specific volume (m^3/kg)

B. Gorai (✉) · S. Sahoo
Department of Mechanical Engineering, Indian Institute of Technology (Indian School of Mines), Dhanbad, Jharkhand 826004, India

P. Mahanta et al. (eds.), *Advances in Thermofluids and Renewable Energy*, Lecture Notes in Mechanical Engineering, https://doi.org/10.1007/978-981-16-3497-0_8

Subscripts

AC	Air cooled
amb	Ambient
comp	Compressor
cond	Condenser
diff	Diffuser
EVC	Evaporative cooling
evp	Evaporator
in	Inflow
out	Outflow
p	Primary
ref	Refrigerant
s	Secondary

1 Introduction

Considerable exergy consumption in CO_2 refrigeration systems is a major concern for its high operating pressure in gas cooler. Nowadays, researchers are showing their keen interest in using CO_2 in various refrigeration systems especially in dual systems where both cooling and heating are required, mobile air conditioning as well as in supermarkets. CO_2 is an environment friendly (ODP = 0 & GWP = 1), non-toxic and less expensive refrigerant. So, the application of CO_2 as refrigerant is both economical as well as environmental benign. In addition to that CO_2 is non-flammable, having excellent potential for charge reduction and has better transport properties. Moreover, unlike NH_3, it has no compatibility issue with copper and other alloys mainly used as the tubing material in refrigeration units. Considering the above factors, CO_2 can be considered as a suitable refrigerant for dairy and automotive air conditioning applications. On the other hand, the major issues with CO_2 as a refrigerant, such as low critical temperature (31.4 °C) and high critical pressure (73.84 bar) are coming in the way of its large-scale commercial use. Because of these adverse critical properties in most cases, the heat rejection takes place in the super-critical zone based on the availability of heat sink temperature. Because of its high critical pressure, the operating pressure of these systems is quite high resulting in increased power consumption as well as more robust design. Hence, the initial as well as running cost of these systems are quite high. In order to address these issues and make these systems commercially viable, many researchers have suggested various system modifications as well finding simultaneous cooling and heating applications, where these systems can best fit. Liu et al. [1] did a comparative study on NH_3 and CO_2 heat pump systems for both heating and cooling applications. The annual cost shaving for different food processes and for different locations is estimated. Omer et al. [2] have explained the best use of heat pump for milk pasteurisation system.

Robinson and Groll [3] have established for a given gas cooler exit temperature the COP of the transcritical CO_2 system is a strong function of the gas cooler pressure, and the maximum COP is achieved at the optimum gas cooler pressure. The study also demonstrated that the maximum irreversibility associated with expansion valve resulted in lower 2nd law efficiency. Ahammed et al. [4] have analysed a conventional transcritical CO_2 system for simultaneous cooling and heating application for a dairy plant. The performance of four different configurations with and without ejectors employing multiple gas coolers and an internal heat exchanger is compared. The result clearly demonstrates that the position of the IHX plays a vital role, and the best performance was realised when the IHX was placed between the gas coolers. Deng et al. [5] described the usefulness of ejectors in CO_2-based transcritical refrigeration systems by showing 25% improvement in cooling COP. Experimental study carried out by Hajidavalloo, Eghtedari [6], established the usefulness of evaporative condenser over air cooled condenser in air conditioning applications. The experimental study has demonstrated up to 20% reduction in power consumption and 50% improvement in COP by replacing the air cooled condenser by an evaporative condenser. Additionally, in hot and dry weather conditions, the application of evaporative condenser is commercially recommended. From the survey, it is observed that not much study on evaporative cooling-based transcritical CO_2 systems is available in open literature. Moreover, the exergetic analysis especially systems incorporating evaporative cooling system which handles the air water vapour mixture is scarce in open literature. Hence, a comprehensive thermodynamic comparison both 1st and 2nd law is the prime focus of the present investigation.

2 CO_2 Refrigeration System with Air Cooled and Evaporative Gas Coolers

2.1 Component Modelling

As shown in Fig. 1a, transcritical CO_2-based vapour compression refrigeration system deploying an ejector in place of the expansion valve and an air cooled gas cooler as the heat rejecting unit has been considered for the present analysis. It has been observed from the literature that in the case of CO_2 system, gas cooler rejects heat in the transcritical zone. So the operating pressure of the gas cooler, hence the pressure ratio of the compressor is a strong function of gas cooler exit temperature which mainly depends on the type of cooling medium and the design of the gas cooler. In order to overcome this problem, an attempt has been made to study a modified system in which the air cooled gas cooler is replaced by an evaporative cooling-based gas cooler as shown in Fig. 1b. The refrigerant coming from the gas is accelerated through the primary nozzle, thus creating a suction which sucks the refrigerant vapour from the evaporator through the secondary nozzle (i.e. process 6–10). Both the streams undergo a constant pressure mixing process while passing through

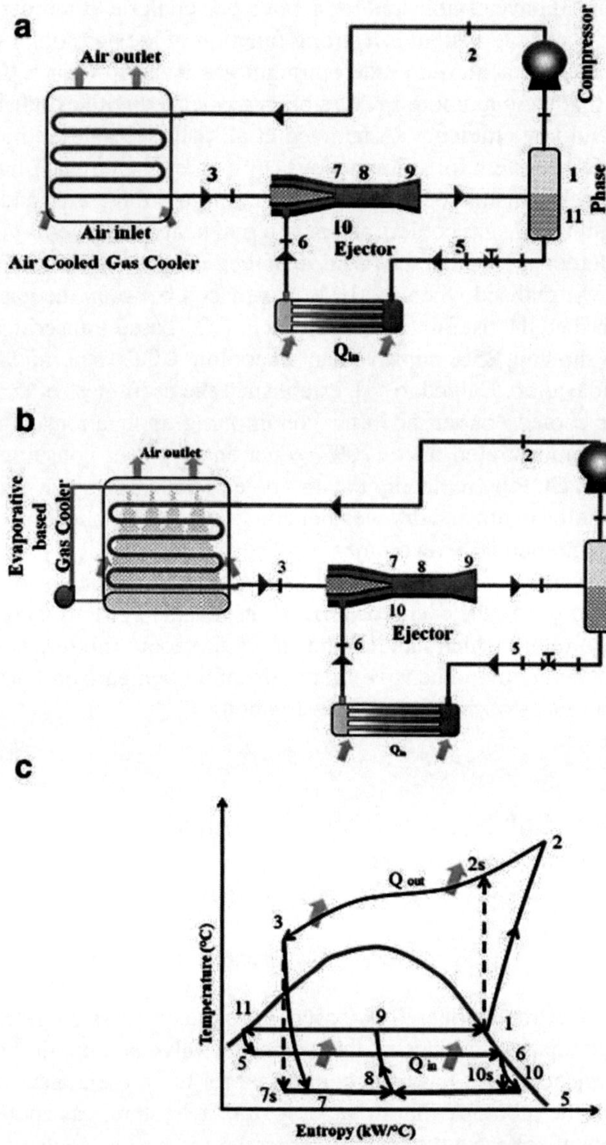

Fig. 1 **a** Schematic of an ejector-based CO_2 refrigeration system with air cooled gas cooler. **b** Schematic of an ejector-based CO_2 refrigeration system with evaporative gas cooler. **c** T-s plot of CO_2 refrigeration cycle with ejector

the mixing zone (i.e. process 8–9). After the mixing zone, the refrigerant passes through a diffusing section to recover the pressure. The phase separator installed at the exit of the diffuser separates the vapour part from that of the liquid part. The liquid part enters into the evaporator to take care of the cooling load, and the gaseous part is compressed by the compressor and then cooled in the gas cooler, completing the cycle. The compressed gas rejects the heat to the cooling medium while passing through the gas cooler, thus completing the cycle.

The analysis is based on the following simplifying assumptions:

- The heat loss from different components to ambient is negligible.
- The pressure drop in various components is negligible.
- Refrigerant flow is steady state and steady flow condition.
- The kinetic energy at the entry of the nozzle and at the exit of the diffuser is negligible
- Energy inputs to the air circulation fan and water pump in the heat rejecting unit are negligible.

The steady-state thermodynamic analyses of both the systems are carried out applying 1st and 2nd law to each individual component. The thermodynamic cycle is pictorially presented on $T{-}s$ plot as given in Fig. 1c.

2.2 Mathematical Equations

Evaporator

Applying steady flow energy equation and exergy balance to the evaporator, the refrigeration capacity and rate of exergy destruction can be expressed as given in Eqs. 1 and 2, respectively. Here $\dot{m}_{\text{ref\_s}}$ is mass flow rate of refrigerant passing through the secondary nozzle and also through the evaporator.

$$\dot{Q}_{\text{ref}} = \dot{m}_{\text{ref\_s}}(h_6 - h_5) \tag{1}$$

where enthalpy of state point 5 and 6 is the functions of evaporator temperature and dryness fraction at the inlet and exit of the evaporator.

$$\begin{aligned} \dot{I}_{\text{evp}} =& \dot{m}_{\text{evp\_s}}((h_5 - h_6) - T_{\text{amb}}(s_5 - s_6)) \\ &+ \dot{Q}_{\text{ref}}\left(1 - \frac{T_{\text{amb}}}{T_{\text{space}}}\right) \end{aligned} \tag{2}$$

where T_{space} is the refrigerated space temperature.

Compressor

The compression process is assumed to be non-isentropic by including the isentropic efficiency, which is given as a function of pressure ratio Robinson et al. [3]

$$\eta_{\text{comp}} = 0.815 + 0.022\left(\frac{P_2}{P_1}\right) - 0.0041\left(\frac{P_2}{P_1}\right)^2 + 0.0001\left(\frac{P_2}{P_1}\right)^3 \quad (3)$$

The power input to the compressor is given by:

$$\dot{W}_{\text{comp}} = \dot{m}_{\text{ref\_p}}(h_2 - h_1) = \frac{\dot{m}_{\text{ref\_p}}(h_{2s} - h_1)}{\eta_{\text{com}}} \quad (4)$$

Exergy destruction of the compressor is given by:

$$\dot{I}_{\text{comp}} = \dot{m}_{\text{ref\_p}}((h_1 - h_2) - T_{\text{amb}}(s_1 - s_2)) + \dot{W}_{\text{comp}} \quad (5)$$

Ejector

Based on the simplifying assumption mentioned earlier, the ejector part which comprises the primary, secondary nozzle, mixing chamber and diffuser is simulated using governing mass, energy and exergy balance equations given below. The ratio of secondary to primary mass flow rate of the refrigerant in secondary and primary nozzles, respectively, is termed as entrainment ratio and is expressed as:

$$\mu_{\text{ejec}} = \dot{m}_{\text{ref\_s}}/\dot{m}_{\text{ref\_p}} \quad (6)$$

Primary Nozzle

$$\eta_{\text{ejec\_np}} = \left(\frac{h_3 - h_7}{h_3 - h_{7s}}\right) \quad (7)$$

$$h_{7s} = f(s_3, P_{10}) \quad (8)$$

$$h_3 = h_7 + \frac{u_7^2}{2} \quad (9)$$

$$\dot{I}_{\text{nozzle\_p}} = \dot{m}_{\text{ref\_p}} T_{\text{amb}}(s_3 - s_7) \quad (10)$$

Secondary Nozzle

$$\eta_{\text{ejec\_ns}} = \left(\frac{h_6 - h_{10}}{h_6 - h_{10s}}\right) \quad (11)$$

$$h_{10s} = f(s_6, P_{10}) \quad (12)$$

$$h_6 = h_{10} + \frac{u_{10}^2}{2} \quad (13)$$

$$\dot{I}_{\text{nozzle\_s}} = \dot{m}_{\text{ref\_s}} T_{\text{amb}} (s_6 - s_{10}) \tag{14}$$

Constant pressure mixing zone

$$\begin{aligned} \dot{m}_{\text{ref\_total}} \left(h_8 + \frac{u_8^2}{2} \right) &= \dot{m}_{\text{ref\_}p} \left(h_7 + \frac{u_7^2}{2} \right) \\ &+ \dot{m}_{\text{ref\_s}} \left(h_{10} + \frac{u_{10}^2}{2} \right) \end{aligned} \tag{15}$$

$$\dot{I}_{\text{mix}} = \left(\dot{m}_{\text{ref\_total}} T_{\text{amb}} s_8 \right) - \left(\dot{m}_{\text{ref\_p}} T_{\text{amb}} s_7 \right) - \left(\dot{m}_{\text{ref\_s}} T_{\text{amb}} s_{10} \right) \tag{16}$$

Diffuser

$$h_9 = h_8 + \frac{u_8^2}{2} \tag{17}$$

$$\eta_{\text{ejec\_d}} = \left(\frac{h_{9s} - h_8}{h_9 - h_8} \right) \tag{18}$$

$$\dot{I}_{\text{dif}} = \dot{m}_{\text{ref\_total}} T_{\text{amb}} (s_9 - s_8) \tag{19}$$

Total exergy destruction of the ejector is given by:

$$\dot{I}_{\text{eje}} = \dot{I}_{\text{dif}} + \dot{I}_{\text{mix}} + \dot{I}_{\text{nozzle\_p}} + \dot{I}_{\text{nozzle\_s}} \tag{20}$$

Phase separator

It separates the saturated liquid and vapour from the vapour and liquid mixture. The isothermal separation is expressed as:

$$\dot{m}_{\text{total}} h_9 = \dot{m}_p h_1 + \dot{m}_s h_{11} \tag{21}$$

Exergy destruction in phase separator is calculated as follows:

$$\dot{\psi}_{\text{in\_1}} = \dot{m}_{\text{ref\_total}} ((h_9 - h_{\text{amb}}) - T_{\text{amb}} (s_9 - s_{\text{amb}})) \tag{22}$$

$$\dot{\psi}_{\text{out,1}} = \dot{m}_{\text{ref\_s}} ((h_{11} - h_{\text{amb}}) - T_{\text{amb}} (s_{11} - s_{\text{amb}})) \tag{23}$$

$$\dot{\psi}_{\text{out\_2}} = \dot{m}_{\text{ref\_}p} ((h_1 - h_{\text{amb}}) - T_{\text{amb}} (s_1 - s_{\text{amb}})) \tag{24}$$

$$\dot{I}_{\text{phs}} = \dot{\psi}_{\text{in\_1}} - \dot{\psi}_{\text{out\_1}} - \dot{\psi}_{\text{out\_2}} \tag{25}$$

Air cooled gas cooler

The heat rejected by the refrigerant is taken by atmospheric air. A constant flow rate of air in the heat exchanger is maintained by installing a fan of required capacity. The refrigerant exit temperature of the gas cooler is fixed by assuming a minimum temperature difference of 15 °C [7] between the coolant (air) and refrigerant. From the energy balance:

$$\dot{Q}_{gc\_AC} = \dot{m}_{ref\_p}(h_2 - h_3) = \dot{m}_{air}.c_{p\_air}.\Delta T_{air} \tag{26}$$

where (m_{air}) mass flow of water through gas cooler and ΔT_{air} is temperature rise of air in the gas cooler.

Exergy destruction on the gas cooler is calculated using following equations:

$$\dot{\psi}_{ref\_in} = \dot{m}_{ref\_p}\left(\left(h_2 - h_{amb\_ref}\right) - T_{amb}\left(s_2 - s_{amb\_ref}\right)\right) \tag{27}$$

$$\dot{\psi}_{ref\_out} = \dot{m}_{ref\_p}\left(\left(h_3 - h_{amb\_ref}\right) - T_{amb}\left(s_3 - s_{amb\_ref}\right)\right) \tag{28}$$

$$\dot{\psi}_{air\_in} = 0 \tag{29}$$

$$\dot{\psi}_{air\_out\_AC} = \dot{m}_{air}\left(\begin{array}{c}\left(h_{air\_out} - h_{air\_amb}\right) - \\ T_{amb}\left(s_{air\_out} - s_{air\_amb}\right)\end{array}\right) \tag{30}$$

$$\dot{I}_{gc\_AC} = \dot{\psi}_{ref\_in} + \dot{\psi}_{air\_in} - \dot{\psi}_{ref\_out} - \dot{\psi}_{air\_out\_AC} \tag{31}$$

Evaporative cooling-based gas cooler

In the evaporative cooler, water is sprayed with the help of a small pump and water sprinkler. A centrifugal fan maintains the air flow inside the evaporative gas cooler. Dry atmospheric air enters into the evaporative gas cooler and becomes cooler by absorbing water particles. This humid air absorbs heat from the refrigerant passing through the gas cooler tubing. Both latent and sensible heat transfer results in improved system performance by maintaining lower gas cooler exit temperature. Based on the data reported in the existing literature, the present analysis assumes that the refrigerant can be cooled up to 11–13 °C above WBT of the incoming air. The following mathematical formulation is used to simulate the evaporative cooling-based gas cooler part.

$$P_{sat,in} = f(T_{in}) \tag{32}$$

$$P_{sat,out} = f(T_{out}) \tag{33}$$

$$P_{v,in} = P_{sat,in}.RH_{in} \tag{34}$$

$$P_{v,out} = P_{sat,out}.RH_{out} \tag{35}$$

$$h_{water,in} = f(T_{water,in}) \tag{36}$$

$$v = f(T_{water,in}) \tag{37}$$

$$\omega_{in} = 0.622\left(\frac{P_{v,in}}{P_{atm} - P_{v,in}}\right) \tag{38}$$

$$\omega_{out} = 0.622\left(\frac{P_{v,out}}{P_{atm} - P_{v,out}}\right) \tag{39}$$

$$\dot{m}_{air} = \left(\frac{\dot{m}_{w\_makeup}}{\omega_{out} - \omega_{in}}\right) \tag{40}$$

Applying the steady flow energy equation to the evaporative cooling-based gas cooler.

$$\begin{aligned}\dot{Q}_{gc\_EVC} &= \dot{m}_{ref\_p}(h_2 - h_3)\\ &= \dot{m}_{air}\left(h_{air\_out} - h_{air\_in}\right) - h_{water\_in}.\dot{m}_{w\_makeup}\end{aligned} \tag{41}$$

The expressions for flow exergy associated with refrigerant in, out and air in are the same as given in Eqs. 27, 28 and 29, respectively.

$$\dot{\psi}_{air\_out\_EVC} = \dot{m}_{air} \left[\begin{array}{c}\left(c_{p\_air} + \omega_{out} c_{p\_water}\right).\left(T_{air\_input}\right)\left[\frac{T_{air\_output}}{T_{air\_input}} - 1 - \ln\left(\frac{T_{air\_output}}{T_{air\_input}}\right)\right] + \\ 0.287.\left(T_{air\_input}\right).\left[\begin{array}{c}(1 + 1.608.\omega_{out}).\ln\left(\frac{(1+1.608.\omega_{in})}{(1+1.608.\omega_{out})}\right) + \\ 1.608.\omega_{in}.\ln\left(\frac{\omega_{out}}{\omega_{in}}\right)\end{array}\right]\end{array}\right] \tag{42}$$

$$\dot{\psi}_{water\_in} = \dot{m}_{w\_makeup} \left[\begin{array}{c}(h_{water\_in} - h_{water\_atm}) - T_{atm}.(s_{water\_in} - s_{water\_atm}) + \\ (p_{atm} - p_{sat\_in}).v_{water} - \left(\frac{8314}{18}\right).T_{air\_input}.\ln\left(Rh_{input}\right)\end{array}\right] \tag{43}$$

$$\begin{aligned}\dot{I}_{gc\_EVC} &= \dot{\psi}_{ref\_in} - \dot{\psi}_{ref\_out} + \dot{\psi}_{water\_in}\\ &+ \left(\dot{\psi}_{air\_in} - \dot{\psi}_{air\_out\_EVC}\right)\end{aligned} \tag{44}$$

Expansion valve

In an ejector-based system, the expansion valve is installed at the exit port of phase separator. While flowing through the expansion valve, no change in enthalpy is observed. However, exergy destruction takes place due to irreversible pressure drop by frictional dissipation and is estimated using the expression given in Eq. 44.

$$\dot{I}_{\text{ev}} = T_{\text{amb}} \dot{m}_{\text{ref\_s}} (s_5 - s_{11}) \tag{45}$$

The total exergy destruction is given by:

$$\dot{I}_{\text{t}} = \dot{I}_{\text{evp}} + \dot{I}_{\text{comp}} + \dot{I}_{\text{eje}} + \dot{I}_{\text{gc\_AC}} / \dot{I}_{\text{gc\_EVC}} + \dot{I}_{\text{phs}} + \dot{I}_{\text{ev}} \tag{46}$$

System COP is calculated considering the cooling effect of the refrigeration system:

$$COP_{\text{system}} = \frac{\dot{Q}_{\text{evp}}}{\dot{W}_{\text{comp}}} \tag{47}$$

3 Result and Discussion

All the simulations are carried out for the input parameters included in Table 1. If nothing is mentioned, the RH is assumed to be 40%.

Table 1 Input parameters for simulation [7–9]

	Range
Cooling load	37 kW
Evaporator temperature	−3 °C
Volume flow rate of air for air cooled gas cooler	169.1 cm
Volume flow rate of air for EVC-based gas cooler	74.5 cm
Air temperature at the inlet gas cooler	30–40 °C
Relative humidity at the inlet to the EVC-based gas cooler	30–60%
Relative humidity at the exit of the EVC-based gas cooler	85%
Nozzle efficiency (both primary and secondary)	85%
Diffuser efficiency	80%
Pressure drop of secondary nozzle	0.3 bar
Pressure drop in gas cooler	1 bar
Water inlet temperature to the evaporative cooler	20 °C

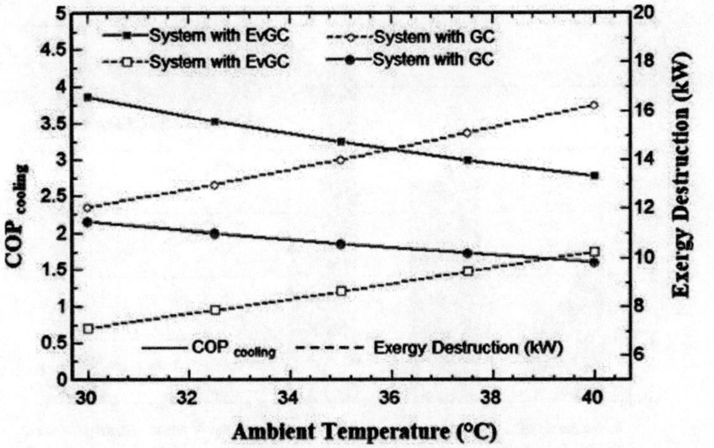

Fig. 2 Effects of ambient temp on system COP and the total irreversibility

Figure 2 presents the effect of ambient temperature on COP and total irreversibility for both the systems. The figure clearly demonstrates a significant improvement in COP and reduction in total irreversibility for EVC-based gas coolers compared that of the air cooled system, and this may be attributed to lower gas cooler exit temperature in the earlier case which results in lower operating pressure. It has been observed that for a given gas cooler exit temperature, there exists an optimum gas cooler pressure at which the transcritical system gives its best performance. Hence, for the present comparison, the operating pressures for both the air cooled and EVC-based gas coolers are fixed at their optimum values corresponding to the respective gas cooler exit temperatures. Because of the reduced operating pressure of the EVC-based gas cooler system, the power input to the compressor decreases resulting in improved COP. However, the difference decreases with increase in the ambient temperature. In case of air cooled systems because of the large temperature difference between the refrigerant and the air and higher heat rejection rate due to high operating pressure significantly increases the system irreversibility.

Figure 3 is a pictorial representation of exergy destruction or irreversibility present in various components. The figure clearly depicts that in case of air cooled systems, the maximum irreversibility is recorded in the ejector; however, in EVC-based systems, the gas cooler overtakes the ejector in terms of exergy destruction. It is noteworthy to mention here that in all the components except the evaporator, the exergy destruction is higher in case of air cooled systems compared to that of EVC-based systems. One of the major reasons behind this large difference in the irreversibility associated with all the components is the high operating gas cooler pressure in case of air cooled systems. High operating gas cooler pressure results in higher work of compression, increased compressor inefficiency, higher heat rejection rate in the gas

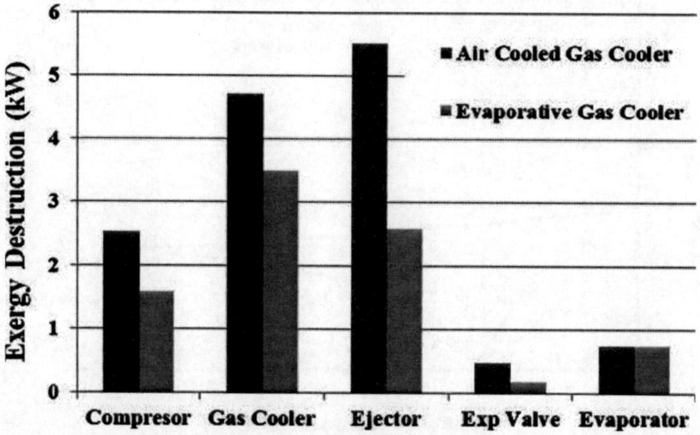

Fig. 3 Component-wise exergy destruction

cooler with larger temperature difference and consequently higher exergy destruction. Similarly for the expansion valve the exergy destruction is higher for systems with air cooled gas coolers due to its higher operating pressure ratio.

Figure 4 indicates the effects of relative humidity on the system COP as well as the makeup water requirement for different ambient conditions. It has been observed that for dry atmosphere, the COP of the EVC-based system is significantly higher compared to the air cooled system. This is due to lower WBT of ambient air which decides the refrigerant exit temperature in case of EVC-based gas coolers. As RH of the ambient air increases, the COP of the EVC systems approaches that of the air

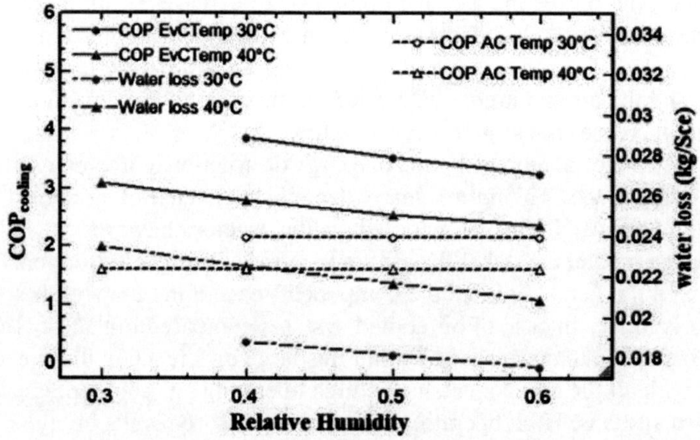

Fig. 4 Variation of system COP and makeup requirement with the relative humidity of the ambient air

cooled system. This is more pronounced at lower DBT of the air. The makeup water requirement reduces with increase in RH and increases with DBT.

4 Conclusion

In the present investigation, a steady-state simulation is carried out to compare the performance of two ejector-based transcritical CO_2 systems: one with air cooled gas cooler and other one with EVC-based gas cooler. The subject work clearly demonstrates a significant improvement in the case of EVC-based systems both from 1st and 2nd law viewpoints. From the parametric study, it is observed that the percentage improvement in COP decreases with increase in ambient temperature and relative humidity. One of the important highlights of the present investigation is the estimation of the component-wise exergy destruction with a special focus on an EVC-based system which deals with binary mixture subjected to both heat and mass transfer. From the analysis, it is observed that the maximum irreversibility is associated with the ejector in case of air cooled system and with the gas cooler for the EVC-based system. Based on this simplified thermodynamic analysis, the study recommends the use of EVC-based gas coolers in CO_2 systems. However, the design and other operating issues can be realised only after a detailed dynamic analysis with experimental validation.

References

1. Liu, Y., Groll, E. A., Yazawa, K., & Kurtulus, O. (2017). Energy-saving performance and economics of CO_2 and heat pumps with simultaneous cooling and heating applications in food processing Case studies. *International Journal of Refrigeration, 73*, 111–124.
2. Omer, C., Yuksel, B., Kara, Y. A., Caglar, A., & Tulek, Y. (1994). Heat pump utilization in milk pasteurization. *Energy Conversion and Management, 35*(2), 91–96.
3. Robinson, D. M., & Groll, E. A. (1998). Efficiencies of transcritical CO_2 cycles with and without an expansion turbine. *International Journal of Refrigeration, 21*(7), 577–589.
4. Ahammed, E., Bhattacharyya, S., & Ramgopal, M. (2018). Analysis of CO_2 based refrigeration systems with and without ejector for simultaneous pasteurization and chilling of milk. *International Journal of Refrigeration, 95*, 61–72.
5. Deng, J., Jiang, P., Lu, T., & Lu, W. (2007). Particular characteristics of transcritical CO_2 refrigeration cycle with an ejector. *Applied Thermal Engineering, 27*, 381–388.
6. Hajidavalloo, E., & Eghtedari, H. (2010). Performance improvement of air-cooled refrigeration system by using evaporatively cooled air condenser. *International Journal of Refrigeration, 33*, 982–988.
7. Arora, R. C. (2010). Refrigeration and air conditioning. PHI Learning Private Limited, p. 507
8. Ahammed, E., Bhattacharyya, S., & Ramgopal, M. (2014). Thermodynamic design and simulation of a CO_2 based transcritical vapour compression refrigeration system with an ejector. *International Journal of Refrigeration, 45*, 177–188.
9. Singh, S., & Dasgupta, M. S. (2017). CO_2 heat pump for waste heat recovery and utilization in dairy industry with ammonia based refrigeration. *International Journal of Refrigeration, 78*, 108–120.

Influence of Pulsating Flow on Thermal Characteristics in a Triangular Sharp-Edged Wavy Channel

Rakesh Kumar, Kapil Dev, Ranjan Kumar, Pulak Sen, and Dipak Sen

Nomenclature

Aw	Amplitude of the wavy wall
Nu	Nusselt number
Pr	Prandtl number
Re	Reynolds number
St	Strouhal number
S_b	Profile of the bottom wavy wall
S_t	Profile of the top wavy wall
T	Temperature
T_{in}	Inlet flow temperature
T_w	Wall temperature of wavy channel
U	Velocity vector
b/L	Horizontal pitch-to-module ratio
n	Power law index
p	Pressure
t	Time
u	X-direction velocity
u_{avg}	Average inlet velocity
u_{in}	Inlet velocity
v	Y-direction velocity

R. Kumar · K. Dev · R. Kumar · D. Sen (✉)
Department of Mechanical Engineering, National Institute of Technology Arunachal Pradesh, Yupia 791112, India

P. Sen
Department of Mechanical Engineering, Techno College of Engineering Agartala, Tripura 799001, India

P. Mahanta et al. (eds.), *Advances in Thermofluids and Renewable Energy*, Lecture Notes in Mechanical Engineering, https://doi.org/10.1007/978-981-16-3497-0_9

1 Introduction

In many applications like heat exchangers, heat transfer enhancement is of great concern. Many ways are being proposed in design of walls. One of these is the corrugation of wavy channel. Due to corrugation, area of contact between fluid and the walls is increased, thereby giving enhanced heat transfer rate and better Nusselt number. Also heat transfer effectiveness is affected by various parameters ranging from geometry of walls and nature of flow to properties of working fluid. Effectiveness of such channel flow can be easily compared with flow through similar flat walled channels. Flow through triangular-shaped channel has been studied by various researchers [1–4]. Wang and Vanka [5] in their study of heat transfer in a periodic wavy channel flow concluded that heat transfer rate was high for high Reynolds number (Re) compared to that of in parallel-plate flow. Wang and Chen [6] in their study of heat transfer in channel having different amplitude–wavelength ratio (**a/λ**) concluded that high local Nusselt number was achieved at higher a/λ around 0.2 and 0.3. It was observed in studies of Fabbri [7] that high heat transfer rate and Nusselt number was achieved at sufficiently high Re and Prandtl number (Pr). Rush et al. [8] studied the effect of phase shift between two wavy plates on various fluid flow parameters. Phase difference (ϕ) 0° was reported to give unsteady flow compared to $\phi =$ 1800. Mohammed et al. [9] studied forced convection in corrugated channels. Hossain and Islam [10] numerically examined unsteady flow in periodic wavy channels having a horizontal pitch at top. Bahaidarah et al. [11] in their comparative study of heat transfer in case of Sine- and arc-shaped channel concluded that the rate of heat transfer increases with increasing Re but at the cost of pressure drop rise. Niceno and Nobile [12] numerically studied the effect of sinusoidal- and arc-shaped corrugation of walls on heat transfer rate. They concluded that heat transfer increases up to three times for Re = 100–200 along with huge pressure drop compared to straight channel. Piao et al. [13] experimentally studied the forced convection in cross-corrugated solar air heaters. Snyder et al. [14] presented heat transfer enhancement in serpentine channel. Zhang et al. [15], both theoretically and practically, studied the effect of corrugation tilt angles (β). Naphon [16] studied two opposite plates with corrugations at 40°–60°. Nishandar and Yadav [17] experimentally investigated the pulsating turbulent flow in pipe. Sorour and Estafanous [18] did a numerical study on effect of pulsating flow between two parallel plates. Ali and Hanaoka [19] experimentally studied the effect of V-shaped corrugation on heat transfer and concluded that effect of local Reynolds number is significant at entrance region. They also concluded that with increasing tilt angles of corrugation from 0° to 60°, there is a significant enhancement of heat transfer ranging from 33 to 67.3%. Ozbolat et al. [20] numerically investigated the flow of water in 2D corrugated channels.

An extensive reviews of pulsating flow in thermal engineering was carried out by several researchers [21–30]. Tutty [31] studied the effect of pulsatile flow in corrugated channel and by Kim et al. [32], and wavy channel was studied by Metwally and Manglik [33]. Mehta and Khandekar [34] studied pulsating laminar flow in a square mini channel with varied frequencies. Xie et al. [35] numerically studied the effects of Reynolds numbers, pulsating amplitude, and frequency on heat transfer enhancement and flow friction. Nishimura et al. [36] both theoretically and experimentally studied the effect of pulsating flow in symmetric wavy channel. Jin et al. [37] experimentally studied the effect of pulsating flow on thermal characteristics fluid flowing through of triangular grooved channel. Bai et al. [38] did Numerical investigation on thermal hydraulic performance of supercritical LNG in sinusoidal wavy channel based printed circuit vaporizer. Shi et al. [39] studied the influence of structure parameters on entropy generation performance in cross wavy channels with fluid–solid coupled heat transfer. Wen et al. [40] did the numerical study on heat transfer behavior of wavy channel supercritical CO_2 printed circuit heat exchangers with different amplitude and wavelength parameters.

After doing a rigorous study, it is found that limited results were found for heat transfer through the channel having horizontal pitch-to-module ratio in case of pulsating flow. In the present study, an attempt has been made to determine optimum channel geometry at a particular Reynolds number for different sets of Strouhal number.

2 Problem Descriptions

An attempt has been made to increase the heat transfer rate by increasing the contact area between working fluid and walls of the channel. The fluid used is incompressible and Newtonian with laminar flow and heat transfer. Specific heat, thermal conductivity and viscosity are independent of temperature and negligible gravity, radiation heat transfer and viscous dissipation. The microchannel used in this study is shown in Fig. 1. The channel consists of two flat-plate sections at its entry and exit and triangular wavy walls bounding the channel. The lengths of all the portions of the microchannel are shown in Fig. 1a, b. In this work, a variation has been given to sharp-edged converging–diverging channels with varying sizes of horizontal pitch which lead to 5 different sets of b/L ratio.

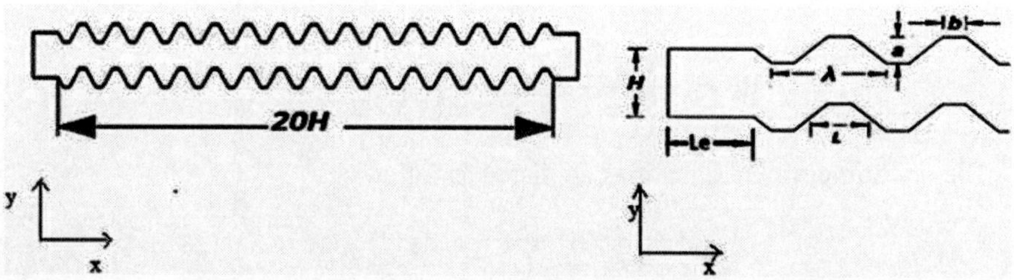

(a) (b)

Fig. 1 **a** Triangular Sharp-Edged Wavy Channel **b** Zoom view of Triangular Sharp-Edged Wavy channel

In the present work, two different a/λ ratios have been used. It has been found in the literature that a higher heat transfer rate is obtained for higher a/λ of 0.2 or 0.3. To get a better picture of optimum channel geometry, five different horizontal pitch to module (b/L) ratios have been used. Used b/L ratios are 0, 0.2, 0.4, 0.6, 0.8, 1, where b/L = 0 resembles completely triangular geometry and b/L = 1 represents channel having square corrugations. For each geometry wavelength of the channel was determined using a/λ = 0.2 or a/λ = 0.3 and fixing a = 0.4.

2.1 Governing Equations

The continuity, momentum and energy equations for unsteady two-dimensional laminar incompressible Newtonian fluid are considered. Using non-dimensional variables, the non-dimensional governing equations are obtained.

Continuity equation:

$$\nabla \mathbf{U} = 0 \tag{1}$$

Momentum equation:

$$\frac{\mathrm{d}U}{\mathrm{d}t} + (\mathbf{U}\cdot\nabla)\mathbf{U} = -\nabla p + \frac{1}{\mathrm{Re}}\nabla\tau_{ij} \quad (i, j) = (x, y) \tag{2}$$

Energy equation:

$$\frac{\mathrm{d}T}{\mathrm{d}t} + (\mathbf{U}\cdot\Delta)T = \frac{1}{\mathrm{Re\ Pr}}\nabla^2 T \tag{3}$$

where $\mathbf{U}$ is the velocity vector,p is the pressure and T is the non-dimensional temperature.

Appearing in Eqs. (2) and (3) Re and Pr are the Reynolds and Prandtl numbers, respectively, which are defined as

$$\mathrm{Re} = \frac{\rho u_{\mathrm{ave}}^{*2-n} L^n}{m}, \mathrm{Pr} = \frac{m c_p}{k}\left(\frac{u_{\mathrm{ave}}^*}{L}\right)^{n-1} \tag{4}$$

The non-dimensional quantities are listed as follows:

$$x = \frac{x^*}{L}, y = \frac{y^*}{L}, A = \frac{A^*}{L}, (x) = \frac{S^*(x^*)}{L},$$
$$u = \frac{u^*}{u_{\mathrm{ave}}^*}, v = \frac{v^*}{v^*}, T = \frac{T^* - T_w^*}{T_{\mathrm{in}}^* - T_w^*}, t = \frac{u_{\mathrm{ave}}^* t^*}{L}, p = \frac{p^*}{u_{\mathrm{ave}}^{*2} L} \tag{5}$$

2.2 *Boundary Conditions*

Both the corrugated walls of the microchannel are at constant temperatures. The straight portions of the wall of the channel are insulated. Both the top and the bottom walls are at a constant temperature of 301 K ($Tw*$). No slip boundary condition is used for both the walls. The velocity and temperature at inlet are given below:

u_{in} = Inlet velocity.
T_{in} = Inlet flow temperature.

3 Numerical Approach

The continuity, momentum and energy equations are solved using commercially available software ANSYS 16.2. Density-based transient energy solver using the SIMPLE scheme has been solved.

Nusselt number for wavy channel is calculated as

$$\text{Local:Nu} = \left(\frac{-L}{T_w^* - T_{\text{in}}^*}\right)\left(\left(\frac{\mathrm{d}T}{\mathrm{d}n}\right)\bigg|_{\text{top wall}} + \left(\frac{\mathrm{d}T}{\mathrm{d}n}\right)_{\text{bottom wall}}\right) \tag{6}$$

$$\text{Time and Space averaged} : < \text{Nu} > \quad = \frac{1}{\mathrm{t_p}} \int \langle Nu \rangle \mathrm{d}t \ \text{over}\ t_p \tag{7}$$

Here, t_p is the dimensionless time period of the inlet flow pulsation and n is the direction normal to the wall.

3.1 *Grid Independency Test*

We have used a mapped mesh in the present problem. The grid independence study has been conducted for the steady inflow with Re = 100, Pr = 6.93 and $a = 0.4$. The below table shows the values of Nusselt number for three different grid sizes on the wavy part of the channel at the peaks and the troughs:

Different nodes and numbered grids were tested and numerical results for Nusselt number were found to be almost independent of grids (as shown in Table 1). The percentage error in variation of Nusselt number between 100,523 nodes and 201,908 nodes is very less. Also, considering 201,908 nodes for simulation would take more run time. So the grid with 100,523 nodes was used in further calculations.

Table 1 Nu variation with number of grids

Number of grid points for straight channel for b/L=0.8, a = 0.4, a/λ = 0.3	Re = 100 Pr = 6.93		
	Nu	% error	
50232	7.693	0.155%	
100523	7.681		0.065%
201908	7.686		

3.2 *Validation*

In order to ensure that the results obtained in the present study are accurate, the local Nusselt number values obtained by Wang and Chen [6] have been reproduced here along with the results for the same problem obtained in the present study. The simulations have been performed for steady Newtonian fluid flow with = 6.93. The amplitude of the wavy wall (A_w) is 0.2. The close agreement of the two results ensures the accuracy of the results obtained in the present study (Fig. 2).

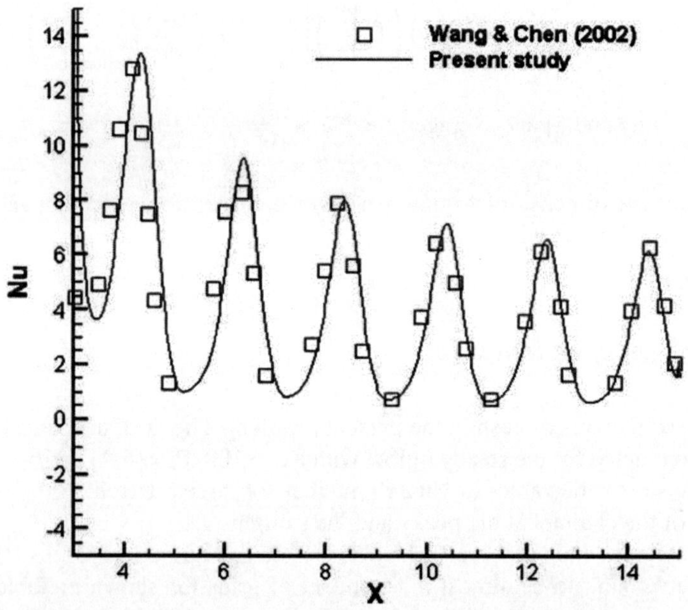

Fig. 2 Comparison between the local Nusselt number distribution along the bottom wavy wall for constant flow at different Re and for Pr = 6.93 and Re = 100 with Wang and Chen [6]

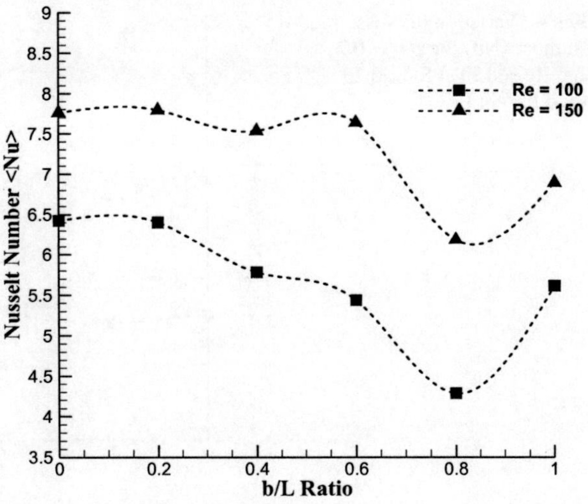

Fig. 3 Variation of Nu for $a/\lambda = 0.2$, $St = 0.25$ and $Pr = 6.93$ at different Re

4 Results and Discussion

4.1 Effect of Reynolds Number on Heat Transfer Rate

Initially for channel geometry with $a/\lambda = 0.2$ and $Pr = 6.93$ applying the boundary conditions problem was solved for $Re = 100$ and $Re = 150$.

Figure 3 shows that heat transfer rate is appreciably higher at higher Reynolds Number ie at $Re = 150$. This is because as the Reynolds number increases there is more turbulence in fluid. This turbulence causes intermixing of fluid particles. Also at higher Reynolds number, there is more velocity gradient in converging section which leads to increase of heat transfer between hot and cold fluid.

4.2 Effect of Prandtl Number on Heat Transfer Rate

Figure 4 shows the variation of heat transfer rate for different b/L ratio at two different $Pr = 4$ and $Pr = 6.93$ at $a/\lambda = 0.2$, $St = 0.25$ and $Re = 150$. In the present study, two different Reynolds numbers and Prandtl numbers have been considered, and their effects have been investigated. Figure 4 explains the effects of higher Prandtl number on the Nusselt number for **$a/\lambda = 0.2$, Re = 150** for different b/L ratios. And it is quite explicit that Nusselt Number is quite higher for $Pr = 6.93$ than that of $Pr = 4.0$. Also, it has been concluded in various literature that higher Reynolds number leads to more heat transfer rate. Hence, for the further studies, $Re = 150$ has been used to determine the optimum channel geometry.

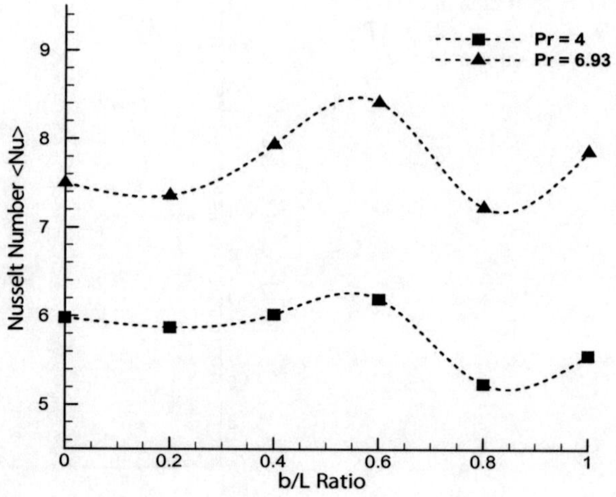

Fig. 4 Variation of Nusselt Number (Nu) for a/λ = 0.2 and Re = 150 at St = 0.25 for different Pr

4.3 *Effect of a/λ Ratio on Heat Transfer Rate*

Figure 5 explains the effect of amplitude-to-wavelength ratio (a/λ) for different b/L ratio at Re = 150 and Pr = 6.93. Heat transfer rate is quite higher in both the cases for a/λ = 0.3 compared to a/λ = 0.2. As a/λ ratio increases, wavelength of the corrugation decreases, and we get more number of corrugations in unit length of the channel. This enhanced number of corrugations gives more surface area per unit length for heat exchange between working fluid and solid walls.

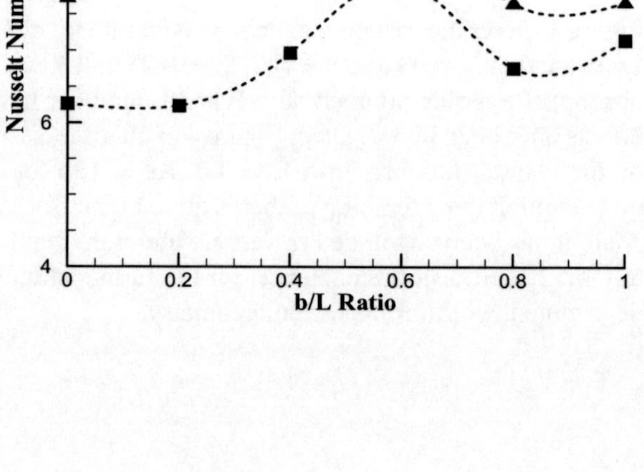

Fig. 5 Comparision of Nu for two different a/λ for Sr = 1, Re = 150, Pr = 6.93

4.4 Comparison at Different Strouhal Number

Figure 6a, b shows the comparison of heat transfer rate for different b/L ratio for Re = 150 and Pr = 6.93 at five different St. Figure 6a shows the variation at a/λ = 0.2 while Fig. 6b describes the same at a/λ = 0.3. For a/λ = 0.2, heat transfer rate is maximum when St = 0.5 while it is minimum when St = 0 ie steady flow. Heat transfer rate increase while Strouhal number increase from St = 0 to St = 0.5 while the heat transfer rate decreases at St = 0.75 and further decreases at St = 1. From Fig. 6b it is evident that heat transfer rate at a/λ = 0.3is maximum in the case of St

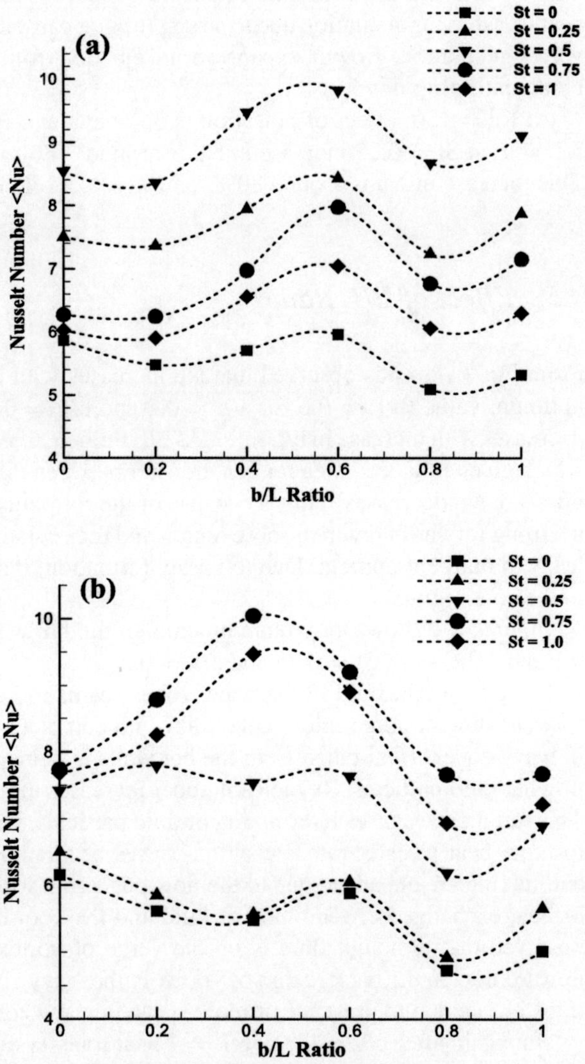

Fig. 6 **a** Variation of Nu for different St at Re = 150, Pr = 6.93, a/λ = 0.2; **b** Variation of Nu for different St at Re = 150, Pr = 6.93, a/λ = 0.3

= 0.75. With increase in Strouhal number from St = 0 to St = 0.75 heat transfer rate increase and decreases at St = 1.

It is intricate to drag an inclusive pattern for the rate of heat transfer in a corrugated channel with the frequency of the pulsating flow. As presented in Fig. 6a, b firstly, the increase of Struohal number causes increase of Nusselt number. Then, after reaching an extremum variable of Nusselt number, further increase of Strouhal number causes the value of Nusselt number to drop. The result shows that the maximum value of Nu depends on a/λ ratio.

Pulsation causes oscillation of flow. Flow disturbance and mixing are enhanced due to the flow pulsation, which leads to the enhancement of heat exchange between the hot fluid and the cool fluid in the wavy zones. The results demonstrate that the heat transfer augmentation due to strong mixing caused by the repeating sequence of vortex generation, growth, expansion and ejection from the groove to the mainstream by the pulsating flow.

At Re = 150, effect of pulsation is quite evident. For $a/\lambda = 0.2$, $b/L = 0.8$ and $b/L = 1$, at St = 0.25 increase in Nu is around 70% to that of steady case (St = 0). This increase in Nu is around 80% for $a/\lambda = 0.3$, $b/L = 0.4$ at St = 0.75.

4.5 *Effect of B/L Ratio*

From Fig. 6a, b, it is observed that Nu increases with increase in b/L ratio up to an optimum value ($b/L = 0.6$ for $a/\lambda = 0.2$ and $b/L = 0.4$ for $a/\lambda = 0.3$) and further decreases with increase in b/L ratio. As b/L ratio increases, horizontal pitch increases which gives enhanced area for heat transfer between fluid and walls. Generally, after b/L=0.6, Nu decreases. This is because of the formation of complete vortex leading to strong forward flow in the core region and recirculating flow in wavy zones which leads to fluid entrapment. Due to vortex formation, fluid is not able to transfer heat to the core region.

Figure 7a–c shows the stream function of fluid flow for different b/L ratios of 0.6, 0.8 and 1 incase of St = 0.75 and $a/\lambda = 0.2$

For $b/L = 0.6$, fluid in the wavy zones seems to be carried away by the strong forward flow of core region. Due to this, no complete vortex formation takes place in wavy zones. Heat taken from the hot walls is being carried away by the forward flowing fluid particles. Reynolds number increases there is more turbulence in fluid. This turbulence causes intermixing of fluid particles. Higher Reynolds number leads to more heat transfer rate. Pulsation causes oscillations of flow. Flow disturbance and mixing are enhanced due to the flow pulsation, which leads to the enhancement of heat exchange between the hot fluid and the cool fluid in the wavy zones. It is observed that although fluid is on the verge of vortex formation, also there is an upward flow in the back region of vortex in the wavy zones which enhances the heat transfer rate due to presence of more space in wavy zones.

But again after $b/L = 0.8$ to $b/L = 1$ heat transfer rate again increases. Figure 7c shows the stream function for geometry with $b/L = 1$. It is observed that although

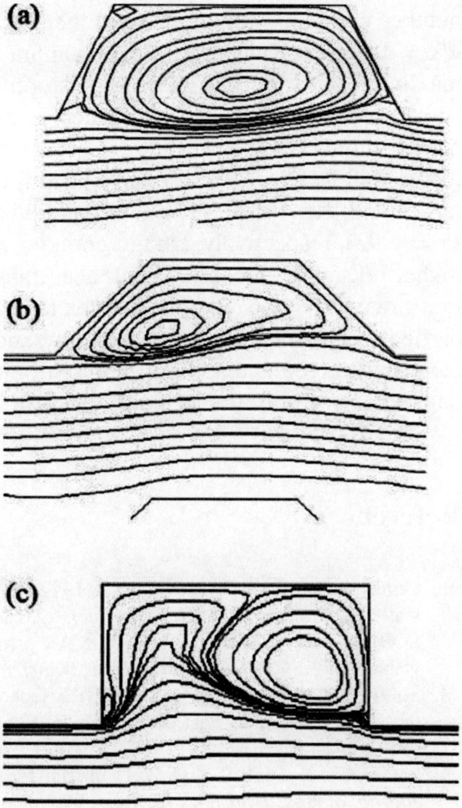

Fig. 7 **a** Stream function distribution at St = 0.75 for a/λ = 0.2, b/L = 0.8; **b** Stream function distribution at St = 0.75 for a/λ = 0.2, b/L = 0.6; **c**: Stream function distribution at St = 0.75 for a/λ = 0.2, b/L = 1

fluid is on the verge of vortex formation, also there is an upward flow in the back region of vortex in the wavy zones which enhances the heat transfer rate due to the presence of more space in wavy zones.

5 Conclusions

In the present study, five different flow pulsation frequencies in the form of St have been used (St = 0, 0.25, 0.5, 0.75, 1.0) along with Re = 100, 150 and Pr = 4, 6.93. From the investigations performed through this study, it can be concluded that heat transfer rate is enhanced at higher Reynolds number and Prandtl number. It is quite tangible that there is a sufficient increase in heat transfer rate with the increase of Re and Pr. While for pulsation, the heat transfer rate may increase or decrease depending upon the channel flow parameters, and hence, no Strouhal number can be considered to be optimum for maximum heat transfer rate as it is quite dependent on a/λ ratio. For different a/λ ratios, heat transfer rate may increase with increase in Strouhal number up to an optimum St, and then, it decreases with increase in St. In case of a/λ = 0.2, heat transfer rate is maximum at St = 0.5, while for a/λ = 0.3, Nu is maximum for St = 0.75. Beyond this optimum value of St (St = 0.5 for a/λ = 0.2 and St = 0.75 for a/λ = 0.3), Nu decreases with increase in St. For a/λ = 0.3, and there is more increase in heat transfer rate than that of a/λ = 0.2. Even 83% and 72% increase in heat transfer rate have been found in the cases of b/L = 0.4, St = 0.75 (a/λ = 0.3, Re = 150, Pr = 6.93) and b/L = 0.4, St = 1.0 (a/λ = 0.3, Re = 150, Pr = 6.93), respectively. Due to complex vortex formations in the wavy zones for higher b/L ratios, like b/L = 0.8, heat transfer rate decreases whereas in lower b/L ratios like b/L = 0.6, complete vortex formation does not take place, hence heat gets carried away with the forward core flow, and as a result heat transfer rate increases. It has also been found that for different Strouhal number, the heat transfer rate changes uniquely since pulsation of fluid flow is different.

References

1. Goldstein, L., Jr., & Sparrow, E. M. (1977). Heat/mass characteristics for flow in a corrugated wall channel. *Journal of Heat Transfer, 99*, 187–195.
2. O' Brien, J. E., & Sparrow, E. M. (1982). Corrugated-duct heat transfer, pressure drop and flow visualization. *ASME Journal of Heat Transfer, 104*, 410–416.
3. Sparrow, E. M., & Comb, J. W. (1983). Effect of interval spacing and fluid flow inlet conditions on a corrugated-wall heat exchanger. *International Journal of Heat and Mass Transfer, 26*, 993–1004.
4. Saniei, N., & Dini, S. (1993). Heat transfer characteristics in a wavy wall channel. *ASME Journal of Heat Transfer, 115*, 788–792.
5. Wang, G., & Vanka, S. P. (1995). Convective heat transfer in periodic wavy passages. *International Journal of Heat and Mass Transfer, 38*, 3219–3230.
6. Wang, C. C., & Chen, C. K. (2002). Forced convection in a wavy-wall channel. *International Journal of Heat and Mass Transfer, 45*, 2587–2595.
7. Fabbri, G. (2000). Heat transfer optimization in corrugated wall channels. *International Journal of Heat and Mass Transfer, 43*, 4299–4310.
8. Rush, T. A., Newell, T. A., & Jacobi, A. M. (1999). An experimental study of flow and heat transfer in sinusoidal wavy passages. *International Journal of Heat and Mass Transfer, 42*, 1541–1553.
9. Mohammed, H. A., Abed, A. M., & Wahid, M. A. (2013). The effects of geometrical parameters of a corrugated channel with in out-of-phase arrangement. *International Communications in Heat and Mass Transfer, 40*, 47–57.
10. Hossain M. Z., & Islam, A. K. M. S. (2004). Numerical investigation of unsteady flow and heat transfer in wavy channels. In: *15th Australasian Fluid Mechanics Conference, the University of Sydney*, Australia, December 13–17.
11. Bahaidarah, H. M., Anand, N. K., & Chen, H. C. (2005). Numerical study of fluid flow and heat transfer over a bank of flat tubes. *Numerical Heat Transfer A, 48*, 359–385.
12. Niceno, B., & Nobile, E. (2001). Numerical analysis of fluid flow and heat transfer in periodic wavy channels. *International Journal of Heat and Fluid Flow, 22*, 156–167.
13. Piao, Y., Hauptmann, E. G., & Iqbal, M. (1994). Forced convective heat transfer in cross-corrugated solar air heaters. *Journal of Solar Energy Engineering, 116*, 212–214.
14. Snyder, B., Li, K. T., & Wirtz, R. A. (1999). Heat transfer enhancement in a serpentine channel. *International Journal of Heat and Mass Transfer, 36*, 2965–2916.

15. Zhang, G., Tian, M., & Zhou, S. (2006). Simulation and analysis of flow pattern in cross-corrugated plate heat exchangers. *Journal of Hydrodynamics, Ser. B, 18*, 547–551.
16. Naphon, P. (2008). Effect of corrugated plates in an in-phase arrangement on the heat transfer and flow developments. *International Journal of Heat and Mass Transfer, 51*, 3963–3971.
17. Nishandar, S. V., & Yadav, R. H. (2015). Experimental investigation of heat transfer characteristics of pulsating turbulent flow in a pipe. *International Research Journal of Engineering and Technology, 02*, 487–492.
18. Sorour, M. M., & Estafanous, S. F. (1988). Heat transfer between two parallel plates with laminar pulsating flow. *International Journal Energy Research, 12*, 11–21.
19. Ali, A. H. H., & Hanaoka, Y. (2002). Experimental study on laminar flow forced-convection in a channel with upper V-corrugated plate heated by radiation. *International Journal of Heat and Mass Transfer, 45*, 2107–2117.
20. Ozbolat, V., Tokgoz, N., & Sahin, B. (2013). Flow characteristics and heat transfer enhancement in 2D corrugated channels. *International Journal of Mechanical, Aerospace, Industrial, Mechatronic and Manufacturing Engineering, 7*, 2074–2078.
21. Carpinlioglu, M., & Gündogdu, M. Y. (2001). A critical review on pulsatile pipe flow studies directing towards future research topics. *Flow Measurement and Instrumentation, 12*, 163–174.
22. Chattopadhyay, H., Durst, F., & Ray, S. (2006). Analysis of heat transfer in simultaneously developing pulsating laminar flow in a pipe with constant wall temperature. *International Communications in Heat and Mass Transfer, 33*, 475–481.
23. Moschandreou, T., & Zamir, M. (1997). Heat transfer in a tube with pulsating flow and constant heat flux. *International Journal of Heat and Mass Transfer, 40*, 2461–2466.
24. Rahgoshay, M., Ranjbar, A. A., & Ramiar, A. (2012). Laminar pulsating flow of nanofluids in a circular tube with isothermal wall. *International Communications in Heat and Mass Transfer, 39*, 463–469.
25. Hemida, H. N., Sabry, M. N., Abdel-Rahim, A., & Mansour, H. (2002). Theoretical analysis of heat transfer in laminar pulsating flow. *International Journal of Heat and Mass Transfer, 45*, 1767–1780.
26. He, Y.-L., WeiWei, Y., Zhao, C.-F., & Tao, W.-Q. (2005). Numerical study of enhancing heat transfer by pulsating flow. *Journal of Engineering Thermophysics, 3*, 041.
27. Kurzweg, U. H. (1985). Enhanced heat conduction in oscillating viscous flows within parallel-plate channels. *ASME Journal Fluid Mechanics, 156*, 291–300.
28. Jafari, M., Farhadi, M., & Kurosh, S. (2013). Pulsating flow effects on convection heat transfer in a corrugated channel: A LBM approach. *International Communications in Heat and Mass Transfer, 45*, 146–154.
29. Nield, D. A., & Kuznetsov, A. V. (2007). Forced convection with laminar pulsating flow in a channel or tube. *International Journal of Thermal Sciences, 46*, 551–560.
30. Scotti, A., & Ugo, P. (2001). Numerical simulation of pulsating turbulent channel flow. *Physics of Fluids, 13*, 1367–1384.
31. Tutty, O. R. (1992). Pulsatile Flow in a Constricted Channel. *Journal of Biomechanical Engineering, 114*, 50–54.
32. Kim, S. Y., Byung, H. K., & Jae, M. H. (1993). Heat transfer in the thermally developing region of a pulsating channel flow. *International Journal of Heat and Mass Transfer, 36*, 4257–4266.
33. Metwally, H. M., & Manglik, R. M. (2000). Numerical solutions for periodically-developed laminar flow and heat transfer in sinusoidal corrugated plate channels with constant wall temperature. In: *Proceedings of Heat Transfer Conference ASME*, NewYork.
34. Mehta, B., & Khandekar, S. (2015). Local experimental heat transfer of single-phase pulsating laminar flow in a square mini-channel. *International Journal Thermal Sciences, 91*, 157–166.
35. Xie, G., et al. (2006). Heat transfer enhancement inside converging-diverging wavy channel by pulsating flow. *Journal Chemical Engineering of Chinese Universities, 20*, 31–35.
36. Nishimura, T., Yoshino, T., & Kawamura, Y. (1987). Numerical flow analysis of pulsatile flow in a channel with symmetric wavy walls at moderate Reynolds numbers. *Journal Chemical Engineering of Japan, 20*, 479–485.

37. Jin, D. X., Lee, Y. P., & Lee, D.-Y. (2007). Effects of the pulsating flow agitation on the heat transfer in a triangular grooved channel. *International Journal of Heat and Mass Transfer, 50*, 3062–3071.
38. Bai, J., Pan, J., He, X., Wang, K., Tang, L., & Yang, R. (2020). Numerical investigation on thermal hydraulic performance of supercritical LNG in sinusoidal wavy channel based printed circuit vaporizer. *Applied Thermal Engineering, 175*, 115379.
39. Shi, X., Wang, Y., Huai, X., & Cheng, K. (2020). Influence of structure parameters on entropy generation performance in cross wavy channels with fluid-solid coupled heat transfer. *Applied Thermal Engineering, 181*, 115882.
40. Wen, Z. X., Lv, Y. G., Li, Q., & Zhou, P. (2020). Numerical study on heat transfer behavior of wavy channel supercritical CO2 printed circuit heat exchangers with different amplitude and wavelength parameters. *International Journal of Heat and Mass Transfer, 147*, 118922.

Numerical Modelling of Porous Square Cavity Heated on Vertical Walls in Presence of Magnetic Field

L. Jino, A. Vanav Kumar, Swapnali Doley, M. Berlin, and P. K. Mohanty

1 Introduction

An investigation of natural convection in a fluid flow via a porous media was increasing because of many engineering applications such as separation process in a chemical industries, crystal growth structure, phase change applications, dispersion of chemical contaminant in a through water and geophysics [1–3]. The natural convective flow through the porous media with the effect of magnetic field is governed by non-dimensional parameters such as Rayleigh number Ra, Darcy number Da, Prandtl number Pr, and Hartmann number Ha. Note that, Ra denotes the natural convective effect, porosity by Da and magnetic field strength by Ha, respectively. A comprehensive study is found in the references [4, 5].

Basak et al. [6, 7] discussed the effect of natural convection in a porous cavity for the various thermal boundaries such as constantly heated, linearly heated and sinusoidally heated. An increase in Ra leads to conversion of conduction dominant flow to the convection dominant flow, and also increase in Da motivates the circulation intensities. Natural convection heat transfer in a nanoparticle into a fluid saturated porous media is presented by Groşan et al. [8]. Sheikholeslami [9, 10] investigated the influence of magnetic field in a different shaped porous cavities filled with nanofluid. The results denote that increase in the nanofluid solid volume fraction ϕ, Ra, Da enhances the heat transfer and increase in the Ha reduces the heat transfer. Also, the

L. Jino · A. Vanav Kumar (✉) · S. Doley
Department of Basic and Applied Science, National Institute of Technology Arunachal Pradesh, Yupia, India

M. Berlin
Department of Civil Engineering, National Institute of Technology Arunachal Pradesh, Yupia, India

P. K. Mohanty
Department of Mechanical Engineering, National Institute of Technology Arunachal Pradesh, Yupia, India

P. Mahanta et al. (eds.), *Advances in Thermofluids and Renewable Energy*, Lecture Notes in Mechanical Engineering, https://doi.org/10.1007/978-981-16-3497-0_10

increase in the Darcy number from 10^{-2} to 10^{2} changes the shape of the streamlines and increases the flow rushing. Rashad et al. [11] investigated the effect of heat generation and magnetic field in a porous cavity saturated with Cu-water nanofluid. The results shows that, increase in the Ha and ϕ reduces the average Nusselt number and increase in Hartmann number increases the maximum temperature while it reduces the flow circulation intensity. Kumar et al. [12, 13] investigated the fluid flow due to natural convection in a porous square cavity with sinusoidally and partially heated boundaries. Jino and Vanav [14] discussed the heat flow, fluid flow and nanoparticle concentration generated due to the magneto-hydrodynamics natural convection in a porous cavity using Carman–Kozeny equation model.

Teamah and El-Maghlany [15] discussed natural convective on magnetic field and heat generation, absorption in a square cavity. Ghasemi et al. [16] in their study concluded that an increase in ϕ causes increase of drop in heat transfer performance depending upon the value of Ha and Ra selected. Rudraiah et al. [17] studied the magnetic field effects on natural convective cavity. The rate of convection is reduced by increasing the Hartmann number and reduces the average Nusselt number in an electrically conducting fluid.

In this study, the effect magnetic field on natural convective porous cavity filled with the Cu-water nanofluid is carried out. The Carman–Kozeny equation model is used to model the fluids passes via porous media.

2 Problem Description

A cavity ($H \times H$) is filled with the Cu-water nanofluid (Pr $= 6.2$) in a two-dimensional cartesian coordinate system with higher wall temperatures in both the vertical walls and lower temperature at the bottom wall. The top side wall temperature of the cavity is maintained adiabatically as shown in Fig. 1.

The nanofluid is assumed to be Newtonian, incompressible and laminar with the validation of Boussinesq approximation. Here, the water is considered as the base fluid containing the Cu nanoparticle. The thermo-physical properties of Cu-water nanofluid is listed in Table 1. The gravity acts parallel to the vertical walls, and the induced magnetic field is negligible when compared to the applied magnetic field.

3 Mathematical Model

The governing equations are made to fluid and heat flow in a cavity by the conservation of mass, momentum and energy with the applicable assumptions made above as,

$$\frac{\partial u}{\partial x} + \frac{\partial v}{\partial y} = 0 \tag{1}$$

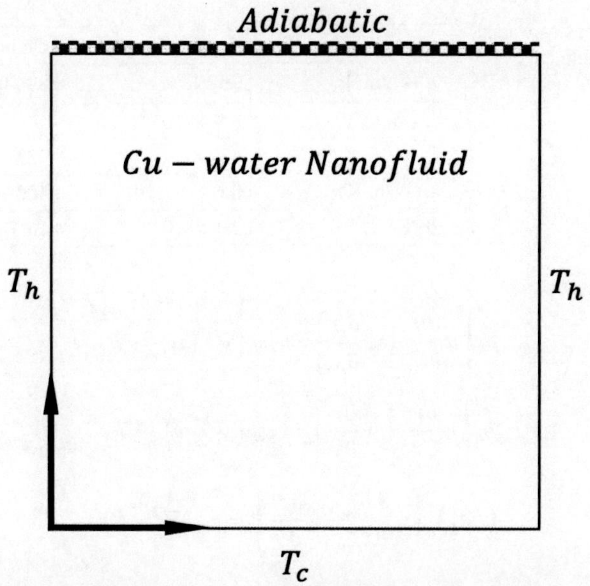

Fig. 1 Physical model

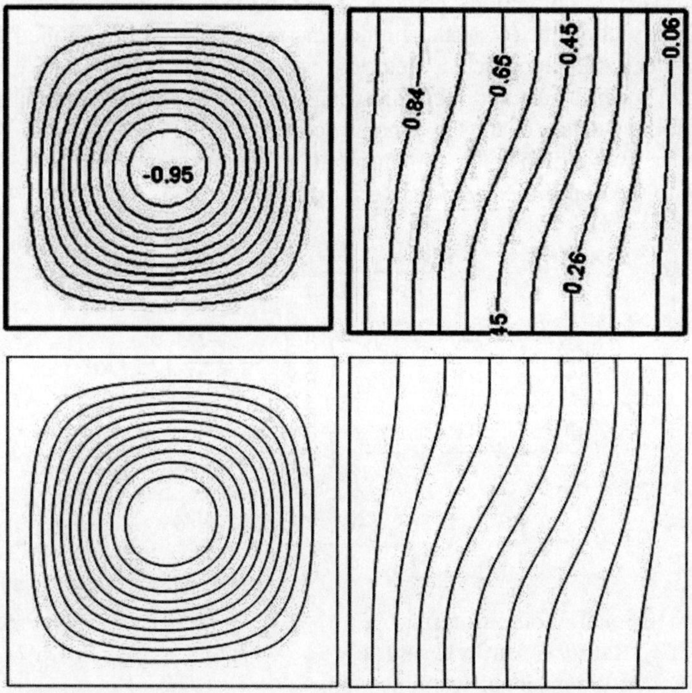

Fig. 2 Code validation of streamlines (left) and isotherm (right) contours of present study (bottom) to that of Ghasemi et al. [16] (top)

Table 1 Properties of water and Cu nanoparticle

	Water	Cu particle
ρ (kg/m^3)	997.1	8954
μ (Pa s)	8.9×10^{-4}	–
c_p (J/kg K)	4179	383
k (W/m K)	0.6	400
β (K^{-1})	2.1×10^{-4}	1.67×10^{-5}

$$\rho_n\left[u\frac{\partial u}{\partial x}+v\frac{\partial u}{\partial y}\right]=\mu_n\left(\nabla^2\cdot \boldsymbol{u}\right)-F_x \tag{2}$$

$$\rho_n\left[u\frac{\partial v}{\partial x}+v\frac{\partial v}{\partial y}\right]=\mu_n\left(\nabla^2\cdot \boldsymbol{v}\right)-F_y \tag{3}$$

$$\left(\rho c_p\right)_n\left[u\frac{\partial T}{\partial x}+v\frac{\partial T}{\partial y}\right]=k_n\left(\nabla^2\cdot \boldsymbol{T}\right) \tag{4}$$

Here $F_y = -\partial p/\partial y - \rho_n[C_m(1-\lambda)]^2/\lambda^3 v - \sigma_n B^2 v + (\rho\beta)_n g(T - T_0)$ and $F_x = -\partial p/\partial x - \rho_n[C_m(1-\lambda)]^2/\lambda^3\ u$. u, v are velocities, p is pressure and T is temperature with the directions x and y. The fluid in a porous media is controlled by the porosity constant (C_m) and liquid fraction (λ). B is the applied magnetic field. The other parameters such as electrical conductivity, density, viscosity, thermal conductivity, specific heat and thermal expansion coefficient depend on the amount of solid volume fraction of the Cu nanoparticle [13].

The above two-dimensional dimensional governing equations Eqs. (1–4) are modified into the non-dimensional governing equation as,

$$\frac{\partial U}{\partial X}+\frac{\partial V}{\partial Y}=0 \tag{5}$$

$$U\frac{\partial U}{\partial X}+V\frac{\partial U}{\partial Y}=\frac{\mu_n}{\rho_n\alpha_{fl}}\left(\frac{\partial^2 U}{\partial X^2}+\frac{\partial^2 U}{\partial Y^2}\right)-F_X \tag{6}$$

$$U\frac{\partial V}{\partial X}+V\frac{\partial V}{\partial Y}=\frac{\mu_n}{\rho_n\alpha_{fl}}\left(\frac{\partial^2 V}{\partial X^2}+\frac{\partial^2 V}{\partial Y^2}\right)-F_Y \tag{7}$$

$$U\frac{\partial \theta}{\partial X}+V\frac{\partial \theta}{\partial Y}=\frac{\alpha_n}{\alpha_{fl}}\left(\frac{\partial^2 \theta}{\partial X^2}+\frac{\partial^2 \theta}{\partial Y^2}\right) \tag{8}$$

Non-dimensional source terms $F_Y = -\partial P/\partial Y - \rho_{fl}(\text{Pr/Da})V + (\rho\beta)_n/(\rho_n\,\beta_{fl})\text{RaPr}\theta - \text{Ha}^2\text{Pr}V$ and $F_X = -\partial P/\partial X - \rho_{fl}(\text{Pr/Da})U$, respectively. The non-dimensional variables are $X = x/H$, $Y = y/H$, $U = (uH)/\alpha_f$, $V = (vH)/\alpha_f$, $P = (pH^2)/\left(\rho_n\alpha_f^2\right)$ and $\theta = (T - T_0)/(T_h - T_0)$. The other non-dimensional numbers such as Prandtl number $\left(\text{Pr} = \nu_f/\alpha_f\right)$, Darcy number $\text{Da} = (\mu_f\, \lambda^3)/(C_m\, (1-\lambda)^2 H^2)$ to control the porosity, Rayleigh number $\text{Ra} = (g\beta_f H^3 (T_h - T_c))/(\nu_f \alpha_f)$ responsible for natural convection effects, and intensity of magnetic field by Hartmann number $\text{Ha} = Bl\sqrt{\sigma_n/(\rho_n \nu_n)}$.

The flow governing equations are solved by the essential boundary conditions such as $(\theta = 1)$ for higher temperature walls, $(\theta = 0)$ for lower temperature wall and adiabatic wall as $(\partial\theta/\partial n = 0)$. The no-slip velocity $(U = V = 0)$ boundaries are imposed on all the walls.

The fluid motion in the cavity is visualized using the streamlines which are obtained from the velocities $U = (\partial\psi/\partial Y)$ and $V = -(\partial\psi/\partial X)$.

The heat transfer enhancement of the convective to the conductive heat transfer at the wall is denoted by the dimensionless quantity local Nusselt number, $\text{Nu} = -(k_n/k_{fl})(\partial\theta/\partial n)$.

4 Numerical Description

The streamlines ψ from the velocities U, V and isotherms θ are solved using the second order of finite different method in a 100×100 constant spaced computational domain. The governing equations Eq. (5–8) are formulated into vorticity–stream-function, which are solved using the respective boundary conditions. The solutions are formed using the iterative process and iteration can be done till the convergence criteria $([(\xi^{n+1} - \xi^n)/\xi^{n+1}\,]| < 10^{-5})$ for temperature and velocities. Here, n represents the iteration number.

The validation of code has been performed for effect of magnetic field on natural convection in a square cavity filled with nanofluid. The results obtained are quite similar to the work of Ghasemi et al. [16]. The grid independence study is carried out (ref Fig. 3), and constant grid of 100×100 is considered in this entire study.

5 Results and Discussion

The numerical result presents the streamlines and isotherms of Cu-water nanofluid for the various values of $10^3 \leq \text{Ra} \leq 10^5$ with the constant volume fraction of the particle $\phi = 0.1$, within the Darcy number $10^{-2} \leq \text{Da} \leq 10^2$. The applied magnetic field strength in a cavity is indicated by Hartmann number $0 \leq \text{Ha} \leq 50$. The local Nusselt number (Nu) is calculated at heated vertical walls for Ra with respect to the Darcy number and Hartmann number.

Figure 4 represents the flow structure of streamlines and isotherms at $\text{Ra} = 10^4$ for various Da and Ha. The flow in a cavity initiates by the temperature difference and has the symmetry flow pattern in both the side vertical orientation with anti-clockwise and clockwise rotation. The variation of rotation is due to the difference in temperature distribution of higher to lower temperature and lower to higher temperature of bottom and vertical walls. At the lower Darcy number, the flow rush is low and increased

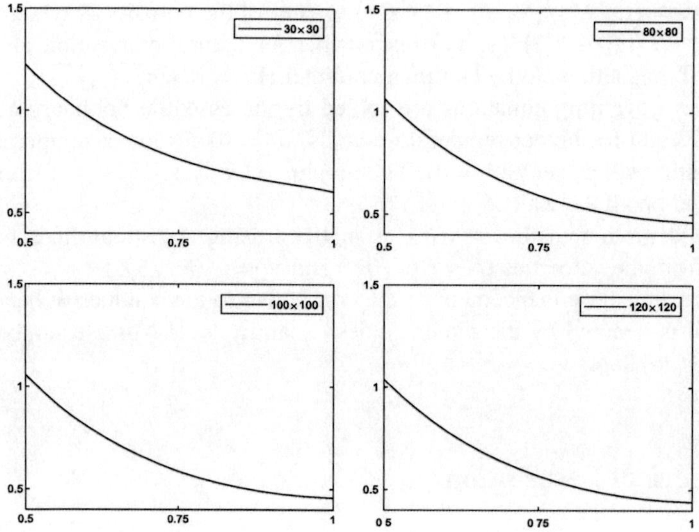

Fig. 3 Grid independency study

for increase in the Da from 10^{-2} to 10^{2}. The isotherm also flows from both the vertical walls symmetrically towards the mid-plane without any collaboration. The higher Darcy number makes increase in flow rush, and the isotherm that moves from the vertical walls is merged and makes the temperature distribution from the lower bottom to top cavity linearly.

The effect of Hartmann number has a very less sense at the lower Da in both the streamlines and the isotherms but at the Da = 10^{2} the increase in the Ha suppresses the flow rush. The isotherm at top separated because of reduction in the heat transfer rate at the Ha = 50.

The Ra = 10^{5} is set, so that the natural convective effect is increased and the flow rush also increases than before and isotherm also merged half and distributed. An increase in the Hartmann number has very minor change in the streamlines and in isotherm, the temperature distribution ($\theta = 0.5$) small elongation towards the adiabatic wall at Da = 10^{-2} as shown in Fig. 5. When increasing the Darcy number, the flow rush is increased more as discussed previously, and the temperature distribution from the vertical wall is more. The increase in the Ha = 50 leads to reduce in flow rush and push towards the bottom. The temperature deviation also reduces as compared to Ha = 0.

The local Nusselt number for the left isothermal wall is shown in Fig. 6. Nu is more for Da = 10^{-2} at lower Rayleigh number and decreases for increasing in the Ra. But Nu is less for same Rayleigh number as Darcy number increased to 10^{2} because at Da = 10^{-2} the conduction effects are more dominant. The increase in the Ra gradually decreases in Nu for smaller Da and increases for higher Da. At when Rayleigh number reaches to 10^{5}, the Nusselt number becomes almost similar for both the Da = 10^{-2} and 10^{2}. The effect of applied magnetic field leads to the slight

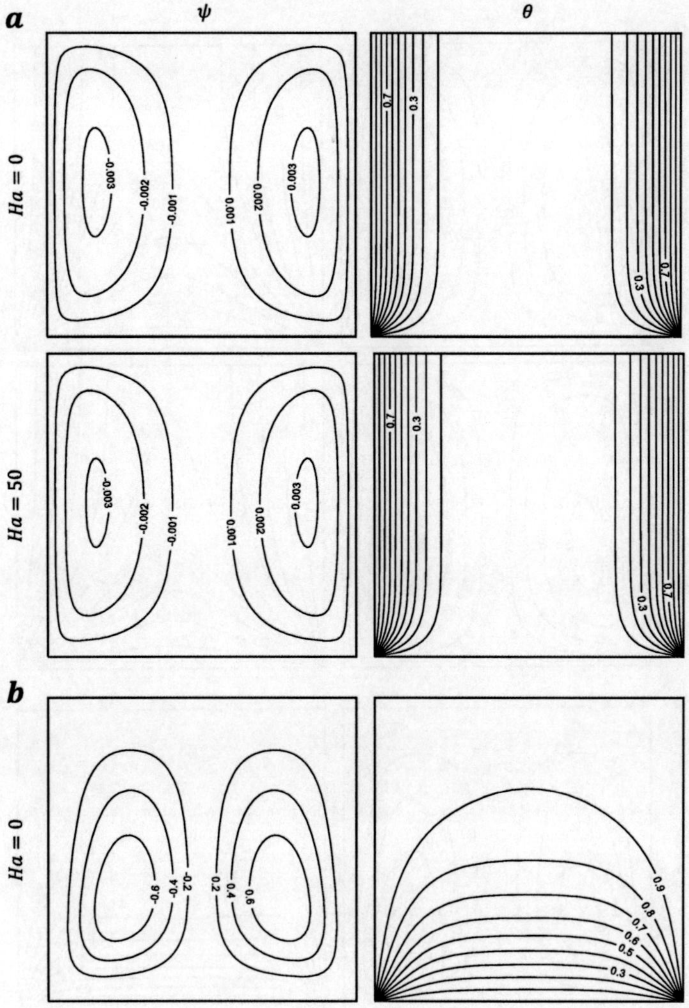

Fig. 4 Streamfunction and temperature contours for Ra = 10^4 with **a** Da = 10^{-2} and **b** Da = 10^2

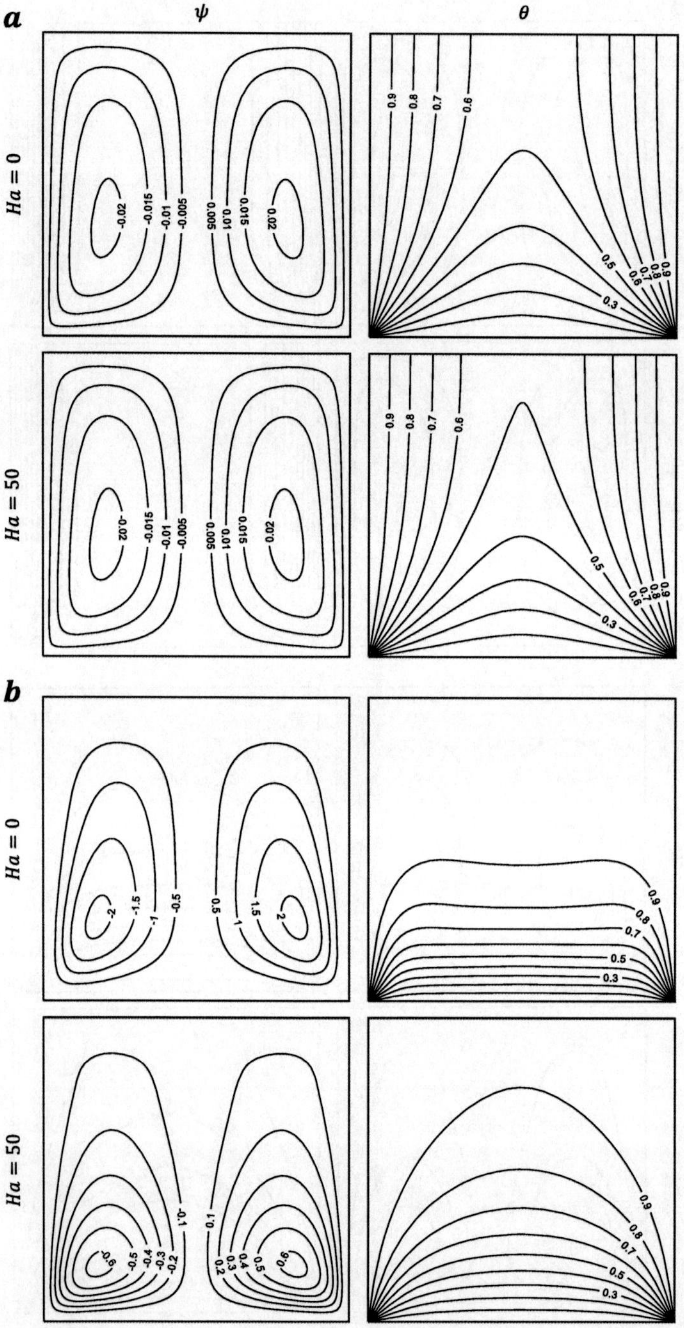

Fig. 5 Streamfunction and temperature contours for Ra $= 10^5$ with **a** Da $= 10^{-2}$ and **b** Da $= 10^2$

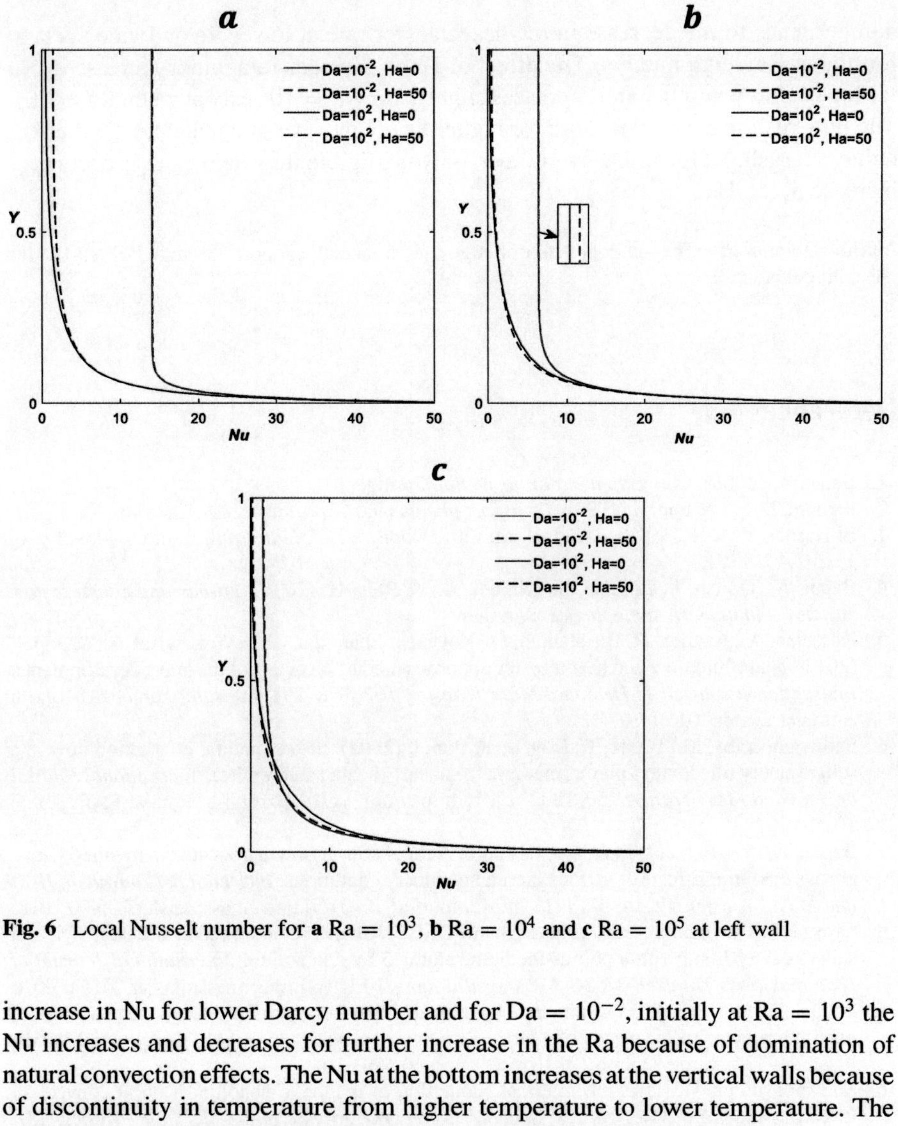

Fig. 6 Local Nusselt number for **a** Ra = 10^3, **b** Ra = 10^4 and **c** Ra = 10^5 at left wall

increase in Nu for lower Darcy number and for Da = 10^{-2}, initially at Ra = 10^3 the Nu increases and decreases for further increase in the Ra because of domination of natural convection effects. The Nu at the bottom increases at the vertical walls because of discontinuity in temperature from higher temperature to lower temperature. The variation in a local Nusselt number is same for both the vertical walls.

6 Conclusion

The flow modelling of Cu-water nanofluid in a porous cavity is studied numerically using the various Darcy number due to natural convection. The magnetic field strength is introduced by varying the Hartmann number from 0 to 50. The increase in the Darcy number leads to the decrease in the heat transfer rate at lower Ra and vice versa at for higher Rayleigh number. The effect of Ha = 50 leads to a minor increase in Nu for lower Darcy number and decreases slightly for Da = 10^2 except when Ra = 10^3. The flow rush in a cavity is more for higher Darcy number and magnetic field effect reduces it rush. At lower Darcy number, only negligible flow field change occurs for increase in the Ha.

Acknowledgements The authors acknowledge the financial support through TEQIP-III for presenting this work.

References

1. Bejan, A. (2006). *Convection in porous media*. Springer.
2. Ingham, D. B., & Pop, I. (1998). *Transport phenomena in porous media*. Elsevier.
3. El Hasadi, Y. M. F., & Khodadadi, J. M. (2013). *Journal of Heat Transfer*. https://doi.org/10.1115/1.4023542
4. Bejan, A., Dincer, I., Lorente, S., Miguel, A., & Reis, H. (2004). *Porous and complex flow structures in modern technologies*. Springer.
5. Kasaeian, A., Azarian, R. D., Mahian, O., Kolsi, L., Chamkha, A. J., Wongwises, S., & Pop, I. (2017). Nanofluid flow and heat transfer in porous media: A review of the latest developments. *International Journal of Heat and Mass Transfer, 107*, 778–791. https://doi.org/10.1016/j.ijheatmasstransfer.2016.11.074
6. Sathiyamoorthy, M., Basak, T., Roy, S., & Pop, I. (2007). Steady natural convection flow in a square cavity filled with a porous medium for linearly heated side wall(s). *International Journal of Heat and Mass Transfer, 50*, 1892–1901. https://doi.org/10.1016/j.ijheatmasstransfer.2006.10.010
7. Basak, T., Roy, S., Paul, T., & Pop, I. (2006). Natural convection in a square cavity filled with a porous medium: Effects of various thermal boundary conditions. *International Journal of Heat and Mass Transfer, 49*, 1430–1441. https://doi.org/10.1016/j.ijheatmasstransfer.2005.09.018
8. Groşan, T., Revnic, C., Pop, I., & Ingham, D. B. (2015). Free convection heat transfer in a square cavity filled with a porous medium saturated by a nanofluid. *International Journal of Heat and Mass Transfer, 87*, 36–41. https://doi.org/10.1016/j.ijheatmasstransfer.2015.03.078
9. Sheikholeslami, M. (2017). Influence of magnetic field on nanofluid free convection in an open porous cavity by means of Lattice Boltzmann method. *Journal of Molecular Liquids, 234*, 364–374. https://doi.org/10.1016/j.molliq.2017.03.104
10. Sheikholeslami, M. (2017). Numerical simulation of magnetic nanofluid natural convection in porous media. *Physics Letters, Section A: General, Atomic and Solid State Physics, 381*, 494–503. https://doi.org/10.1016/j.physleta.2016.11.042
11. Rashad, A. M., Rashidi, M. M., Lorenzini, G., Ahmed, S. E., & Aly, A. M. (2017). Magnetic field and internal heat generation effects on the free convection in a rectangular cavity filled with a porous medium saturated with Cu–water nanofluid. *International Journal of Heat and Mass Transfer, 104*, 878–889. https://doi.org/10.1016/j.ijheatmasstransfer.2016.08.025
12. Kumar, A. V., Jino, L., Berlin, M., & Mohanty, P. K. (2019). Magnetic field effect on nanofluid suspension cavity by non-uniform boundary conditions. *AIP Conference Proceedings AIP Publishing*. https://doi.org/10.1063/1.5120205
13. Jino, L., & Vanav Kumar, A. (2020). Natural Convection of Water-Cu Nanofluid in a Porous Cavity with Two Pairs of Heat Source-Sink and Magnetic Effect. *International Journal of Mechanical and Production Engineering Development, 10*, 14481–14492. https://doi.org/10.24247/ijmperdjun20201378

14. Jino, L., & Vanav Kumar, A. (2020). Cu-water nanofluid natural convective heat and fluid flow in a porous cavity. *International Journal of Mechanical and Production Engineering Research Development, 10*, 13695–13706. https://doi.org/10.24247/ijmperdjun20201305
15. Teamah, M. A., El-Maghlany, W. M. (2012). Augmentation of natural convective heat transfer in square cavity by utilizing nanofluids in the presence of magnetic field and uniform heat generation/absorption. *International Journal of Thermal Sciences, 58*, 130–142. https://doi.org/10.1016/j.ijthermalsci.2012.02.029
16. Ghasemi, B., Aminossadati, S. M., & Raisi, A. (2011). Magnetic field effect on natural convection in a nanofluid-filled square enclosure. *International Journal of Thermal Sciences, 50*, 1748–1756. https://doi.org/10.1016/j.ijthermalsci.2011.04.010
17. Rudraiah, N., Barron, R. M., Venkatachalappa, M., & Subbaraya, C. K. (1995). Effect of a magnetic field on free convection in a rectangular enclosure. *International Journal of Engineering Science, 33*, 1075–1084. https://doi.org/10.1016/0020-7225(94)00120-9

Thermal Effects of CNTs Nanoliquid Film Flow During Spin Coating

Swatilekha Nag and Susanta Maity

Nomenclature

Bi	Biot number
C_0	Constant
C_1	Constant
C_p	Heat capacity at constant pressure (J/kg K)
Fr	Froude number
h	Film thickness (m)
h_0	Initial film thickness (m)
k	Thermal conductivity (W/m K)
p	Pressure of nanoliquid (Pa)
Pr	Prandtl number
Re	Reynolds number
t	Rotational time (s)
t_c	Characteristic time (s)
T	Nanoliquid temperature (K)
T_g	Temperature in gas phase (K)
T_0	Room temperature (K)
We	Weber number

Greek Symbols

α	Thermocapillary parameter
β	Heat transfer coefficient
ϵ	Aspect ratio
μ	Dynamic viscosity (kg/m s)

S. Nag · S. Maity (✉)
Department of Basic and Applied Science, National Institute of Technology, Arunachal Pradesh, India
e-mail: susanta@nitap.ac.in

P. Mahanta et al. (eds.), *Advances in Thermofluids and Renewable Energy*, Lecture Notes in Mechanical Engineering, https://doi.org/10.1007/978-981-16-3497-0_11

ν Kinematic viscosity (m^2/s)
Ω Angular velocity of the disk (s^{-1})
ϕ Nanoparticle volume fraction
ϕ_1, ϕ_2 Dimensionless constant
ψ_1 Radial velocity component (m/s)
ψ_2 Cross-radial velocity component (m/s)
ψ_3 Velocity component perpendicular to disk (m/s)
ρ Density (kg/m^3)
σ Surface tension (kg/s^2)
σ_0 Surface tension at initial time (kg/s^2)

Subscript

bl Base liquid
nl Nanoliquid
CNTs Nanoparticle

1 Introduction

The spin coating is a process by which a very thin uniform film layer (layers of paint, polymer, lacquer, etc.) is developed over the surface of a horizontal rotating substrate/disk with the help of centrifugal force. The thin film development over a spinning disk is very important in microelectronic industries. Nowadays, we are living in the electronic age and using electronic gadgets frequently in our daily life. Many of these gadgets have the hard disk, screen, electronic circuit, etc. In manufacturing of these gadgets, the coating is required in many intermediate steps. This technique is also broadly used in magnetic disk for data storage, small-scale gadget industry to fabricate the incorporated circuits, optical mirrors and color television screens, etc. The investigation of thin Newtonian liquid film over a spinning disk was first theoretically initiated by Emslie et al. [1]. Acrivos et al. [2] and Jenekhe and Schuldt [3] extended the analysis to think about the stream of a non-Newtonian power-law liquid film flow over a spinning disk. In spin coating process, the effect of solvent evaporation was first introduced by Myerhofer [5]. Middleman [6] examined the impact of air shear at the liquid–air interface on the development of liquid film through the spin coating process. Higgins [4] analyzed the unsteady liquid film flow on a rotating disk by match asymptotic methods by considering full set of momentum equations. Dandapat and Ray [7, 8] explored the influence of heating/cooling of the disk surface on thin liquid film development over a spinning disk. Dandapat et al. [9] studied the two-layer thin film flow on a rough spinning disk in the presence of air shear at the liquid–air interface. Dandapat and Maity [10] investigated the impacts of external air flow on a thin film flow over a rotating annular disk with the deformable free surface. Dandapat et al. [11] analyzed the thermal effects of thin liquid film flow on a spinning disk.

In recent year, nanofluid has gained huge attentions of the researchers due to its increased heat transfer performance. Nanofluid is the colloidal suspension of nanometer size metallic or nonmetallic particles (1–100 nm) with the conventional heat transfer fluids like water, ethylene glycol, engine oil, etc. In the presence of tiny-sized particles, the thermal conductivity of the conventional fluids enhances anomalously. The nanofluid was first discovered by Choi [12] and showed the anomalous increase of thermal conductivity of the conventional fluids. Buongiorno [13] investigated the laminar flow of nanofluids by taking effects of thermophoresis and Brownian motion. Carbon nanotubes-based nanofluids (CNTs) are also very important due to its unique properties such as high electrical conductivity, high mechanical strength, and high thermal conductivity. Akbar and Butt [14] considered the peristaltic flow of CNTs suspended nanofluid over a curved channel. Hussanan et al. [15] studied the magnetohydrodynamic flow of CNTs nanofluid flow on a stretching sheet. Recently, Maity [16] investigated water-based thin nanoliquid film development during spin coating process.

In this article, we have modeled the thin CNTs nanoliquid film development over a spinning horizontal disk. It is assumed that the nanoliquid film is deformable. The effects of both single-walled carbon nanotubes (SWCNTs) and multi-walled carbon nanotubes (MWCNTs) are considered for investigation.

2 Formulation of the Problem

We have considered an axisymmetric, viscous, incompressible water-based CNTs nanoliquid film flow on a rotating disk. We choose the cylindrical polar coordinate system with the origin at center of the disk, and the z-axis is vertically upward. Initially, the disk is at rest at time $t = 0$ and it starts to rotate with angular velocity Ω about the z-axis. At the begin, the framework is at room temperature T_0. Let (ψ_1, ψ_2, ψ_3) and T be the velocity components along (r, θ, z) direction and temperature of the nanoliquid, respectively. The axisymmetric equation of continuity and momentum and energy equations can be written as:

$$\frac{\partial \psi_1}{\partial r} + \frac{\psi_1}{r} + \frac{\partial \psi_3}{\partial z} = 0, \tag{1}$$

$$\rho_{\mathrm{nl}}\left[\frac{\partial \psi_1}{\partial t} + \psi_1 \frac{\partial \psi_1}{\partial r} + \psi_3 \frac{\partial \psi_1}{\partial z} - \frac{{\psi_2}^2}{r}\right] = -\frac{\partial p}{\partial r}$$
$$+\mu_{\mathrm{nl}}\left[\frac{\partial^2 \psi_1}{\partial r^2} + \frac{\partial}{\partial r}\left(\frac{\psi_1}{r}\right) + \frac{\partial^2 \psi_1}{\partial z^2}\right], \tag{2}$$

$$\rho_{\text{nl}}\left[\frac{\partial \psi_2}{\partial t}+\psi_1\frac{\partial \psi_2}{\partial r}+\psi_3\frac{\partial \psi_2}{\partial z}+\frac{\psi_1\psi_2}{r}\right]=$$
$$\mu_{\text{nl}}\left[\frac{\partial^2 \psi_2}{\partial r^2}+\frac{\partial}{\partial r}\left(\frac{\psi_2}{r}\right)+\frac{\partial^2 \psi_2}{\partial z^2}\right], \tag{3}$$

$$\rho_{\text{nl}}\left[\frac{\partial \psi_3}{\partial t}+\psi_1\frac{\partial \psi_3}{\partial r}+\psi_3\frac{\partial \psi_3}{\partial z}\right]=-\frac{\partial p}{\partial z}$$
$$+\mu_{\text{nl}}\left[\frac{\partial^2 \psi_3}{\partial r^2}+\frac{1}{r}\frac{\partial \psi_3}{\partial r}+\frac{\partial^2 \psi_3}{\partial z^2}\right], \tag{4}$$

$$\left(\rho C_p\right)_{\text{nl}}\left[\frac{\partial T}{\partial t}+\psi_1\frac{\partial T}{\partial r}+\psi_3\frac{\partial T}{\partial z}\right]=k_{\text{nl}}\left[\frac{\partial^2 T}{\partial r^2}\right.$$
$$\left.+\frac{1}{r}\frac{\partial T}{\partial r}+\frac{\partial^2 T}{\partial z^2}\right], \tag{5}$$

where p, ρ_{nl}, μ_{nl}, $(\rho C_p)_{\text{nl}}$, and k_{nl} are, respectively, the pressure, density, dynamic viscosity, heat capacity, and thermal conductivity of the water-based CNTs nanoliquid. These are defined as follows:

$$\rho_{\text{nl}}=(1-\phi)\rho_{\text{bl}}+\phi\rho_{\text{CNTs}}, \tag{6}$$

$$\mu_{\text{nl}}=\frac{\mu_{\text{bl}}}{(1-\phi)^{\frac{5}{2}}}, \tag{7}$$

$$(\rho C_p)_{\text{nl}}=(1-\phi)(\rho C_p)_{\text{bl}}+\phi(\rho C_p)_{\text{CNTs}}, \tag{8}$$

$$k_{\text{CNTs}}=k_{\text{bl}}\left(\frac{(1-\phi)+\frac{2\phi k_{\text{CNTs}}}{k_{\text{CNTs}}-k_{\text{bl}}}\log\left(\frac{k_{\text{CNTs}}+k_{\text{bl}}}{2k_{\text{bl}}}\right)}{(1-\phi)+\frac{2\phi k_{\text{bl}}}{k_{\text{CNTs}}-k_{\text{bl}}}\log\left(\frac{k_{\text{CNTs}}+k_{\text{bl}}}{2k_{\text{bl}}}\right)}\right), \tag{9}$$

where suffixes *bl* and CNTs stand for the properties of the base liquid and carbon nanotubes, respectively, and ϕ is the volume fraction of CNTs.

The corresponding boundary and initial conditions are given below: at the surface of the rotating disk ($z=0$),

$$\psi_1 = 0, \psi_2 = r\Omega, \psi_3 = 0, T = \theta(r), \tag{10}$$

at the liquid–air interface $z = h(r, t)$,

$$-p + \frac{2\mu_{\mathrm{nl}}}{1 + {h_r}^2}[{h_r}^2 \psi_{1_r} - h_r(\psi_{1_z} + \psi_{3_r}) + \psi_{3_z}] = \frac{\sigma}{r}\left[\frac{h_r(1 + {h_r}^2) + rh_{\mathrm{rr}}}{(1 + {h_r}^2)^{3/2}}\right], \tag{11}$$

$$\frac{\mu_{\mathrm{nl}}}{(1 + {h_r}^2)^{1/2}}[2h_r(\psi_{1_r} - \psi_{3_z}) + ({h_r}^2 - 1)(\psi_{3_r} + \psi_{1_z})] = -(\sigma_r + h_r\sigma_z), \tag{12}$$

$$\psi_{2z} - rh_r\left(\frac{\psi_2}{r}\right)_r = 0, \tag{13}$$

$$h_t + \psi_1 h_r = \psi_3, \tag{14}$$

$$-k_{\mathrm{nl}}[T_z - h_r T_r] = \beta(T - T_0)\sqrt{1 + {h_r}^2}, \tag{15}$$

the initial condition at time $t = 0$,

$$\psi_1 = 0, \psi_2 = 0, \psi_3 = 0, h(r, 0) = \delta(r), T = T_0, \tag{16}$$

where σ is the surface tension which varies linearly with the temperature T as $\sigma = \sigma_0 - \gamma(T - T_0)$ and γ is a positive constant. β is the heat transfer coefficient. The temperature depended physical properties of the water and CNTs are given in Table 1.

Table 1 Physical properties of base liquids and CNTs (see Hussanan et al. [15], Krishanan et al. [17])

	$\rho\left(\frac{\mathrm{kg}}{\mathrm{m}^3}\right)$	$C_p\left(\frac{\mathrm{J}}{\mathrm{kgK}}\right)$	$k\left(\frac{\mathrm{W}}{\mathrm{mK}}\right)$
Water	997.1	4179	0.613
SWCNT	2600	425	6600
MWCNT	1600	796	3000

To find the non-dimensional form of the governing set of Eqs. (1)–(5), the following dimensionless variables are introduced as

$$\hat{t} = \frac{t}{t_c}, \hat{r} = \frac{r}{R}, \hat{z} = \frac{z}{h_0}, \hat{\delta} = \frac{\delta}{h_0}, \hat{h} = \frac{h}{h_0},$$
$$\hat{\psi}_1 = \frac{\psi_1}{U_0}, \hat{\psi}_2 = \frac{\psi_2\sqrt{\epsilon \mathrm{Re}}}{U_0}, \hat{\psi}_3 = \frac{\psi_3}{\epsilon U_0}, \hat{p} = \frac{{h_0}^2 t_c}{\nu_{\mathrm{bl}} R^2 \rho_{\mathrm{bl}}} p,$$
$$U_0 = \frac{R}{t_c}, \hat{T} = \frac{T - T_0}{T_{d_0} - T_0}, t_c = \frac{\nu_{\mathrm{bl}}}{(\Omega h_0)^2},$$

where $\mathrm{Re} = U_0 h_0/\nu_{\mathrm{bl}}$ is the Reynolds number, h_0 is the average value of initial film profile, $\epsilon = h_0/R(\ll 1)$ is the aspect ratio, R is the radius of the disk, and T_{d0} is imposed temperature at the center of the disk. Finally, we get the non-dimensional governing set of equations as:

$$\frac{\partial \hat{\psi}_1}{\partial \hat{r}} + \frac{\hat{\psi}_1}{\hat{r}} + \frac{\partial \hat{\psi}_3}{\partial \hat{z}} = 0, \tag{17}$$

$$\epsilon \mathrm{Re}\phi_1 \left[\frac{\partial \hat{\psi}_1}{\partial \hat{t}} + \hat{\psi}_1 \frac{\partial \hat{\psi}_1}{\partial \hat{r}} + \hat{\psi}_3 \frac{\partial \hat{\psi}_1}{\partial \hat{z}} \right] = -(1-\phi)^{2.5} \frac{\partial \hat{p}}{\partial \hat{r}}$$
$$+\phi_1 \frac{\hat{\psi}_2^{\,2}}{\hat{r}} + \epsilon^2 \left[\frac{\partial^2 \hat{\psi}_1}{\partial \hat{r}^2} + \left(\frac{\hat{\psi}_1}{\hat{r}} \right)_{\hat{r}} \right] + \frac{\partial^2 \hat{\psi}_1}{\partial \hat{z}^2}, \tag{18}$$

$$\epsilon \mathrm{Re}\phi_1 \left[\frac{\partial \hat{\psi}_2}{\partial \hat{t}} + \hat{\psi}_1 \frac{\partial \hat{\psi}_2}{\partial \hat{r}} + \hat{\psi}_3 \frac{\partial \hat{\psi}_2}{\partial \hat{z}} + \frac{\hat{\psi}_1 \hat{\psi}_2}{\hat{r}} \right] =$$
$$\left[\epsilon^2 \left(\frac{\partial^2 \hat{\psi}_2}{\partial \hat{r}^2} + \left(\frac{\hat{\psi}_2}{\hat{r}} \right)_{\hat{r}} \right) + \frac{\partial^2 \hat{\psi}_2}{\partial \hat{z}^2} \right], \tag{19}$$

$$\epsilon^3 \mathrm{Re}\phi_1 \left[\frac{\partial \hat{\psi}_3}{\partial \hat{t}} + \hat{\psi}_1 \frac{\partial \hat{\psi}_3}{\partial \hat{r}} + \hat{\psi}_3 \frac{\partial \hat{\psi}_3}{\partial \hat{z}} \right] = -(1-\phi)^{2.5} \frac{\partial \hat{p}}{\partial \hat{z}}$$
$$+\epsilon^2 \left[\epsilon^2 \left(\frac{\partial^2 \hat{w}}{\partial \hat{r}^2} + \frac{1}{\hat{r}} \frac{\partial \hat{w}}{\partial \hat{r}} \right) + \frac{\partial^2 \hat{w}}{\partial \hat{z}^2} \right] - \epsilon \mathrm{Fr}\phi_1, \tag{20}$$

$$\epsilon \text{Re Pr}\phi_2\left[\frac{\partial \hat{T}}{\partial \hat{t}} + \hat{\psi}_1\frac{\partial \hat{T}}{\partial \hat{r}} + \hat{\psi}_3\frac{\partial \hat{T}}{\partial \hat{z}}\right] =$$

$$\frac{k_{\text{nf}}}{k_f}\left[\epsilon^2\left(\frac{\partial^2 \hat{T}}{\partial \hat{r}^2} + \frac{1}{\hat{r}}\frac{\partial \hat{T}}{\partial \hat{r}}\right) + \frac{\partial^2 \hat{T}}{\partial \hat{z}^2}\right], \tag{21}$$

where $\text{Fr} = gh_0^2/(U_0\nu_{\text{bl}})$ is the modified Froude number, $\text{Pr} = \frac{\nu_f}{\alpha_f}$ is the Prandtl number, $\phi_1 = (1-\phi)^{2.5}[(1-\phi) + \phi\frac{\rho_s}{\rho_f}]$, $\phi_2 = (1-\phi) + \phi\frac{(\rho c_p)_s}{(\rho c_p)_f}$.

The non-dimensional boundary and initial conditions are given below:
at $\hat{z} = 0$,

$$\hat{\psi}_1 = 0, \hat{\psi}_2 = \hat{r}, \hat{\psi}_3 = 0, \hat{T} = \hat{\theta}(\hat{r}), \tag{22}$$

at $\hat{z} = \hat{h}(\hat{r}, \hat{t})$,

$$-p + \frac{2\epsilon^2(1-\phi)^{2.5}}{(1+\epsilon^2\hat{h}_{\hat{r}}^2)}[\epsilon^2\hat{h}_{\hat{r}}\hat{\psi}_{1\hat{r}} - \hat{h}_{\hat{r}}(\hat{\psi}_{1\hat{z}} + \epsilon^2\hat{\psi}_{3\hat{r}})$$

$$+\hat{\psi}_{3\hat{z}}] = \frac{\epsilon}{\hat{r}}[\text{We} - \epsilon^2\alpha\hat{T}][\frac{\hat{h}_{\hat{r}}(1+\epsilon^2\hat{h}_{\hat{r}}^2) + \hat{r}\hat{h}_{\hat{r}\hat{r}}}{(1+\epsilon^2\hat{h}_{\hat{r}}^2)^{3/2}}], \tag{23}$$

$$2\epsilon^2\hat{h}_{\hat{r}}(\hat{\psi}_{3\hat{z}} - \hat{\psi}_{1\hat{r}}) + (1-\epsilon^2\hat{h}_{\hat{r}}^2)(\epsilon^2\hat{\psi}_{3\hat{r}} + \hat{\psi}_{1\hat{z}})$$

$$= -(1-\phi)^{2.5}\epsilon\alpha(\hat{T}_{\hat{r}} + \hat{h}_{\hat{r}}\hat{T}_{\hat{z}})\sqrt{1+\epsilon^2\hat{h}_{\hat{r}}^2}, \tag{24}$$

$$\hat{\psi}_{2\hat{z}} - \epsilon^2\hat{r}\hat{h}_{\hat{r}}\left(\frac{\hat{\psi}_2}{\hat{r}}\right)_{\hat{r}} = 0, \tag{25}$$

$$\frac{k_{\text{nf}}}{k_f}(\hat{T}_{\hat{z}} - \epsilon^2\hat{h}_{\hat{r}}\hat{T}_{\hat{r}}) = -Bi\hat{T}\sqrt{1+\epsilon^2\hat{h}_{\hat{r}}^2}, \tag{26}$$

$$-\frac{\partial \hat{h}}{\partial \hat{t}} - \hat{\psi}_1\frac{\partial \hat{h}}{\partial \hat{r}} + \hat{\psi}_3 = 0, \tag{27}$$

at $\hat{t} = 0$,

$$\hat{\psi}_1 = 0, \hat{\psi}_2 = 0, \hat{\psi}_3 = 0, \hat{h} = \hat{\delta}(\hat{r}), \hat{T} = \hat{\theta}(\hat{r}), \tag{28}$$

where $\mathrm{We} = \epsilon^2\sigma_0/(\nu_{\mathrm{bl}}U_0)$ is the Weber number, $\alpha = \gamma(T_{d_0} - T_0)/(\nu_{\mathrm{bl}}U_0)$ is the thermocapillary parameter, and $\mathrm{Bi} = \beta h_0/k_{\mathrm{bl}}$ is the Biot number.

3 Asymptotic Solution

From this section onward, we have dropped the hat over the dimensionless variables for simplicity. To find the analytical solution of the coupled nonlinear system of partial differential Eqs. (17)–(21), we expanded the dependent variables in power of ϵ as

$$(\psi_1, \psi_2, \psi_3, p) = (\psi_{10}, \psi_{20}, \psi_{30}, p_0) \\ +\epsilon(\psi_{11}, \psi_{21}, \psi_{31}, p_1) + O(\epsilon^2), \tag{29}$$

We substituted Eq. (29) upon guiding Eqs. (17)–(21) and the boundary conditions (22)–(26) and compared the coefficients of like order term of ϵ from both sides. We have obtained set of partial differential equations and solved these equations up to the first order of ϵ as

$$\psi_1 = \phi_1(h - \frac{z}{2})rz + \epsilon\mathrm{Re}{\phi_1}^2 r\Bigg[h_t\left(\frac{z^3}{6} - \frac{h^2 z}{2}\right) \\ +\phi_1\left(\frac{z^6}{360} - \frac{hz^5}{60} + \frac{2\,h^3z^3}{9} - \frac{3\,h^5 z}{5} + \frac{rhh_r}{6}\left(\frac{z^4}{4} - h^3 z\right)\right)\Bigg] \\ +\epsilon\left[\mathrm{Fr}h_r\phi_1 - \mathrm{We}(1-\phi)^{2.5}\left(\frac{1}{r}(rh_r)_r\right)_r\right]\left(\frac{z^2}{2} - hz\right) \\ +\epsilon K\alpha(1-\phi)^{2.5}\left(\frac{\theta}{(K + Bih)}\right)_r z, \tag{30}$$

$$\psi_2 = r + \epsilon 2\mathrm{Re}{\phi_1}^2 r\left(\frac{hz^3}{6} - \frac{z^4}{24} - \frac{h^3 z}{3}\right), \tag{31}$$

$$\psi_3 = \phi_1\left(\frac{z^3}{3} - hz^2\right) - \frac{\phi_1 rh_r z^2}{2} + \epsilon \mathrm{Re}{\phi_1}^2\left[\left(h_t + \frac{rh_{rt}}{2}\right)\right.$$
$$\times\left(\frac{h^2z^2}{2} - \frac{z^4}{12}\right) + \frac{rhh_rh_t}{2} + \phi_1\left[\frac{-z^7}{1260} + \frac{hz^6}{180} + \frac{rh_rz^6}{360}\right.$$
$$-\frac{rhh_rz^5}{40} - \frac{r^2h_r^2z^5}{120} - \frac{r^2hh_{rr}z^5}{120} - \frac{h^5z^4}{9} - \frac{rh^2h^rz^4}{6}$$
$$\left.\left.+\frac{3}{5}h^5z^2 + \frac{7}{4}rh^4h_rz^2 + \frac{r^2h_r^2h^3z^2}{3} + \frac{r^2h^4h_{rr}z^2}{12}\right]\right]$$
$$-\frac{\epsilon}{r}\left[\mathrm{Fr}\phi_1(rh_r)_r - \mathrm{We}(1-\phi)^{2.5}\left\{r\left(\frac{1}{r}(rh_r)\right)_r\right\}_r\right]$$
$$\times\left(\frac{z^3}{6} - \frac{hz^2}{2}\right) + \epsilon\left[\mathrm{Fr}\phi_1 h_r - \mathrm{We}(1-\phi)^{2.5}\left(\frac{1}{r}(rh_r)_r\right)_r\right]$$
$$\times\frac{h_rz^2}{2} - \epsilon K\alpha(1-\phi)^{2.5}\frac{1}{r}\left[r\left(\frac{\theta}{K+\mathrm{Bi}h}\right)_r\right]_r\frac{z^2}{2}, \tag{32}$$

$$T = \frac{K+\mathrm{Bi}(h-z)}{K+\mathrm{Bi}h}\theta(r) + \epsilon\left(\frac{\mathrm{Re}\,\mathrm{Pr}\phi_2}{K}\right)\left[\frac{\mathrm{Bi}^2h_t\theta}{(K+\mathrm{Bi}h)^2}\right.$$
$$\times\left(\frac{z^3}{6} - \frac{h^2}{6}\left(\frac{3K+\mathrm{Bi}h}{K+\mathrm{Bi}h}\right)z\right) + r\phi_1\theta_r\left(\frac{hz^3}{6} - \frac{z^4}{24}\right.$$
$$\left.-\frac{h^3}{24}\left(\frac{8K+3\mathrm{Bi}h}{K+\mathrm{Bi}h}\right)z\right) - r\phi_1\mathrm{Bi}\left(\frac{\theta}{K+\mathrm{Bi}h}\right)_r\left(\frac{hz^4}{12}\right.$$
$$\left.-\frac{z^5}{40} - \frac{h^4}{120}\left(\frac{25K+7\mathrm{Bi}h}{K+\mathrm{Bi}h}\right)z\right) - \frac{\mathrm{Bi}\theta\phi_1}{K+\mathrm{Bi}h}\left(\frac{z^5}{60}\right.$$
$$\left.-\frac{hz^4}{12} - \frac{h^4}{60}\left(\frac{15K+4\mathrm{Bi}h}{K+Bih}\right)z\right) + \frac{1}{24}\phi_1\mathrm{Bi}$$
$$\left.\times\left(\frac{rh_r\theta}{K+\mathrm{Bi}h}\right)\left(z^4 - h^3\left(\frac{4K+\mathrm{Bi}h}{K+\mathrm{Bi}h}\right)z\right)\right], \tag{33}$$

where $K = \frac{k_{\mathrm{nf}}}{k_{\mathrm{bl}}}$. It is to be mentioned here that the solutions given by Eqs. (30)–(33) do not satisfy the initial condition (28). The film evolution equation is derived by integrating the kinematic condition (27) with respect to z as

$$h_t + \frac{1}{r}\left[\frac{\phi_1r^2h^3}{3} - \epsilon\mathrm{Re}{\phi_1}^2r^2\left(\frac{5}{24}h_th^4 + \phi_1\left(\frac{311}{1260}h^7\right.\right.\right.$$
$$\left.\left.+\frac{3}{40}rh^6h_r\right)\right) - \frac{\epsilon rh^3}{3}\left\{Fr\phi_1h_r - We(1-\phi)^{2.5}\left(\frac{1}{r}(rh_r)_r\right)_r\right\}$$
$$\left.+\frac{\epsilon K\alpha r}{2}(1-\phi)^{2.5}\left(\frac{\theta}{k+Bih}\right)_r h^2\right]_r = 0. \tag{34}$$

To find the solution of Eq. (34), we expanded $h(r, t)$ by following the asymptotic procedure (29). Finally, collecting the coefficients of different powers of ϵ in Eq. (34) we get Eq. (35) for coefficient of ϵ^0 as

$$h_{0t} + \phi_1 r {h_0}^2 h_{0r} = -\frac{2}{3}\phi_1 {h_0}^3, \tag{35}$$

while coefficient of ϵ^1 gives Eq. (36).

$$\begin{aligned} h_{1t} + \phi_1 r {h_0}^2 h_{1r} &= -2\phi_1({h_0}^2 + r{h_0}^2 h_{0r} h_1) \\ &-\frac{1}{r}[\mathrm{Re}{\phi_1}^2 r^2(-\frac{5}{24}h_{0t}{h_0}^4 + \phi_1(-\frac{311}{1260}{h_0}^7 \\ &-\frac{3}{40}r{h_0}^6 h_{0r})) - \frac{1}{3}Fr\phi_1 r h_{0r}{h_0}^3 + \frac{1}{3}We(1-\phi)^{2.5} r{h_0}^3 \\ &\times\left(\frac{1}{r}(rh_{0r})_r\right)_r + k\alpha r(1-\phi)^{2.5}\left(\frac{\theta}{k + Bih_0}\right)_r \frac{{h_0}^2}{2}]_r. \end{aligned} \tag{36}$$

It is followed from Eq. (35) that

$$\frac{\mathrm{d}}{\mathrm{d}t}h_0(r(t), t) = -\frac{2}{3}\phi_1 {h_0}^3(r(t), t), \tag{37}$$

along the characteristics curve $r(t)$ satisfying

$$\frac{\mathrm{d}}{\mathrm{d}t}r(t) = \phi_1 r(t){h_0}^2(r(t), t). \tag{38}$$

Integrating above two equations, we get Eqs. (39) and (40), respectively, as

$$h_0(r(t), t) = C_0\chi^{-1/2}, \tag{39}$$

$$r(t) = C_1\chi^{\frac{3}{4}}, \tag{40}$$

where $\chi = 1 + (4/3)\phi_1 {C_0}^2 t$. It also follows from Eqs. (39) and (40) that

$$r(t){h_0}^{3/2}(r(t), t) = {C_0}^{3/2}C_1 = \text{Constant}. \tag{41}$$

Solution of Eq. (36) along the characteristic curve (40) is obtained as

$$
\begin{aligned}
h_1(r(t),t) = {} & \frac{K\alpha}{2}(1-\phi)^{2.5}\phi^{-1}{C_1}^{-1}(\theta_r\zeta^{-1}\chi^{-3/4} \\
& +\frac{2}{3}\theta(r)\mathrm{Bi}C_0{C_1}^{-1}\zeta^{-2}\chi^{-2})+\frac{62}{315}\mathrm{Re}{\phi_1}^2{C_0}^5\chi^{-5/2} \\
& -\frac{2}{9}\mathrm{Fr}{C_0}^2{C_1}^{-2}\chi^{-5/2}+\frac{32}{81}\mathrm{We}{\phi_1}^{-1}(1-\phi)^{2.5}{C_0}^2{C_1}^{-4}\chi^{-4} \\
& +C_2\chi^{-1/2},
\end{aligned} \tag{42}
$$

where $\zeta = K + \mathrm{Bi}C_0\chi^{-1/2}$.

It is to be mentioned here that the above solutions do not satisfy the initial condition and these solutions are known as the longtime solution. To find the solutions that satisfy the initial condition (known as short-time solution), we have stretched the time coordinate as $\tau = t/\epsilon$ and keeping other variables same as before. Following the asymptotic expansion (29), we have obtained the short-time film thickness that satisfies the initial condition.

Now, both the solutions are matched by the following Van Dyke [18] matching principle under the assumption that flow is continuous from the start of the spinning of the disk to all succeeding time as

$$
\lim_{\tau\to\infty}\overline{h}(r,\tau) = \lim_{t\to 0} h(r,t), \tag{43}
$$

where $\overline{h}(r,\tau)$ is the short-time film thickness.

We have obtained the integrating constants by using equation (43) in the expression of film thickness in both the timescales as $C_1 = r(0) = \xi(\text{say})$, $C_0 = h_0(\xi,0) = \delta(\xi)$ and

$$
\begin{aligned}
C_2 = {} & -\frac{K\alpha}{2}(1-\phi)^{2.5}\phi^{-1}\xi^{-1}(\theta_\xi{\zeta_0}^{-1}+\frac{2}{3}\theta(\xi)\mathrm{Bi}\delta(\xi)\xi^{-1}{\zeta_0}^{-2}) \\
& +\frac{2}{9}\mathrm{Fr}\xi^{-2}\delta(\xi)^2-\frac{32}{81}\mathrm{We}(1-\phi)^{2.5}\phi^{-1}\xi^{-4}\delta(\xi)^2 \\
& +0.462594\mathrm{Re}\phi_1^2\delta^5+1.83188\mathrm{Re}\phi_1^2\xi\delta^4\delta_\xi,
\end{aligned} \tag{44}
$$

where $\zeta_0 = 1 + Bi\delta(\xi)$. The composite film thickness that is valid for all timescales is given by

$$
h^c(r,t) = \overline{h}(r,t/\epsilon) + h(r,t) - h_{\mathrm{com}}(r,t), \tag{45}
$$

where $h_{\mathrm{com}}(r,t)$ is the common solution.

4 Results and Discussion

In this paper, we have analyzed the development of water-based SWCNTs and MWCNTs nanoliquid film during the spin coating process. The analytical expression for the velocity, temperature field, and film thickness is obtained by the perturbation technique. The film height is computed numerically by using the initial non-uniform film profile and disk temperature distribution, respectively, as $\delta(r) = 0.47(r + 0.1)^2 e^{-r^{0.76}}$ and $\theta(r) = e^{-r^2}$. Figure 1 depicts the variation of the composite film thickness h^c with time t for different values of ϕ with SWCNTs nanoliquid, and it is observed from the figure that the film thinning rate diminishes for the increasing values of ϕ. Due to addition of the SWCNTs on the base liquid water, the viscosity of the water increases.

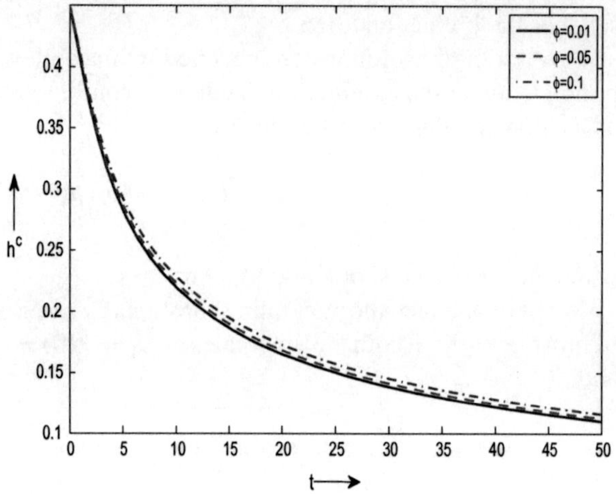

Fig. 1 Variation of film thickness h^c with time t for water-based SWCNTs nanoliquids with different values of ϕ. Here, $\text{Re} = 1.0$, $\text{Fr} = 0.1$, $\text{We} = 0.1$, $\epsilon = 0.01$, $\alpha = 0.1$, $\text{Bi} = 0.1$, $r = 4$, $\delta(r) = 0.47(r + 0.1)^2 e^{-r^{0.76}}$, $\theta(r) = e^{-r^2}$

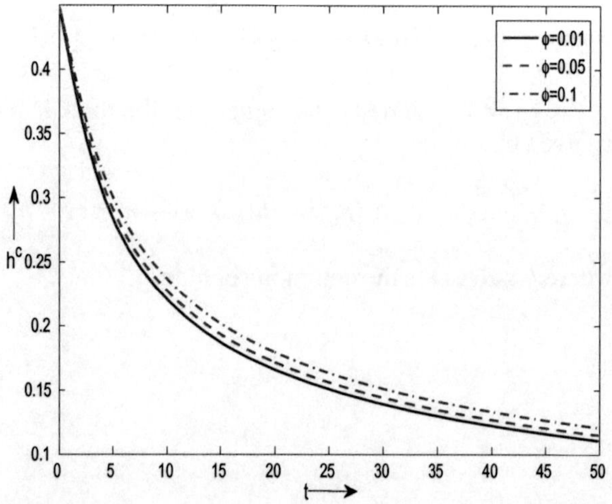

Fig. 2 Variation of film thickness h^c with time t for water-based MWCNTs nanoliquids with different values of ϕ. Here, $\text{Re} = 1.0$, $\text{Fr} = 0.1$, $\text{We} = 0.1$, $\epsilon = 0.01$, $\alpha = 0.1$, $\text{Bi} = 0.1$, $r = 4$, $\delta(r) = 0.47(r + 0.1)^2 e^{-r^{0.76}}$, $\theta(r) = e^{-r^2}$

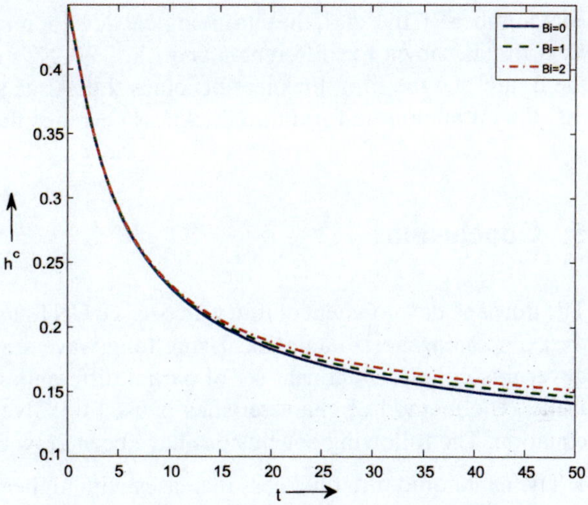

Fig. 3 Variation of film thickness h^c with time t for water-based MWCNTs nanoliquids with different values of Biot number Bi. Here, $\mathrm{Re} = 1.0$, $\mathrm{Fr} = 0.1$, $\mathrm{We} = 0.1$, $\epsilon = 0.1$, $\alpha = 1$, $\phi = 0.05$, $r = 4$, $\delta(r) = 0.47(r + 0.1)^2 \mathrm{e}^{-r^{0.76}}$, $\theta(r) = \mathrm{e}^{-r^2}$

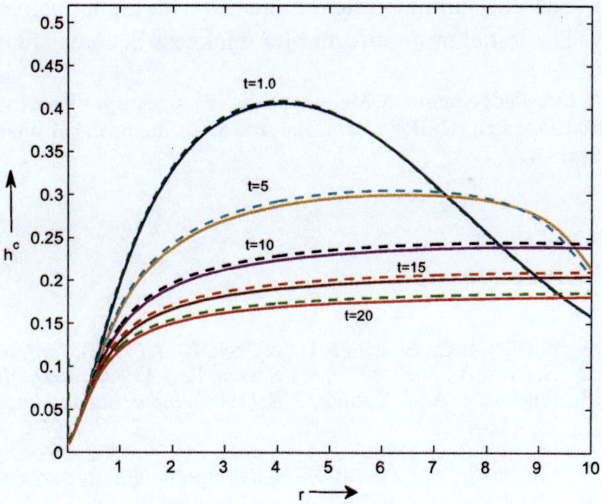

Fig. 4 Variation of film thickness h^c with r for water-based SWCNTs nanoliquids with different ϕ and t. Here, continuous line for $\phi = 0.05$ and dashed line for $\phi = 0.1$ when $Re = 1.0$, $\mathrm{Fr} = 0.1$, $\mathrm{We} = 0.1$, $\epsilon = 0.01$, $\alpha = 0.1$, $\mathrm{Bi} = 0.1$, $\delta(r) = 0.47(r + 0.1)^2 \mathrm{e}^{-r^{0.76}}$, $\theta(r) = \mathrm{e}^{-r^2}$

As a result, the radial velocity decreases for higher values of ϕ and the film thinning rate reduces. In Fig. 2, we have plotted the variation of h^c with time t for different ϕ and MWCNTs nanoliquid. It is found the rate of film thinning decreases with rising values of ϕ.

The effects of the Biot number Bi on the film thinning process are presented in Fig. 3 for MWCNTs nanoliquid. It is clear from the figure that the rate of film thinning decreases for rising values of Bi. In general, the Biot number represents the amount of heat transfer from the SWCNTs nanoliquid to the surrounding fluid medium in comparison with heat transfer within the nanoliquid. For this reason, the thermocapillary action diminishes and the film thinning rate diminishes for larger

Biot number Bi. In Fig. 4, the chronological development of composite film thickness h^c with r is shown for different ϕ and t in SWCNTs nanoliquid. It is evident from the figure that the film thickness becomes thinner as time increases. It is also clear that the initial non-uniform film thickness becomes flat at large time.

5 Conclusions

The flow and development of thin water-based CNTs nanoliquid film are investigated over a spinning horizontal disk. Using long wave singular perturbation technique, governing coupled nonlinear set of partial differential equations are solved analytically. The method of characteristics is used to solve the nonlinear film evolution equation. The following conclusions have been drawn from this study.

- The nanoliquid film thickness increases with higher values of CNTs volume fraction ϕ for both SWCNTs and MWCNTs.
- The film thinning rate diminishes with higher values of the Biot number Bi.
- The initial non-uniform film thickness becomes planar at large time.

Acknowledgements S. Maity expresses his sincere gratitude to the Council of Scientific and Industrial Research (CSIR), New Delhi, India, for the financial support with file number 25(0267)/17/EMR-II.

References

1. Emslie, A. C., Bonner, F. D., & Peck, L. G. (1958). *Journal of Applied Physics, 29*, 858–862.
2. Acrivos, A., Shah, M. J., & Petersen, E. E. (1960). *Journal of Applied Physics, 31*, 963–968.
3. Jenekhe, S. A., & Schuldt, S. B. (1984). *Industrial and Engineering Chemistry Fundamentals, 23*, 432–436.
4. Higgins, B. G. (1986). *Physics of Fluids, 29*, 3522–3529.
5. Myerhofer, D. J. (1978). *Journal of Applied Physics, 49*, 3993–3997.
6. Middleman, S. (1987). *Journal of Applied Physics, 62*, 2530–2532.
7. Dandapat, B. S., & Roy, P. C. (1994). *Journal of Physics D: Applied Physics, 27*, 2041–2045.
8. Dandapat, B. S., & Roy, P. C. (1993). *International Journal of Non-Linear Mechanics, 28*, 489–501.
9. Dandapat, B. S., & Singh, S. K. (2015). *Applied Mathematics and Computation, 258*, 545–555.
10. Dandapat, B. S., & Maity, S. (2009). *International Journal of Non-Linear Mechanics, 44*, 877–882.
11. Dandapat, B. S., Santra, B., & Kitamura, A. (2005). *Physics of Fluids, 17*.
12. Choi, S. U. S., & Proc., (1995). *ASME International Mechanical Engineering Congress and Exposition, FED231/MD66* (pp. 99–105). ASME.
13. Buongiorno, J. (2006). *Journal of Heat Transfer, 128*, 240–250.
14. Akbar, N. S., & Butt, A. W. (2015). *Applied Mathematics and Computation, 259*, 231–241.
15. Hussanan, A., Khan, I., Gorji, M. R., & Khan, W. A. (2019). *Bio Nano Science, 9*, 21–29.
16. Maity, S. (2017). *International Nano Letters, 7*, 9–23.
17. Krishanan, R., Maity, S., Singh, S. K., Dandapat, B. S. (2020). *International Journal of Applied and Computational Mathematics, 6*, 147(1–19).
18. Van Dyke, M. (1964). *Perturbation Methods in Fluid Mechanics*. Academic Pssress.

Performance Improvement Techniques in Shell-and-Tube Type of LHS Unit

Berihu Gebreyohannes Abreha, Pinakeswar Mahanta, and Gaurav Trivedi

Nomenclature

A_{Mush}	Mushy region constant value
Cp	Specific heat capacity (J kg^{-1} K^{-1})
F	Body force (N m^{-3})
g	Acceleration due to gravity (m s^{-2})
h	Specific enthalpy (J/kg)
L	Latent heat of fusion (J kg^{-1})
k	Thermal conductivity (W m^{-1} K^{-1})
P	Pressure (Pa)
S	Darcy law's source term
T	Temperature (K)
v	Velocity (m/s)

Greek Symbols

ρ	Density (kg/m^3)

B. G. Abreha
Center for Energy, Indian Institute of Technology Guwahati, Guwahati, Assam 781039, India

P. Mahanta (✉)
Department of Mechanical Engineering, Indian Institute of Technology Guwahati, Guwahati, Assam 781039, India
e-mail: pinak@iitg.ac.in

National Institute of Technology Arunachal Pradesh, Yupia, Arunachal Pradesh 791112, India

G. Trivedi
Department of Electronics and Electrical Engineering, Indian Institute of Technology Guwahati, Guwahati, Assam 781039, India

P. Mahanta et al. (eds.), *Advances in Thermofluids and Renewable Energy*, Lecture Notes in Mechanical Engineering, https://doi.org/10.1007/978-981-16-3497-0_12

β	Coefficient of thermal expansion (1/K)
μ	Dynamic viscosity (Ns/m^2)
θ	Melt fraction
φ	Nanoparticles volume fraction

Subscripts

ini	Initial
l	Liquid
np	Nanoparticle
nPCM	Nanophase change material
s	Solid
Ref	Reference

1 Introduction

Fossil fuel is the primary energy source to fulfill the ongoing increase in global energy demand. Increasing the share of renewable energy, such as solar energy in the energy mix is a very important step to alleviate environmental hazards and energy security problems. However, renewable energy sources are intermittent and unpredictable in nature. Thus, storing solar energy is necessary to maintain a continuous supply of energy and to satisfy the demand of end users [1].The difference between energy supply and demand can be minimized by developing effective energy storage technologies. Generally speaking, thermal energy storage systems are of three kinds, viz. sensible heat storage (SHS), latent heat storage (LHS), and chemical heat storage system [2]. Latent heat type storage device is considered more efficient and highly compacted, and due to these attractive properties, LHS is preferred frequently [3].

However, the latent thermal energy system suffers from poor thermal conductivity, and this is a critical problem that results in low charging and discharging rates of PCM [4]. Consequently, several effective thermal conductivity enhancement techniques have been practiced including embedding fins into the PCM [5], inserting metal matrix or carbon foams [6, 7], dispersing nanoparticles and carbon-based particles embedded in PCM [8], etc.

A comprehensive examination of the heat transfer characteristics between HTF and PCM is essential to design and develop the most effective storage unit. For the selection of the most efficient and economically affordable TES unit, conducting optimization techniques is important. This is a difficult task considering the cost and performance criteria of the storage unit. For optimal design and material selection of TES numerical modeling of the LHS system is a perfect solution to overcome the above mentioned problem. From the aforementioned reviewed literature, it is observed that there is a lack of investigations undertaken on a comparison study of various performance enhancement techniques of shell-and-tube LHS unit. Hence, the principal objective of this paper is to perform a performance comparative study of three shell-and-tube type LHS units using erythritol as PCM.

2 Numerical Formulation

In this section, the geometric characteristics of the selected LHS units are described. Moreover, the governing equations that are used to solve the phase transition characteristics of the PCM are formulated, and the material properties of the domains considered in the numerical investigation are also defined. In addition, the grid size and boundary conditions are explained in this section.

2.1 Model Description

Figure 1 represents three physical modeling diagram of shell-and-tube LHS units considered for this study. With the help of an optimization technique, 19 tubes are selected with an internal and external diameter of 17 mm and 20 mm, respectively. The shell of the LHS unit illustrated in Fig. 1a is filled with PCM. In the second case, the LHS model is also modeled without fins, but the phase changing materials (PCM) are mixed with 5% volume fraction of Cu nanoparticles and is exhibited in Fig. 1b. In the third case, each HTF tube is supported by four longitudinal fins to improve the heat transfer penetration from the hot walls into PCM and is presented in Fig. 1c. In order to maintain the same volume of PCM in all models, the container diameter of the LHS models is kept different. The internal shell diameter of the storage unit in the first case is 235 mm. On the other hand, the internal shell diameters of the second and third models of LHS units are the same and it is set to be 240 mm. In case two, 5% of the volume of the shell is contained by the Cu nanoparticles and this is equal to the volume of the fins in case three. Because of superior thermal conductivity, HTF pipes and fins are considered to be of copper material and, width and thickness of each fin are chosen as 12 mm and 2 mm, respectively. Commercial sugar alcohol known as erythritol ($C_4H_{10}O_4$) is considered as PCM. The thermal properties of PCM and copper are presented in Table 1.

2.2 Thermal Properties of Materials

The thermal properties of the nano-PCM (a mixture of PCM and nanoparticle) are derived from the respective volume fraction of nanoparticle and PCM considering

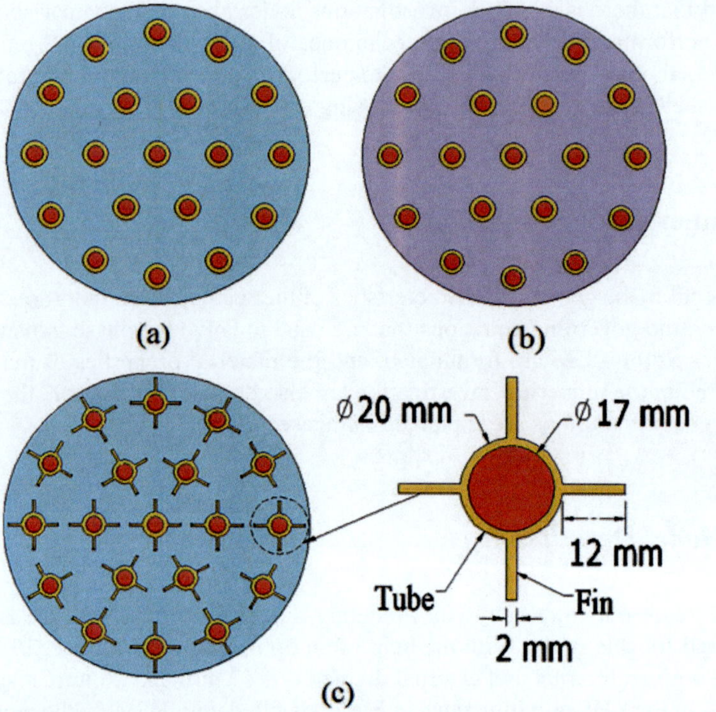

Fig. 1 Physical modeling diagram of shell-and-tube LHS units consider for this study

Table 1 Thermophysical properties of erythritol and copper [7, 10–12]

Properties	Erythritol	Copper
$\rho\ (\text{kg/m}^3)$	1450 (S)	8978
	1280 (L)	
C_p(J/kgK)	1680 (S)	381
	2660 (L)	
K (W/m K)	0.73 (S)	387.6
	0.33 (L)	
L (kJ/kg)	354.7	
T_m (°C)	118	
μ(kg/ms)	1.6×10^{-2}	
$\beta(\text{K}^{-1})$	2.94×10^{-5}	

the theoretical mixing and are listed below [9]. The thermosphysical properties of nano-PCM are also listed in Table 1.

Density

$$\rho_{n,\text{PCM}} = \varphi_n \rho_{\text{np}} + (1 - \varphi_n)\rho_{\text{PCM}} \tag{1}$$

Specific heat capacity

$$C_{P,\text{npcm}} = \frac{\varphi_n\left(\rho_{\text{np}} C_{P,\text{np}}\right) + (1-\varphi_n)(C_P\rho)_{\text{PCM}}}{\rho_{n,\text{PCM}}} \tag{2}$$

Latent heat

$$L_{n,\text{PCM}} = \frac{(1-\varphi_n)(\rho L)_{\text{PCM}}}{\rho_{n,\text{PCM}}} \tag{3}$$

Thermal expansion coefficient

$$\beta_{n,\text{PCM}} = \frac{\varphi_n(\rho\beta)_{\text{np}} + (1-\varphi_n)(\beta\rho)_{\text{PCM}}}{\rho_{n,\text{PCM}}} \tag{4}$$

The effective dynamic viscosity of nano-PCM can be calculated as follows [13].

$$\mu_{n,\text{PCM}} = 0.983\text{e}^{(12.959\varphi_n)}\mu_{\text{PCM}} \tag{5}$$

The effective thermal conductivity of the mixture of nanoparticle and PCM is described with the help of Maxwell's theory and Brownian motion [14].

$$k_{n,\text{PCM}} = \frac{k_{\text{np}} + 2k_{\text{PCM}} - 2 * \varphi_n\left(k_{\text{PCM}} - k_{\text{np}}\right)}{k_{\text{np}} + 2k + \varphi_n\left(k_{\text{PCM}} - k_{\text{np}}\right)} k_{\text{PCM}} \tag{6}$$

2.3 Thermal Modeling

In this study, a two-dimensional CFD model has been built to analyze the melting performance of LHS units using computational fluid dynamics (CFD) ANSYS Fluent software. For modeling the phase transition of PCM, the commonly used enthalpy–porosity technique is applied, and this method helps to combine enthalpy with the temperature jointly as a function presented below. The numerical model is developed using Navier–Stokes equations that define the conservation of mass, momentum, and energy. For the buoyancy entity inside the momentum equation, density gradients are handled using Boussinesq approximation [15].

The next assumptions are taken into account in the numerical formulation to evaluate the performance of storage unit. The PCM and nanoparticles are mixed uniformly; the flow is assumed to be laminar, incompressible, and unsteady. The viscous and dissipation term is also assumed to be negligible. The PCM and HTF tubes have steady reference temperature, and the storage system is assumed to be thermally insulated to ensure that there is no heat loss toward external wall of the PCM.

Continuity equation

$$\frac{\partial \rho}{\partial t} = \nabla.\left(\rho \vec{V}\right) \tag{7}$$

Momentum conservation equation

$$\rho\frac{\partial \vec{V}}{\partial t} + \rho\vec{V}.\left(\nabla.\vec{V}\right) = -\nabla P + \mu\nabla^2\vec{V} + \rho\beta\vec{g}(T - T_{\text{ref}}) + \vec{S} \tag{8}$$

Energy conservation equation

$$\frac{\partial}{\partial t}(\rho h) + \nabla.\left(\rho.\vec{V}h\right) = \nabla(k\nabla T) \tag{9}$$

The overall enthalpy (H) is equal to the sum of sensible enthalpy (h) and latent heat (ΔH), which is resulted due to phase conversions.

$$H = h + \Delta H \tag{10}$$

where,

$$h = h_{\text{e}} + \int\limits_{T_{\text{ref}}}^{T} C_{\text{p}}\text{d}T \tag{11}$$

$$\Delta H = \theta L \tag{12}$$

The movement of PCM solid–liquid interface is tracked by enthalpy–porosity method. To monitor liquid part in the mushy region, a liquid fraction parameter is introduced in the enthalpy–porosity technique. The liquid fraction (θ) of each cell represents porosity of that given cell. The solid region has no porosity and its liquid fraction is zero. The melting front of PCM is considered as a mushy region having a liquid fraction value that spans between 0 and 1. Normally, melting fraction (θ) of PCM can be expressed as Eq. (13).

$$\theta = \frac{\Delta H}{L} = \begin{cases} 0 & \text{if } T < T_{\text{s}} \\ \frac{T - T_{\text{s}}}{T_l - T_{\text{s}}} & \text{if } T_{\text{s}} < T < T_l \\ 1 & \text{if } T > T_1 \end{cases} \tag{13}$$

The source term $\vec{S}$, which is added in the momentum conservation equation, is defined as:

$$\vec{S} = \frac{(1-\theta)^2}{(\theta^3 + \varepsilon)} A_{\text{Mush}}\vec{V} \tag{14}$$

where, ε is an insignificant quantity (10^{-3}) placed to avoid division by zero. The mushy region's constant value lies between 10^4 and 10^7, and is frequently applied for computations, and the value for this study is set equal to 105 [16]. All the cases of LHS are tested with three different meshes sizes to examine dependency on grid size and time step size. This helps in deciding an optimal size considering both accuracy and time. The variations between finer and extra fine mesh outcomes are insignificant. Hence, finer mesh has been selected for the simulation of all the LHS models. The mesh elements size of first, second, and third LHS models are 11,031, 15,728, and 20,931, respectively. In all the cases, initial temperature of the entire system is 20 °C, and there is no heat loss from external walls. The inlet temperature of the HTF is set as 138 °C.

3 Results and Discussion

The governing equations that are formulated in Section 2 are solved using CFD software. The transient temperature and liquid fraction parameters are employed to evaluate the performances of selected storage units. Due to the symmetrical half of LHS, models are considered for the numerical analysis. The time taken for the entire melting of PCM inside the storage unit is related to the rate of temperature increase. In order to ensure that the numerical results are correct and reliable, the experimental study of Agyenim et al. [17] has been solved numerically using the same numerical setup, which is utilized for the current study.

The same properties of PCM and, similar initial and boundary conditions of the experimental study are employed for the validation. The timewise temperature parameter during the charging process is used for the comparison of these two studies.

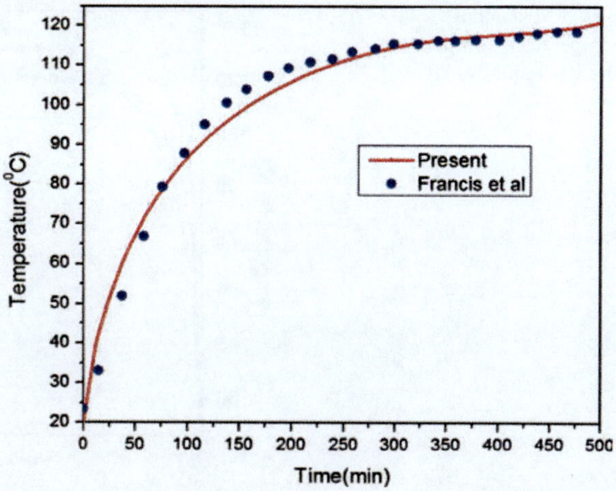

Fig. 2 Comparison of present study versus Agyenim et al. [17] results

Figure 2 illustrates the numerical and practical results, and a good agreement is observed between them.

Figure 3 depicts the variations of average temperature versus time of three LHS units during the charging process. It can be noticed that the temperature curves of all the models are becoming steeper up to 10 min without significant difference among them. This could be resulted due to the dominant conduction heat transfer in all the cases near the beginning of the process. Later, the rate of heat addition in the finned LHS unit is found to be higher than the other two. This is because of the better heat transfer in the finned LHS unit due to the extended surface of the fins. Due to the uniformly distributed fins, the temperature expansion is more uniform and highest as compared with LHS models without fins and the LHS unit with nano-PCM. Even though the thermal conductivity of PCM in the second LHS model is improved by adding 5% volume fraction Cu nanoparticle, the heat transfer rate is even slower than the first model. The internal shell diameter of the first model is smaller as compared to the other models. Hence, the outward heat transfer from the wall of HTF tubes of the LHS model with nano-PCM is delayed as compared to the first model. As it is explained above, the temperature of finned LHS unit increases more quickly, and since its magnitude is reached 137.8 °C at at 80 min, which is almost same as the inlet temperature of HTF tubes.

The liquid fraction variation as a function of time for all LHS units is shown in Fig. 4. It can be seen from this figure that the finned LHS has a constant increase in the melt fraction than the LHS model without fins. As it is expected, the melting progress of the PCM in the finned LHS unit is faster. From the beginning of the charging process, there is a noticeable difference in the melting progress in the finned LHS among all the storage systems. The presence of fins in the HTF tubes augments the heat transfer across the storage unit and improves thermal conductivity of the PCM.

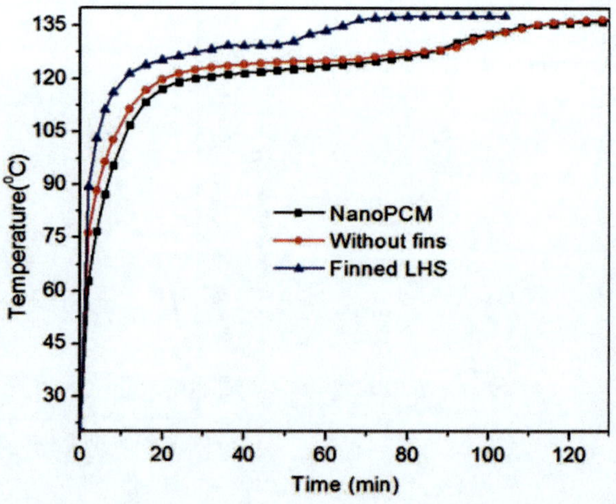

Fig. 3 Transient temperature comparison of the LHS units

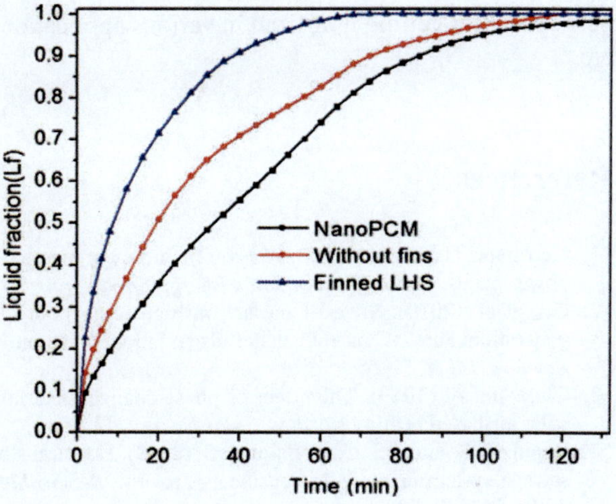

Fig. 4 Transient liquid fraction comparison of the LHS units

In this context, the total time taken for the complete melting of PCM in the finned LHS unit is 80 min, whereas the LHS models in the first and second cases consume 120 and 136 mins, respectively. As a result, the third LHS model supported by fins has been found the best option to be used in various applications.

4 Conclusion

This study presents a comparative study of the thermal performances of three shell-and-tube LHS models. With the help of an optimization technique, 19 tubes are selected with an internal and external diameter of 17 mm and 20 mm, respectively for the proposed study. In the first case, LHS is considered without fins.

In the second case, LHS is also modeled without fins, but the phase changing materials (PCM) are mixed with 5% volume fraction of Cu nanoparticles. In the third case, every HTF tube is supported by four fins. In order to maintain the same volume of PCM in all the models, the shell diameter of the first LHS model is 235 mm and, for the second and third cases, the diameter is set to be 240 mm. Erythritol ($C_4H_{10}O_4$) is considered as PCM in this study. The thermal performance of LHS is evaluated numerically and charging time, liquid fraction, transient temperature, and stored energy are estimated. The buoyancy effect of liquid PCM is modeled using Boussinesq approximation method in the proposed study.

The outcome of numerical analysis is validated with the experimental results to ensure reliability of the proposed work. It is observed that the temperature of finned LHS unit increases quickly. The LHS in the third case gets fully charged after 80 min, whereas LHSs in the first and second cases take 120 and 136 min, respectively, for complete melting. This exhibits that the third LHS model supported by fins is the best option for getting it utilized in various applications, such as hybrid solar cooker etc.

References

1. Herrmann, U., & Kearney, D. W. (2002). Survey of thermal energy storage for parabolic trough power plants. *The Journal of Solar Energy Engineering Transfer ASME, 124*(2), 145–152.
2. Gil, et al. (2010). State of the art on high temperature thermal energy storage for power generation. Part 1-Concepts, materials and modellization. *Renewable and Sustainable Energy Reviews, 14*(1), 31–55.
3. Ghoneim, A. (1989). The effect of phase-change material properties on the performance of solar air-based heating systems. *Solar Energy, 42*(6).
4. Regin, F., Solanki, S. C., & Saini, J. S. (2008). Heat transfer characteristics of thermal energy storage system using PCM capsules: A review. *Renewable and Sustainable Energy Reviews, 12*(9), 2438–2458.
5. Niyas, H., Prasad, S., & Muthukumar, P. (2017). Performance investigation of a lab–scale latent heat storage prototype—Numerical results. *Energy Conversion and Management, 135*, 188–199.
6. Zhao, Y., Lu, W., & Tian, Y. (2010). Heat transfer enhancement for thermal energy storage using metal foams embedded within phase change materials (PCMs). *Solar Energy, 84*(8), 1402–1412.
7. Wang, W., He, S., Guo, S., Yan, J., & Ding, J. (2014). A combined experimental and simulation study on charging process of Erythritol-HTO direct-blending based energy storage system. *Energy Conversion and Management, 83*, 306–313.
8. Owolabi, L., Al-Kayiem, H. H., & Baheta, A. T. (2016). Nanoadditives induced enhancement of the thermal properties of paraffin-based nanocomposites for thermal energy storage. *Solar Energy, 135*, 644–653.
9. Khan, Z., Khan, Z., & Tabeshf, K. (2016). Parametric investigations to enhance thermal performance of paraffin through a novel geometrical configuration of shell and tube latent thermal storage system. *Energy Conversion and Management, 127*, 355–365.
10. Nomura, T., Okinaka, N., & Akiyama, T. (2009). Impregnation of porous material with phase change material for thermal energy storage. *Materials Chemistry and Physics, 115*(2–3), 846–850.
11. Fasina, O. O., & Colley, Z. (2008). Viscosity and specific heat of vegetable oils as a function of temperature: 35 °C to 180 °C. *International Journal of Food Properties, 11*(4), 738–746.
12. Al-Abidi, A., Mat, S., Sopian, K., Sulaiman, M. Y., & Mohammad, A. T. (2013). Numerical study of PCM solidification in a triplex tube heat exchanger with internal and external fins. *International Journal of Heat and Mass Transfer, 61*(1), 684–695.
13. Vajjha, R. S., Das, D. K., & Namburu, P. K. (2010). International Journal of Heat and Fluid Flow Numerical study of fluid dynamic and heat transfer performance of Al_2O_3 and CuO nanofluids in the flat tubes of a radiator. *International Journal of Heat and Fluid Flow, 31*(4), 613–621.
14. Mahdavi, M., Tiari, S., & Pawar, V. (2020). A numerical study on the combined effect of dispersed nanoparticles and embedded heat pipes on melting and solidification of a shell and tube latent heat thermal energy storage system *27*(September 2019).
15. Prakash, & Voller, V. (1989). On the numerical solution of continuum mixture model equations describing binary solid-liquid phase change. *Numerical Heat Transfer, Part B: Fundamentals, 15*(2), 171–189.
16. Hong, Y., Ye, W., Du, J., & Huang, S. (2019). International journal of heat and mass transfer solid-liquid phase-change thermal storage and release behaviors in a rectangular cavity under

the impacts of mushy region and low gravity. *International Journal of Heat and Mass Transfer, 130*, 1120–1132.
17. Agyenim, F., Eames, P., & Smyth, M. (2009). A comparison of heat transfer enhancement in a medium temperature thermal energy storage heat exchanger using fins. *Solar Energy, 83*(9), 1509–1520.

Thermal Response of Nanoparticle Infused Tissue Phantoms During Microwave Heating—A Comparative Study

Abishek Sarkar, Mukesh Mehta, Shubham Gupta, and Anup Paul

Nomenclature

E Electric field (V/m)
f Frequency (s^{-1})
H Magnetic field intensity (T)
Q Rate of heat generation (W)

Greek Symbols

σ Effective conductivity (S/m)
μ Permeability of free space
χ Susceptibility of the magnetic nanoparticle

Subscripts

h Heat
0 Free space

A. Sarkar · M. Mehta · S. Gupta · A. Paul (✉)
Department of Mechanical Engineering, NIT Arunachal Pradesh, Yupia, Andhra Pradesh 791112, India

P. Mahanta et al. (eds.), *Advances in Thermofluids and Renewable Energy*, Lecture Notes in Mechanical Engineering, https://doi.org/10.1007/978-981-16-3497-0_13

1 Introduction

Cancer is the second leading cause of death in the world, with around 1 out of every 6 persons dying because of it. Low-and-middle class family are the ones who are affected the most, accounting for 70% of total deaths because of cancer. Cancer has very adverse effects on the economy. The year of 2010 saw a total expenditure of approximately US$ 1.16 trillion. Surgery, radiation therapy, chemotherapy, immunotherapy, targeted therapy, hormone therapy, stem cell transplant and precision medicine are some of the methods used for cancer treatment. But surgery is sometimes difficult to perform, and it comes along with various post-surgical complications. Therefore, in recent years, researchers have given more attention towards non-surgical procedures and found out different ways to improve their tumour destructing capacity. Hyperthermia is a non-invasive method of increasing tumour temperature to stimulate blood flow, increase oxygenation and make tumour cells more sensitive to radiation. Depending upon the extent of the area being treated, hyperthermia treatment can be classified as local, regional, or whole-body hyperthermia. For heating the area, various forms of energy such as radio waves, micro waves, ultrasound waves, etc., can be used. The process of using microwaves to achieve hyperthermia and destroying cancerous cells is called microwave ablation. Microwave ablation is one of the newer methods in the domain of tumour ablation. Percutaneous, laparoscopic and access through open surgery are some of the approaches for treatment by this technique. The operating principle comprises of a microwave generator which emits electromagnetic waves through the bare non-insulated portion of the antenna. Microwaves agitate water molecules in the surrounding tissue producing friction and heat, and thus causing death of cells by coagulation necrosis. But it is important to know that complete destruction of the tumour is necessary to prevent the regeneration of cancerous cells. The microwave technology stands out to be advantageous over other existing thermal ablation technologies because of its consistently higher intratumoural temperatures, larger ablation volumes, faster ablation periods and an improved convection profile [1].

1.1 Microwave Heating Mechanism and Characteristics

Dielectric hysteresis is the process by which heat is generated by the microwaves. Polar molecules (e.g. water) try to continuously realign with an applied electromagnetic field that alternates polarity billions of times per second. When the molecules begin to lag behind the changes of the alternating field, some of the microwave energy is absorbed by the material and converted to heat. The rate of heat generation (Q_h) is proportional to the square of the applied electric field magnitude (E) [1], or

$$Q_{\mathrm{h}} = \sigma |E|^2$$

The most prominent characteristic of microwave heating is volumetric heating, which is quite different from conventional techniques [2]. Conventional heating depends upon the processes of conduction and convection for heat transfer, which requires relatively longer period of time, whereas microwave heating has the potential to deliver heat instantly throughout the product due to volumetric heat generation [3]. Higher the temperature and longer the heating periods, higher is the thermal damage to the product. As the heat is generated throughout the material, faster heating rates are achieved [4]. One of the major drawbacks of microwave heating is the existence of hot spots in several zones depending on product geometry [5], which sometimes becomes a matter of concern for the safety of food products, as sometimes at a cold spot, the food may not be cooked to an extent to kill the microbes present in the food [6]. And because of these cold spots, microwave ovens are designed with turntables, so that each part of the food passes through at least one hot spot. The existence of these hot and cold spots is mainly because of two reasons: First is because of the standing waves, and the second is due to the rapid decay of microwaves [4].

1.2 *Probe Versus Oven*

Though in our experiment, we have used a microwave oven to carry out the ablation on the tissue mimic; but in practical medical applications, a microwave antenna is used to supply the microwaves and ablate the affected zone. In the case of microwave oven, whole of the sample is kept inside the oven cavity. There is no scope of focussing the heat to the sample or the tumour, and therefore, there is a high non-uniformity in temperature distribution within the sample. But in the case of the probe, it is directly inserted into the tumour, which helps in focussing the radiation to the target zone, and thus, it gives a comparatively more uniform temperature distribution. Since, both in the oven and the probe, the heating depends upon the water content and the dielectric properties of the sample, and their heating characteristics tends to show similarity in many cases. Both have a parabolic temperature profile, which tends to vary with the duration of heating [7]. The heating sees a sudden jump from the initial temperature, and the temperature rises following a parabolic profile. Because of the existence of standing waves, presence of hot and cold spots is inevitable for the sample to be heated in the microwave oven [8]. Though the physics of heating of the probe slightly varies from oven, scientists and researchers have observed the presence of hot and cold spots in the case of probe too [9].

1.3 *Use of Nanoparticles in Microwave Heating*

The recent trends in the field of microwave ablation involve the use of nanoparticles to focus the radiation to the affected zone. Heavy damage to the healthy tissues surrounding the tumour is minimized by localizing the nanoparticles within or at proximity to the affected tissues [10]. Magnetic nanoparticles, such as iron oxide nanoparticles are introduced into the tumour to reduce the time of exposure to electromagnetic radiation. Iron oxide nanoparticles are biocompatible, non-toxic to humans and degradable, useful for in vivo applications and have good absorption at microwave frequencies [11]. Magnetic properties of the tumour are intensified because of the insertion of nanoparticles into it. Therefore, in addition to the effect of the electric field, the effect of magnetic field also has to be considered for the net heat generated by the nanoparticles.

$$Q = \pi \mu_0 \chi f |H|^2$$

Though nanoparticles seem to have a promising future in the field of cancer therapy, it is very important to implement the right nanoparticle in the therapeutic procedures. The particle chosen should not only give higher temperature readings, but also should be biocompatible. Biocompatibility of the nanoparticle plays a significant role in treatment of cancer as implementation of non-biocompatible particle may result in undesirable complexities which may turn out to be fatal. Biocompatibility of nanoparticles depends on various factors such as—shape, size, structure and the surface properties [12, 13]. These nanoparticles are inserted into the human tissue via two schemes: Intravenous (I.V.) and Intratumoural (I.T.). The I.V. scheme has the helps in covering the entire tumour, but lacks the ability to generate high temperatures because of lacking enough concentration of nanoparticles to focus the radiation efficiently. I.T. schemes effectively attain higher temperatures at the locations where nanoparticles are injected, but lack the ability to effectively cover the entire tumour [14].

2 Materials and Methods

2.1 Experimental Setup

In our experiments, we used a domestic microwave oven within which the sample is heated for 10 seconds, at a power rating of 900 W. The complete experimental setup is shown in Fig. 1. It consists of a 400 ml borosilicate beaker having a diameter of 75 mm, which contained the agar gel-based tissue mimic, J-type thermocouples for measuring temperatures, a data logger (KEYSIGHT 34980A) for measuring real-time temperatures, a microwave oven (max. output = 1500W), ultra-sonicater, magnetic stirrer and a computer to record the results.

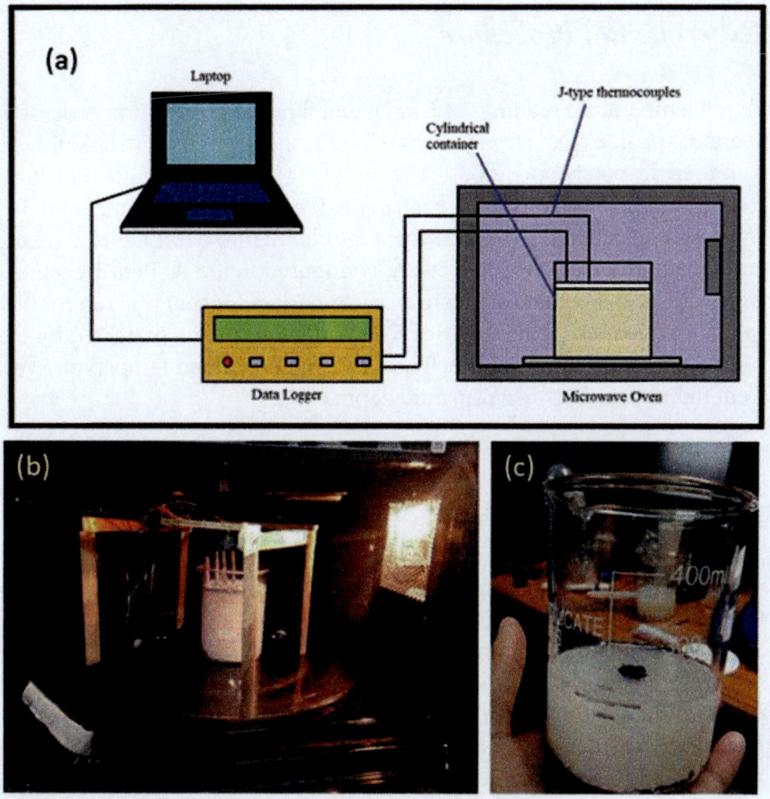

Fig. 1 **a** Schematic diagram of experimental setup, **b** Thermocouples inserted at radial distances and **c** Nanoparticle and agar gel mixture filled in the 8 mm × 10 mm slot

2.2 *Sample Preparation*

At first, 400 ml of warm distilled water was taken on a sauce pan and then, the weighed 10.4 gm of agar powder was poured slowly into the saucepan. The mixture was then stirred continuously and was simultaneously heated till a clear solution was formed. It was ensured that no bubbles were formed. The solution was then allowed to cool for some time and then was poured slowly in a borosilicate beaker which already had its base covered with aluminium foil. For preparation of tissue mimic with nanoparticles, first, 200 ml agar gel was prepared and then poured it into the beaker. Then the nanoparticle solution was prepared using an ultra-sonicator. Nanoparticle solution was added in the 10 ml beaker containing 3 ml agar gel, and the solution was then mixed using a magnetic stirrer. The mixture obtained was filled in the 8 mm × 10 mm slot made in the tissue mimic using a spatula. The second layer of agar gel was then poured till the 400 ml mark.

2.3 Experimental Procedure

We obtained temperature readings at 6 axial and 4 radial positions in order to obtain the temperature profiles. An arrangement of PTFE sheet was used to hold the thermocouples in their respective positions. The circular PTFE lid contained drilled holes at different radial positions from its centre (i.e. 0 mm, 10 mm, 20 mm and 30 mm). The sample was placed in the oven cavity at the centre of the turntable. To extract the temperature data, the thermocouples were configured in the Agilent Benchlink data logger software, and then data acquisition system (data logger) was run for the same duration as of the experiment, i.e. 10 s. The temperature data extracted by the data logger was obtained as an excel file. The excel file contained temperature readings at different radial positions for a specific depth.

3 Result and Discussions

In our study on microwave ablation, we used microwave oven to generate microwaves and ablate the tissue mimic by achieving hyperthermia. The discussion in this section is based on biocompatible nanoparticles, whether they focus the microwave radiation to the region where they infused or not, so that the nearby healthy cells are protected from the heating effect of microwaves. Various temperature profiles have been presented to understand nanoparticles' nature and trend in the presence of microwave radiation. During the experiment, each sample (tissue phantom made by agar powder inside which nanoparticles injected at a depth of 30–40 mm) was kept inside the microwave oven (provided the environment like microwave ablation procedure) for 10–15 s to heat. Data was collected from experimental methods and used to assess the temperature profile. A test sample without nanoparticles (WNPs) was also heated separately in the microwave for the same duration, so that the effect of the presence of microwaves can be assessed.

Figure 2 depicts the collation of temperature profiles of tissue phantom infused with and without nanoparticles at various radial positions for an axial depth of 30 mm and 40 mm, respectively. In this profile, samples with nanoparticles show a higher temperature near the region where nanoparticles are injected (near the central core of the example). It shows the tendency of the nanoparticles to focus microwave radiation towards itself and thus increasing the safety of nearby healthy cells surrounding the tumour. It also displays better uniformity near the nanoparticles' zone because of less deviation from mean temperature. This result efficiently applies on AlNPs & IONPs. CuNPs show deviation from the mentioned results as seen in (Fig. 2). Some irregularities also exist in above graphical representations due to non-uniform heating nature of the microwave oven; it happens due to generation of hot and cold spots inside microwave oven. This observation is similar to Fakhouri et al. [6].

Figure 3 shows the temperature profile for the representation of temperature versus axial depth for different radial positions. This profile indicates that the temperature

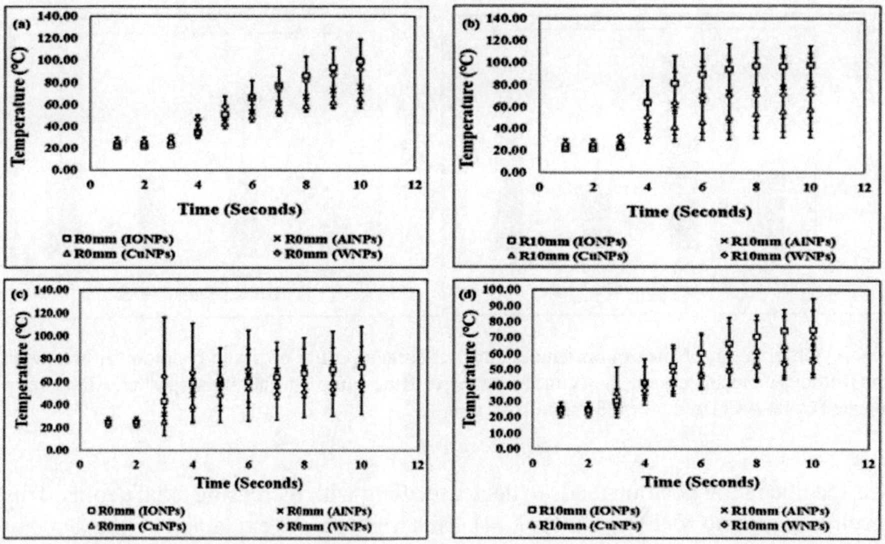

Fig. 2 Collation of temperature profile of tissue phantom infused with and without nanoparticles at the central core part of profiles the sample where tissue nanoparticles have been injected, viz. axial depth of (**a, b**) 30 mm; (**c, d**) 40 mm and various radial positions, viz. (**a, c**) at the centre (0 mm); (**b, d**) 10 mm away from the centre

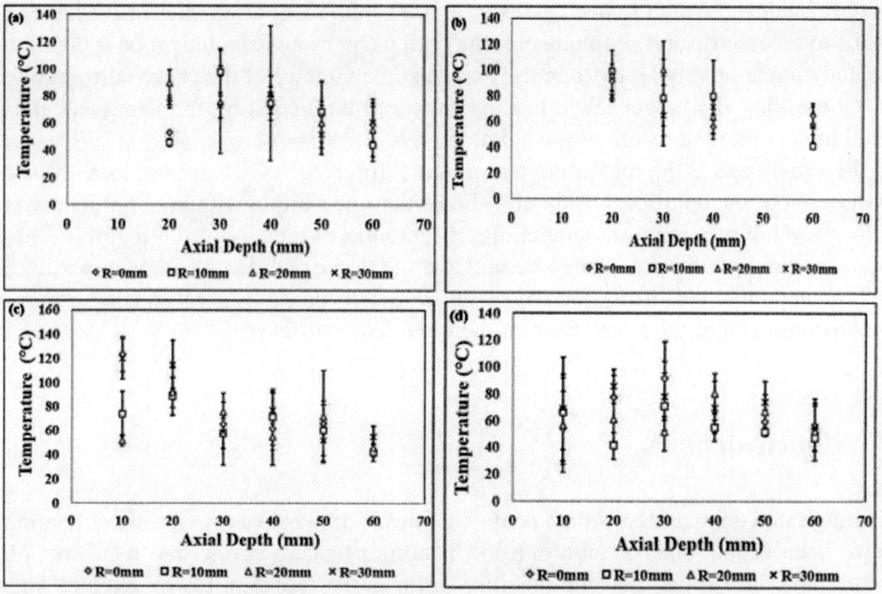

Fig. 3 Representation of temperature profile versus various axial depths (i.e. 10 mm to 60 mm) of tissue phantom infused with and without nanoparticles for different radial positions. **a** Tissue phantom with IONPs; **b** Tissue phantom with AlNPs; **c** Tissue phantom with CuNPs; **d** Tissue phantom WNPs

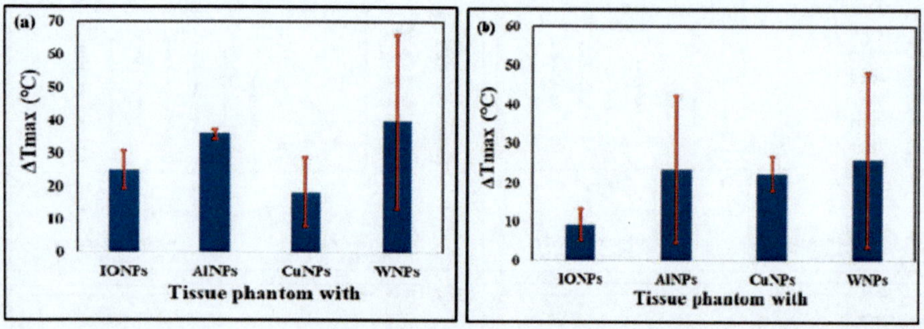

Fig. 4 Observation of maximum temperature difference profile of tissue phantom infused with and without nanoparticles at the central core part of the sample (where nanoparticles have been injected), viz. Axial depth of **a** 30 mm; **b** 40 mm

at a specific radial position tends to decrease along with increasing axial depths. This result is similar to Vadivambal et al. [4]. This trend can be explained by the fact that the upper surface of the cylindrical sample is closer to the waves emanating from the waveguide of the microwave oven. The upper surface tends to interfere more with the microwaves and thus gets more heated. In the case of microwave ablation by a probe, the heated volume is axially concentrated around the probe which helps in focussing the heat to the tumour and thus destroying it effectively. The heating pattern achieved from the oven gives us an idea that, though the microwave heating tends to volumetric and simultaneous, the regions far away from the probe is ought to heated comparatively lesser than the tissues at the vicinity of the probe surrounding it. Controlling the power level, heating time and the operating frequency are thus vital in deciding the extent of the ablation area.

Figure 4 depicts the maximum temperature difference of tissue phantom infused with/without nanoparticles. This plot shows that the sample with nanoparticles has a lower value of maximum temperature difference, especially in the region of 30–40 mm. Through this data, it can be said that the tissue phantoms with nanoparticles gets heated more uniformly and tend to show less deviation from mean temperatures when compared to the tissue phantom without nanoparticles.

4 Conclusions

Through this comparative study, some significant drawbacks in microwave heating have been found, such as non-uniform heating of microwaves (due to formation of hot and cold spots) and the inability to focus the radiation to the sample. This inefficiency of microwaves cannot be removed entirely, but it can be reduced to some extent. Especially during the experiment, some results show irregularity due to non-uniform heating of microwaves. But the characteristics of volumetric heating and high heating rates makes it a useful method for the ablation process. There might be a chance of damaging the healthy cells surrounding the tumour because of its rapid, volumetric heating nature. Therefore, biocompatible nanoparticles like IONPs are very helpful to concentrate the microwave radiation near the vicinity of where it is injected. The experimental results also indicate that the region near the nanoparticles shows higher rates of heating and ability of the nanoparticles to focus microwave radiations. And also the nanoparticle injected region indicates less deviation from its mean temperature that signifies the strength of nanoparticles to achieve uniform heating inside the sample. (All the mentioned scenarios have been found true nearby the region where nanoparticles injected.)

References

1. Brace, C. L. (2010). *Microwave ablation technology: What every user should know* (Vol. 38, Issue 2, pp. 61–67).
2. Lee, B., & Cho, Y. *The principal of microwave oven and microwave heating*.
3. Mullin, J. (1995). Microwave processing. In G. W. Gould (Ed.), *New methods of food preservation* (pp. 112–134). Bishopbriggs, UK: Blackie Academic and Professional.
4. Vadivambal, R., & Jayas, D. S. (2010). Non-uniform temperature distribution during microwave heating of food materials—A Review. *Food and Bioprocess Technology, 3*, 161–171.
5. Ho, Y. C., & Yam, K. L. (1992). Effect of metal shielding on microwave heating uniformity of a cylindrical food model. *Journal of Food Processing and Preservation, 16*(5), 337–359.
6. Fakhouri, M. O., & Ramaswamy, H. S. (1993). Temperature uniformity of microwave heated foods as influenced by product type and composition. *Food Research International, 26*(2), 89–95.
7. Dolande, J., & Datta, A. (1993). Temperature profile in microwave heating of solids: A systematic study. *Journal of Microwave Power and Electromagnetic Energy, 28*(2), 56–57.
8. Funawatashi, Y., & Suzuki, T. (2003). Numerical analysis of microwave heating of a dielectric. *Heat Transfer-Asian Research, 32*(3), 227–236.
9. James, B. J., Strohbehn, J. W., Mechling, J. A., & Trembly, S. (1989). The effect of insertion depth on the theoretical SAR patterns of 915 MHz dipole antenna arrays for hyperthermia. *International Journal of Hyperthermia, 5*(6), 733–747.
10. Cherukuri, P., Glazer, E. S., & Curley, S. A. (2010). Targeted hyperthermia using nanoparticles. *Advanced Drug Delivery Reviews, 62*, 339–345.
11. Suseela, S., Urdaneta, M., & Wahid, P. (2015). Use of nanoparticles in microwave ablation. In *IEEE 16th Annual Wireless and Microwave Technology Conference*.
12. Xie, W., Guo, Z., Gao, F., Gao, Q., Wang, D., Liaw, B. S., Cai, Q., Sun, X., Wang, X., & Zao, L. (2018). Shape-, size- and structure, controlled synthesis and biocompatibility of iron oxide nanoparticles for magnetic theranostics. *Theranostics, 8*(12), 3284–3307.
13. Feng, Q., Liu, Y., Huang, J., Chen, K., Huang, J., & Xiao, K. (2018). *Uptake, distribution, clearance and toxicity of iron oxide nanoparticles with different sizes and coatings*. Scientific report, Nature (Vol. 8).
14. Hainfeld, J. F., & Huang, H. S. (2013). Intravenous magnetic nanoparticle hyperthermia. *International Journal of Nanomedicine, 8*(1), 2521–2532.

Study of Bubble Dynamics in Aerated Liquid Storage Tanks Opened to Atmosphere

Raj Sarath and J. S. Jayakumar

Nomenclature

D_b	Equivalent bubble diameter (mm)
D_{bav}	Average equivalent bubble diameter (mm)
D_o	Orifice diameter (mm)
D_t	Detachment time (s)
g	Acceleration due to gravity (m s^{-2})
p	Specified total pressure (Pa)
V_g	Velocity of gas (ms^{-1})
Y_p	Vertical location of the bubble from column bottom (m)

Greek Symbols

α	Volume fraction (dimensionless)
ρ	Density (kgm^{-3})
σ	Surface tension, (Nm^{-1})
μ	Dynamic viscosity (Pa s)

Subscripts

1, 2	Fluid 1, Fluid 2
av	Average
b	Bubble/Body

R. Sarath · J. S. Jayakumar (✉)
Department of Mechanical Engineering, Amrita Vishwa Vidyapeetham, Amritapuri, India

P. Mahanta et al. (eds.), *Advances in Thermofluids and Renewable Energy*, Lecture Notes in Mechanical Engineering, https://doi.org/10.1007/978-981-16-3497-0_14

g Gas phase
l Liquid phase
o Orifice
p Position

1 Introduction

We know that gas bubble liquid columns are widely used as multiphase heat exchangers, contactors and reactors in chemical, biochemical, petrochemical industries. It has numerous applications in various other engineering fields due to the main advantages such as, high heat and mass transfer rates, compactness, low operating cost, high and accurate interactions at the interphase. Even if large number of studies exists in the area of gas bubble liquid columns, the studies are not well established because, most of them concentrate only on one phase that is either in gas or in liquid. Since the ratio of density and viscosity is higher, the tracking of bubble motion within the bubble column is very complicated and there is no accurate mathematical model which can be used for estimating various bubble behaviors by considering different fluid parameters.

In the past numerous time period, wide variety of numerical and experimental techniques was used for finding the bubble behaviors. Celata et al. [1] executed experiments to study the heat transfer enhancement rate due to the injection of air in heated pipe water up flow system. To visualize the bubble movement and slip ratio they used Plexiglass and high-speed cameras. From the analysis they concluded that due to the injection of air bubbles heat transfer rate increased. Delaure et al. [2] carried out experimental analyses to study heat transfer phenomenon due to the movement of bubbles on a heated surface dipped in water at 20 °C. Also they numerically investigated the variations in heat transfer coefficient and the effect of inlet gas velocity on heat transfer rates along the heated surface. Bubble velocity at different time was estimated with PIV and found that heat transfer rates were more at regions were bubble showed a zigzag motion. Hetsroni et al.[3] did experiments in order to compare the bubble growth behavior in water and in a surfactant solution. Two heat flux values were considered for the experiments. Experiment with pure water at heat flux 10 kW/m^2 showed that bubbles showed a shape which is axially symmetric and vertically non-symmetric. But boiling case of surfactant solution at 50 kW/m^2 bubble showed a vigorous motion, and bubbles were clustered.

Senthil Kumar [4] did 2D numerical analysis for studying the effect of inclination of heated surface on sliding bubble behavior. For tracking the interface they used VOF method in CFX software. They used SIMPLE algorithm for pressure velocity coupling. For interface reconstruction they used Young's PLIC method. They found out that convective heat transfer increased due to the disturbances created with the bubble flow within the bulk liquid. By using VOF and Level Set Method of ANSYS-Fluent Gupta et al. [5] analyzed the bubble behavior of Taylor flow region in a micro channel. For the analyses they used Re (Reynolds number) value of 280 and a Ca (Capillary number) value of 0.006. On comparing results with single-phase condition showed that presence of gas phase in the system increased the average Nusselt number (Nu) values 2.5 times higher. Using VOF method in `interFoam` Sarath Raj and Jayakumar [6] numerically studied the variations in different bubble characteristic by varying geometric and fluid parameters in a 2D rectangular domain. For the simulation authors used Open Source Software `OpenFOAM`.

In the present numerical simulation, the VOF [7] method in `OpenFOAM` is utilized for studying the various bubble behaviors within a 2D rectangular domain having single and double gas (air) inlets. The influence of density and viscosity of bulk liquid on bubble diameter, detachment time and bubble agglomeration are numerically studied. Details regarding interaction of wakes during the time of bubble coalescence with single and double inlet cases are described in details and to the best knowledge of authors it is not recorded anywhere. In this work, numerical analyses were performed using Open Source CFD code `OpenFOAM` (Open Field Operation And Manipulation), which is a C++ library consisting of numerous solvers. Usage of commercial CFD code for analyzing multiphase problem may not be a good choice because they are very expensive and incorporation of suitable models are difficult. Using of an Open Source Software, (OSS) `OpenFOAM` is an alternative way, as it is economical.

2 Governing Equations

For any two phase air water 2D system the conservation equations for mass and momentum are written as follows [6, 8].

$$\frac{\partial \rho}{\partial t} + \nabla \cdot (\rho \vec{v}) = 0 \tag{1}$$

$$\frac{\partial (\rho \vec{v})}{\partial t} + \nabla \cdot (\rho \vec{v} \vec{v}) = -\nabla p\_rgh + \nabla \left[\mu (\nabla \vec{v} + \vec{v}^T) \right] + F_s + \rho \vec{g} \tag{2}$$

In Eq. (2), p_rgh is pseudo or hydrostatic pressure and can be written as follows.

$$p_{\_rgh} = p - \rho \vec{g} \cdot \vec{y} \tag{3}$$

In Eq. (3), p represents the specified total pressure and $\rho \vec{g} \cdot \vec{y}$ is the static pressure in Pascal. $\vec{y}$ represents the position vector.

2.1 *Volume of Fluid Model and Reconstruction Technique at the Interphase*

In comparison with other front tracking models in computational fluid dynamics (CFD), VOF model has less complexity, relatively the model is highly accurate, simple and easy for understanding. The model utilizes single set of conservation equation for momentum and equation for volume fraction, therefore each cell is traced out throughout the given domain. The method works on the principle that the gas and the liquid phase do not interpenetrate each other and uses a function parameter α [8] whose value ranges from zero to one. For those cells filled with pure water α has a value of one and those cells filled with pure air α value will be zero. At the interphase region the α value will be between zero and one. Equations (4) and (5) show the representations of function α.

$$\alpha = \frac{\text{Fluid volume in unit cell}}{\text{Unit cell volume}} \tag{4}$$

$$\alpha = \begin{cases} 0 & \text{Cells filled with air} \\ 0 < \alpha < 1 & \text{Water Air Interface} \\ 1 & \text{Cells filled with water} \end{cases} \tag{5}$$

Even though various methods such as Donor–Acceptor, FLAIR and SLIC [6, 8] methods are available for interphase reconstructions, but for the current numerical analysis PLIC (Piecewise Linear Interface Calculation) [6, 8] is used for interphase reconstruction. In the present work, for finding the surface tension force at the air–water interphase continuum surface model (CSF) [6, 8] was used and the equation is given in (6)

$$F_{\text{vol}} = \sigma_{12} \frac{\alpha_1 \rho_1 k_2 \nabla \alpha_2 + \alpha_2 \rho_2 k_1 \nabla \alpha_1}{0.5(\rho_1 + \rho_2)} \tag{6}$$

where, $k_l = \frac{\nabla \alpha_l}{|\nabla \alpha_l|}$ and $k_v = \frac{\nabla \alpha_v}{|\nabla \alpha_{vl}|}$ are the curvature at the liquid and gas phase [7–9].

3 Validation

For validating the methodology used for the present numerical analyses, the works of Ma et al. [8] were taken. By changing different fluid and geometric parameters they investigated the variation of different bubble parameters. For the validation purpose, the variation of bubble diameter at different inlet gas velocity for an inlet diameter (D_o) 0.3 mm was considered. The comparison of variation of bubble diameter with inlet gas velocity is shown in Fig. 1. The numerical results (Fig. 1) showed that

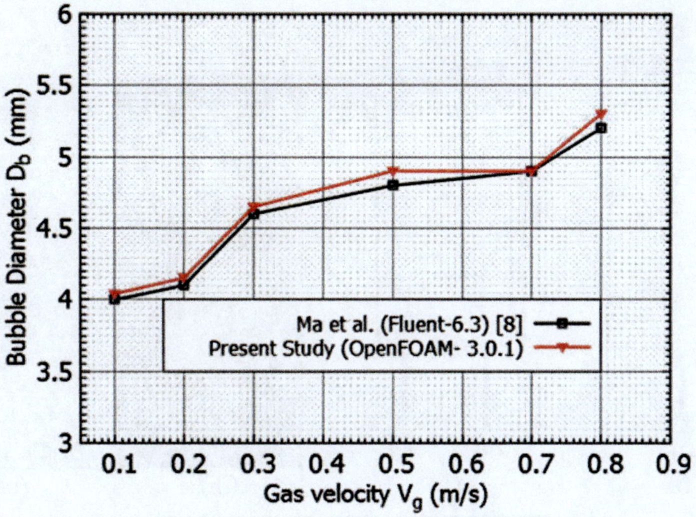

Fig. 1 Variation of average bubble diameter with inlet gas velocity, Orifice Diameter $D_o = 0.3$ mm

values of bubble diameter (D_b) get increased with inlet gas velocity (V_g). The graph shows (Fig. 1) that the results of Ma et al. [8] are in good agreement with the present analyses with `OpenFOAM`. The relative error of validation was of the order of less than 5%.

4 Analysis

For the present numerical simulation, a 2D rectangular container with 15% of free space is considered. The length and width of the domain are 100 and 50 mm, respectively. Figure 2a, b shows the dimensional details of the computational domain with single and double inlets. The `OpenFOAM` representation and the enlarged view at the orifice are shown in Fig. 2c, d, respectively. For discretization of the governing equations, that is, converting partial differential equations (PDE) to system of algebraic equation, finite volume method (FVM) is utilized [6, 8]. Pressure velocity coupling is done with PIMPLE algorithm. PLIC technique [6, 9] is used for interphase reconstruction. The numerical computations were done with multiphase solver interFoam of `OpenFOAM`.

A small time step of the order of 0.1 ms was considered for the analyses. The pressure gradients normal to the walls are calculated from velocity boundary condition predicted boundary flux (Φ). Side walls and bottom walls (except inlet) are consigned as no-slip boundary condition. At the top boundary velocity gradient is set as zero in the direction normal to the boundary for out flow ($\Phi > 0$), and velocity is set as zero for inflow ($\Phi < 0$) condition. Air inlet (orifice) is defined as velocity inlet

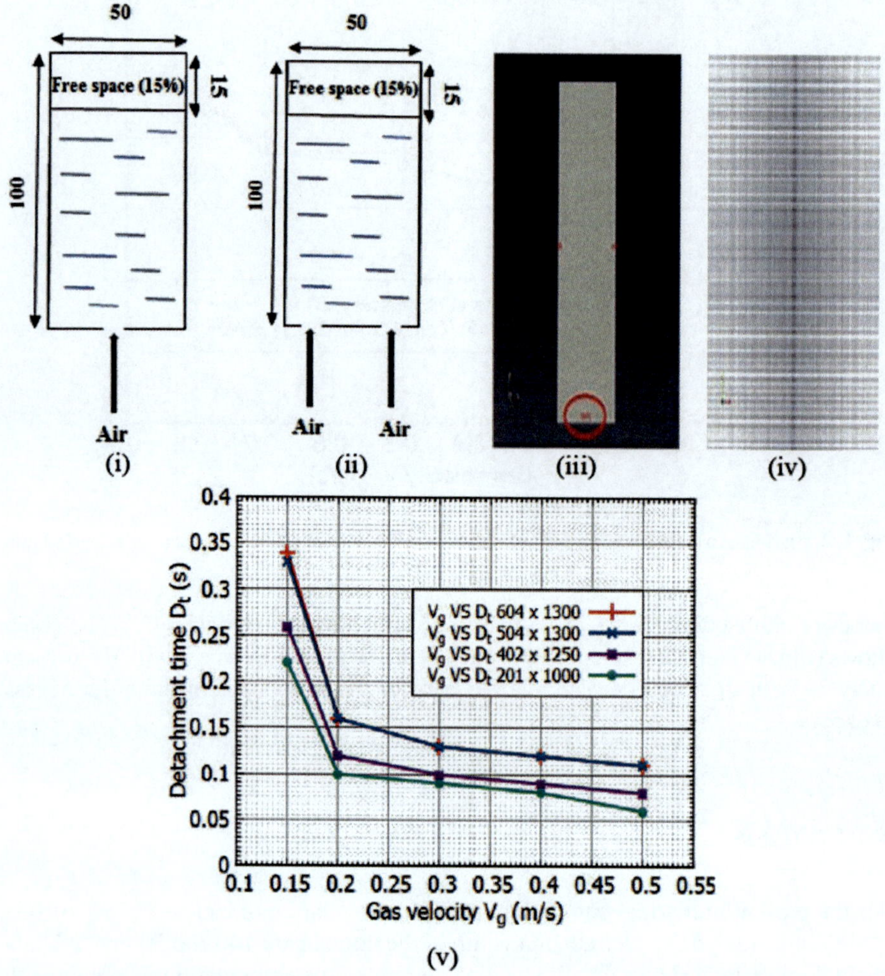

Fig. 2 (i) 2D rectangular container with single inlet. (ii) 2D rectangular container with double inlets. (iii) OpenFOAM representation of the 2D domain. (iv) 2D grid system. (v) Mesh independency study, (Variations in bubble detachment time with inlet gas velocity)

for single and double inlet cases. Since air at a particular velocity is passing through the inlet, the void fraction at inlet is set as 1. Gradient of void fraction at the walls (side walls and the bottom wall except inlet) is set as zero in the direction normal to the wall. At the top the gradient of void fraction of air is set to one for outflow and inflow condition. Hence, a mixed boundary condition is applied on the top boundary. The top boundary is provided with atmospheric pressure condition permitting both inflow and outflow and at remaining all boundaries are set to zeroGradient. The static pressure at the top boundary is deduced by subtracting hydrodynamic pressure from the specified total pressure. Initial conditions are specified as follows. Both gas and liquid phase velocity are consigned as zero ($u = v = 0$). Since the current solver `interFoam` solves void fraction for air (alpha.air), the value for alpha.air is specified as 1 (i.e., filled with air) at the space opened to atmosphere and remaining portion alpha air is specified as 0 (i.e., filled with water).

Analyses were done for four different density values of bulk liquid viz., 998.2, 1500, 2000 and 2500 kg/m3. For that dynamic viscosity of bulk liquid and the surface tension values was kept constant at 0.001005 Pa s and 0.00728 N/m, respectively. Four viscosity values viz., 349.4e-05, 720.3e-05, 1169.8e-05 and 1501.0e-05 Pa s are considered for studying the effect of liquid viscosity on various bubble behaviors. The simulations are carried out at a constant surface tension value of 0.0728 N/m. Simulations with different densities and viscosity values are carried out by keeping the transport properties of gas phase at constant value. The values considered are as follows $\mu_g = 1.7894 \times 10^{-5}$ Pa s, $\rho_g = 1.225$ kg/m^3, $\mu_l = 0.001005$ Pa s, $\rho_l =$ 998.2 kg/m^3, $\sigma = 0.0728$ N/m.

In order to check the dependency of number of mesh in the present work mesh independency check was done with four ranges of mesh numbers. Figure 2(v) shows the variations of bubble detachment time with different inlet gas velocity by considering four ranges of mesh numbers. From figure, it is clear that for 201 × 1000 and 402 × 1250 the graphs are not over lapping, i.e., there is a difference in values of detachment time. For mesh number 504 × 1300 and 604 × 1300 the graphs for bubble detachment time overlaps. The mesh independency precision range will not be increased even if the cell number increased beyond 604 × 1300. Therefore for single inlet simulations are done with a cell number of 504 × 1300.

5 Results and Discussions

In this section, the effect of variations in density and viscosity of bulk liquid in various bubble behavior are discussed. For all the simulations, the red portion and the blue portion within the volume fraction contours represents the air bubble and the water phase, respectively.

5.1 Effect of Liquid Density

By varying density values of continuous phase, we have analyzed variations in different bubble behaviors such as detachment time, equivalent bubble diameter and the wakes created in the bulk liquid due to bubble movement. For the analysis, four density values of liquid viz., 998.2, 1500, 2000 and 2500 kg/m^3 are considered. Analyses are carried out with an inlet gas velocity of 0.35 m/s and orifice diameter 0.3 mm. Figure 3(i–iv) illustrate effect of density on different bubble behaviors. From the analysis, it was found that changing the density of liquid from 998.2 to 2500 kg/m^3 has substantial effect in the various bubble behavior and free liquid

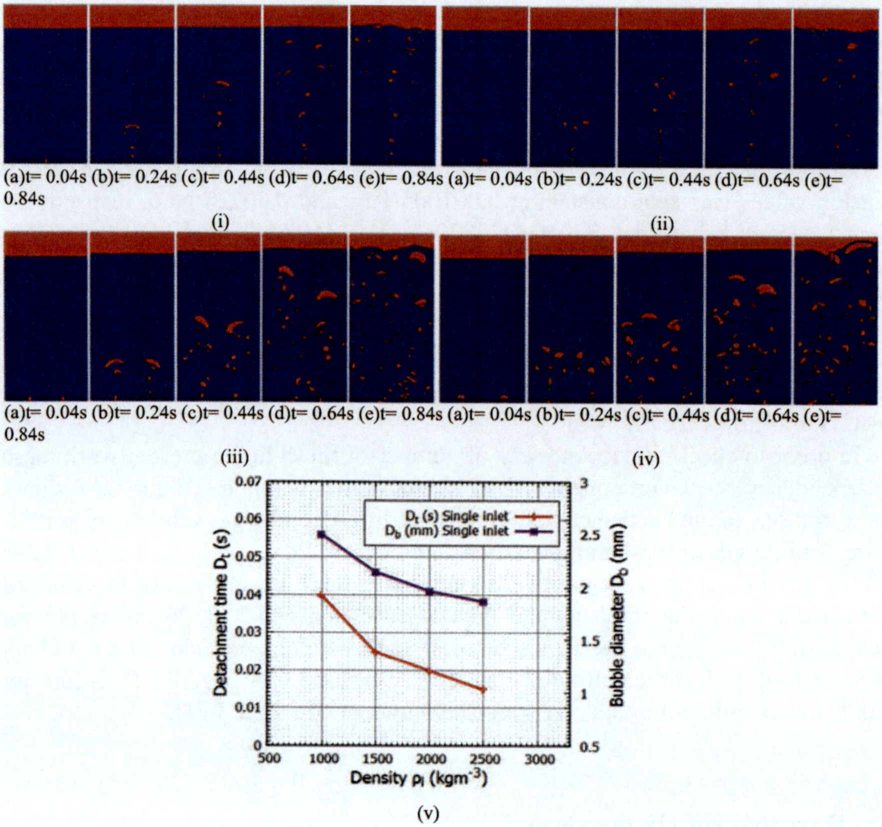

Fig. 3 Void fraction contours at different time instant, single and double inlet cases. (i) and (iii). $\rho_l = 998.2$ kg/m^3, (ii) and (iv). $\rho_l = 2500$ kg/m^3. (v) Variation of bubble detachment time and equivalent bubble diameter

surface deformation due to bubble bursting. The void fraction contours show various stages of bubble growth, detachment, bubble bursting and subsequent free liquid deformation at the free liquid surface. On comparing the void fraction contours, it is seen that for the same time instant when density value increased from 998.2 to 2500 kg/m^3 bubbling frequency get increased. For single inlet case with $\rho_l =$ 998.2 kg/m^3 at t = 0.24 s four bubbles are generated, whereas at the same time instant for $\rho_l = 2500$ kg/m^3 five bubbles are generated. In the case of double inlet case at t = 0.24 s, for $\rho_l = 998.2$ kg/m^3 seven bubble are generated, but for $\rho_l =$ 2500 kg/m^3 ten bubble are generated [See Fig. 3(iii) (b) and (iv) (b)]. On comparing other time instant for single and double inlet case the same trend can be seen for ρ_l = 998.2 and 2500 kg/m^3. Thus, it is inferred that for both single and double inlet case bubble formation frequency increased with increase in density values. When the buoyancy force on the bubble is predominant, bubble get detached from the orifice and moves upward. Figure 3(v) shows the variation in bubble detachment time and equivalent bubble diameter at different values of liquid densities. From graphs, it is clear that for single inlet case at a particular inlet gas velocity, the detachment time and bubble diameter show a downward trend with the increase in bulk liquid density. It is mainly due to the increment in bubble buoyancy with the increase in liquid density. On comparing Fig. 3(i) and (ii)–(a) it is seen that for higher density value since D_t value is less. Therefore, bubble after detachment has risen appreciably. The same observation is seen for double inlet cases. With the increment in density the bubble buoyancy increased therefore bubbles detaches from the orifice with less time and therefore less amount of air get entrapped within the bubble. Hence, equivalent bubble diameter gets reduced. From the analyses it is seen that since the orifice diameter (D_o) and inlet gas velocity (V_g) values are constant the same values are obtained for D_b and D_t for double inlet case. For double inlet case D_b is calculated in terms of average equivalent bubble diameter (D_{bav}) values.

From the numerical analysis, it is observed that the wakening effect created by the leading bubbles enhances the velocity of trailing bubbles by which bubble coalescence happens. The black encircled portions of Fig. 4(i) and (ii) shows wakes created by the leading bubble and the subsequent accelerated motion of trailing bubble under the action of wake. Since the bubble buoyancy is more for higher density value, more number of bubbles will be formed. Therefore the case with density value 2500 kg/m^3 and at velocity 0.35 m/s the wakening effect became more prominent [See Fig. 4(i) and (ii)-(e)]. In order to study, the interaction of bubbles from two adjacent orifice double inlet cases is considered. The highlighted black circles in Fig. 4 shows wakes created due to the movement of bubble from same orifice. The highlighted square portion shows the wake interaction point due to the movement of bubbles from the adjacent orifice. Due to this wake, interaction bubbles repel with each other and moves towards the wall side away from the domain axis. At time instant, Fig. 4 (iii) $t = 0.64$ s and 0.84 s for $\rho_l = 2500$ kg/m^3, due to the vigorous movements of upper leading bubbles create wakes which carries the bubbles from right orifice towards left side. The highlighted rectangular portions show the movement of generated bubbles from right orifice toward left side wall and subsequent bubble agglomeration between bubbles from adjacent orifices. These observations are not seen in double inlet cases with $\rho_l = 998.2$, 1500 and 2000 kg/m^3.

Figure 4(iv) and (v) shows the bubble positions of first detached leading and trailing bubbles approaching for bubble agglomeration process for single inlet case considering $\rho_l = 998.2$ and 2500 kg/m^3. From figure, it is clear that for single inlet case with $\rho_l = 998.2$ kg/m^3 first detached leading and trailing bubble approach for coalescence between time interval 0.1 to 0.13 s. Due to the reduction in time for bubble detachment for $\rho_l = 2500$ kg/m^3 above-mentioned observation is seen between time interval 0.9 to 0.12 s. It is seen that for both density values first detached bubbles approaching for agglomeration occurs at shorter heights [See Fig. 4(iv) and (v)]. In the case of $\rho_l = 998.2$ kg/m^3 first detached bubbles approaching for agglomeration is observed at a height of 0.0098 m, whereas for of $\rho_l = 2500$ kg/m^3 this observation is seen approximately at a height of 0.0089 m form the bottom wall of the computational domain [See Fig. 4(iv) and (v)]. Since the time for bubble detachment reduces with increase in density values, the height at which first detached bubble

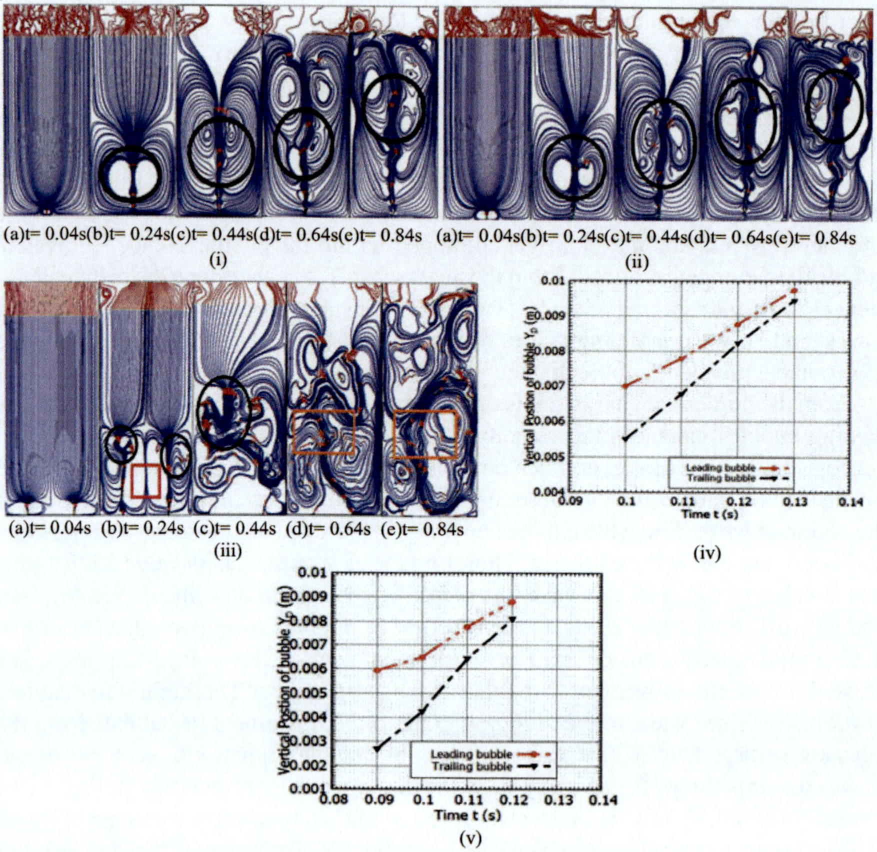

Fig. 4 (i) and (ii). Stream line plots single inlet, (i) $\rho_l = 998.2$ kg/m^3. (ii) $\rho_l = 2500$ kg/m^3. (iii) Stream line plots double inlet case $\rho_l = 2500$ kg/m^3. (iv) and (v) Positions of bubbles approaching for coalescence, (iv) $\rho_l = 998.2$ kg/m^3, (v) $\rho_l = 2500$ kg/m^3

approaching for coalescence get reduced. For the present work write interval of 0.01 s is utilized. For double inlet case since bubble movement is vigorous compared to sing inlet case, the position of first detached bubble approaching for coalescence cannot be tracked under the write interval 0.01 s.

5.2 *Effect of Dynamic Viscosity*

By varying the dynamic viscosity of liquid phase (μ_l = 349.4e-05, 720.3e-05, 1169.8e-05 and 1501.0e-05 Pa.s) and by considering single and double gas inlets, simulations are done for analysing the variation in detachment time and bubble diameter. From the numerical simulations, it was observed that for single inlet case at a constant inlet gas velocity the detachment time and equivalent bubble diameter values increased slightly with the increase in viscosity of liquid. That is the numerical results showed that variation in liquid viscosity values have no considerable effects on D_t and D_b. From the analyses it is seen that D_t and D_b values become constant at high values of viscosity [See Fig. 5(i)]. The same trend is seen for double inlet cases. For double inlet case D_b is calculated as average equivalent bubble diameter D_{bav}.

Figure 5(ii–v) shows the void fraction contours of movement of detached bubbles toward the free liquid space at different time instant using single and double inlets and considering two viscosity values (720.3e-05 and 1501.0e-05 Pa s). From Fig. 5(ii) and (iv) at t = 0.11 s the bubble at the verge of detachment. But for viscosity value 1501.0e-05 Pa.s for single and double inlet case thickness of bubble neck is more, i.e., bubbles are not under the verge of detachment [See Fig. 5(iii) and (v)–(a). For single and double inlet case the reported time of bubble detachment at viscosity value 1501.0e-05 Pa s is 0.13 s. Due to this delay in bubble detachment, compared to viscosity value 720.3e-05 Pa s in cases with viscosity value 1501.0e-05 Pa s more amount air get entrapped and hence D_b values are more for cases with viscosity value 1501.0e-05 Pa s [See Fig. 5(i)]. Numerical analyses showed that for double inlet cases with $\mu_l = 720.3\text{e-}05$ and 1501.0e-05 Pa.s, above-mentioned similar bubble behaviors are seen at the time of bubble detachment. In single inlet case on comparing Fig. 5(ii) and (iii) at $t = 0.21$ s one detached bubble is seen. At time instant, Fig. 5(ii) and (iii) $t = 0.31$ s the number of detached bubbles are two each for viscosity values 720.3e-05 and 1501.0e-05 Pa s. For double inlet cases on comparing Fig. 5(iv) and (v) at $t =$ 0. 21 s two detached bubbles are seen. At time instant Fig. 5(iv) and (v) $t = 0.31$ s the number of detached bubbles are four each for viscosity values 720.3e-05 and 1501.0e-05 Pa.s for double inlet cases. Therefore, present simulations with different μ_l values showed that for both single and double inlet cases at a given inlet gas velocity bubbling frequency is not affected with the increment in viscosity values.

In order to study the bubble behavior at the time of bubble coalescence void fraction contors and corresponding stream lines contours at different time instant are plotted for single inlet case with $\mu_l = 1501.0\text{e-}05$ Pa.s [See Fig. 5(vi) and (vii)]. From the void fraction contours [See Fig. 5(ii) and (iii)] it is clearly seen that for single inlet case at different μ_l values the bubble positions are orientations are same up to the time instant $t = 0.41$ s. But after the agglomeration process under the effect of wakes created by the movement of bubbles towards the free space, for $\mu_l = 1501.0\text{e-}05$ Pa s [See Fig. 5(vi) and (vii)] it is seen that the agglomerated bubble divided into two bubbles and moves toward the wall side. From the stream line plot (Fig. 5vii) the highlighted circular portions show the wakes which are generated by the movement of leading bubble. The highlighted rectangular portions show the wakes on the bubbles which are under development stage. This wakes are created from already detached bubbles from both the orifices. These wakes act as a driving force for the formation and detachment of newly formed bubbles from the orifices which are at the verge of detachment. Wakes created during the time of bubble movement in double inlet

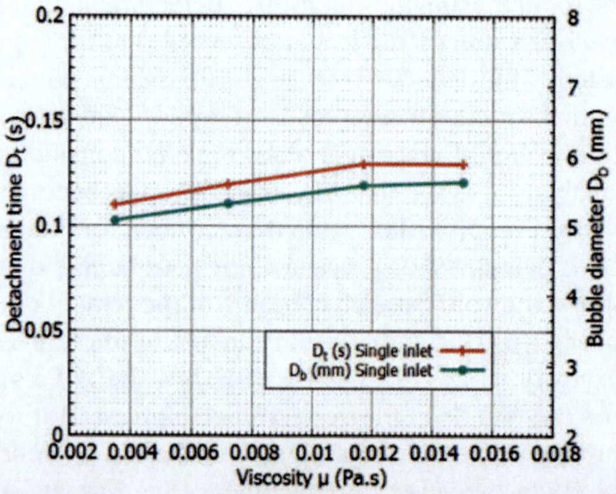

(i)

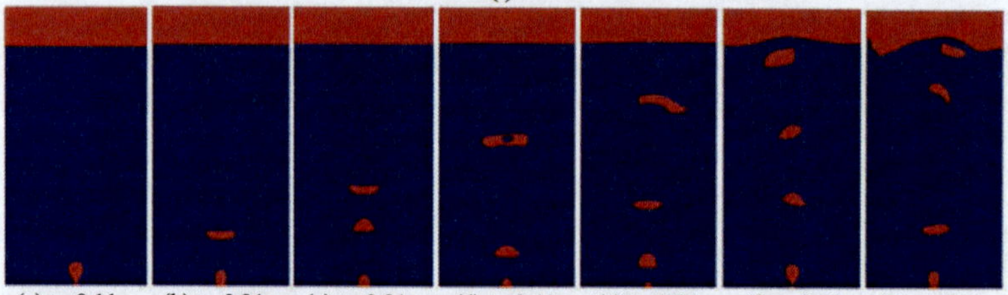

(a) t= 0.11s (b) t= 0.21s (c) t= 0.31s (d) t= 0.41s (e) t= 0.51s (f) t= 0.61s (g) t= 0.71s

(ii)

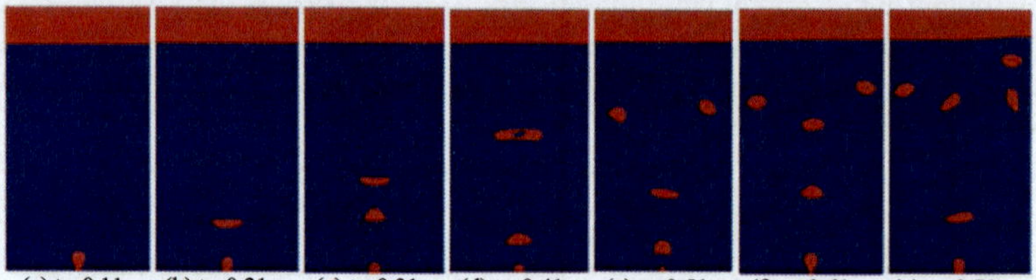

(a) t= 0.11s (b) t= 0.21s (c) t= 0.31s (d) t= 0.41s (e) t= 0.51s (f) t= 0.61s (g) t= 0.71s

(iii)

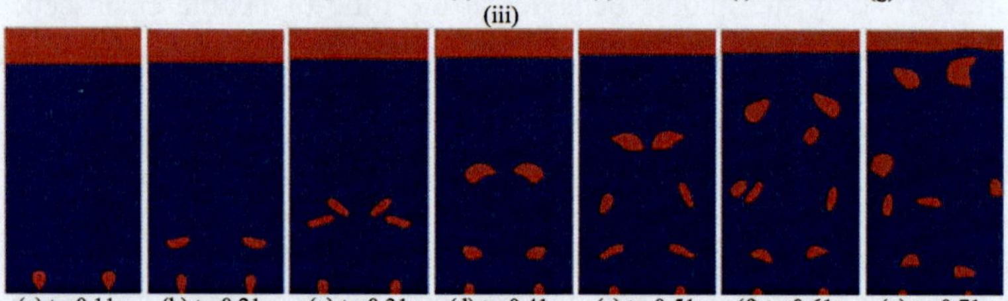

(a) t= 0.11s (b) t= 0.21s (c) t= 0.31s (d) t= 0.41s (e) t= 0.51s (f) t= 0.61s (g) t= 0.71s

(iv)

Fig. 5 (i) Variation of bubble detachment time and equivalent bubble diameter. (ii) and (iv) Void fraction contours at different time intstat, Single and Double inlet case, μ_l = 720.3e-05. (iii) and (v) Single and Double inlet case, μ_l = 1501.0e-05 Pa s. (vi) Bubble agglomeration in single inlet case, μ_l = 1501.0e-05 Pa s. (vii) Wake effect during the time of bubble agglomeration single inlet case, μ_l = 1501.0e-05 Pa.s. (viii) Wake effect during the time of bubble agglomeration double inlet case, μ_l = 1501.0e-05 Pa s. (ix) and (x) Positions of bubbles approaching for coalescence, for μ_l = 1501.0e-05 Pa s, (ix) Single inlet. (x) Double inlet case

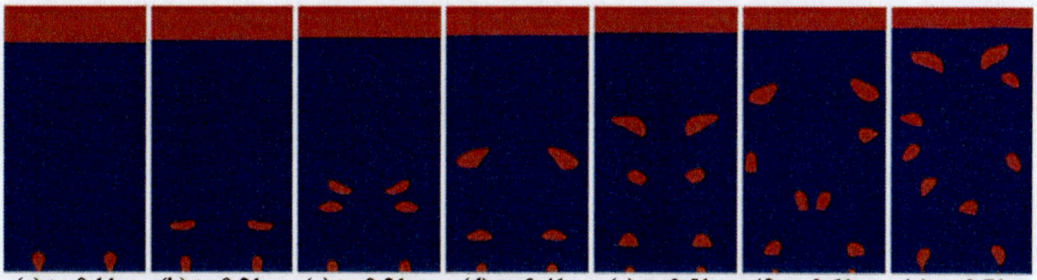

(a) t= 0.11s (b) t= 0.21s (c) t= 0.31s (d) t= 0.41s (e) t= 0.51s (f) t= 0.61s (g) t= 0.71s

(v)

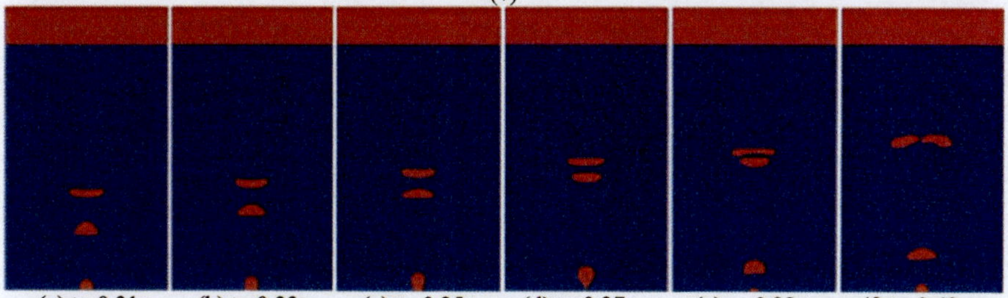

(a) t= 0.31s (b) t= 0.33s (c) t= 0.35s (d) t= 0.37s (e) t= 0.39s (f) t= 0.42s

(vi)

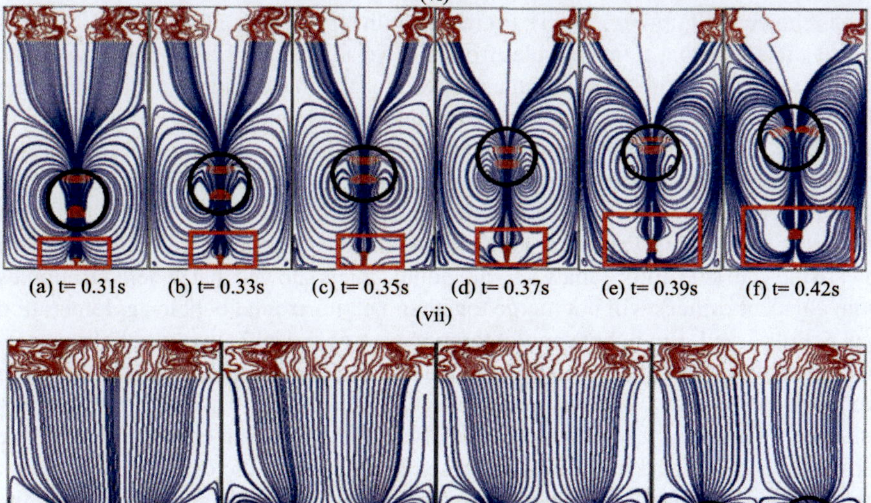

(a) t= 0.31s (b) t= 0.33s (c) t= 0.35s (d) t= 0.37s (e) t= 0.39s (f) t= 0.42s

(vii)

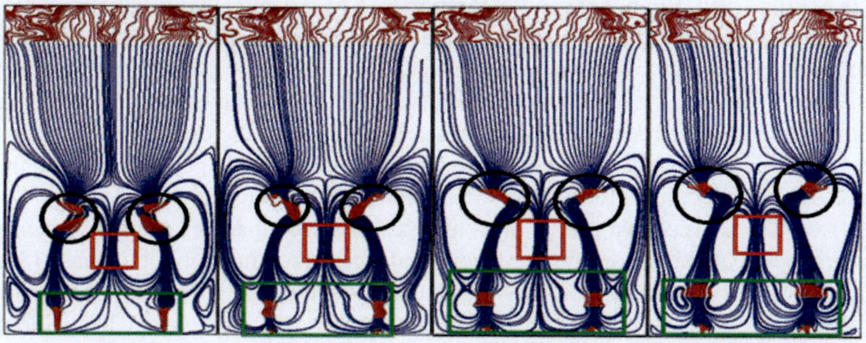

(a) t= 0.31s (b) t= 0.32s (c) t= 0.33s (d) t= 0.34s

(viii)

Fig. 5 (continued)

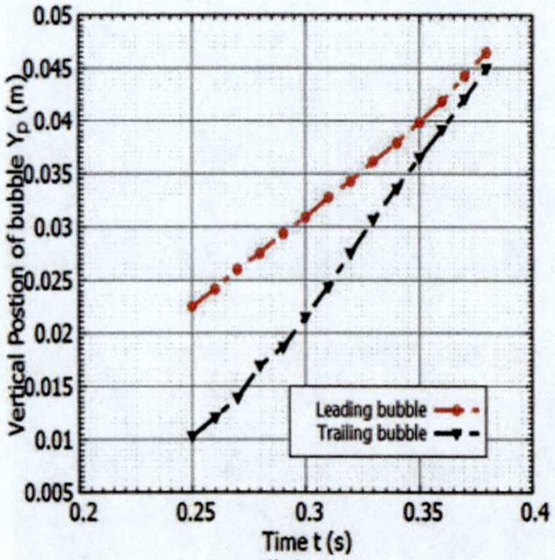

(ix)

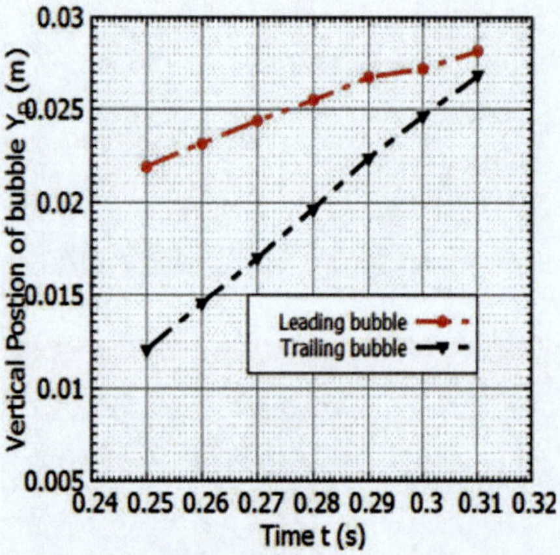

(x)

Fig. 5 (continued)

case is shown in figure Fig. 5(viii). From streamline plots [Fig. 5(viii), $\mu_l = 1501.0$e-05 Pa s], it is seen that at first bubbles from the two adjacent orifices moves toward the domain axis. The highlighted circular portions in stream line plot show the wakening effect created by displaced liquid and the sequence of bubble agglomeration process of bubbles generated from the same orifice. After the bubble agglomeration the newly formed bubbles moves away from the domain axis. The highlighted square portions show the points of wake interaction created due to the movement of bubbles from two inlets. Due to this wake interaction points bubbles repels horizontally and move toward the wall side. Entire analyses with double inlets showed that generated bubbles from adjacent orifices will not merge together, i.e., horizontal bubble agglomeration was not observed. But bubble agglomerations are observed between bubbles generated from the same orifice. The highlighted rectangular portions show the wakes on the bubbles which are in development stage. This wakes are created from already detached bubbles from both the orifices. These wakes act as a driving force for the formation and detachment of newly formed bubbles from the orifices which are at the verge of detachment. The quantitative variations of bubble positions during the time of agglomeration process for single and double inlet case considering $\mu_l =$ 1501.0e-05 Pa.s is shown in Fig. 5(ix) and (x). It is seen that for both the viscosity values identical bubbles are formed from both the orifice in double inlet case [See Fig. 5(iv) and (v)], hence bubble movements approaching for bubble agglomeration from both the orifices showed the same trend. In order to plot the change in positions of bubbles during the time of bubble coalescence, bubbles detached from right side orifice in considered in double inlet case. From graph, it is clear that due the movement of upper leading bubble a lower drag is created on the trailing bubble. This will accelerate the trailing bubbles. Hence, the distance between bubbles reduces. Subsequently bubbles merge together. From the numerical analyses with viscosity values 349.4e-05 and 1169.8e-05 Pa.s considering single and double inlet cases, above explained same of trends bubble behaviors (Variations in bubble detachment time and equivalent bubble diameter, bubble detachment frequency, positions and height of first detached bubbles approaching for coalescence) were observed.

.

6 Concluding Remarks

The gas bubble bursting and ultimate free liquid surface deformation has an important role on mass transfer between different phases in stratified layers of two phase flows. In the present work, volume of fluid (VOF) method of OpenFOAM is utilized for investigating the bubble dynamics of air bubbles in isothermal water pool. From the investigation, the following conclusions are drawn.

i. For single and double inlet cases it is observed that the equivalent bubble diameter and the average equivalent bubble diameter get decreased with the increase in density values.
ii. Detachment time noted for bubbles at the verge of detachment for both single and double inlet cases get reduced with the increase in liquid density.
iii. It is observed that for single and double inlet cases at a constant inlet gas velocity the detachment time, equivalent and average equivalent bubble diameter values increased slightly with the increase in viscosity of liquid. That is the numerical results showed that variation in liquid viscosity values have no considerable effects on D_t and D_b.
iv. For both single and double inlet cases bubble detachment frequency get increased with increase in density.
v. Present simulations with different viscosity values showed that for both single and double inlet cases at a given inlet gas velocity bubbling frequency is not affected with the increment in viscosity values.
vi. The time when first two detached bubbles begin to agglomerate get reduced with the increase density.
vii. For all the cases with different viscosity values, for double inlet cases it is seen that due to the interaction of wakes generated by the movement of bubbles side wise bubble coalescence are not observed. But for higher density value due to the vigorous movements of upper leading bubbles create wakes which carries the bubbles from right orifice toward left side and subsequent bubble agglomeration can be seen between bubbles from adjacent orifices.

References

1. Cellata, G. P., Chiaradia, A., Cumo, M., & Annibale, F. D. (1999). *Int. J. Multi-phase Flow, 25*, 1033–1052.

2. Delaure, Y. M. C., Chan, V. S. S., & Murray, D. B. (2003). *Exp. Ther. Fluid Sci., 27*, 911–926.
3. Hetsroni, G., Mosyak, A., Pogrebnyak, E., Sher, I., & Segal, Z. (2006). *Int. J. Multi-phase Flow, 32*, 159–182.
4. Senthil, K. (2009). PhD thesis, Dublin City University. http://doras.dcu.ie/14862/1/thesis_S_Senthil_kumar_2009.pdf.
5. Raghvendra, G., David, F. F., & Brian, S. H. (2010). *Chemical Engineering Science, 65*, 2097–2107.
6. Sarath, R., & Jayakumar, J. S. (2017). *Proc. of Indian Society for Heat and Mass Transfer Digital Library*, IHMTC-2017, 2089–2096. https://doi.org/10.1615/IHMTC-2017.2930.
7. Hirt, C. W., & Nichols, B. D. (1981). *Journal of Computational Physics, 39*, 201–225. https://doi.org/10.1016/0021-9991(81)90145-5.
8. Ma, D., Mingyan, L., Yonggui, Z., & Can, T. (2012). *Chemical Engineering Science, 72*, 61–77.
9. Raj, S., & Jayakumar, J. S. (2019). *Advances in Interdisciplinary Engineering*, Lecture Notes in Mechanical Engineering. Springer. https://doi.org/10.1007/978-981-13-6577-5_75

Unsteady Casson Fluid Flow Past a Stretching Sheet Subject to Non Linear (Quadratic) Free Convection Along with Suction

R. Balamurugan and A. Vanav Kumar

Nomenclature

t'	Time
x'	Horizontal axis
y'	Vertical axis
u'	Horizontal velocity
v'	Vertical velocity
T'	Transient temperature
T'_w	Wall temperature
T'_∞	Ambient fluid temperature
g	Gravitational acceleration
C_p	Specific heat at constant pressure
Nu_X	Local Nusselt number
SP	Suction parameter
t	Dimensionless time
x	Dimensionless horizontal axis
y	Dimensionless vertical axis
u	Dimensionless horizontal velocity
v	Dimensionless vertical velocity
T	Dimensionless transient temperature
Pr	Prandtl number
Gr	Grashof number
$Gr*$	Quadratic free convection parameter
β	Heat transfer coefficient
β^*	Non linear volumetric thermal expansion coefficient
Π	Product of the component of deformation rate with itself

R. Balamurugan · A. Vanav Kumar (✉)
Mathematics Division, Department of Basic and Applied Science, National Institute of Technology, Yupia, Andhra Pradesh 791112, India

P. Mahanta et al. (eds.), *Advances in Thermofluids and Renewable Energy*, Lecture Notes in Mechanical Engineering, https://doi.org/10.1007/978-981-16-3497-0_15

κ	Thermal conductivity of fluids
ρ	Density of the fluid
σ	Electrical conductivity of the fluid
υ	Kinematic viscosity of the fluid
τ_{ij}	Rheological equation of state
τ_x	Local skin friction
μ_B	plastic dynamic viscosity
β	Casson fluid parameter
c'	Stretching parameter

1 Introduction

Fluid flow of Casson type past a stretching sheet has been a major source of investigation. Casson fluid past an stretching sheet along with temperature effects has been reported widely. With advent of computation powers and better mathematical techniques effects of non linear free convection has also been included in the research topics of late frequently. Shawky [1] investigated the heat and mass transfer in Casson fluid on porous medium over a stretching sheet with transverse magnetic field. The governing equations so obtained were solved by using Kummers function. Mukhopadhyay [2] studied Casson fluid flow and its heat transfer on unsteady suction stretching sheet under the circumstance of thermal radiation and they utilized shooting technique to solve the equations. Shehzad [3] studied the effects of chemical reaction and suction on MHD Casson flow over a stretching sheet and series solution method has been implemented to solve the problem. Casson flow on a non linear stretching sheet with heat source/sink was examined by Sumalatha et al. [4] and the partial differential equations converted into ordinary differential equation by similarity transformation and the results are found by Keller Box method. The heat and mass transfer effects over the boundary layer flow of MHD Casson fluid flow with heat source/ absorbtion was employed by Runge-Kutta-Fehlberg fourth-fifth order method by Gireesha [5]. Reddy [6] investigated convective boundary layer Casson fluid flow with variable thermal conduction and they involved shooting method for solving the problem. Benazir et al. [7] studied the unsteadiness of MHD Casson fluid over the Geometry vertical cone and plate with heat source/sink. It is found that the heat source playing important role on temperature. The effect of Soret, radiation and heat source on three-dimensional Casson fluid flows was studied by Sulochana et al. [8]. Reddy [9] investigated MHD Casson flow under radiation, chemical reaction over an exponentially inclined geometry. Animasaun [10] et al. studied temperature variation on Casson fluid flow over an exponential stretching sheet with heat generation and used homotopy analysis method to get the results.

The present study devotes to non linear free convection of fluid flow of Casson fluid past a stretching sheet. Such type of study was carried out by Shaw et al. [11]. They involve convective boundary conditions in addition to viscous dissipation along with Soret and Dufour effects. Kumar and Sood [12] carry out the study of non linear convective fluid flow past a stretching sheet. They investigate the stagnation point flow in two dimension. Keller box scheme is employed to obtain the results. RamReddy et al. [13] studied the non Newtonian Power law fluid flow past an inclined plate saturated with an Darcy porous medium. The study include non linear natural convection and thermal convective boundary conditons. Ibrahim and Gamachu [14] studied the non linear effects and electric field effect on the Williamson nanofluid past a radially stretching sheet. Mandal and Mukhopadhyay [15] investigated the effects of non linear convection of the micropolar fluid past an exponentially stretching sheet with free stream moving exponentially. Non linear convection with cross diffusion was again researched upon by RamReddy et al. [16]. Mahanthesh et al. [17] reported the effect of non linear convection on flow of Oldroyd-B fluid. It may be also noted that Mahanthesh et al. [18] report the effect of quadratic (non linear) free convection in their study carried out on dusty Carreau and dusty Casson fluids past a stretching surface. Apart from this, permeability case was also considered in [19–21] over a Casson fluid flow.

In the present study, Casson fluid flow subject to nonlinear (quadratic) convection, suction is studied.

The study differs from the previous investigation by the fact that the fluid is already flowing with a non zero velocity and is subject to a temperature field which follows an hypergeometric function. The authors believe that such a report has not been mentioned in the literature, hence, the present study is carried out.

2 Mathematical Formulation of Governing Equations

Consider a Casson fluid flow which is incompressible, laminar moving through a stretching sheet. The fluid is flowing with a initial velocity with horizontal velocity and vertical velocity given by u' and v' along the x' and y' which are in the horizontal and vertical directions, respectively. The fluid is maintained at a velocities $u' = x'e^{-\sqrt{\frac{c'}{\upsilon}}y'}$ and $v' = -\sqrt{\upsilon c'}\left(1 - e^{-\sqrt{\frac{c'}{\upsilon}}y'}\right)$, respectively, for time given by $(t' < 0)$. The letters c', υ represent the stretching parameter, and the kinematic viscosity. Also, the temperature of the ambient fluid is maintained at T'_∞ and the wall temperature, T'_w, of the sheet is given by expression $T' - T'_\infty = (T'_w - T'_\infty)\frac{e^{-\sqrt{\frac{c'}{\upsilon}}y'}\mathrm{H}\left(\mathrm{Pr}, 1+\mathrm{Pr}, -\mathrm{Pr}\, e^{-\sqrt{\frac{c'}{\upsilon}}y'}\right)}{\mathrm{H}(\mathrm{Pr}, 1+\mathrm{Pr}, -\mathrm{Pr})}$. during the time, $t' < 0$. H(*,*,*) represents the hypergeometric function. Pr represents the Prandtl number.

The sheet is stretched with a velocity given by by $c'x'$, where c' is the proportionality constant (stretching constant), with wall temperature changing to $T' - T'_\infty = T'_w - T'_\infty$ at $t' \geq 0$. The distance x' is measured from the origin. The fluid is subject to suction along with quadratic free convection. The fluid being of Casson fluid type, the rheological equation of state for an isotropic and incompressible flow of a Casson fluid is given by [13],

$$\tau_{ij} = \begin{cases} 2\left(\mu_B + \frac{P_y}{\sqrt{2\prod_c}}\right)E_{ij}, \prod > \prod_c, \\ 2\left(\mu_B + \frac{P_y}{\sqrt{2\prod_c}}\right)E_{ij}, \prod < \prod_c, \end{cases} \tag{1}$$

where, τ_{ij}, μ_B and P_y are the (i, j)-th component of the stress tensor, the plastic dynamic viscosity of the non-Newtonian fluid, and yield stress of the fluid. Further, details of this equation can be had from reference given in literature (Animasaun [10]). Invoking the Boussinesq's approximation, the model for the fluid flow obeying the above conditions are given by the governing equations,

$$\frac{\partial u'}{\partial x'} + \frac{\partial v'}{\partial y'} = 0 \tag{2}$$

$$\begin{aligned} \frac{\partial u'}{\partial t'} + u'\frac{\partial u'}{\partial x'} + v'\frac{\partial u'}{\partial y'} = \upsilon\left(1 + \frac{1}{\gamma}\right)\frac{\partial^2 u'}{\partial y'^2} \\ + \rho g\beta\left(T' - T'_\infty\right) \\ + \rho g\beta^*\left(T' - T'_\infty\right)^2, \end{aligned} \tag{3}$$

$$\frac{\partial T'}{\partial t'} + u'\frac{\partial T'}{\partial x'} + v'\frac{\partial T'}{\partial y'} = \frac{\kappa}{\rho C_p}\frac{\partial^2 T'}{\partial y'^2}, \tag{4}$$

with initial and boundary conditions given by,

$$t' < 0,\ y' = 0:$$

$$u' = c'x' \exp\left(-\sqrt{\frac{c'}{\upsilon}}y'\right), \tag{5}$$

$$v' = -\sqrt{c'\upsilon}\left(1 - \exp\left(-\sqrt{\frac{c'}{\upsilon}}y'\right)\right), \tag{6}$$

$$\frac{T' - T'_\infty}{T'_w - T'_\infty} = \frac{e^{-\sqrt{\frac{c'}{\upsilon}}y'} H\left(Pr, Pr + 1, -Pre^{-\sqrt{\frac{c'}{\upsilon}}y'}\right)}{H(Pr, Pr + 1, -Pr)}, \tag{7}$$

$$t' \geq 0,\ \forall y' = 0, x \geq 0:$$

$$u' = c'x', \tag{8}$$

$$v' = -v_0, \tag{9}$$

$$T' - T'_\infty = T'_w - T'_\infty, \tag{10}$$

$$\forall y' \to \infty, x' \geq 0:$$

$$u' \to 0, \tag{11}$$

$$T' \to T'_\infty, \tag{12}$$

Using the dimensionless variables

$$t = t'c'x = x'\sqrt{\frac{c'}{v}}, y = y'\sqrt{\frac{c'}{v}}, u = \frac{u\prime}{\sqrt{c'v}},$$

$$v = \frac{v'}{\sqrt{c'v}}, T = \frac{T' - T'_\infty}{T'_w - T'_\infty}, \gamma = \mu_B\sqrt{2\prod_c / P_y},$$

$$Gr = \frac{\rho g\beta\left(T'_w - T'_\infty\right)}{c'\sqrt{c'v}}, Gr^* = Gr\beta^*\left(T'_w - T'_\infty\right),$$

$SP = -\frac{v_0}{\sqrt{v\beta}}$ and $\Pr = \frac{\mu C_P}{\kappa}$ non dimensionalizing (2)–(4) along with the initial conditions and boundary conditions (5)–(12), we get the dimensionless governing equations given by

$$\frac{\partial u}{\partial x} + \frac{\partial v}{\partial y} = 0, \tag{13}$$

$$\frac{\partial u}{\partial t} + u\frac{\partial u}{\partial x} + v\frac{\partial u}{\partial y} = \left(1 + \frac{1}{\gamma}\right)\frac{\partial^2 u}{\partial y^2} + GrT + Gr^*T^2, \tag{14}$$

$$\frac{\partial T}{\partial t} + u\frac{\partial T}{\partial x} + v\frac{\partial T}{\partial x} = \frac{1}{\Pr}\frac{\partial^2 T}{\partial y^2} \tag{15}$$

along with dimensionless initial and boundary conditions given by,

$$t < 0, \forall\, x, y > 0:$$

$$u = x\exp(-y), \tag{16}$$

$$v = -(1 - \exp(-y)), \tag{17}$$

$$T = \frac{e^{-y} H\left(Pr, Pr + 1, -Pre^{-y}\right)}{H(Pr, Pr + 1, -Pr)}, \quad (18)$$

$$t \geq 0, \; y = 0 :$$

$$u = x, \quad (19)$$

$$v = SP, \quad (20)$$

$$T = 1, \quad (21)$$

$$\forall y \to \infty, x \geq 0 :$$

$$u \to 0, \quad (22)$$

$$T \to T_{\infty}, \quad (23)$$

where, *SP*, *Gr*, *Gr** are the Suction parameter, Grashof number, and Quadratic free convection parameter.

3 Computational Approach

The non-dimensional equations Eqs. (13–15) along with non-dimensional initial and boundary conditions (16) to (23) are solved implementing the finite difference method which follows the Crank-Nicolson scheme of implicit type. Forward difference for $\frac{\partial(.)}{\partial t}$, derivative terms with respect to time, backward difference for $\frac{\partial(.)}{\partial x}$, central difference for the $\frac{\partial(.)}{\partial y}$, $\frac{\partial^2(.)}{\partial y^2}$ are applied for discretization. The method is uncondtionally stable [22]. The computation is carried out for the domain $0 \leq x \leq 1$ and $0 \leq y \leq 25$. The grid size is kept at $x = 0.002$, $y = 0.01$ and $t = 0.001$ along the x, y axis and time, t, respectively. This was achieved by performing numerical experiments by changing the values of parameters. The domain which is fixed here are obtained after observing that there is no more significant changes and are enough to capture the profiles smoothly.

Iterations would stop when the absolute difference between the consecutive values of dependent variable is less than or equal to 5×10^{-5}.u, T, are the dimensionless horizontal velocity and the dimensionless temperature field.

Figure 1 (Table 1) shows the comparision curves of the local Skin friction for $Gr = 0.0$, $Gr^* = 0.0$, $SP = 0.0$ and $\beta = 0.1$. From figure, it is clear that they are overlapping,

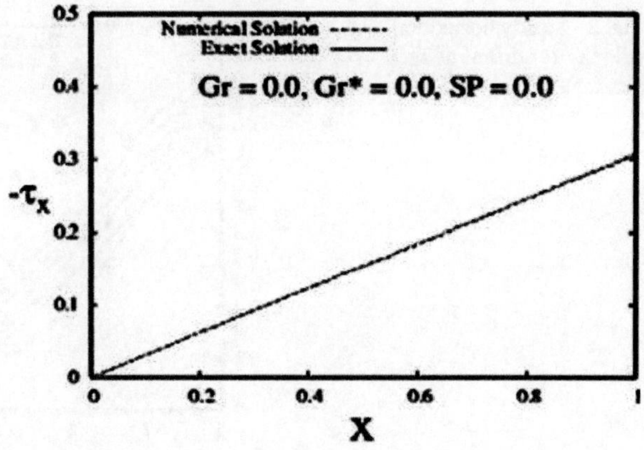

Fig. 1 Comparison between Computed and Exact local Skin friction, $-\tau_x$

Table 1 Values of the exact and computed local skin friction for different x

x	$-\tau_x$	
	Exact	Computed
0.0	0.0	0.0
0.2	0.060	0.061
0.4	0.121	0.122
0.6	0.181	0.184
0.8	0.241	0.245
1.0	0.301	0.306

which shows the our computations are in order and are in good agreement with the exact solutions, $-\tau_x = \frac{\partial u}{\partial y} = \frac{x}{\sqrt{1+\frac{1}{\beta}}}$.

4 Results and Discussion

In the present sections, we describe the plots and discuss the effects of the parameters namely the Casson fluid parameter (β), Suction parameter SP, Pr (=0.71) and Quadratic free convective parameter (Gr^*).

The plots in Figs. 2, 3, 4, 5, 6, 7 and 8 have been described for $\beta = 0.2, 0.8$, $SP = -1.0, -0.5, -0.25, 0.0$,$Gr^* = 0.1$, and the Grashof number, $Gr = 5.0$.

Figures 2, 3 and 4, describe the steady state velocity and temperatures. One can observe that the velocity shows a increase with respect to suction parameter as the values changes (increases) from -1.0 to 0.0. The continuous line represent $\beta = 0.2$ and the dotted line represent $\beta = 0.8$ in Figs. 2, 3, 4, 5 and 6. Observing the horizontal velocity in Fig. 2, it can be concluded that the increase in the Casson fluid parameter leads to decreases in the horizontal velocity as $y \geq 1.5$.

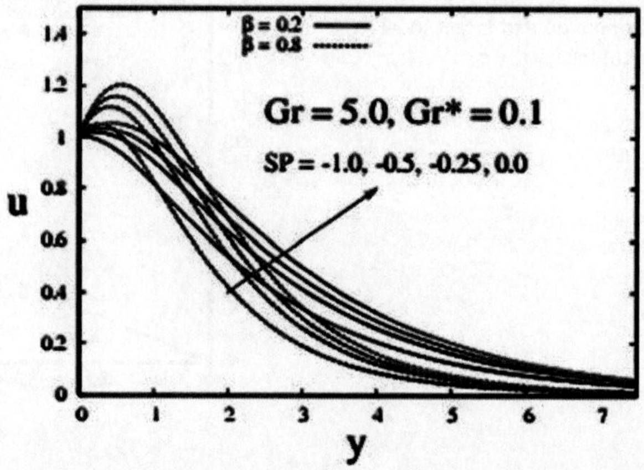

Fig. 2 Steady horizontal velocity for different suction parameter, *SP*

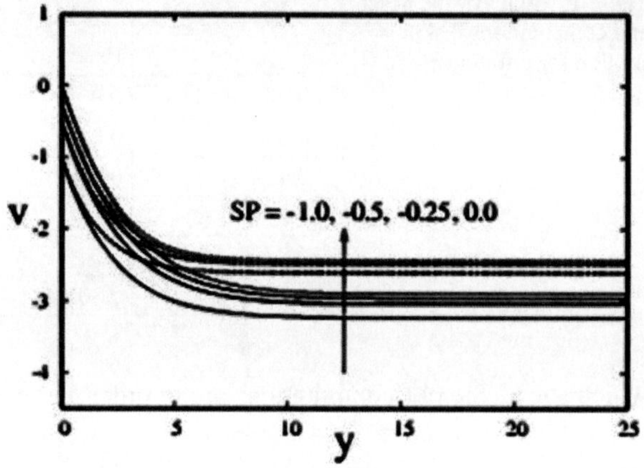

Fig. 3 Steady vertical velocity for different suction parameter, *SP*

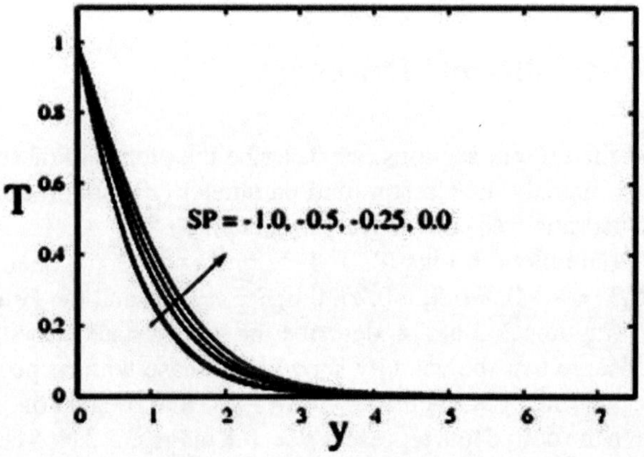

Fig. 4 Steady temperature for different suction parameter, *SP*

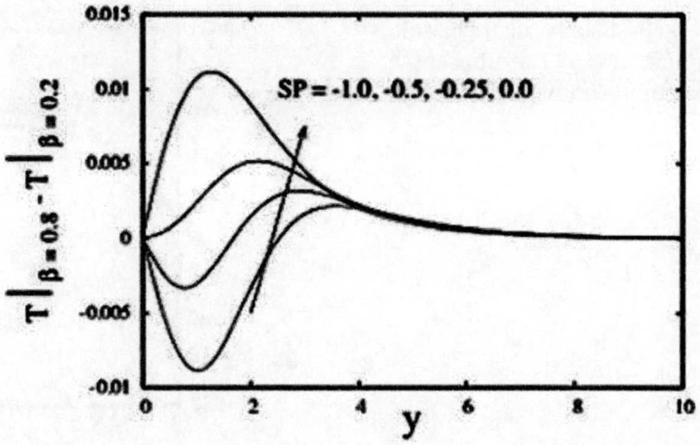

Fig. 5 Difference of steady temperature for Casson fluid parameter, β between 0.2 and 0.8, for different suction parameter, *SP*

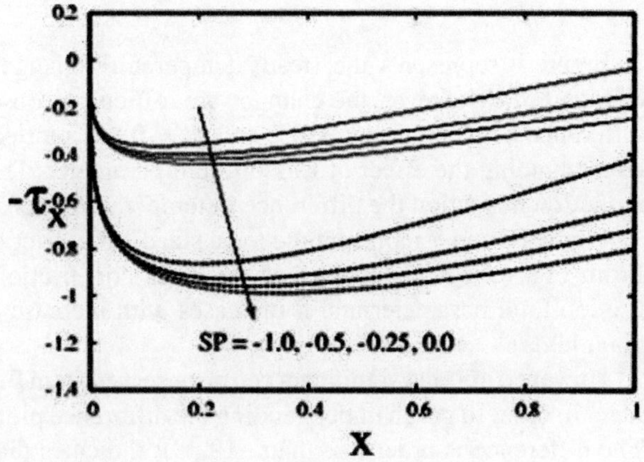

Fig. 6 Steady skin friction, $-\tau_x$ for different suction parameter, *SP* and β

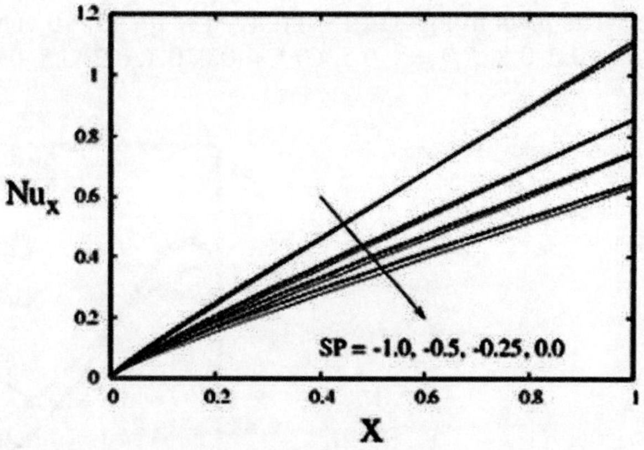

Fig. 7 Steady local Nusselt number, Nu_x for different suction parameter, *SP* and β

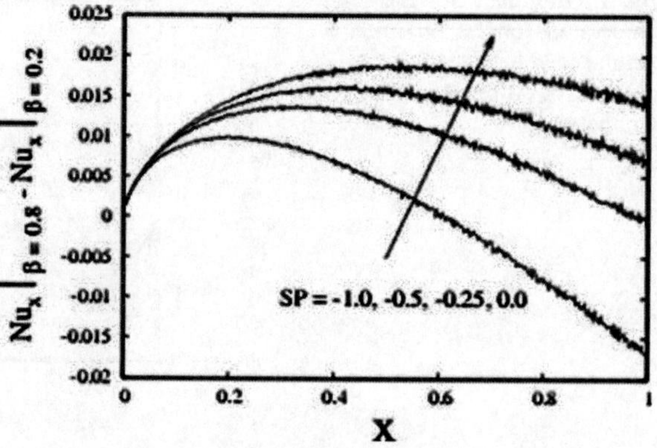

Fig. 8 Steady local Nusselt number, Nu_x for different suction parameter, *SP* and β

Figure 4 represents the steady temperature change when the $\beta = 0.2, 0.8$ for different *SP*. However, the changes are difficult to observe, hence, the temperature difference between the $\beta = 0.2$ and $\beta = 0.8$ is plotted in the Fig. 5. This helps in understanding the effect of Casson fluid parameter. The plot clearly shows that the reduction in suction the difference in temperature increases.

Figures 6 and 7 represent the local steady skin friction and local Nusselt number. From Fig. 6, it is observed that the local skin friction decreases with increase in Casson fluid parameter and it increases with increase in suction value (increase in magnitude).

However, nothing significant can be observed from Fig. 7 for local Nusselt number, Nu_X. In order to get an understanding the difference plot has been displayed in Fig. 8. The difference is order less than ~10^{-3}. It indicates that as the suction inceases the steady local Nusselt number decreases.

The plots in Figs. 9, 10, 11, 12, 13, 14, 15, 16 and 17 have been described for $\beta = 0.2, 0.8$, $SP = -0.5$, $Gr^* = 0.0, 0.1, 0.2, 0.5, 0.8$ and $Gr = 5.0$.

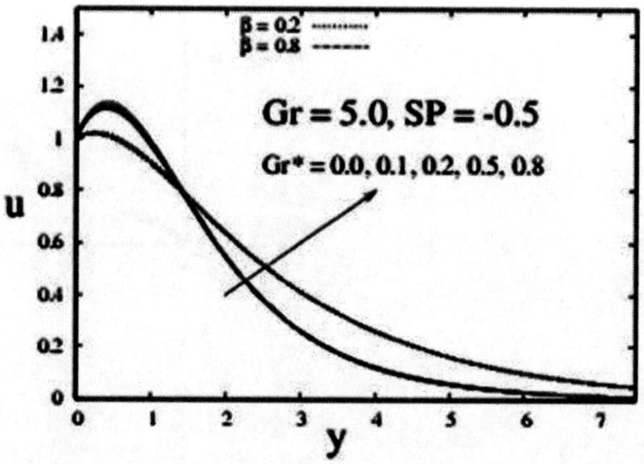

Fig. 9 Steady horizontal velocity for different Quadratic convective parameter, Gr^*

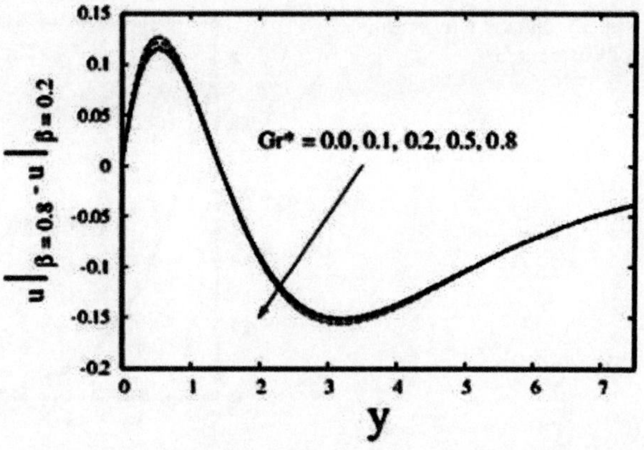

Fig. 10 Difference of steady horizontal velocity for different, *Gr**

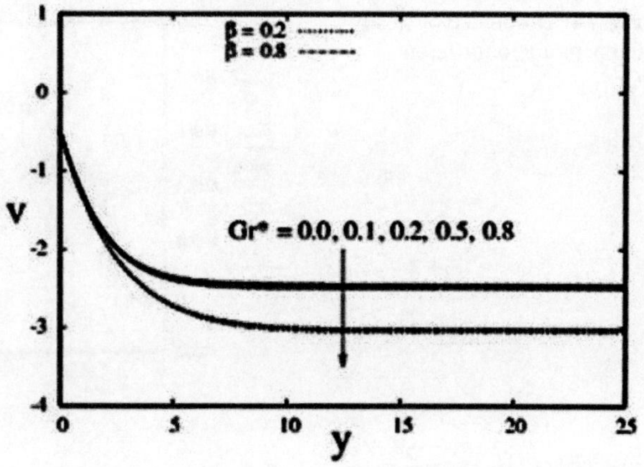

Fig. 11 Steady vertical velocity for different *Gr**

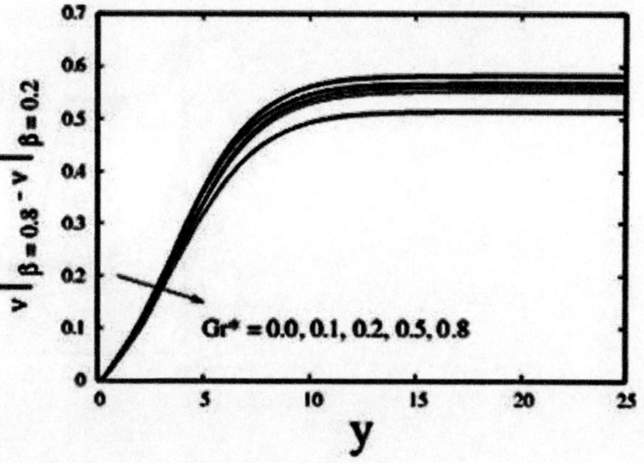

Fig. 12 Difference of steady vertical velocity for different, *Gr**

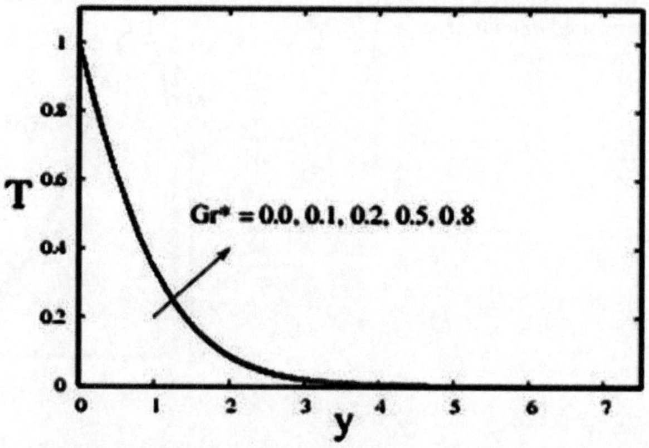

Fig. 13 Steady temperature for different Gr*

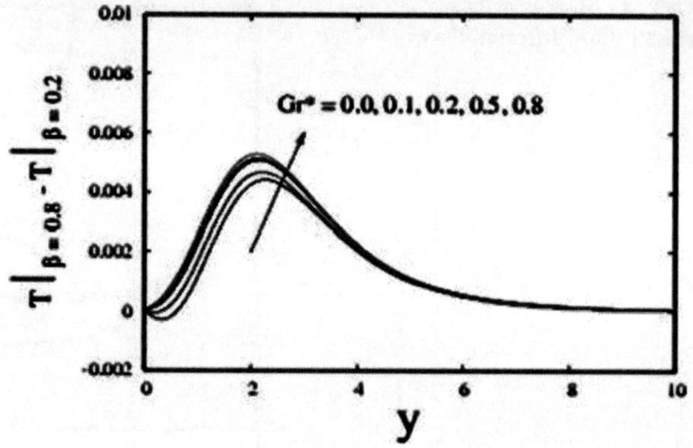

Fig. 14 Difference of steady temperature for different, Gr^*

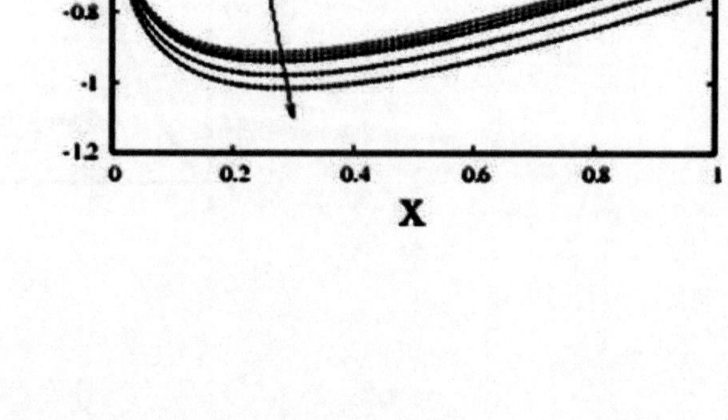

Fig. 15 Steady skin friction, $-\tau_x$ for different Gr^*

Fig. 16 Steady Nusselt number, Nu_x for different Gr^*

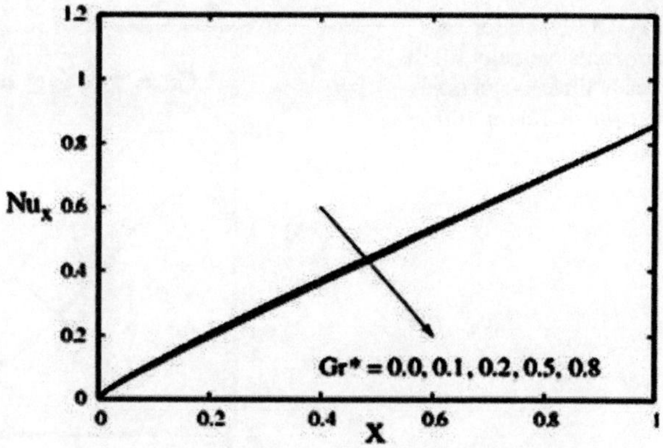

Fig. 17 Difference between Steady Nusselt number, Nu_x for different Gr^*

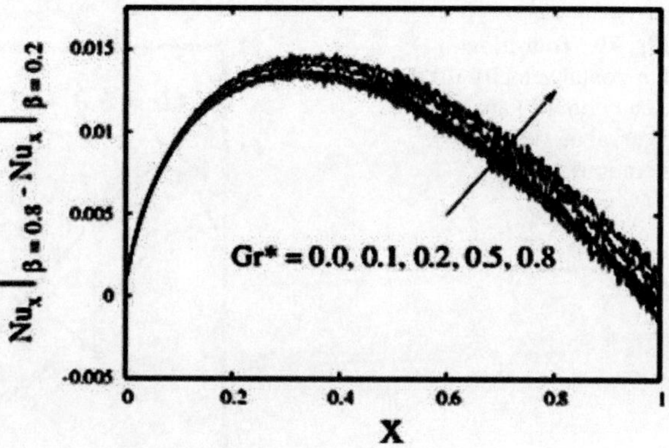

In this graphs, dotted lines are used to represent the curves for $\beta = 0.2$, and dashed lines are used to represent $\beta = 0.8$

Figures 9, 11 and 13 describe the steady horizontal, vertical velocities and temperature for the parameter values mentioned previously.

Figures 10, 12 and 14 describe the difference graphs.

It must be noted that the difference values between $\beta = 0.2$ and $\beta = 0.8$ are of the order less than ~10^{-1} for the velocity and is of order less then ~10^{-3} for the temperature field.

Figures 15 and 16 describe the local skin friction and Nusselt number. Figure 17 describes the difference plots for local Nusselt number (~10^{-2}).

The time evolutions are plotted from Figs. 18, 19, 20, 21, 22 and 23. It is seen that application of suction actually decreases the horizontal velocity. As the velocity dominates, the same is reflected in the temperature field which shows the decrease in the temperature profiles. The change in values of Gr* from 0.1 to 0.8 is also of order less then ~10^{-3}.

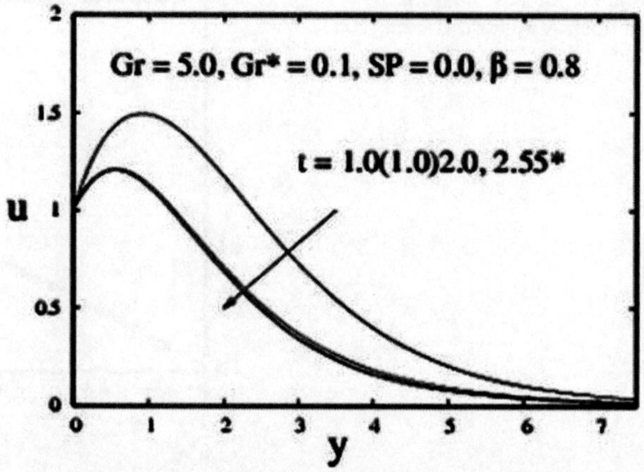

Fig. 18 Transition of horizontal velocity till the steady time(*) at a time interval of 1.0 (or 100 iterations)

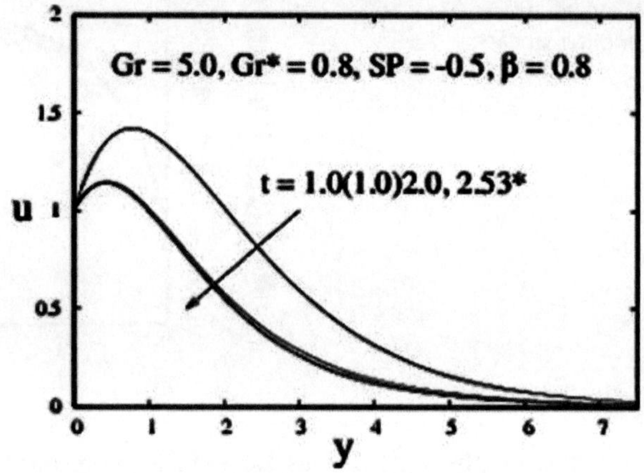

Fig. 19 Transition of horizontal velocity till the steady time(*) at a time interval of 1.0 (or 100 iterations)

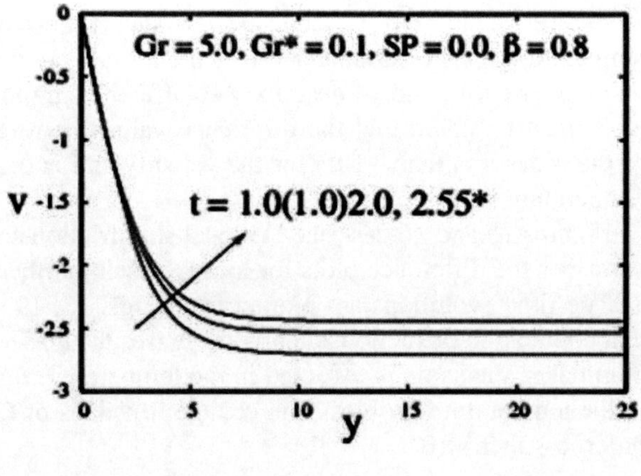

Fig. 20 Transition of vertical velocity till the steady time (*) at a time interval of 1.0 (or 100 iterations)

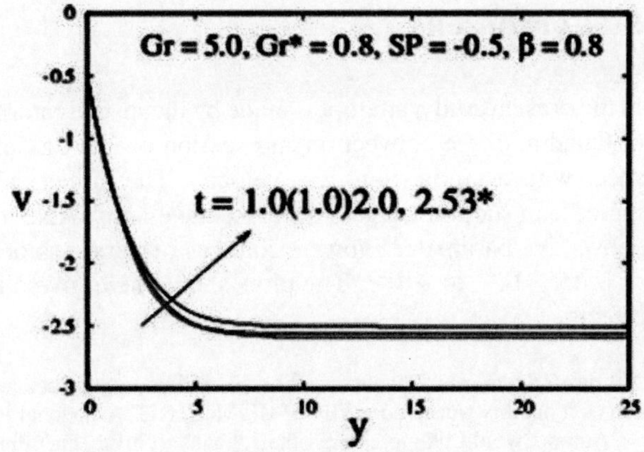

Fig. 21 Transition of vertical velocity till the steady time(*) at a time interval of 1.0 (or 100 iterations)

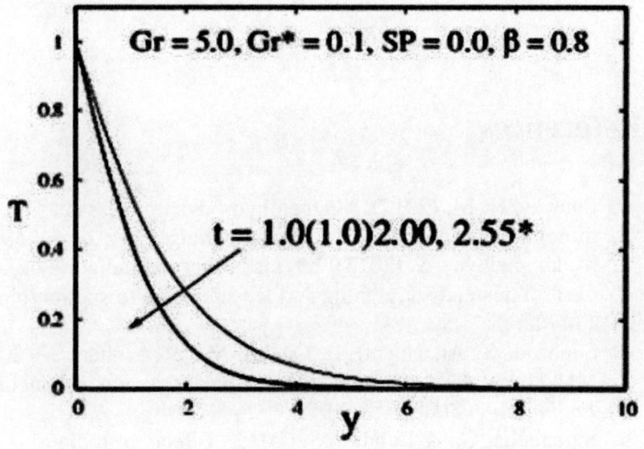

Fig. 22 Transition of temperature velocity till the steady time(*) at a time interval of 1.0 (or 100 iterations)

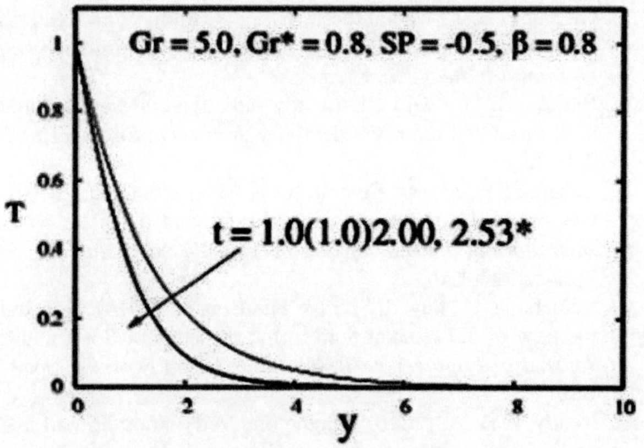

Fig. 23 Transition of temperature velocity till the steady time(*) at a time interval of 1.0 (or 100 iterations)

5 Conclusions

In the present study, attempt is made by the investigators to study the combined effect of Quadratic free convection and suction over a Casson fluid flow past a stretching sheet with an intial non zero velocity. The numerical experiments show that the increase in suction leads to decrease in velocity. Also, the changes in Qaudratic free convective parameter shows a changes of very less order and there contribution is of order $\sim 10^{-1}$ to $\sim 10^{-3}$. The plots shown here give a better understanding of their effects.

Acknowledgements The first author (R. Balamurugan) acknowledge the financial support for carrying out this work under TEQIP-III, MoE, NIT, Arunachal Pradesh.

Authors would like to express their thanks to Prof. (Dr.) Pinakeswar Mahanta, Director, NIT, Arunachal Pradesh for encouraging as to take up this work.

References

1. Shawky, H. M. (2012). MagnetohydrodynamicCasson fluid flow with heat and mass transfer through a porous medium over a stretching sheet. *Journal of Porous Media, 15*(4).
2. Mukhopadhyay, S. (2013). Effects of thermal radiation on Casson fluid flow and heat transfer over an unsteady stretching surface subjected to suction/blowing. *Chinese Physics B, 22*(11), 114702.
3. Shehzad, S. A., Hayat, T., Qasim, M., & Asghar, S. (2013). Effects of mass transfer on MHD flow of Casson fluid with chemical reaction and suction. *Brazilian Journal of Chemical Engineering, 30*(1), 187–195.
4. Sumalatha, C., & Bandari, S. (2015). Effects of radiations and heat source/sink on a Casson fluid flow over nonlinear stretching sheet. *World Journal of Mechanics, 5*(12), 257.
5. Gireesha, B. J., Mahanthesh, B., & Rashidi, M. M. (2015). *MHD boundary layer heat and mass transfer of a chemically reacting Casson fluid over a permeable stretching surface with non-uniform heat source/sink.*
6. Reddy, M. G. (2015). Unsteady radiative-convective boundary-layer flow of a Casson fluid with variable thermal conductivity. *Journal of Engineering Physics and Thermophysics, 88*(1), 240–251.
7. Jasmine Benazir, A., Sivaraj, R., & Makinde, O. D. (2016). Unsteady magnetohydrodynamic Casson fluid flow over a vertical cone and flat plate with non-uniform heat source/sink. In *International Journal of Engineering Research in Africa* (Vol. 21, pp. 69–83). Trans Tech Publications Ltd.
8. Sulochana, C., Payad, S. S., & Sandeep, N. (2016). Non-uniform heat source or sink effect on the flow of 3D Casson fluid in the presence of Soret and thermal radiation. In *International Journal of Engineering Research in Africa* (Vol. 20, pp. 112–129). Trans Tech Publications Ltd.
9. Reddy, P. B. A. (2016). Magnetohydrodynamic flow of a Casson fluid over an exponentially inclined permeable stretching surface with thermal radiation and chemical reaction. *Ain Shams Engineering Journal, 7*(2), 593–602.
10. Animasaun, I. L., Adebile, E. A., & Fagbade, A. I. (2016). Casson fluid flow with variable thermo-physical property along exponentially stretching sheet with suction and exponentially decaying internal heat generation using the homotopy analysis method. *Journal of the Nigerian Mathematical Society, 35*(1), 1–17.

11. Shaw, S., Mahanta, G., & Sibanda, P. (2016). Non-linear thermal convection in a Casson fluid flow over a horizontal plate with convective boundary condition. *Alexandria Engineering Journal, 55*, 1295–1304.
12. Kumar, R., & Sood, S. (2016). Interaction of magnetic field and nonlinear convection in the stagnation point flow over shrinking sheet. *Journal of Engineering*, Article ID 6752520, 8 pages.
13. RamReddy, Ch., Naveen, P., & Srinivasacharya, D. (2019). Influence of non-linear Bousinsinesq approximation on natural convective flow of a power-law fluid along an inclined plate under convective thermal boundary condition. *Nonlinear Engineering, 8*(1), 94–106.
14. Ibrahim, W., & Gamachu, D. (2019). Non linear convection flow of Williamson nanofluid past a radially stretching surface. In *AIP Advances*, 9, 085026.
15. Mandal, I. C., & Mukhopadhyay, S. (2019). Nonlinear convection in micropolar fluid flow past an exponentially stretching sheet in an exponentially moving stream with thermal radiation. *Mechanics of Advanced Materials and Structure, 26*(24), 2040–2046.
16. RamReddy, C., Naveen, P., & Srinivasacharya, D. Effects of nonlinear convection and cross-diffusion for the flow of Darcy-Forchheimer model micropolar fluid with convective boundary condition. *Computational Thermal Sciences: An International Journal, 11*(3), 205–218.
17. Mahanthesh, B., Gireesha, J., Shehzad, S. A., Abbasi, F. M., & Gorla, R. S. R. (2017). Nonlinear three-dimensional stretched flow of an Oldroyd-B fluid with convective condition, thermal radiation, and mixed convection *38*(7), 969–980.
18. Mahanthesh, B., Animasaun, I. L., Rahimi-Gorji, M., Alarifi, I. M. (2019). Quadratic convective transport of dusty Casson and dusty Carreau fluids past a stretched surface with nonlinear thermal radiation, convective condition and non-uniform heat source/sink. *Physica A, 535*, 122471.
19. Raju, C. S. K., Sandeep, N., Sugunamma, V., Babu, M. J., & Reddy, J. R. (2016). Heat and mass transfer inmagnetohydrodynamic Casson fluid over an exponentially permeable stretching surface. *Engineering Science and Technology, An International Journal, 19*(1), 45–52.
20. Ibrahim, S. M., Lorenzini, G., Kumar, P. V., & Raju, C. S. K. (2017). Influence of chemical reaction and heat source on dissipative MHD mixed convection flow of a Cassonnanofluid over a nonlinear permeable stretching sheet. *International Journal of Heat and Mass Transfer, 111*, 346–355.
21. Konda, J. R., Madhusudhana, N. P., & Konijeti, R. (2018). MHD mixed convection flow of radiating and chemically reactive Cassonnanofluid over a nonlinear permeable stretching sheet with viscous dissipation and heat source. *Multidiscipline Modeling in Materials and Structures.*
22. Tannehil, J. C., Anderson, D. A., & Phetcher, R. H. (1997) *Computational fluid mechanics and heat transfer* (2nd Ed.), Taylor and Francis.

Bio Fuel, Combustion, Aerodynamics

Wind Tunnel Experiments to Estimate the Performance of a Novel Arc-Elliptical-Bladed Savonius Wind Rotor

Parag K. Talukdar, Nur Alom, Umang H. Rathod, Vinayak Kulkarni, Palash Saikia, and Deva Kanta Rabha

Nomenclature

A	swept area of rotor (m^2)
AR	aspect ratio (dimensionless)
C_P	power coefficient (dimensionless)
C_T	torque coefficient (dimensionless)
d	blade/bucket diameter (m)
D	diameter of rotor (m)
D_0	diameter of end plate (m)
H	height of rotor (m)
N	rotational speed (rpm)
$P_{\text{available}}$	power available in the wind (W)
P_{turbine}	power output of the rotor (W)
R	radius of rotor (m)
Re	Reynolds number (dimensionless)
r_p	radius of the braking pulley fitted on rotor shaft (m)
S	tension in slack side (kg)
T	torque generated by the rotor (Nm)
V	wind velocity at rotor inlet (m/s)
W	tension in the tight side (kg)
β	overlap ratio (dimensionless)

P. K. Talukdar (✉) · P. Saikia · D. K. Rabha
Department of Mechanical Engineering, Jorhat Engineering College, Jorhat 785007, India

N. Alom
Department of Mechanical Engineering, National Institute of Technology Meghalaya, Meghalaya 793003, India

U. H. Rathod · V. Kulkarni
Department of Mechanical Engineering, Indian Institute of Technology Guwahati, Assam 781039, India

P. Mahanta et al. (eds.), *Advances in Thermofluids and Renewable Energy*, Lecture Notes in Mechanical Engineering, https://doi.org/10.1007/978-981-16-3497-0_16

ρ density of wind (kg/m^3)
ω rotational speed (rad/s)
λ tip-speed ratio (dimensionless)

1 Introduction

Energy plays an essential part in the socio-economic development of a nation. The rapidly growing urbanization and industrialization, the consumption rate of per capita resources, is increasing day by day [1]. Due to depletion of the non-renewable energy sources and their harmful effects, the researchers are keen on finding alternative sources of energy that are easily available, clean and benign to living beings. In this sense, wind energy is one of the most promising green energy resources which can play a crucial role in meeting potential energy needs. The importance of wind as an important resource for generating electricity has been recognized by both developed and developing countries, and the requisite steps are being taken worldwide to harness the energy for its efficient use in power production.

While ongoing work is underway and significant progress has been made with regard to large wind turbines, considerable research still needs to be done with regard to small turbines [2]. In the late 1920s, Finnish engineer S. J Savonius invented the drag-based vertical-axis Savonius rotor composed of two or more semicircular blades with a slight overlap between them. The traditional Savonius rotor consists of two half-cylinders mounted horizontally so that when seen from the top, the cross section resembles the letter ‘S’ [3, 4]. The Savonius rotor’s aerodynamic performance is inferior to that of lift-type wind rotors, but this rotor is simple in structure, omni-directional, and have good starting torque characteristics [5]. It is also regarded as the quietest wind turbine as it can deliver power at low wind speed [6, 7]. A Savonius rotor can be developed easily using nearby materials. Further, it runs at lower rotational speed and is thus prone to reduced vibration.

A two-bladed Savonius rotor indicating advancing blade, returning blade and end plates is shown in Fig. 1. It rotates and extracts power from wind primarily because of the drag force between the concave and convex portions of the blades. Thus, drag is the main driving force for this kind of rotor [8, 9]. In addition, to produce a positive torque, the advancing blade is exposed to more drag force as opposed to the return blade. The negative torque caused by the returning blades of Savonius rotor contributes towards dropping of performance. However, a little lift force also contributes towards overall torque generation by the Savonius rotors [2]. Savonius had conducted a series of experiments on around 30 different rotors to find the optimum design configuration and he obtained maximum or peak power coefficient ($C_{P\max}$) of 0.31. However, an experimental study done by Simonds and Bodek [10] in the Brace Research Institute revealed a $C_{P\max}$ of 0.14 at tip-speed ratio (λ) of 0.70.

Aldoss [11] proposed a novel swinging blade in 1984 to eliminate this drawback. In that context, Aldoss [12] investigated several designs of Swinging blades at an inlet

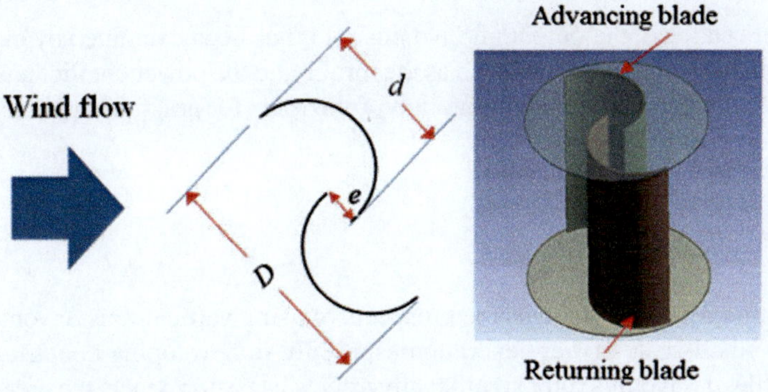

Fig. 1 Schematic diagram of vertical axis Savonius wind turbine

wind speed (V) of 18 m/s and reported a 23.5% increment of CPmax. Ushiyama et al. [13] compared the output of the Savonius rotors with Bach-type (profile composed of an arc and a straight line) and conventional semicircular type blade profiles. The Bach-type rotor yielded maximum gain of peak power coefficient ($C_{P\max}$) in the range of overlap ratio (β) = 0–0.30, and it demonstrated a $C_{P\max}$ of 0.23 which is 9% higher than that of a semicircular-bladed rotor. Sheldahl et al. [14] investigated the performance of two- and three-bladed rotors and revealed 1.5 times greater $C_{P\max}$ for two-bladed rotor as compared to three-bladed rotor. Roy et al. [15] investigated the influence of the arc angle (90°–165°) on the Bach rotor. The numerical study revealed a drop in performance for arc angle is greater than 135°, and the $C_{P\max}$ of 0.34 is obtained at $\lambda = 0.80$. Sanusi et al. [16] experimentally compared the performance of semicircular-bladed and elliptical-bladed Savonius rotors. At $\lambda = 0.80$, the $C_{P\max}$ of 0.26 and 0.24 have been obtained for elliptical-bladed and semicircular-bladed rotors, respectively. Rathod et al. [17] estimated the performance of the semicircular-bladed rotor with and without vent. The study indicated that the vents decreased the positive torque and increased the negative torque which in turn results in inferior performance of the Savonius rotor with capped vent as compared to the rotor having blades without capped vent.

The design parameters that influence the performance of Savonius rotor are aspect ratio, number of blades, overlap ratio, blade shape, use of end plates, use of power augmenting devices, tip speed ratio and Reynolds number. These parameters are investigated by several researchers over a period of time through computational and experimental investigations.

The blade spacing (e) and overlap ratio (β) as shown in Fig. 1 significantly influence the rotor's performance. Most studies conclude that the rotor maintaining the $\beta = 0.15$ between the blades indicates superior performance [18, 19]. Furthermore, the Savonius rotor with endplates showed higher C_P than that of the rotor without endplates [4, 20]. As per the literature [4, 8, 15], the optimal size of end plate diameter (D_o) is equal to 1.1 times the rotor diameter (D).

In the recent past, the output of Savonius rotor has been examined by calculating the performance characteristics such as the torque and the power coefficients in order to optimize the geometrical configuration of the rotor for practical application.

1.1 Present Objective

The literature review enlightens the prospects of using vertical-axis Savonius rotors for future small-scale power generation especially in developing countries. As the blade profile of Savonius rotor significantly affects its performance, the present study is devised to compare the performance of the semicircular-bladed and newly developed arc-elliptical-bladed Savonius rotors. Wind tunnel testing of the rotors is carried out at different loading conditions and the variation of power coefficient at various tip-speed ratios are analysed.

2 Design and Performance Calculation

Design specifications of the Savonius wind rotors are listed in Table 1. The cross-sectional area or swept area (A) of the rotors is set at 0.0448 m^2. The end plates with diameter ($D_p = 1.1 \times D$) of 222.2 mm are used to keep the blades in position. The arc-elliptical profile has drawn by considering the arc section in the leading side and elliptical section in the trailing side of the blade. In case of arc-elliptical-blade, the maximum force is expected to shift towards the leading edge, whereas, in semicircular blade, it is located centrally, and hence, higher moment arm and higher performance can be anticipated in the former one. Further, the 1.0-mm-thick mild steel sheets are used to fabricate the blades. Figure 2 shows the CAD model of the two blade shapes considered in the present study.

The output characteristics of a Savonius rotor is given in the form of power coefficient (C_P) torque coefficient (C_T) and tip-speed ratio (λ). The following calculations are used to estimate those parameters [18]:

Table 1 Design specifications of Savonius rotors

Parameter	Specification
Blade profile	Semicircular and Arc-elliptical
Number of blades (n)	2
Rotor diameter (D)	0.202 m
Rotor height (H)	0.222 m
Overlap ratio (β)	0.15
Aspect ratio (H/D)	1.1
Chord length (d)	0.110 m

Fig. 2 2D top view of Savonius rotor designs, **a** semicircular-bladed shape, **b** Arc-elliptical-bladed shape

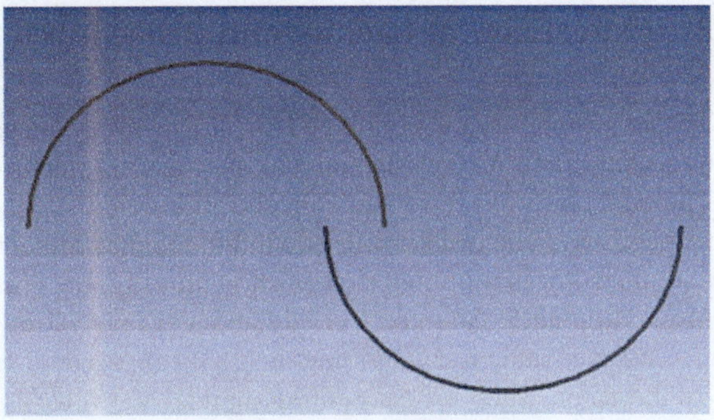

a) Semicircular-bladed shape

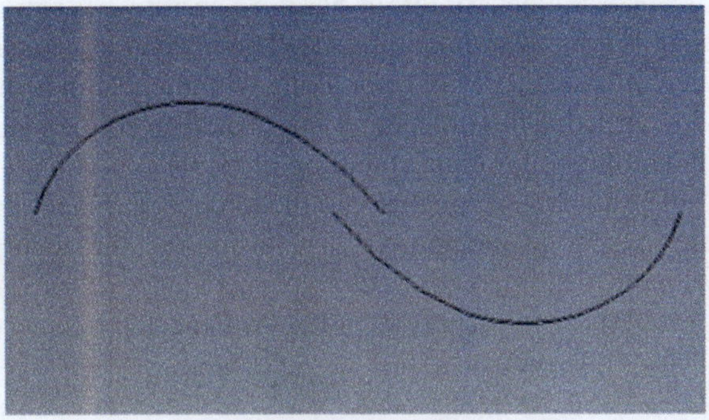

b) Arc-elliptical-bladed shape

$$C_P = \frac{P_{\text{turbine}}}{P_{\text{available}}} = \frac{T\omega}{\frac{1}{2}\rho A V^3} \quad (1)$$

$$T = 9.81 \times (W - S) \times r_p \quad (2)$$

$$C_T = \frac{2T}{\rho A V^2 R} = \frac{C_P}{\lambda} \quad (3)$$

$$\lambda = \frac{\omega R}{V} \quad (4)$$

where A is the cross-sectional or swept area of rotor (m^2), $P_{\text{available}}$ is the power available at rotor inlet, P_{turbine} is the output power of rotor (Watt), R is the radius of rotor (m), r_p is the radius of the braking pulley fitted to the rotor shaft (m), S is the tension in the slack side (kg), T is the torque produced by rotor (N m), V is the incoming wind velocity (m/s), W is the tension in the tight side (kg), ρ is the density of wind (kg/m^3) and ω is the angular speed of the rotor (rad/s).

3 Experimental Facility and Procedure

The experiments are conducted in an open-circuit subsonic wind tunnel (closed test section) set-up of Mechanical engineering department at IIT Guwahati, India. The concentration ratio and concentration length of wind tunnel are 9:1 and 1.8 m, respectively. The test section of wind tunnel has dimensions of 0.6 m × 0.6 m × 2 m and the rotor is mounted in it as shown in Fig. 3. The maximum operational speed of wind tunnel is 50 m/s.

The mechanical dynamometer has been used to measure the torque produced by the rotor. The vertical central shaft of the rotor is mounted in bearings on top and bottom plate of wind tunnel. The dynamometer mainly comprises of spring balance, weighing pan, braking pulley mounted on the rotor shaft and nylon string. The measurement of torque has been performed using methods reported by Talukdar et al. [4] and Kamoji et al. [21]. The mechanical loads are gradually increased, and thereafter, the RPM of rotor along with the readings displayed on the spring balance is recorded. This procedure is repeated from the state of no-load condition to the maximum braking load (stalling of rotor) condition.

The uncertainty analysis for the rotor performance has been calculated out by using the prescribed methods [22]. Considering the accuracies/uncertainties associated with the manometer (±0.3%), spring balance (±0.2%) and tachometer (±0.5%), the overall measurement of λ and C_P is expected to be accurate within ± 0.58% and ± 0.858%, respectively.

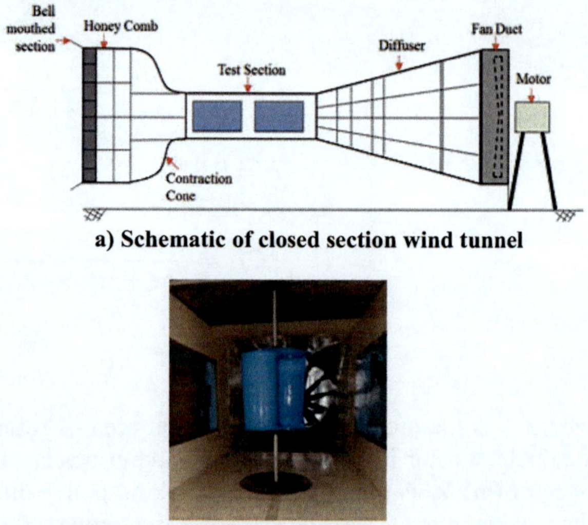

a) Schematic of closed section wind tunnel

b) Mounted rotor

Fig. 3 Wind tunnel facility at IIT Guwahati, **a** Schematic of closed section wind tunnel, **b** Mounted rotor

4 Results and Discussion

The vertical-axis Savonius wind rotor can play a vital role in rural and urban electrification. In the present study, a novel arc-elliptical blade shape has been developed, and its output characteristics have been compared with the conventional semicircular-bladed rotor. The wind tunnel testing has been performed at inlet $V = 6$ m/s ($Re =$ 87,039).

Figure 4 demonstrates the results of wind tunnel experiments in the form of performances indices such as torque coefficient, power coefficient and tip-speed ratio. It can be observed from this figure that the C_P value initially increases with an increase in λ till an inversion point, and beyond which, it starts decreasing with a corresponding increase in λ for both the rotors. The trend of C_P variation with λ is observed to similar for both the rotors. The arc-elliptical-bladed rotor indicated $C_{P\max}$

Fig. 4 Performance characteristics, **a** C_P versus λ, **b** C_T versus λ

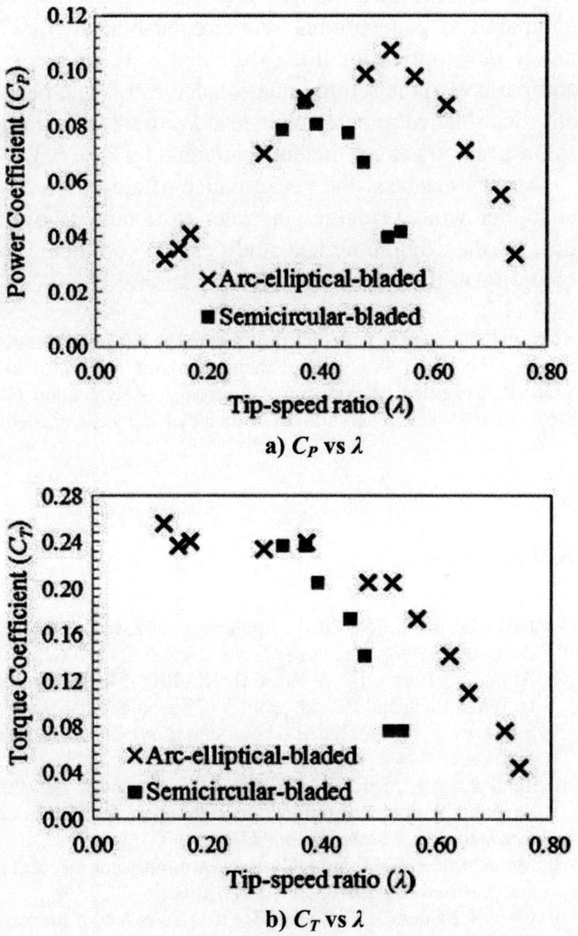

of 0.11 at $\lambda = 0.52$, whereas the conventional semicircular-bladed rotor showed a $C_{P\max}$ of 0.08 at $\lambda = 0.37$. Furthermore, the arc-elliptical-bladed rotor is observed to be operated at higher λ values as compared to the semicircular-bladed rotor.

5 Conclusions

In this present analysis, an effort has been made to enhance the performance of Savonius wind rotor by introducing a new blade profile. A novel arc-elliptical-shaped blade has been developed, and its performance is then compared with the conventional semicircular-bladed Savonius wind rotor. The testing has been conducted in a subsonic wind tunnel having closed test section at varying mechanical loads at wind velocity of 6 m/s.

The arc-elliptical-bladed rotor showed a 37% improvement performance as compared to conventional semicircular-bladed rotor. It is also observed that the newly developed arc-elliptical-bladed rotor operates at higher tip-speed ratios in comparison to the semicircular-bladed rotor. The peak power coefficient of the arc-elliptical-bladed rotor is obtained at $\lambda = 0.52$, whereas in case of semicircular-bladed rotor, peak power coefficient is attained for $\lambda = 0.37$.

As a future task, the performance of arc-elliptical-bladed rotor can be checked at higher wind velocities in order to obtain its overall performance characteristics. Further, the numerical study can be conducted to obtain the flow physics and compliment the experimental observation.

Acknowledgements The study documented here forms a part of the TEQIP Collaborative Research Scheme of National Project Implementation Unit (NPIU), India supported research project (ID: 1-5738545131) titled "Aerodynamic Performance Evaluation of Savonius Vertical-axis Rotor for Small-scale Power Generation". The financial support extended is gratefully acknowledged.

References

1. Doso, O., & Gao, S. (2020). Application of Savonius rotor for hydrokinetic power generation. *Journal of Energy Resources Technology, 142*(1), 1–6.
2. Alom, N., Borah, B., & Saha, U. K. (2018). Insight into the drag and lift characteristic of modified Bach and Benesh profiles of Savonius rotor. *Energy Procedia, 44*, 50–56.
3. Savonius, S. J. (1931). The S-rotor and its applications. *Mechanical Engineering, 53*(5), 333–338.
4. Talukdar, P. K., Sardar, A., Kulkarni, V., & Saha, U. K. (2018). Parametric analysis of model Savonius hydrokinetic turbines through experimental and computational investigations. *Energy Conversion and Management, 158*, 36–49.
5. Menet, J. L. (2004). A double-step Savonius rotor for local production of electricity: A design study. *Renewable Energy, 29*, 1843–1862.
6. Chen, L., Chen, J., & Zhang, Z. (2018). Review of the Savonius rotor's blade profile and its performance. *Journal of Renewable and Sustainable Energy, 10*, 1–20.

7. Fujisawa, N. (1992). On the torque mechanism of Savonius rotors. *Journal of Wind Engineering and Industrial Aerodynamics, 40*, 277–292.
8. Akwa, J. V., Vielmo, H. A., & Petry, A. P. (2012). A review on the performance of Savonius wind turbines. *Renewable and Sustainable Energy Reviews, 16*, 3054–3064.
9. Roy, S., & Saha, U. K. (2013). Review on the numerical investigations into the design and development of Savonius wind rotors. *Renewable and Sustainable Energy Reviews, 24*, 73–83.
10. Simonds, M. H., & Bodek, A. (1964). *Performance test of a Savonius rotor*. Quebec, Canada: Brace Research Institute McGill University.
11. Aldos, T. (1984). Savonius rotor using swinging blades as an augmentation system. *Wind Engineering, 8*, 214–220.
12. Aldoss, T., & Najjar, Y. (1985). Further development of the swinging-blade Savonius rotor. *Wind Engineering, 9*, 165–170.
13. Ushiyama, I., Nagai, H., & Shinoda, J. (1986). Experimentally determining the optimum design configuration for Savonius rotors. *Bull. JSME, 29*(258), 4130–4138.
14. Sheldahl, R. E., Feltz, L. V., & Blackwell, B. F. (1978). Wind tunnel performance data for two and three-bucket Savonius rotors. *AIAA Journal of Energy, 2*(3), 1–105.
15. Roy, S., & Saha, U. K. (2014). An adapted blockage factor correlation approach in wind tunnel experiments of a Savonius-style wind turbine. *Energy Conversion and Management, 86*, 418–427.
16. Sanusi, A., Soeparman, S., Wahyudi, S., & Yuliati, L. (2016). Experimental study of combined blade Savonius wind turbine. *International Journal of Renewable Energy Research, 6*(2), 614–619.
17. Rathod, U. H., Talukdar, P. K., Kulkarni, V., & Saha, U. K. (2019). Effect of capped vents on torque distribution of a semicircular-bladed Savonius wind rotor. *ASME Journal of Energy Resource Technology, 141*(10), 1–15.
18. Talukdar, P. K., Alom, N., Rathod, U. H., & Kulkarni, V. (2021). Alternative blade profile based on savonius concept for effective wind energy harvesting. *Journal of Energy Resources Technology*, 1–32. https://doi.org/10.1115/1.4051675.
19. Nasef, M. H., El-Askary, W. A., Abdel-Hamid, A. A., & Gad, H. E. (2013). Evaluation of Savonius rotor performance: Static and dynamic studies. *Journal of Wind Engineering and Industrial Aerodynamics, 123*, 1–11. https://doi.org/10.1016/j.jweia.2013.09.009.
20. Talukdar, P. K., Kulkarni, V., & Saha, U. K. (2018). Performance estimation of Savonius wind and Savonius hydrokinetic turbines under identical power input. *Journal of Renewable and Sustainable Energy, 10*(6), 064704. https://doi.org/10.1063/1.5054075.
21. Kamoji, M. A., Kedare, S. B, & Prabhu, S. V. (2009). Experimental investigations on single stage modified Savonius rotor. *Applied Energy, 86*(7–8), 1064–1073. https://doi.org/10.1016/j.apenergy.2008.09.019.
22. Moffat, R. J. (1988). Describing the uncertainties in experimental results. *Experimental Thermal and Fluid Science, 1*(1), 3–17. https://doi.org/10.1016/0894-1777(88)90043-X.

Design and Performance Analysis of Hydrokinetic Turbine with Aerodynamic Stall Model

Mahendra Kumar Gupta and P. M. V. Subbarao

Nomenclature

A	Rotor swept area (m^2)
a	Induced axial flow (dimensionless)
a'	Induced tangential flow (dimensionless)
B	Number of blades (dimensionless)
C_d	Coefficient of drag (dimensionless)
C_l	Coefficient of lift (dimensionless)
C_p	Rotor power coefficient (dimensionless)
C_t	Coefficient of thrust (dimensionless)
c	Blade section chord length (m)
F	Prandtl's tip loss correction factor (dimensionless)
k	Blade aspect ratio (dimensionless)
N	Number of blade elements (dimensionless)
P	Rotor power (W)
R	Radius of a rotor (m)
Re	Reynolds Number (dimensionless)
r	Radial position of blade element (m)
U_a	Axial fluid velocity (m/s)
U_t	Tangential fluid velocity (m/s)
V_o	Free stream fluid velocity (m/s)
V_{rel}	Relative velocity (m/s)
α	Angles of attack (deg)
α_s	Stall angle (deg)
λ	Tip speed ratio (dimensionless)
Ω	Rotational speed of a rotor (rad/s)

M. K. Gupta (✉) · P. M. V. Subbarao
Department of Mechanical Engineering, Indian Institute of Technology Delhi, New Delhi 110016, India

P. Mahanta et al. (eds.), *Advances in Thermofluids and Renewable Energy*, Lecture Notes in Mechanical Engineering, https://doi.org/10.1007/978-981-16-3497-0_17

Ø Flow angle (deg)
ρ Density of the fluid (kg/m^3)
σ Solidity (dimensionless)
θ Local pitch angle (degree)
θ_p Blade pitch angle (degree)
i ith blade section
j jth iteration
s the starting point at stall angle (degree)

1 Introduction

Conventional hydropower plant requires a high head of water to drive turbines, but what if, there are zero head or little head of water. A feasible solution to this lies in hydrokinetic turbines (HKT). HKT systems are electromechanical energy converters that convert the kinetic energy of river streams, tidal currents and human-made water channels into other usable forms of energy.

1.1 *Hydrokinetic Turbine's Principle*

The energy conversion principle of HKT and wind turbines is the same. So for the study of free flow hydrokinetic turbines, the corresponding parameters of wind turbines can be used from Khan et al. [1]. Comprehensive reviews on HKT including detailed fundamental studies and turbine classification can be found in Khan and Iqbal [2] and Khan et al. [3].

The amount of hydrokinetic power available for extraction depends on the free stream fluid velocity (V_o), rotor swept area (A) and density of the fluid (ρ).The power coefficient (C_p) determines the performance of HKT as expressed in Eq. 1, Swenson [4]. The typical overall efficiency for a hydrokinetic turbine with low-mechanical losses is approximately 30%, Tiago [5]. The performance of HKT depends primarily on these design parameters, tip speed ratio (TSR), solidity (σ) and blade pitch angle (θ_p). The TSR (λ) expressed in Eq. 2 is defined as the ratio of the blade tip speed to the fluid velocity (V_o). The solidity is related to the radius (R), blade chord length (c), and some blades (B), and it is the ratio of the planform area of the blades to the swept area as expressed in Eq. 3.

Generally, a low-solidity axial flow turbine turns at a higher TSR than a high solidity turbine, results in higher rotational speed for the same diameter and current speed. However, solidity should be kept around 30% to produce enough starting torque, as demonstrated by the Brazilian and Australian experiences, Swenson [4], Tiago [5].

Hydrokinetic energy converters can be divided into lift based and drag based machines. The maximum power coefficient, which can be achieved in a drag based device, is 0.29, which is significantly lower than the Betz limit of 0.59, Bahaj et al. [6]. Hence, all modern hydrokinetic turbines are being designed on lift principles, using airfoils which should have a significant lift to drag ratio to achieve a maximum coefficient of performance (C_p). Here, author focuses on horizontal axis hydrokinetic turbine (HAHkT) among other HKT because it can self-start in lower flow velocities, and the horizontal axis of rotation allows each blade to produce a continuous torque throughout one rotation, outputting a smooth and constant power, Batten et al. [7].

$$C_p = \frac{P}{(1/2)\rho A \mathbf{V}_{\mathbf{o}}^3} = \frac{\text{Rotor Power}}{\text{Power in the water current}} \tag{1}$$

$$\lambda = \frac{\Omega R}{\mathbf{V}_{\mathbf{o}}} \tag{2}$$

$$\sigma = \frac{Bc}{2\pi R} \tag{3}$$

For predicting the performance, and designing of HAHkT, the most employed method is the blade element momentum theory (BEM), Bahaj and Myers [8], Batten [9], and Massouh and Dobrev [10]. The starting point for a BEM model is the knowledge of the reliable two-dimensional (2D) aerodynamic coefficient; lift (C_l) and drag (C_d), but these coefficients are often available only for small angles of attack(α) (attached regime), while a large range of α (stalled regime) appears on the blade element of the rotor when HAHkT operates at high speed.

To overcome this problem of non- availability of C_l and C_d in the stalled regime of the blade element, the mathematical models for estimation of lift and drag coefficients in this region have been proposed by Viterna and Corrigan(VC), Viterna and Corrigan [11]. Prandtl's tip loss correction factor (F) to account the finite-size effect of a rotor blade, Manwell and McGowan [12] and the VC model is incorporated with original BEM model so that improved BEM model can predict rotor performance correctly for a wide range of α.

1.2 Blade Design and Modeling

In BEM theory, the rotor blade is divided into some elemental sections along the span, and both the linear and angular momentum changes of flowing water through the rotor disk are balanced with the axial force and torque generated on the rotor blades, respectively, Manwell and McGowan [12].

Figure 1 shows the velocity diagram for the current configuration, where V_{rel} is the relative velocity; the combination of axial fluid velocity (U_a), and tangential fluid velocity due to the rotation (U_t). The angle between chord and rotor plane, between

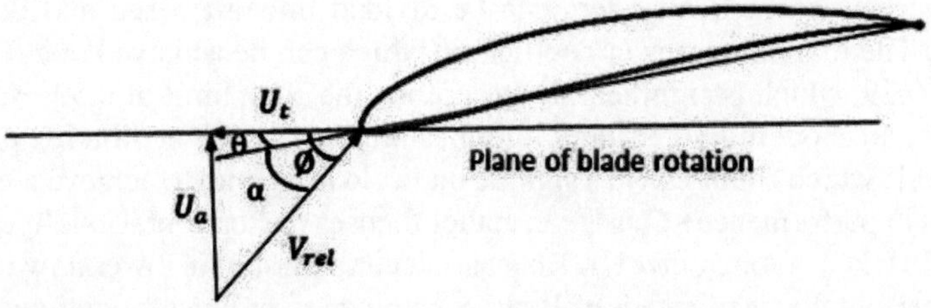

Fig. 1 Velocity diagram for airfoil

the rotor plane and V_{rel} and between chord and V_{rel} are local pitch angle (θ), flow angle (Ø) and angle of attack (α), respectively.

2 Mathematical Model for the Aerodynamic Characteristics of an Airfoil in Stall Regimes

Comparisons among power curve of different horizontal axis wind turbine (HAWT) have been shown for different aerodynamic models and conclude that VC model predicts close to experimental data, David [13]. The VC model plays an important role in predicting the peak power of a fixed pitch turbine.

In Viterna–Corrigan stall model lifting line theory has been used for prediction of airfoil properties in attached regime only ($\alpha < \alpha_s$), and finite-size effects have been considered by this theory in this VC model. For blade element under the stalled regime ($\alpha > \alpha_s$), Viterna–Corrigan developed the following empirical relations to predict C_l and C_d, depending on the blade aspect ratio (k).

$$C_l = \frac{1}{2} C_{D,\max} \sin 2\alpha + A_2 \cos^2 \alpha / \sin \alpha \tag{4}$$

$$C_d = C_{d,\max} \sin^2 \alpha + B_2 \cos \alpha \tag{5}$$

where

$$A_2 = \frac{\left(C_{L,s} - C_{D,\max} \sin \alpha_s \cos \alpha_s\right) \sin \alpha_s}{\cos^2 \alpha_s} \tag{6}$$

$$B_2 = \left(C_{D,s} - C_{D,\max} \sin^2 \alpha_s\right) / \cos \alpha_s \tag{7}$$

$$C_{D,max} = \begin{cases} 1.11 + 0.018k, & k \leq 50 \\ 2.01 & k > 50 \end{cases} \tag{8}$$

The subscript s denotes the constant at stall angle (α_s). $C_{D,\max}$ is the maximum drag coefficient at an angle of attack of 90°.

3 Modified BEM Model

For predicting the performance of low-aspect ratio rotor blade, the original BEM theory can be validated with an experiment in attached regime only. Tests on the NASA Mod-0 100 kW wind turbine indicated that for finite-sized rotor blade operating at higher α.BEM theory was incapable of predicting rotor performance in stalled regime, Viterna and Corrigan [11].

So BEM theory is improved with the inclusion of Prandtl's tip loss correction factor to account for losses due to fluid flow from the pressure side to suction side at the blade's tip, Manwell and McGowan [12] and VC model described above.

In BEM theory, blade is divided into N elements (Fig. 1), and rotor theory is applied to each element to estimate the shape of each element. Due to the presence of rotor, induced axial and tangential flow field at the rotor plane is expressed as a and $a\prime$, respectively. At each rotor blade element (r_i), after initialization for $a_{i,1}$ and $a\prime_{i,1}$, the iterative solution procedure for the j-th iteration is as follows, Manwell and McGowan [12].

$$\tan \emptyset_{i,j} = \frac{1 - a_{i,j}}{\left(1 + a'_{i,j}\right)\lambda_{r,i}} \tag{9}$$

$$F_{i,j} = \left(\frac{2}{\pi}\right)\cos^{-1}\left[\exp\left(-\left\{\frac{(B/2)[1 - (r_i/R)]}{(r_i/R)\sin \emptyset_{i,j}}\right\}\right)\right] \tag{10}$$

$$a_{i,j+1} = \frac{1}{\left(1 + 4F_{i,j}\sin^2 \emptyset_i/\left(\sigma_i' C_{l,i}\cos \emptyset_i\right)\right)} \tag{11}$$

$$a\prime_{i,j+1} = \frac{1}{\left(4F_{i,j}\cos \emptyset_i/\sigma_i' C_{l,i}\right) - 1} \tag{12}$$

4 Results and Discussion

4.1 Predicted Aerodynamic Coefficients in the Stalled Regime

Among other airfoils designed specially for HAWT, the S809 airfoil has significant lift-to-drag ratio under an operating range of Re from 10^5 to 10^6, which also comes

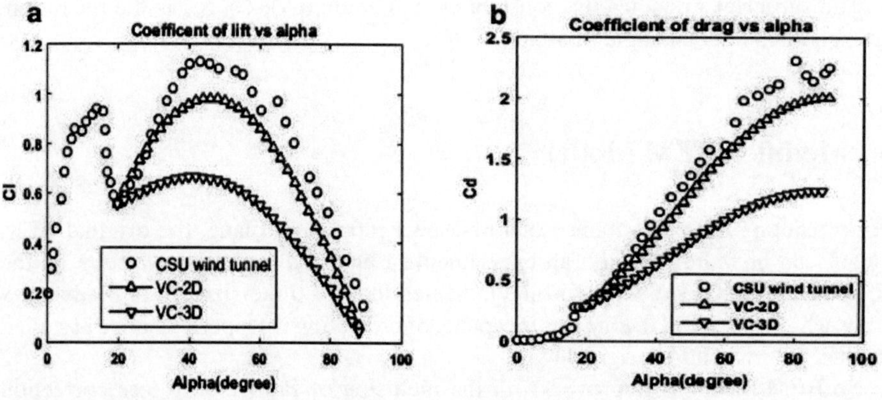

Fig. 2 **a, b** Measured versus predicted aerodynamic coefficients for S809 at Re 650,000

under our working conditions. This airfoil was designed with an objective of insensitivity toward leading edge (LE) roughness so that the value of the maximum lift coefficient does not decline due to debris over LE of an airfoil, Somers and Dan [14].

For testing the VC model against measured data, the measured aerodynamic data for the whole range of α up to 90° is available only at the Reynolds number of 650,000 (CSU) for S809 airfoil. Martínez et al. [15] showed that predicted 2D data using the VC model is close to measured 2D data if α_s is chosen corresponding to the local $C_{l,\min}$ rather than $C_{l,\max}$ which was suggested by the original Viterna–Corrigan model.

Many authors found that in the attached regime, the performance of the rotor can be predicted well using 2D aerodynamic coefficient, so 2D and 3D aerodynamic data are predicted using the VC model for stall regime only.

Predicted 2D aerodynamic coefficients are compared with measured data (CSU wind tunnel) in Fig. 2. It can be seen that predicted 2D aerodynamic coefficients in the stall regime using VC-2D ($k > 50$) are close to 2D experimental data as compare to VC-3D (for $k < 50$).

4.2 Rotor Performance Predictions with Modified BEM

The fixed point iterative numerical approach was used in MATLAB to calculate optimized blade profile, and C_p. Blade designed data are shown in Table 1.

Axial induction factor(a) distribution along the blade for pitch angle of 0°, 1°, 3° and 5° are shown in Fig. 3. It can be seen that for pitch angle of zero and one degree and at tip speed ratio (TSR) of 6.28, a is close to Betz limit of (1/3) and results in a maximum coefficient of power (C_p) at this position.

Table 1 Blade designed data

Design parameters	Values
Rated power (W)	800
Nominal water current velocity (m/s)	2
Design TSR (λ)	4
Number of blades	3
Airfoils	S809

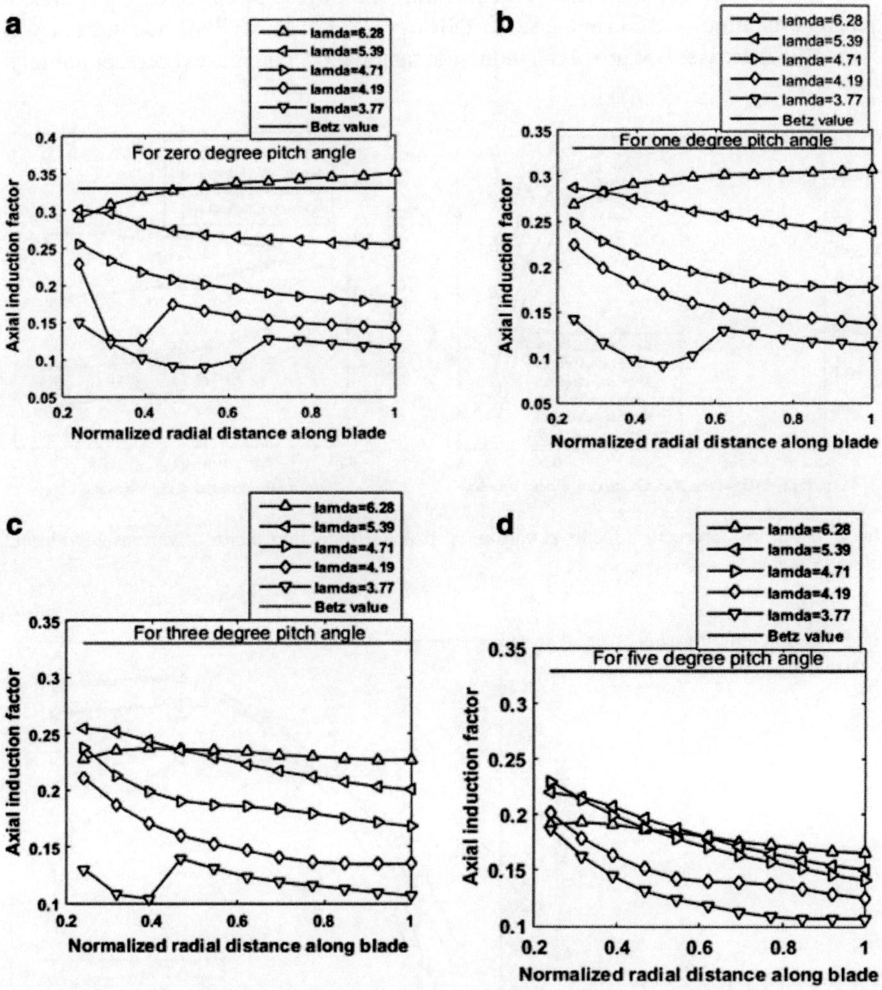

Fig. 3 **a–d** Axial Induction factor at different-2 TSR along the blade for pitch angle of 0°, 1°, 3° and 5°

Figure 4a has shown tip loss correction factor (F) distribution along the blade. The figure shows as tip speed ratio (TSR) decreases, the effect of tip loss increases at tip considerably. The effect of tip loss covers approximately 30% of the blade.

From the tip, Figure 4b has shown axial induction factor for pitch angle of 1° with and without considering F. It also indicates that from root to 70% of the blade axial induction factor is same along the blade.

Figure 5 shows the coefficient of power (C_p) versus tip speed ratio (λ) for the pitch angle of 0°, 1°, 3° and 5°. For zero degree and one-degree pitch angle, C_p is almost same and maximum (0.51) at the same TSR of 6.28. For lower TSR, C_p increases as pitch angle increases because axial induction factor and C_l increases correspondingly,

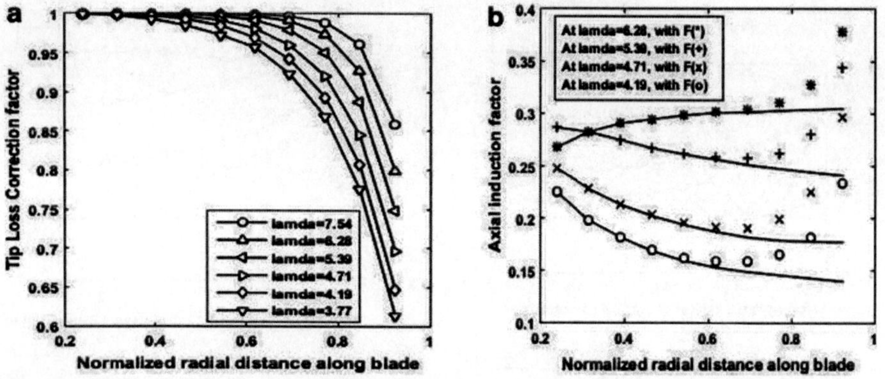

Fig. 4 **a** Tip loss correction factor at various λ, **b** Axial Induction factor at various λ with and without F

Fig. 5 Coefficient of power at various pitch angle versus TSR

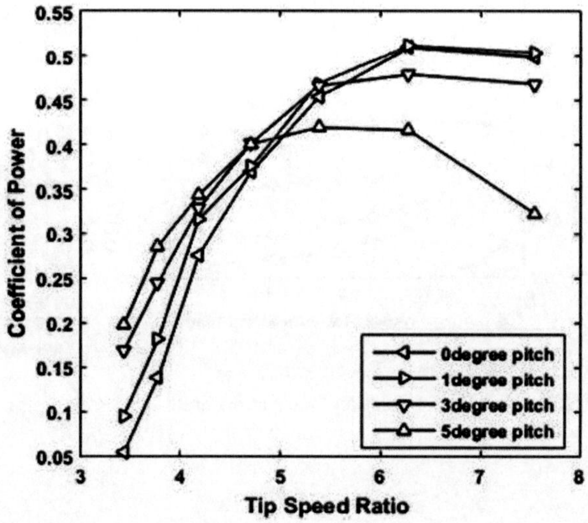

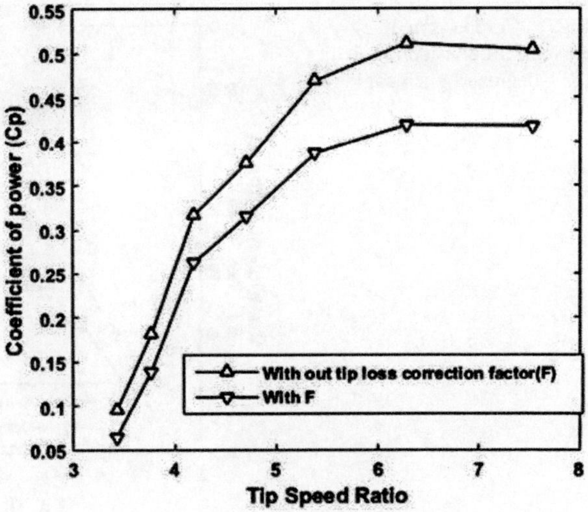

Fig. 6 Comparison of power versus TSR considering *F* for 1-degree of pitch angle only

while for higher TSR axial induction factor and C_l decreases shown in Fig. 3, results in a decrement in C_p as pitch angle increases Fig. 5.

Figures 6 and 7 show C_p versus TSR and Coefficient of thrust (C_t) versus TSR for pitch angle of 1-degree considering tip loss correction factor. It can be seen that C_p is 0.42 almost same at $\lambda = 6.28$. C_t increases as TSR increases, but the rate of growth decrease.

5 Conclusions

It has been shown that predicted 2D aerodynamic coefficients in the stall regime using VC-2D ($k > 50$) are close to 2D experimental data as compare to VC-3D (for $k < 50$). It is found that maximum $C_{p,\max}$ is 0.50 at optimum λ_{opt} of 6.3 for pitch angle of one degree. With Prandtl's tip loss correction factor (F), it was showed that thirty percent of the blades from the blade's tip is affected due to fluid flow from the pressure side to the suction side, and the effect is significantly higher at large TSR (Figs. 6 and 7).

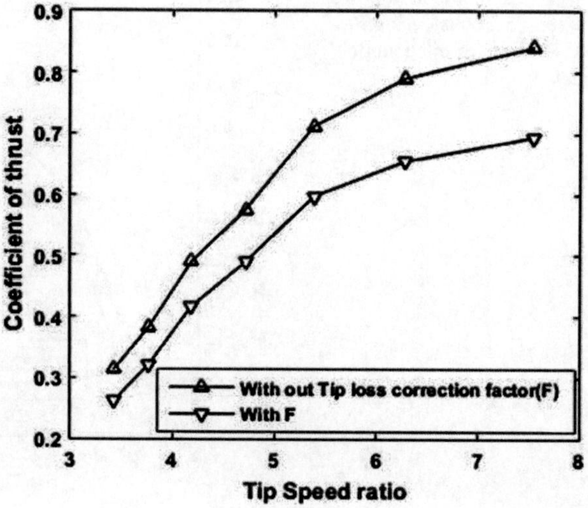

Fig. 7 Comparison of C_t versus TSR considering F for 1-degree of pitch angle only

Acknowledgements The author wishes to thank his colleagues and supervisor for his invaluable guidance.

References

1. Khan, M. J., Iqbal, M. T., & Quaicoe, J. E. (2008). River current energy conversion systems: Progress, prospects and challenges. *Renewable and Sustainable Energy Reviews, 12*(8), 2177–2193.
2. Khan, M. J., Iqbal, M. T., & Quaicoe, J. E. (2006). *A technology review and simulation-based performance analysis of river current turbine systems.* Ottawa: IEEE CCECE/CCGEI.
3. Khan, M. J., Bhuyan, G., Iqbal, M. T., & Quaicoe, J. E. (2009). Hydrokinetic energy conversion systems and assessment of horizontal and vertical axis turbines for river and tidal applications: A technology status review. *Applied Energy, 86*(10), 1823–1835.
4. Swenson, W. J. (1999). *The evaluation of an axial flow, lift type turbine for harnessing the kinetic energy in the tidal flow.* Darwin, Australia: Australian Universities Power Engineering Conference AUPEC.
5. Tiago, G. L. (2003). *The state of the art of hydrokinetic power in Brazil.* Waterpower XIII, Buffalo New York.
6. Bahaj, A. S., & Myers, L. E. (2003). Fundamentals applicable to the utilization of marine current turbines for energy production. *Renewable Energy, 28*, 2205–2211.
7. Batten, W. M. J., Bahaj, A. S., Molland, A. F., & Chaplin, J. R. (2006). Hydrodynamics of marine current turbines. *Renewable Energy, 31*(2), 249–255.
8. Bahaj, A. S., & Myers, L. E. (2006). Power output performance characteristics of a horizontal axis marine current turbine. *Renewable Energy, 31*, 197–208.
9. Batten, W. M. J., Bahaj, A. S., Molland, A. F., & Chaplin, J. R. (2008). The prediction of hydrodynamics of marine current turbines. *Renewable Energy, 33*, 1085–1096.
10. Massouh, F., & Dobrev, I. (2007). Exploration of the vortex wake behind of wind turbine rotor. *J Phys Conf Ser, 75*, 1–9.

11. Viterna, L. A., & Corrigan, R. D. (1981). Fixed pitch rotor performance of large horizontal axis wind turbines. In *DOE/NASA Workshop on Large Horizontal Axis Wind Turbines*, Cleveland, OH (pp. 69–85).
12. Manwell, J. F., & McGowan, J. G. (2002). *Wind energy explained: Theory, design and application*. New York: John Wiley and Sons.
13. David, A., & Spera (2009). *Wind turbine technology: Fundamental concepts in wind turbine engineering*. New York: ASME.
14. Somers, D. M. (1997). *Design and experimental results for the S809 airfoil*. Airfoils, Inc.
15. Martínez, J., et al. (2005). An improved BEM model for the power curve prediction of stall-regulated wind turbines. *Wind Energy, 8*, 385–402.

Energy and Exergy Analyses of a Gas Turbine and Reheat-Regenerative Steam Turbine Integrated Combined Cycle Power Plant

J. Nondy and T. K. Gogoi

Nomenclature

$\dot{E}$	Exergy rate (kW)
e	Specific exergy (kJ/kg)
h	Specific enthalpy (kJ/kg)
$\bar{h}$	Molar specific enthalpy (kJ/kmol)
$\overline{LHV}$	Molar lower heating value (kJ/kmol)
$\dot{m}$	Mass flow rate (kg/s)
$\dot{n}$	Molar flow rate (kmol /s)
P	Pressure (kPa)
$\dot{Q}$	Heat transfer rate (kW)
$\bar{R}$	Universal gas constant (kJ/kmol-K)
r_p	Pressure ratio (dimensionless)
T	Temperature (K)
$\dot{W}$	Power (kW)

Greek Symbols

η	Energy efficiency (%)
ε	Exergy efficiency (%)

J. Nondy (✉) · T. K. Gogoi
Department of Mechanical Engineering, Tezpur University, Napaam, Tezpur 784028, India
e-mail: joy_mtm16@agnee.tezu.ernet.in

P. Mahanta et al. (eds.), *Advances in Thermofluids and Renewable Energy*, Lecture Notes in Mechanical Engineering, https://doi.org/10.1007/978-981-16-3497-0_18

Superscripts

CH	Chemical
PH	Physical

Subscripts

a	Air
AC	Air compressor
D	Destruction
e	Exit
f	Fuel
FP	Feedwater pump
g	Combustion gas
GT	Gas turbine
k	Component
i	Inlet
s	Isentropic
ST	Steam turbine
tot	Total
w	Water
0	Dead state

1 Introduction

The global energy demand is rising rapidly due to ever-increasing population growth, urbanisation and industrialisation. It also increased the environmental issues such as global warming and climate change. This situation has brought many researchers attention to develop efficient energy generation systems. The combined cycle power plant (CCPP) is one such area that has emerged as an attractive power generation option due to their relatively high efficiency and low-greenhouse gas emission.

Several studies have been carried out in the last two decades to propose an efficient CCPP. For instance, Ersayin and Ozgener [1] presented a performance analysis of a gas turbine (GT)-steam turbine (ST) integrated CCPP based on energy and exergy analyses. They found that the energy and exergy efficiency of the CCPP are 56% and 50.04%, respectively. They also reported that about 63% of the total exergy destruction occurs in the combustion chamber. In another study, Ameri et al. [2] investigated Neka CCPP using exergy analysis and evaluated irreversibility of each component. They observed that the combustion chamber, gas turbine and heat recovery steam generator (HRSG) are the major sources of irreversibilities, collectively contributing more than 83% to the total exergy destruction. They also recommended, optimising the HRSG for reducing exergy losses. Likewise, Abuelnuor et al. [3] performed the exergy analysis on the Garri "2" CCPP and found similar results. Ahmadi et al. [4] investigated the effect of supplementary firing on the performance of the CCPP. They observed that with the increase in the degree of supplementary firing, the exergy efficiency of the CCPP increases, mainly due to the increase in the bottoming cycle work output.

Many studies have focused on improving the performance of the bottoming cycle to increase the efficiency of the CCPP. For instance, Gogoi and Talukdar [5] presented the energy and exergy analyses of a reheat-regenerative ST cycle. They observed that with the increase in boiler pressure and the boiler temperature, both the energy and the exergy efficiency of the ST cycle increases. They also reported that the maximum irreversibility occurs at the boiler. Ganjehkaviri et al. [6] performed comprehensive thermodynamic modelling of a CCPP and discussed the importance of steam turbine outlet quality. The results showed that the system with the steam turbine outlet quality of 88% is the most realistic. Sharma and Singh [7] carried out a thermodynamic analysis for the optimal performance of HRSG using different layouts. They reported that the dual pressure HRSG provides 10–15% higher reduction in stack temperature when compared to the single-pressure HRSG. In another study, Mansouri et al. [8] investigated the effect of HRSG pressure levels on the performance of the CCPP. They found that with the increase in the number of pressure level, the exergy efficiency increases and the total exergy destruction rate decreases.

In this study, a CCPP is proposed that comprises of a recuperative GT cycle integrated with a reheat-regenerative ST cycle. The feasibility of the proposed CCPP is evaluated based on energy and exergy analyses using an in-house MATLAB based code. Also, a parametric study is conducted to reveal the effect of some important process parameters on the overall system performance.

2 System Description

The schematic of the proposed CCPP is shown in Fig. 1. It comprises a recuperative GT cycle and a reheat-regenerative ST cycle integrated through a single-pressure heat recovery steam generator (HRSG). The GT cycle includes an air compressor (AC), where the ambient air is compressed from state 1–2, followed by an air preheater (APH) which is employed to preheat the compressed air by recapturing heat from the hot GT outlet exhaust gas. Fuel is fed to the combustion chamber (CC) where it burns in the presence of compressed air to produce the hot combustion gas that enters the gas turbine. After expansion at the GT and producing the power, the combustion gas leaves the GT at state 5 and enters the APH. The GT exhaust gas leaving the APH is then passed through the HRSG. The HRSG further recaptures the thermal energy from the exhaust flue gas which otherwise would have wasted into the atmosphere. The HRSG comprises reheater (REH), economizer (ECO), evaporator (EVA),and superheater (SUP). The water fed to the HRSG leaves at a superheated vapour state

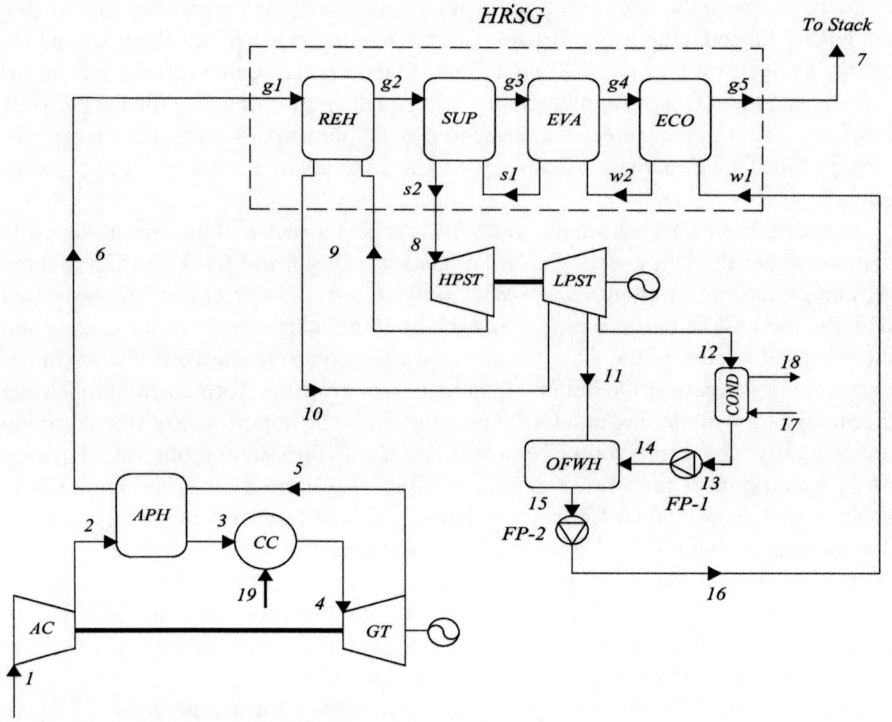

Fig. 1 Schematic of the proposed CCPP

with the same inlet pressure and serves the purpose of driving the high-pressure steam turbine (HPST). The steam leaving the HPST is reheated before sending to the low-pressure steam turbine (LPST). Then, some amount of steam is extracted from the LPST at an appropriate pressure and used to preheat the boiler feedwater exiting the condenser, at the open feedwater heater (OFWH).

3 Modelling

This section first discusses the assumptions considered for carrying out the modelling, followed by the methodologies used for the energy and exergy analyses.

3.1 Assumptions

The following assumptions were made for modelling the proposed system:

Table 1 The assumptions considered for the present study

Description	Unit	Values
r_p	–	10
$\eta_{s,AC}$	%	86
$\eta_{s,GT}$	%	86
T_3	K	850
T_4	K	1520
$\dot{W}_{net,GT}$	MW	50
T_8	K	693.15
P_8	bar	40
$\eta_{s,ST}$	%	85
$\eta_{s,FP}$	%	80
P_{12}	kPa	6

- Air and flue gas are considered as an ideal gas mixture.
- Ambient temperature and pressure are 298.15 K and 101.3 kPa.
- Natural gas (100% Methane) is used as a fuel in the combustion chamber of the gas turbine cycle with a molar lower heating value $(\overline{LHV})$ of 802,361 kJ/kmol, as reported by Khalijani et al. [9].
- The volumetric composition of inlet air is considered as 0.778 N_2, 0.2059 O_2, 0.0003 CO_2 and 0.019 H_2O, as reported by Bejan et al. [10].
- From the combustion chamber, 2% of the fuel lower heating value is considered lost as heat.
- Air preheater pressure drop of 3% on the gas side and 5% on the air-side are considered.
- Inlet temperature and pressure of fuel injected to the combustion chamber are 298.15 K and 1200 kPa as reported by Nondy and Gogoi [11].

The other parameters and their corresponding values are given in Table 1.

3.2 Energy Analysis

The thermodynamic modelling of the proposed CCPP is performed based on mass and energy balance equations as given by:

$$\sum \dot{m}_i = \sum \dot{m}_e \tag{1}$$

$$\sum (\dot{m}h)_i + \sum \dot{Q} = \sum (\dot{m}h)_e + \dot{W} \tag{2}$$

where, h is the specific enthalpy, $\dot{Q}$ and $\dot{W}$ are the heat and work transfer rate.

The enthalpy and entropy of air and the combustion product at each state point of the CCPP are evaluated considering the ideal gas mixture principle as described by Bejan et al. [10]. The fuel–air ratio (λ) was calculated using the energy balance equation and combustion stoichiometry. The HRSG components; reheater (REH), economiser (ECO), evaporator (EVA) and superheater (SUP) were modelled using the pinch point temperature difference (PPTD) and the approach point temperature difference (APTD) of 20 K and 10 K, respectively, using the procedure presented by Ganapathy [12]. The enthalpy and entropy of water at various states of HRSG and ST cycle were calculated using the REFPROP 9.0 database provided by Lemmon et al. [13]. The steam generation rate and the unknown temperatures corresponding to the various points of the HRSG were calculated iteratively solving the set of energy balance equations for the economiser, evaporator, superheater and reheater as reported by Nondy and Gogoi [14]. Besides, the HRSG outlet temperature at the gas side is limited to 400 K in order to avoid the formation of corrosive acid due to the condensation of water vapour.

In the steam turbine cycle, the steam is expanded in a series of high-pressure and low-pressure stages. This approach was adopted for reheating the steam to limit the steam quality at the turbine exhaust above 88% as per Nag [15]. For optimum cycle efficiency, the reheat pressure ratio of 0.20 was chosen for this study, referring Pandey and Gogoi [16] and the operating pressure of OFWH was evaluated following the condition given by Gogoi and Talukdar [5]. The OFWH was employed for improving the cycle efficiency and also to remove the gases dissolved in feed water to prevent boiler corrosion.

3.3 Exergy Analysis

The exergy rate at a point is the sum of the physical exergy and the chemical exergy. Physical exergy is the maximum useful work that a system generates when it undergoes a reversible process from the specified initial state to the dead state. The specific physical exergy is given by Eq. (3):

$$e^{PH} = (h - h_0) - T_0(s - s_0) \tag{3}$$

The chemical exergy, on the other hand, is the maximum work obtained when limited dead state of the system reaches the real dead state. The chemical exergy of water/steam and methane was calculated using the standard chemical exergy table provided by Bejan et al. [10]. The chemical exergy of combustion gas mixtures was calculated using the Eq. (4).

$$\bar{e}^{CH} = \sum_k x_k\left[\bar{e}_k^{CH} + \bar{R}T_0 \times \ln(x_k)\right] \tag{4}$$

where, x_k is the mole fraction of the gaseous component k, $\bar{R}$ is the universal gas constant, and $\bar{e}_k^{CH}$ is the chemical exergy per mole of gas k.

The values of the standard chemical exergy per mole of the combustion gases are also taken from Bejan et al. [10]. Further, the exergy flow rate across all the components of the CCPP is evaluated solving the exergy balance equation as given by Eq. (5).

$$\dot{E}_F = \dot{E}_P + \dot{E}_L + \dot{E}_D \tag{5}$$

where, $\dot{E}_F$, $\dot{E}_P$, $\dot{E}_D$ and $\dot{E}_L$ denotes the rate of fuel exergy, product exergy, exergy destroyed and the exergy loss, respectively. The desired result produced by the system is considered as the product exergy and the resources utilised to produce that result is represented as the fuel exergy, whilst the exergy carried away by the flue gas, away from the system is considered as the exergy loss. The remaining exergy is considered as destroyed due to the presences of system irreversibilities.

The exergy efficiency and the exergy destruction ratio are important parameters that are used to evaluate the exergetic performance of the component as well as the overall system. Exergy efficiency gives the percentage of the fuel exergy that has been converted into the product exergy in a component. On the other hand, the exergy destruction ratio gives the comparison of exergy destroyed in a component to the exergy destroyed by the combined system. The expressions for the exergy efficiency (ε) and exergy destruction ratio (Y_D) are given by Eqs. (6) and (7), respectively.

$$\varepsilon = \frac{\dot{E}_P}{\dot{E}_F} \tag{6}$$

$$Y_D = \frac{\dot{E}_{D,k}}{\dot{E}_{D,tot}} \tag{7}$$

The governing energy relations, the exergy balance equations and the exergy efficiency expressions for each component of the proposed CCPP are shown in Appendix.

4 Results and Discussion

This section first presents the validation of the GT and ST models, followed by the discussion on the findings of the energy and exergy analyses.

4.1 Validation

The simulation models used for the topping gas turbine cycle and the bottoming Rankine cycle in this study is already validated in one of the previous papers of the corresponding authors in Nondy and Gogoi [17] and therefore in this paper, the validation part is not repeated.

4.2 Energy Results

The state properties, mass flow rate and exergy rate of the proposed system at each state point are shown in Table 2. For a 50 MW gas turbine cycle, the fuel–air ratio of 0.0321 and a fuel flow rate of 2.74 kg/s was obtained. The heat generation rate at the combustion chamber was 136.87 MW and the first law efficiency for the gas turbine was found to be 36.53%. After integrating the steam turbine cycle to the gas turbine, the overall thermal efficiency of the combined system becomes 49.37%. The

Table 2 Thermodynamic properties, mass flow rate and exergy rate at each state point of the proposed CCPP

State	T (K)	P (kPa)	$\dot{m}$ (kg/s)	$\dot{E}$ (MW)
1	298.15	101.3	152.13	0.00
2	603.54	1013	152.13	45.89
3	850	962.35	152.13	69.90
4	1520	914.23	154.87	171.06
5	1006	109.93	154.87	66.32
6	776.6	106.63	154.87	37.38
7	488.36	101.30	154.87	4.53
8	693.15	4000	15.48	19.00
9	505.5	800	15.48	13.04
10	693.15	760	15.48	16.00
11	465.62	92.96	1.48	0.77
12	309.3	6.0	14.0	1.22
13	309.3	6.0	14.0	0.045
14	309.32	92.96	14.0	0.047
15	370.73	92.96	15.48	0.53
16	371.25	4040	15.48	0.59
17	298.15	101.30	780.35	1.95
18	308.15	101.30	780.35	2.48
19	298.15	1200	2.74	141.67

net power produced by the steam turbine was 17.58 MW with a thermal efficiency of 35.02%. Therefore, the total power yield of the proposed CCPP is 67.58 MW.

4.3 Exergy Results

The results obtained from the exergy analysis of the proposed CCPP are summarised in Table 3. The components are listed in order of decreasing exergy destruction rate. Also, the exergy destruction ratio for each component of the CCPP is shown in Fig. 2. It can be seen that the combustion chamber is the major site of exergy destruction, accounting 63.30% of the total exergy destruction rate. The next major component contributing to the overall exergy destruction is the HRSG, followed by gas turbine, air preheater and air compressor. The exergy destruction in a system occurs due to the irreversibilities associated with chemical reaction, heat transfer and fluid friction. The combustion chamber is accounted for most exergy destruction due to the presence of all the three principal irreversibilities. In the steam turbine cycle, the low-pressure steam turbine shows the highest exergy destruction, followed by reheater. Moreover, about 6.8% (9567.10 kW) of the fuel exergy is carried out of the system with the flue gas at the HRSG outlet which is considered as exergy loss.

The gas turbine and the air compressor are the two most efficient components with exergy efficiency of 94.91% and 92.86%, respectively. However, the condenser is the least efficient component with the exergy efficiency of 45.57% followed by OFWH and the HRSG with the respective exergy efficiency of 64.42% and 76.82%.

For the clear visualisation of exergy destruction throughout the CCPP, the exergy flow diagram is presented in Fig. 3. It can be observed that the total fuel exergy $\left(\dot{E}_1 + \dot{E}_f\right)$ input to the system is 141.7 MW, and the total product exergy is

Table 3 Summary of results obtained from exergy analysis of the CCPP

Components	$\dot{E}_D$ (kW)	$\dot{E}_L$ (kW)	ε (%)
CC	40,509	–	80.85
HRSG	6444.6	9567	76.82
GT	5329.20	–	94.91
APH	4928.50	–	82.97
AC	3525.70	–	92.86
LPST	1725.10	–	87.67
COND	641.74	–	45.47
HPST	578.55	–	90.33
OFWH	294.33	–	64.42
FP–I	12.76	–	83.93
FP–II	0.2985	–	80.72
CCPP	63,990	9567	47.70

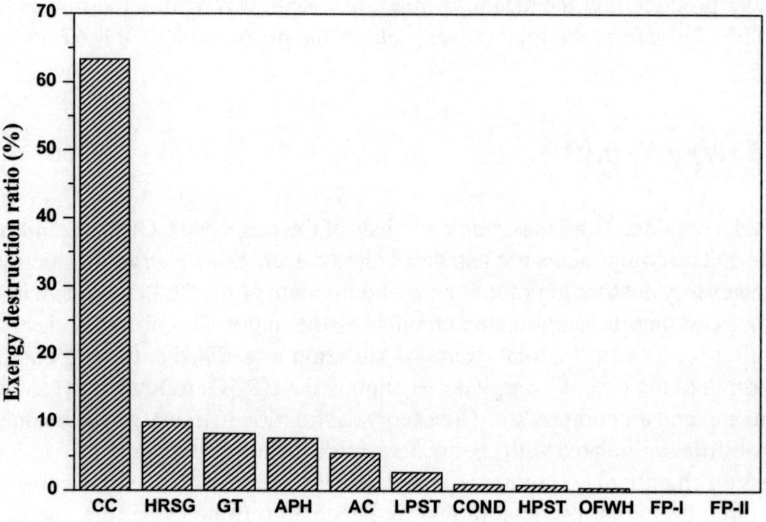

Fig. 2 Exergy destruction ratio of all the CCPP components

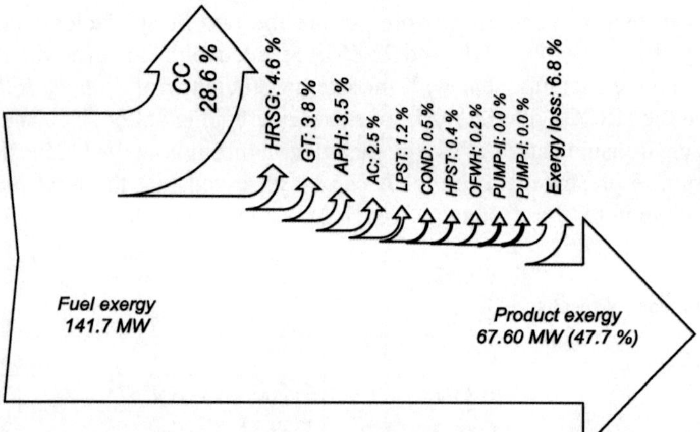

Fig. 3 Exergy flow diagram for the CCPP

67.60 MW. Amongst all the components, 28.60% of the fuel exergy is lost as exergy destruction in the combustion chamber.

The exergy efficiency for the overall CCPP is found to be 47.70% with 63.99 MW of total exergy destruction. Meanwhile, 82.13 % of the exergy is destroyed in the topping GT cycle whilst the remaining 17.87 % of exergy is destroyed in the ST cycle. Moreover, in terms of power output, ST cycle accounts for 35.16% of the total power produced from the CCPP and the remaining 64.84% of power is provided by the topping gas turbine cycle. Therefore, this shows that with the integration of the bottoming cycle, higher efficiency and power output are obtained from the proposed system in comparison to the stand-alone GT plant.

4.4 Parametric Study Results

In this section, a parametric study is performed to investigate the effects of pinch point temperature difference, boiler pressure, air preheater outlet temperature and gas turbine inlet temperature on the energy and exergy efficiencies (η and ε) of the GT cycle and the GT-ST cycle (CCPP). It also shows the effect of variation in the selected process parameters on the total exergy destruction rate $\left(\dot{E}_{D,tot}\right)$ of the overall combined cycle.

The effect of pinch point temperature difference (PPTD) is shown in Fig. 4. With the increase in PPTD, the mass flow rate of steam decreases and consequently the net ST power reduces resulting energy and exergy efficiencies of the combined cycle to decrease slightly. However, PPTD is the property of the HRSG, therefore, the performance of the topping GT plant remains unaffected and that is well evident from the Fig. 4. Also, the exergy destruction rate is found to be reduced from 64.05 MW to 63.99 MW due to the reduction in the irreversibilities associated with the components of the ST cycle owning to decrease in the mass flow rate of steam.

As seen from Fig. 5, it was observed that the GT efficiencies (energy and exergy) are independent of the boiler pressure which is reasonable because boiler pressure is

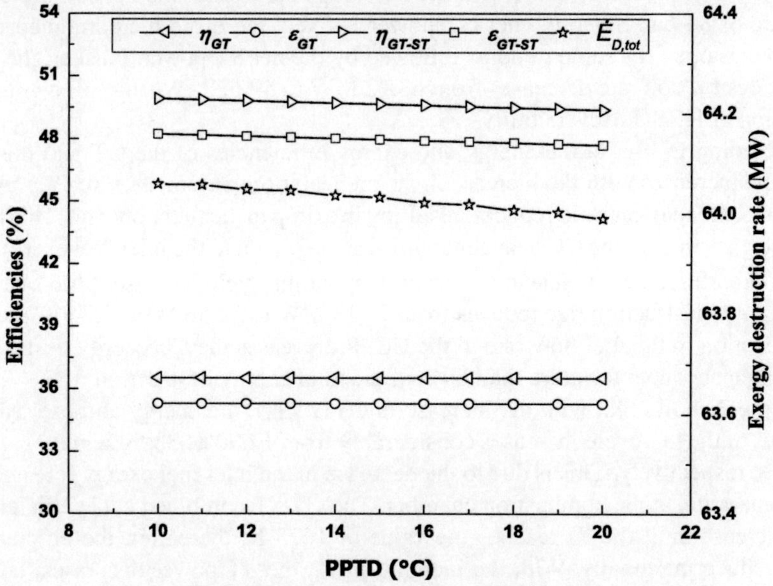

Fig. 4 Effect of pinch point temperature difference (PPTD) on the CCPP performance

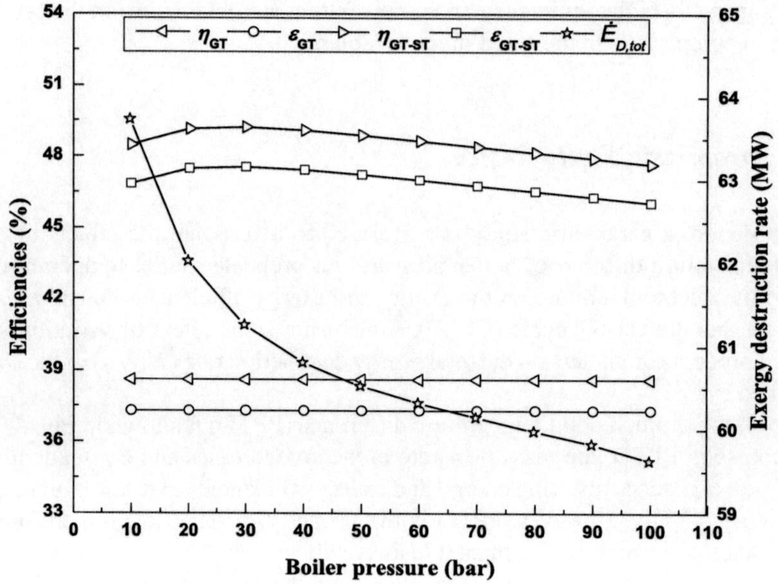

Fig. 5 Effect of boiler pressure on the CCPP performance

only associated with the bottoming ST cycle. However, it was found that the HPST power reduces and the LPST power increases resulting in a net increase in ST power with the rise in boiler pressure. But the net ST power again falls beyond the boiler pressure of 30 bar. Since the net GT power is fixed, therefore the combined cycle efficiencies obey the same trend as followed by the net ST power. Further, the total exergy destruction rate decreases from 63.72 MW to 59.62 MW (6.43%) owing to a reduction of HRSG irreversibility.

According to Fig. 6, the energy and exergy efficiencies of the GT and the GT-ST cycle increases with the increase in air preheater outlet temperature (T_3). As T_3 increases, the fuel–air ratio reduces, resulting in a drop in the fuel flow rate. This leads to a decrease in both the CC heat generation rate $\left(\dot{Q}_{CC}\right)$ and the inlet fuel exergy rate $\left(\dot{E}_f\right)$. Therefore, the efficiencies of the corresponding cycle increase. Moreover, the total exergy destruction rate reduces from 73.13 MW to 54.56 MW (25.40%) due to the reduction in the fuel flow rate to the CC. It decreases the CC exergy destruction rate which accounts for more than 60% of the total exergy destruction.

Figure 7 shows that with the increase in GTIT (T_4), the energy and exergy efficiencies of the GT cycle increases considerably from 12.90%–38.59% and 12.47%–37.28%, respectively. This is due to the decrease in the inlet fuel exergy rate and the heat generation at the combustion chamber. The GT-ST combined cycle efficiencies also increase until the T_4 reaches the value of 1475 K, thereafter, the efficiencies again reduce marginally. With the increase in T_4, the ST power decreases, but the rate of decrease in ST power is less than that of $\dot{E}_f$ and $\dot{Q}_{CC}$, therefore, the GT-ST efficiencies increase at the beginning. However, the rate of decrease in ST power

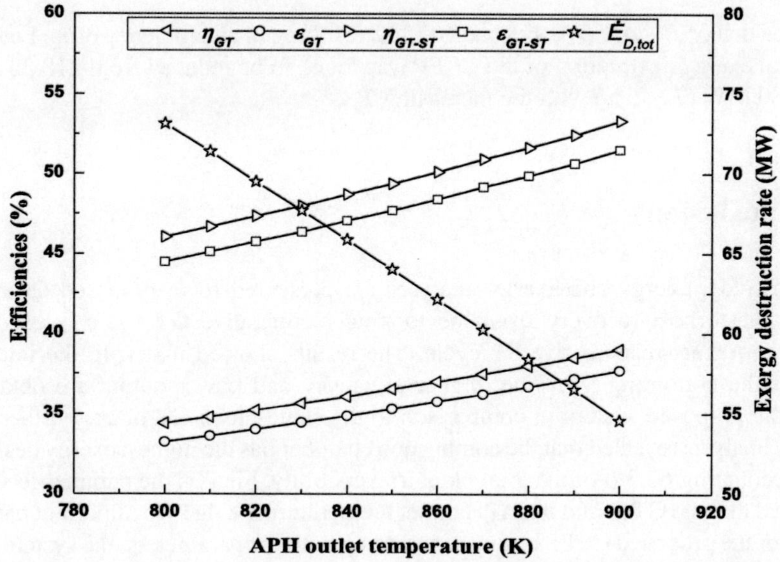

Fig. 6 Effect of APH outlet temperature on CCPP performance

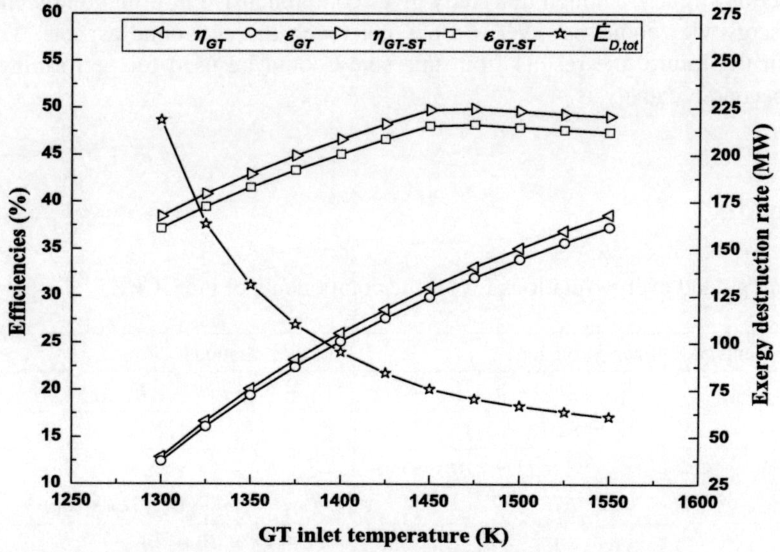

Fig. 7 Effect of GT inlet temperature on the CCPP performance

exceeds that of $\dot{E}_f$ and $\dot{Q}_{CC}$ beyond 1475 K, resulting in an efficiency drop. Further, the total exergy destruction of the CCPP was found to be reduced from 218.38 MW to 60.80 MW (72.15%) with the increasing T_4.

5 Conclusions

In this study, energy and exergy analyses are presented for a proposed CCPP in which waste heat recovery from the topping recuperative GT cycle is used for driving a reheat-regenerative ST cycle. The results showed that with the integration of the bottoming ST cycle, higher efficiency and power output are obtained from the proposed system in comparison to the stand-alone GT plant. Further, the exergy analysis revealed that the combustion chamber has the highest exergy destruction accounting 63.3% of the total plant irreversibility. Finally, the parametric study reported that the GTIT and the APH outlet temperature are the two important parameters for the proposed CCPP because increasing both the parameters, the system efficiencies increases and the total exergy destruction decreases significantly. Besides, the boiler pressure and the PPTD reduces both the efficiencies and the total exergy destruction slightly. Though this study gives comprehensive information on energy and exergy viewpoint, however, it does not touch the economic aspects. Therefore, in the future, the results from this study could be used for performing the exergoeconomic analysis.

Appendix

The energy and exergy relations for all the components of the CCPP.

Components	Energy relations	Exergy relations
AC	$\eta_{s,AC} = (\bar{h}_{2s} - \bar{h}_1)/(\bar{h}_2 - \bar{h}_1)$	$\dot{E}_1 + \dot{W}_{AC} = \dot{E}_2 + \dot{E}_{D,AC}$
	$\dot{W}_{AC} = \dot{n}_a(\bar{h}_2 - \bar{h}_1)$	$\varepsilon_{AC} = (\dot{E}_2 - \dot{E}_1)/\dot{W}_{AC}$
APH	$(\bar{h}_3 - \bar{h}_2) = (1+\lambda)(\bar{h}_5 - \bar{h}_6)$	$\dot{E}_2 + \dot{E}_5 = \dot{E}_3 + \dot{E}_6 + \dot{E}_{D,APH}$
	$\lambda = \dot{n}_f/\dot{n}_a$	$\varepsilon_{APH} = (\dot{E}_3 - \dot{E}_2)/(\dot{E}_5 - \dot{E}_6)$
CC	$-0.02\lambda\,\overline{LHV} + \bar{h}_3 + \lambda\,\bar{h}_f =$ $(1+\lambda)\,\bar{h}_4$	$\dot{E}_f + \dot{E}_3 = \dot{E}_4 + \dot{E}_{D,CC}$
	$(1+\lambda) = \dot{n}_g/\dot{n}_a$	$\varepsilon_{CC} = \dot{E}_4/(\dot{E}_3 + \dot{E}_f)$
GT	$\eta_{s,GT} = (\bar{h}_4 - \bar{h}_5)/(\bar{h}_4 - \bar{h}_{5s})$	$\dot{E}_4 = \dot{W}_{GT} + \dot{E}_5 + \dot{E}_{D,GT}$
	$\dot{W}_{GT} = \dot{n}_g(\bar{h}_4 - \bar{h}_5)$	$\varepsilon_{GT} = \dot{W}_{GT}/(\dot{E}_4 - \dot{E}_5)$
HRSG	$\dot{n}_g(\bar{h}_{g2} - \bar{h}_{g4}) =$ $\dot{m}_w(h_{s2} - h_{w2})$	$\dot{E}_6 + \dot{E}_9 + \dot{E}_{16} =$ $\dot{E}_7 + \dot{E}_8 + \dot{E}_{10} + \dot{E}_{D,HRSG}$

(continued)

(continued)

Components	Energy relations	Exergy relations
	$\dot{n}_g(\bar{h}_{g2} - \bar{h}_{g3}) = \dot{m}_w(h_{s2} - h_{s1})$	$\varepsilon_{HRSG} = (\dot{E}_8 - \dot{E}_{16} + \dot{E}_{10} - \dot{E}_9)/(\dot{E}_6 - \dot{E}_7)$
HPST	$\eta_{s,HPST} = (h_8 - h_9)/(h_8 - h_{9s})$	$\dot{E}_8 = \dot{E}_9 + \dot{W}_{HPST} + \dot{E}_{D,HPST}$
	$\dot{W}_{HPST} = \dot{m}_w(h_8 - h_9)$	$\varepsilon_{HPST} = \dot{W}_{HPST}/(\dot{E}_9 - \dot{E}_8)$
LPST	$\dot{W}_{LPST} = \dot{m}_w(h_{10} - h_{11}) + \dot{m}_{w1}(h_{11} - h_{12})$	$\dot{E}_{10} = \dot{E}_{11} + \dot{E}_{12} + \dot{W}_{LPST} + \dot{E}_{D,LPST}$
	$\dot{m}_{w1} = \dot{m}_w(h_{11} - h_{15})/(h_{11} - h_{14})$	$\varepsilon_{LPST} = \dot{W}_{LPST}/(\dot{E}_{10} - \dot{E}_{11} - \dot{E}_{12})$
OFWH	$\dot{m}_{w2}h_{11} + \dot{m}_{w1}h_{14} = \dot{m}_w h_{15}$	$\dot{E}_{11} + \dot{E}_{14} = \dot{E}_{15} + \dot{E}_{D,OFWH}$
	$\dot{m}_{w2} = \dot{m}_w - \dot{m}_{w1}$	$\varepsilon_{OFWH} = \dot{E}_{15}/(\dot{E}_{11} + \dot{E}_{14})$
COND	$\dot{m}_{w1}(h_{12} - h_{13}) = \dot{m}_w(h_{18} - h_{17})$	$\dot{E}_{12} + \dot{E}_{17} = \dot{E}_{13} + \dot{E}_{18} + \dot{E}_{D,COND}$
	$\dot{Q}_{COND} = \dot{m}_w(h_{18} - h_{17})$	$\varepsilon_{COND} = (\dot{E}_{18} - \dot{E}_{17})/(\dot{E}_{13} - \dot{E}_{12})$
FP–I	$\eta_{s,FP-I} = (h_{14s} - h_{13})/(h_{14} - h_{13})$	$\dot{E}_{13} + \dot{W}_{FP-I} = \dot{E}_{14} + \dot{E}_{D,FP-I}$
	$\dot{W}_{FP-I} = \dot{m}_{w1}(h_{14} - h_{13})$	$\varepsilon_{AFP-I} = (\dot{E}_{14} - \dot{E}_{13})/\dot{W}_{FP-I}$
FP–II	$\eta_{s,FP-II} = (h_{16s} - h_{15})/(h_{16} - h_{15})$	$\dot{E}_{15} + \dot{W}_{FP-II} = \dot{E}_{16} + \dot{E}_{D,FP-II}$
	$\dot{W}_{FP-II} = \dot{m}_w(h_{16} - h_{15})$	$\varepsilon_{AFP-II} = (\dot{E}_{16} - \dot{E}_{15})/\dot{W}_{FP-II}$

References

1. Ersayin, E., & Ozgener, L. (2015). Performance analysis of combined cycle power plants: A case study. *Renewable and Sustainable Energy Reviews, 43*, 832–842.
2. Ameri, M., Ahmadi, P., & Khanmohammadi, S. (2008). Exergy analysis of a 420 MW combined cycle power plant. *International Journal of Energy Research, 32*(2), 175–183.
3. Abuelnuor, A. A. A., Saqr, K. M., Mohieldein, S. A. A., Dafallah, K. A., Abdullah, M. M., & Nogoud, Y. A. M. (2017). Exergy analysis of Garri "2" 180 MW combined cycle power plant. *Renewable and Sustainable Energy Reviews, 79*, 960–969.
4. Ahmadi, P., Dincer, I., & Rosen, M. A. (2011). Exergy, exergoeconomic and environmental analyses and evolutionary algorithm based multi-objective optimization of combined cycle power plants. *Energy, 36*(10), 5886–5898.
5. Gogoi, T., & Talukdar, K. (2014). Thermodynamic analysis of a combined reheat-regenerative thermal power plant and water-LiBr vapour absorption refrigeration system. *Energy Conversion and Management, 78*, 595–610.
6. Ganjehkaviri, A., Jaafar, M. M., & Hosseini, S. (2015). Optimization and the effect of steam turbine outlet quality on the output power of a combined cycle power plant. *Energy Conversion and Management, 89*, 231–243.

7. Sharma, M., & Singh, O. (2019). Thermodynamic study of multi-pressure HRSG in gas/steam combined cycle power plant. *Journal of the Institution of Engineers (India): Series C, 100*(2), 361–369. https://doi.org/10.1007/s40032-017-0423-5.
8. Mansouri, M. T., Ahmadi, P., Kaviri, A. G., & Jaafar, M. N. M. (2012). Exergetic and economic evaluation of the effect of HRSG configurations on the performance of combined cycle power plants. *Energy Conversion and Management, 58*, 47–58.
9. Khalijani, M., Saray, R. K., & Bahlouli, K. (2015). Comprehensive analysis of energy, exergy and exergoeconomic of cogeneration of heat and power in a combined gas turbine and organic Rankine cycle. *Energy Conversion and Management, 97*, 154–165.
10. Bejan, A., Tsatsaronis, G., & Moran, M. (1995). *Thermal design and optimization*. John Wiley and Sons.
11. Nondy, J., & Gogoi, T. K. (2020). A comparative study of metaheuristic techniques for the thermoenvironomic optimization of a gas turbine-based benchmark combined heat and power system. *Journal of Energy Resources Technology*. https://doi.org/10.1115/1.4048534.
12. Ganapathy, V. (2014). *Steam generators and waste heat boilers: For process and plant engineers*. CRC Press.
13. Lemmon, E., Huber, M., & McLinden, M. (2010). *NIST standard reference database 23, reference fluid thermodynamic and transport properties (REFPROP)*, version 9.0, National Institute of Standards and Technology. R1234yf.d_file.
14. Nondy, J., & Gogoi, T. K. (2019). ASME gas turbine India conference-ASME GTINDIA'19. IIT Madras, Chennai, India, 83525 V001T02A002. https://doi.org/10.1115/GTINDIA2019-2351.
15. Nag, P. (2002). *Power plant engineering*. Tata McGraw-Hill Education.
16. Pandey, M., & Gogoi, T. (2013). Energy and exergy analysis of a reheat regenerative vapour power cycle. *International Journal of Emerging Technology and Advanced Engineering, 3*(3), 427–434.
17. Nondy, J., & Gogoi, T. K. (2020). Comparative performance analysis of four different combined power and cooling systems integrated with a topping gas turbine plant. *Energy Conversion and Management, 223*, 113242. https://doi.org/10.1016/j.enconman.2020.113242.

Comparative Analysis of Savonius Type Ultra-Micro Hydrokinetic Turbine of Experimental and Computational Investigation

Thochi Seb Rengma and P. M. V. Subbarao

Nomenclature

C_p	Power coefficient (Dimensionless)
C_m	Torque coefficient (Dimensionless)
A	Projected area (m^2)
U	Free stream velocity (m/s)
H	Height of the turbine (m)
D	Diameter of the turbine (m)
e	Overlap ratio (Dimensionless)

Greek Symbols

ω	Angular velocity (rad/s)
λ	Tip-speed ratio (TSR) (Dimensionless)
ρ	Density (kg/m^3)

1 Introduction

Over the years, vertical axis wind turbines have fascinated researchers, thus contributing to sufficient advancements of this technology to become a prominent energy harvester in different wind terrains. However, research on this technology for river streams harvesting has probably picked up from the last decade. Savonius

T. S. Rengma (✉) · P. M. V. Subbarao
Department of Mechanical Engineering, Indian Institute of Technology, Delhi, New Delhi, India

P. Mahanta et al. (eds.), *Advances in Thermofluids and Renewable Energy*, Lecture Notes in Mechanical Engineering, https://doi.org/10.1007/978-981-16-3497-0_19

hydrokinetic turbine is one of the emerging technologies for power generation which has gained considerable interest among the researchers in recent times.

Vertical axis Savonius rotor was initially developed by S. J. Savonius in the late 1920s. The idea of the conventional Savonius rotor is based on cutting a cylinder into two halves along the central plane and then moving the two sideways along the cutting plane, so that the cross section resembles the letter S. A schematic of this design is shown in Fig. 1. The rotation of the turbine is due to the drag difference between the advancing and the returning blades. The net driving force can be enhanced by increasing the positive force on the advancing blade or by reducing the opposite force on the returning blade [3]. The rotation rate of the turbine is measured by angular velocity (ω). Since, the flow of water is only in one direction, therefore, the rotor spins around its axis and the change of coefficient of drag are experienced in both the advancing and the returning blades, resulting in the change in torque produced by a rotor at constant ω. The change in torque cyclically varies during the rotation of the rotor. Hydrokinetic turbines are those that convert the kinetic energy of flowing water current into mechanical power. Hydrokinetic turbines are also known as ultra-low-head hydro turbines and zero head water turbines [4]. These turbines offer many advantages like simple construction, low noise, reduced wear on moving parts, better self-starting, etc. [5]. However, the major problem associated with these turbines is their low efficiency, although its performance in water application is comparatively better than in wind applications due to the greater weight density of water [6, 7]. Some researchers have attempted to improve the turbine efficiency by modifying the original design using blade twists, blade overlap, multi-staging, deflector plates, etc.

Kumar and Saini [8] analysed the performance of a modified Savonius water turbine with twisted blades using computational CFD Fluent Simulation. A maximum power coefficient (C_p) of 0.39 was reported at a tip speed ratio of 0.9 for an optimum twist angle of 12.5°. The performance study of a Savonius turbine using two deflector plates was carried by Kailash et al. [9]. They found that the power coefficient increased to 0.35 at TSR 1.08 by placing a deflector plate and advancing plate at

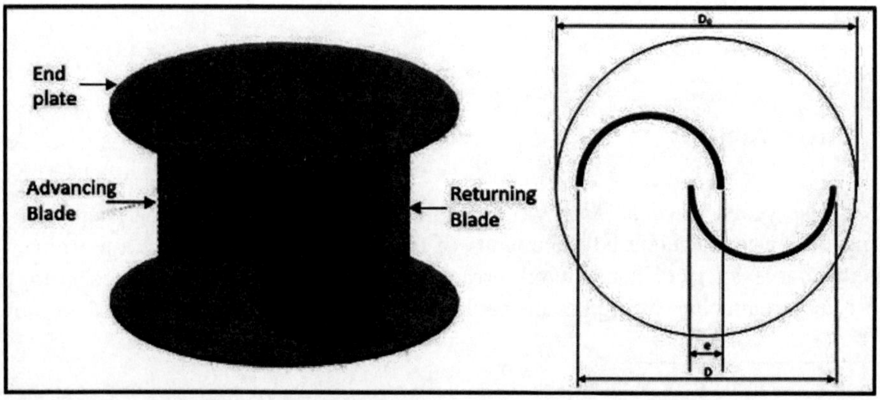

Fig. 1 Geometrical parameters of SHKT

an angle of 50° at a distance of 1.8R. Again, Kumar and Saini [10] attempted to improve the performance of a single stage Savonius water turbine by modifying the blade shape. The effects of two parameters, namely, the blade arc angle and blade shape factor were studied. A maximum C_p of 0.426 was reported for blade arc angle of 150° and blade shape factor of 0.6, which was obtained at a tip speed ratio of 0.9 and flow velocity of 2 m/s. Patel et al. [11] investigated the effect of overlap ratio on the performance of a Savonius water turbine and obtained the highest hydrodynamic torque at an overlap ratio of 20%.

2 Design and Performance Parameter

Theoretical power of Savonius hydrokinetic turbine can be calculated as [12]:

$$P_{\text{Theory/in}} = \frac{1}{2}\rho AU^3 = \frac{1}{2}\rho HDU^3 \tag{1}$$

In the above expression, the term 'ρ'is the density of water, 'A' is the projected area, and 'U' is the free stream velocity, 'H' is the height of the turbine, and 'D' is the diameter of the turbine. The entire energy impacting the turbine does not contribute to the rotation of the turbine; a fraction is inevitably lost because of the blade profile and large flow domain. Thus, only a fraction of kinetic energy (KE), known as power coefficient (C_p), can be captured in the turbine cross section. The actual power captured by the turbine is given as:

$$P_{\text{Turbine}} = C_P\frac{1}{2}\rho AU^3 \tag{2}$$

The power coefficient of hydrokinetic turbines depends on the tip speed ratio (TSR), which is defined as the speed of the blade at the tip to the speed of water [13]. The expression is given as:

$$\lambda = \frac{\text{speed of blade at tip}}{\text{speed of flowing water}} = \frac{\omega D}{2U} \tag{3}$$

The expression for outlet power extraction (P_{out}) is given as:

$$P_{\text{out}} = T\omega = \frac{2\pi NT}{60} \tag{4}$$

The torque coefficient (C_m) is generally expressed as:

$$C_m = \frac{T}{\frac{1}{2}\rho AU^2R} = \frac{2T}{\frac{1}{2}\rho HD^2U^2} \tag{5}$$

The power coefficient (C_p) is defined as the ratio of power output given by the turbine to power input on the turbine. The expression is given as:

$$C_p = \frac{P_{\text{out}}}{P_{\text{in}}} = \frac{T\omega}{\frac{1}{2}\rho H D U^3} = C_m \lambda \tag{6}$$

3 Computational Methodology

The flow field around the turbine is complex due to flow separation and wake formation. The computation fluid dynamics (CFD) is an important tool to solve a wide range of fluid flow problems. ANSYS workbench allows processing geometry modeling, mesh generation, fluent solver and CFD post for results. For the present study, unsteady Reynolds-Averaged Navier–Stokes (URANS) solver with K-ω SST turbulence model was used for numerical analysis.

3.1 Computational Domain

2D and 3D geometry were created in Ansys Design Modeler, as per the parameter given in Table 1. A cylinder enclosure was created around the turbine to capture the angular velocity of the turbine. The computational domain was divided into two zones (i.e., stationary zone and rotating zone) to calculate the rotation of the turbine. The circumference of the cylinder was set as an interface to maintain continuity of the flow. Figure 2 represents the 3D open channel computational domain.

Table 1 Geometric parameters of SHKT

Parameter	2D	3D
Number of blades	2	2
Aspect ratio (H/D)	1	0.86
Overlap ratio (e)	0	0
Turbine diameter (D), mm	250	152
Turbine height (H), mm	250	132
Endplate diameter (D_O), mm	275	167.2

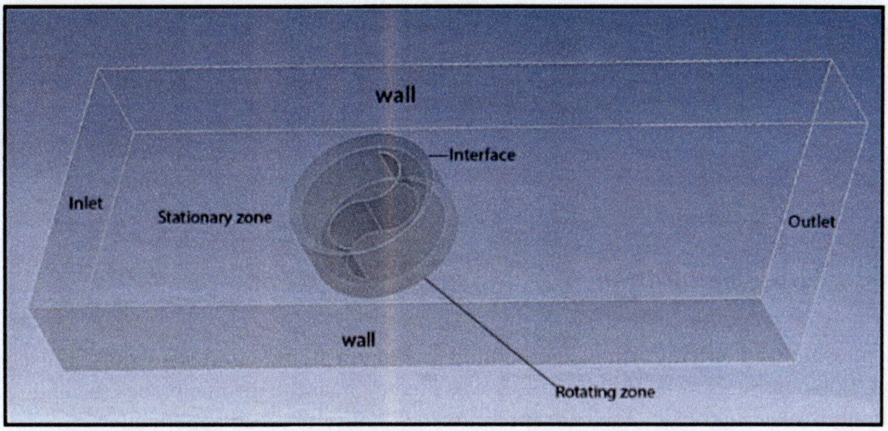

Fig. 2 Open channel of 3D computational domain

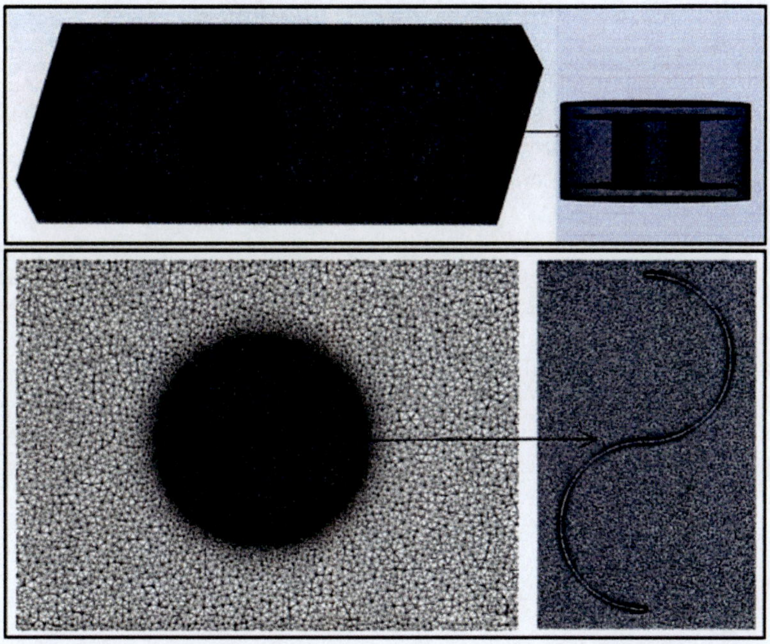

Fig. 3 Meshing of the model channel

Table 2 Mesh details

Quality aspect	2D	3D
No.of elements	310,721	9,813,699
Skewness	0.60	0.87
Orthogonality	0.40	0.78
Aspect ratio	17	39

3.2 Mesh Generation

In meshing, a non-conformal unstructured grid having tetrahedral elements was used as shown in Fig. 3. The inflation layer was created near the turbine blades to capture the flow near the blade. Table 2 represents the quality of mesh generated.

3.3 Boundary Condition and Solver Simulation

Boundary conditions used for numerical analysis are given in Table 3. Inlet boundary condition was assigned as velocity inlet (Dirichlet boundary condition) which corresponds to free steam velocity. The outlet boundary condition was set to be outflow. The top of the channel was assigned as symmetry. Side and bottom wall boundaries were set as wall.

A finite volume method (FVM) multiple reference frame (MRF) was applied to a rotating zone and set at an angular velocity to resolve the transient flow problem. Second order Upwind scheme was applied to discretise convective terms and the pressure was interpolated by applying linear interpolation scheme. The diffusive terms were solved using central difference scheme. Pressure velocity coupling was solved using semi-implicit methods for pressure-linked equation (SIMPLE). Convergence criteria were obtained at a residual value of 10^{-4} for each time step. Unsteady Navier Averaged Stokes equation (URANS) was used as the flow solver which is written in the following form.

Table 3 Boundary conditions

Name	2D		3D	
	Boundary type	Boundary condition	Boundary type	Boundary condition
Inlet	Velocity inlet	0.46 m/s	Velocity inlet	1 m/s
Outlet	Outflow	Outflow	Outflow	Outflow
Channel Top surface	–	–	Symmetry	Symmetry
Side and bottom wall	Wall	Wall	Wall	Wall
Turbine	No slip	Rotates at rpm	No slip	Rotates at rpm

$$\frac{\partial u_i}{\partial x_i} = 0 \tag{7}$$

$$\rho\left(\frac{\partial u_i}{\partial t} + \frac{\partial}{\partial x_i}(u_i u_j)\right) = -\frac{\partial p}{\partial x_i} + \frac{\partial}{\partial x_i}(-\rho \bar{u}'_i \overline{u'}_j) + \frac{\partial}{\partial x_j}\left[\mu \frac{\partial u_i}{\partial x_j} + \frac{\partial u_j}{\partial x_i}\right];$$
$$-\rho \bar{u}'_i \overline{u'}_j = \mu_t\left(\frac{\partial u_i}{\partial x_j} + \frac{\partial u_j}{\partial x_i}\right) - \frac{2}{3}(\rho k)\delta_{ij} \tag{8}$$

where = (*u*, *v*, *w*); = (*x y z*).

3.4 Turbulence Model

The selection of the turbulence model depends on Reynolds number and flow geometry. In earlier studies, standard k-ϵ model, realizable k-ϵ model, RNG k-ϵ model, standard k-ω model and K-ω SST model have been used by different researchers. Accordingly, k-ω SST model has been selected in the present study as it is a combination of standard k-ϵ model and standard k-ω. The standard k-ϵ model captures boundary treatment, whereas the standard k-ω model is better to resolve free stream treatment.

The transport equations for k-ω in k-ω SST model are given as

$$\frac{\partial(pk)}{\partial t} + \frac{\partial}{\partial x_i}(pku_i) = \frac{\partial}{\partial x_j}\left(\Gamma_{\mathrm{k}}\frac{\partial \mathrm{k}}{\partial \mathrm{x_j}}\right) + G_k - Y_k \tag{9}$$

$$\frac{\partial(p\omega)}{\partial t} + \frac{\partial}{\partial x_i}(p\omega u_i) = \frac{\partial}{\partial x_j}\left(\Gamma_{\omega}\frac{\partial \omega}{\partial \mathrm{x_j}}\right) + G_\omega - Y_\omega + D_\omega \tag{10}$$

where, Γ_k and Γ_ω are the effective diffusivity for k and ω. G_k is the generation of turbulence kinetic energy due to mean velocity gradients, and G_ω represents the generation of ω. Y_k and Y_ω represent the dissipation of k and ω due to turbulence, while D_ω is the cross-diffusion term.

4 Results and Discussion

In this study, various simulations have been carried out for different TSR ranging from 0.5 to 1.2. It is seen that the quasi-steady time convergence is found after four revolutions in both 2D and 3D cases. Figure 4 shows the time convergence in 3D

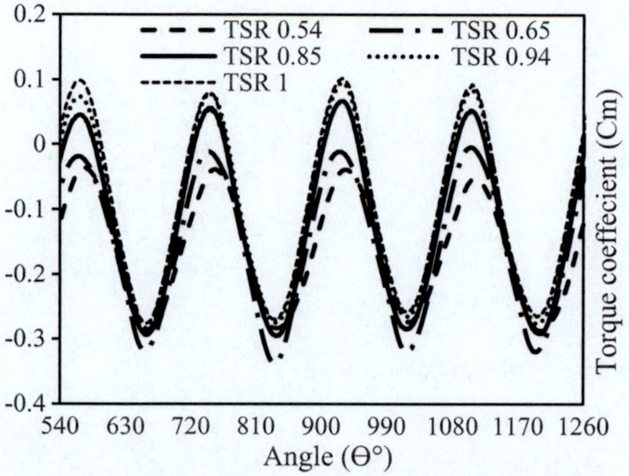

Fig. 4 Quasi-steady convergence

computations. For both 2D and 3D cases, flow distribution is shown corresponding to maximum C_p for all values of parameters considered.

4.1 Velocity Contours, Pressure Contours and Streamlines

Pressure and velocity contour plots predicting the variation of pressure and velocity, respectivey, in various regions near the blades within the flow domain are shown in Fig. 5a, b.

The pressure contour shows that the pressure exerted on the concave side is more prominent than the convex side. The velocity contour shows the velocity gradient near the turbine. The velocity wake is created when the fluid pass a turbine.

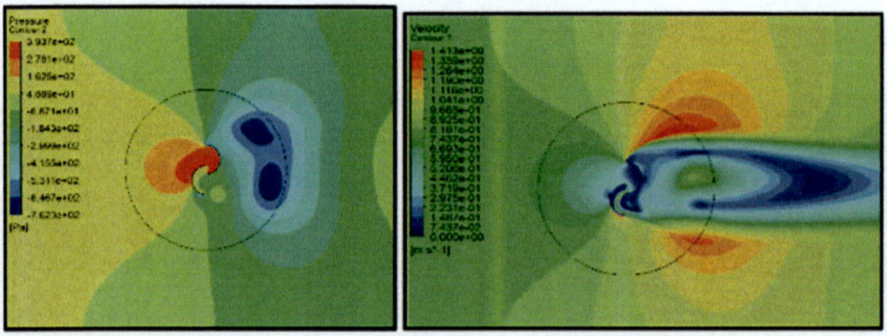

(a) (b)

Fig. 5 **a** Pressure contour and **b** velocity contour of two-bladed Savonius turbine

4.2 Performance Characteristics

In the current work, the computational study of the performance of SHKT is carried out for both 2D and 3D cases and an attempt has been made to compare the power coefficient obtained with the published result. For CFD analyses, the turbine and blade particulars were taken from and compared with the work of Talukdar et al. [1] and Vimal et al. [2] for 2D and 3D computations, respectively. The comparison is shown in Fig. 6a, b. In both cases, the trend of the computed values seems to be following that of the experimental. In 2D, the results appear to be underestimated in 2D computations in comparison with the corresponding experimental values. The deviation may be because in 2D simulation, the effect of height was not taken into consideration. However, in the 3D computations, the results in terms of C_p obtained computationally appear to be slightly higher than the corresponding experimental results. The results illustrate that the flow physics and flow behavior can be captured in the computational simulation.

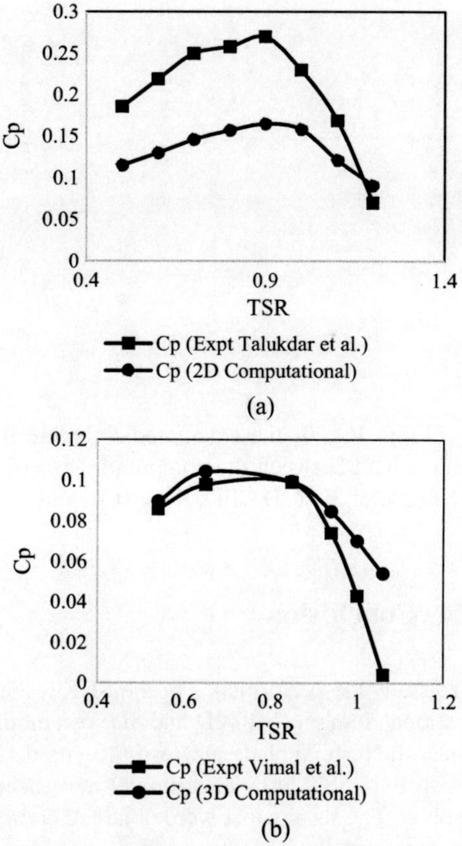

Fig. 6 Comparison between experimental and computational results: **a** 2D; **b** 3D

Fig. 7 Corelation coefficient: **a** 2D; **b** 3D

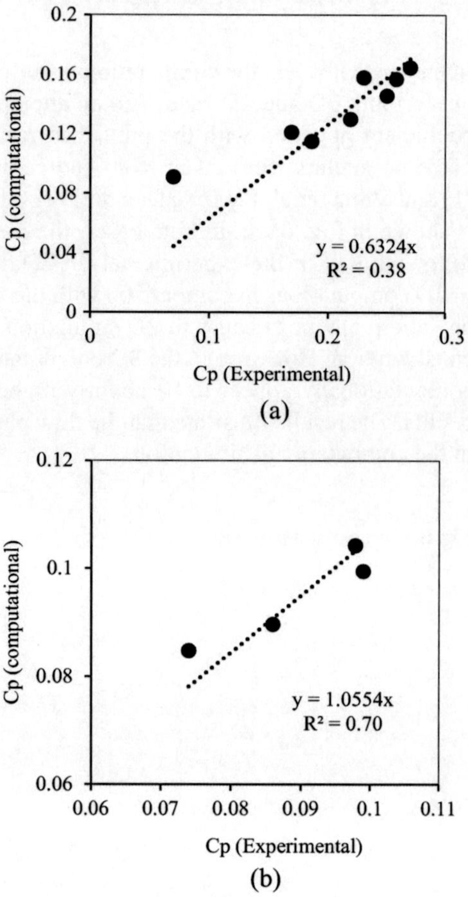

From Fig. 7, the calculated R^2 value further supports that there exists a good agreement between the computational and the experimental results, particularly in the case of 3D (3D CFD, $R^2 = 0.70$ and 2D CFD, $R^2 = 0.38$).

5 Conclusion

To study the performance of the SHKT, the simulation is carried out in the computational domain (both 2D and 3D) resembling an open channel. To capture the flow near the turbine blade, unsteady Reynolds-Averaged Navier–Stokes equation along with K-ω SST turbulence model have been solved through commercial FLUENT solver. The results that were obtained computationally were close to those obtained experimentally. The agreement is more prominent in the case of 3D computations.

The findings of this work suggest that the flow physics and flow behavior can be captured and studied through computational simulation.

References

1. Talukdar, P. K., Sardar, A., Kulkarni, V., & Saha, U. K. (2018). Parametric analysis of model Savonius hydrokinetic turbines through experimental and computational investigations. *Energy Conversion and Management, 158*, 36–49.
2. Patel, V., Bhat, G., Eldho, T. I., & Prabhu, S. V. (2017). Influence of overlap ratio and aspect ratio on the performance of Savonius hydrokinetic turbine. *International Journal of Energy Research, 41*(6), 829–844.
3. Kamoji, M. A., Kedare, S. B., & Prabhu, S. V. (2008). Experimental investigations on single stage, two stage and three stage conventional Savonius rotor. *International Journal of Energy Research, 32*(2008), 877–895.
4. Khan, M. J., Bhuyan, G., Iqbal, M. T., & Quaicoe, J. E. (2009). Hydrokinetic energy conversion systems and assessment of horizontal and vertical axis turbines for river and tidal applications: A technology status review. *Applied Energy, 86*(10), 1823–1835.
5. Gupta, R., Biswas, A., & Sharma, K. K. (2008). Comparative study of a three-bucket Savonius rotor with a combined three-bucket Savonius-three-bladed Darrieus rotor. *Renewable Energy, 33*(9), 1974–1981.
6. Sarma, N. K., Biswas, A., & Misra, R. D. (2014). Experimental and computational evaluation of Savonius hydrokinetic turbine for low velocity condition with comparison to Savonius wind turbine at the same input power. *Energy Conversion and Management, 83*, 88–98.
7. Mabrouki, I., Driss, Z., & Abid, M. S. (2014). Experimental investigation of the height effect of water Savonius rotors. *International Journal of Applied Mechanics, 4*(1), 8–12.
8. Kumar, A., & Saini, R. P. (2017). Performance analysis of a Savonius hydrokinetic turbine having twisted blades. *Renewable Energy, 108*, 502–522.
9. Kailash, G., Eldho, T. I., & Prabhu, S. V. (2012). Performance study of modified savonius water turbine with two deflector plates. *International Journal of Rotating Machinery, 2012.*
10. Kumar, A., & Saini, R. P. (2017). Performance analysis of a single stage modified Savonius hydrokinetic turbine having twisted blades. *Renewable Energy, 113*, 461–478.
11. Patel, C. R., Patel, V. K., Prabhu, S. V., & Eldho, T. I. (2013). Investigation of overlap ratio for Savonius type vertical axis hydro turbine. *2*, 379–383.
12. Pham, L., & Member, S. (2014). Riverine hydrokinetic technology : A review (pp. 1–6). Oregon Tech—Ree516 Term Pap.
13. Jones, B. (1889). *Elements of aerodynamics*. New York: J. Wiley.

Effect of Biomass Blending Ratio and Catalysts in the Kinetic Modelling of Coal and Duckweed Co-Pyrolysis

Munmi Bhattacharyya, Pinakeswar Mahanta, and Kaustubha Mohanty

Nomenclature

E_a	Activation energy (kJ/mol)
A.C.	Ash content
BBR	Biomass blending ratio
m_t	Corresponding mass at time (t) (mg)
daf	Dry-ash-free basis
m_f	Final mass (mg)
F.C.	Fixed carbon
m_o	Initial mass (mg)
mt	Metric tonnes
M.C.	Moisture content
n	Order of reaction
A	Pre-exponential factor (1/min)
k(T)	Rate constant at temperature (T)
R^2	Regression coefficient
T	Temperature (K)
t	Time

M. Bhattacharyya · K. Mohanty
School of Energy Science and Engineering, Indian Institute of Technology Guwahati, Guwahati 781039, India

P. Mahanta (✉)
Department of Mechanical Engineering, Indian Institute of Technology Guwahati, Guwahati 781039, India
e-mail: pinak@iitg.ac.in

NIT Arunachal Pradesh, Arunachal Pradesh, Papum Pare, Yupia 791112, India

K. Mohanty
Department of Chemical Engineering, Indian Institute of Technology Guwahati, Guwahati 781039, India

P. Mahanta et al. (eds.), *Advances in Thermofluids and Renewable Energy*, Lecture Notes in Mechanical Engineering, https://doi.org/10.1007/978-981-16-3497-0_20

R Universal gas constant
V.M. Volatile matter content
w/w Weight/Weight basis

Greek Symbols

$f(\alpha)$ Differential form of reaction model
α Extent of conversion
β Heating rate, K/min
$g(\alpha)$ Integral form of reaction model

Subscripts

onset Initital devolatilization temperature
offset Final devolatilization temperature
max Peak DTG temperature

1 Introduction

Biomass valorization presents as a vital cue to the energy economy since it is carbon neutral, addresses greenhouse gas mitigation if harvested sustainably and casts limited effect on ozone layer due to carbon offsetting. Aquatic biomass confers as an expedient alternative energy source since it is devoid of arable land usage, recovers from extreme temperature, and from nutrient loadings. It also escapes nutrient balance and pH in the environment. Majority of the focus is invested upon algal research but this ponders upon a heavy challenge on high energy requirement for harvestation and its extraction from aqueous medium in dry form. Members of the family *Lemnaceae* include the world's smallest angiosperms (commonly known as duckweeds) render less mechanical separation process for its harvestation, thus proving it to be less energy-intensive. These plants also sequester CO_2 from the atmosphere through consumption of CO_2 to the water through a closed-loop process making it a potent biofuel. There are 37 species of duckweed identified as bioenergy crop worldwide, where majority are found in South-East Asian countries. In the Indian subcontinent, even though duckweed is abundantly available in Odisha, Madhya Pradesh, Maharashtra, Kerela, Thailand, Myanmar and Malaysia, in Assam there are four species (*Landoltia punctata, Lemna perpusilla, Wolffia arrhiza, Spirodela polyrrhiza*) from Kamrup, Barak valley and Darrang districts. In this case, one of them is rare specimen.

Coal is extensively produced in the world with leading producers being India, USA, Australia, China and Indonesia with coal production increased globally by 1.9 (%) in 2018. India being the fifth largest coal reserve in the world has many domestic reserves of coal in the states of Jharkhand (26.06%), Odisha (24.86%), Chattisgarh, West-Bengal and few of the north-eastern states. North-Eastern India contributes 105 mt of Gondwana coal and 1492 mt of tertiary coal reserves [1]. 73% of the tertiary coal reserves is enundated from Meghalaya and Assam coal mining fields [1]. But the coal reserve in the country comes with a heavy price of high ash content and high sulphur content. In such cases, coal washing, coal dewatering and coal demineralization are effective for its usage. Installed coal washing capacity in the country is about 145 mt/annum and is projected to increase to 350 mt by 2022. However, in the state of Meghalaya, the total reserve is 576.48 mt. It is characterized by high ash content (~45%) and high sulphur content (~2–8%), high volatile matter, low ash fusion temperature and high calorific value in comparison to other Indian coals. Co-conversion of such low-grade coal with biomass leads to an energy efficient pathway for production of fuel-pyro gas and by-product char.

In this paper, we propose a kinetic modelling-based (model-fitting) strategy for the thermochemical conversion of coal-duckweed blends. Recent developments in process synthesis strategies for hybrid coal and biomass blends have shown that onerous and computationally efficient methodologies exist to analyse thousands of process reactions simultaneously to extract the optimal process reaction(s) in a multi-step kinetic pathway. Iso-conversional model-fitting determination of activation energy, pre-exponential factors through mathematical function that elucidates the reaction mechanism of coal-duckweed blends by characteristic compositions in a thermal decomposition process is the prime essence of this paper. Coats and Redfern method is used for the elucidation of such kinetic values [2]. The main advantage and significance of using this method is the estimation of order of reaction and activation energy (kJ/mol) at a single stage of reaction at a single heating rate by keeping the heating rate constant. However, precision of this method at low-temperature integral values renders certain relative errors in the decomposition process [3]. Feedstock type and feedstock composition (biomass blending ratio) in coal and duckweed blends are considered measuring parameters to study synergistic interactions amongst the blends and also to establish a parallel link for product distribution profile from co-pyrolysis studies. As an addendum, integration of catalysts such as dolomite and zeolite also pose as an implausible factor for enhancement of product gas yield and bio-oil when further scaled up in a pyrolyzer unit.

The novelty of the paper instils the utilization of a biomass blending strategy to coal and duckweed feedstocks and its comparison in product manifestation through degradation profile studies. Dolomite and zeolite catalysts when integrated into coal-duckweed blends might enhance the product gas yield which has not been studied so far in the literatures as reviewed. Kinetic modelling and optimization through a model-fitting Coats and Redfern method for estimation of activation energy for both non-catalytic and catalytic co-pyrolysis of coal-duckweed blended feedstocks are studied which gives excellent fits for non-isothermal data. However, precision of this particular method for determination of apparent activation energy values without neglecting the low-temperature integral in the above thermal decomposition process is yet to be studied.

2 Materials and Methods

2.1 Materials

In this study, coal and duckweed are the raw materials used for process intensification studies. The coal collected from Meghalaya is of drift origin and sub-bitumunous rank while *Lemna minor* (common name: Duckweed) is procured from water bodies of Amingaon, North Guwahati. Both the coal sample and duckweed are oven-dried at 378 K for 24 hour overnight, and the dried sample is ground and ball milled and sieved in a mortar and pestle to lesser than 300 μm for efficient heat and mass transfer degradation studies. Mixtures of coal-duckweed with biomass blending ratio of (BBR) 0, 33.7, 50, 66.3 and 100 (%) are prepared. Zeolite and dolomite catalysts are also procured commercially for catalytic degradation studies. Both the catalysts are grounded by a mortar and pestle and sieved to lesser than 300 μm particle size. The ultimate analysis (carbon, hydrogen, sulphur and nitrogen) is ascertained using CHNS Elemental Analyzer (Make: Eurovector, Model: Eurovector: EA3000). However, the oxygen content is evaluated by difference as received-basis (Eq. 1)

Oxygen weight (%) as received basis as given in Eq. (1).

$$(\%)\mathrm{O} = 100-(\%\mathrm{C} + \%\mathrm{H} + \%\mathrm{N} + \%\mathrm{S}) \tag{1}$$

2.2 Experimental Methods

To evaluate the pyrolytic behaviour of the raw samples, a Netzsch STA thermogravimetric analyzer (TGA) (Make: Netzsch, Model: STA449F3A00) is employed as shown in Fig. 1. In each pyrolysis experiment, approximately 8.13 mg of the raw samples are loaded into an alumina ceramic crucible and heated from ambient temperature to 1173 K at a single heating rate 10 K min^{-1}, respectively. On the basis of previous literature, a constant Argon flow rate of 60 mL min^{-1} is maintained in this study to avoid the influence on the weight measurement, buoyancy and heat transfer limitation. However, consequently 5% (w/w) dolomite and 5% (w/w) Zeolite is impregnated to coal-duckweed blend with 50 (%) BBR to estimate catalytic degradation pattern. Prior to the thermogravimetric experiments, the catalysts are calcined at 873 K for 4 hours in a muffle furnace to activate (reduce) its metal active sites.

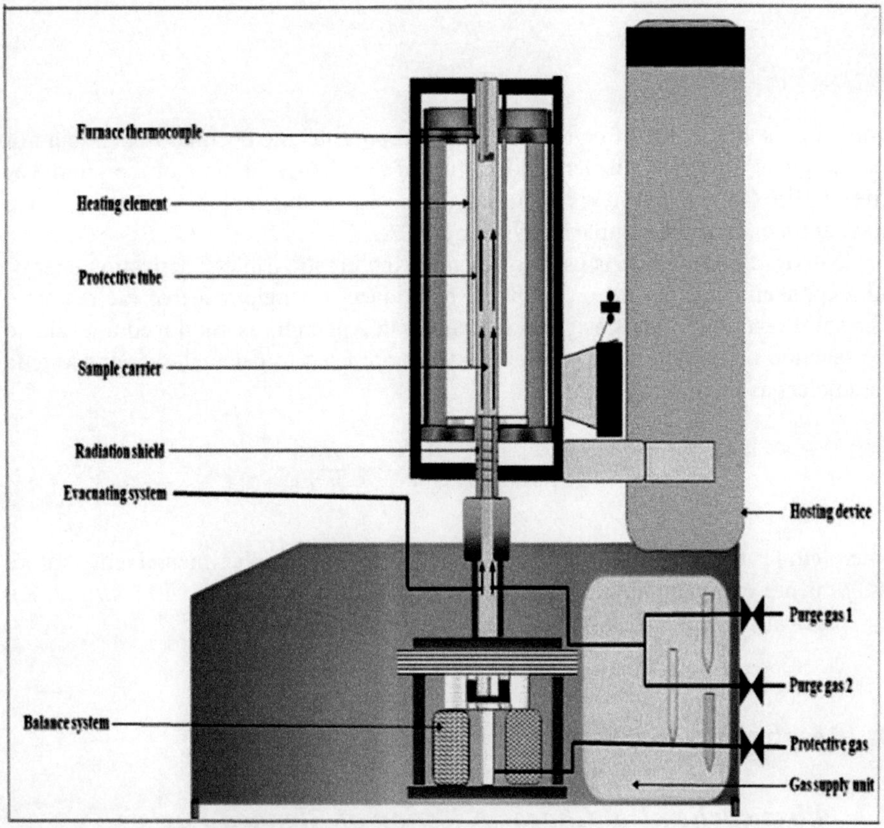

Fig. 1 NETSZCH-STA thermogravimetric analyzer

2.3 Kinetic Theory

For the study of heterogeneous reaction kinetics, Arrhenius expression is used to correlate the rates of mass loss with temperature as given in Eqs. (2, 3).

$$k\,(T) = Ae^{\frac{-Ea}{R.T}} \tag{2}$$

$$\ln k = \ln A - \frac{Ea}{R.T} \tag{3}$$

where E_a is the activation energy (kJ/mol), A is the pre-exponential factor (1/min), $g(\alpha)$ and $f(\alpha)$ are functions of (α) depending on the reaction mechanism, k is the rate constant at temperature T (K), which generally obeys the Arrhenius equation.

$$\propto = \frac{m_0 - m_t}{m_0 - m_f} \tag{4}$$

where (α) is the degree of conversion which represents the decomposed amount of the sample at time *t*, and is defined in terms of the change in mass of the sample as given in Eq. (4): where (m_0) is the initial mass, (m_f) is the final mass and (m_t) is the mass at time (*t*) of the sample analysed by TGA.

Coats-Redfern method is used to calculate the kinetic triplets: activation energy, pre-exponential factor and order of the reaction in a single stage of the reaction. The solid-reaction models as given in Table 6 in Appendix is employed to evaluate the reaction order. The following expressions are used to calculate various kinetic parameters as mentioned in Eq. (5).

$$\ln \frac{g(\alpha)}{T^2} = \ln \frac{A.R}{\beta.\,Ea} - \frac{Ea}{R.T} \tag{5}$$

where $g(\alpha)$ is the integral form of equation, *T* is the mean experimental temperature (K), *A* is pre-exponential factor (1/min), *R* is gas constant (8.314×10^{-3} kJ/mol. K), E_a is activation energy (kJ/mol).

3 Results and Discussion

3.1 Physicochemical Characterization of Blended Fuels

The physicochemical characterization of the blended fuels is mentioned in Table 1 where the ultimate analysis is given. It is observed that coal has highest carbon content 61.945 (%) followed by the blend with 33.7 (%) BBR which is 41.29 (%). Blend with 50 (%) BBR has a carbon content of 39.4 (%) and blend with 66.3% BBR has carbon content of 35.6 (%). Duckweed has the lowest carbon content 29.789 (%). The hydrogen content of the feedstocks is higher for blend with 50% BBR which is 4.65%, followed by duckweed 4.397% and blend with 33.7 % BBR which is 3.231%. However, blend with 66.3% BBR has a hydrogen content of 2.62% followed by coal

Table 1 Ultimate analysis of the blended fuels

Feedstock	C (%)	H (%)	N (%)	S (%)	O (%)
Coal	61.9	0.4	1.46	1.4	34.75
Blend (33.7% BBR)	41.2	3.2	0.97	–	54.5
Blend (50% BBR)	39.4	4.6	0.65	–	55.29
Blend (66.3% BBR)	35.6	2.6	0.75	–	61.02
Duckweed (100% BBR)	29.7	4.3	2.21	2.1	61.44

with 0.411%. Nitrogen content in coal is 1.469 (%) while duckweed is 2.215 (%), while for the blends is (blend with 33.7 (%) BBR: 0.979 (%), blend with 50 (%) BBR: 0.656 (%), blend with 66.3 (%) BBR: 0.759 (%)). Sulphur content is seen in coal 1.425 (%) and duckweed 2.151 (%). Oxygen is the difference of the rest of the elements from 100, and it increases linearly with the blending ratio with the highest value for duckweed. Oxygen content is 34.75 (%) for coal, blend with 33.7 (%) BBR is 54.5 (%), blend with 50% BBR is 55.29 (%), blend with 66.3 (%) BBR is 61.02 (%) and duckweed with 100 (%) BBR is 61.44 (%).

3.2 *Influence of Biomass Blending Ratio*

The proportion of biomass in the blend has a significant influence on product distribution of solid, and gas [4]. With the increase of biomass blending ratio, the char yield decreases, while the yields of gas increase [5]. Typical TG and DTG curves for duckweed, coal and their mixtures are presented in Figs. 2 and 3. Char yield of the fuels is (Coal: 85.3 (%), Coal-duckweed blend with 33.7 (%) BBR: 79.54 (%), Coal-duckweed blend with 50% BBR: 75.35 (%), Coal-duckweed blend with 66.3 (%) BBR:48.35 (%) and Duckweed: 35.21 (%)), while gas yields are (Coal: 10.09 (%), Coal-duckweed blend with 33.7 (%) BBR: 20.46 (%), Coal-duckweed blend with 50 (%) BBR: 24.65 (%), Coal-duckweed blend with 66.3 (%) BBR: 33.7 (%)

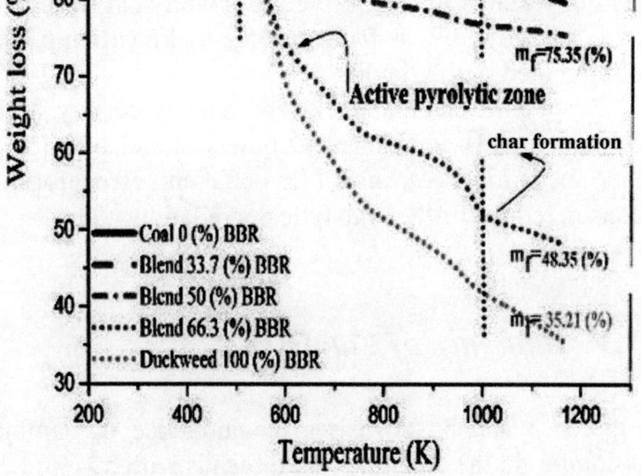

Fig. 2 TGA curve of coal-duckweed blends with variation in biomass blending ratio

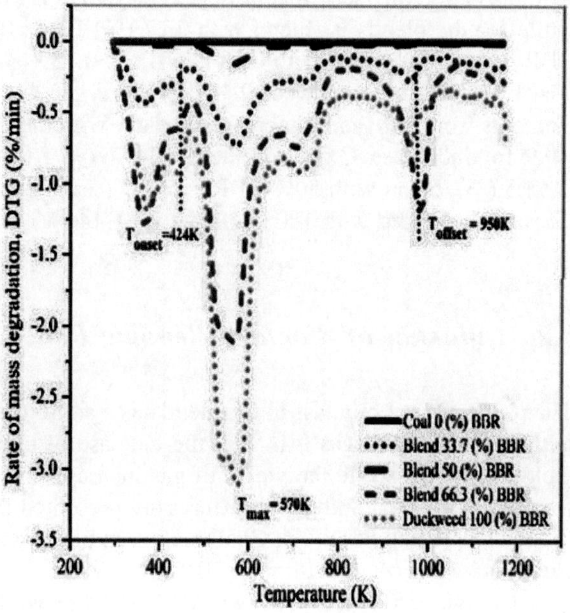

Fig. 3 DTG curve of coal-duckweed blends with variation in biomass blending ratio

and Duckweed: 64.79 (%)). As observed from Fig. 3, coal has very less thermal degradation curve (DTG curve) due to its higher density, higher ash content and less volatile matter while duckweed has a predominant degradation zone due to its higher volatile matter content. This correlates to a higher thermal stability of coal than duckweed biomass. This states that the biomass blending strategy in coal renders a higher degradation pattern in the feedstocks by decreasing the thermal stability of the coal. The DTG curves evaluate that T_{onset} is 424 K and T_{offset} is 950 K for blends with coal, where T_{onset} is temperature of biomass devolatilization and T_{offset} is temperature of coal devolatilization. Thus, increasing biomass content in blends decreases the per cent of residual mass. Co-pyrolysis with coal would also generate better quality of pyrolytic oil to use as fuel since the hydrocarbon present in coal ensures formation of upgraded pyrolytic oil.

Despite the fact that duckweed is a low-density, high volatile matter, less sulphur content, high H-content and CO_2 neutral fuel, addition of more biomass would make thermochemical conversion less efficient. Henceforth, the blend with 50% BBR is considered for further catalytic degradation studies.

3.3 Influence of Catalysts

Figures 4 and 5 showcases the influence of commercial catalysts (zeolite and dolomite) in the coal-duckweed blends with 50 (%) BBR via TG-DTG curves. The

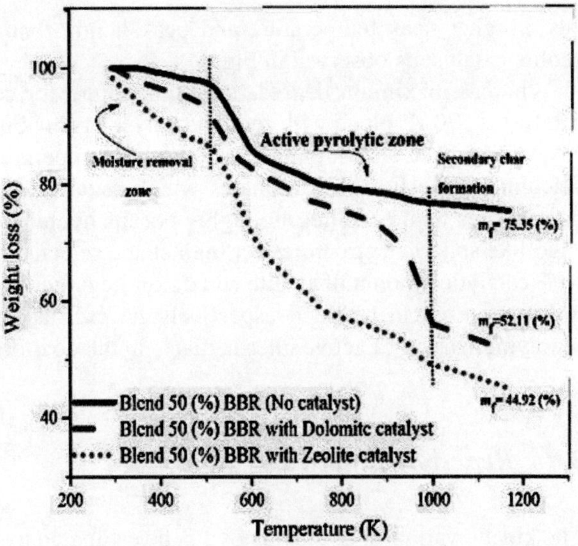

Fig. 4 TGA curve of coal-duckweed blend embedded with dolomite and zeolite catalyst

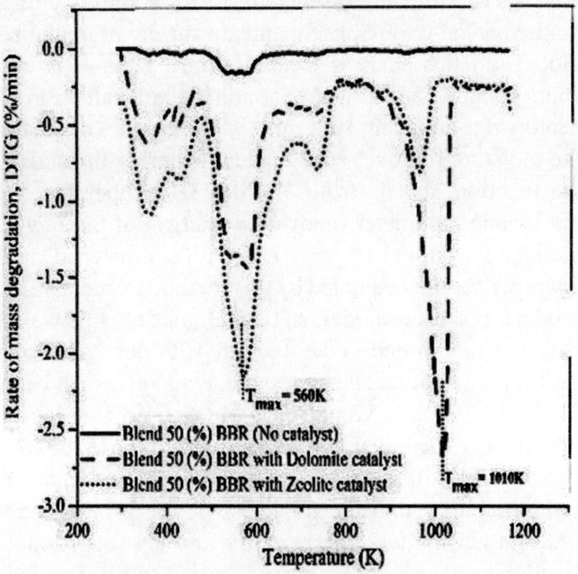

Fig. 5 DTG of Coal-duckweed blend embedded with dolomite and zeolite catalyst

char yields in fuels are Coal-duckweed blend with 50 (%) BBR: 75.35 (%), Coal-duckweed blend with 5 w/w (%) dolomite: 52.21 (%), Coal-duckweed blend with 5 w/w (%) Zeolite: 44.98 (%) while gas yields are Coal-duckweed blend with 50 (%) BBR: 24.65 (%), Coal-duckweed blend with 5 (%) dolomite: 47.79 (%), Coal-duckweed blend with 5% Zeolite: 55.02 (%). The blend impregnated with dolomite has a higher peak temperature and peak height than the blend impregnated with zeolite catalyst as observed in Fig. 5.

Whereas maximum degradation temperature for coal-duckweed blend with no catalyst is 570 K, blend with zeolite catalyst has maximum degradation temperature (T_{max}) of 560 K and dolomite has a higher temperature of 1010 K. Dolomites are calcium magnesium bi-carbonates which require calcination for activation of its porous sites while zeolites are highly porous hydrated aluminosilicates which have cage-like spherical structures for high shape selectivity. An optimum presence of 5 wt% catalytic amount of zeolite and dolomite indicates catalytic cracking of various linkages present in the blend respectively. It would also optimize the side cracking and repolymerization of active intermediates in the coal-duckweed blend respectively.

3.4 Reaction Kinetics

The kinetic variables of the blends can be estimated by kinetic modelling where iso-conversional techniques can be employed to model the co-pyrolysis process. Kinetic analysis captures the essence of the exceedingly complex reaction set in a tractable mathematical way. Consequently, a variety of mathematical techniques, often called global kinetics such as Coats-Redfern method is used to calculate the activation energy values and pre-exponential factor values in a single stage by keeping the heating rate constant, 10 K min^{-1}. The kinetics of the feedstocks is deduced excluding the moisture removal zone, and considering the active pyrolytic zone till the end of the reaction, that is from 424–1173K temperature. Tables 2, 3, 4 and 5 features the kinetic parameter (activation energy) of the raw fuels, blended feedstocks, and blended feedstocks impregnated with commercial catalyst of 5 wt (%) dolomite and 5 wt (%) Zeolite evaluated by this method, respectively, for all the solid-state reaction models. For each model, statistical goodness fit was determined by the correlation value, as mentioned in the respective Tables 2, 3, 4 and 5.

Tables 2, 3, 4 and 5 suggests that coal gives the best fit of 0.7739 regression value for 1-D Diffusion Model of 20.48 (kJ/mol) while duckweed gives the best fit of 0.9629 regression value for Mampel first order of 22.53 (kJ/mol) activation energy values. 1-D Diffusion Model suggests that the rate of product formation decreases proportionally with the thickness of the product barrier layer as seen in coal. In order-based models, the reaction rate is proportional to concentration, amount or fraction remaining of reactant(s) raised to a particular power (integral or fractional) [6]. Mampel first order is a special case of Avrami-Erofeev model where $n = 1$ while $n = 2$ for second order model [6]. The blended feedstocks render regression coefficient values of 0.9792 for 33.7 (%) BBR coal-duckweed blend, 0.9675 for 50 (%) BBR coal-duckweed blend and 0.9629 for 66.3 (%) BBR for coal-duckweed blend with reaction orders of Mampel first order for both 33.7 (%) BBR blend and 66.3 (%) BBR blended feedstock and second order for 50 (%) BBR blend, respectively. Activation energy values are 17.6 (kJ/mol) and 18.22 (kJ/mol) for 33.7 (%) BBR blend and 66.3 (%) BBR blend, respectively, for 50 (%) BBR blend it is 37.67 (kJ/mol) which is

Table 2 Estimation of E_a (kJ/mol) for coal and duckweed raw feedstock

Reaction Model	Coal		Duckweed	
	E_a (kJ/mol)	R^2	E_a (kJ/mol)	R^2
P2, $n = 1/2$	10.24	0.7739	16.63	0.8950
P3, $n = 1/3$	10.24	0.7739	16.63	0.8950
P4, $n = 1/4$	10.24	0.7739	16.63	0.8950
A4 Model	12.44	0.6954	22.53	0.9629
A3 Model	12.44	0.6954	22.53	0.9629
A2 Model	12.44	0.6954	22.53	0.9629
1-D Model	20.48	0.7739	33.20	0.9023
3-D Model	−11.72	0.2850	3.473	0.9325
Crank Model	1.125	0.2988	−2.499	0.7668
Anti-Jander equation	–	–	–	–
Z-L-T equation	3.04	0.4398	-5.649	0.8956
F0 Model	10.24	0.7739	16.63	0.8950
F1 Model	12.44	0.6954	22.53	0.9629
F2 Model	15.57	0.5914	32.06	0.9356
R2 Model	1.777	0.5335	2.985	0.94012
R3 Model	1.015	0.5123	1.736	0.9325

Table 3 Estimation of E_a (kJ/mol) for coal-duckweed blends- 33.7 (%) and 66.3 (%) BBR

Reaction Model	Blend 33.7 (%) BBR		Blend 66.3 (%) BBR	
	E_a (kJ/mol)	R^2	E_a (kJ/mol)	R^2
P2, $n = 1/2$	13.25	0.9446	13.48	0.9328
P3, $n = 1/3$	13.25	0.9446	13.48	0.9328
P4, $n = 1/4$	13.25	0.9446	13.48	0.9328
A4 Model	17.6	0.9792	18.22	0.9740
A3 Model	17.6	0.9792	18.22	0.9740
A2 Model	17.6	0.9792	18.22	0.9740
1-D Model	26.51	0.9446	26.48	0.9342
3-D Model	−23.08	0.6658	−25.82	0.6303
Crank Model	−2.34	0.7489	−2.45	0.7484
Anti-Jander equation	–	–	–	–
Z-L–T equation	17.6	0.9792	−5.30	0.8931
F0 Model	13.25	0.9446	13.48	0.9328
F1 Model	17.6	0.9792	18.22	0.9740
F2 Model	24.27	0.9266	25.90	0.9041
R2 Model	2.67	0.949	2.69	0.9515
R3 Model	1.56	0.9394	1.58	0.9420

Table 4 Estimation of E_a (kJ/mol) for coal-duckweed blend with 50 (%) BBR

Reaction Model	Blend 50 (%) BBR	
	E_a (kJ/mol)	R^2
P2, $n = 1/2$	18.14	0.8837
P3, $n = 1/3$	18.14	0.8837
P4, $n = 1/4$	18.14	0.8837
A4 Model	25.39	0.9644
A3 Model	25.39	0.9644
A2 Model	25.39	0.9644
1-D Model	35.53	0.8310
3-D Model	−30.99	0.7395
Crank Model	−2.83	0.8163
Anti-Jander equation	–	–
Z-L-T equation	−6.14	0.8954
F0 Model	18.14	0.8837
Fl Model	25.39	0.9644
F2 Model	37.67	0.9675
R2 Model	3.18	0.91562
R3 Model	1.85	0.9117

slightly higher than prior two. 50 (%) BBR blend impregnated with dolomite catalyst decomposition profile features a goodness fit of 0.9398, whereas blend with zeolite catalyst gives 0.9786, both imbibing zero-order reaction mechanism for the best statistical goodness fit. Therefore through model-fitting procedure, implementation of catalyst dolomite into 50 (%) BBR blend gives an activation energy of 16.31 (kJ/mol) while zeolite gives 10.31 (kJ/mol) values which are slightly lower than the non-catalytic pyrolysis feedstock. The lower values of activation energy for catalytic pyrolysis of coal-duckweed blend might be due to a vivid mechanistic pathway in pyrolysis due to the presence of the catalyst. Zeolite catalyst favors catalytic cracking reactions under pyrolysis conditions in addition to the reactions that are occurring when the catalyst is absent.

4 Conclusions

It can be concluded that coal-duckweed blend proves to be a viable feedstock for co-pyrolysis. The blends with various biomass blending ratios depict decrease in carbon content with increase in blending along with no sulphur content in the blends except coal and duckweed raw feedstock. Coal render reaction mechanism as 1-D diffusion model and duckweed as a first-order reaction when best statistical fit is

Table 5 Estimation of E_a (kJ/mol) for coal-duckweed blend with 50 (%) BBR integrated with dolomite and zeolite catalyst

Reaction Model	Blend with 50 (%) BBR with dolomite catalyst		Blend with 50 (%) BBR with zeolite catalyst	
	E_a (kJ/mol)	R^2	E_a (kJ/mol)	R^2
P2, $n = 1/2$	16.31	0.9398	−7.54	0.3073
P3, $n = 1/3$	16.31	0.9398	−37.13	0.5539
P4, $n = 1/4$	16.31	0.9398	51.39	0.9699
A4 Model	20.93	0.9366	−9.78	0.3329
A3 Model	20.93	0.9366	−6.22	0.3558
A2 Model	20.93	0.9366	60	0.9711
1-D Model	33.12	0.9331	102.79	0.9699
3-D Model	−21.66	0.4415	0.57	0.0018
Crank Model	−1.92	0.4921	0.49	0.0687
Anti-Jander equation	–	–	–	–
Z-L-T equation	−4.68	0.7318	−5.29	0.9016
F0 Model	16.31	0.9398	10.3	0.9786
F1 Model	20.93	0.9366	15.14	0.9768
F2 Model	28.44	0.8130	23.10	0.8774
R2 Model	13.59	0.8933	7.61	0.9385
R3 Model	1.49	0.8462	1.53	0.9427

obtained through Coats-Redfern method. However, the blended feedstocks with 33.7 (%) BBR and 66.3 (%) BBR shows first order, except 50 (%) BBR blend displays second order reaction mechanism for the best statistical fit analysis. While blends with catalyst dolomite and zeolite show zero-order kinetics. Coal and duckweed have similar activation energy values while 50 (%) BBR has the highest activation energy value for the best regression fit curve. Blend with 50 (%) BBR when impregnated with dolomite and zeolite catalyst results in further lower activation energy values. However, increasing the biomass blending ratio and integrating catalysts into the mixture enhances its product yields by maximizing its gas production.

Acknowledgements The authors acknowledge the help received by the Central Instrumentation Facility, School of Energy Science and Engineering, Department of Mechanical Engineering and Department of Chemical Engineering, Indian Institute of Technology Guwahati and Guwahati Biotech Park for carrying out various experiments and analyses.

Appendix

See Table 6.

Table 6 Representation of solid-state kinetic models [6]

No.	Reaction model	Differential form $f(\alpha) = 1/k\ dR/dt$	Integral form $g(\alpha) = kt$
1	*Power Law Model (Exponential nucleation)*		
(i)	Power Law Model, P2, $n = 1/2$	$2\alpha^{1/2}$	$\alpha^{1/2}$
(ii)	Power Law Model, P3, $n = 1/3$	$3\alpha^{2/3}$	$\alpha^{1/3}$
(iii)	Power Law Model, P4, $n = 1/4$	$4\alpha^{3/4}$	$\alpha^{1/4}$
2	*Avrami-Erofeev Model (Random nucleation and nuclei growth)*		
(i)	Avrami-Erofeev, A4	$4(1 - \alpha)\ [-\ln(1 - \alpha)]^{3/4}$	$[-\ln(1 - \alpha)]^{1/4}$
(ii)	Avrami-Erofeev, A3	$3(1 - \alpha)\ [-\ln(1 - \alpha)]^{2/3}$	$[-\ln(1 - \alpha)]^{1/3}$
(iii)	Avrami-Erofeev, A2	$2(1 - \alpha)\ [-\ln(1 - \alpha)]^{1/2}$	$[-\ln(1 - \alpha)]^{1/2}$
3	*Diffusion Model*		
(i)	1-D Diffusion Model	$(½)\ \alpha^{-1}$	α^2
(ii)	3-D Diffusion Model (Janders)	$2(1 - \alpha)^{2/3}[1 - (1 - \alpha)^{1/3}] - 1$	$[1 - (1 - \alpha)$^1/3$]^2$
(iii)	Diffusion Control (Crank)	$3/2[((1 - \alpha)^{-1/3}) - 1]^{-1}$	$1 - 2\alpha/3 - (1 - \alpha)^{2/3}$
iv)	Anti−Jander equation	$3/2(1 - \alpha)^{4/3}[(1 - \alpha)^{1/3} - 1]^{-1}$	$3/2(1 + \alpha)^{2/3}[(1 + \alpha)^{1/3} - 1]^{-1}$
(v)	Z-L-T equation	$[(1 - \alpha)^{-1/3} - 1]^2$	$[(1 + \alpha)^{1/3} - 1]^2$
4	*Reaction Order*		
(i)	Zero Order, F0	1	α
(ii)	Mampel First Order, F1	$(1 - \alpha)$	$-\ln(1 - \alpha)$
(iii)	Second Order, F2	$(1 - \alpha)^2$	$((1 - \alpha)^{-1}) - 1$
5	*Geometrical Contraction Model (Limiting surface reaction between two phases)*		
(i)	Contracting Cylinder, R2	$2(1 - \alpha)^{1/2}$	$(1 - (1 - \alpha))^{1/2}$
(ii)	Contracting Sphere, R3	$3(1 - \alpha)^{2/3}$	$(1 - (1 - \alpha)^{1/3}$

References

1. Chabukdhara, M., & Singh, O.P. (2016). *International Journal of Coal Science & Technology, 3*, 87–96.
2. Vyazovkin, S., & Wight, C. A. (1999). *Thermochimica Acta, 340–341*, 53–68.
3. Cai, J., & Bi, L. (2008). *Energy and Fuels, 22*, 2172–2174.
4. Aboyade, A. O., Carrier, M., Meyer, E. L., Knoetze, H., & Görgens, J. F. (2013). *Energy Conversion and Management., 65*, 198–207.
5. Guo, M., & Bi, J. C. (2015). *Fuel Processing Technology., 138*, 743–749.
6. Khawam, A., & Flanagan, D. R. (2006). *The Journal of Physical Chemistry B, 110*(35), 17315–17328.

Integral Sliding Mode Controller Design for Francis Turbine Electrohydraulic IGV System

J. Vinod and Bikash Kr. Sarkar

1 Introduction

Hydro turbine power generation system is most widely used among all other renewable sources due to its flexibility to load variation, minor impact on environment and low cost by high-value power etc. Hydropower potential is abundant and can replace as an alternative instead of the shortage and high-polluting nonrenewable resources. Francis turbines are widely employed compare to other hydroturbines as it suits efficiently on the wide range of water flow and head, Vinod et al., Vinod et al. and Liang et al. [1–3]. Hydro turbine system consists of headrace, penstock, casing, turbine runner, draft tube and tail race, through water flows from high head, Vinod et al. [1] to low-head converting water energy to rotating mechanical energy. The rotating mechanical energy is converted to electrical energy through an alternator or generator, which has been connected to turbine shaft. The turbineinlet guide vane (IGV) along with actuating system is responsible to regulate amount of water flow through the turbine, Vinod et al., Vinod et al. and Liang et al. [1–3]. The amount of water flow through the turbine controls the mechanical power generation, which should be equivalent to the demand electrical power to maintain the quality of the generated power at desired frequency.

The hydro turbine governing system is quite complex due to non-minimal phase characteristic, complex coupling between hydraulic, mechanical and electrical system, Jiangbin et al., Yang et al., Zhang et al., Li et al. and Wu et al. [4–8]. A precise modeling of the hydro turbine governing system is important as it relates the system stability. A gate opening transfer fuction model with one pole and one zero has been reported in most of the previous researches without a detailed modeling of the IGV, Vinod et al., Vinod et al. and Liang et al. [1–3] actuating system, as the IGV blade opening and closing for the corresponding water flow dynamics. Stability

J. Vinod · B. Kr. Sarkar (✉)
Department of Mechanical Engineering, NIT Meghalaya, Shillong, India
e-mail: bikash.sarkar@nitm.ac.in

P. Mahanta et al. (eds.), *Advances in Thermofluids and Renewable Energy*, Lecture Notes in Mechanical Engineering, https://doi.org/10.1007/978-981-16-3497-0_21

of the hydro turbine governing system is another important concern. Some of the recent reported research concentrated on the stability analysis of the hydro turbine governing system, Yang et al., Huang et al. and IEEE Working Group [9–11].

Various controller design approaches are implemented to improve the turbine system performance. Starting from conventional proportional integral derivative (PID) controller, Wang et al. [12] with droop setting for hydro turbine governing system various controller design approach has been implemented for turbine governing system. Recently, various nonlinear controller design approach have been implemented for turbine IGV system. Fuzzy, Chamorroa et al. [13] based controller, artificial neural network (ANN), Dasmahapatra et al. [14] based controller, fixed time sliding mode controller, Yang et al., Huang et al. and IEEE Working Group [9–11] has been implemented for hydro turbine governing system. Researchers have attempted the controller design approach through turbine modeling, neglecting the detailed dynamic model of the IGV actuating system, which is really an integrated part of the IGV ring gate.

Implementation of the controller design through a detailed dynamic modeling of the actuating system is quite justified as it is directly responsible for IGV angular position, Vinod et al., Vinod et al. and Liang et al. [1–3]. The electrohydraulic IGV actuating system have nonlinear characteristic with parametric uncertainty, unknown external disturbance etc. Robust controller design approach has become quite essential for turbine IGV system for save operation Vinod et al., Vinod et al. and Liang et al. [1–3].

In the present study, a detailed modeling of the electrohydraulic IGV actuating system has been utilized forturbine governor control application. The robust integral sliding mode controller has been proposed for the turbine IGV position control application. The controller performance has been studied with the sinusoidal tracking demand of the IGV angular position with unknown disturbance in water flow over IGV, internal parametric uncertainty. The simulation study has been carried out in MatlabSimulink environment.

2 Francis Turbine IGV System Dynamic Modeling

Schematic diagram of the Francis turbine IGV system for closing and opening condition has been shown in Fig. 1a, b, respectively. The IGV blades are connected over the ring, which has been connected with a double acting, single rod linear hydraulic actuator. The linear movement of the actuator provides angular movement of the ring, finally provides angular movement of the IGV blade. The angular position of the IGV blades maintains the projected passage area of water flow. The variation of the IGV angular position, which regulates the flow area, i.e., flow rate through the turbine.

The mathematical modeling of the electrohydraulic system present as, Sarkar et al. and Turton [15, 16]

Fig. 1 Schematic diagram of the Francis turbine control by electrohydraulicactuating system, IGV a closed condition b open condition

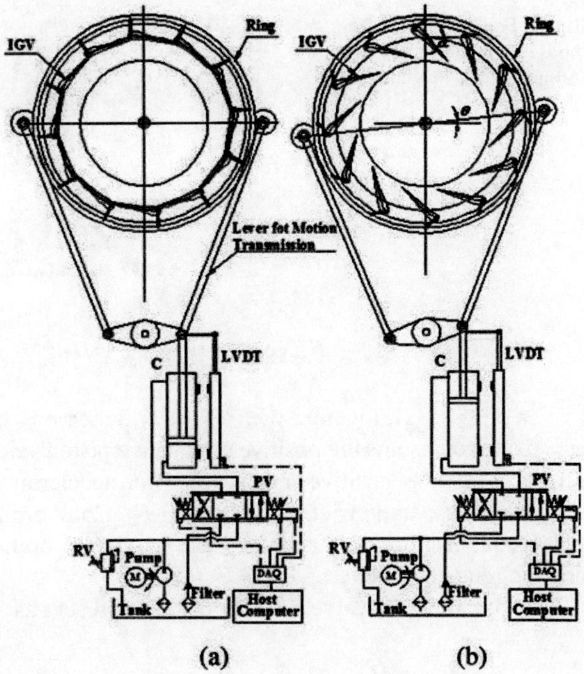

$$\dot{y} = v \tag{1a}$$

$$\dot{v} = a \tag{1b}$$

$$\dot{v} = (P_1 A_{a1} - P_2 A_{a2} - F_f - F_l - F_i)/m_a \tag{2}$$

$$\dot{P}_1 = \beta(Q_1 - A_{a1} v)/(V_{10} + A_{a1} y) \tag{3a}$$

and

$$\dot{P}_2 = \beta(A_{a2} v - Q_2)/(V_{20} - A_{a2} y) \tag{3b}$$

where, y and v are the piston displacement and velocity, respectively, P_1 and P_2 are the pressure, A_{a1} and A_{a2} are the effective piston area, m_a is the moving mass, F_f is the piston friction, β is the oil compressibility, V_{10} and V_{20} are the initial volume of chamber, subscripts 1 and 2 represent for the chambers of the hydraulic cylinder (Fig. 2).

The piston friction characteristic, Sarkar [15] of the cylinder represent as,

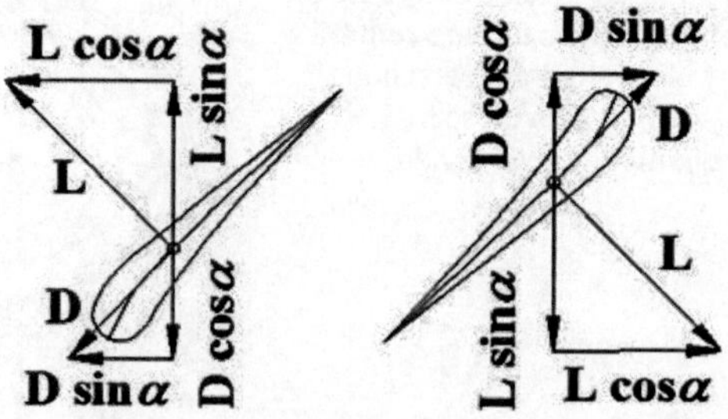

Fig. 2 Forces diagram of the IGV of the Francis turbine [2, 16, 17]

$$F_f = F_{c\varphi}^{\gamma} + (F_{b\varphi}^{\gamma} - F_{c\varphi}^{\gamma}) \exp\{-(v - v_{b\varphi}^{\gamma})^2/(v_{s\varphi}^{\gamma} - v_{b\varphi}^{\gamma})^2\} + \alpha_{v\varphi}'^{\gamma}(v - v_{0\varphi}^{\gamma}) \quad (4)$$

For $|v| \geq \left|v_{0\varphi}^{\gamma}\right|$, together with $F_f = F_t$, when $v = 0$ and $|F_t| \leq \left|F_{0\varphi}^{\gamma}\right|$. Subscript $\phi = \text{p}$ or n represents the positive or negative piston velocity, respectively, and superscripts $\gamma = +$ or $-$ positive or negative piston acceleration, respectively. The terms are represented as, static friction or stiction F_0, Coulomb friction F_c, Stribeck velocity v_s, viscous friction coefficient α_v, the maximum boundary lubrication friction, F_b and the related velocity v_b.

The dynamics of valve flow for piston represent as

$$Q_1 = (\bar{C}_{v1}e + C_l)\sqrt{P_P - P_1} - C_l\sqrt{P_1 - P_T}, \quad \text{for } e > e_0 \quad (5a)$$

$$Q_2 = (\bar{C}_{v4}e + C_l)\sqrt{P_2 - P_T} - C_l\sqrt{P_P - P_2}), \quad \text{for } e > e_0 \quad (5b)$$

$$Q_1 = (\bar{C}_{v2}e + C_l)\sqrt{P_1 - P_T} - C_l\sqrt{P_1 - P_T}), \quad \text{for } e < e_0 \quad (5c)$$

$$Q_2 = (\bar{C}_{v3}e + C_l)\sqrt{P_P - P_2} - C_l\sqrt{P_P - P_2}), \quad \text{for } e < e_0 \quad (5d)$$

where C_l is the leakage coefficient and $\bar{C}_{v1}, \bar{C}_{v2}, \bar{C}_{v3}$ and $\bar{C}_{v4}$ are the valve coefficients.

The resultant flow force of the IGV, Vinod et al. and Sarkar et al. [2, 15] will produce torque on the ring gate during steady state condition and transient condition. To neutralize that torque, an opposite torque should be generated by the hydraulic cylinder. Force required by each cylinder to hold the ring gate is,

$$F_l = 6D_{gc}\rho V^2 cbC_L[\sin\alpha - 0.005\cos\alpha]/2D_a, \quad (6a)$$

$$C_D/C_L = 0.05 \quad (6b)$$

Considering the aerofoil section of the IGV and ring gate with 6 blades. To simplify the analysis, consider $C_L = 1.1$ (4% camber).

Force due to the Inertia of the Ring Gate:

$$T = I\alpha \tag{7a}$$

$$F_i = I\ddot{\theta}/R_a \tag{7b}$$

Considering the linear relation between the rotation of the ring gate, θ and the IGV angle, α hence

$$\alpha = k\theta \tag{8}$$

Relation between piston movement and the ring gate rotation, θ is

$$y = \theta(D_a/2) \quad \dot{y} = \dot{\theta}(D_a/2) \quad \ddot{y} = \ddot{\theta}(D_a/2) \tag{9}$$

$$y = \theta(D_a/2) \quad \dot{y} = \dot{\theta}(D_a/2) \quad \ddot{y} = \ddot{\theta}(D_a/2) \tag{10}$$

$D_{a/2}$ is the distance between piston head to the center of the ring gate.
Equations (1)–(10) together with we can write,

$$\begin{aligned}\ddot{\alpha} = 2/Dka[(P_1A_{a1} - P_2A_{a2} - F^{\gamma}_{c\varphi} + (F^{\gamma}_{b\varphi} - F^{\gamma}_{c\varphi})\\ \exp\{-(\dot{\alpha}kDa/2 - v^{\gamma}_{b\varphi})^2/(v^{\gamma}_{s\varphi} - v^{\gamma}_{b\varphi})^2\}\\ + \alpha'^{\gamma}_{v\varphi}(\dot{\alpha}kDa/2 - v^{\gamma}_{0\varphi}) - F_l - F_i)/m_a]\end{aligned} \tag{11}$$

3 Controller Design

Proposed controller block diagram has been shown in Fig. 3. The controller has been constructed based on the definition of sliding surface and Lyapnov function. The

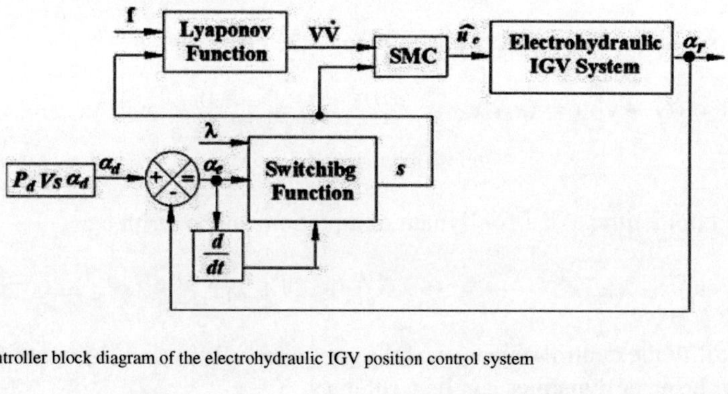

Fig. 3 Controller block diagram of the electrohydraulic IGV position control system

stability of controller ia ensured by the definition of Lyaponov function and consideration of bounded disturbance. The dynamic equation of IGV actuating system Eq. (11) can be represented, Yang et al., Huang et al., IEEE Working Group and Turton [9–11, 16] as,

$$x^{(n)} = g(x, e, t) + d(x, t) \tag{12a}$$

involving a state vector

$$x(t) = (x_1 \quad x_2 = x_1^{(1)} \ldots x_i = x_1^{(i-1)} \cdots \ x_1^{(n-2)} \quad x_n = x_1^{(n-1)})^T \tag{12b}$$

Single input voltage e, single output

$$\alpha = x_1 \tag{12c}$$

a nonlinear function g, a disturbance d and superscripts indicating the order of derivative with respect to time t. Further, assuming that part of knowledge of the function g allows the construction of an input $\hat{u}$ providing an input-linearized form of the system dynamics as, Yang et al., Huang et al., IEEE Working Group and Turton [9–11, 16],

$$x^{(n)} = f(x, t) + \hat{u}(t) + d(x, t) \tag{13a}$$

where,

$$g(x, e, t) = f(x, t) + \hat{u}(t) \tag{13b}$$

Corresponding to a demand vector,

$$x_d(t) = (x_{d1} \quad x_{d2} = x_{d1}^{(1)} \ldots x_{di} = x_{d1}^{(i-1)} \ldots x_{d1}^{(n-2)} \quad x_{dn} = x_{d1}^{(n-1)})^T \tag{14}$$

Of course, the control objective is to reduce the error vector

$$x_e(t) = x_d(t) - x(t) = (x_{e1}\ x_{e2} = x_{e1}^{(1)} \ldots x_{ei} = x_{e1}^{(i-1)} \ldots x_{e1}^{(n-2)} x_{en} = x_{e1}^{(n-1)}) \tag{15}$$

to zero in a finite time. An error dynamics equation can be written as,

$$x_{e1}^{(n)} + \lambda_{n-1} x_{e1}^{(n-1)} + \ldots + \lambda_i x_{e1}^{(i)} + \ldots + \lambda_1 x_{e1}^{(1)} + \lambda_0 x_{e1} = 0 \tag{16}$$

Wouldfulfill the control objective, if the coefficients $\lambda_0, \lambda_1, \ldots, \lambda_{n-1}$ are Hurwitz. Further the error dynamics can be written as,

$$x_d^{(n)} = f(x,t) + u(t) + d(x,t) - (\lambda_{n-1}x_{e1}^{(n-1)} + \ldots + \lambda_1 x_{e1}^{(1)} + \lambda_0 x_{e1}) \tag{17}$$

With the consideration of the above equation we can write the expression for sliding surface with integral form as, Yang et al., Huang et al., IEEE Working Group and Turton [9–11, 16],

$$s(t) = -x_1^{(n-1)}(t) + \int_0^t (x_{d1}^{(n)} + \lambda_{n-1}x_{e1}^{(n-1)} + \ldots + \lambda_1 x_{e1}^{(1)} + \lambda_0 x_{e1})\mathrm{d}t \tag{18a}$$

Getting derivative of Eq. (18a) we can write,

$$\dot{s}(t) = -x_1^n(t) + (x_{d1}^{(n)} + \lambda_{n-1}x_{e1}^{(n-1)} + \ldots + \lambda_1 x_{e1}^{(1)} + \lambda_0 x_{e1}) \tag{18b}$$

With the consideration of tracking error reduction sliding mode controller can be defined as [9–11, 16],

$$\hat{u}_c(s) = -u_c \mathrm{sgn}(s) \tag{19}$$

In order to construct the control parameter u_c, a Lyapunov function is introduced as

$$V = s^2/2 \tag{20}$$

Getting differentiation of Eq. (20) we can write as,

$$\dot{V} = s\dot{s} \tag{21}$$

Now the controller parameter can be obtained with the consideration of,

$$\dot{V} \le |s|(L - u_c) \tag{22}$$

where L is the bounded disturbance.

4 Results and Discussion

Performance of the proposed integral sliding mode controller has been shown in Fig. 4 for sinusoidal 0.05 Hz frequency and 15° amplitude IGV angular position demand. To study the robustness of the controller, controller performance has been verified with sinusoidal IGV angular position demand. As the hydro turbine governing system, quite complex controller performances have been studied with 0.05 and 0.1 Hz frequency IGV position demand.The controller response depicts clearly the reaching

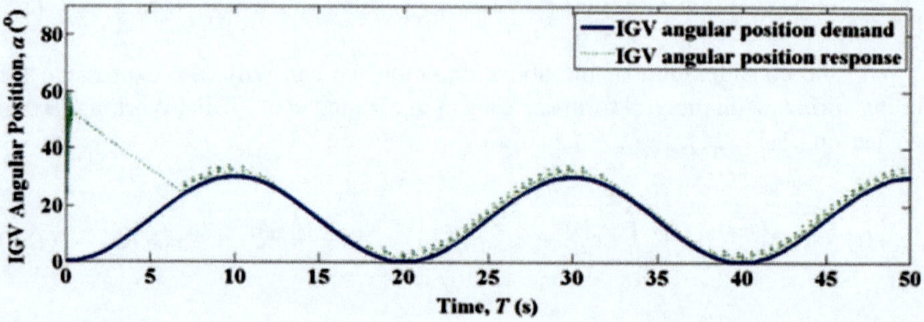

Fig. 4 Performance of the proposed controller for sinusoidal 0.05 Hz frequency IGV position demand

phase and sliding phase. The position response that it flows the angular position demand as requires after reaching phase, i.e., in the sliding phase.

Figure 5 has been shown the IGV angular position response with respect to 0.1 Hz and 8° amplitude sinusoidal position demand. The control voltage signal with respect to the response of Fig. 5 has been shown in Fig. 6. The voltage signal has been

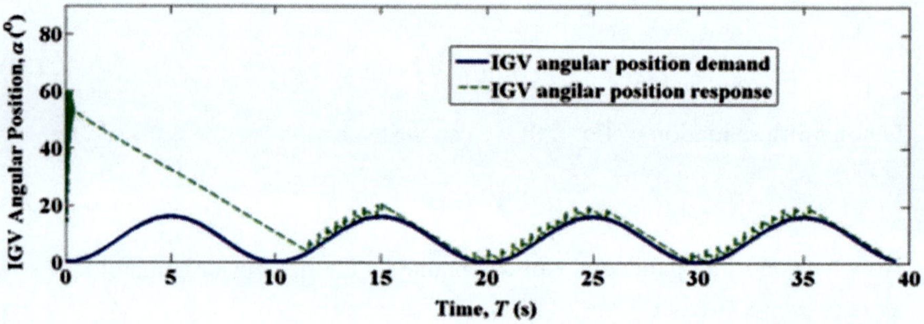

Fig. 5 Performance of the proposed controller for sinusoidal 0.1 Hz frequency IGV position demand

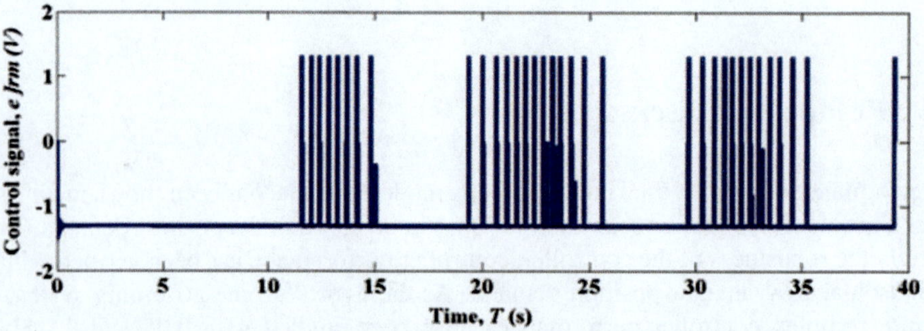

Fig. 6 Control signal for sinusoidal 0.1 Hz frequency IGV position demand

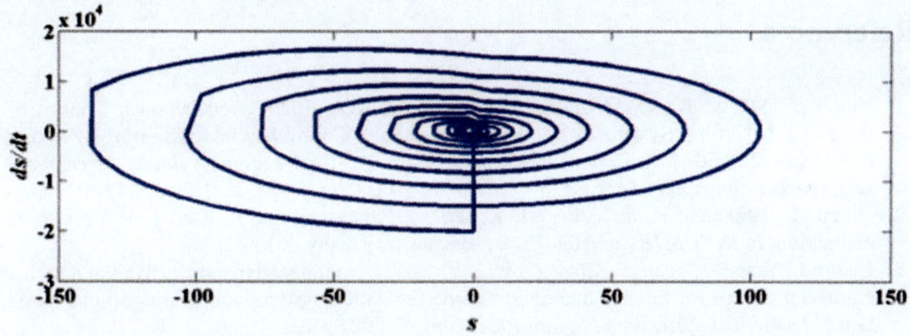

Fig. 7 Trajectory of the sliding variable

shown consistent with respect to the reaching phase and sliding phase in Fig. 5. The IGV angular position response in Figs. 4 and 5 has been shown as some chattering characteristic, is quite as the proposed controller has been designed based on SMC-I criteria and there is no adaption.

The initial part of the control action is quite complex as the hydro turbine poses non-minimal phase characteristic. The controller poses reaching phase movement toward sliding phase with constant control signal. While in the sliding phase control signal varies from the positive to negative with IGV position response flows demand.

Trajectory of the sliding variable s and $\dot{s}$ has been shown in Fig. 7. The trajectory has been converging to the origin. The trajectory ensures stability of the controller. The stability has been ensured with the external disturbances as water flow velocity, piston friction and valve dead band discontinuous.

5 Conclusion

Electrohydraulic turbine IGV system has been modeled for simulation study of the controller performance. Integral sliding mode controller has been designed for Francis turbine IGV position control. The simulation study has been done in Matlab Simulink. The solver configuration has been considered as RungeKutta 4 with time step 10^{-4}. The controller performance has been studied 0.05 and 0.1 Hz frequency IGV angular position demand. The tracking performance of the controller has been found satisfactory. The controller performance has been obtained with reaching phase and sliding phase depicted in the controller performance.The trajectory of the sliding surface depicts the convergence of the proposed controller.

Acknowledgements We sincerely acknowledge the Department of Science and Technology (DST), Science and Engineering Research Board (SERB), for their support to conduct this research. The grand number is ECR/2015/000077.

References

1. Vinod, J., Sarkar, B. K., Mookherjee, S., & Sanyal, D. (2020). Active Power Control of the Francis Turbine System by Model-Free Adaptive Controller. In B. Biswal, B. Sarkar & P. Mahanta (Eds.), *Advances in Mechanical Engineering. Lecture Notes in Mechanical Engineering*. Singapore: Springer. https://doi.org/10.1007/978-981-15-0124-1_142.
2. Vinod, J., Venkaiah, P., & Sarkar. B. K. (2018). Francis turbine IGV control under force estimation. In *INCOM18* (pp. 769–772), Jadavpur University, Kolkata.
3. Liang, J., Yuan, X., Yuan, Y., Chen Z., & Li. Y. (2017). Nonlinear dynamic analysis and robust controller design for Francis hydraulic turbine regulating system with a straight-tube surge tank. *Mechanical Systems and Signal Processing, 85*, 927–946.
4. Jiangbin, Z., Hua, L., & Huiping, X. (2017). Analysis and improvement on primary frequency regulation control system of hydro generator units. *Journal of Hydroelectric Engineering, 28*, 206–213.
5. Yang, W., Yang, J., Guo, W., & Norrlund, P. (2016). Response time for primary frequency control of hydroelectric generating unit. *International Journal of Electrical Power and Energy Systems, 74*, 16–24.
6. Zhang, G., Cheng, Y., Lu, N., & Guo, Q. (2016). Research of hydro-turbine governor supplementary control strategy for islanding AC grid at sending terminal of HVDC system. *IEEE Transactions on Energy Conversion, 31*, 1229–1238.
7. Li, H., Chen, D., Zhang H., Wang, F., & Ba. D. (2016). Nonlinear modeling and dynamic analysis of a hydro-turbine governing system in the process of sudden load increase transient. *Mechanical Systems and Signal Processing, 80*, 414–428.
8. Wu, F., Ding, J., & Wang. Z. (2016). Robust Finite-Time terminal sliding mode control for a Francis hydroturbine governing system. *Journal of Control Science and Engineering*, Article ID 2518734, 8. https://doi.org/10.1155/2016/2518734.
9. Yang, T., Wang, B., & Chen. P. (2020). Design of a Finite-Time terminal sliding mode controller for a nonlinear Hydro-Turbine governing system. Energies, *13*(3), 634. https://doi.org/10.3390/en13030634.
10. Huang, S., Zhou, B., Bu, S., Li, C., Zhang, C., Wang, H., & Wang, T. (2019). Robust fixed-time sliding mode control for fractionalorder nonlinear hydro-turbine governing system. *Renewable Energy, 139*, 447–458.
11. IEEE Working Group. (1992). Hydraulic turbine and turbine control models for system dynamic studies. *IEEE Transactions on Power Systems, 7*, 167–179.
12. Wang, B., Xue, J., Wu, F., & Zhu. D. (2016). Robust Takagi-Sugeno fuzzy control for fractional order hydro-turbine governing system. *ISA Transactions, 65*, 72–80.
13. Chamorroa, H. R., Sanchezb, A. C., Pantojac, A., Zelinkad, I., Longatte, F. G., & Sood, V. K. (2019). A network control system for hydro plants to counteract the nonsynchronous generation integration. *Electrical Power and Energy Systems, 105*, 404–419.
14. Dasmahapatra, S., Sarkar, B. K., Chatterjee, A., Mookherjee, S., & Sanyal. D. (2015). Design of an adaptive fuzzy-bias SMC and validation for a rugged electrohydraulic system. *IEEE/ASME Transactions on Mechatronics, 20*(6), 2708–2715.
15. Sarkar, B. K., Das, J., Saha, R., Mookherjee, S., & Sanyal. D. (2013). Approaching servoclass tracking performance by a proportional valve-controlled system. *IEEE/ASME Transactions on Mechatronics, 18*, 1425–1430.
16. Turton, R. K. (2012). *Principle of turbomachinary* (2nd ed.). Netherlands: Springer.
17. Vinod, J., & Sarkar, B. K. (2021). Francis turbine electrohydraulic inlet guide vane control by artificial neural network 2 degree-of-freedom PID controller with actuator fault. *Proc IMechE Part I: J Systems and Control Engineering, 235*(8), 1494–1509.

Bio-Waste Fired Gas Turbine and Transcritical Co_2 Cycle Based Combined Power Plant: Thermodynamic, Economic and Environmental Performance Assessment

P. Mondal, S. Samanta, S. Ghosh, and S. N. Barman

Nomenclature

r_p	Pressure ratio (–)
TIT	Turbine inlet temperature (°C)
HETD	Hot end temperature difference (°C)
LHV	Lower heating value (kJ/kg)
SI	Sustainability index

Greek Symbols

η	Efficiency (dimensionless)

Subscripts

b	biomass
C	Cost

P. Mondal · S. N. Barman
Department of Mechanical Engineering, NIT Sikkim, Ravangla 737139, India

S. Samanta (✉)
School of Mechanical Engineering, KIIT, Bhubaneswar 24, India

S. Ghosh
Department of Mechanical Engineering, IIEST Shibpur, Howrah 711103, India

P. Mahanta et al. (eds.), *Advances in Thermofluids and Renewable Energy*, Lecture Notes in Mechanical Engineering, https://doi.org/10.1007/978-981-16-3497-0_22

1 Introduction and Perspective

Exploration and shifting toward the new and renewable energy sources are being the primary concerns for the energy researchers, since past two decades. Energy generation from different available renewable sources and using different routes with different analysis strategies have been carried out and reported by numerous researchers, in the literature. Among different renewable sources, biomass is considered as one of the promising sources as India being the agriculture based country. However, efficient energy generation from biomass in economic and environment friendly manner is still in the focus area of the energy scientist. There are different types of biomass waste like agricultural crop wastes and residues, wood wastes and residues, different plants residue, household waste, bio-waste from industries, other waste materials including some municipal wastes etc. All these bio-waste are hazardous in nature, creates environmental pollution and creates others problems to human lives. On the other hand, these bio-wastes are good source of energy which are unutilized and not explored properly. So, use of these bio-wastes to generate energy would solve the triangular problem of energy demand and environmental pollution and waste management. In India, only power generation or combined heat and power (CHP) generation through bio-waste incineration/combustion have been proposed by various earlier researchers [1–4]. Thermal energy generation either in the form of power or heat from bio-waste incineration/combustion has been proposed by various researchers throughout the world [5–8]. In all the above-mentioned cases, it is observed that the power is generated by using conventional gas turbines (GT). After the combustion of the bio-waste, the hot flue gas is allowed to expand in the GT only after proper gas cleaning. But, till the conventional GT blades faces corrosion and particulate matter deposition related problems. Instead of conventional GT, the use of indirectly fired GT/externally fired GT/air turbine gives the system more stability or reliability in terms of turbine life as the harmful corrosive hot flue gas does not come in direct contact of the turbine blades. Many previous research groups have done the energy, exergy, economic analyses of biomass fuelled power generation system constituted with the indirectly heated air turbine or externally fired air turbine [9–12].

Waste heat recovery is found to be useful in every power generation cycle. It helps to enhance the overall power output as well the efficiency of any plant which in turns, help to improve the economy of the plant. Among various alternative options, the supercritical CO_2 (s-CO_2) cycle as bottoming cycle for waste heat recovery has been recommended by various researchers. Thermodynamic and economic analyses have also been conducted by different group of researchers [13–15] to show the ability of integration of s-CO2 cycle for downstream heat recovery options. To cross-check the feasibility of any newly developed plant, thermodynamic, economic and environmental analyses are highly essential. Various researchers have conducted thermo-economic analysis of biomass fuelled power generation system to estimate the energy and exergy efficiency as well as cost of electricity delivered from plant [16–18]. Based on the above considerations, present study is intended to design and analysis a novel and efficient biomass based combined cycle plant for generation of power in off-grid manner. Energetic, exergetic, environmental and economic performances of the plant at varying topping compressor pressure ratios ($r_p = 4$–16), air turbine inlet temperatures (TIT = 900, 1000 and 1100 °C) and terminal temperature difference of hot end of the CCHR unit (70–150 °C) are evaluated and reported in this paper.

2 Plant Configuration

Schematic diagram of the proposed plant is shown in Fig. 1. Solid biomass after drying and segregation (D&S) enters the combustion chamber (CC) of the CCHR unit. Then, solid biomass gets combusted with the help of the exhaust air, coming out from the air turbine (T1). Then, the burnt flue gas enters the shell side of the topping heat exchanger (HX1) of the CCHR unit. Atmospheric air after being compressed in the topping compressor (C1) enters the tube side of the HX1. Then, pressurized air gets heated during the heat exchanging process in HX1 and then enters the T1. Hot and pressurized air after getting expanded in the T1, enters the CC to assist the combustion of solid biomass in the CC. Finally the flue gas, after passing through the heat recovery unit (HX2) gets exhausted through the stack. An induced drought compressor is connected in between the HX2 and stack to recover the pressure drop in the flue gas line. Carbon-di-oxide (CO_2) acts as working fluid for the bottoming cycle. Supercritical CO_2 after passing through the tube side of the HX2 enters the

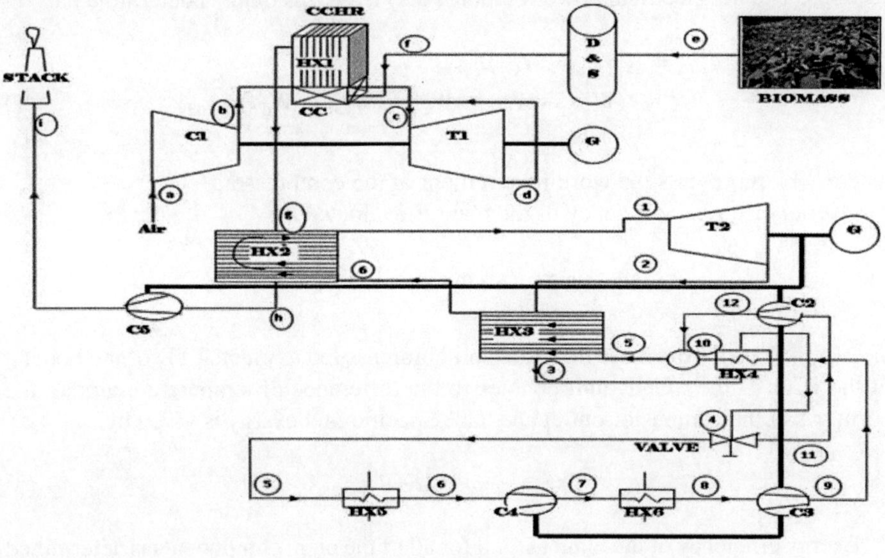

Fig. 1 Schematic diagram of the proposed plant

CO2-turbine (T2). T2 exhaust then enters the internal recuperator (HX3) and after then passes through the three compressors (C2, C3 and C4) and intercoolers (HX4, HX5 and HX6). Pressurized CO_2 stream returns back to the HX3 and flows through the tube side of it. CO_2 stream finally enters the tube side of the heat recovery unit and flow to the T2, and thus, the cyclic process is executed.

3 Model Equations

3.1 Energy and Exergy Analysis

Necessary model equations as well as parametric assumptions for carrying out the energetic and exergetic performance analysis of the plant are presented in this section. Assumptions made for the analysis are presented as bellow:

- Woody biomass is fed to plant having the proximate analysis: C-46.28%; H-5.40%; N-0.28%; O-36.97%; H_2O-8%; Ash-3.09% and LHV-16935 kJ/kg.
- Topping GT cycle produces fixed electrical output of 500 kW.
- No heat loss assumed across any component.

Component based assumptions are shwon in Table 1. Component wise model equations of the plant are considered in line with the works of Mondal and Ghosh [10] and Roy et al. [18]. However, necessary model equations for carrying out the energetic and exergetic performance analyses of the plant are presented in the following sections. The net electrical power output (Pnet) from this plant is determined as:

$$P_{\text{net}} = \{(W_{T1} - W_{C1})\eta_{gen1} + (W_{T2} - W_{C2} - W_{C3} - W_{C4} - W_{C5})\eta_{\text{gen2}}\} \tag{1}$$

where "W" represents the work requirement of the compressors.

Overall electrical efficiency of the plant is evaluated as:

$$\eta_{\text{overall}} = P_{\text{net}}/m_{\text{biomass}}.\text{LHV}_{\text{biomass}} \tag{2}$$

m_{biomass} represents the mass flow rate solid biomass fed to plant. TTD of the hot end of the HX1 of the CCHR unit denotes to the difference of temperature among the burnt gas at the combustor outlet and TIT. Specific fuel exergy is given by:

$$\text{Ex}_{\text{biomass}} = \beta.\text{LHV}_{\text{biomass}} \tag{3}$$

Exergy efficiency of the plant as well for all of the plant components is determined as:

Table 1 Assumptions made for the analysis

Component	Parameter/s	Value	Ref
CC of CCHR	Operating pressure (bar) Pressure drop (bar)	1.013 0.05	Mondal and Ghosh [10] Datta et al. [9]
HX of DCHR	Pressure drop at air side Pressure drop at gas side	3% of inlet pressure 1.5% of inlet pressure	Datta et al. [9]
C1	Isentropic efficiency (%)	90	Mondal and Ghosh [10]
C5	Isentropic efficiency (%)	90	
	Outlet pressure (bar)	1.1	
T1	Isentropic efficiency (%) Mechanical Efficiency (%)	90 95	
T2	Isentropic efficiency (%) Inlet pressure (bar)	90 200	Saha et al. [18]
C2, C3, C5	Isentropic efficiency (%)	90%	Mondal and Ghosh [10]

$$\eta_{ex} = \frac{Ex_{pt}}{Ex_{fl}} \tag{4}$$

Ex_{pt} and Ex_{fl} refer the product and fuel exergy of the plant as well as of the components, respectively.

3.2 Environmental and Economic Analysis

Necessary model equations as well as parametric assumptions for energetic and exergetic performance analyses of the plant are presented in this section. Assumptions made for the analysis are presented bellow:

Useful life of plant is taken as 20 years and annualized interest rate on plant components is taken as 10% [10]. Capital utilization factor (CUF) is taken to be 0.5 as the plant is designed to operate 12 h a day basis. Cost of CO_2 emission is taken as 0.0145 $/kg [18].

Environmental performance of the modeled plant is assessed via determining the sustainability index of the plant (SI). SI of a plant is evaluated to evaluate how much a newly developed plant is sustainable. SI of the plant is calculated as:

$$SI = \frac{1}{(1 - \eta_{ex,plant})} \quad (5)$$

where $\eta_{ex,plant}$ denotes the exergetic efficiency of the entire plant. Total environmental benefit cost (ENBC) due to savings in CO_2 emission because of the establishment of the bio-waste based power plant is given by the following equation:

$$ENBC = CO_2\_savings.C\_CO_2.8760.CUF \quad (6)$$

where C_CO_2 denotes the environmental damage cost due to the emission of CO_2 into the atmosphere. Economic performance of the modeled plant is evaluated via determining the unit cost of electricity delivered by the plant. Unit cost of delivered electricity is determined from the annualized capital cost of the plant components, annualized O&M cost as well annualized fuel cost. Capital cost of the individual plant component is taken from the work of Mondal and Ghosh [10] and Roy et al. [18].

Fixed capital cost (CZF) of the bio-waste based plant is calculated as:

$$CZF = \sum_{i=1}^{n} Z_i + Z_{CW,DN} \quad (7)$$

Z represents the capital cost for the respective plant components as well as for the civil works and distribution network. Capital cost associated with civil works and distribution network is calculated as:

$$Z_{CW,DN} = \sum_{i=1}^{n} Z_i.(1 + f) \quad (8)$$

The value of f in the above expression is taken as 0.09 [18]. Ministry of New and Renewable Energy (MNRE), Government of India provides subsidy on the establishment of new renewable energy based plants. Capital subsidy (in INR) of the plant is calculated as:

$$Subsidy = 2.10^6.P_{annualized,electrical}^{0.646} \quad (9)$$

Annualized capital cost (AC_Z) of the bio-waste based plant is calculated by considering the recovery factors of the individual plant components and is presented as follows:

$$AC_Z = (CZF - Subsidy).RF \quad (10)$$

Recovery factor (RF) in the above expression is derived as:

$$RF = \frac{d(1+d)^T}{(1+d)^T - 1} \tag{11}$$

d is discount rate, T is the useful life of the plant.

Therefore, annualized O&M cost for the plant is determined as:

$$AC_{OM,p} = CZF.m_{OM} + 8760.CUF.C_{mp}.n_{mp} \tag{12}$$

In the above expression, m_{OM} stands for the multiplication factor of capital cost for calculation of OM cost of the respective plant components and the value is taken as 2.8% of total capital cost [18]. C_{mp} represents the salary of the plant personals in hourly basis and n_{mp} signifies the numbers of plant operator.

Annualized fuel cost is calculated as:

$$AC_F = C_{biomass}.m_{biomass}.8760.CUF \tag{13}$$

Finally, delivered unit cost of electricity (DUCE) in INR/kWh is determined as:

$$DUCE = \frac{AC_{C,p} + AC_{OM,p} + AC_F}{P_{annualized,electrical}} \tag{14}$$

Annualized electricity delivered by the plant (in kWh) is calculated as:

$$P_{annualized,electrical} = W_{net,electrical} \times CUF \times 8760 \times (1 - L_{DN}) \tag{15}$$

Loss is distribution network (L_{DN}) is taken as 5%.

4 Results and Discussions

4.1 Energy and Exergy

Base case performance of the plant is shown in Table 2. For the base case analysis, thermodynamic state points of the plant are considered to be $r_p = 4$, TIT = 1000 °C and TTD = 70 °C. It is observed from Table 2 that the plant delivers about 545 kW electricty with an overall efficiency value of about 44%. Furthermore, it is also observed from the table that specific biomass consumption is about 0.5 kg/kWh at the base case.

Table 2 Base case performance of the plant

Parameter	Unit	Value
Net T1 output	kW	500
Net Plant output	kW	545.51
Net electrical efficiency	%	43.63
Electrical specific biomass consumption	kg/kWh	0.52
Exergy efficiency of the CC of the CCHR unit	%	64.64
Exergy efficiency of the CC of the CCHR unit	%	92.94

Overall exergy balance of the plant at the base case is shown in Fig. 2. It is observed from figure that maximum exergy destruction occurs at the combustion chamber of the CCHR and at the heat exchanger of the CCHR unit. About 42% of the input exergy is destructed at the CCHR unit. This is due to the occurrence of chemical reaction at the combustion chamber of the CCHR unit as well as the heat exchanging process across the topping heat exchager at the high temperature. Furthermore, it is also observed from the plot that, traceable amonut of exergy is destructed at the turbines, compressors and other heat exchangers.

Figure also indicates that CC and HX1 are the most important component of plant, whose exergetic performance at different thermodynamic state points may influence the performance of the plant. Based on this consideration, exergy efficiencies of these two components at different r_p and TITs (keeping TTD fixed at 70 °C) are shown in Fig. 3. It is seen from Fig. 3 that with the increase in r_p, exergy efficiency of the CC decreases (at moderate rate) with, while that for the HX1 increses (at higher rate). Figure also helps to identify that higher TIT yields in better exergetic performance by these two components. This discussion generates an idea that the energetic performance of the plant needs to be evaluted at the considered ranges of r_p and TIT (keeping TTD fixed at 70 °C). Variations in overall electrical efficincy and net electrical ouput from the plant with the change in r_p and TIT are shown in Fig. 4a,

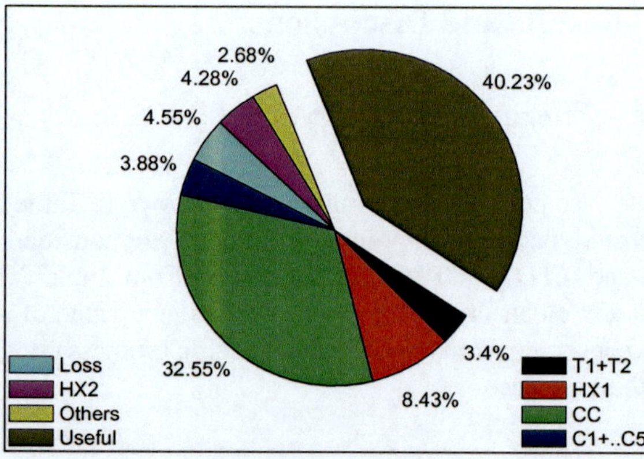

Fig. 2 Exergy balance of the plant at the base case

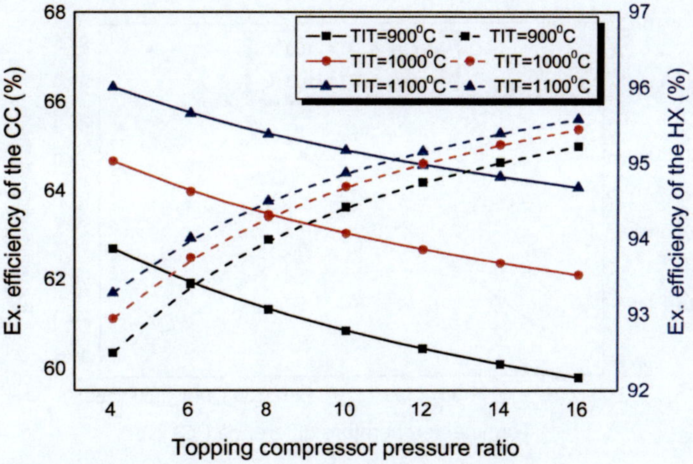

Fig. 3 Variations in exergy efficiency of the CC and HX1 with r_p and TIT

Fig. 4 **a** Variation in electrical efficiency with r_p and TIT. **b** Variation in net electrical output with r_p and TIT

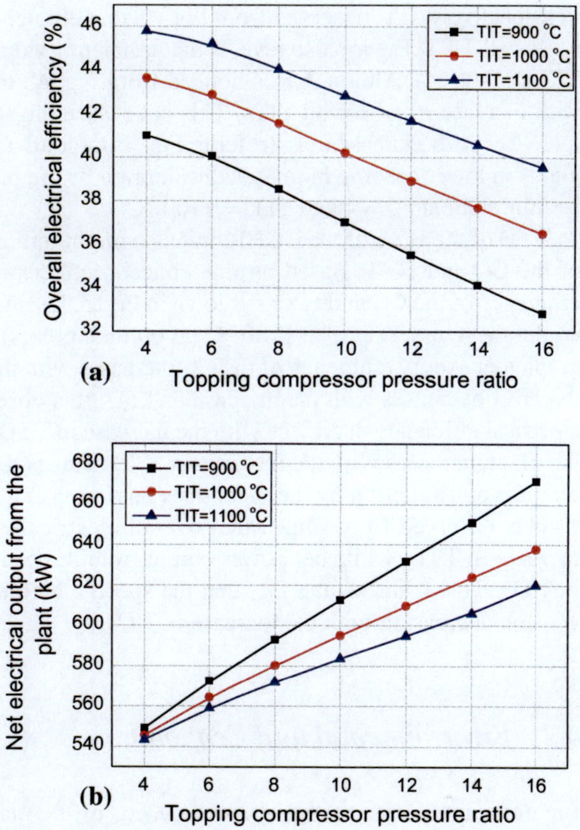

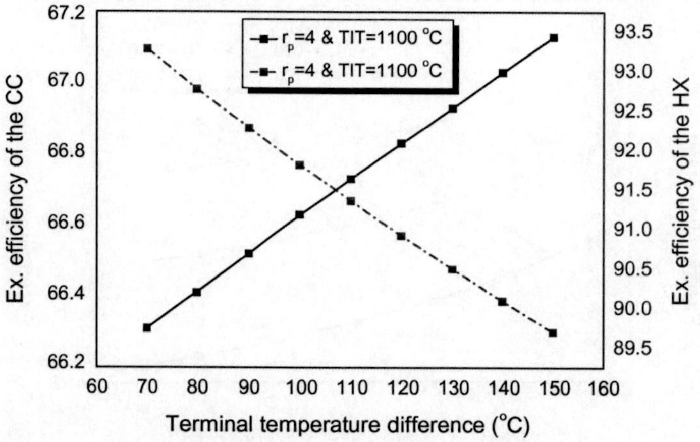

Fig. 5 Variations in exergy efficiency of the CC and HX1 with TTD

b, respectively. It is observed from Fig. 4a that efficiency decreases with increase in r_p, for all TITs. Figure also reveals that efficiency value is maximized at $r_p = 4$ and TIT = 1100 °C. Although it is noticed from Fig. 4b that net plant output increases with increase in r_p, for all TITs. This is because the specific biomass consumption increases with increase in r_p (refer to Fig. 7a), for all TITs. Furthermore, higher TIT yields in lower specific biomass consumption by the plant. Power output is found to be minimum at $r_p = 4$ and TIT = 1100 °C.

TTD of the HX1 of the CCHR unit also might influence the exergy performances of the CC and HX1. Based on this concept, effect of varying TTD on the exergy efficiency of the CC and HX1 is shown in Fig. 5. During TTD variation, r_p is considered to be 4 and TIT to be 1100 °C (at optimal points). It is noticed from the above graph that exergy efficiency of the CC increases with increase in TTD, while that for the HX1 decreases with the increase in TTD. It is observed from Fig. 6 that overall electrical efficiency decreases with the increase in TTD while net power output from the plant increases with the increase in TTD. This is due to the fact that the specific biomass consumption by the plant goes on decreasing with increase in TTD.

It is observed from Fig. 6 that overall electrical efficiency decreases with the increase in TTD, while net power output from the plant increases with the increase in TTD. This is due to the fact that the specific biomass consumption by the plant goes on decreasing with the increase in TTD.

4.2 Environmental and Economic

Environmental and economic performances of the plant at base case ($r_p = 4$, TIT = 1000 °C and TTD = 70 °C) are shown in Table 3. SI value is found to be 1.63 at the base case. Annually, about 30,500 USD is saved due to less CO_2 emission

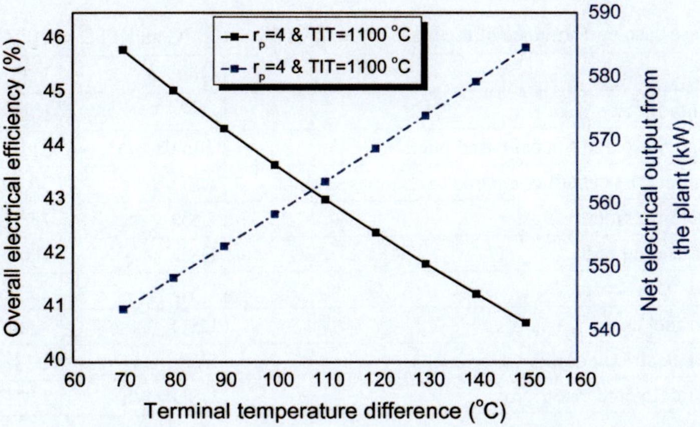

Fig. 6 Variations in overall electrical efficiency and power output of the plant with TTD

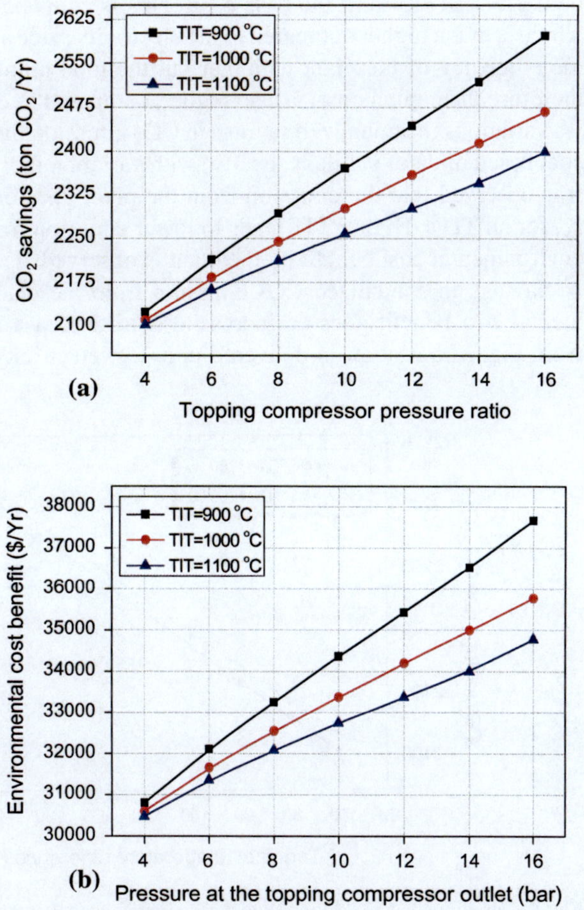

Fig. 7 **a** Variation in annualized CO_2 savings with r_p and TIT. **b** Variation in environmental cost benefit with r_p and TIT

Table 3 Base case performance of the plant ($r_p = 4$, TIT = 1100 °C and TTD = 70 °C)

Parameter	Unit	Value
Sustainability index	–	1.63
CO_2 savings compared to coal based plant	Ton CO_2/Y	2110
Environmental cost benefit compared to coal based plant	USD/Y	30,593
Total capital investment	USD	7.98E+06
Annualized capital cost	USD	3.98E+05
Annualized O&M cost	USD	2.23E+05
Annualized fuel cost	USD	3.50E+04
Annualized electricity delivered	kWh	2.33E+06
Unit cost of delivered electricity	USD/kWh	2.81E−01

from the plant. Furthermore, the unit cost of delivered electricity price is found to be 0.218 USD/kWh at the base case. This is equivalent to about INR 18 per kWh which is much higher compared to the electricity price at the Indian market. Although the efficiency of the plant is higher but the total number of plant component (and therefore the capital cost) is also higher, compared to other such kind of plants.

Variations in annualized savings in CO_2 emission by the plant (when compared to coal based plant of similar capacity) with variation in r_p and TIT is shown in Fig. 7a. It is observed that the emission from the plant goes on increasing with increase in r_p, for all TITs. Higher TIT leads to lower emission from the plant. Similar trend of environmental cost benefit by the plant is observed, as seen from Fig. 7b.

Savings in annualized CO_2 emission from the plant and therefore the environmental cost benefit goes on increasing with increase in TTD of the plant, as seen from Fig. 8. Variation in delivered unit cost electricity (DUCE) with the variation

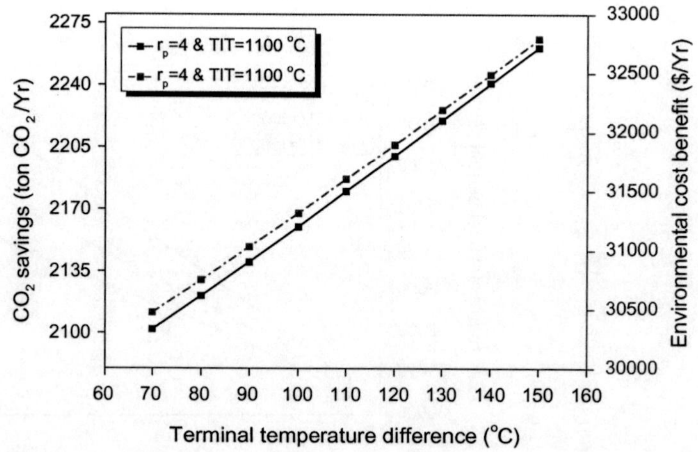

Fig. 8 Variations in CO_2 savings and environmental cost benefit with TTD

Fig. 9 **a** Variation in DUCE with r_p and TIT. **b** Variation in DUCE with TTD

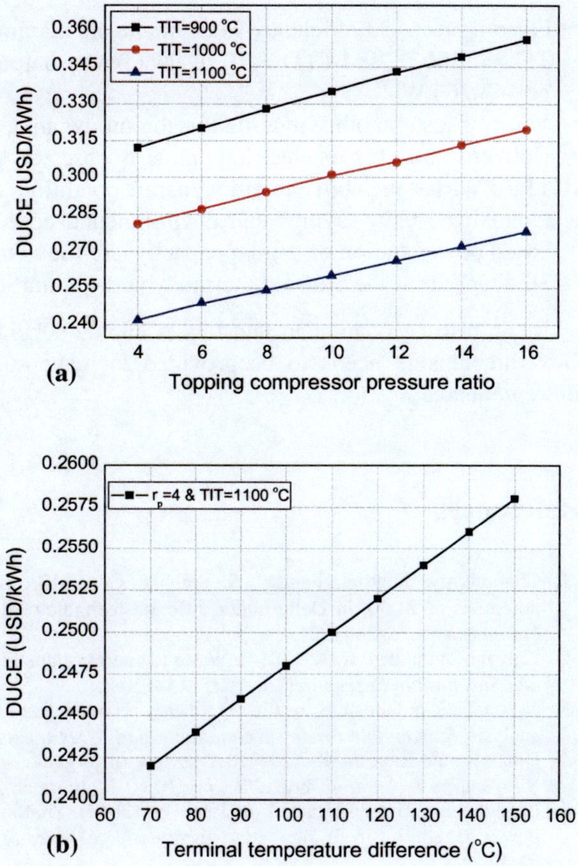

in r_p and TIT is shown in Fig. 9a. It is observed from figure that cost of electricity value goes on increasing with increase in r_p, for all TITs. However, higher TIT yields in lower DUCE value. This is because, with increase in TIT, biomass and air flow rate through the topping cycle go on decreasing and therefore the capital cost of the components become lower. For given r_p and TIT value, increase in TTD value of the CCHR unit results in higher biomass consumption and therefore higher capital cost of plant components. This ultimately affects the DUCE value to increase with increase in TTD value, as seen from Fig. 9b.

5 Conclusions

After carrying out the 4-E analysis of the plant, following conclusions can be drawn:

- Overall electrical efficiency value attains its maximum value (45.8%) at $r_p = 4$, TIT = 1100 °C and TTD = 70 °C. Net power output at this condition is found to be 543.26 kW.
- Value of sustainability index at this thermodynamic condition is found to be 1.67.
- Deliverd unit cost of electricty value is observed to be USD 0.242/kWh (INR 15.7/kWh) at the above thermodynamic condition.
- Annualized CO_2 savings and environmental cost benifit, compared to a coal-based power station of similar capacity, are found to be 2100 Ton/Year and USD 30,473/Year at the pre-said thermodynamic condition.

It can also be concluded that higher capital subsidy on such projects from the Government side needs to be provided to make such projects attractive to the Entrepreneures.

References

1. Chakraborty, Sharma, Pandey, & Prabhat, G. (2013). Assessment of energy generation potentials of MSW in Delhi under different technological options. *Energy Conversion and Management*, *75*, 249–255.
2. Kalyani, & Pandey, K. K. (2014). Waste to energy status in India: A short review. *Renewable and Sustainable Energy Reviews*, *31*, 113–120.
3. Nixon, Dey, & Ghosh, S. K. (2017). Energy recovery from waste in India: An evidence-based analysis. *Sustainable Energy Technologies and Assessments*, *21*, 23–32.
4. Dhar, K., & Rakesh, K. (2017). A review on organic waste to energy systems in India. *Bioresource Technology*. https://doi.org/10.1016/j.biortech.2017.08.159
5. Abdulhakim, Hamad, Hamad, & John, W. (2014). Sheffield. Study of energy recovery and power generation from alternative energy source. *Case Studies in Thermal Engineering*, *4*, 92–98.
6. Tan, Ho, Hashim, Lee, Taib, & Ho, C. S. (2015). Energy, economic and environmental (3E) analysis of waste-to-energy (WTE) strategies for municipal solid waste (MSW) management in Malaysia. *Energy Conversion and Management*, *102*, 111–120.
7. Azamia, Taheri, Pourali, & Farschad, T. (2018). Energy and exergy analyses of a mass-fired boiler for a proposed waste-toenergy power plant in Tehran. *Applied Thermal Engineering*, *140*, 520–530.
8. Ayodele, O., & Alao, M. A. (2018). Economic and environmental assessment of electricity generation using biogas from organic fraction of municipal solid waste for the city of Ibadan, Nigeria. *Journal of Cleaner Production*. https://doi.org/10.1016/j.jclepro.2018.08.282.
9. Datta, Ganguly, & Sarkar, L. (2010). Energy and exergy analyses of an externally fired gas turbine (EFGT) cycle integrated with biomass gasifier for distributed power generation. *Energy*, *35*, 341–350. https://doi.org/10.1016/j.energy.2009.09.031.
10. Mondal, & Ghosh, S. (2017). Exergo-economic analysis of a 1 MW biomass-based combined cycle plant with externally fired gas turbine cycle and supercritical organic Rankine cycle. *Clean Technologies and Environmental Policy*, *19*, 1475–1486. https://doi.org/10.1007/s10098-017-1344-y.
11. Soltani, Mahmoudi, Yari, & Rosen, M. A. (2013). Thermodynamic analyses of an externally fired gas turbine combined cycle integrated with a biomass gasification plant. *Energy Conversion and Management*, *70*, 107–115. https://doi.org/10.1016/j.enconman.2013.03.002.
12. Al-attab, & Zainal, Z. A. (2010). Turbine startup methods for externally fired micro gas turbine (EFMGT) system using biomass fuels. *Applied Energy*, *87*, 1336–1341. https://doi.org/10.1016/j.apenergy.2009.08.022.

13. Liu, Zhang, Ma, & Yan, J. (2018). Thermo-economic analyses on a new conceptual system of waste heat recovery integrated with an S-CO_2 cycle for coal-fired power plants. *Energy Conversion and Management*, *161*, 243–253. https://doi.org/10.1016/j.enconman.2018.01.049.
14. Villafana, Bueno, V. M., & Pablo, J. (2019). Thermoeconomic and environmental analysis and optimization of the supercritical CO_2 cycle integration in a simple cycle power plant. *Applied Thermal Engineering*, *152*, 1–12. https://doi.org/10.1016/j.applthermaleng.2019.02.052.
15. Jokar, Ahmadi, Sharifpur, Meyer, Pourfayaz, & Ming, T. (2017). Thermodynamic evaluation and multi-objective optimization of molten carbonate fuel cell-supercritical CO_2 Brayton cycle hybrid system. *Energy Conversion and Management*, *153*, 538–556. https://doi.org/10.1016/j.enconman.2017.10.027.
16. Mamaghani, Najafi, Shirazi, & Rinaldi, F. (2015). Exergetic, economic, and environmental evaluations and multi-objective optimization of a combined molten carbonate fuel cell-gas turbine system. *Applied Thermal Engineering*, *77*, 1–11. https://doi.org/10.1016/j.applthermaleng.2014.12.016.
17. Soltani, Mahmoudi, Yari, & Rosen, M. A. (2013). Thermodynamic analyses of a biomass integrated fired combined cycle. *Applied Thermal Engineering*, *59*, 60–68. https://doi.org/10.1016/j.applthermaleng.2013.05.018.
18. Roy, Samanta, Ghosh, S. (2019). Thermo-economic assessment of biomass gasification-based power generation system consists of solid oxide fuel cell, supercritical carbon dioxide cycle and indirectly heated air turbine. *Clean Technologies and Environmental Policy*, 21, 827–845. https://doi.org/10.1007/s10098-019-01671-7.

Production of Hydrogen Gas from Al-Water Reaction in the Presence of Aqueous NaOH—An Experimental Approach

Biswajyoti Das, Pinakeswar Mahanta, and P. S. Robi

Nomenclature

J	Joule
K	Kelvin
L	Litre
M	Molarity
m	Meter
T	Temperature

Greek Symbols

μ	Micron

1 Introduction

In the era of green technology, the energy carrier should not contribute anything to the deterioration of the environment directly or indirectly. Hydrogen is lightweight, non-toxic, and high-energy density material [1]. Various methods are available for the generation of hydrogen viz. water splitting under the solar energy with the help of some catalyst [2], partial oxidation of hydrocarbons [3], steam reforming of hydrocarbons [3], by reacting chemical hydrides with water [4], by the chemical method in

B. Das · P. Mahanta (✉) · P. S. Robi
Department of Mechanical Engineering, Indian Institute of Technology, Guwahati, Assam, India
e-mail: pinak@iitg.ac.in

P. Mahanta et al. (eds.), *Advances in Thermofluids and Renewable Energy*, Lecture Notes in Mechanical Engineering, https://doi.org/10.1007/978-981-16-3497-0_23

the presence of aqueous alkalis [5]. The production of hydrogen gas from Al-water reaction is a promising technique in which the by-product is only bayerite, which can be recycled [6]. However, Al is the most abundant metal on the earth's crust. Scrap Al is available in households and it is cheap as well (3 dollars/kg approximately) [7]. The Al-water reaction is given by the following chemical equation [8]:

$$2Al + 6H_2O \overset{yields}{\rightarrow} 2Al(OH)_3 + 3H_2 \tag{1}$$

The Al-water reaction is dependent on the initial temperature and concentration of the NaOH solution. However, the reaction follows first-order kinetics. It was reported that 1 g of Al when reacts completely with water under ambient condition yields 1360 ml of hydrogen gas [9]. However, the reaction slows down after some time due to the formation of a passive oxide layer over Al surface [10]. Hence, an aqueous NaOH solution is introduced in the reaction. It acts as a catalyst. The size of the Al flakes also determines the rate of the reaction. Al flakes with lower specific surface areas are found to be more profound to the liberation of hydrogen gas [11].

1.1 Storage

The storage of hydrogen gas is a very challenging task due to its low volumetric energy density. Moreover, the ignition temperature of hydrogen is very low (0.03 mJ). That means a small spark is enough to catch fire [12]. Chemical hydrides are employed to store hydrogen. But it is also having some demerits.

2 Materials and Method

2.1 Fabrication of Hydrogen Gas Generator

A hydrogen gas generator was fabricated out of an acrylic cylindrical tube with 9.7 L of the inner volume. The pressure sensor is connected with the data acquisition system and computer to record the data. A schematic diagram of the setup is shown in Fig. 1

The raw material for the reaction is scrap Al. The bars are sheared according to the following sizes (25 × 25) mm, (25 × 37.5) mm, and (25 × 75) mm. The thickness of the flakes is 1 mm. NaOH pellets of normal grade were procured from the market. Distilled water and tap water are used for the reaction have a resistivity of 0.5 MΩ.cm and 0.003 MΩ cm, respectively.

Fig. 1 Schematic diagram of the hydrogen generation setup with other instruments

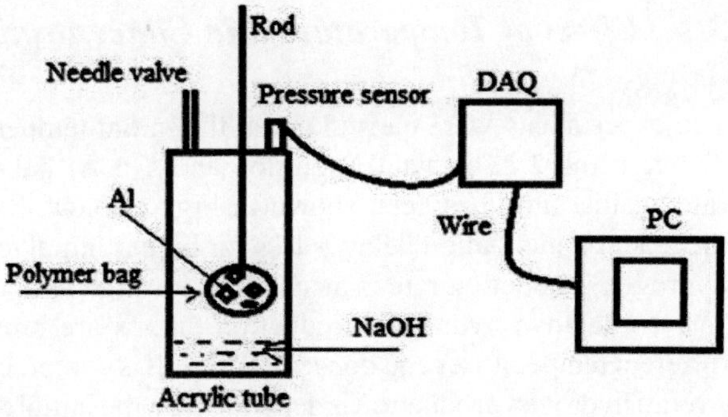

2.2 Experimental Procedure

The experiments were carried out according to the following experimental matrix for three different cases:

(a) water type- distilled water and size of flakes is (25 × 25 × 1) mm
(b) water type- distilled water and size of flakes is (25 × 37.5 × 1) mm
(c) water type- tap water and size of flakes is (25 × 25 × 1) mm.

The reaction is started by dipping the Al flakes into the NaOH solution. Hydrogen gas bubbles are observed and the increase in the volume of hydrogen gas per unit time is recorded. Van der Wall's equation is employed to find the amount of hydrogen gas evolved.

3 Results and Discussions

The effects of temperature, the concentration of NaOH solution, the size of Al pellets, and the type of water used, on the evolution of H_2 gas, were studied. The cumulative production of hydrogen gas is found to be 1186 ml/g of Al which is 87.21% of the theoretical yield of 1360 ml/g [9]. The fact of not having 100% yields can be attributed to the un-reacted layer of some other material over the Al surface.

Table 1 Experimental matrix

M	1 M	2 M	3 M	4 M	5 M
T					
303 K					
313 K					
323 K					
333 K					

3.1 Effect of Temperature and Concentration of NaOH

The experiments were carried out at the initial temperatures of 303, 313, 323, and 333 K using 2 L of NaOH solution and 5 g Al flakes. The hydrogen generation rate against time has been shown in Fig. 2 below. It is observed from figure that the reaction incubation delay is less for higher initial temperature and the maximum hydrogen production rate is also higher for high temperature, which was predicted. The cumulative hydrogen production curves are represented in Figs. 3 and 4 at different temperatures and concentrations. It is comprehended from the figure that the overall hydrogen production is dependent on the initial temperature and concentration of the NaOH solution.

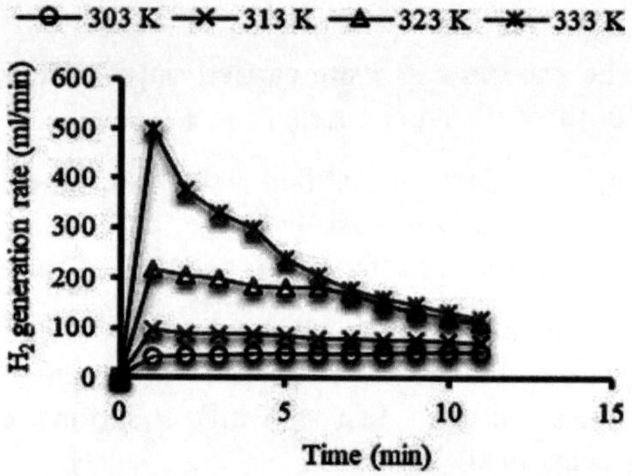

Fig. 2 H_2 generation rate versus time at different temperatures

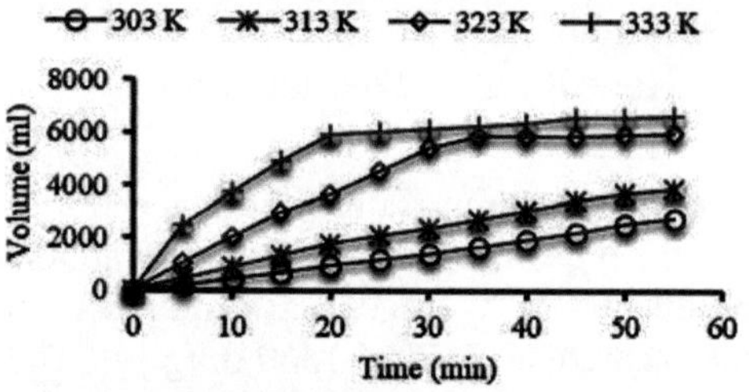

Fig. 3 H_2 production versus time at different temperatures

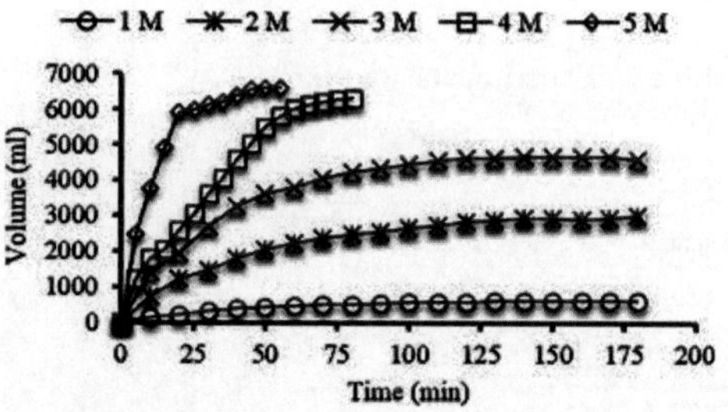

Fig. 4 H_2 production versus time at different concentrations of NaOH solution

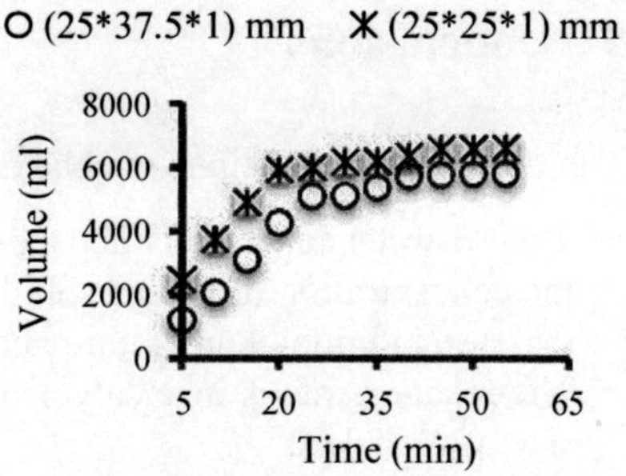

Fig. 5 Production of H_2 gas versus time for two different sizes of Al flakes

3.2 *Effect of Size of Al Flakes*

Two different sizes of Al flakes viz. (25 × 25) mm and (25 × 37.5) mm were used in the reaction to evaluate the effect of flake size on the production of hydrogen gas. Figure 5 shows the hydrogen generation plot for Al flakes of two different sizes. It is obvious from figure that the reaction rate is faster for Al flakes with a smaller surface area of 625 mm^2 than 937.5 mm^2. The reaction for smaller particle size reaches the maximum value of 1186 ml/g of Al within 20 min, whereas; the other reaction reaches the maximum value in 55 min.

3.3 *Effect of Type of Water*

A solution of NaOH of concentration 5 M was prepared in 2 L of distilled water and tap water to conduct two tests to observe the effect of type of water on the hydrogen generation rate. The results are shown in Fig. 6. The rate of the reaction is faster and yields a maximum of 1186 ml/g of Al in the first 20 min. Whereas, for tap water, the maximum yield is obtained as 1070 ml/g of Al in 55 min. The results can be validated with Wang et al. [13]. The amount of salt content in the water affects the rate of the reaction. The more concentrated the salt, the faster is the reaction [14].

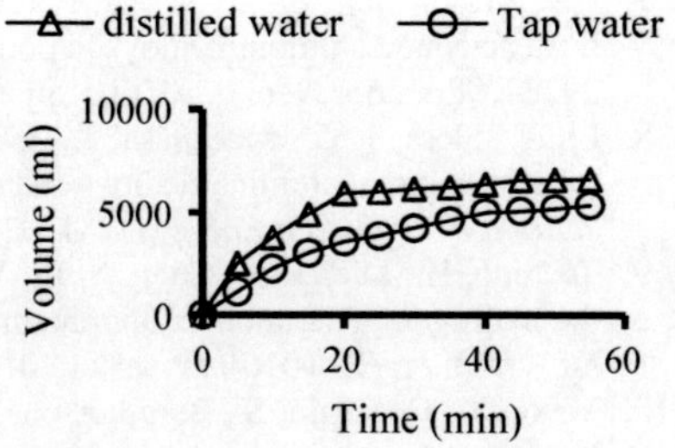

Fig. 6 Production of hydrogen gas versus time for two different types of water

4 Conclusions

The following conclusions may be derived from the above experiments:

- The Al-water reaction is dependent on the initial temperature of the water and the concentration of NaOH solutions. The rate of the reaction increases with the increase in initial temperature and concentration of NaOH solution.
- The reaction rate is inversely proportional to the resistivity of the water and the size of Al flakes.
- The optimum result was obtained at 5 M concentration of NaOH and at a temperature of 333 K when the NaOH solution was prepared in distilled water with Al pellets size used was (25 × 25) mm. The maximum volume of hydrogen gas obtained in the aforesaid condition was evaluated to be 1186 ml/g Al. Which is 87.21% of the total theoretical yield from the stoichiometry of the reaction.

References

1. Rotegard, D. (2002). Tomorrow's energy: Hydrogen, fuel cells, and the prospects for a cleaner planet. Peter Hoffmann. 2001. MIT Press, Cambridge, MA. 289 pp. $32.95 hardcover., Environ. Pract. https://doi.org/10.1017/s1466046602251261v.
2. Kitano, M., Tsujimaru, K., & Anpo, M. (2008). Hydrogen production using highly active titanium oxide-based photocatalysts. *Topics in Catalysis*. https://doi.org/10.1007/s11244-008-9059-2
3. Brown, L. F. (2001). A comparative study of fuels for on-board hydrogen production for fuel-cell-powered automobiles. *International Journal of Hydrogen Energy*. https://doi.org/10.1016/S0360-3199(00)00092-6
4. Wee, J. H. (2006). Which type of fuel cell is more competitive for portable application: Direct methanol fuel cells or direct borohydride fuel cells? *Journal of Power Sources*. https://doi.org/10.1016/j.jpowsour.2006.07.032
5. Belitskus, D. (1970). Reaction of aluminum with sodium hydroxide solution as a source of hydrogen. *Journal of the Electrochemical Society*. https://doi.org/10.1149/1.2407730
6. Wang, W., Chen, D. M., & Yang, K. (2010). Investigation on microstructure and hydrogen generation performance of Al-rich alloys. *International Journal of Hydrogen Energy, 35*(21), 12011–12019. https://doi.org/10.1016/j.ijhydene.2010.08.089
7. Fan, M. Q., Sun, L. X., & Xu, F. (2010). Experiment assessment of hydrogen production from activated aluminum alloys in portable generator for fuel cell applications. *Energy, 35*(7), 2922–2926. https://doi.org/10.1016/j.energy.2010.03.023
8. Lluís Solera, J. C., Macanása, J., & Muñoza, M. (2007). Synergistic hydrogen generation from aluminum, aluminum alloys and sodium borohydride in aqueous solutions. *International Journal of Hydrogen Energy, 32*, 4702–4710.
9. Te Teng, H., Lee, T. Y., Chen, Y. K., Wang, H. W., & Cao, G. (2012). Effect of $Al(OH)_3$ on the hydrogen generation of aluminum-water system. *Journal of Power Sources, 219*, 16–21. https://doi.org/10.1016/j.jpowsour.2012.06.077
10. Yavor, Y., Goroshin, S., Bergthorson, J. M., Frost, D. L., Stowe, R., & Ringuette, S. (2013). Enhanced hydrogen generation from aluminume water reactions. *International Journal of Hydrogen Energy, 38*(35), 14992–15002. https://doi.org/10.1016/j.ijhydene.2013.09.070.

11. Ivanov, V. G., Safronov, M. N., & Gavrilyuk, O. V. (2001). Macrokinetics of oxidation of ultradisperse aluminum by water in the liquid phase. *Combustion, Explosion, and Shock Waves*. https://doi.org/10.1023/A:1017505709456
12. Tzimas, E., Filiou, C., Peteves, S. D., & Veyret, J. (2003). Hydrogen storage : State-of-the-art and future perspective.
13. Wang, H. W., Chung, H. W., Te Teng, H., & Cao, G. (2011). Generation of hydrogen from aluminum and water—Effect of metal oxide nanocrystals and water quality. *International Journal of Hydrogen Energy, 36*(23), 15136–15144. https://doi.org/10.1016/j.ijhydene.2011.08.077
14. Soler, L., Candela, A. M., Macanás, J., Muñoz, M., & Casado, J. (2009). Hydrogen generation by aluminum corrosion in seawater promoted by suspensions of aluminum hydroxide. *International Journal of Hydrogen Energy, 34*(20), 8511–8518. https://doi.org/10.1016/j.ijhydene.2009.08.008

Thermal Cracking of High-Density Polyethylene Wastes and Production of Liquid Hydrocarbon

Awinash Kumar, Vikeshachand Prasad Singh, and Pradip Lingfa

Nomenclature

BSFC	Brake-specific fuel consumption
cm	Centimeter
°C	Degree centigrade
EDAX	Energy dispersive X-ray analysis
EVA	Ethylene-vinyl acetate
FTIR	Fourier Transform Infrared Spectroscopy
GCMS	Gas Chromatography Mass Spectroscopy
gm	Gram
g/m	Gram per milliliter
HDPE	High Density Polyethylene
Kv	Kilovolt
LDPE	Low Density Polyethylene
mA	Milliampere
ml	Milliliter
mm	Millimeter
PVC	Poly-Vinyl Chloride
SEM	Scanning Electron Microscopy
WPF	Waste Plastic Fuel
WPPO	Waste Plastic Product Oil

A. Kumar (✉) · V. P. Singh · P. Lingfa
Department of Mechanical Engineering, North Eastern Regional Institute of Science and Technology, Nirjuli, Arunachal Pradesh 791109, India

P. Mahanta et al. (eds.), *Advances in Thermofluids and Renewable Energy*, Lecture Notes in Mechanical Engineering, https://doi.org/10.1007/978-981-16-3497-0_24

1 Introduction

Plastic was invented in 1860, but it is widely used in the past 30 years because of its durability and lightweight nature, cheap, and easily availability. In 1907, the first plastic synthetic in nature discovered by Leo Baekeland in New York. Plastic has been named after the Greek word "Plastikos" concerning molded or shaped behavior. It is the part of organic polymer which contains high-molecular mass. It is obtained from petrochemicals. As plastics are products of petroleum derivatives, therefore it contains hydrocarbons, but despite hydrocarbons, it also contains antioxidants, colorants, and stabilizers. Plastics have high-molecular mass. Besides polymers, it also contains other substances to improve its performance and reduce its cost. These polymers are made up of monomers due to which it can be molded and shaped to various desired formed. Two types of plastics are mainly focused: thermoplastics and thermosetting polymers. If enough heat is applied thermoplastics can repeatedly soften and hardened on cooling so that they can be used as new plastics goods. Examples are PVC, polyethylene, and polystyrene. Thermosetting plastics can take shape once. They are not suitable for repeated heat treatments; therefore, after they have solidified, they stay solid. Phenol–formaldehyde and urea–formaldehyde are good examples of thermosets. The use of plastic is increasing day-by-day and, it is a huge concern for the entire world as it is non-biodegradable. The consumption of plastic in the world per annum has been increased 20 times from 20 million in 1950 to 100 million tons [1]. The plastic is used as it is cheaper and easily available. As more plastics are used, more waste will be generated. To protect the environment, these plastic wastes are to be utilized. The best way to utilize it to convert it into fuel as plastic polymers were originated from petroleum to meet future fuel demands. According to the fact sheet on plastic waste in India, 40% of the total waste plastics may be caused for environmental pollution. Most of the plastic products have lead and cadmium pigments are hazardous and directly take part to create an unhealthy environment for at least 267 species worldwide [2]. Several researchers have been studied different types of plastic wastes based on reuse, recycle, and recovery methods for better utilization and eco-friendly alteration. Different types of catalysts have been used to improve the pyrolysis process during the alteration of plastic wastes such as sodium bentonite, zeolites, and kaolin. Khan et al. [7] reported high-density polyethylene (HDPE) for generating oil through pyrolysis. They compared the oil produced with the properties of diesel. The plastic was pyrolyzed for 2–3 h up to 330–490°C. About 38.5% of WPPO was recovered at a temperature of 330 °C. The oil percentage increased constantly to 76.0% at 425 °C. Only about 10.0% of distilled WPPO was achieved at the temperature of 116 °C. However, the percentage of WPPO increased constantly to 73.5% at a temperature of 264 °C from 10% at a temperature of 116 °C [2–7]. Syamsiro et al. [8] produced fuel oil from municipal waste by pyrolysis and catalytic reforming process. Results showed that types of feedstock strongly affect product yields and the quality of liquid and solid products. HDPE waste produced the highest liquid fraction. The presence of catalyst reduced the liquid fraction and increased the gaseous fraction. Furthermore, municipal plastic wastes pyrolysis produced higher heating value solid than those of biomass and low-rank coal. It has been studied that thermal pyrolysis (without catalyst) produced the highest liquid fraction. Singh et al. [9] converted waste plastic to high-quality oil by the method of pyrolysis and studied blends with diesel. High-quality alternative fuel was used in CI engine. Fast pyrolysis method was reported for the production of high-quality hydrocarbons. LDPE (low-density polyethylene) and EVA (Ethylene–vinyl acetate) were used as feedstocks. The study explored that the BSFC was increased with load with respect to WPPO 50, WPPO 70, and diesel fuel. The reading of BSFC decreases for all fuel blends due to the load increases. At full load WPPO blends show the specific fuel consumption is higher than diesel. It was concluded that waste plastic oil is a good energy resources as the future transportation fuel. Kumar et al. [10] evaluated the waste plastic oil and its blend in diesel engines. Physicochemical properties like FTIR and GCMS were investigated for waste plastic fuel and waste plastic fuel blends with diesel. Performance and emission characteristics have been studied with the single-cylinder engine and compared with the diesel. It was observed that the engine can operate with neat waste plastic fuel and its blends without any modification and can be used as an alternative fuel for the diesel engine. It was found that WPF10D90 (10% waste plastic oil and 90% diesel fuel) shows similar results as compared to diesel. Rao et al. [11] experimented with detecting the properties of the pyrolysis oil. A mixture of one ton of plastic waste is subjected to produce oil. The process involves washing waste several times to ensure dust elimination. After which the waste is dried up to room temperature to remove moisture content. The reaction is carried out in the presence of catalyst 'Aluminum silicate'. The volume of catalyst is kept to 2.5% of the mass of the waste. The mixture is heated at 603–723 K. As a result of this pyrolysis, three products are obtained pyrolysis oil (60–70%), gas (15–20%), and carbon black (20–30%). Further various mixture containing 40% oil + 60% petrol, 10% oil + 90% diesel, 30% oil + 70% diesel and 50% oil + 50% diesel are subjected to various characteristic tests and are found similar to that of pure fuel oil. Increasing resin wastes may be bad concerns for the climate change. It has been found that about 60% of solid plastic waste is decomposed in open space or land filled around the world. Municipal wastes are heterogeneous unlike the industrial wastes. It is difficult to recover the energy resources from heterogeneous wastes as mixture of resins is present. Therefore, a specific, beneficial, ecofriendly, and economic method is required. It is good approach for waste to wealth is recycling of waste plastics in a specific way. Available alternate recycling investigations are used to explain wide range of methods such as pyrolysis, methanolysis, gasification, hydrolysis and hydrogenation. The proper disposal and incineration have too much cost as plastics wastes are prodigious than other organic residues and occupy mountainous space in landfills. Also, incineration of plastic wastes is caused for different environmental problems due to continuous emission of hazardous components. In the present study, conversion of waste HDPE into useful hydrocarbon has been focused on and compared with the transport fuel. Different types of components present in the waste HDPE have been reported by SEM/EDAX analysis.

2 Experimental

2.1 Materials and Methods

The waste HDPE plastics were collected from various waste sites. These plastics were washed properly to remove dirt and then dried. After drying, plastics were crushed into smaller pieces. These plastics are more tightly and dense and hence the name was given "High-Density Polyethylene". 10 gm of the sample was taken for each triplicate experiment, while for collection of the main volume of liquid product for further study, 460 gm was taken. Thermo-electric degradation process has been carried out in the stainless-steel reactor with height; 24 cm inner diameter; 180 mm and a wall thickness of 2.5 mm. A bulb condenser has been used for condensation. The delivery pipe of the reactor is connected with the bulb condenser leading to the receiver as shown in Fig. 1. The outcomes of the reaction have been found as non-condensable gaseous products, wax, solid residue, and condensable oil. The yellowish color of the liquid sample was recovered. The total percentage of conversion has been followed by:

$$\text{Conversion}(\%) = \left(\left[\text{Weight of HDPE sample} - \text{Weight of Residue}\right] \times 100\right) / (\text{Weight of HDPE sample})$$

Physical parameters of the oil sample have been followed by ASTM methods. Viscosity was determined by viscometer and specific gravity was evaluated mathematically. The gross calorific value of the oil sample was reported using Parr 6200 (Model: A1290DDEE) Bomb Calorimeter. Abbe's Refractometer (NAR-IT ATAGO CO., TOKYO, Japan) was used for calculating the refractive index. 460 gm of the HDPE sample was heated within the temperature ranges from 300 to 450 °C for five hours. Most of the researchers have not considered the effect of reaction time

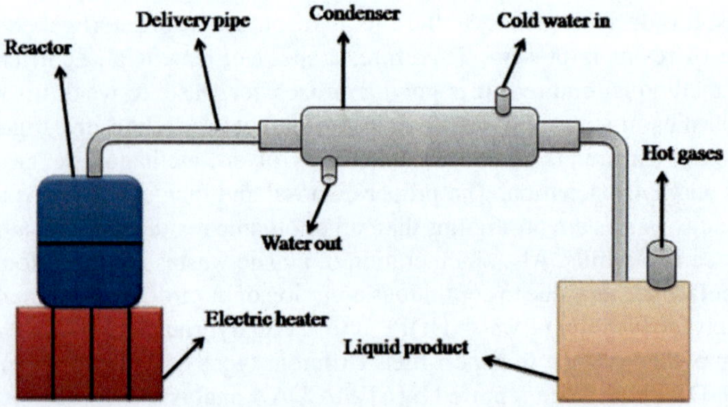

Fig. 1 Systematic diagram of the experimental setup

concerning the thermal cracking of waste plastics. Seo et al. [12] and Miskolezi et al. [13] studied the catalytic deformation of HDPE for about one hour. Fixed process time of 160 min has been considered by Manos et al. [14].

After that the weight of the wax including with gases have been calculating as 365 gm. The conversion percentage of HDPE waste is calculated as 20.65%.

2.2 *Scanning Electron Microscope/Energy-Dispersive X-ray Spectroscopy Analysis of HDPE*

A morphological investigation has been carried out as shown in Fig. 2. SEM analysis was recorded by LEO S430 scanning electron microscope attached with energy dispersive X-ray analyzer model OXFORD LINK ISIS. Thread type structure has been observed and focused at spectrum 1 shown in Fig. 2. Table 1 shows the weight % and atomic % of the elements present in the HDPE sample. Arc discharge method was used for carbon coating on the samples for SEM/EDAX. The sputter current was set to 20 mA and 59 s for coal coating. 5 and 20 kV electrons high tension was used for SEM and EDAX analysis. Chamber pressure was recorded as 1.0 e^{+3} mbar. Less atomic weight of FeS_2 has been found about to 0.06 percent which is nearby half of its weight percentage.

The micrograph shows broken edges in the SEM images and a large number of particles are spotted with Fe, and Ti [15]. It has been observed that titanium and iron

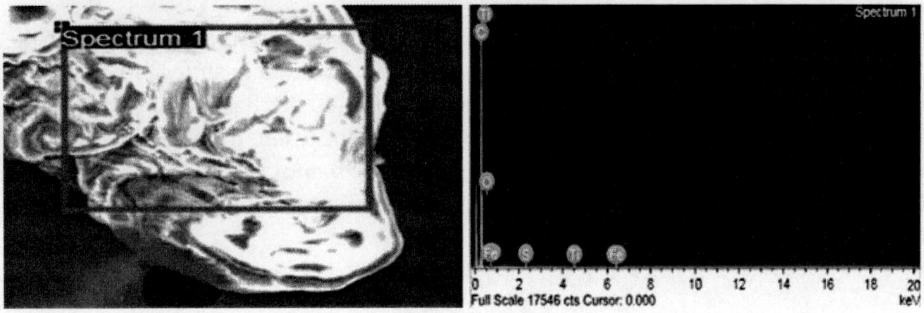

Fig. 2 SEM image and EDAX view of HDPE sample

Table 1 Weight % and Atomic % of HDPE components traced by EDAX

HDPE sample					
Percentage	$CaCO_3$	SiO_2	FeS_2	Ti	Fe
Weight	98.71	0.81	0.15	0.13	0.21
Atomic	99.25	0.61	0.06	0.03	0.04

Table 2 Comparison of physical parameters of waste HDPE oil with standard fuel [16–21]

Physical properties				
Waste HDPE Oil		Standard values		
		Diesel	Kerosene	Gasoline
Density (g/mL)	0.798	0.83–0.85	0.78–0.82	0.72–0.74
Refractive index	1.296	1.484	1.440	1.434
Specific gravity	0.765	0.80–0.85	0.70–0.80	0.71–0.73
Viscosity (mm^2/s)	3.75	1.9–4.5	1.1–1.9	0.775–0.840
Flashpoint (ºC)	48.0	52.0–60.0	50.0–55.0	37.8–42.0
Pour point (ºC)	20	6	–	–
Cal.Value (MJ/kg)	41.7	43–44	42.5–47.0	40.0–250.0

have less atomic percentage 0.03 and 0.04, respectively. High weight and atomic percentage of calcium carbonate have been found 98.71 and 99.25, respectively.

3 Results and Discussions

3.1 Physical Properties Calculations of Waste HDPE Plastic Oil

The density reading was taken at the stable stage of the hydrometer. 30 ml of the sample was put into 375 ml of measuring cylinder. A closed flash tester of Pensky-Martens Kehler-Model K-16270 was used to determine the flashpoint of the liquid sample. The dried test sample was put into the cup of the tester, followed by regulated heating temperature. The temperature at which a large flame occurred was recorded as the flashpoint of the liquid sample. Different physical properties have been reported in Table 2.

3.2 Comparative Exploration of Waste HDPE Oil with Standard Economical Fuels

It has been studied that the density of the recovered waste HDPE oil is nearby the density ranges from 0.83 to 0.85 g/mL of diesel fuel available in the market. Thermo-electric waste HDPE oil has a less refractive index in comparison with others. The viscosity has been calculated at 40 °C and it is within the range of standard diesel. The flashpoint of obtained oil is greater than the gasoline. An effective calorific value close to gasoline has been reported. Calorific value of the HDPE oil was found 41.7 MJ/kg which is a relevant result and shows that the multipurpose use of the

Fig. 3 Waste HDPE oil and wax

product oil in different sectors like transportation, industry, and agree-machinery etc. Waste HDPE oil has relevant potential as a new energy.

Resource for the transportation fuel industry. It was observed that most of the polymer portion has been converted into wax (79.34%) with gaseous products as hydrocarbons. Jan et al. [22–27] studied that in catalytic degradation amount of wax is reduced and the oil yield has been increased by 4.8%.

4 Conclusion

Recycling procedure of waste HDPE has been studied in this research, and it has been explored that the easy conversion of waste HDPE into useful hydrocarbons. The recovery percentage of the liquid product is 20.65%. SEM and EDAX analysis of the HDPE sample shows that the minor (atomic %) presence of iron and titanium. The high-calorific value of the liquid product has been recovered which has similar physical properties of transportation fuel. The liquid product is more viscous in comparison with the other fuels. Lower operating temperature is caused for the good yield of liquid products. The color of the thermo-electric waste HDPE oil is yellowish and similar to petrol or diesel. The wax obtained may be used as a binder in different cement works or civil works. This study concluded that the production of transportation type fuel is less cost-effective, eco-friendly, and a part of the methodology of recycling plastic wastes. A simple and easy method of recycling plastic wastes has been explored by comparative study with commercial fuels. Blending of HDPE plastic waste oil may be used for transportation industry. Wax recovered by this process may be used in civil works like construction of roads, and smooth flexible surfaces. The study has a path of three '*R*' reduce, reuse and recycle of plastic wastes. An exploration and comparison of product samples with the standard fuel show that the liquid may serve as alternative resources of energy demand in future.

References

1. Kumar, A., Dash, S. K., Ahamed, M. S., & Lingfa, P. (2020). Study on conversion techniques of alternative fuels from waste plastics. In S. Gosh (Ed.), *Energy Recovery Processes from Wastes* (Vol. 1, pp. 213–224). Springer. https://doi.org/10.1007/978-981-32-9228-4_18.
2. (2018) Fact sheet on plastic waste in India (pp. 1–4). The Energy and Resources Institute.
3. Kumar, A., & Lingfa, P. (2019). Sodium bentonite and kaolin clays, comparative study on their FT-IR, XRF, and XRD. *Materials Today: Proceedings, 22*, 737–742. https://doi.org/10.1016/j.matpr.2019.10.037
4. Kumar, A., & Lingfa, P. (2020). Physico-chemical characterization of hydrochloric acid-treated kaolin clay: An industry approach as a potential catalyst. *Combinatorial Chemistry & High Throughput Screening, 23*, 205–213. https://doi.org/10.2174/1386207323666200219123459
5. Kumar, A., & Lingfa, P. (2019). Physicochemical characterization of sodium bentonite clay and its significance as a catalyst in plastic wastes valorization. In *3rd International Conference on Electronics, Materials Engineering & Nano-Technology (IEMENTech)* (pp. 1–4). IEEE Explore, Kolkata, India. https://doi.org/10.1109/IEMENTech48150.2019.8981195.
6. Kumar, A., & Lingfa, P. (2019). An exploration of waste plastic oil and different hydrocarbon fuels derived from plastic wastes. *Emerging Renewable Energy Technology: Assam, India, 2019*, 1–13.
7. Khan, M. Z. H., Sultana, M., Al-Mamun, M. R., & Hasan, M. R. (2016). Pyrolytic waste plastic oil and its diesel blend: Fuel characterization. *Journal of Environmental and Public Health, 2016*. Article ID 7869080. https://doi.org/10.1155/2016/7869080.
8. Syamsiro, M., Saptoadi, H., Norsujianto, T., Noviasri, P., Cheng, S., Alimuddin, Z., & Yoshikawa, K. (2014). Fuel oil production from municipal plastic wastes in sequential pyrolysis and catalytic reforming reactors. *Energy Procedia, 47*, 180–188. https://doi.org/10.1016/j.egypro.2014.01.212
9. Singh, D., Soni, S. L., Sharma, D., & Kumari, D. (2017). A review on use of waste plastic oil as alternative fuel in CI engine. *International Research Journal of Engineering and Technology, 4*, 1042–1048.
10. Kumar, R., Mishra, M. K., Singh, S. K., & Kumar, A. (2016). Experimental evaluation of waste plastic oil and its blends on a single cylinder diesel engine. *Journal of Mechanical Science and Technology, 30*, 4781–4789. https://doi.org/10.1007/s12206-016-0950-7
11. Rao, L. N., Jayanthi, J. L., & Kamalakar, D. (2015). Conversion of watste plastics into alternative fuel. *International Journal of Engineering Sciences and Research Technology, 4*, 195–201.
12. Seo, Y. H., Lee, K. H., & Shin, D. H. (2003). Investigation of catalytic degradation of high-density polyethylene by hydrocarbon group type analysis. Journal of Analytical and Applied Pyrolysis, *70*, 383–398.
13. Miskolezi, N., Bartha, L., Deak, G., Jover, B., & Kallo, D. (2004). Thermal and thermo-catalytic degradation of high-density polyethylene waste. *Journal of Analytical and Applied Pyrolysis, 72*, 235–242.
14. Manos, G., Garforth, A., & Dwyer, J. (2000). Catalytic degradation of high-density polyethylene over different zeolite structures. *Industrial and Engineering Chemistry Research, 39*, 1198–1202.
15. Kumar, A., & Lingfa, P. (2020). Acid-activated sodium bentonite and kaolin clay: Comparative study by physicochemical properties. *Combinatorial Chemistry & High Throughput Screening, 23*, 433–445. https://doi.org/10.2174/1386207323666200311114349
16. García, R. A., Serrano, D. P., & Otero, D. (2005). Catalytic cracking of HDPE over hybrid zeolitic-mesoporous materials. *Journal of Analytical and Applied Pyrolysis, 74*, 379–386.
17. Sarker, M., Kabir, A., Rashid, M. M., Molla, M., & Din Mohammad, A. S. M. (2011). Waste polyethylene terephthalate (PETE-1) conversion into liquid fuel. *Journal of Fundamentals of Renewable Energy and Applications*, (1), 1–5.
18. Manickaraja, E., & Tamilkolundu, S. (2014). Catalytic degradation of waste PVC into liquid fuel using $BaCO_3$ as catalyst and its blending properties with diesel fuel. *Discovery, 23*, 74–78.

19. Desai, S. B., & Galage, C. K. (2015). Production and analysis of pyrolysis oil from waste plastic in Kolhapur city. *International Journal of Engineering Research and General Science, 3*, 590–595.
20. Pinto, F., Costa, P., Gulyurtlu, I., & Cabrita, I. (1998). Pyrolysis of plastic wastes. 1. Effect of plastic waste composition on product yield. *Journal of Analytical and Applied Pyrolysis, 51*, 39–55.
21. Blazso, M. (2006). Composition of liquid fuels derived from the pyrolysis of plastics. In J. Scheirs & W. Kaminsky (Eds.), *Feedstock recycling and pyrolysis of waste plastics: Converting waste plastics into diesel and other fuels* (pp. 315–342). Wiley.
22. Jan, M. R., Shah, J., & Gulab, H. (2010). Catalytic degradation of waste high-density polyethylene into fuel products using $BaCO_3$ as a catalyst. *Fuel Processing Technology, 91*, 1428–1437.
23. Kaminsky, W., & Zorriqueta, I.-J.N. (2007). Catalytical and thermal pyrolysis of polyolefins. *Journal of Analytical and Applied Pyrolysis, 79*, 368–374. https://doi.org/10.1016/j.jaap.2006.11.005
24. Mastral, J. F., Berrueco, C., Gea, M., & Ceamanos, J. (2006). Catalytic degradation of high-density polyethylene over nanocrystalline HZSM-5 zeolite. *Polymer Degradation & Stability, 91*, 3330–3338. https://doi.org/10.1016/j.polymdegradstab.2006.06.009
25. Park, D. W., Hwang, E. Y., Kim, J. R., Choi, J. K., Kim, Y. A., & Woo, H. C. (1999). Catalytic degradation of polyethylene over solid acid catalysts. *Polymer Degradation & Stability, 65*, 193–198.
26. Huang, W.-C., Huang, M.-S., Huang, C.-F., Chen, C.-C., & Ou, K.-L. (2010). Thermochemical conversion of polymer wastes into hydrocarbon fuels over various fluidizing cracking catalysts. *Fuel, 89*, 2305–2316. https://doi.org/10.1016/j.fuel.2010.04.013
27. Demirbas, A. (2004). Pyrolysis of municipal of plastic wastes for recovery of gasoline-range hydrocarbons. *Journal of Analytical and Applied Pyrolysis, 72*, 97–102. https://doi.org/10.1016/j.jaap.2004.03.001.

A Review on Indigenous Freshwater Microalgae Isolated from Natural Habitats of Arunachal Pradesh, India, as a Biodiesel Source

Nabam Hina Papu and Pradip Lingfa

1 Introduction

With increase in global energy demand due to change in lifestyle, heavy industrial development and increase in population, the dependence on fossil fuels is rising exponentially. Fossil fuels are unsustainable energy sources and the widescale use of fossils might exhaust world fossil oil reserves [1, 2]. Also as per European Environmental Agency (EEA), the energy and transport sectors are responsible for around 80% of the greenhouse gas emission in the European continent [2]. And as per International Energy Agency (IEA) in the future, the countries with the fastest-growing major economy such as India and China will further raise energy consumption which may lead to major release of greenhouse gases [3]. Greenhouse gas contributes to climate changes, global warming, respiratory diseases and changes in pH level of ocean, affecting the safety of both environment and human life [4]. Therefore, due to energy crisis coupled with environmental pollution has forced several nations to consider alternate source of energy which is clean and renewable [5].

1.1 Energy Demand of India

India is highly dependent on imports of crude oil, the country imported an estimated 80.6, 81.7, 82.69 and 83.3% of oil it consumes in year 2015–16, 2016–17, 2017–18 and 2018–2019, respectively [6]. As per IEA 2007, India's crude oil consumption may rise to 6 million barrels/day (2.2 billion barrels/year) [3]. Apart from energy insecurity, India has to deal with environmental safety which is caused by pollution generated by fossil fuels. Therefore, the need of sustainable fuel supplies and rise of crude oil prices have led the Indian government and other economic experts to search

N. H. Papu (✉) · P. Lingfa
Department of Mechanical Engineering, NERIST, Nirjuli, Arunachal Pradesh 790102, India

P. Mahanta et al. (eds.), *Advances in Thermofluids and Renewable Energy*, Lecture Notes in Mechanical Engineering, https://doi.org/10.1007/978-981-16-3497-0_25

for an alternative source of fuel supply. Presently, the Government of India found their interest in biofuel as an alternative source of energy and issued a comprehensive national policy on biofuels created by the Ministry of New and Renewable Energy (MNRE). In 2003, The National Policy on Biofuels encouraged the blending of 5% ethanol with petrol in 9 states of India and later included 20 states in year 2006. And by year 2017, they called for blending at least 20% of biofuels as an additive with petrol and diesel fuel [6].

1.2 Biofuel Feedstock

Biofuel is derived from biomass which can be replenished readily like plant, animal waste and algae material. It is considered to be the largest source of renewable energy and has potential to replace petroleum-based oil. The most common biofuel is bio-ethanol, biodiesel, methane, biogas, etc. Depending upon feedstock from which biofuel is obtained, they are divided into four generations (Fig. 1). First-generation biofuels are derived from food-based feedstock like corn, sugarcane, soybean, and animal fat. Second-generation biofuels are obtained from non-edible oil-based feedstock, agricultural waste, forest residue and plant waste biomass like Jatropha, Nahar, rubber seed, castor, and waste cooking oils. Third-generation biofuels are produced from algae (popular algae species like *Chlorella vulgaris, Botryococcus Braunii,* etc.). And fourth-generation biofuels are just extensions of third-generation biofuels where algae are genetically modified to alter its cellular metabolism for higher CO_2 capture ability and higher lipid production rate [7]. Currently, most of biofuel production comes from first-generation feedstock except for Jatropha. However, extensive use of first-generation feedstock may raise a scarcity of food for human beings,

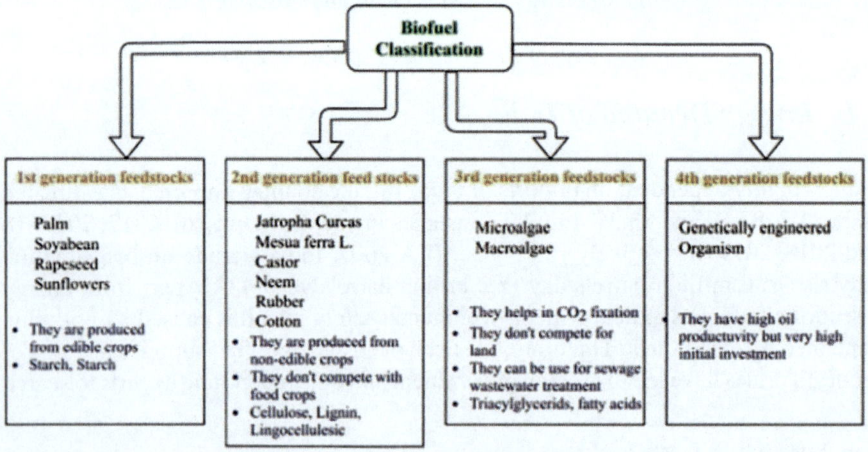

Fig. 1 Biofuel classifications

and it will be very expensive compared to conventional diesel fuel. First-generation feedstock may dominate biofuel production in the future, since industrial production is well-established for them but for developing countries like India, Indonesia and Malaysia producing biofuel from edible crops are not economically feasible [8]. To avoid competition with food, biofuel policy of India (2018) has encouraged the production of biofuel from second-generation feedstock [9]. Nevertheless, the available quantities of second-generation feedstock are not enough to meet the demand of biofuel and growth of plant-based crops needs large amount of land which can lead competition with land and has bad impact on biodiversity [10]. Therefore, recently many researchers have shown interest in third-generation feedstock, algae, because of their ability to remove heavy metals and harness nutrients from the wastewater, ability to capture CO_2 from atmosphere, can be grown in wastelands, high oil content and required less land [7].

1.3 Algae: As Potential Source for Biofuel Production

The use of algae for wastewater treatment using solar energy through photosynthesis subside harmful heavy metal and nutrients, also entrap CO_2 emissions of fossil fuels that have powered machineries like IC engine and power plant. The algae have higher oil yield per acre of land than most terrestrial crop plants, i.e. around 500 to 1000 times more that crop-based plants. Some of algae have high oil content which is around 25–50% of dry weight, and most of them have high triglycerides which is currently thought to be feasible and the higher values for projecting a possibility in the future as a potential source of biofuel production [11]. However, there are severable shortcoming, production of biofuel from algae is not cheap and its production cost surpasses the price of conventional fuel price by vast margin. Use of fertilizers and harvesting accounts for 60% of total production cost, and microalgae cultivation is a water-intensive process which required around 11–13 million litre of water per hectare of land for cultivation in open ponds [12]. The loss of freshwater is a huge economical drawback, also the excess unfiltered fertilizer in a wastewater production from algae cultivation leads to bioaccumulation in the ecosystem harming the surviving flora and fauna. Considering the scope and issue related to microalgae biodiesel production, this review focuses its attention on biodiesel production from naturally available wild algae strains and stresses out its importance [13].

Present study also focuses on scope of freshwater microalgae growth in North-eastern region of India, including their biomass growth, biomass productivity, lipid content, and compared with other microalgae species which are common and not detected in Arunachal Pradesh. The review also discuss performance, emission and combustion characteristics of diesel engine fuelled with microalgae species which were isolated from local water bodies.

2 Biodiesel Productions from Microalgae

Many studies have found that triglycerides have potential to replace conventional diesel fuel in diesel engine and vegetable oils primarily consist of triglycerides. However, the direct use of vegetable oils or its blends can lead to injector choking, plugging and gumming of filter and fuel lines because of unfavourable properties like high viscosity, free fatty acid, density and acid composition of such oils. To overcome the hindrance caused by direct use of vegetable oils in diesel engine, a considerable amount of effort has been made by researchers to develop a technology that could bring properties and performance of vegetable oils similar to conventional diesel fuel. Till now four main processes have been investigated to overcome the drawbacks of vegetable oils, and they are micro-emulsification, pyrolysis, dilution and transesterification [14]. As per Murugesan et al. 2009 and Fukuda et al. 14, transesterification is the most promising technology available right now for biodiesel production and biodiesel derived from third-generation feedstock is one of the best possible options because of its high yield productivity per hectare, CO_2 absorbing capability and act as wastewater purifier [15].

2.1 Recent Development on Microalgae Biofuel

With increase in popularity of microalgae as a source of renewable energy, researchers have underwent several studies, and recently, they have shown their interest towards algae which can grow in dark. Algae which can be cultivated under heterotrophic and mixotrophic modes have higher lipid content than in a photoautotrophic mode of cultivation; though they have lower biomass productivity than latter one, the overall energy consumption of heterotrophic and mixotrophic was observed to be less than phototrophic [16]. Lately, microalgae–microbial fuel cell technology (MMFC) piques the interest of scientist as they can be utilized for bioelectricity production without damaging environment unlike lithium-based battery [17]. There are many studies where cultivation algae in a wastewater were carried, and it was observed that a very few algae species were able to thrive on it. *Chlorella sp.* was able to survive in biochar which was derived from residue of hematite and produce perceptible amount of biomass which can be used for biofuel production [18]. Rasouli et al. cultivated co-culture of green microalgae and methanotrophs in an industrial wastewater and were able to recover nutrient and heavy metal, which aided the cleaning of wastewater [19]. Zainan et al. studied the quality of bio-oil of microalgae *C. vulgaris* which underwent catalytic pyrolysis process and found that the quality was significantly improved compared to other extraction processes [20]. Karthikeyan et al. improved the diesel engine emission by adding Ni-doped ZnO and ZrO_2 nanoparticles in a *B. braunii* [21].

2.2 Prospects of Biodiesel Production from Algae in India

India's dependency on imported crude oil is country's greatest economic vulnerability but despite being energy deficit, many researchers have claim that India has potential to develop promising sustainable energy sources. Researchers claimed that southern part of Asia mainly India is suitable for mass algae cultivation due to its tropical climate which naturally serves as advantages over other countries and vast availability of usable wasteland in India can be utilized for algae biomass production with reasonable effort [22]. Developing countries like India lack proper technique to treat wastewater, and at the same time, there is a growing disquietude on the depravity of wastewater generation. Study showed that the concentration of nutrient level in both municipal and agriculture wastewater was high, and algae grow rapidly in nutrient-rich environment which makes it most suitable for low-cost wastewater treatment [23]. Raceway open pond algae cultivation is considered as satisfactory for tropical countries like India due to its cost-effectiveness and superior flexibility [24]. Back on days India has been mass cultivating *Spirulina* algae in a raceway pond for food and other medical supplements now several institutions and organization have carried work on various aspects of algae [25]. University of Madras, Chennai, cultivated *Sargassum, B.Braunii* microalgae species in an open raceway pond for biogas production, *C. Vulgaris* were extensively studied and cultivated to estimate its biomass production and its resilience in Indian tropical condition by Central Rice Research Institute (CCRI), Cuttack [26]. Life cycle analysis (LCA) of several diatoms species and its potential as an additional biofuel sources were research by Vivekananda Institute of Algal Technology (VIAT), Chennai [27]. VIAT is also developing a technology to treat industrial wastewater using microalgae [28]. Alternate hydro-energy centre, IIT Roorke, successfully converted microalgae oil into biodiesel using several conversion methods [29].

3 Feasibility of Growing Freshwater Microalgae in North-Eastern Region of India

Factors which are associated to selection of site for mass microalgae cultivation are:

(a) *Availability of land for algae cultivation*
With growing population and high demand of fertilized land, it has become impractical to use fertilized land for algae biomass cultivation. Though algae have high oil yield per area as compared to terrestrial crops, still it is not economically viable to compete for fertilized land with food crops especially for developing countries like India. Therefore, wasteland is considered to be attractive options for algae cultivation. The GOI classifies 90 million hectares as wasteland which can be utilized for commercial production of algae biomass. Arunachal Pradesh accounts for total 21.7% of land as wasteland (18,175.95

km^2) which if taken advantage of than it can help out reduce CO_2 from the atmosphere, reduces current diesel consumption and creates job for native [30].

(b) *Availability of water*

The main obstacle for cultivating freshwater microalgae is vast usage of freshwater, as per study 12 million litre of water is used for cultivation of one hectare of land which and around 80% of water are wasted during microalgae cultivation [12]. It is a hefty price to cultivate microalgae when usage of water is considered. Nevertheless, the average rainfall recorded for Arunachal Pradesh is 450 cm, ranging from 100 to 990 cm. The vast majority of rainfall is at mid altitude to low altitude. Arunachal Pradesh has large freshwater sources for mass cultivation and usage of freshwater won't be foremost problem for domestic consumption [31].

(c) *Nutrient sources and supply*

Arunachal Pradesh is filled with vast amount of biomass where animals and birds excreta and fallen leaves can be used as a bio-fertilizer for algae cultivation. Using biological wastages will also help cleaning of environmental.

(d) *Wastewater composition*

Wastewater has lot of nutrients, and few of them can thrive on wastewater. According to Debnath, composition of waste found in Arunachal Pradesh is mostly organic which was 53% in average and can be used for heterotrophic and mixotrophic cultivation as those mode of cultivation tend to deliver higher lipid productivity than photoautotrophic mode [32].

3.1 List of Indigenous Algae Strains Detected in Arunachal Pradesh

Selecting suitable microalgae species as an ideal source of biodiesel is one of the most crucial steps. The important characteristics of selected microalgae species are to have high biomass growth rate, high lipid content with desirable fatty acids, robust in nature, ability to compete against other microorganisms. A very few researchers have studied the diversity of algae in natural habitats of north-eastern region of India, Das and Adhikary documented total of 86 freshwater species which includes 8 Cyanophyta, 30 Chlorophyta, 8 Euglenophyta, 39 diatoms and 1 Xanthophyta from different locations of Arunachal Pradesh [33]. Out of 86 freshwater species which were documented by Das and Adhikary only 13 of them were found to be studied by researchers for their cell growth, lipid content and their potential as a feedstock for biofuel (Table 1).

Table 1 List of Indigenous algae reported in the literature

Freshwater algae species	Ref
Cynophyta (Blue-green algae)	
Oedogonium sp	[34]
Spirogyra sp	[34]
Botryococcus Braunii	[12,35–39]
Scenedesmus Bijugatus	[40 41]
Scenedesmus quadricauda	[42–45]
Euglenophyta	
Euglena sanguinea	[46]
Diatoms (Bacillariophyta)	
Tabellaria fenestrate	[47]
Fragilaria capucina	[48]
Synedra ulna	[49, 50]
Diadesmis confervacea	[51]
Navicula sphaerophora	[52]
Nitzschia species	[47]
Navicula sp	[47]

4 An Overview of Growth Dynamics and Lipid Content of Freshwater Microalgae Found in Arunachal Pradesh

The biomass productivity, specific growth rate, oil content and culture medium used for the growth of same algae species found in local water bodies are collected from several literatures and placed in Tables 2 and 3. From this study, we will get the general idea of growth behaviour, nature and biofuel potential of locally available algae strains. Since every algae is different and using fourth-generation algae is presently not practical; therefore, opting for naturally available biomass is economical viable for now. Oedogonium as a genus of filamentous green algae are generally found in freshwater pond and lakes. Lawton et al. (2013) studied three types of freshwater macroalgae of genus *Oedogonium, Spirogyra* and *Cladophora.* The biomass productivity of Oedogonium was at least 20% higher than *Spirogyra* and *Cladophora* when they are treated with low air supply and no external CO_2 supply. *Oedogonium* was able to dominate in mixed species culture and grew exponentially within a week [53]. *Oedogonium* were able to withstand varying range on environmental conditions and demonstrated that natural occurring species of *Oedogonium* can be domesticated which provides first crucial steps to determine suitability of algae as an energy source [54].

B. braunii strain is a green, pyramid-shaped planktonic microalgae which grow in cluster and can be found in tropical and temperate lakes and estuaries [55]. It was claimed by many researchers as one of the most promising feedstock out of all microalgae and macroalgae because of their high lipid content and high biomass

Table 2 Biomass productivity of freshwater algae [12, 34–51]

Algae species	Nutrient	Biomass productivity/ Biomass yield*
Mycrocystis sp	Modified Murashige and Skoog Basal Medium	20.30 ± 1.79 mg/L*
Oedogonium species	BB medium	0.478 g/L*
Botryococcus braunii	Chu13	0.026 mg/L/d
	na	0.035 mg/L/d
	Modified Chu13	0.026 mg/L/d
	BG11	0.037 mg/L/d
	Secondary domestic wastewater	0.034 mg/L/d
	na	0.19 mg/L/d
	na	0.26 mg/L/d
	na	0.345 mg/L/d
	BG11	0.03 mg/L/d
	Modified Chu13	0.114 mg/L/d
Scenedesmus bijugatus	BG11 medium	28.67 mg/L/d
	BBM	0.26 g/L/d
Scenedesmus quadricauda	Aquaculture wastewater	4.85 ± 0.02 mg/L/h 10.44 ± 0.14 mg/L/h
	BG11medium	68.92 mg/L/d
	na	0.19 g/L/d
	BG11 medium	81.79 mg/L/d
	Sewage wastewater	79.17 mg/L/d
	BBM	0.19 g/L/d
Euglena sanguinea	0.5 mL seaweed + 0.25 mL poultry extract + 19 NPK fertilizer	0.927 g/L/d
Tabellaria sp	Modified Murashige and Skoog Basal Medium	16.90 ± 0.80 mg/L
Fragilaria capucina	Modified WC medium (NaCL & Si added)	0.61/ d
Synedra crystalline		
Synedra species	Modified Murashige and Skoog Basal Medium	21.30 ± 1.53 mg/L
Synedra ulna	na	0.75/ d
	sD11 medium with 540 μg $NaSiO_3$	0.64/d (6.4 g lipid/ml)
	Modified Csi medium	0.245/ d
Diadesmis confervacea	modified f/2 media	
Navicula sphaerophora	Chu13 medium	0.508 g/L

(continued)

Table 2 (continued)

Algae species	Nutrient	Biomass productivity/ Biomass yield*
Navicula species	Modified Murashige and Skoog Basal Medium	38.20 ± 1.13* mg/L
	CSi medium	0.17–0.221/ d
Nitzschia species	Modified Murashige and Skoog Basal Medium	20.50 ± 0.36* mg/L
Spirogyra species	BG medium	0.469 g/L

productivity. Some algae strain of *B.braunii* were able to produce about 86% of hydrocarbons [40, 42, 43, 56]. *B.braunii is* lipid composed of saturated and polysaturated fatty acids and also triacylglycerols (TAGs) which makes it suitable for biodiesel production [56].

Scenedesmus is a genus of green algae which grow into colony and can exist in unicells. Many researchers have studied the species of *Scenedesmus* for wastewater treatment, biomass estimation and their potential as a feedstock for biodiesel. Among several species of genus *Scenedesmus*, *Scenedesmus quadricauda* and *Scenedesmus bijugatus* were reported in Arunachal Pradesh, India [33]. Han et al. were able to obtain 28.67 mg/L/d of algae biomass productivity when *S. bijugatus* were cultivated with BG11 culture medium; he also reported that *S. bijugatus* content 24.3% of total lipid dry weight basis (DWB) which makes it good competitor as a biodiesel feedstock [40]. Ashok et al. reported that they were able to obtain 0.26 g/L/d of *S. bijugatus* biomass productivity. Halfhide et al. cultivated *S. quadricauda* freshwater algae in a filtered wastewater. He used two different culture environment, axenic culture and non-axenic culture. They obtained maximum biomass productivity of 4.85 ± 0.02 mg/L/h for axenic culture and 10.44 ± 0.14 mg/L/h for non-axenic culture [42]. Ahmad et al. reported 0.19 g/L/d biomass productivity with 18.4% of total lipid content on DWB [43]. Han et al. (2015) studied the survivability and change in chemical composition of *S. quadricauda* in different environments. They cultivated *S. quadricauda* in two media: sewage water and BG culture medium. They reported that by using sewage water as a nutrient medium, they were able to obtain 79.17 mg/L/d biomass productivity with total lipid content 27.36% DWB, and by using BG culture medium, they obtained 81.79 mg/L/d biomass productivity with total lipid content 31.31% DWB [44].

Euglena Sanguinea is unicellular species which belongs to family of *Euglenaceae* and order of *Euglenales*. They are motile in nature and their shape changes depending upon motion. They are extremely robust in nature and can survive in sewage wastewater. They are also good in breaking down the nutrient and can be used to minimizing the harmful chemicals in wastewater. They have biomass productivity of 0.927 g/L/d and can store 25% DWB lipid on their cell [45].

Tabellaria is a genus of freshwater diatoms or Bacillariophyta. They have cuboid shape and siliceous cell walls are attached at the corners due to which colonies are

Table 3 Freshwater algae lipid content

Algae species	Lipid content (% DWB)	Ref
First-generation biodiesel feedstock		
Soyabean	18	[60]
Sunflower	40	[60]
peanut	45–52	[60]
Rapeseed	41	[60]
Oil palm	40–52	[60]
groundnut	66	[60]
Second-generation biodiesel feedstock		
Jatropha curcas	20–60	[61]
Rubber	40–60	[61]
Cotton	17–25	[61]
Neem	25–45	[61]
Mahua	35–50	[61]
Pongamia Pinnata	27–39	[61]
Castor	45–50	[61]
Tung	35–40	[61]
Mesua ferra L	58–75	[61]
Soapnut	51.8	[61]
Third-generation biodiesel feedstock		
Mycrocystis sp	16.5	[47]
Oedogonium species	9.2	[62]
	11.4	[34]
Botryococcus braunii	25.7	[35]
	13.5	[12]
	36.14	[37]
	18.4	[38]
	21.1	[38]
	17	[38]
	40	[39]
	19	[63]
Scenedesmus bijugatus	24.3	[40]
Scenedesmus quadricauda	18.4	[43]
	31.31	[44]
	27.36	[44]
Tabellaria sp	10.90 ± 1.71	[47]
Fragilaria capucina	14 (30 Low N)	[48]

(continued)

Table 3 (continued)

Algae species	Lipid content (% DWB)	Ref
Synedra crystalline	na	
Synedra species	20.30 ± 2.01	[47]
Synedra ulna	6.4 * (g lipid/mL)	[50]
	29	[50]
Diadesmis confervacea	14.6	[51]
Navicula sphaerophora	13.8	[52]
Navicula species	27.20 ± 1.37	[47]
	24	[50]
Nitzschia species	16.50 ± 1.24	[47]
Spirogyra species	7.3	[62]
	4.93	[34]
Euglena sanguinea	25	[46]

shaped in zig-zag order [57]. Saputro et al. isolated total 8 freshwater microalgae from Wonorejo River and Surabaya River of Indonesia. They cultivated *Tabellaria sp* using modified Murashige and Skoog Basal culture medium and obtained 16.90 ± 0.80 mg/L total biomass yield. They also reported that *Tabellaria sp* content 10.90 ± 1.71% of total lipid on dry weight basis [47].

Fragilaria capucina is filamentous freshwater diatoms/ Bacillariophyta. *F. capucina* can accumulate lipid even at high concentration level of N and Si. Justin et al. reported that the maximum specific growth rate of *F. capucina* was 0.61 per day, and total lipid content at low N level was more than 30% DWB [48]. *Synedra ulna* comes under family of Fragilariaceae. Specific growth rate of *S. ulna* is reported to be 0.245 per day to 0.75 per day and has oil content around 29% DWB [34, 58, 59]. Saputro et al. studied the specific growth rate and lipid content of 8 isolated algae using modified Murashige and Skoog Basal culture medium and reported that Synedra species have total biomass growth 21.30 ± 1.53 mg/L and content 20.30 ± 2.01% of oil in its cell on DWB [47].

Navicula is a genus of boat-like diatoms, and they have been identified as high lipid-producing algae. They have biomass productivity of around 38.2 mg/L/d and specific growth rate around 0.17–0.22 per day. As per report, they are able to produce 27.2–24% of lipid [47]. Vinayak reported that they have discovered a strain of diatom *Diadesmis confervacea* which can accumulate lipid in its cell up to 14.6% DWB and also extract oil naturally when cells attain maturity in 31st day of culturing in vivo [51].

5 Performance and Emission Analysis of Diesel Engine Fuelled with Algae Biodiesel

There are various physicochemical properties of biodiesel which contribute to any changes in performance and emission characteristics of a diesel engine. Biodiesel is an oxygenated fuel, and it often contains extra oxygen atoms which influence performance, emission and combustion significantly. Most of the cases, use of biodiesel in diesel engine results in reduction of unburned hydrocarbons (UHC), Carbon monoxide (CO) and particulate matter (PM) as shown in Table 5. Brake-specific fuel consumption (BSFC) and brake thermal efficiency (BTE) are two important parameters to study the performance characteristics of an engine. BTE is a measure of % of chemical energy in a fuel which transform into useful work. There are numerous publications on engine analysis using biodiesel and its blend as shown in Table 4. For most of the crop-based biodiesel, the BTE decrease with increase in volume percentage of biodiesel in a blend which is due to low energy density of biodiesel fuel as compared to diesel fuel [64]. BSFC and BTE have an inversely relationship as because mathematical formula of BTE is almost reciprocal of BSFC mathematical formula. Therefore, brake-specific fuel consumption (BSFC) is generally higher for biodiesel and its blends than conventional diesel fuel as the quality of biodiesel depends upon the composition of oil. *C. vulgaris* biodiesel and its blends increase the BTE of diesel engine and similar reports were found for *B. Braunii* and *Scenedesmus Obliquus* biodiesel [65, 58]. This may be due to higher cetane number of algae biodiesel which improves the ignition delay and enhance the combustion process, thus lowering the fuel consumption or due to higher oxygen molecules in biodiesel which help burning the fuel efficiently. The result of *C. vulgaris, B. Braunii* and *Scenedesmus Obliquus* displayed the potential of algae biodiesel as an alternative fuel which even improves the engine performance.

NO_x is formed due to ignition delay, oxygen content in the fuel and temperature of combustion chamber. Unsaturated fatty acids and longer hydrocarbon chain in biodiesel are correlated to the formation of NO_x. Except for *Scenedesmus Obliquus,* all other algae biodiesel tested in diesel engine increase the NO_x emissions which could be result of extra oxygen atoms which improves combustion and thus increases the in cylinder temperature. Subsequently, it enhances The NOx formation [59] and CO are formed due to incomplete combustion of air and fuel mixture, this happens due to insufficient oxygen supply to complete the combustion reaction. The ability to provide extra oxygen atoms helps in peroxidising the air-fuel mixture and biodiesel being an oxygenated fuel reduces the harmful emission of HC and CO. Patel et al. also reported the reduce in particulate matter (PM) when using *C. vulgaris* biodiesel in a diesel engine [65]

Table 4 Performance of diesel engine using crop-based and algae biodiesel [12, 52, 58, 59, 61, 64, 65]

Biodiesel	Testing unit	Engine performance	
		BTE	BSFC
Crop-based biodiesel			
Soyabean	na	↓	↑
Sunflower	na	↓	↑
Waste cooking oil	na	↓	↑
Mesua ferra L	Single cylinder, water cooled (6.5 hp)	↓	↑
Algae biodiesel			
Chlorella Vulgaris	Single cylinder, water cooled (5 hp)	↑	↓
Crypthecodinium cohnii	Four cylinder, turbocharged, intercooled, (134 hp)	↓	↑
*B.braunii**	Single cylinder, air cooled	↑	↓
*Scendesmus Obliqus**	Single cylinder, water cooled (6.5 hp)	↑	↓
Navicula sphaerophora *	Single cylinder, water cooled (4.69 hp)	↓	↑
*Euglena sanguinea**	Single cylinder, water cooled (4.69 hp)	↓	↑

*Algae species found in Arunachal Pradesh

6 Combustion Analysis of Diesel Engine Fuelled with Algae Biodiesel

In Table 6, all the report on combustion analysis of microalgae biodiesel in diesel engine stated and compared with each other. Islam et al. reported that cylinder pressure (CP) of *Crypthecodinium cohnii* biodiesel was slightly less than diesel fuel at 25% engine load and after further increase in load there was no significant difference between blends and diesel fuel. The decrease in cylinder presser (CP) at lower engine load may be because of high viscosity of biodiesel [59]. Heaik et al. tested *Ankistrodesmus braunii* and *Nannochloropsis sp* biodiesel in single cylinder Ricardo engine and observed that slight in-cylinder pressure, Rate of pressure rise (RPR) and net heat release rate NHRR [35] and reported just opposite result of Islma et al.

Table 5 Emission of diesel engine while running using crop-based biodiesel and algae biodiesel [52, 58, 59, 64, 65]

Biodiesel	Testing unit	Emissions				
		NO_x	HC	CO	PM	EGT
B.braunii	1-cylinder, air cooled	↑	↓	↓	na	na
Scendesmus Obliqus*	1-cylinder, water cooled (6.5 hp)	↓	↓	↓	na	na
*Navicula sphaerophora**	1-cylinder, water cooled (4.69 hp)	↑	↓	↓	na	na
*Euglena sanguinea**	1-cylinder, water cooled (4.69 hp)	↑	↓	↓	na	↑
Chlorella Vulgaris	1-cylinder, water cooled (5 hp)	↑	↓	↓ (↑at max load)	↓	na
Crypthecodinium cohnii	1-cylinder, turbocharged, intercooled, (134 hp)	↑	↓	na	na	na

*Algae species found in Arunachal Pradesh

Table 6 Combustion characteristic of diesel engine fuelled with algae biodiesel [35, 59]

Biodiesel	Test unit	Combustion		
		CP	RPR	NHRR
Ankistrodesmus braunii	Single cylinder Ricardo-E6 engine (12 hp)	↑	↑	↑
Nannochloropsis sp	Single cylinder Ricardo-E6 engine (12 hp)	↑	↑	↑
Crypthecodinium cohnii	Four cylinder, turbocharged, intercooled, (134 hp)	↓ (same after 25% load)	↑ (↓ after 50% load)	na

*Algae species found in Arunachal Pradesh

7 Conclusions

To find an ideal sustainable source for biodiesel production, the feedstock should have sufficient amount of oil content, high productivity and suitable fatty acid chains for having good biodiesel properties. Most of recent researches on algae species were on *C. vulgaris* and *Chlamydomonas reinhardtii*, now recently many researchers have shifted their attention towards locally available wild algae species. This review described the potential of 13 indigenous algae species which were found in Arunachal Pradesh and all algae species have found to yield biomass productivity and content high % of oil (DCW). Four wild freshwater algae species *B. braunii, Scenedesmus obliquus, C. vulgaris and E. sanguinea found to be most favourable among other species,* these have ability to utilize for wastewater treatment and carry out nutrient breakdown as well as aid the production of algae biodiesel. By using sewage wastewater for algae biomass production is not only advantageous for environmental safety and CO_2 reduction from atmosphere but also lower the cost of biodiesel production by avoiding fertilizers.

7.1 Scope for Future Research Work

1. Chemical and physical composition changes for algae depending upon their environment. Therefore, lipid content for the same species differs as per report because researcher has used different environment for algae cultivation and lipid content generally varies with several different parameters like temperature, culture medium, pH value, nutrient level of N, K, Ni, P, CO_2 concentration level and air turbulence. Thus, there is a scope to further study the suitability of algae species for more biomass growth rate and higher lipid productivity.
2. Microalgae biodiesel generally depends upon TAGs and free fatty acids of lipid. Not many reports have been found on fatty acid composition of microalgae and the report varies for same species depending upon extraction method, culture medium and cultivating environment. Therefore, there is a scope to further study the range of fatty acid compositions of algae lipid
3. Many publications have reported the potential of algae biodiesel as a replacement of conventional diesel fuel in a diesel engine but a very few report have been found studying the actual engine characteristics when fuelled with algae biodiesel due to the limited amount of fuel. A complete mapping of engine performance, emission, and combustion characteristics with microalgae biodiesel needs to be done.

References

1. Chandrasekhar, K., Lee, Y.-J., & Lee, D.-W. (2015). Biohydrogen production: Strategies to improve process efficiency through microbial routes.*International Journal Molecular Science, 16*, 8266–8293.
2. European Environmental Agency (EEA) .(2004). Greenhouse gas emission trends and projections in Europe 2004: Progress by the EU and its Member States towards achieving their Kyoto Protocol targets. Report N85.
3. International Energy Agency (IEA). World Energy Outlook (2007). China and India Insights, Paris, France.

4. Ormerod, W. G., Freund, P., Smith, A., & Davison, J. (2002). Ocean storage of CO_2. IEA greenhouse gas R&D programme. International Energy Agency
5. Ministry of Petroleum and Natural Gas, Government of India. (2020). http://petroleum.nic.in.
6. Basavaraj, G., Rao, P., Reddy, C., Ashok, A., Pinnamaneni, S. R., & Reddy, B. V. S. (2012). A review of national biofuel policy in India: A critique need for promotion of alternative feedstock. *Journal Biofuels, 3*(2), 65–78.
7. Kasturi, D., Achlesh, D., & GawLin, J. (2014). Evolution retrospective for alternative fuels: First to fourth generation. *Renewable Energy, 69*, 114–122.
8. Ravindranath, N. H., Lakshmi, C. S., Manuvie, R., & Balachandra, P. (2011). Biofuel production and implications for land use, food production and environment in India. *Energy Policy, 39*(10), 5737–5745.
9. Ministry of Petroleum and Natural Gas. (2018). National Policy of Biofuel, New Delhi.
10. Reinhardt, G., Rettenmaier, N., Ko¨ppen, S. (2008). How sustainable are biofuels for transportation? In *Bioenergy: Challenges and opportunities. International Conference and Exhibition on Bioenergy*. ISBN 978–92–5–107414–5.
11. Lundquist, T. L., Woertz, I. C., Quinn, N. W. T., & Benemann, J. R. (2010). A realistic technology and engineering assessment of algae biofuel production. *EBI, 2010*, 1–178.
12. Chinnasamy, S., Bhatnagar, A., Hunt, R. W., & Das, K. C. (2010). Microalgae cultivation in a wastewater dominated by carpet mill effluents for biofuel applications. *Bioresource Technology, 101*, 3097–3105.
13. Mahapatra, D. M., Chanakya, H. N., & Ramachandra, T. V. (2011). Assessment of treatment capabilities of Varthur Lake, Bangalore India. *International Journal Environment, 14*, 84–102.
14. Fukuda, H., Kondo, A., & Noda, H. (2001). Biodiesel fuel production by transesterification of oils. *Journal of Bioscience and Bioengineering, 92*, 405–416.
15. Srivastava, A., & Prasad, R. (2000). Triglycerides-based diesel fuels. *Renewable Sustainable Energy Review, 4*(2), 111–133.
16. Papu, N. H., Lingfa, P., Dash, S. K. (2020). *Euglena Sanguinea* algal biodiesel and its various diesel blends as diesel engine fuels: A study on the performance and emission characteristics. *Energy Sources, Part*. https://doi.org/10.1080/15567036.2020.1798566
17. Costa, C., & Hadiyanto, A. (2018). Bioelectricity production from microalgae-microbial fuel cell technology (MMFC). *MATEC Web Conference, 156*, 2–4.
18. Ashokkumar, V., Chen, W.-H., Kamyab, H., Kumar, G., Al-Muhtaseb, A. H., & Ngamcharussrivichai, C. (2019). Cultivation of microalgae *Chlorella sp.* in municipal sewage for biofuel production and utilization of biochar derived from residue for the conversion of hematite iron ore (Fe_2O_3) to iron (Fe)– Integrated algal biorefinery. *Energy, 189*, 116–128.
19. Rasouli, Z., Valverde-Pérez, B., D'Este, M., De Francisci, D., & Angelidaki, I. (2018). Nutrient recovery from industrial wastewater as single cell protein by a co-culture of green microalgae and methanotrophs. *Biochemical Energy Journal, 134*, 129–133.
20. Zainan, N. H., Srivatsa, S. C., Li, F., & Bhattacharya, S. (2018). Quality of bio-oil from catalytic pyrolysis of microalgae *Chlorella vulgaris. Fuel, 223*, 12–19.
21. Karthikeyan, S., & Prabhakaran, T. D. (2018). Emission analysis of *Botryococcus braunii* algal biofuel using Ni-Doped ZnO nano additives for IC engines. *Energy Sources, Part A433A: Recovery, Utilization, and Environmental Effects+A465, 40*, 1060–1067.
22. Beer, L. L., Boyd, E. S., Peters, J. W., Posewitz, M. C. (2009). Engineering algae for biohydrogen and biofuel production. *Current Opinion Biotechnology, 20*(3), 264–271.
23. Chanakya, H. N., Karthick, B., Ramachandra, T. V. (2006) .Nitrogen and carbon flows through Bellandur lake-role of Bellandur lake as a natural wetland treating Bangalore wastewater. http://wgbis.ces.iisc.ernet.in/energy/lake2006
24. Pulz, O. (2001). Photobioreactors: production systems for phototrophic microorganisms. *Applications Microbiology Biotechnology*, 287–293.
25. Venkatamaran, L. V. (1986). Blue-green algae as biofertilizer. In A. Richmond (Ed.), *CRC Handbook of microalgal mass culture* (pp. 455–471). Boca Raton: CRC Press.
26. Chinnasamy, S., Ramakrishnan, B., Bhatnagar, A., & Das, K. C. (2009). Biomass production potential of a wastewater alga *Chlorella vulgaris* ARC 1 under elevated levels of CO_2 and temperature. *International Journal of Molecular Sciences, 10*(2), 518–532.

27. Ramachandra, T. V., Durga, M. M., & Karthic, B. K. (2009). Milking diatoms for sustainable energy: Biochemical engineering versus gasoline-secreting diatom solar panels. *Industrial and Engineering Chemistry Research, 48*(19), 8769–8788.
28. Sivasubramanian, V., Subramanian, V. V., Raju, P. A., Muthukumaran, M. (2009). In *Phycoremediation of Oil Drilling Waste at Kakinada International Conference on Algal Biomass,Resources and Utilization*, Stella Maris College, Chennai.
29. Rajvanshi, S., & Sharma, M. P. (2012). Micro Algae: A potential source of biodiesel. *Journal of Sustainable Bioenergy Systems, 02*(03).
30. Government of India .(2011). Wasteland atlas of India, Arunachal Pradesh. https://dolr.gov.in/documents/wasteland-atlas-of india.
31. Dhar, O. N., & Nandargi, S. (2004). Rainfall distribution over the Arunachal Pradesh Himalayas . *Weather, 59*(6), 155–157.
32. Debnath, A., & Mitra, A. (2013, December 4–8). Challenges of solid waste accumulation and management in a growing urban area: A study of Itanagar town of Arunachal Pradesh. *Indian Society for Ecological Economics-Global Change, Ecosystems, Sustainability*.
33. Das, S. K., & Adhikar, S. P. (2012). Diversity of freshwater algae in Arunachal Pradesh and their distribution in different altitudes. *Journal Indian Bot Society, 91*(1–3), 160–182.
34. Munir, N., Imtiaz, A., Sharif, N., & Naz, S. (2015). Optimization of growth conditions of different algal strains and determination of their lipid contents. *Journal Animals Plant Science, 25*(2), 546–553. ISSN: 1018–7081
35. Haik, Y., Selim, Md. Y. E., & Abdulrehman, T. (2010). Combustion of algae oil methyl ester in an indirect injection diesel engine. *Energy, 36*, 1827–1835.
36. Lee, J. Y., Yoo, C., Jun, S. Y., Ahn, C. Y., Oh, H. M. (2010). Comparison of several methods for effective lipid extraction from microalgae. *Bioresourcing Technology, 101*(1), S75–S77.
37. Sydney, E. B., da Silva, T. E., Tokarski, A., Novak, A. C., de Carvalho, J. C., & Wojciechowski, A. L. (2011). Screening of microalgae with potential for biodiesel production and nutrient removal from treated domestic sewage. *Applied Energy, 18*(10), 3291–3294.
38. Ahmad, A. L., Yasin, N. H. M., Derek, C. J. C., & Lim, J. K. (2011). Microalgae as a sustainable energy source for biodiesel production: A review. *Renewable Sustainable Energy Review, 15*, 584–593.
39. Abomohra, A.E.-F., Wagner, M., El-Sheek, M., & Hanelt, D. (2012). Lipid and total fatty acid productivity in photoautotrophic fresh water microalgae: Screening studies towards biodiesel production. *Journal of Applied Phycology, 25*(4), 931–936.
40. Han, L., Pei, H., Hu, W., Han, F., Song, M., & Zhang, S. (2014). Nutrient removal and lipid accumulation properties of newly isolated microalgal strains. *Bioresourcing Technology, 165*(C), 38–41.
41. Ashokkumar, V., Salam, Z., Tiwari, O. N., Chinnasamy, S., Mohammede, S., & Ani, F. N. (2015). An integrated approach for biodiesel and bioethanol production from *Scenedesmus bijugatus* cultivated in a vertical tubular photobioreactor. *Energy Conversation Management, 101*, 778–786.
42. Halfhide T, Åkerstrøm, A., Lekang, O. I., Gislerød, H. R., & Ergas, S. J. (2014). Production of algal biomass, chlorophyll, starch and lipids using aquaculture wastewater under axenic and non-axenic conditions. *Algal Research, 6*(B), 152–159.
43. Ahmad, A. L., Yasin, N. H. M., Derek, C. J. C., & Lim, J. K. (2011). Microalgae as a sustainable energy source for biodiesel production: A review. *Renewable Sustainable Energy Review, 15*, 584–593.
44. Han, L., Pei, H., Hu, W., Jiang, L., Ma, G., Zhang, S., & Han, F. (2015). Integrated campus sewage treatment and biomass production by *Scenedesmus quadricauda* SDEC-13. *Bioresource Technology, 175*, 262–268.
45. Rodolfi, L., Zittelli, G. C., Bassi, N., Padovani, G., Biondi, N., & Bonini, G. (2009). Microalgae for oil: Strain selection, induction of lipid synthesis and outdoor mass cultivation in a low-cost photo bioreactor. *Biotechnology and Bioengineering, 102*, 100–112.
46. Kings, A. J., Raj, R. E., Miriam, L. R. M., & Visvanathan, M. A. (2017). Growth studies on microalgae *Euglena sanguinea* in various natural eco-friendly composite media to optimize the lipid productivity. *Bioresource Technology, 244*, 1349–1357.

47. Saputro, T. B., Purwani, K. I., Ermavitalini, D., & Saifullah, A. F. (2019). Isolation of high lipid content microalgae from wonorejo river, Surabaya, Indonesia and its identification using rbcL marker gene. *Biodiversitas, 20*(5), 1380–1388.
48. Chaffin, J. D., Mishra, S., Kuhaneck, R. M., Heckathorn, S. A., & Bridgeman, T. B. (2012). Environmental controls on growth and lipid content for the freshwater diatom, *Fragilaria capucina*: A candidate for biofuel production. *Journal of Applied Phycology, 24*, 1045–1051.
49. Tilman, D., Mattson, M., & Langer, S. (1981). Competition and nutrient kinetics along a temperature gradient: An experimental test of a mechanistic approach to niche theory. *Limnology and Oceanography, 26*(6), 1020–1033.
50. Graham, J. M., Graham, L. E., Zulkifly, S. B., Pfleger, B. F., Hoover, S. W., & Yoshitani, J. (2011). Freshwater diatoms as a source of lipids for biofuels. *J. Ind., 39*, 419–428.
51. Vinayak, V., Gordon, R., Gautam, S., Rai, A. (2014). Discovery of a diatom that oozes oil. *Advanced Science Letters, 20*(7–9), *Special Issue: National Conference on Nanotechnology and Renewable Energy*, New Delhi: Jamia Millia Islamia.
52. Papu, N. H., & Lingfa, P. (2018). Isolation biomass estimation and characterization of the biofuel potential of diatom *Navicula Sphaerophora. AIP Conference Proceedings, 10*(1063/1), 5032022.
53. Lawton, R. J., de Nys, R., & Paul, N. A. (2013). Selecting reliable and robust freshwater Macroalgae for biomass applications. *PLoS ONE, 8*(5), e64168.
54. Lawton, R. J., de Nys, R., Skinner, S., & Paul, N. A. (2014). Isolation and identification of *Oedogonium species* and strains for biomass applications. *PLoS ONE, 9*(3), e90223.
55. Metzger, P., & Largeau, C. (2005). *Botryococcus braunii*: A rich source for hydrocarbons and related ether lipids. *Applications Microbiology Biotechnology, 66*(25), 486–496.
56. Kalacheva, G. S., Zhila, N. O., & Volova, T. G. (2002). Lipid and hydrocarbon compositions of a collection strain and a wild sample of the green microalga Botryococcus. *Aquatic Ecology, 36*(2), 317–331.
57. Hoek, C., Mann, D. G., Martin, J. H. (1995). *Algae: An introduction to phycology*. Cambridge University Press. ISBN 0-521-31687-1.
58. Farouk, K. El-B., Gad, M. S., Abdoc, S. M., Abed, K. A., Matte, I. A. (2016). Performance and exhaust emissions of a diesel engine burning algal biodiesel blends. *International Journal Mechanics Mechatronics, 16*(3), 151–158. 165803–9797-IJMME-IJENS
59. Islam, Md. A., Rahman, M. M., Heimann, K., Nabi, Md. N., Ristovski, Z. D., Dowell, A., Thomas, G., Feng, B., Alvensleben, N., & Brown, R. J. (2014). Combustion analysis of microalgae methyl ester in a common rail direct injection diesel engine. *Fuel, 143*, 351–360.
60. Hoekman, S. K., Broch, A., Robbins, C., Ceniceros, E., & Natarajan, M. (2012). Review of biodiesel composition, properties, and specifications. *Renewable Sustainable Energy Review, 16*(1), 143–169.
61. Dash, S. K., & Lingfa, P. (2018). Performance evaluation of Nahar oil–diesel blends in a single cylinder direct injection diesel engine. *International Journal of Green Energy, 15*(6), 400–405.
62. Sharif Hossain, A. B. M., Salleh, A., Boyce, A. N., Chowdhury, P., & Naqiuddin, M. (2008). Bodiesel fuel production from algae as renewable energy. *American Journal of Biochemistry and Biotechnology, 4*(3), 250–254.
63. Ashokkumar, V., & Rengasamy, R. (2012). Mass culture of B braunii Kutz. Under open raceway pond for biofuel production. *Bioresource Technology, 104*, 394–399.
64. Jahirul, M. I., Brown, R. J., Senadeera, W., O'Hara, I. M., Ristovski, Z. D. (2013). The use of artificial neural networks for identifying sustainable biodiesel feedstocks. *Energies, 6*(8), 3764–3806.
65. Patel, J. S., Kumar, N., Deep, A., Sharma, A., Gupta, D. (2014). Evaluation of emission characteristics of blend of algae oil methyl ester with diesel in a medium capacity diesel engine. SAE Technical Paper 2014–01–1378.

Experimental Investigation of Feasibility of Using Biodiesel Produced from Nahar, Castor, and Rice Bran Seeds in CI Engine

Animesh Das, Nibedita Das, and Manjula Das Ghatak

1 Introduction

Fossil fuels are produced from the anaerobic decomposition of buried dead organism, and thus, millions of years are required for the production of fossil fuels [1]. Therefore, the rate of production of fossil fuels is quite slow as compared to its consumption due to drastically increasing number of automobiles. Thus, the faster-growing rate of automobiles has a major contribution for polluting the environment [2]. So it is very important to develop some ecofriendly and convenient way which can possibly diminish the scarcity of fossil as well as to save the non-renewable resources to some extent for the future. Biodiesel is a cost-effective way which can behave as a fuel for diesel engine with little or no engine modification [3]. The raw materials that are used for the production of biodiesel are usually taken from two sources, i.e., animal source and plant source [4].

Current work mainly focuses on experimental investigation of feasibility of using biodiesel produced from nahar oil, castor oil, and rice bran oil, respectively, in CI engine [5–10]. The prime objective of the work is to run a CI engine with the biodiesel blends B10, B20, and B30 at a fixed compression ratio by varying load to compare the engine performance parameters and combustion parameters with a diesel run engine. The performance parameters chosen for the experimental investigation are brake thermal efficiency and specific fuel consumption (SFC) [11–16].

A. Das · N. Das · M. Das Ghatak (✉)
Department of Mechanical Engineering, National Institute of Technology, Yupia, Arunachal Pradesh, India

P. Mahanta et al. (eds.), *Advances in Thermofluids and Renewable Energy*, Lecture Notes in Mechanical Engineering, https://doi.org/10.1007/978-981-16-3497-0_26

2 Methodology

The very first step for the production of biodiesel is the collection of seeds. Nahar (Mesuaferrea) seeds are collected from some nearby villages of Arunachal Pradesh as well as from some districts of Assam. Other two kinds of seeds (castor and rice bran) are also collected from different other sources. After collection of seeds, the seeds are cleaned and dried in the sunlight for 7 days to remove the moisture content from the seeds. After removal of the moisture content from the seeds, oil is extracted using Mahindra 6 V mechanical oil expeller. After extraction of oil from the collected seeds, the raw oil is filtered with the help of filter paper to remove the impurities present in the raw oil (Figs. 1, 2, and 3).

2.1 Transesterification

This process is carried out for the pretreatment of free fatty acid (FFA) in the biodiesel production. By definition, esterification is the conversion of carboxylic acid to ester

Fig. 1 Nahar seed

Fig. 2 Castor seed

Fig. 3 Rice bran

Fig. 4 Nahar oil methyl ester, castor oil methyl ester, and rice bran oil methyl ester

using acid and alcohol [17–20]. In this esterification process, the methanol-to-oil ratio is taken as 3:1(v/v ratio). After that, sulfuric acid (H_2 SO_4) is added as a catalyst in the methanol oil mixture. This mixture is heated at 70 °C in a magnetic stirrer at a speed of 400 rpm for 2.5 h. During the heating of the solution, the methanol tends to evaporate from the flask.

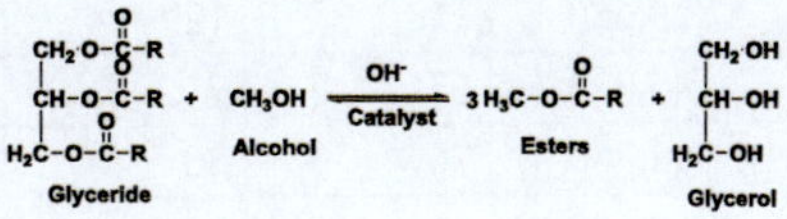

After the esterification is completed, the solution is kept in a separating funnel to settle for 24 h. After settling for 1-day formation of two layers in the separating funnel is observed. The bottom layer is glycerol, and the upper layer is ester. The glycerol layer is separated out; the upper layer of ester is taken out in a beaker; and the remaining volume of the solution is noted. The solution is washed with warm distilled water, and this process is repeated for 6 times until the cloudy-like liquid in the bottom of the funnel is observed to clear and transparent. Normally, water wash is carried out to remove the unreacted excess methanol and glycerol from the ester water; thus, the ester is obtained (Fig. 4).

3 Results and Discussion

Before conducting engine performance test of Nahar oil methyl ester (NOME), Castor oil methyl ester (COME) and rice bran oil methyl ester (RBOME) significant properties like calorific value, viscosity, flash point, and pour point of three blends of biodiesel (B10, B20, and B30) for all the above mentioned methyl esters along with pure diesel and pure biodiesel (B100) which is shows in Table 1 were tested. Equipments used for characterization of nahar oil methyl ester, castor oil methyl ester, and rice bran oil methyl ester are bomb calorimeter, density meter, flash point and fire point measuring apparatus and cloud point and pour point measuring apparatus.

In order to study the performance and combustion parameters using the lab prepared three testing fuels, i.e., NOME, COME, and RBOME, the experiment was carried out in a four-cylinder four-stroke with eddy current dynamometer diesel engine as shown in Fig. 5. The full engine specification is given in Table 2.

The experiment was carried out in a four -cylinder four-stroke water-cooled diesel engine. The engine performance parameters such as brake thermal efficiency, specific fuel consumption, and combustion parameters such as pressure angle and heat release rate were obtained with the help of a software called as engine soft. Normally, this engine is run by high-speed diesel. 1500 ml of biofuel blend is required to run the engine. Biodiesel blends are prepared manually by mixing of B100 and

Table 1 Fuel characterization

Fuels	Density(Kg/m^3)	Calorific value (KJ/Kg)	Flash point (°C)	Fire point (°C)	Cloud point (°C)	Pour point (°C)
Diesel	830	45,220	65	78	5	−8.5
Naliar oil	932.1	3,6870	178	189	11	5
NOME	900.05	38,650	118	129	8	3
Nahar B10	825	45,030	63	71	8.5	−8.1
Nahar B20	833	42,380	76	80	9.8	−7.3
Nahar B30	845	40,670	83	86	10	−5.3
Castor oil	948	37,400	290	314	10.8	−3.2
COME	910	38,150	180	213	7.5	−5.5
Castor B10	840	41,000	79	84	8.1	−7.2
Castor B20	850	39,500	83	87	8.9	−8.1
Castor B30	864	37,700	90	93	9.6	−8.9
Rice bran oil	895	40,171	300	332	1 1	7
RBOME	873	37,200	195	213	7.7	4
Rice bran B10	865	41,500	81	86	9.1	2.9
Rice bran B20	870	39,000	85	89	10.7	2.1
Rice bran B30	888	36,790	91	93	11.9	−1.2

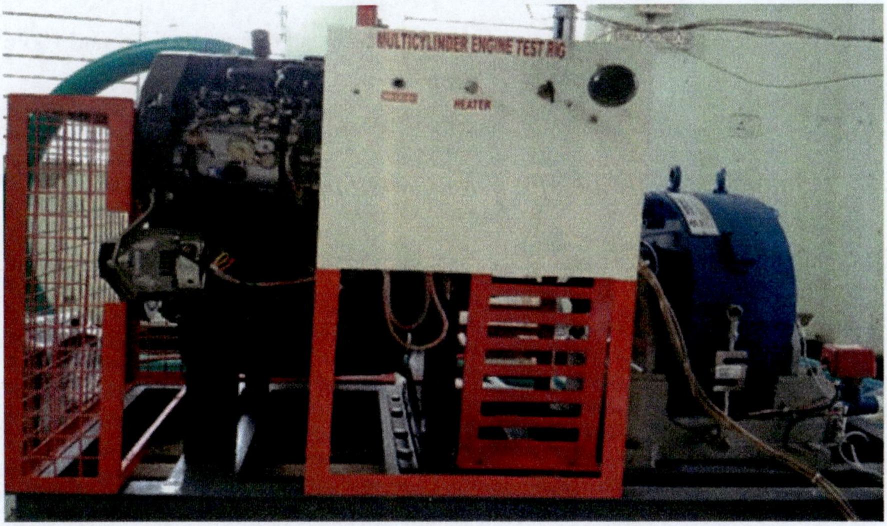

Fig. 5 The test setup

Table 2 Values of fuel characterization

Parameters	Specification
Make	Finite technologies
Model	ICE 115
Type	EDDY CURRENT
Year of manufacturing	2017/18
At rpm	1500/3000
No. of cylinders	Four
No. of strokes	Four
Fuel	diesel
Rated power	3.5 KW at 1500 rpm
Cylinder diameter	87.5 mm
Stroke length	110 mm
Connecting rod length	234 mm
Compression ratio vary12 to 18:1	12 to 18:1
Orifice diameter	20 mm
Dynamometer arm length	185 mm

diesel so as to prepare B10, B20, and B30 blends, respectively, and poured through the engine fuel tank. The engine performance parameters such as brake thermal efficiency, specific fuel consumption, brake power, indicated power, frictional power have been determined at 25%, 50%, 75%, and 100% load, respectively. The 25%

load represents 3 kg, 50% load represents 6 kg, 75% load represents 9 kg, and 100% load which is full load represents 12 kg.

3.1 Comparison of Specific Fuel Consumption

Figure 6 shows the comparison of fuel consumption for B10 blends of three different fuels. Figures CB10, NB10, and RBB10 represent 10% blend of bio-diesel of castor, nahar, and rice bran, respectively, with pure diesel. It is clearly visible from the trend that all the three different biodiesel blends have slightly more fuel consumption rate than pure diesel. But the interesting fact is that with the increase of load the SFC is lowering for all three different fuel.

Figure 7 shows the fuel consumption for B20 of castor, nahar, and rice bran with load. From Fig. 7 , it is clearly visible that though the fuel consumption rate is high in the lower loads but with the increase of load fuel consumption is getting lowered.

Figure 8 indicates the variation in fuel consumption for B30 of castor, nahar, and rice bran with load. With the increase in percentage of biodiesel in the fuel blend up to 30%, the consumption is further getting lowered not only that with the increase in load the fuel consumption is even better than pure diesel in higher load conditions. Especially for nahar oil biodiesel (NB30), it is clearly visible that the consumption rate is lower for throughout the load variation.

For all three types of blend percentages of biofuels, initially fuel consumption is much higher than pure diesel due to the higher density, high viscosity, and low calorific value of biofuel than pure diesel. With the increase in blend percentage density and viscosity of blended increase, and at the same time, calorific value

Fig. 6 Fuel consumption for B10 of castor, nahar, and rice bran with load

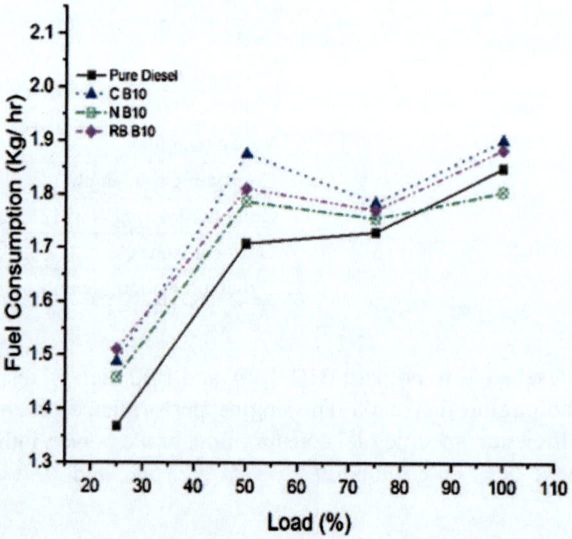

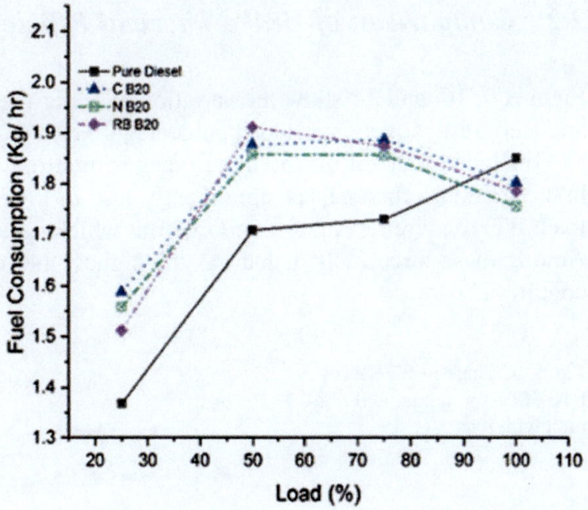

Fig. 7 Fuel consumption for B20 of castor, nahar, and rice bran with load

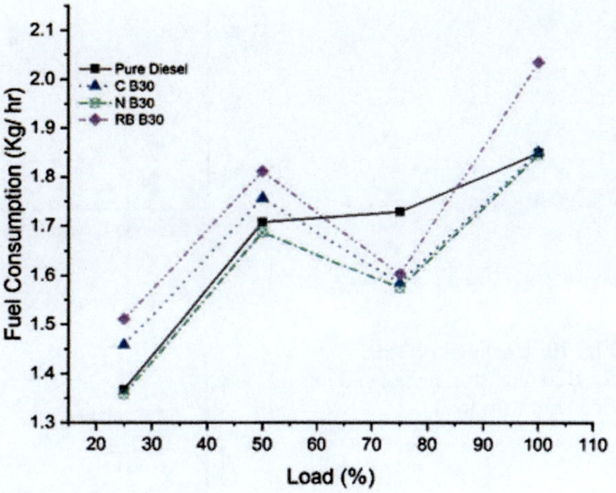

Fig. 8 Fuel consumption for B30 of castor, nahar, and rice bran with load

decreases due to that engine consumes much fuel than that of pure diesel. This trend is continued for almost throughout the curve with increase in load percentage except for very higher loads. And the interesting fact is that fuel consumption is getting lowered for very higher loads. This may be in higher loads small percentage blends of biofuel with pure diesel gives proper combustion, and also, the ignition lag is reduced so that engine consumes less fuel than pure diesel.

3.2 Comparison of Brake Thermal Efficiency

Figures 9, 10, and 11 show the variation in break thermal efficiency for B10, B20, and B30, respectively, with load percentage. From the trend of all the figures, it can be concluded that for all the three types of fuel (castor, nahar, and rice bran), the brake thermal efficiency is significantly less in lower loads. With the increase of load, BTHE increases up to a certain limit which again lowering in very high loads. Among these three fuels, nahar has the highest thermal efficiency in medium load condition.

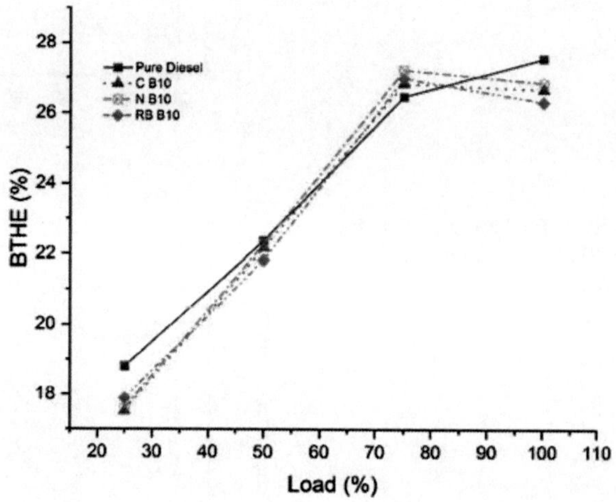

Fig. 9 Change in BTHE for B10 of castor, nahar, and rice bran with load

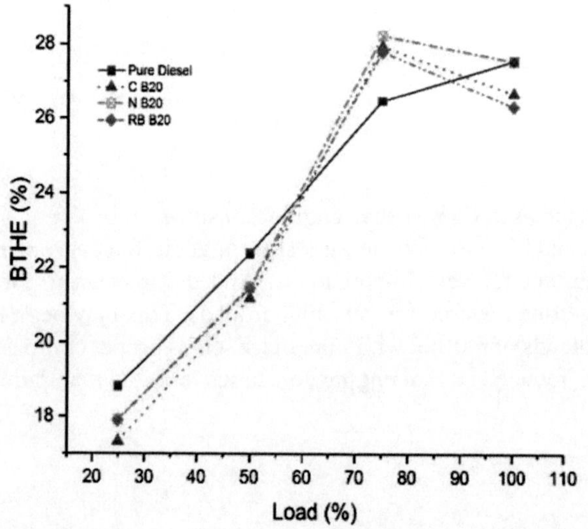

Fig. 10 Change in BTHE for B20 of castor, nahar, and rice bran with load

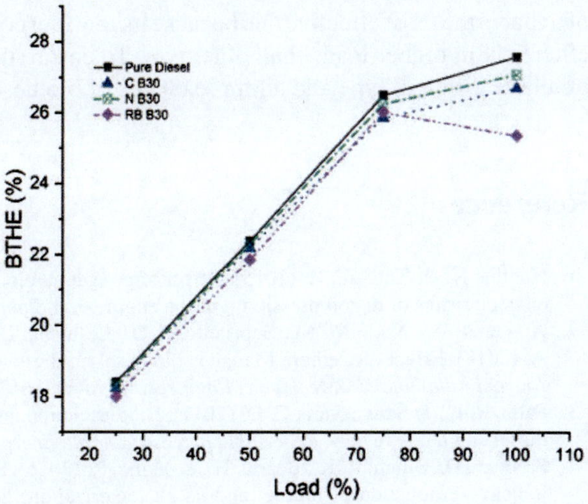

Fig. 11 Change in BTHE for B30 of castor, nahar, and rice bran with load

Thermal efficiency for B30 sample of all three kinds of fuel is almost same from starting with no-load condition to almost 80% load. Although thermal efficiency of B30 sample of rice bran oil decreases rapidly than other two oil after a certain amount of load, B20 sample of nahar oil has the highest thermal efficiency from the mid-range of load to almost 80% load. From the above three figures, it is clearly visible that up to 20% blend of biofuel with diesel break thermal efficiency increases significantly for mid-range of load. But particularly for the higher load condition, BTHE decreases with the increase in blend percentage. Though BTHE increases almost linearly for all types of blends up to 80% of load, it decreases significantly for higher loads. The higher viscosity of biofuel BTHE is less than pure diesel but with the increase in load and with increase in cylinder temperature the droplets of biofuel atomizes properly and BTHE increases. At the same time, at very high load condition may be due to the lack of oxygen in combustion chamber otherwise due to the improper air fuel ratio BTHE decreases slightly.

4 Conclusion

All the three biodiesels from three different sources castor, nahar, and rice bran oil can be a very effective alternative of diesel fuel considering all the engine parameters as well as any other factors like availability, cost, and health issue. Although pure diesel is the most efficient one, the biodiesel from vegetable sources is not far behind. The experiments are done without any modification in conventional diesel engine so the efficiency may be increased with some little modifications. Also, the fuel parameters may be better by adding of some additives which can also help to increase engine performances. Among all three different fuel and its different blend percentage B30 of nahar is the most effective one because its low fuel consumption and higher thermal efficiency in higher loads than other two. To ensure the practical use, emission test of all the different type and all three blends of biodiesel is also needed.

References

1. Nantha, K., & Thundil, R. (2015). Effect of pongamia biodiesel on emission and combustion characteristics of di compression ignition engine. *Ain Shams Engineering Journal, 6*, 1.
2. Anggono, W., Noor, M. M., Suprianto, F. D., Lesmana, L. A., Gotama, G. J., & Setiyawan, A. (2018). Effect of Cerbera Manghas biodiesel on diesel engine performance. *International Journal Automotation Mechanics Engineering, 15*(3), 5667–5682.
3. Patel, R. L., & Sankhavara, C. D. (2017). Biodiesel production from Karanja oil and its use in diesel engine: A review. *Renewable and Sustainable Energy Reviews, 71*.
4. Pradhan, D., Singh, R. K., Bendu, H., & Mund, R. (2016). Pyrolysis of Mahua seed (Madhuca indica)—Production of biofuel and its characterization. *Energy Conversation Management, 108*.
5. Dharmadhikari, H. M., Kumar, P. R., & Rao, S. S. (2012). Performance and emissions of C.I. engine using blends of biodiesel and diesel at different injection pressures. *Interantional Journal Applications Research Mechanics Engineering*, 1–6.
6. Nithyananda, B. S., Anand, A., & Prakash, G. V. N. (2013). Performance study on diesel engine using different blends of neem biodiesel. *International Journal Engineering Research Applications, 3*(4), 1778–1781.
7. Demirbas, A. (2005). Biodiesel production from vegetable oils via catalytic and non-catalytic supercritical methanol transesterification methods.*Programming Energy Combustion Science, 31*(5–6), 466–487.
8. Dorado, M. P., Ballesteros, E., Arnal, J. M., Gómez, J., & López, F. J. (2003). Exhaust emissions from a diesel engine fueled with transesterified waste olive oil. *Fuel, 82*(11), 1311–1315.
9. Wood, B. M., Kirwan, K., Maggs, S., Meredith, J., & Coles, S. R. (2015). Study of combustion performance of biodiesel for potential application in motorsport. *Journal of Cleaner Production, 93*, 167–173.
10. Wirawan, S. S., Tambunan, A. H., Djamin, M., & Nabetani, H. (2008). The effect of palm biodiesel fuel on the performance and emission of the automotive diesel engine. *Agricultural Engineering International CIGR Journal*, 1–13.
11. Chuah, L. F., Aziz, A. R. A., Yusup, S., Klemeš, J. J., & Bokhari, A. (2009). 2016, Waste cooking oil biodiesel via hydrodynamic cavitation on a diesel engine performance and greenhouse gas footprint reduction. *Chemical Engineering Transactions, 50*, 301–306.
12. Raheman, H., Jena, P. C., & Jadav, S. S. (2013). Performance of a diesel engine with blends of biodiesel (from a mixture of oils) and high-speed diesel. *International Journal of Energy and Environmental Engineering, 4*(1), 1–9.
13. Mofijur, M., Rasul, M. G., Hassan, N. M. S., & Nabi, M. N. (2019). Recent development in the production of third generation biodiesel from microalgie. *Energy Procedia, 156*, 53–58.
14. Atmanli, A. (2016). Comparative analyses of diesel-waste oil biodiesel and propanol, n-butanol or 1-pentanol blends in a diesel engine. *Fuel, 176*, 209–215.
15. Damanik, N., Ong, N. C., & Tong, C. W. (2018). A review on the engine performance and exhaust emission characteristics of diesel engines fuelled with biodiesel blends. *Environmental Science and Pollution Research, 25*, 15307–15325.
16. Banapurmath, N. R., Tewari, P. G., & Hosmath, R. S. (2008). Performance and emission characteristics of a DI compression ignition engine operated on Honge, Jatropha and sesame oil methyl esters. *Renewable Energy, 33*(9), 1982–1988.

17. Randazzo, M. L., & Sodré, J. R. (2011). Exhaust emissions from a diesel powered vehicle fuelled by soybean biodiesel blends (B3–B20) with ethanol as an additive (B20E2-B20E5). *Fuel, 90*(1), 98–103.
18. Altin, R., Çetinkaya, S., & Yücesu, H. S. (2001). Potential of using vegetable oil fuels as fuel for diesel engines. *Energy Conversation Management*, *42*(5), 529–538.
19. Al-Widyan, M. I., Tashtoush, G., & Abu-Qudais, M. (2002). Utilization of ethyl ester of waste vegetable oils as fuel in diesel engines. *Fuel Processing Technology*, *76*(2), 91–103.
20. Huzayyin, A. S., Bawady, A. H., Rady, M. A., & Dawood, A. (2004). Experimental evaluation of diesel engine performance and emission using blends of jojoba oil and Ddiesel fuel. *Energy Conversation Management*, *45*(13–14), 2093–2112.

Effect of EGR on Performance and Emission Characteristics of a Diesel Engine Fueled with Yellow Oleander Seed Oil Biodiesel

Mridul Deka, Pradeep Kumar Mahanta, and Nabajit Dev Choudhury

1 Introduction

Thevetia peruviana or cascabela thevetia, commonly known as yellow oleander is a tropical shrub. The oil content of yellow oleander seeds is as high as 67%, Adamu et al. [1]. But the seed is known for its toxicity in many parts of the world. Ingestion of thevetia seeds may be fatal and causes severe cardiovascular effect with rhythmic disorder as these seeds contain three types of poisonous chemicals, namely glycosides-thevetin, thevetoxin, and peruvoside, Bose et al. [2]. As these chemicals have detrimental effect on the cardiovascular system, they are known as cardiac glycosides. Because of this toxicity, these seeds are not at all edible although medicinal use of the extract of the plant, leaves, and seeds has been reported, Zibbu and Batra [3]. Because of the inedibility and high oil content of the seeds, they may be used as a source of fuel. Fatty acid methyl esters of lipids show synonymity of its fuel properties with petroleum diesel, and this is the main cause of its study of its uses in I.C. engines.

In C.I. engines, introduction of biodiesel promisingly reduces the emission of CO and production of unburned hydrocarbons, but at the same time, it is seen that there is a drastic increase in NOx emission with increase in percentage of biodiesel present in the diesel fuel, Hoekman and Robbins [4]. To reduce the production of NOx emission, EGR may be introduced, which undoubtedly reduces the production of NOx by reducing the oxygen content of the incoming air into the combustion chamber along with reducing the maximum combustion temperature, Zeng et al. [5]. It is also seen that with the increase in percentage of the EGR, there is an increase

M. Deka (✉)
Department of Mechanical Engineering, JIST, Jorhat, Assam 785010, India

P. K. Mahanta
Department of Mechanical Engineering, JEC, Jorhat, Assam 785007, India

N. D. Choudhury
Department of Energy, ASTU, Guwahati, Assam 781019, India

P. Mahanta et al. (eds.), *Advances in Thermofluids and Renewable Energy*, Lecture Notes in Mechanical Engineering, https://doi.org/10.1007/978-981-16-3497-0_27

Table 1 Fuel properties of thevetia biodiesel and petroleum diesel

Parameters	ASTM standard	Petroleum diesel	TPOME
Density at 15 °C kg/m^3	D4052	830	889
Kinematic Viscosity at 40 °C cSt	D445	2.2	5
Flash point (closed cup), °C, minimum	D93	66	84
Calorific value MJ/kg	D240	46	40
Pour point °C	D97	−11	1
Cloud point °C	D2500	−9	13
Acid value mg of KOH/ g of FAME	D6751	–	0.27

in the production of CO, HC, and particular mass. With further increase in EGR, the performance of the engine starts to behave abnormally. The experimentations have been limited to 10, 20, and 30% blend ratios of biodiesel to petroleum diesel (B10, B20, and B30), as the higher percentages of biodiesel in the fuel blend result in very high emission of the NOx. The various fuel properties of thevetia oil methyl esters (TOME) have been determined and tabulated in Table 1.

1.1 NOx Production

NOx production can be categorized into three different types, viz. thermal NOx, prompt NOx, and fuel NOx. At a temperature as high as 1700 K, thermal reaction of NO production takes place, Varatharajana and Cheralathan [6, 7]. At this temperature, the atmospheric nitrogen reacts with oxygen to produce a huge quantity of NO. NO is the predominant component of engine NOx. Prompt NOx is another source of production of NOx during combustion as described by *Fenimore*, and such types of NOx have got the name *Fenimore* NOx. The combustion of hydrocarbons plays an important role in the production of such types of NOx, Harrington et al. [8]. The reaction for the same may be as summarized below, *Fenimore* [9].

$$CH + N_2 \leftrightarrow HCN + N$$

$$N + O_2 \leftrightarrow NO + O$$

$$HCN + OH \leftrightarrow CN + H_2O$$

$$CN + O_2 \leftrightarrow NO + CO$$

If any trace of nitrogen is present in the fuel, it may generate NOx by the reaction of the nitrogen with oxygen. Such types of NOx production are known as fuel NOx. As diesel fuels and biodiesel contains no or very less amount of such nitrogen, therefore, it may be noted that during combustion of diesel and biodiesel blends, only prompt NOx and thermal NOx may be considered, Fernando et al. and Moser et al. [10, 11].

2 Extraction and Transesterification of the Oil

Fresh handpicked thevetia seeds have been collected, and the outer covers have been removed. The inner soft kernels are taken and sun dried. The sun-dried kernels are then grounded. In a soxhel apparatus, using solvent extraction method, oil has been extracted. n-hexane is the solvent used during the oil extraction process. After degumming of the oil, acid value of the oil has been evaluated. The acid value has been found to be 1.9 mg of KOH per gram of the oil although after transesterfiction, it is seen that the acid value of the oil drops down to 0.27 mg of KOH per gram of the oil. The oil has been directly sent for methanolysis as the acid value of raw oil is found to be less than 2 mg of KOH per gram of the oil [12]. During this process a NaOH pellets of weight percentage 0.5% have been dissolved in methanol, Canakci and Gerpen [13]. This solution of NaOH in methanol was poured into the oil in a conical flask, while stirring at 700 rpm and the temperature was kept at 65 °C for 2 h. Excess methanol has been removed in rotavapor, and the rest of the product is allowed to settle in a separatory funnel till the layers of biodiesel over glycerol have been visible. After about 8 h, these two layers are distinctly visible. And the glycerol has been allowed to flow through the bottom of the separatory funnel, and the biodiesel has been recovered from the top.

3 Experimental Setup

IC engine setup under test is research diesel by Apex innovation Pvt. Ltd.

Various engine specifications are as follows: (Fig. 1)

No. of cylinders	1
No. of strokes	4
Cylinder diameter	87.5 mm
Stroke length	110 mm
Connecting rod length	234 mm
Orifice diameter	20 mm
Dynamometer arm length	185 mm
Power	3.5 kW
Speed	1500 RPM

Various measuring parameters are:

Fig. 1 Experimental setup

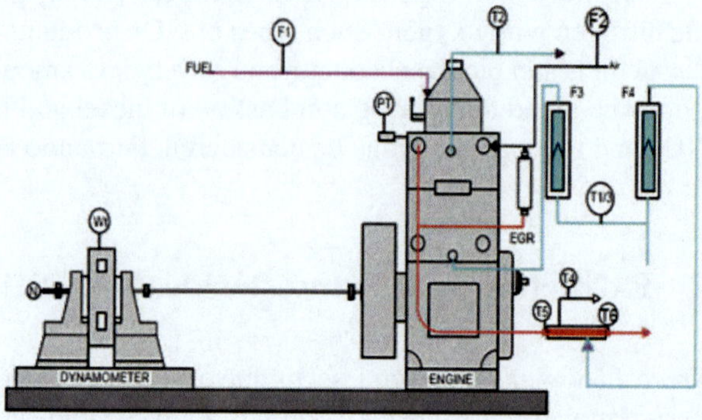

T_1 and T_3 are temperature of cooling water through engine and calorimeter, respectively.

T_2 Fuel inlet temperature
T_5 Flue gas temperature
T_4 Exhaust gas cooling water inlet temperature
T_6 Exhaust gas cooling water outlet temperature
F_1 Fuel flow rate
F_2 Air flow rate
F_3 Cooling water flow rate through engine
F_4 Water flow rate through dynamometer

The engine used in this study is a water-cooled single-cylinder CRDI engine. An eddy current dynamometer has been used for changing the load. The engine can handle a maximum load of 12.3 kg for which the power output is approximately 3.5 kW. Compression ratio 18 has been taken in this study. Exhaust gas recirculation system diverges a part of the exhaust gas to the intake air going into the cylinder. The EGR percentages considered are 0, 10, and 15%, respectively. Exhaust gas going into the engine has been cooled before it enters the engine by a cooling water circulation system. EGR percentage may be calculated by using the following formula:

$$\text{EGR}(\%) = \frac{\text{Air flow rate without EGR} - \text{Air flow rate with EGR}}{\text{Air flow rate without EGR}} \times 100$$

An electronic control unit (ECU) provides various measuring data through computer interface. NIRA RAK software is used for measuring various data from the ECU, and software IC Engine Soft is used for processing of the data. For each set of the experiments, the loads have been changed as shown in Table 2.

The engine exhausts have been analyzed by TESTO 350 flue gas analyzer (Fig. 2).

Table 2 Load used during experimentations:

Load (kg)	Power in (kW)	Percentage load
1	0.28	8
3	0.85	24.29
5	1.42	40.57
7	1.99	56.86
9	2.56	73.14
11	3.13	89.43
12.3	3.50	100

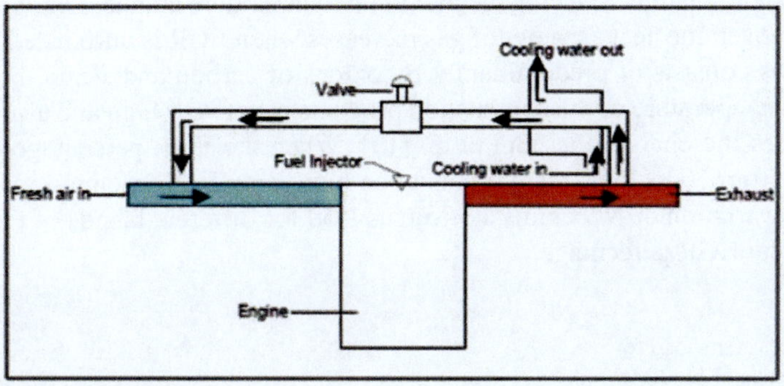

Fig. 2 Line diagram of EGR

4 Results and Discussion

Experimentations have been performed at a constant rpm of 1500, and various performance parameters have been analyzed during the experiments, such as.

BTE, BSFC exhaust gas temperature, etc., also various exhaust gas emissions such as NOx, CO, and UHC have been analyzed.

4.1 Exhaust Emission

It has been well established that the use of EGR is highly fruitful in reducing the NOx emission, Agarwala et al. and Hira et al. [14, 15]. In this study, it is seen that there has been a drastic rise in NOx emission as soon as B10 blend of the fuel has been used instead of petroleum diesel. The NOx emission increases from a range of 263–600 ppm at a load condition of 8% as soon as petroleum diesel has been replaced by B10 of TPME, and similarly at full load condition, the NOx emission increases from 409 to 831 ppm. This increase in NOx may be attributed by increase in adiabatic flame temperature inside the combustion chamber, Tamilselvana et al.

[16]. This increase in flame temperature is because of the presence of oxygen present in the fuel itself leading to better combustion and temperature rise and increase NOx emission, Chen et al. [17]. With higher percentage of biodiesel present in the fuel, the NOx emission is also noted to be increased. As soon as 10% cold EGR is introduced with the intake air, it is seen that the emission of NOx decreases rapidly. At 10% EGR for B10 of the fuel, the NOx value reduces down to 315 ppm from a value of as high as 586 ppm at part load and the same decreases to 449 ppm from 841 ppm at peak load. This decrease in NOx emission may be attributed by the fact that, with addition of exhaust gas to the intake air, there is a drop in intake oxygen and nitrogen inside the cylinder, and as there is decrease in oxygen, the combustion remains incomplete to some extent, thereby reducing the maximum temperature attained, Vallapudi et al. [18]. Moreover, the heat capacity of gas increases when EGR is introduced since the exhaust gas consists of predominantly by oxides of carbon and steam, because of which the temperature of the combustion products is not very high and also the inert EGR dilutes the charge, Pierpont et al. [19]. When the EGR percentage has been increased from 10 to 15%, there seems to be more drop in NOx emission. Figure 3 shows the variation of NOx emission versus load for different blends of the biofuel and different EGR percentage.

Fig. 3 NOx versus load for different blends of biodiesel with and without EGR

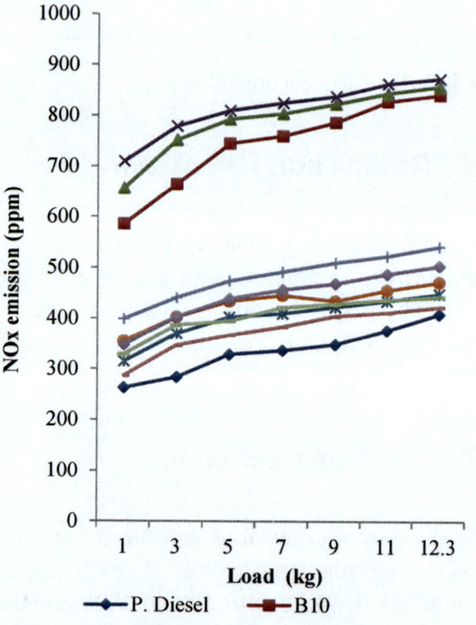

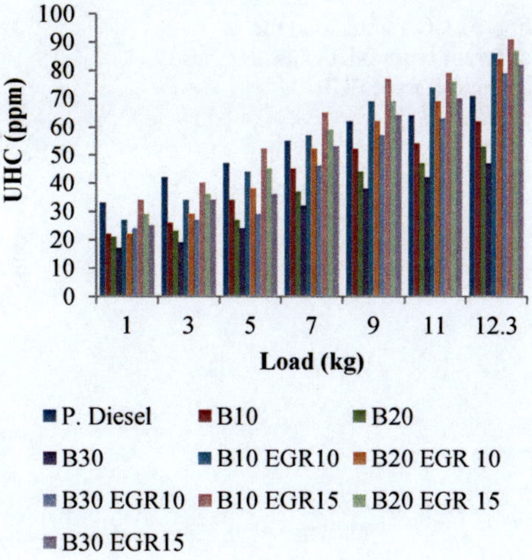

Fig. 4 UHC versus load for different types of fuel used and at different EGR percentage

4.2 Unburned Hydrocarbons

In Fig 4, UHC versus load is shown. It is seen that at least load condition, the amount of UHC is less and it increases with increase in load. As the percentage of biodiesel is increased in the fuel, the production of UHC is decreased. This may be attributed by the additional oxygen present in the fuel which is liable to improve the combustion of the fuel, and hence, the production of UHC decreases with increase in the percentage of biodiesel in the fuel, Mahalingam et al. [20]. When the percentage of EGR is increased, it is seen that the emission of UHC increases. This is attributed by the lack of complete combustion of the fuel because of the lack of oxygen in the combustion chamber, Nabi et al. [21]. Maximum emission of UHC has been noted at the peak load condition with B10 of the biodiesel with an EGR percentage of 15%. It is also seen that the emission of UHC decreases slowly with increase in the percentage of biodiesel, since the addition of biodiesel increases the percentage of oxygen in the fuel.

4.3 Carbon Monoxide

Oxygen content of biodiesel plays an important role in production of CO. Biodiesel contains 10–12% oxygen by weight, Chauhan et al. and Xue et al. [22, 23]. When petroleum diesel is blended with biodiesel, the CO emission reduces as oxygen is intact in biodiesel and combustion is enhanced, Rasheda et al. [24]. When petroleum diesel is replaced by B30, the emission of CO has been decreased by approximately 26%, at full load condition. Similarly at part load conditions also appreciable

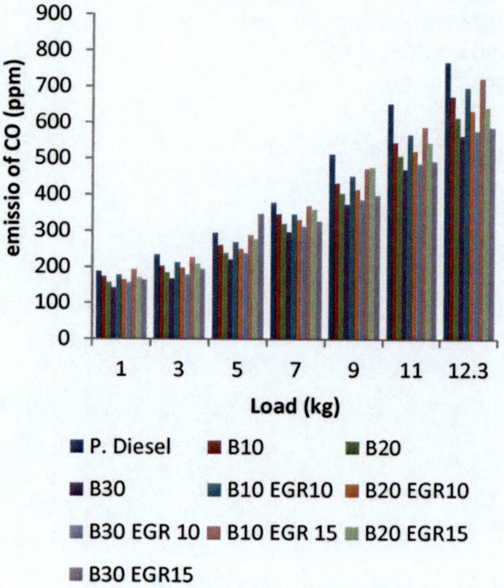

Fig. 5 CO versus load for different types of fuel used and at different EGR percentage

decrease in CO may be noted. But as soon as EGR has been introduced to the experiments, emission of CO increases. The reason of increase in CO is similar to that for UHC; when EGR is introduced, there is a decrease in oxygen inside the combustion chamber, leading to incomplete combustion of the fuel, and hence, the emission of CO and UHC increases. For B10 of the fuel at peak load condition, when the EGR increases from 0 to 15%, the CO emission is seen to increase approximately by 7%. Similarly for other blending percentage of the fuel, there is a similar range of increase in CO emission when the percentage of EGR increases (Fig. 5).

4.4 Exhaust Gas Temperature

Exhaust gas temperature versus load for petroleum diesel and B10 of the fuel have been shown in Fig. 6 for illustration of the variation in temperature with addition of biodiesel in the fuel and also the variation of the temperature when EGR is used. It is seen that the temperature of the exhaust gas increases when biodiesel is added in petroleum diesel. This is due to the fact that additional oxygen present in the biodiesel gets added to the fuel, and as a result, a better combustion environment is created and hence there is an increase in the temperature of the combustion products. When EGR has been introduced, there is a decrease in oxygen content on the combustion chamber, and as a result, there is a decrease in temperature of the engine exhaust. More the percentage of EGR more is the decrease in the temperature of the exhaust. At peak load condition, the temperature of the exhaust decreases from 534 to 488 °C for B10 of the fuel and for B10 along with 25%EGR.

Fig. 6 Exhaust gas temperature versus load for different types of fuel used and at different EGR percentage

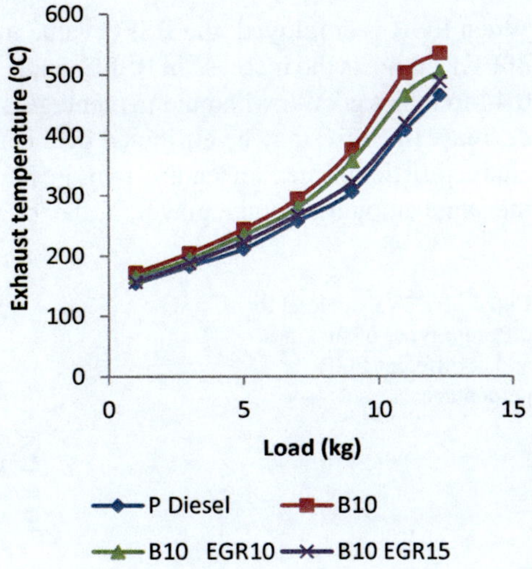

4.5 BTE

It is seen that the BTE increases with the increase in load and it is maximum at the peak load. B10 blend of the fuel shows better result than that shown by petroleum diesel; this is due to the fact that as biodiesel contains oxygen atoms intact in its molecules, better combustion occurs, Ghazali et al. [25]. BTE shown by B20 blend is also comparable to that with the BTE shown by petroleum diesel. When EGR valve is opened it is seen that there is a slight decrease in the BTE of the engine. The BTE for B10 changes from 21.79 to 21.12% at full load condition without EGR and with a maximum of 15% EGR, respectively. Therefore, it may be said an EGR of 15% does not change the BTE of the engine too adversely. Similarly for other blends of the biodiesel also, the BTE is not affected too much by up to an EGR of 15%.

4.6 BSFC

For any combination of compression ratio and blending of fuel, the BSFC value is seen to decrease sharply with the increase in load. The main reason behind this phenomenon is that the percentage increase in brake power is more as compared to the percentage increase in fuel burned and relatively the losses after combustion are less at higher loads, Raheman and Ghadge [26]. For B10 there is a slight decrease in the BSFC value as the excess oxygen present in the fuel itself causes the complete combustion of the fuel. But a higher percentage of biodiesel in the fuel is detrimental to the BSFC value since the calorific value of biodiesel is less than the petroleum diesel, Raheman and Ghadge [26]. The BSFC value decreases from 0.4 to 0.36 kg/kWh at peak load condition when petroleum diesel is replaced by B10.

When EGR is employed, the BSFC value increases slightly. More is the percentage of EGR; more is the increase in BSFC value. For B30, the BSFC value increases from 0.44 to 0.53 kg/kWh when the percentage of EGR is increased from 0 to 15%. This decrease in BSFC may be attributed by the adverse effect of EGR on the combustion characteristics of the fuel leading to more amount of fuel consumption for developing the same amount of brake power, Yadav et al. [27] (Figs. 7 and 8).

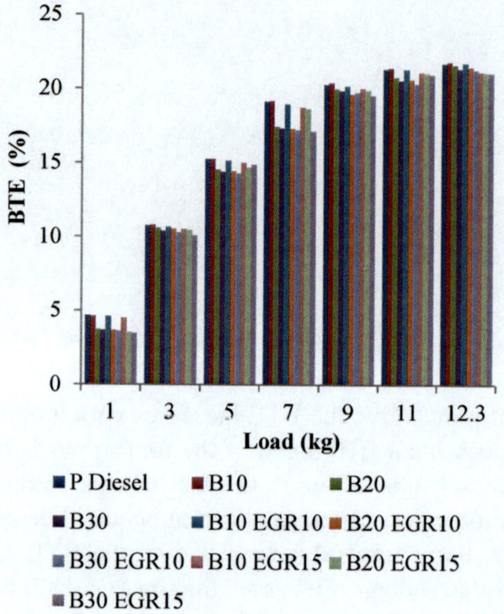

Fig. 7 BTE versus load for different types of fuel used and at different EGR percentage

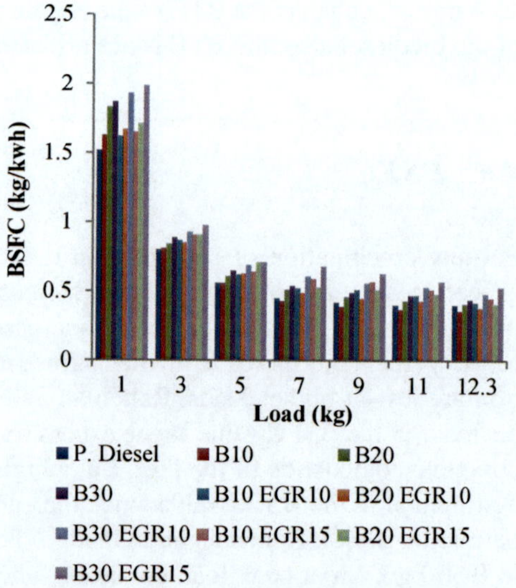

Fig. 8 BSFC versus load for different types of fuel used and at different EGR percentage

5 Conclusions

A single-cylinder C.I. research engine has been operated with petroleum diesel and B10, B20, and B30 of TPME. Cold EGR has been employed to the engine with an intension to reduce the NOx emission from the engine. Investigations related to the performance and emission characteristics of the engine have been made under the application of EGR when the engine is fueled with various blends of TPME blends. The results have been as summarized below:

- NOx emission of the engine increases with the increase in the percentage of biodiesel in the fuel. Since, the oxygen present in biodiesel enhances combustion process which further increases the combustion temperature leading to more production of thermal and prompt NOx. This NOx emission may be decreased from 841 ppm to 423 ppm when the FGR percentage is increased from 0 to 15 % for Bl0 of the fuel at peak load condition.
- Thevetia peruviana oil has been extracted and methyl ester of the oil has been prepared using NaOH as a catalyst.
- Engine performance parameters such as BTE, BSFC, and exhaust gas temperature have been investigated under different load conditions and with and without EGR. It is seen that the BTE shows better result when B10 of the fuel is used than that when petroleum diesel is used. B20 blend of the fuel also shows BTE comparable to the BTE when petroleum diesel has been used. When EGR is used the BTE values deteriorates slightly. For B10 the BTE changes from 21.9 to 21.12% for 0% EGR and 15% EGR, respectively.
- BSFC value initially decreases when the B10 of the fuel is used. Further increase in the percentage of biodiesel increases the BSFC value, since the calorific value of biodiesel is less than petroleum diesel. The BSFC values further increases when EGR is used. The value changes from 0.4 to 0.53 kg/kWh for petroleum diesel and for B30 with 15% EGR.
- The exhaust gas temperature shows a rise when the percentage of biodiesel is more in the fuel. This increase in the temperature is the main cause of increase in NOx when biodiesel is used. This temperature shows a decreasing pattern with increase in the percentage of EGR.
- CO and UHC both decreases when the percentage of biodiesel in the fuel is decreased, since biodiesel enhances combustion process because of its higher oxygen content. CO and UHC emission further increases when EGR percentage starts to increase.

References

1. Adamu, F. A., Jahun, B. G., & Daniel, C. (2013). Assessment and optimization of energy use of yellow oleander (Thevetia Peruviana) for biodiesel blends in Nigeria. *International Journal of Science and Reseorch (USR).*

2. Bose, T. K., Basu, R. K., Biswas, B., De J. N., Majumdar, B. C., & Datta, S. (1999, October). Cardiovascular effects of yellow oleander ingestion. *Journal of the Indian Medical Association, 97*(10), 407–410. PMID: 10638101.
3. Zibbu, G., & Batra, A. (2011). Thevetia peruviana (Pers) Schum: A plant with enormous therapeutic potential. *Journal of Pharmacy Research, 4*(12), 4461–4464.
4. Hoekman, S. K., & Robbins, C. (2012). Review of the effects of biodiesel on NOx emissions. *Fuel Processing Technology, 96*, 237–249.
5. Zheng, M., Reader, G. T., & Hawley, J. G. (2004). Diesel engine exhaust gas recirculation—A review on advanced and novel concepts. *Energy Conversion and Management, 45*, 883–900.
6. Varatharajana, K., & Cheralathan, M. (2012). Influence of fuel properties and composition on NOx emissions from biodiesel powered diesel engines: A review. *Renewable and Sustainable Energy Reviews, 16*, 3702–3710.
7. Varatharajana, K., & Cheralathan, M. (2013). Effect of aromatic amine antioxidants on NOx emissions from a soybean biodiesel powered DI diesel engine. *Fuel Processing Technology, 106*, 526–532.
8. Harrington, K. J. (1986). Chemical and physical properties of vegetable oil esters and their effect on diesel fuel performance. *Biomass, 9*, 1–17.
9. Fenimore, C.P. (1971). Formation of nitric oxide in premixed hydrocarbon flames. In Thirteenth Symposium (International) on Combustion (Vol. 13, Issue 1, pp. 373–380).
10. Fernando, C., Hall, C., & Jha, S. (2006). NOx reduction from biodiesel fuels. *Energy and Fuels, 20*, 376–382.
11. Moser, B. R., Williams, A., Haas, M. J., & McCormick, R. L. (2009). Exhaust emissions and fuel properties of partially hydrogenated soybean oil methyl esters blended with ultra low sulfur diesel fuel. *Fuel Processing Technology, 90*, 1122–1128.
12. Canakci, M., & Gerpen, J. V. (2001). Biodiesel production from oils and fats with high free fatty acids. *446*, 1429–1436.
13. Sut, D., Chutia, R. S., Bordoloi, N., Narzari, R., & Kataki, R. (2016). Complete utilization of non-edible oil seeds of Cascabela thevetia through a cascade of approaches for biofuel and by-products. *Bioresourcing Technology*, 111–120.
14. Agarwala, D., Sinhab, S., & Agarwal, A. K. (2006). Experimental investigation of control of NOx emissions in biodiesel-fueled compression ignition engine. *Renewable Energy, 31*, 2356–2369.
15. Hira, J., Sharma, R., Kamboj, K., Kumar, V., & Shama, P. (2019). NOx emission reduction in diesel engine through developed cooled EGR setup. In M. Kumar, R. Pandey, &V. Kumar (Eds.), *Advances in interdisciplinary engineering. Lecture Notes in Mechanicol Engineering*. Springer, Singapore.
16. Tamilselvana, P., Nallusamy, N., & Rajkumar, S. (2017). A comprehensive review on performance, combustion and emission characteristics of biodiesel fuelled diesel engines. *Renewable and Sustainable Energy Reviews, 79*, 1134–1159.
17. Chen, H., Xie, B., Ma, J., & Chen, Y. (2018). NOx emission of biodiesel compared to diesel: Higher or lower? *Applied Thermal Engineering, 137*, 584–593.
18. Vallapudi, D. R., Makineni, H. K., Pisipaty, S. K., & Vernu, H. (2018). Combined impact of EGR and injection presure in perfomarce improvement and NOx cotrol of a DI diesel engine powered with tamarind seed biodiesel blend. *Environmental Science and Pollution Research.*
19. Pierpont, D. A., Montgomery, D. T., & Reitz, R. D. (1995). Reduing particulate and NOx using mukipe injections and EGR in a D.I. diesel. SAE Technical Paper 950217.
20. Mahalingam, A., Munuswamy, D. B., Devarajan, Y., & Radhakrishnan, S. (2018). Emission and performance analysis on the effect of exhaust gas recirculation in alcohol-biodiesel aspirated research diesel engine. *Environmental Science and Pollution Research, 25*, 12641–12647.
21. Nabi, M. N., Akhter, M. S., & Shahadat, M. M. Z. (2006). Improvement of engine emissions with conventional diesel fuel and diesel–Biodiesel blends. *Bioresource Technology, 97*, 372–378.
22. Chauhan, B. S., Kumar, N., & Cho, H. M. (2012). A study on the performance and emission of a diesel engine fueled with Jatropha biodiesel oil and its blends. *Energy, 37*, 616–622.

23. Xue, J., Grift, T. E., & Hansen, A. C. (2011). Effect of biodiesel on engine performances and emissions. *Renewable and Sustainable Energy Reviews, 15*, 1098–1116.
24. Rasheda, M. M., Kalama, M. A., Masjukia, H. H., Mofijurb, M., Rasulb, M. G., & Zulkifli, N. W. M. (2016). Performance and emission characteristics of a diesel engine fueledwith palm, jatropha, and moringa oil methyl ester. *Industrial Crops and Products, 79*, 70–76.
25. Ghazali, W. N. M. W., Mamat, R., Masjuki, H. H., & Najaf, G. (2015). Effects of biodiesel from different feed stocks on engine performance and emissions: A review. *Renewable and Sustainable Energy Reviews, 51*, 585–602.
26. Raheman, H., & Ghadge, S. V. (2008). Performance of diesel engine with biodiesel at varying compression ratio and ignition timing. *Fuel, 87*, 2659–2666.
27. Yadav, V.S., Soni, S. L., & Sharma, D. (2012). Performance and emission studies of direct injection C.I. engine in duel fuel mode (hydrogen-diesel) with EGR. *International Journal of Hydrogen Energy, 37*(4) 3807–3817.

Synthesis of a Low Cost, Natural Base Solution for CO_2 Scrubbing from Biogas and its Comparison with KOH

Deep Bora, Lepakshi Barbora, and Pinakeswar Mahanta

Nomenclature

CO_2	Carbon dioxide
CH_4	Methane
KOH	Potassium hydroxide
BPS	Banana pseudostem
MBP	*Musa balbisiana Peel*
MAP	*Musa acuminata Peel*
NBS	Natural base solution
M	Molar
L	Litre
HCV	Higher calorific value (MJ/m^3)
LCV	Lower calorific value (MJ/m^3)

D. Bora · L. Barbora (✉)
Centre for Energy, Indian Institute of Technology Guwahati, Guwahati, Assam 781039, India
e-mail: lopab@iitg.ac.in

P. Mahanta (✉)
Department of Mechanical Engineering, Indian Institute of Technology Guwahati, Guwahati, Assam 781039, India
e-mail: pinak@iitg.ac.in

Department of Mechanical Engineering, National Institute of Technology Arunachal Pradesh, Yupia, Arunachal Pradesh 791112, India

P. Mahanta et al. (eds.), *Advances in Thermofluids and Renewable Energy*, Lecture Notes in Mechanical Engineering, https://doi.org/10.1007/978-981-16-3497-0_28

1 Introduction

Biogas is a mixture of different gases, which are produced from biodegradation of organic matters in the absence of oxygen and the presence of anaerobic microorganisms [1]. Biogas, typically a mixture of 45–60% methane (CH_4), 40–55% carbon dioxide (CO_2) and a trace amount of other gases like hydrogen (H_2), nitrogen (N_2), and hydrogen sulfide (H_2S), is recognized as a potential form of renewable energy today alongside solar energy, wind energy, and biofuels [2, 3]. The CH_4 in biogas is measured as a high valued source of energy, while CO_2 has zero energy yield through combustion and significantly reduces the overall heating value due to its high concentration per volume of biogas [4]. The calorific value of biogas with CO_2 varies from 18.7 to 26 MJ/m^3 and that without CO_2 is between 33.5 and 35.3 MJ/m^3 [5]. Thus, removal of CO_2 and other impurities from biogas enhances the fuel efficiency which could serve as a source of immense energy that can be used effectively for different applications like cooking [6], vehicle fuel, and power generation [7–9]. After scrubbing, the biogas can be compressed and stored in gas cylinder and transported to the desired location for utilization. Additionally, the scrubbed biogas reduces greenhouse gas emissions [5]. Numerous basic mechanisms are employed to attain selective separation of gas constituents from biogas, such as membrane separation, cryogenic separation, adsorption on solid surface, physical or chemical absorption, pressure swing adsorption (PSA), and so on [10–14]. Cryogenic separation and membrane separation give high purity but they are economically not viable for pilot plant as they are energy intensive and consume relatively more power per unit of gas produced [15]. Chemical adsorption on solid surface is effective as well as economical, has good moisture removal capacities, and is simple in design. Nevertheless, due to the compactness of the packing material, a very high pressure for chemical regeneration is required [16]. Physical scrubbing of biogas using water is simple and one of the cheapest methods to remove CO_2 and H_2S simultaneously. However, the requirement of high pressure poses as a constraint. Chemical absorbents comparatively are considered to be more efficient as they do not require high pressure to act as absorbents. Besides, they store CO_2 permanently in a solid form by converting them into alkali metal carbonates [17, 18]. The most commonly used chemical bases for biogas scrubbing are NaOH, $Ca(OH)_2$, and KOH [19]. These are no doubt of commonly available chemicals but use of these solvents in rural areas for biogas scrubbing becomes impractical owing to technical and cost issues. In this context, this work proposes the use of a natural base solution (NBS) targeting the rural households of the northeastern region of India. Locally referred to as kola khar, the NBS is prepared from the ash of banana peels and is very popular among the rural communities of Northeast India [20]. For the present study, natural bases prepared from banana pseudostem (BPS), *Musa balbisiana peel* (MBP), and *Musa acuminata peel* (MAP) have been tested for CO_2 scrubbing from biogas. To assess the potential of the NBS as a CO_2 absorbent, the performance results have been compared with potassium hydroxide (KOH) bases of varying concentrations. Application of NBS for CO_2 scrubbing from biogas is envisioned to be advantageous over other conventional chemicals used for pretreatment as because the source material is abundantly available, has almost zero production cost, process of preparation is easy, and can be easily adopted in the rural areas.

2 Materials and Methods

2.1 Preparation of Natural Base Solution (NBS)

The banana peels were obtained from local farmers and were sun dried for a week, till it lost its yellow color. The peels were then burnt to ash in air on a wood flame. The banana peel ash formed thereafter was crushed into a fine powder and stored in dry conditions at 25 °C. To prepare the NBS, ash was taken and mixed with distilled water followed by checking of pH, which was measured using a pH meter (Make: Mettler Toledo, Model: 30,266,889). The process was repeated till no further increase in pH was observed even after addition of ash. Banana peels are natural sources of potassium, and the maximum pH that could be achieved of the solution prepared from ash of *Musa balbisiana Peel* (MBP) was 12.77. The solution was allowed to settle and the supernatant liquid was filtered with the help of a filter paper to get the base solution.

2.2 Base Solutions for CO_2 Scrubbing

Five different molar concentrations of potassium hydroxide (KOH) solution of 0.1, 0.3, 0.5, 0.7, and 1.0 M with pH of 12.68, 13.14, 13.30, 13.34, and 13.42 were prepared. Accordingly, three natural base ash solutions with saturated pH 11.48, 12.77, and 11.63 were prepared from different parts of banana plant, viz. banana pseudostem (BPS), *Musa balbisiana Peel* (MBP), and *Musa acuminata Peel* (MAP), respectively. These eight types of base solutions were used for CO_2 scrubbing from biogas and their performance compared.

2.3 Description of the Scrubber

The scrubber was fabricated using PVC pipe having dimension of 88 mm inner diameter and height of 328 mm with a volume of 2 L. The gas inlet is connected to the bottom of the scrubber while the gas outlet is placed at the top. Another two openings are located at the top and side lower portion of the scrubber for feeding and extraction of the saturated scrubbing agent, respectively.

2.4 Experimental Procedure

The scrubber was filled with each type of scrubbing agent and the biogas was directly fed from a 3m^3 biogas plant (Deenbandhu model) installed at Aaoniati Satra, North Guwahati, Assam. The flow rate was controlled at around 200 ml/min in order to measure the CO_2 absorption and saturation rate of the scrubbing agents. The treated and untreated biogas samples were collected in tedlar bags at 5 min time interval and were analyzed for CH_4 and CO_2 concentration by gas chromatograph (Thermo Fisher Scientific Trace GC 1110 Gas Chromatograph, USA). All experiments were conducted at ambient temperature (303 K), and the biogas pressure was maintained slightly above the atmospheric level (gauge pressure nearly 20 kPa) throughout the experiment with the help of a micropressure gauge. The pH of the solutions was measured using a digital pH meter (Make: Mettler Toledo, Model: 30,266,889) at regular time intervals.

2.5 Kinetics of CO_2 Absorption in Base Solutions

The rate of CO_2 absorption in base solution is obtained as per the procedure reported by Tippayawong and Thanompongchart [21]. The fraction of CO_2 in biogas that is absorbed by the base solution at time t is represented by Eq. 1.

$$A = 1 - \frac{C_o}{C_i} \tag{1}$$

where

A Fraction of CO_2 absorbed from the gas mixture after time t.
C_o outlet CO_2 concentration (% v/v) after time t.
C_i inlet CO_2 concentration (% v/v).

The removal rate of CO_2 from biogas is momentary for this particular scrubbing setup. The rate of declining CO_2 absorption is assumed to be proportional to the CO_2 fraction that was absorbed in the scrubbing column and the CO_2 fraction that passed through. It is expressed using the following equations reported by the authors [21–23].

$$\frac{dA}{dt} = -kA(1 - A) \tag{2}$$

After integration the above equation

$$\ln\left(\frac{A(1 - A_0)}{A_0(1 - A)}\right) = k(\tau - t) \tag{3}$$

After rearranging Eq. 3, we get:

$$t = \frac{1}{k}\ln\left(\frac{C_o}{C_i - C_o}\right) + \tau \quad (4)$$

where

- k Absorption constant.
- A_0 Fraction of CO_2 absorbed from the gas mixture after τ.
- τ Characteristic absorption time when 50% absorption of CO_2 occurs. This indicates that the solutions in the scrubbing column should be completely saturated after elapsed time of 2τ.
- t Experimental time.

2.6 Calorific Value Analysis Method

Higher calorific values (HCV) and lower calorific values (LCV) of enriched biogas were calculated according to the formula reported by Li et al. [24].

$$\mathrm{HCV}_{\mathrm{biogas}} = 0.3989 \times \mathrm{MC} + 0.0213\left(R^2 = 1\right) \quad (5)$$

$$\mathrm{LCV}_{\mathrm{biogas}} = 0.3593 \times \mathrm{MC} + 0.0192\left(R^2 = 1\right) \quad (6)$$

where MC is the methane content in biogas (%) and R^2 is the square of the correlation. The HCV and LCV of pure methane were 39.82 and 35.87 MJ/m^3, respectively [25].

3 Results and Discussions

3.1 CO_2 Absorption Capacity of the Base Solutions

Figure 1 shows the CO_2 absorption capacity of the three types of NBS and KOH of varying concentrations as a function of pH. It is observed that the absorption capacity of the scrubbing agent is a function of its pH value. Higher the pH value of the solvent, greater is its CO_2 absorption capacity. Thus, the result shows that the concentration of scrubbing agent plays an important role in the CO_2 absorption. The pH of the solvent decreased on CO_2 absorption, which may be attributed to the increase in concentration of H_2CO_3 and K_2CO_3 in the absorption system due to chemical reactions that take place during the absorption of CO_2 with base solutions as described by following chemical equilibrium reactions.

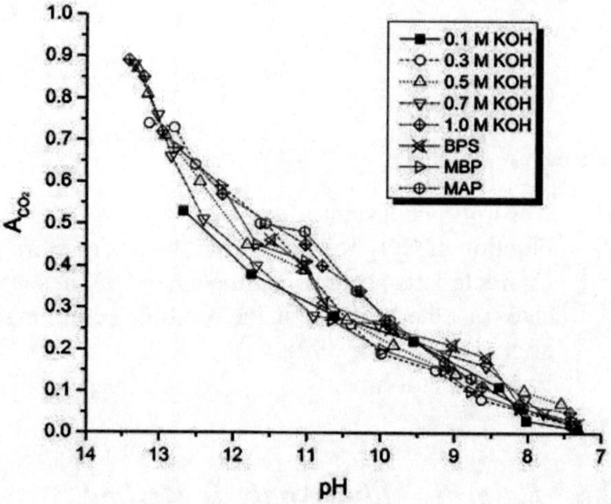

Fig. 1 CO_2 absorption capacity of scrubbing agents with respect to pH

$$H_2O\ (aq) + CO_2(g) \leftrightarrow H_2CO_3(aq) \tag{7}$$

$$2KOH\ (aq) + CO_2(g) \leftrightarrow K_2CO_3(aq) + H_2O \tag{8}$$

From the graph it is observed that base solutions with a pH of 11 and above had a capacity of approximately 40% CO_2 scrubbing from the fed raw biogas.

3.2 Variation of CO_2 and CH_4 in Treated Biogas with Time

Figure 2 compares the outlet CO_2 concentration in the biogas samples after passing through various scrubbing agents with respect to time. The increase in CO_2 concentration in the treated biogas samples with time was due to the steady saturation of the scrubbing agent, which also increases the solution viscosity, which simultaneously decreases the diffusion coefficient and hence the absorption rate [26]. For every solvent, after the respective saturation point CO_2 started desorbing and evolved in the outlet gas stream. It is observed from the graph that among the chemical bases 0.5, 0.7, and 1.0 M KOH solution were able to reduce the CO_2 concentration of the inlet biogas by approximately similar amount (~88%) followed by 0.3 M KOH (74%) and among the natural bases, MBP could reduce the maximum amount of CO_2 by 68% approximately. At lower concentration of scrubbing agents (especially 0.1 M KOH, MAP, and BPS) the ability to absorb CO_2 is less and hence outlet CO_2 concentration (C_o/C_i) is found more.

Figures 3 and 4 show the variation of CH_4 content in the treated biogas samples and the variation of pH of the scrubbing agents with respect to time. It was observed that with increase in KOH concentration and at higher value of pH the ability of the

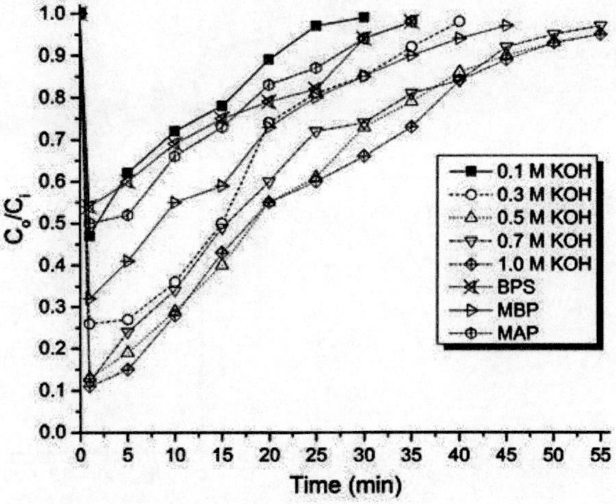

Fig. 2 Outlet CO_2 concentration with respect to time

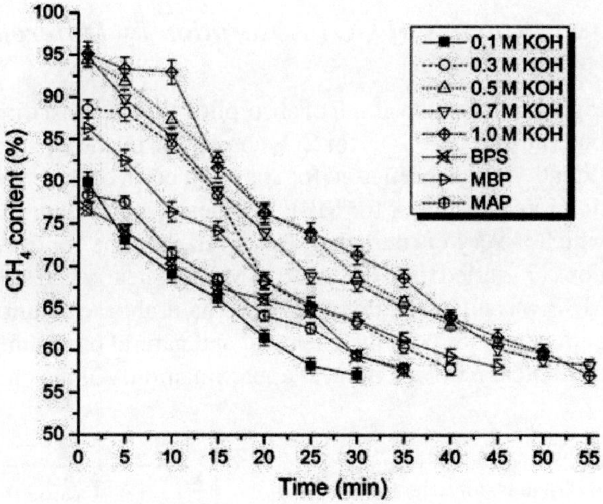

Fig. 3 Variation of CH_4 content with respect to time

scrubbing agents to absorb CO_2 increases thereby enriching CH_4 content in the outlet during the initial period of time. The process performance is fully reliant on factors such as dimensions of the scrubber, concentration of scrubbing agents, raw biogas composition, gas pressure, and flow rates. Accordingly, depending on the need one can calculate the replenishment time of the scrubbing agent or the number of cycles the biogas needs to pass through or the volume of the scrubbing agent required to attain the desired CH_4 content.

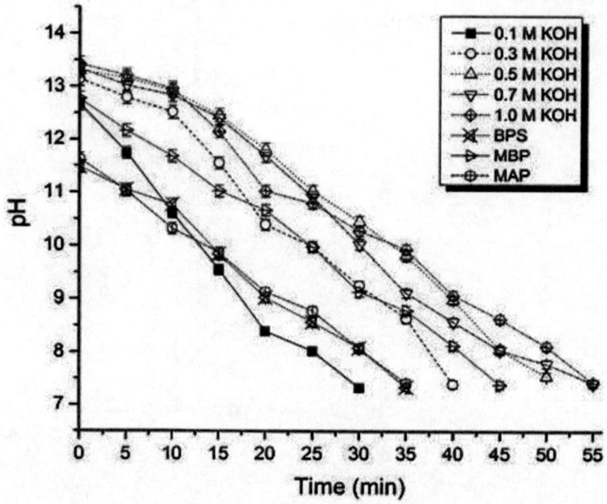

Fig. 4 Variation of pH with respect to time

3.3 Kinetics of CO_2 Absorption by Different Solvents

Figure 5 shows the graph of absorption time against dimensionless biogas concentration ($\ln[C_o/(C_i - C_o)]$) for CO_2 absorption by the chemical and natural base solutions. Equation 4 has been used for approximation of the curves to straight lines. The reaction rate constant k for MBP was nearly equivalent to that of 0.5 and 0.7 M KOH, which is very encouraging. However, the time for 50% saturation of the base solutions (τ) varied; 0.5, 0.7, and 1.0 M KOH nearly had similar τ values. The τ value for MBP was highest among the three natural base solutions. Table 1 shows the model parameters for both the chemical and natural base solutions and also the correlation coefficient R^2 of each linear approximation. All the chemical as well as natural base

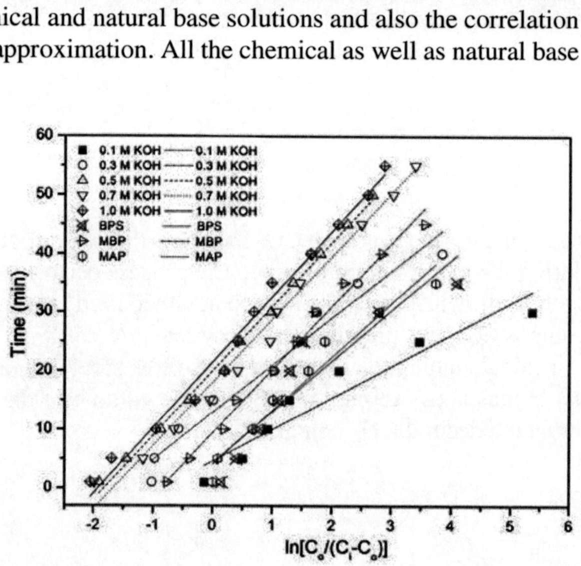

Fig. 5 Kinetics of CO_2 absorption by different molar concentration of KOH and natural base at 303 K

Table 1 Kinetic parameters for CO_2 absorption by different molar concentration of KOH and natural base solutions

Scrubbing agent	k (min^{-1})	τ(min)	R^2
0.1 M KOH	5.29	4.89	0.91
0.3 M KOH	7.96	13.06	0.96
0.5 M KOH	10.88	20.07	0.99
0.7 M KOH	10.68	18.12	0.98
1.0 M KOH	11.08	21.57	0.98
BPS	8.53	4.62	0.87
MBP	10.67	9.61	0.99
MAP	9.03	4.32	0.96

solutions excepting 0.1 M KOH and BPS exhibited R^2 values of greater than 0.95 and thereby reflecting good linear relationships.

3.4 Methane Enrichment by Chemical and Natural Base Solutions

The volume percentage of methane content (%) after every five minutes of time interval was analyzed with the help of gas chromatograph. Figure 6 compares the methane enrichment that could be achieved with the different scrubbing agents. The methane content of the untreated biogas was 56.9%. KOH of 0.5 M concentration could enrich the methane content upto 94.4% followed by 0.3 M KOH (88.6%), MBP (86.3%), 0.1 M KOH (79.9%), MAP (78.4%), and BPS (76.8%). The findings clearly indicate that natural base solutions have almost equivalent potential like chemical base solutions for scrubbing of CO_2 from biogas. Besides, the natural base ash

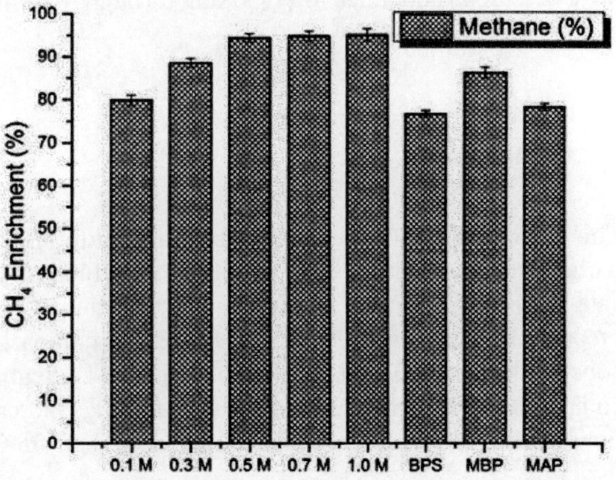

Fig. 6 Methane enrichment by different scrubbing agents

Table 2 Calorific value of raw and enriched biogas treated with scrubbing agents

Scrubbing agents	HCV (MJ/m^3)	LCV (MJ/m^3)
Raw	22.72	20.46
0.1 M KOH	31.89	28.73
0.3 M KOH	35.36	31.85
0.5 M KOH	37.68	33.94
0.7 M KOH	37.84	34.08
1 M KOH	37.96	34.19
BPS	30.66	27.61
MBP	34.45	31.03
MAP	31.29	28.19

solutions are environment friendly, disposal is hassle free, and process of production is inexpensive and simple. Moreover, the raw materials are easily available, which makes the solution cost effective and can be rendered as an appropriate technology for rural applications.

3.5 HCV and LCV of Enriched Biogas

HCV and LCV of inlet biogas (raw) and enriched biogas were theoretically calculated using Eqs. 5 and 6, respectively. Table 2 shows the calorific value of raw and enriched biogas treated with different scrubbing agents. The HCV (22.72 MJ/m^3) and LCV (20.46 MJ/m^3) of untreated biogas were significantly less before the treatments. The HCV and LCV of scrubbed biogas obtained after passing through KOH solutions of 0.5, 0.7, and 1.0 M concentrations have almost equivalent HCV and LCV of 37.82 and 34.06 MJ/m^3, respectively. However, with the natural base solutions, the scrubbed biogas obtained after passing through MBP had the highest HCV of 34.45 and LCV of 31.03 MJ/m^3.

4 Conclusions

This work reports the use of natural base ash solutions (NBS), prepared from different parts of banana plant for CO_2 scrubbing from biogas. The experimental results indicate that the NBS possess varying pH, and the CO_2 absorption capacity of NBS prepared from *Musa balbisiana Peel* (MBP) is comparable to KOH solution of 0.3 M concentration. MBP could enrich the methane content in biogas from 56.9 to 86.3% compared to 88.6% by 0.3 M KOH. The findings indicate the potential of NBS solutions to be usable as substitute of chemical solvents for CO_2 scrubbing with the added advantage of these being environment friendly, abundant resource availability, easy preparation process, and cost effective.

Acknowledgements The authors gratefully acknowledge the financial support of the Ministry of New and Renewable Energy, Government of India for the above work (256/3/2017-BIOGAS).

References

1. Mamun, M. R. A., Karim, M. R., Rahman, M. M., Asiri, A. M., & Torii, S. (2016). Methane enrichment of biogas by carbon dioxide fixation with calcium hydroxide and activated carbon. *Journal of the Taiwan Institute of Chemical Engineers, 58*, 476–481.
2. Mann, G., Schlegel, M., Schumann, R., & Sakalauskas, A. (2009). Biogas-conditioning with microalgae. *Agronomy Research, 7*(1), 33–38.
3. Karki, A. B., Shrestha, J. N., & Bajgain, S. (2005). Biogas: As renewable source of energy in Nepal theory and development. *BSP-Nepal*. 1–189.
4. Abatzoglou, N., & Boivin, S. (2008). A review of biogas purification processes. *Biofuels Bioproducts and Biorefining, 3*(1), 42–71.
5. Monnet, F. (2003, November). An introduction to anaerobic digestion of organic waste, Final report, Remade Scotland.
6. Ray, N. H. S., Mohanty, M. K., & Mohanty, R. C. (2016). Biogas compression and storage system for cooking applications in rural households. *International Journal of Renewable Energy Research, 6*(2), 593–598.
7. Abushammala, M. F. M., Qazi, W. A., Azam, M. H., Mehmood, U. A., Al-Mufragi, G. A., & Alrawahi, N. A. (2016). Generation of electricity from biogas in Oman. In *3rd MEC International Conference on Big Data and Smart City*.
8. Sun, Q., Li, H., Yan, J., Liu, L., Yu, Z., & Yu, X. (2015). Selection of appropriate biogas upgrading technology-a review of biogas cleaning, upgrading and utilisation. *Renewable and Sustainable Energy Reviews, 51*, 521–532.
9. Zhao, Q., Leonhardt, E., MacConnell, C., Frear, C., & Chen, S. (2010). Purification technologies for biogas generated by anaerobic digestion. In *Climate friendly farming, compressed biomethane* (pp. 1–24). CSANR Research Report, [Chapter 9].
10. Osorio, F., & Torres, J. C. (2009). Biogas purification from anaerobic digestion in a wastewater treatment plant for biofuel production. *Renewable Energy, 34*, 2164–2171.
11. Farooq, M., Chaudhry, I. A., Hussain, S., Ramzan, N., & Ahmed, M. (2012). Biogas upgradation for power generation applications in Pakistan. *Journal Quality Technology Management, 8*(2), 107–118.
12. Yang, H., Xu, Z., Fan, M., Gupta, R., Slimane, R. B., & Bland, A. E. (2008). Progress in carbon dioxide separation and capture: A review. *Journal of Environmental Sciences, 20*, 14–27.
13. Gomes, V. G., & Hassan, M. M. (2001). Coal seam methane recovery by vacuum swing adsorption. *Separ Purif Techn, 24*, 189–196.
14. Ryckebosch, E., Drouillon, M., & Vervaeren, H. (2011). Techniques for transformation of biogas to biomethane. *Biomass Bioenergy 1, 35*(5), 1633–1645.
15. Hagen, M., & Polman, E. (2001). Adding gas from biomass to the gas grid. *Final report submitted to Danish Gas Agency* (pp. 26–47).
16. Wise, D. L. (1981). Analysis of systems for purification of fuel gas. In *Fuel gas production from biomass* (Vol 2). CRC Press.
17. Baciocchi, R., Costa, G., Gavasci, R., Lombardi, L., & Zingaretti, D. (2012). Regeneration of a spent alkaline solution from a biogas upgrading unit by carbonation of APC residues. *Chemical Engineering Journal, 179*, 63–71.

18. Baciocchi, R., Carnevale, E., Corti, A., Costa, G., Lombardi, L., Olivieri, T., Zanchi, L., & Zingaretti, D. (2013). Innovative process for biogas upgrading: Results from pilot plant operation. *Biomass and Bioenergy, 53*, 128–137.
19. Malla, F. A., Khan, S. A., & Gupta, N. (2016). Application of alkaline chemicals for biogas methane enrichment. *Economics Environment and Conservation, 22*(2), 475–484.
20. Hemanta, M. R., Mane, V. K., & Bhagwat, A. (2014). Analysis of traditional food additive kolakhar for its physico-chemical parameters and antimicrobial activity. *European Academic Research, 2, 8*, 10531–10536.
21. Tippayawong, N., & Thanompongchart, P. (2010). Biogas quality upgrade by simultaneous removal of CO_2 and H_2S in a packed column reactor. *Energy, 35*, 4531–4535.
22. Leonzio, G. (2016). Upgrading of biogas to bio-methane with chemical absorption process: Simulation and environmental impact. *Journal of Cleaner Production, 131*, 364–375.
23. Aboudheir, A., Tontiwachwuthikul, P., Chakma, A., & Idem, R. (2003). Kinetics of the reactive absorption of carbon dioxide in high CO_2-loaded, concentrated aqueous monoethanolamine solutions. *Chemical Engineering Science, 58*, 5195–5210.
24. Li, Y., Liu, H., Yan, F., Su, D., Wang, Y., & Zhou, H. (2017). High-calorific biogas production from anaerobic digestion of food waste using a two-phase pressurized biofilm (TPPB) system. *Bioresource Technology, 224*, 56–62.
25. Li, Y., Zhang, R., He, Y., Zhang, C., Liu, X., Chen, C., & Liu, G. (2014). Anaerobic co-digestion of chicken manure and corn stover in batch and continuously stirred tank reactor (CSTR). *Bioresource Technology, 156*, 342–347.
26. Zanfir, M., & Gavriilidis, A. (2005). Carbon dioxide absorption in a falling film microstructured reactor: Experiments and modeling. *Industrial and Engineering Chemistry Research, 44*, 1742–1751.

Vegetable Oil as Fuel in Ci Engine with and Without Exhaust Gas recirculation—A Review

S. S. Mohapatra, S. Choudhury, and Binayak Pattanayak

1 Introduction

The fossil fuel reserve throughout the world is decreasing day by day. To fulfill the energy demands of the world, an alternate energy source is required. Particularly in compression ignition engines, vegetable oils can be used as a replacement of fossil fuel. Vegetable oils are non-toxic, renewable, and bio degradable. The emission from the combustion of vegetable oil is lower as compared to fossil fuel [1– 3]. The first diesel engine was developed by Rudolf Diesel, and he ran it with groundnut oil as fuel [4]. But use of vegetable oil in diesel engine has some drawbacks. The drawbacks arise because of low volatility, high viscosity, and low heating values of vegetables oils [5]. The direct use of straight vegetable oil in compression ignition engine may lead to the problems like poor fuel atomization, heavy smoke emission, lube oil dilation, high carbon deposits, injection nozzle failure, etc. Using single stage trans-esterification or two stage esterification and transesterification process, vegetable oils can be transformed to corresponding vegetable oil esters. These vegetable oil esters are otherwise known as biodiesels. With the help of the above processes, the viscosity of vegetables can be reduced and its volatility can be increased. This may lead to proper combustion of the vegetable oil in CI engines. The combustion of biodiesel in CI engine greatly reduces CO, CO_2, HC, and PAH emission [5]. But NOx formation increases by the use of biodiesel because of its higher oxygen content and proper combustion characteristic [6]. The NOx emission from CI engines can be reduced by the use of techniques like water injection, recirculation of exhaust gas, etc. In exhaust gas recirculation, some portion of the engine exhaust coming out of the combustion chamber is mixed with the fresh air. This mixture is supplied the combustion chamber. This mixture has high heat capacity and moisture content.

S. S. Mohapatra · S. Choudhury · B. Pattanayak (✉)
Department of Mechanical Engineering, Siksha 'O' Anusandhan, Deemed to be University, Bhubaneswar, India
e-mail: binayakpattanayak@soa.ac.in

P. Mahanta et al. (eds.), *Advances in Thermofluids and Renewable Energy*, Lecture Notes in Mechanical Engineering, https://doi.org/10.1007/978-981-16-3497-0_29

This combined effect reduces the flame temperature of combustion [7]. The reduced flame temperature causes reduction in NOx emission from the engine.

2 Fuel Properties

The properties of vegetable oil such as density, viscosity, cloud point, calorific value, pour point, flash point, fire point and cetane number, etc., are discussed in this section.

2.1 *Straight Vegetable Oil (SVO)*

Vegetable oils are plant oils which are produced from the resin or plant seeds. All vegetable oils primarily consist of triglycerides. The viscosity and volatility of vegetable oil are larger than that of fossil fuel due to the molecular size of triglycerides [8]. Flash point signifies the flammability of vapors produced from a fuel, and pour point signifies the flow characteristic of the oil at cold condition. Vegetable oil has larger flash point than diesel fuel. Pour point of edible plant oils are lower than diesel; however, for non-edible plant oils, it is greater than diesel [9]. Various properties of different vegetables oils are given in Table 1.

Table 1 Properties of some vegetable oils [5]

Vegetableoil	Kinematic viscosity at 38^0C (cm^2/s)	Cetane no	Heating value (MJ/kg)	Cloud point (K)	Pour point (K)	Flash point (K)	Density (kg/mm^3)	Reference
Corn	0.349	37.6	39.5	271.9	233	550	909.5	Murugesan et al. [5]
Cottonseed	0.335	41.8	39.5	274.7	258	507	914.8	
Crambe	0.536	44.6	40.5	283	260.8	547	904.8	
Linseed	0.272	34.6	39.3	274.7	258	514	923.6	
Peanut	0.39	41.8	39.8	285.8	266.3	544	902.6	
Rapeseed	0.37	37.6	39.7	269.1	241.3	519	911.5	
Safflower	0.313	41.3	39.5	291.3	266.3	533	914.4	
Sesame	0.355	40.2	39.3	269.1	263.6	533	913.3	
Soya bean	0.326	37.9	39.6	269.1	260.8	527	913.8	
Sunflower	0.339	37.1	39.6	280.2	258	547	916.1	
Palm	0.396	42.0	–	304	–	540	918.0	
Babassu	0.303	38	–	293	–	423	946.0	

2.2 Biodiesel

For reduction of the viscosity and specific gravity of vegetable oil, they are converted to their respective biodiesels. The term biodiesel has no unambiguous definition. Biodiesel are basically alkyl esters of different vegetable oils. Biodiesel is produced by a procedure known as transesterification. From the study of fuel properties of different biodiesels like polanga oil methyl ester and karanja oil methyl ester, it has been found that the property of biodiesel remains close to diesel fuel [10 ,11]. Various properties of biodiesel of different oils are given in Table 2.

3 Experiment

In this section, the experiments conducted by various researchers were discussed using biodiesel with and without exhaust gas recirculation.

3.1 Biodiesel Mode

Puhan et al. [12] conducted various tests on a constant speed direct injection (DI) type diesel engine having variable injection pressure. The experiment compared the diesel with linseed oil methyl ester (LOME) at different injection pressure of 200 bar, 220 bar, and 240 bars. The experiments were used to particularly study the emission and performance characteristics. Banapurmath et al. [13] conducted experiments on a four stroke direct injection single cylinder CI engine operated on neat diesel, Jatropha oil methyl ester (JOME), Honge oil methyl ester (HOME), and sesame oil methyl ester (SOME) at optimum injection pressure and injection timing. Gürü et al. [14] investigated on the use of different blends of tall oil methyl ester (TOME). They choose a direct injection type diesel engine, diesel engine with single cylinder. Al-Widyan et al. [15] conducted experiments using different blends of biodiesel. They used the waste vegetable oil for biodiesel production. Experiments were conducted on a direct injection, naturally aspirated, water cooled, and single cylinder variable compression engine at 1000, 1200, 1450, and 1800 rpm at 20o BTDC injection time and 18:1 compression ratio. Purushothaman and Nagarajan [16] used experimental method to study the emission, combustion, and performance characteristic of orange oil with diethyl ether (DEE). Godiganur, Murthy and Reddy [17] conducted various experiments on a three cylinder, four strokes, direct injection, and air-cooled CI engine using different blendings of fish oil methyl ester (FOME). Experiments are conducted using several blends of fish oil methyl ester. Borgelt et al. [18] did research on a engine which is operated on different blends of Soybean oil methyl ester and diesel fuel. They used B0, B10, B20, B30, and B40 for their investigation. Raheman and Ghadge [19] conducted experiments on a water cooled, four stroke,

Table 2 Properties of some biodiesels

Fuel type	Calorific value (KJ/kg)	Density (kg/m^3)	Viscosity (mm^2/s)	Cetane number	Flash point °C	Reference
Sunflower oil	40,579	878	10.3 (27 °C)	45.5	–	Ramadhas et al. [27]
Cottonseed oil	40,580	874	11.1 (27 °C)	45.5	–	Ramadhas et al. [27]
Soybean oil	39,760	872	11.1(27 °C)	37.0	178	Ramadhas et al. [27]
Karanja oil	36,120	876	9.60 (27 °C)	52.8	187	Murugesan et al. [5]
Pungam oil	39,150	892	5.40 (27 °C)	57	156	Venkatraman and Deveradjane [28]
Jatropha oil	38,450	880	5.65 (27 °C)	50	170	Narayan and Ramesh [29]
Honge oil	36,010	890	5.6 (27 °C)	45	163	Banapurmath and Tewari [30]
Jojoba oil	39,862	920	52 (40 °C)	55	186	Shehata et al. [31]
linseed oil	40,759	865	4.2 (40 °C)	48	161	Puhan et al. [12]
sesame oil	38,836	882	5.34 (40 °C)	–	170	Banapurmath and Tewari [30]
Polanga oil	41,397	869	3.99 (40 °C)	–	140	Sahoo and Das [32]
Tobacco oil	39,811	886.8	3.5 (40 °C)	51	–	Usta [33]
Palm oil	37,130	871.6	4.73 (40 °C)	62	–	Agudelo and Agudelo [34]
Tall oil	40,023	922	7.1 (40 °C)	54	89	Gürü et al. [14]
Fish oil	42,241	880	4.0 (40 °C)	–	176	Godiganur et al.[17]
Animal fat oil	39,858	877	5.072 (40 °C)	58.8	30	Oner and Altun [35]

single cylinder Ricardo E6 engine taking high speed diesel (HSD), mahua biodiesel (B100), and various blendings of methyl ester of Mahua oil (MOME). Aydin and Bayinder [20] conducted various experiments on a single cylinder, four strokes, and air-cooled diesel engine using various blends of cotton seed methyl ester (COME). They compared the performance and emission characteristic of various blends of cotton seed oil methyl ester with diesel fuel at various speed of the engine. Karabektas et al. [21] conducted various experiments on a single cylinder type naturally aspirated, diesel engine of variable speed using pre-heated biodiesel, and diesel fuel. The biodiesel is pre-heated up to various ranges of temperatures which are 30, 60, 90, 120 °C. Tests are performed at full load condition at various engine speeds. Raheman and Phadatare [22] conducted experiments on a four stroke, water cooled diesel engine. The engine was having single cylinder with direct injection. They used different blending of Karanja oil methyl ester for experimentation.

3.2 Biodiesel with EGR Mode

Tsolakis et al. [23] conducted experiments to study the combined effects of EGR with Rapeseed oil methyl ester (RME) on a naturally aspirated, air-cooled, single cylinder, and direct injection diesel engine. They compared results regarding the emission along with the performance of engine obtained with neat and two different blends RME blended with diesel. They investigated at 0, 10, and 20% EGR rate. Rajan and Senthilkumar [24] investigated the effect of EGR with sunflower oil methyl ester blends and diesel fuel on the emission and performance characteristic of a naturally aspirated, twin cylinder, four stroke, water cooled, and DI type diesel engine. Agarwal et al. [6] investigated on the use of EGR with methyl ester of rice bran oil and neat diesel on a two cylinders, constant speed, air-cooled, four stroke, and DI type diesel engine. Pradeep and Sharma [25] studied the implications of hot EGR with diesel and Jatropha methyl ester on a water cooled, DI type four stroke, and one cylinder diesel fuelled engine. They optimized that 15% EGR rate to be suitable for all emission and performance characteristic. Saleh [26] experimented on a diesel engine using blends of methyl ester of Jojoba oil with EGR. They used a two cylinder, DI type engine with natural aspiration.

4 Results and Discussion

4.1 Biodiesel Mode

Engine performance, emission, and combustion of using biodiesel in C.I engine are represented in this section.

4.1.1 Engine Performance

Puhan et al. [12] found that the BSFC of the methyl ester of Linseed oil, under all operating conditions is more as compared to that of diesel. This arises as Linseed oil methyl ester (LOME) is denser than diesel with lesser heating value. At 240 bars operating condition with the use of LOME at full load condition, the BTE is observed to be greater than diesel. This occurs because at higher pressure and full load condition, there is improved atomization of fuel which helps in proper combustion. Banapurmath et al. [13] found that BTE is lowest for JOME and highest for diesel. The increasing order of BTE with different fuels is found to be JOME, HOME, SOME, and diesel. The probable cause of this is the high amount of viscosity, less volatility, and higher density of biodiesel in comparison to diesel which leads to the poor mixture formation. Gürü et al. [14] found that there are no noticeable differences between blends of biodiesel of tall oil methyl ester and diesel in engine power and torque at low speed. The reason behind is this is that the blended fuels have higher cetane number. MI Al-Widyan et al. [15] found that BTE, brake power, and temperature of exhaust were higher for all blends of biodiesel in comparison with the baseline fuel at all speeds. This is explained by the better combustibility nature of biodiesel due to its inherent oxygen content. But the BSFC for all biodiesel blends are found lower than the baseline fuel due to its high density and viscosity. Purushothaman and Nagarajan [16] observed that at all loads, the mixture of orange oil with 36 mg/s of diethyl ether shows maximum BTE. Godiganur et al. [17] found that the BSFC for all blends of FOME is more than that of diesel due to higher mass of consumption of FOME. The BTE for blends of FOME is less than diesel except B20 due to higher mass of blended fuel consumption. Borgelt et al. [18] found that peak torque was not affected by blends of biodiesel, but exhaust gas temperature shows significant increase along with the increase in BTE. Raheman and Ghadge [19] found that when percentage of mahua biodiesel increases in the fuel, the BSFC also increases. Aydin and Bayinder [20] found that the torque along with power were less when the percentage of cotton seed oil methyl ester (COME) increases in its blend. The reason attributed to high amount of viscosity and less heating value of COME. Karabektas et al. [21] found that brake power reduces and BTE improves with use of COME in comparison to diesel in pre-heated condition because of decrease in its viscosity and density. Raheman and Phadatare [22] found that the complete combustion karanja biodiesel leads to higher torque as compared to diesel. The 20% and 40% blend increases the torque by 0.1–1.3%, respectively. But in case of B60, B80, and B100, it decreased due to a reduction in calorific value of fuel.

4.1.2 Combustion

Banapurmath et al. [13] observed shorter ignition delay for diesel as compared to biodiesels. In between SOME, HOME, and JOME, the shortest ignition delay is observed for SOME. The diffusion of combustion phase is longer in case of esters as compared to diesels which leads to higher combustion duration. Purushothaman and Nagarajan [16] found that peak cylinder pressure is higher for diethyl ether with orange oil and lower for diesel fuel. The existence of oxygen content leads to proper combustion hence more cylinder pressure in case of biodiesel and its blends.

4.1.3 Exhaust Gas Emissions

Puhan et al. [12] found that the CO and smoke emission is considerably low for Linseed oil methyl ester as it contains oxygen which helps in proper combustion and reduction in CO, smoke emission which results in higher CO2 emission. NOx emission considerably increases by using Linseed oil methyl ester as compared to diesel because of the high viscosity and oxygen content of LOME. Banapurmath et al. [13] found that the smoke emission level along with HC and CO emission is higher with the 100% biodiesels as compared to standard diesel. The poor atomization, high amount of viscosity, and low volatility are the reason behind it. JOME having highest viscosity has maximum emission, whereas SOME having lowest viscosity has minimum emission. Gürü et al. [14] found that CO emission reduced for blended fuel at all speeds due to the oxygen content of biodiesel. But the presence of higher amount of oxygen content in the blends of biodiesel leads to proper combustion which in turn leads to more NOx emission. Purushothaman and Nagarajan [16] found that as compared to diesel, blends of orange oil and neat orange oil lower NOx emission are obtained from the mixture of DEE with orange oil. The evaporation of DEE lowers the charge temperature which reduces in cylinder gas temperature hence NOx level. The orange oil and its blend with diesel have higher amount of oxygen content. This leads to proper combustion of the fuel and more amount of NOx emission. The emission level of CO and HC is highest for the mixture of orange oil with DEE operation. The reason behind this is incomplete combustion of orange oil due to energy shared by burning of DEE. But for 100% orange oil and blends CO and HC emission is low due to oxygen content of orange oil. The different blends of orange oil have higher oxygen content which leads to proper combustion and in turn low smoke emission. Godiganur et al. [17] found that CO and HC emission level decreases by the use of blends of FOME due to its proper combustion characteristic which attributes to presence of oxygen content in the biodiesel chain. All the blends of biodiesel produce more NOx emission than that of diesel due to proper combustion characteristics of it. Borgelt et al. [18] found that HC and CO emission are reduced due to proper combustion of biodiesel. The presence of oxygen in the fuel is supposed to promote complete combustion. They found reduction in NOx emission by delaying injection timing of 3°. Raheman and Ghadge [19] found that the smoke level and CO emission decreases with increasing concentration of Mahua biodiesel due to its high oxygen content which also increases the NOx emission. Karabektas et al. [21] found that CO emission is lower for COME than diesel due to its oxygen content. But NOx emission is higher for COME at all temperature level than diesel due to the oxygen content of COME and higher combustion temperature in the cylinder. Raheman and Phadatare [22] found that the emission of CO, smoke, and NOx is considerably lower than neat diesel for all the blends of karanja oil methyl ester at any loading condition.

4.2 *Biodiesel with EGR Mode*

The engine performance, combustion, and engine emissions using biodiesel with exhaust gas recirculation in C.I engine are discussed in this part of the article.

4.2.1 Engine Performance

Tsolakis et al. [23] found that reduced efficiency and increased fuel consumption resulted by the use of EGR in the engine. Rajan and Senthilkumar [24] found that BSFC and BTE are increased at part loading condition, but remained unaffected at full loading condition with EGR. The increase in BTE and BSFC are due to the reburning of hydrocarbons circulated in EGR. They also found that there is a lowering of temperature of exhaust gas with diesel/EGR than biodiesel/EGR. Agarwal et al. [6] found that BTE increases and BSFC decreases slightly with EGR at part load condition due to reburning of partially burnt hydrocarbons in the combustion chamber and remains unaffected at full loading condition for neat diesel and various other fuel blends. Pradeep and Sharma [25] found that with 15% EGR, BTE is 30.1% and 32.4% for Jatropha biodiesel and diesel, respectively. When EGR increased more than 15%, BTE is decreased significantly for both fuels as EGR causes dilution of charge. Saleh [26] found that when EGR percentage is increased, the temperature of exhaust gas considerably reduced due to the reduction in flame temperature.

4.2.2 Combustion

Tsolakis et al. [23] observed that by using EGR in an engine fuelled with biodiesel leads to shifting of the starting and ending of the combustion process toward the later part of compression, rise of the ignition delay period. The heat release rate of biodiesel with 15% EGR was found to be lower as compared to without EGR. Heat release rate is higher for engine fuelled with biodiesel without EGR. This occurs because of dynamic and static advance of injection apart from higher presence of oxygen in biodiesel.

4.2.3 Exhaust Gas Emissions

Tsolakis et al. [23] observed that by using EGR, NOx formation decreases but smoke, HC, and CO emission increases for all the fuel blends and diesel. The availability of lesser amount of oxygen for combustion increases smoke, HC, and CO emission, whereas reduced flame temperature decreases NOx formation. Rajan and Senthilkumar [24] found that with EGR use CO and HC emission increased because of the availability of lesser amount of oxygen and hence causes improper combustion.

But NOx emission decreases due to reduced flame temperature and low oxygen available for combustion. Agarwal et al. [6] found that with EGR the smoke opacity, HC, and CO emission increases as compared to without EGR due to reduced oxygen availability for combustion with exhaust gas recirculation. But NOx formation decreases with EGR as compared to without EGR due to lower oxygen availability and lower flame temperature. Pradeep and Sharma [25] observed that the smoke, HC, CO emission is lower for Jatropha methyl ester at full load condition in comparison to diesel. Saleh [26] found that NOx formation decreases, but emission of HC and CO increases with increase in EGR rate for Jojoba oil methyl ester. This occurs mainly because of lesser amount of oxygen availability for combustion and low flame temperature.

5 Conclusions

A number of research activities have been conducted on the combustion, emission, and performance aspect of CI engine fuelled with vegetable oil. When vegetable oils are used as fuels in the CI engine, their low volatility and high viscosity leads to improper combustion. The viscous nature of vegetable oils can be altered in different ways like emulsification, pyrolysis or cracking, dilution with hydrocarbon or blending, and transesterfication. Out of all these methods, transesterification is widely used, and the products of transesterification, which are mono-alkyl esters of different vegetable oils, are known as biodiesel. From various experiments conducted, it has been found that all the emission parameters like HC, CO, and soot emission decreases by the use of biodiesel, but NOx emission increases due to the high oxygen availability and proper combustibility of biodiesel. Exhaust gas recirculation technique has been implemented by many researchers to reduce the NOx emission (Tables 1, 2).

References

1. Srivastava, A., & Prasad, R. (2000). Triglycerides based diesel fuels. *Renewable and Sustainable Energy Reviews, 4*, 111–133.
2. Ma, F., & Hanna, M. F. (1999). Biodiesel production: A review. *Bioresource Technology, 70*, 1–15.
3. Williamson, A. M., & Badr, O. (1998). Assessing the viability of using rape methyl ester (RME) as an alternative to mineral diesel fuel powering road vehicles in the UK. *Applied Energy, 59*, 189–214.
4. Bijalwan, A., Sharma, C. M. and Kediyal, V. K. (2006) 'Biodiesel revolution', Science reporter, pp.14–17.
5. Murugesan, A., Umarani, C., Subramanian, R., & Nedunchezhian, M. (2009). Biodieel as an alternative fuel for diesel engine—A review. *Renewable and Sustainable Energy Reviews, 13*, 653–662.
6. Agarwal, D., Sinha, S., & Agarwal, A. K. (2006). Experimental investigation of control of NOx emissions in biodiesel fueled compression ignition engine. *Renewable Energy, 31*, 2356–2369.

7. Agarwal, D., Singh, S. K., & Agarwal, A. K. (2011). Effect of exhaust gas recirculation (EGR) on performance emissions, deposits and durability of a constant speed compression ignition engine. *Applied Energy, 88*, 2900–2907.
8. Hossain, A. K., & Davis, P. A. (2010). Plant oils as fuels for compression ignition engines: A technical review and life cycle analysis. *Renewable Energy, 35*, 1–13.
9. Ziejewski, M., & Gottler, H. J. (1992). *Design modifications for durability improvements of diesel engine operating on plant oil fuels, SAE 9210630*. USA: SAE.
10. Sahoo, P. K., Das, L. M., Babu, M. K. G., & Naik, S. N. (2007). Biodiesel development from high acid value polanga seed oil and performance evaluation in a CI engine. *Fuel, 86*, 448–545.
11. Naik, M., Meher, L. C., Naik, S. N., & Das, L. M. (2008). Production of biodiesel from high free fatty acid karanja (Pongamia pinnata) oil. *Biomass and Bioenergy, 32*, 354–357.
12. Puhan, S., Jegan, R., Balasubramanian, K., & Nagarajan, G. (2009). Effect of injection pressure on performance, emission and combustion characteristic of high linolenic linseed oil methyl ester in a DI diesel engine. *Renewable Energy, 34*, 1227–1233.
13. Banapurmath, N. R., Tewari, P. G., & Hosmath, R. S. (2008). Performance and emission characteristics of a DI compression ignition engine operated on Honge Jatropha and sesame oil methyl esters. *Renewable Energy, 33*, 1982–1988.
14. Altiparmak, D., Keskin, A., Koca, A., & Gürü, M. (2007). Alternative fuel properties of tall oil fatty acid methyl ester-diesel fuel blends. *Bioresource Technology, 98*, 241–246.
15. Al-Widyan, M. I., Tashtoush, G., & Abu-Qudais, M. (2002). Utilization of ethyl ester of waste vegetable oils as fuel in diesel engine. *Fuel Processing Technology, 76*, 91–103.
16. Purushothaman, K., & Nagarajan, G. (2009). Experimental investigation on a C.I. engine using orange oil and orange oil with DEE. *Fuel, 88*, 1732–1740.
17. Godiganur, S., Murthy, C. S., & Reddy, R. P. (2010). Performance and emission characteristics of a Kirloskar HA394 diesel engine operated on fish oil methyl esters. *Renewable Energy, 35*, 355–359.
18. Schumacher, L. G., Borgelt, S. C., Fosseen, D., Goetz, W., & Hires, W. G. (1996). Heavy duty engine exhaust emission tests using methyl ester soybean oil/diesel fuel blends. *Bioresource Technology, 57*, 31–36.
19. Raheman, H., & Ghadge, S. V. (2007). Performance of compression ignition engine with mahua (Madhuca indica) biodiesel. *Fuel, 86*, 2568–2573.
20. Aydin, H., & Bayinder, H. (2010). Performance and emission analysis of cottonseed oil methyl ester in a diesel engine. *Renewable Energy, 35*, 588–592.
21. Karabektas, M., Ergen, G., & Hosoz, M. (2008). The effects of preheated cottonseed oil methyl ester on the performance and exhaust emissions of a diesel engine. *Applied Thermal Engineering, 28*, 2136–2143.
22. Raheman, H., & Phadatare, A. G. (2004). Diesel engine emissions and performance from blends of karanja methyl ester and diesel. *Biomass and Bioenergy, 27*, 393–397.
23. Tsolakis, A., Megaritis, A., Wyszynski, M. L., & Thennoi, K. (2007). Engine performance and emissions of a diesel engine operating on diesel-RME (rapeseed methyl ester) blends with EGR (exhaust gas recirculation). *Energy, 32*, 2072–2080.
24. Rajan, K., & Senthilkumar, K. R. (2009). Effect of exhaust gas recirculation (EGR) on the performance and emission characteristics of diesel engine with sunflower oil methyl ester. *Jordan Journal of Mechanical and Industrial Engineering, 3*, 306–311.
25. Pradeep, V., & Sharma, R. P. (2007). Use of HOT EGR for NOx control in a compression ignition engine fuelled with bio-diesel from Jatropha oil. *Renewable Energy, 32*, 1136–1154.
26. Saleh, H. E. (2009). Effect of exhaust gas recirculation on diesel engine nitrogen oxide reduction operating with jojoba methyl ester. *Renewable Energy, 34*, 2178–2186.
27. Ramadhas, A. S., Jayaraj, S., & Muraleedharan, C. (2004). Use of vegetable oils as I.C. engine fuels—A review. *Renewable Energy, 29*, 727–742.
28. Venkatraman, M., & Deveradjane, G. (2010). Experimental investigation of performance and emission characteristics of diesel-pungam oil methyl esters diesel blends fueled DI diesel engine at optimum engine operating parameters. *International Journal of Green energy and environment, 1*, 7–12.

29. Narayan, R. J., & Ramesh, A. (2010). Experimental investigations on compression ignition engine fuelled with bio-diesel of Jatropha oil. *International Journal of Green energy, energy and environment, 1*, 1–6.
30. Banapurmath, N. R., & Tewari, P. G. (2009). Comparative performance studies of a 4-stroke CI engine operated on dual fuel mode with producer gas and Honge oil and its methyl ester (HOME) with and without carburetor. *Renewable Energy, 34*, 1009–1015.
31. Shehata, M. S., Razek, S., & Abdel, M. (2011). Experimental investigation of diesel engine performance and emission characteristics using jojoba/diesel blend and sunflower oil. *Fuel, 90*, 886–897.
32. Sahoo, P. K., & Das, L. M. (2009). Combustion analysis of Jatropha, Karanja and Polanga based biodiesel as fuel in a diesel engine. *Fuel, 88*, 994–999.
33. Usta, N. (2005). Use of tobacco seed oil methyl ester in a turbocharged indirect injection diesel engine. *Biomass and Bioenergy, 28*, 77–86.
34. Benjumea, P., Agudelo, J., & Agudelo, A. (2009). Effect of altitude and palm oil biodiesel fuelling on the performance and combustion characteristics of a HSDI diesel engine. *Fuel, 88*, 725–731.
35. Oner, C., & Altun, S. (2009). Biodiesel production from inedible animal tallow and an experimental investigation of its use as alternative fuel in a direct injection diesel engine. *Applied Energy, 86*, 2114–2120.

Extraction of Silica (SiO_2) from Coal Fly Ash by Leaching and Sintering Technology

Rajaram Mohanty, Shantanu Kumar Mishra, Siba Shankar Mohapatra, Sasanka Choudhury, and Binayak Pattanayak

Nomenclature

Greek Symbols

η Efficiency (dimensionless)

Subscripts

CFA Coal fly ash
FP Final product

1 Introduction

Combustion of coal transforms it to residue termed ash. There are two types of ash one is bottom ash and other fly ash. The heavier ash around 20% falls below is known as bottom ash and lighter particles around 80% carried away by flue gas called fly ash. Due to increase in demand of power, coal power plants are growing rapidly as coal is abundantly found in parts of India. Increased burning of coal increases ash generation. At present, nearly 780 million tons of CFA is generated per annum worldwide, while India contributes 220 million tons per annum. The utilisation of

R. Mohanty · S. K. Mishra · S. S. Mohapatra · S. Choudhury · B. Pattanayak (✉)
Department of Mechanical Engineering, ITER, S'O'A Deemed to be University, Bhubaneswar, Odisha 751030, India
e-mail: binayakpattanayak@soa.ac.in

P. Mahanta et al. (eds.), *Advances in Thermofluids and Renewable Energy*, Lecture Notes in Mechanical Engineering, https://doi.org/10.1007/978-981-16-3497-0_30

entire quantity of generated ash is a must for environment protection. Ash is used in brick making, manufacture of cement, concrete ceramic products, building materials, road construction, waste land reclamation, mine void filling, and in agriculture. Even then the national level utilisation of coal ash is around 54% and rest are stored in ash ponds/ mounds which can cause environmental pollution, require huge land for storage besides non-compliance of the fly ash notification of Govt. of India.

To overcome above stated disadvantages, the waste product ash can be transformed to value added product (engineered sand). The CFA contains a considerable percentage of SiO_2 ranging from 35 to 70% depending on source of the coal. The ash obtained from JSL contains 63% of silicon dioxide and, hence, suitable to extract silicon dioxide from the coal fly ash. As CFA has 35–70% silicon dioxide and 10–35% aluminium oxide [1], research has been made to extract silicon dioxide and aluminium oxide in last several years. Acid leaching and alkaline leaching technique are mostly used compared to other methods like line stone sintering and ammonium sulphate sintering method.

Wang et al. [2] studied extraction of silicon dioxide and aluminium oxide from fly ash. They extracted aluminium oxide slag from fly ash by leaching FA with NH_4HSO_4. By the effect of leaching aluminium oxide, extracted slag was formed in which silicon dioxide content increased from 45 to 90%. Then again the aluminium oxide extracted slag was leached with sodium hydroxide. The leaching rate of silicon dioxide further increased to 95% under the optimised condition. Nayak and Panda [3] leached CFA by sulphuric acid, and they concluded that it is not enough to extract alumina from CFA by direct leaching at low acid concentration and at atmospheric temperature. In order to improve the aluminium extraction efficiency, various researches have been developed. Li et al. [4] achieved an alumina extraction efficiency of 87% by sulfuric acid leaching method with leaching temperature of 200-210 °C, leaching time of 80 min and acid to CFA ratio of 5:1. Jiang et al. [5] studied the reaction behaviour of Al_2O_3 and SiO_2 in high alumina coal fly ash during alkali hydrothermal process. In this process, CFA was leached with NaOH with varying concentration of NaOH, leaching time, and temperature. The result shows that the mass ratio of Al_2O_3:SiO_2 increases from 1.25:1 to 2.19:1. Sangita et al. [6] leached the CFA of a thermal power plant by sulphuric acid. They conclude that the aluminium extraction efficiency depends upon varied solid to liquid ratio and particle size of ashes, and they found an aluminium extraction efficiency of 84.14%. Tripathy et al. [7] conducted a sodium fluoride assisted acid leaching technique. The effect of addition of sodium fluoride (NaF) on leaching of alumina was examined at varied acid concentrations, temperatures, and time of leaching. A maximum alumina leaching recovery of 91% can be achieved with 15% sulphuric acid and 3 g NaF, while leaching 10 g alkali treated fly ash residue at a temperature of 90 °C for 2 h. Wu et al. [8] use a differènt leaching technique,where they leached CFA with sulphuric acid in a high pressure reaction kettle and found an aluminium extraction efficiency of 82.4% at temperature 180 °C and reaction time of 4 h.

All of the above studies aimed at finding alternate routes to extract alumina and SiO_2.However, these processes have their own disadvantages for commercial viability. The present investigation is an attempt to convert a bulk amount of CFA

Table 1 Chemical analysis of CFA

Components	%
SiO_2	65.42
Al_2O_3	20.88
Fe_2O_3	4.65
CaO	1.3
MgO	0.12
TiO_2	1.6
LOI	1.4

into environment friendly material-sand for use inconstructionactivity, replacing the natural river sand and making it commercially viable.

2 Experimental Method

2.1 Materials and Instrument

The fly ash generated from captive power plant (CPP) of Jindal Stainless Limited (JSL) Odisha was collected, as collected CFA was tested to find its physical and chemical properties. The chemical composition of the CFA was tested by X-ray fluorescence (XRF-Axio-MAX, PANalytical, Netherland) analyser. The bulk density and compressive strength of the CFA was also tested. All chemicals are taken of commercial grade. Distilled water was taken for leaching (Table 1).

2.2 Experimental Procedure

On long trial of variables, the experiment appeared satisfactory, conducted taking 100 g of fly ash, mixed with 40% NaOH solution. The 40% NaOH solution was prepared by taking 40 g of NaOH pallets added with 100 ml of water. The mixture of CFA and NaOH was leached at 95 °C for one hour. A continuous stirring of the mixture was made for one hour for better leaching efficiency. After one hour of leaching, the mixture formed was like slurry structure. Then the mixture was washed with hot water so that the added NaOH gets separated from the leached fly ash. To ensure that all the NaOH solution has been separated, the pH value of the fly ash was tested after every wash of the mixture. The washing process was continued till the pH value of the mixture became 7. After this, the fly ash was dried by heating technique to evaporate themoisture in the fly. After drying, the sample was collected and details examined (Figs. 1 and 2).

Fig. 1 Leaching of CFA

Fig. 2 Sample collected after drying

The sample was tested in Nalco Research and Technology centre, Gothapatana, Bhubaneswar. The sample was subjected to laser-microtac to find its grain size, XRF-Axio-Max to find the composition of the sample, SEM to know the microscopic structure of the sample. Also the bulk density and compressive strength of the sample were tested.

The alkali dissolution leaching efficiency was obtained by the following formulas:

$$\eta(Al_2O_3) = \frac{m_{CFA}(Al_2O_3) - m_{FP}(Al_2O_3)}{m_{CFA}(Al_2O_3)} \times 100 \quad (1)$$

$$\eta(SiO_2) = \frac{m_{CFA}(SiO_2) - m_{FP}(SiO_2)}{m_{CFA}(SiO_2)} \times 100 \quad (2)$$

where η is the leaching efficiency, $m_{CFA}(Al_2O_3)$ and $m_{RE}(Al_2O_3)$ denote the mass of Al_2O_3 in the CFA and final product, $m_{CFA}(SiO_2)$ and $m_{RE}(SiO_2)$ donate the mass of SiO_2 in the CFA and final product, respectively.

3 Results and Discussion

Leaching process and controlling step for reaction rate the main chemical reaction that occurs between Al_2O_3 extracted slag and NaOH is as follows:

$$nSiO_2 + 2NaOH \rightarrow Na_2O \cdot nSiO_2 + H_2O \tag{3}$$

Due to the existence of a small amount of Al_2O_3, the following reaction will take place in the leaching process:

$$Al_2O_3 + 2NaOH + 3H_2O \rightarrow 2NaAl(OH)_4 \tag{4}$$

With the increase of Al content, the reaction between $NaAl(OH)_4$ and $Na_2O{\cdot}nSiO_2$ will occur, which can be written as

$$Na_2O \cdot nSiO_2 + 2NaAl(OH)_4 \rightarrow Na_2O \cdot Al_2O_3 \cdot nSiO_2 + 3H_2O + 2NaOH \tag{5}$$

The XRF report shows the different compounds contain the CFA in percentage and the result is shown in Table 2.

From this result, it is found that Al_2O_3 percentage increased. To extract SiO_2, the Al_2O_3 percentage has to be decreased. To decrease the Al_2O_3 percentage, the leaching temperature is to be increased. So to leach the CFA at higher temperature, a temperature controlled oil bath was fabricated.

From the report of laser-microtac, the grain size of (d50) is found to be 64.46 micron. And the result is shown in Table 3 and (Fig. 3).

The scanning electron microscopy (SEM) of the after leaching CFA is shown in the figure. The SEM report results the surface structure of the leached CFA. Due to the effect of leaching the surface of the leached CFA became rougher than before

Table 2 Compounds of leached CFA

Compounds	(%)
SiO_2	49.316
Al_2O_3	39.214
Na_2O	1.655
Fe_2O_3	4.426
TiO_2	1.689
Others	3.70

Table 3 Grain size of CFA

%Tile	Size (μm)
10.00	30.50
20.00	39.43
30.00	47.43
40.00	55.53
50.00	64.46
60.00	75.01
70.00	88.66
80.00	108.4
90.00	144.0
95.00	183.8

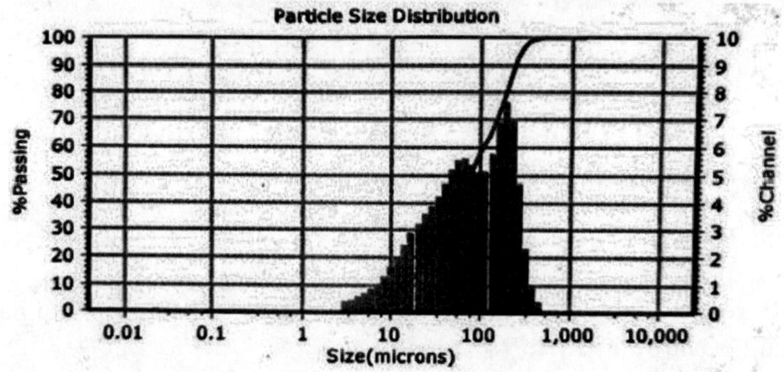

Fig. 3 Shows that the grain size verses % passing

the leaching of CFA. From the microscopic structure, we can see that the bigger size particles break in to smaller particles due to the effect of leaching (Fig. 4).

4 Conclusions

The experiment was conducted by leaching the CFA with NaOH where the concentration of the NaOH was 40% and the mass ratio of CFA to NaOH was 1:1. The leaching process was conducted for 45 min at 95 °C with continuous stirring of the mixture of CFA and NaOH. Due to the effect of leaching, the particles of CFA break into smaller size particles which results that the extraction efficiency of SiO_2 found to be 80%.

In the leaching effect, concentration of NaOH has a major role. It is found that with increasing in concentration, the leaching effect increases. The extraction efficiency

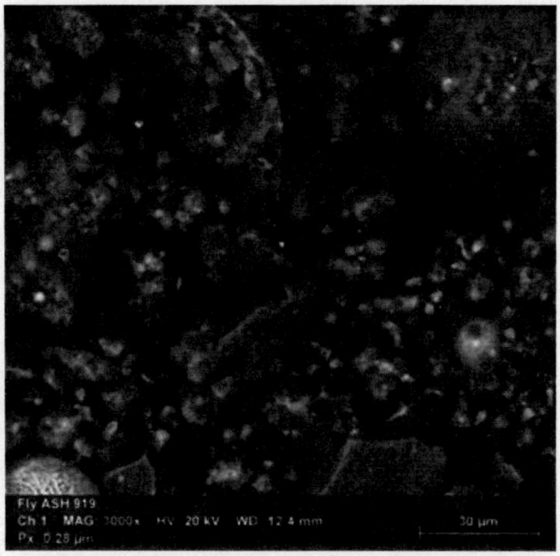

Fig. 4 SEM of CFA after leaching

of SiO_2 can be improved by increasing the temperature of leaching and by increasing the time of leaching.

By increasing the stirring speed and proper separation of NaOH from CFA after leaching can improve the extraction efficiency.

References

1. Behera, D. K. (2019). Annual (2018–19) audited report of SPCB. Odisha on Fly ash generation and utilisation. https://missionenergy.org/flyash2019/presentations/OPCB.pdf
2. Wang, R., Zhai, Y., & Ning, Z. (2014). Kinetics of SiO_2 leaching from Al_2O_3 extracted slag of fly ash with sodium hydroxide solution . *Transactions of Nonferrous Metals Society of China, 24*(6), 1928–1936.
3. Nayak, N., & Panda, C. R. (2010). Aluminium extraction and leaching characteristics of Talcher thermal power station fly ash with sulphuric acid . *Fuel, 89*(1), 53–58.
4. Li, L., Wu, Y., Liu, Y., & Zhai, Y. (2011). Extraction of alumina from coal fly ash with sulfuric acid leaching method. *The Chinese Journal of Process Engineering, 11*(2), 2011.
5. Jiang, Z., Jing, Y. A. N. G., Le, W. A. N. G., & Xi, M. A. (2015). Reaction behaviour of Al_2O_3 and SiO_2 in high alumina coal fly ash during alkali hydrothermal process . *Transactions of Nonferrous Metals Society of China, 25*(6), 2065–2072.
6. Sangita, S., Nayak, N., & Panda, C. R. (2017). Extraction of aluminium as aluminium sulphate from thermal power plant fly ashes . *Transactions of nonferrous metals society of China, 27*(9), 2082–2089.
7. Tripathy, A. K., Behera, B., Aishvarya, V., Sheik, A. R., Dash, B., Sarangi, C. K., Tripathy, B. C.,Sanjay, K., & Bhattacharya, I. N. (2019).Sodium fluoride assisted acid leaching of coal fly ash for the extraction of alumina. *Minerals Engineering, 131*,140—145.
8. Wu, C., Yu, H., & Zhang, H. (2012). Extraction of aluminum by pressure acid-leaching method from coal fly ash . *Transactions of nonferrous metals society of China,* 22(9), 2282–2288.

Extraction of SIO_2 from Coal Fly Ash by Using Sodium Meta Silicate Leaching and Sintering Technology

Suryaranjan Samantaray, Shantanu Kumar Mishra, Siba Shankar Mohapatra, and Binayak Pattanayak

1 Introduction

Fly ash is a solid residual waste material of coal fly ash from coal fired power plants. Nowadays, a huge amount of coal is used in thermal power plant and many other plants for thermal energy. Presently, the ash produced in India is around 220 million tons per annum in India. Due to use of a huge amount of coal, the production of fly ash is also being more. So disposal of the fly ash is being difficult for the requirement of ahuge amount of disposal area, and it brings huge pollution to environment. Generally, coal fly ash is used for production of construction material like break, AAC, and the filling of the land voids of mines. Many thermal power plants are giving their attention towards the problem of disposal of fly ash and trying to find out the way of better utilization of fly ash. Many thermal power plants also give penalty for disposal of fly ash. Fly ash causes environmental pollution as well as creates many respiratory diseases. Low utilization of fly ash requires huge amount of land for disposal. So proper utilization of fly ash is necessary for both environmental and economic point of view.

Utilization of fly ash can be possible by extracting silica as its percentage is about to 50–60% in fly ash. Generally, fly ash consists of some staple components like silicon dioxide, aluminium dioxide, ferrous oxides, calcium oxide, and some amount of other components.

Many researchers and scientists have done many experiments about the extraction of silicon dioxide and aluminium oxide from coal fly ash by using alkali and heat treatment. Gao et al. [1] studied the extraction of silicon dioxide from fly ash by using acetic acid leaching. The leaching temperature is 25 °C, and the corresponding time period is 1 h. The ratio of fly ash and acetic acid is 1:3. Then it is reacted with

S. Samantaray · S. K. Mishra · S. S. Mohapatra · B. Pattanayak (✉)
Department of Mechanical Engineering, ITER, S'O'A Deemed to be University, Bhubaneswar, Odisha 751030, India
e-mail: binayakpattanayak@soa.ac.in

P. Mahanta et al. (eds.), *Advances in Thermofluids and Renewable Energy*, Lecture Notes in Mechanical Engineering, https://doi.org/10.1007/978-981-16-3497-0_31

15% concentrated NaOH solution at 100 °C for 2 h. The final yield strength of silica is 51%, and purity of silica is 98.65%.

Yan et al. [2] studied about the optimization of parameter of acid leaching with fly ash. Here the aluminum leaching rating is about 90.68%, when they took 6 g Sodium carbonate (Na_2CO_3), 100 g of fly ash, leaching time was about 1.5 h, 4 mol/l of HCl, and the corresponding temperature was 100 °C. The concentration of Fe^{2+} is 30.67 gL^{-1}, and concentration of Al^{3+} is 31.21%.

Xu et al. [3] studied the extraction of SiO_2 and Nano-α-Al_2O_3 when the experiment was done under low temperature condition. They took fly ash and sodium carbonate (Na_2CO_3) with ratio 1:1 and roasted at 800 °C for 2 h. The SiO_2 and alumina percentage is more when the HCl concentration is about 3.5 mol/L. The purity percentage is about 99.9%.

Zhou et al. [4] studied about the production of synthetic zeolite from secondary coal fly ash. The parameter which influence the synthesis of zeolite are like the molar ratio of silicon dioxide and aluminium oxide, alkalinity, crystallization time, and the experimental temperature. In this experiment, the percentage of silica is increased from 29.47 to 65.39%, and the percentage of alumina is decreased from 51.72 to 26.25%.

Prasad et al. [5] studied the characteristics and the process of synthesis of zeolite from Indian fly ash. They took NaOH as reactant, the experiment temperature is about 100 °C and the conversion time of 24 h. The sample ratio is 18:1, and the concentration variations are 0.5, 1.0, 2.0, and 3.0 M. The silica availability is about 51.5–64.0%.

Su et al. [6] studied the process of preparation of ultrafine aluminium hydroxide from coal fly ash by using alkali solution. In this experiment, fly ash and 8 M of sodium hydroxide were reacted each other. The experimental temperature is 95 °C, and the corresponding time period is 150 min after leaching and filtration. In the second step, fly ash leached with 20 M sodium hydroxide solution at temperature 260 °C for 1 h to get $NaAlO_2$ solution.

- In first step, the leaching efficiency of SiO_2 was 40% by alkali dissolution.
- In the second step, the leaching efficiency of Al_2O_3 was 89% by alkali dissolution.

2 Experimental Method

2.1 *Materials And Equipments*

The staple equipment is used in this experiment; sodium silicate (Na_2SiO_3), fly ash, and water are the major materials that used in the experiment. Induction heater, temperature measurement device, beaker (500 ml), conical flask, steel utensils, graduated cylinder, and gloves are the equipment used in the experiment (Figs. 1 and 2; Table 1).

Fig. 1 Sodium meta silicate

Fig. 2 Fly ash

Table 1 Chemical composition of coal fly ash

Components	Percentage
SiO_2	65.42
Al_2O_3	20.89
Fe_2O_3	4.65
CaO	1.3
MgO	0.11
TiO_2	1.6
LOI	1.4

Fig. 3 Experimental set up

2.2 *Experimental Procedure*

This experiments used 50 ml of 20, 30, and 40% concentrated sodium silicate (Na_2OSiO_3) and 50gm of fly ash. The first step contains mixing of fly ash with the appropriate amount (1:1) of sodium silicate solution. After mixed up, the heating process is done by the induction heater. The temperature is maintained up to 90 °C for 45 min. To increase the percentage of silica in the sample, the pH of the sample should be '7'. So to maintain the pH as 7, the sample was washed repeatedly with water. Then filter it by filter paper and allowed it to dry. Then the desired product is obtained (Figs. 3 and 4).

The sample was tested in NALCO research and technology center, Gothapatana, Bhubaneswar. The sample was subjected to laser microtac to find out the grain size of the particles present in the sample.

XRF-Axio-Max to find out the amount of components present in the sample. Also the bulk density and the compressive strength of the particles present in the sample were tested.

3 Results and Discussion

The XRF report shows the different compounds contain the CFA in percentage, and result is shown in Table 2.

From the above result, it is shown that the percentage of SiO_2 is 48.328, and Na_2O is 8.656. By maintaining the pH as 7, we can reduce the amount of sodium oxide, and to reduce the percentage of aluminium, the leaching temperature should be high.

Fig. 4 Final product in powder form

Table 2 Compounds of leached CFA before maintaining the pH as 7

Compounds	Concentration (%)
SiO_2	48.328
Al_2O_3	32.246
Na_2O	8.656
Fe_2O_3	4.426
TiO_2	1.676
Others	4.668

The grain size of the sample is determined by the report of laser microtac, and the percentage of grain size is shown by Table 3.

Table 3 Percentage of grain size of the sample

% Tile	Size
10	22.33
20	29.35
30	36.65
40	44.78
50	54.10
60	65.37
70	79.91
80	101.0
90	139.1
95	180.9

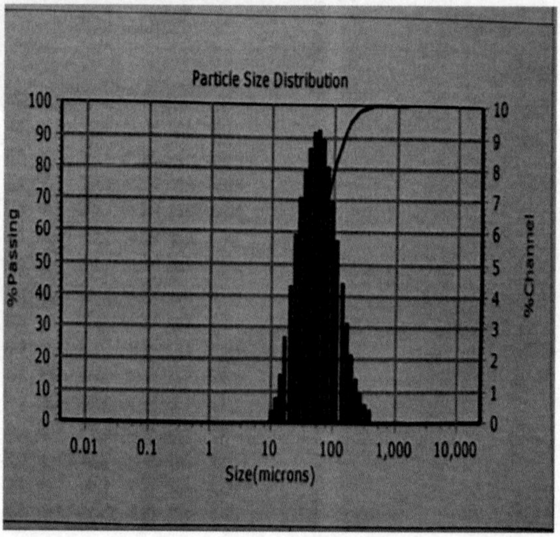

The graph shows the grain size verses % passing.

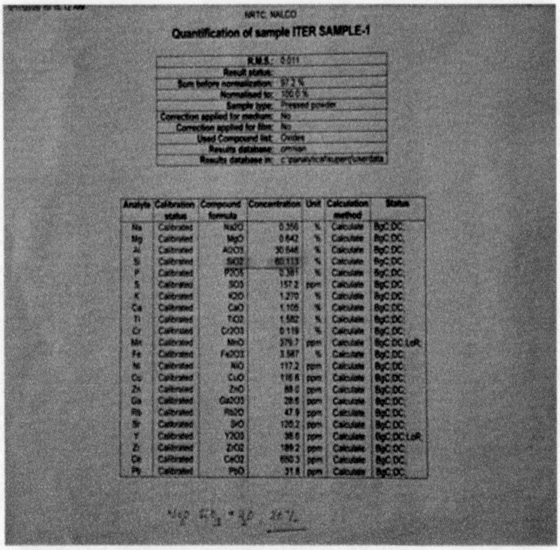

NRTC, NALCO

Quantification of sample ITER SAMPLE-1

R.M.S.:	0.011
Result status:	
Sum before normalization:	97.2 %
Normalised to:	100.0 %
Sample type:	Pressed powder
Correction applied for medium:	No
Correction applied for film:	No
Used Compound list:	Oxides
Results database:	omnian
Results database in:	c:\panalytical\superq\userdata

Analyte	Calibration status	Compound formula	Concentration	Unit	Calculation method	Status
Na	Calibrated	Na2O	0.356	%	Calculate	BgC,DC,
Mg	Calibrated	MgO	0.642	%	Calculate	BgC,DC,
Al	Calibrated	Al2O3	30.646	%	Calculate	BgC,DC,
Si	Calibrated	SiO2	60.113	%	Calculate	BgC,DC,
P	Calibrated	P2O5	0.381	%	Calculate	BgC,DC,
S	Calibrated	SO3	157.2	ppm	Calculate	BgC,DC,
K	Calibrated	K2O	1.270	%	Calculate	BgC,DC,
Ca	Calibrated	CaO	1.105	%	Calculate	BgC,DC,
Ti	Calibrated	TiO2	1.582	%	Calculate	BgC,DC,
Cr	Calibrated	Cr2O3	0.119	%	Calculate	BgC,DC,
Mn	Calibrated	MnO	379.7	ppm	Calculate	BgC,DC,LoR,
Fe	Calibrated	Fe2O3	3.587	%	Calculate	BgC,DC,
Ni	Calibrated	NiO	117.2	ppm	Calculate	BgC,DC,
Cu	Calibrated	CuO	116.6	ppm	Calculate	BgC,DC,
Zn	Calibrated	ZnO	88.0	ppm	Calculate	BgC,DC,
Ga	Calibrated	Ga2O3	28.6	ppm	Calculate	BgC,DC,
Rb	Calibrated	Rb2O	47.9	ppm	Calculate	BgC,DC,
Sr	Calibrated	SrO	120.2	ppm	Calculate	BgC,DC,
Y	Calibrated	Y2O3	38.6	ppm	Calculate	BgC,DC,LoR,
Zr	Calibrated	ZrO2	189.2	ppm	Calculate	BgC,DC,
Ce	Calibrated	CeO2	650.3	ppm	Calculate	BgC,DC,
Pb	Calibrated	PbO	31.6	ppm	Calculate	BgC,DC,

Report-1

Report-1 is the result of sample-1 in which the sodium silicate percentage is about 20. This is a tabular representation of percentage of each componentthat present in the sample. Here the percentage of silica is about 60.113.

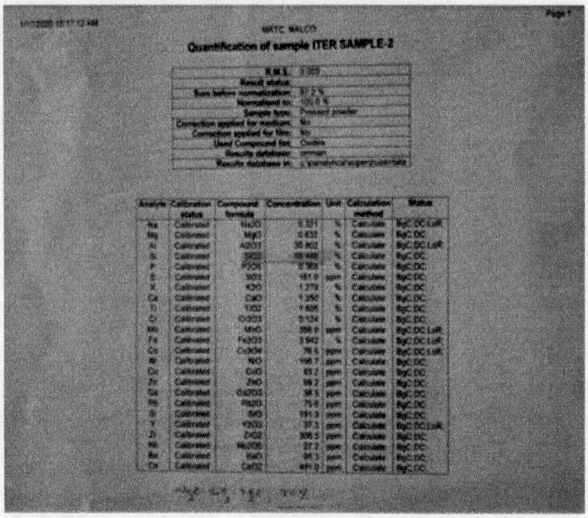

NRTC, NALCO

Quantification of sample ITER SAMPLE-2

R.M.S.:	[illegible]
Result status:	
Sum before normalization:	97.2 %
Normalised to:	100.0 %
Sample type:	Pressed powder
Correction applied for medium:	No
Correction applied for film:	No
Used Compound list:	Oxides
Results database:	omnian
Results database in:	c:\panalytical\superq\userdata

Analyte	Calibration status	Compound formula	Concentration	Unit	Calculation method	Status
Na	Calibrated	Na2O	0.321	%	Calculate	BgC,DC,LoR,
Mg	Calibrated	MgO	0.632	%	Calculate	BgC,DC,
Al	Calibrated	Al2O3	30.802	%	Calculate	BgC,DC,LoR,
Si	Calibrated	SiO2	59.448	%	Calculate	BgC,DC,
P	Calibrated	P2O5	0.365	%	Calculate	BgC,DC,
S	Calibrated	SO3	161.9	ppm	Calculate	BgC,DC,
K	Calibrated	K2O	1.279	%	Calculate	BgC,DC,
Ca	Calibrated	CaO	1.250	%	Calculate	BgC,DC,
Ti	Calibrated	TiO2	1.406	%	Calculate	BgC,DC,
Cr	Calibrated	Cr2O3	0.124	%	Calculate	BgC,DC,
Mn	Calibrated	MnO	356.9	ppm	Calculate	BgC,DC,LoR,
Fe	Calibrated	Fe2O3	3.942	%	Calculate	BgC,DC,LoR,
Co	Calibrated	Co3O4	76.5	ppm	Calculate	BgC,DC,LoR,
Ni	Calibrated	NiO	106.7	ppm	Calculate	BgC,DC,
Cu	Calibrated	CuO	93.2	ppm	Calculate	BgC,DC,
Zn	Calibrated	ZnO	98.2	ppm	Calculate	BgC,DC,
Ga	Calibrated	Ga2O3	38.5	ppm	Calculate	BgC,DC,
Rb	Calibrated	Rb2O	75.6	ppm	Calculate	BgC,DC,
Sr	Calibrated	SrO	151.3	ppm	Calculate	BgC,DC,
Y	Calibrated	Y2O3	37.3	ppm	Calculate	BgC,DC,LoR,
Zr	Calibrated	ZrO2	306.5	ppm	Calculate	BgC,DC,
Nb	Calibrated	Nb2O5	37.2	ppm	Calculate	BgC,DC,
Ba	Calibrated	BaO	95.3	ppm	Calculate	BgC,DC,
Ce	Calibrated	CeO2	441.0	ppm	Calculate	BgC,DC,

Report-2

It is the result of sample-2 in which the sodium silicate percentage is 30, and it consists silica percentage about 59.448.

Table 4 Different concentration of sodium meta silicate

Concentration of sodium meta silicate (Na_2OSiO_3) (%)	Silica percentage (SiO_2) (%)
20	60.113
30	59.448
40	58.385

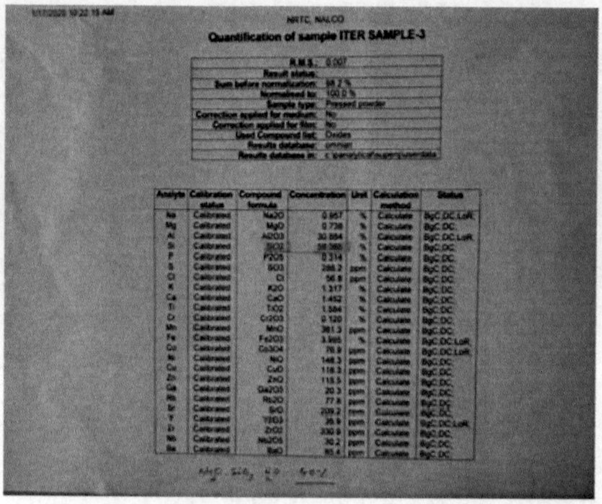
NRTC, NALCO

Quantification of sample ITER SAMPLE-3

R.M.S.:	0.007
Result status:	
Sum before normalization:	98.7 %
Normalised to:	100.0 %
Sample type:	Pressed powder
Correction applied for medium:	No
Correction applied for film:	No
Used Compound list:	Oxides
Results database:	omnian
Results database in:	c:\panalytical\superq\userdata

Analyte	Calibration status	Compound formula	Concentration	Unit	Calculation method	Status
Na	Calibrated	Na2O	0.957	%	Calculate	BgC,DC,LoR,
Mg	Calibrated	MgO	0.738	%	Calculate	BgC,DC,
Al	Calibrated	Al2O3	30.884	%	Calculate	BgC,DC,LoR,
Si	Calibrated	SiO2	58.38[illegible]	%	Calculate	BgC,DC,
P	Calibrated	P2O5	0.314	%	Calculate	BgC,DC,
S	Calibrated	SO3	288.2	ppm	Calculate	BgC,DC,
Cl	Calibrated	Cl	56.8	ppm	Calculate	BgC,DC,
K	Calibrated	K2O	1.317	%	Calculate	BgC,DC,
Ca	Calibrated	CaO	1.452	%	Calculate	BgC,DC,
Ti	Calibrated	TiO2	1.584	%	Calculate	BgC,DC,
Cr	Calibrated	Cr2O3	0.120	%	Calculate	BgC,DC,
Mn	Calibrated	MnO	381.3	ppm	Calculate	BgC,DC,
Fe	Calibrated	Fe2O3	3.985	%	Calculate	BgC,DC,LoR,
Co	Calibrated	Co3O4	76.9	ppm	Calculate	BgC,DC,LoR,
Ni	Calibrated	NiO	148.3	ppm	Calculate	BgC,DC,
Cu	Calibrated	CuO	118.3	ppm	Calculate	BgC,DC,
Zn	Calibrated	ZnO	115.5	ppm	Calculate	BgC,DC,
Ga	Calibrated	Ga2O3	20.3	ppm	Calculate	BgC,DC,
Rb	Calibrated	Rb2O	77.8	ppm	Calculate	BgC,DC,
Sr	Calibrated	SrO	209.2	ppm	Calculate	BgC,DC,
Y	Calibrated	Y2O3	36.9	ppm	Calculate	BgC,DC,LoR,
Zr	Calibrated	ZrO2	330.9	ppm	Calculate	BgC,DC,
Nb	Calibrated	Nb2O5	30.2	ppm	Calculate	BgC,DC,
Ba	Calibrated	BaO	85.4	ppm	Calculate	BgC,DC,

Report-3

It is the result of sample-3 in which the sodium silicate percentage is 40, and it consists silica percentage about 58.385.

The following results are obtained by taking different concentration of sodium meta silicate (Table 4).

4 Conclusions

The aim of this experiment was to extract silica from fly ash for the better utilization of fly ash. Extraction of silica can be done by using sodium meta silicate (Na_2OSiO_3). From this above experiment, we concluded that maintaining the PH value of the sample as '7,' the silica availability is about 58–60% in each sample. From fly ash, we can extract fine grained SiO_2 with high purity of 98.65%, and the yield strength is 51% through organic acid/inorganic alkali and ultrasonic process. The optimal condition for the experiment is determined by various factors like the leaching temperature, leaching timing, and concentration of acid. We can extract SiO_2 from fly ash, and we can test the various properties of SiO_2 by varying other parameters and using sintering technology.

References

1. Gao, M., Ma, Q., Lin, Q., Chang, J., & Ma, H. (2017). A novel approach to extract SiO2 from fly ash and its considerable adsorption properties. *Materials and Design, 116*, 666–675.
2. Yan, L., Shang, J., Wang, Y., Li, J., Liu, H., Qu, T., & Zhang, C. (2016). Experimental parameter optimization study on the acid leaching of coal fly ash. Desalin. *Water Treatment, 57*(23), 10894–10904.
3. Xu, Z., Zhang, M., Zhu, J., & Min, F. (2013). Extraction of Nano-α-Al_2O_3 and SiO_2 from Fly Ash at low temperature conditions. *Integrated Ferroelectrics,147*(1), 8–16.
4. Zhou, C., Yan, C., Zhou, Q., Wang, H., & Luo, W. (2016). Producing a synthetic zeolite from secondary coal fly ash. *Environmental Technology,37*(22), 2916–2923 (2016).
5. Prasad, B., Sudip, M., Sangita, K., Mahato, A. K., & Mortimer, R. J. G. (2012). Studies on synthesis and characteristics of zeolite prepared from Indian fly ash. *Environmental Technology,33*(1), 37–50.
6. Su, S.Q., Yang, J., Ma, H.W., Jiang, F., Liu, Y.Q., & Li, G. (2011). Preparation of ultrafine aluminum hydroxide from coal fly ash by alkali dissolution process. *Integrated Ferroelectrics, 128*(1), 155–162.
7. Annual. (2017–2018). Audited report of SPCB Odisha on Fly ash generation and utilization.

Effect of Obstruction Blockage Ratio in a Suddenly Expanded Micro-Combustor

Arees Qamareen, Shah Shahood Alam, and Mubashshir Ahmad Ansari

1 Introduction

The micro-power generation units [1, 2] based on micro-combustion have the advantages of higher energy density as compared to the conventional batteries used in micro-devices. The batteries pose several limitations when used in micro-systems such as low energy density, bulky structure, lower working duration, and long recharging time. Combustion-based micro-power sources offer the advantages of higher energy density from combustion of hydrocarbons and longer working hours apart from the advantage of simple design and no moving components. However, micro-combustors tend to exacerbate several challenges due to the linearly increased surface area-to-volume ratio due to downscaling of combustor dimensions which leads to heat losses through the walls and radical quenching. At higher fuel flow rate, the residence time for the reactant mixture reduces, causing a decrease in combustion efficiency.

Various strategies are proposed and employed in order to improve the energy conversion efficiency and obtain high and uniform wall temperature. Advanced fabrication methods permit novel designs for the combustors to improve their efficiency by adding catalyst [3, 4], porous media [5, 6] and flow modulating structures such as backward-facing step [7–9], baffles [10–12], bluff bodies [13–15], and cavity [16–18].

In the presented work, the suddenly expanded micro-combustor has been inserted with a triangular obstruction along the centerline and a comparative study has been made with a simple combustor having a backward-facing step and no block insert. The blockage ratio of the obstruction and the inlet mixture velocity was varied and studies on combustor performance in the form of outer wall temperature were conducted.

A. Qamareen (✉) · S. S. Alam · M. A. Ansari
Department of Mechanical Engineering, Aligarh Muslim University, Aligarh 202002, India

P. Mahanta et al. (eds.), *Advances in Thermofluids and Renewable Energy*, Lecture Notes in Mechanical Engineering, https://doi.org/10.1007/978-981-16-3497-0_32

2 Geometrical and Mathematical Model

2.1 Structural Design of Micro-Combustor

The schematic diagram of the structural design of micro-combustor is shown in Fig. 1, and the various geometrical parameters have been listed in Table 1. The height and length of the block insert are a variable with the blockage ratio (BR) defined as the ratio of height of the block to the height of combustion chamber. The block aspect ratio (block length-to-height ratio) is kept as constant and equals to 1.

$$\text{BR(Blockage Ratio)} = b/(h_i + h_s) \quad (1)$$

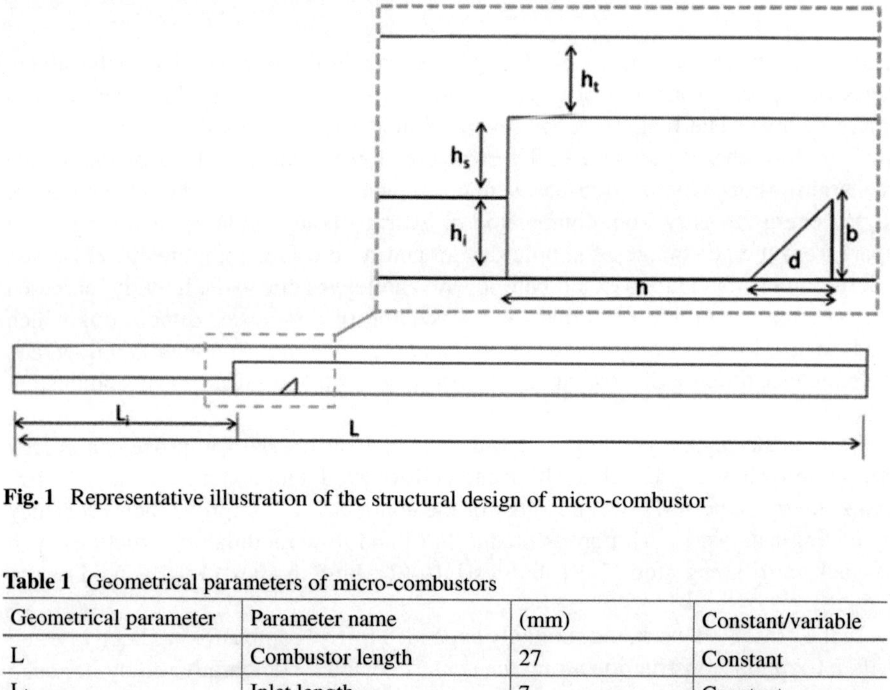

Fig. 1 Representative illustration of the structural design of micro-combustor

Table 1 Geometrical parameters of micro-combustors

Geometrical parameter	Parameter name	(mm)	Constant/variable
L	Combustor length	27	Constant
L_i	Inlet length	7	Constant
h	Block location	2	Constant
h_i	Inlet height	0.5	Constant
h_s	Step height	0.5	Constant
h_t	Combustor wall thickness	0.5	Constant
b	Block height	0.3, 0.4, 0.5, 0.6	Variable
d	Block length	0.3, 0.4, 0.5, 0.6	Variable

2.2 Mathematical and Computational Model

The governing equations for continuity, momentum conservation, energy conservation, and species transport can be found in the literature Li et al. [19] as used by the commercial CFD software ANSYS Fluent 16.0. In the 2D axisymmetric computational domain of the micro-combustor, steady-state combustion of H_2–air mixture occurs. Dufour effects, gas radiation, and work done by viscous and pressure forces are neglected.

Finite rate/eddy dissipation model (FR/EDM) is used for modeling chemical reaction rates. This model utilizes both the Arrhenius and the eddy dissipation expressions for reaction rates [20]. A minimum value of the net reaction rate is considered in the simulation. For the global reaction mechanisms, it is convenient to use FR/EDM which can predict the reaction rates quite accurately. After the combustion gets initiated in the simulation, the eddy dissipation rate is usually smaller than the Arrhenius reaction rate, and hence, the reactions become mixing dependent.

Fluid properties, such as specific heat, are dependent on temperature while thermal conductivity and viscosity have been taken as constant. Mixture density has been calculated using the incompressible ideal gas equation, specific heat from mixing law, thermal conductivity, and viscosity for the mixture from mass-weighted-mixing law while mixture viscosity from kinetic theory. The gas-phase chemistry is taken as the default global H_2–air reaction mechanism as found in the ANSYS Fluent software. Steel is taken as the combustor and block material, and the properties are taken from the reference Akhtar et al. [21].

At the fluid–solid interfaces, no-slip boundary conditions and zero-flux of species are assumed. Coupled boundary conditions for conduction and convection heat transfer are considered at the gas–solid boundaries such as combustor inner walls and block surface. Since both radiation and convection modes of heat transfer are assumed at the outer surface of the combustor solid wall, mixed boundary condition is considered. At the fuel–air inlet, the mixture velocities (uniform) and species concentration (calculated for $\varphi = 0.8$) are defined for the simulation runs. Zero gauge pressure at outlet is considered. The inlet and outlet side walls are set as adiabatic. Convergence criterion for all the equations is 10^{-6}, and the equations are discretized using second-order upwind scheme. SIMPLE pressure–velocity coupling algorithm is employed. To model turbulence, standard k-ε model with non-equilibrium wall condition was considered.

3 Results and Discussion

The released heat from the combustion process is transferred through the solid wall of the combustor, and this thermal energy is emitted from the outer surface of the combustor. This radiated energy is the primary concern for the use of micro-combustor as an emitter in MTPV generator.

Hence, in the work undertaken, the outer wall temperature distribution has been reported for the varying blockage ratio of blocks inserted in the micro-combustors with different inlet velocities. After the grid independency test (coarse, medium, fine mesh), a mesh density of 1.1×10^5 was chosen. Multi-zone quadrilateral grid system was generated using ANSYS Workbench meshing module and is shown in Fig. 2.

Verification of the numerical results was done (see Fig. 3.) with reference to the experimental results of Li et al. [7] for inlet velocity as 12 m/s, equivalence ratio of 0.8, and external surface emissivity as 0.6 for the micro-combustor without any block insertion. A comparison of outer wall temperature distribution was made, and the trend of the numerical result curve was seen to be almost similar. The difference in the two curves is caused due to the global H_2–air reaction mechanism chosen. The global mechanisms generally do not predict temperatures precisely and report a

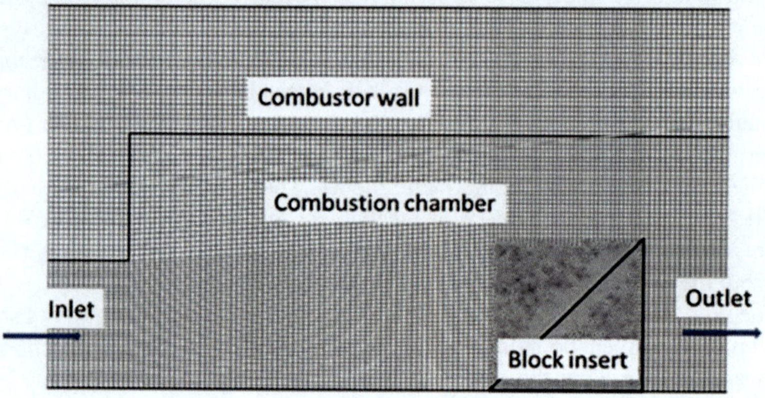

Fig. 2 Multi-zone computational grid generated

Fig. 3 Validation of the numerical results with the experimental results of Li et al. [7]

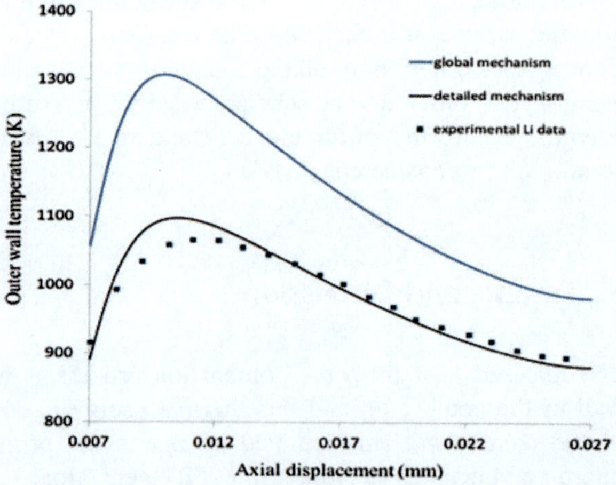

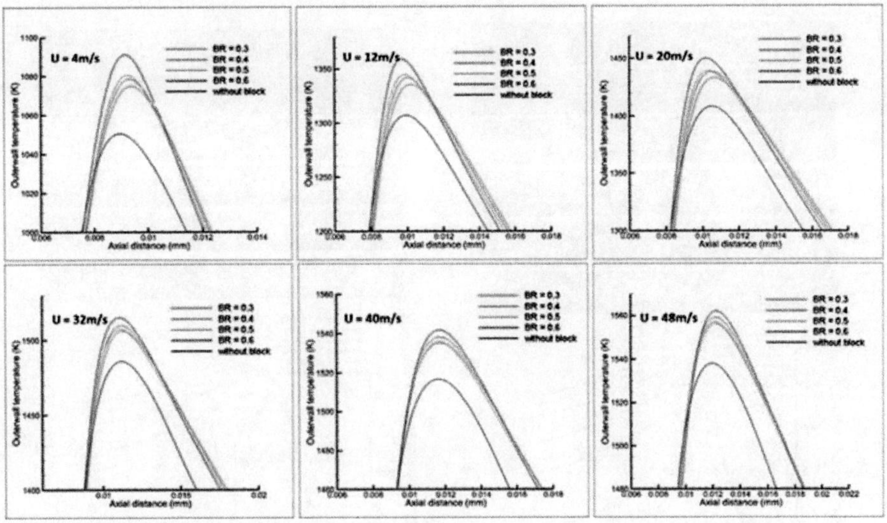

Fig. 4 Comparative representation of the micro-combustors at different inlet velocities

higher value. Work is being done on the same problem with 9 species and 19 reactions mechanisms [21].

In Fig. 4, the maximum temperature zones for the different micro-combustor geometries are compared at different inlet velocities. A jump of about 50 K is observed for the micro-combustor with BR = 0.6 block as compared to the case when no block is present. For lower velocities (i.e., 4, 12, and 20 m/s), it is seen that as block BR is increased there is a clear increment in the temperature distribution in this zone. But for higher velocities, an overlap of temperature curves for BR = 0.3, 0.4, and 0.5 is observed but BR = 0.6 stands out having the maximum temperature peak. Throughout the velocity range, the case without any block displays the worst performance and it can be concluded that an obstruction can significantly enhance the performance of the micro-combustor in terms of the outer wall temperature but the size of the block only slightly affects the temperature field.

Figure 5 shows the temperature contour plots for different cases besides the outer wall temperature distribution along the axial length of the micro-combustor. The contours clearly show the shifting of flame downstream as the velocity is increased.

In Fig. 6, the streamlines formed are shown overlapping the temperature contour plots for different micro-combustors with an inlet velocity of u = 48 m/s. In the micro-combustor without any obstruction, only one recirculation zone is formed that is just after the backward-facing step while for the block inserted combustors two recirculation zones are visible. An increase in BR slightly increases the area covered by the recirculation zone thus causing a larger flame attachment behind the block which increases the outer wall temperature but this increase is not very significant. The reason is that the temperature fields for the four micro-combustors are almost

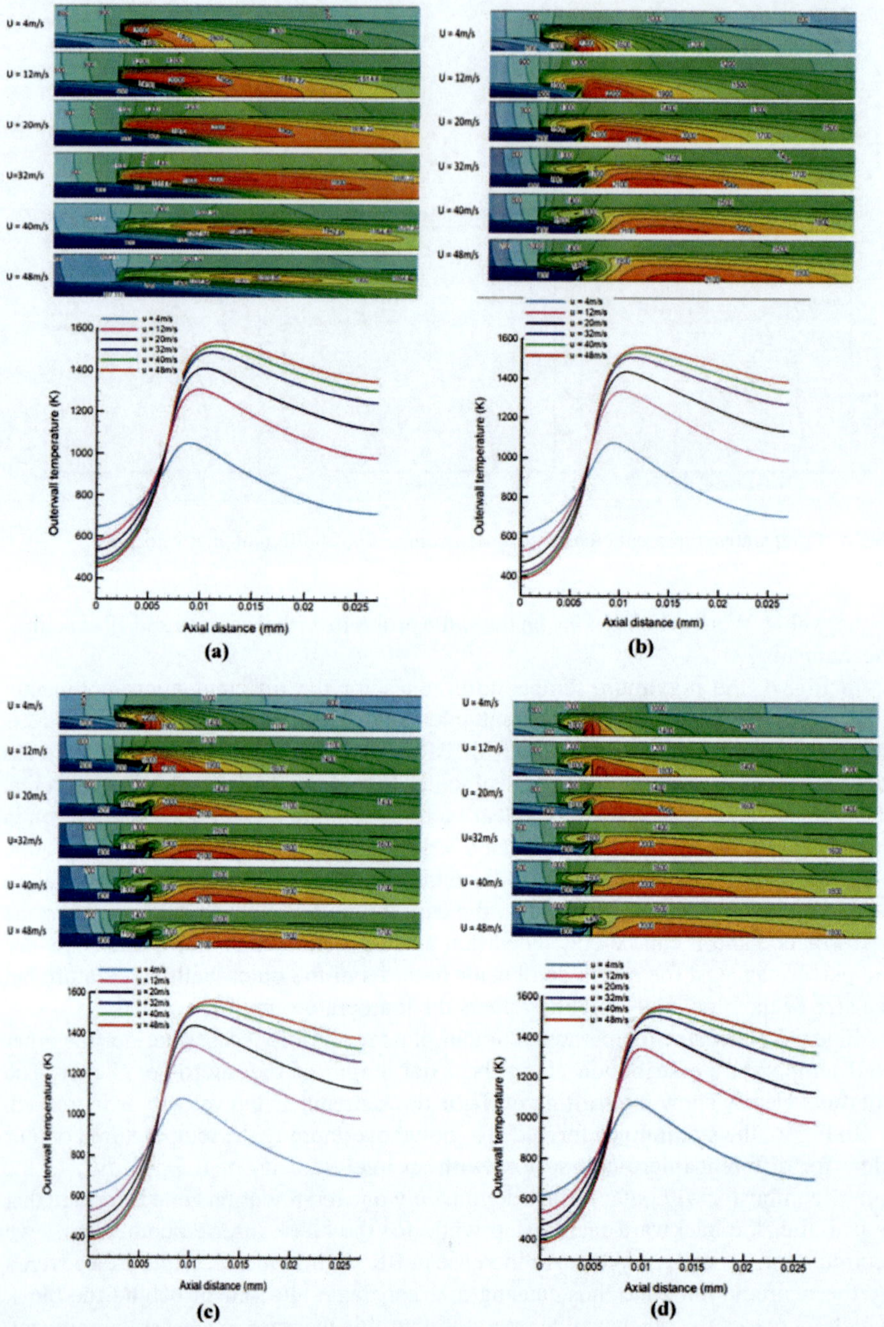

Fig. 5 Contour plots and temperature distribution curves along the axial length of the micro-combustors **a** without block, **b** with block BR = 0.3, **c** with block BR = 0.4, **d** with block BR = 0.5, and **e** with block BR = 0.6

Fig. 5 (continued)

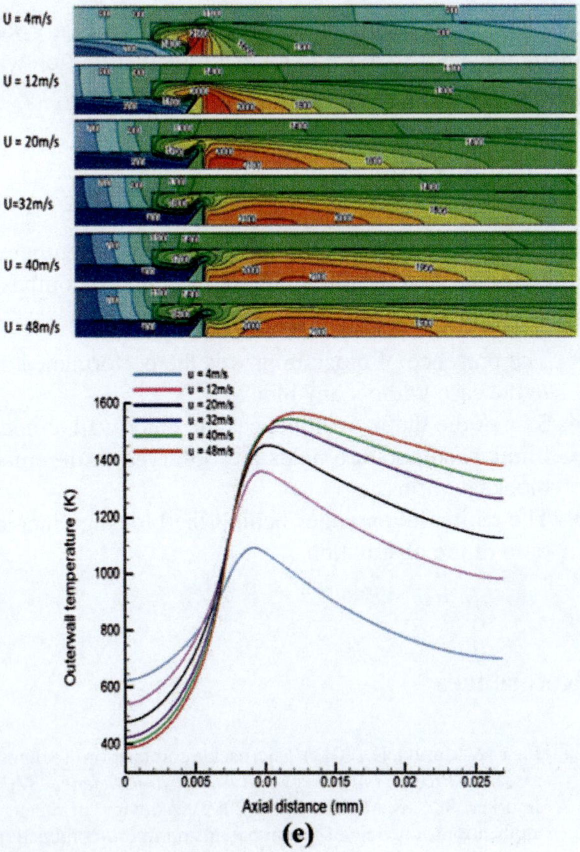

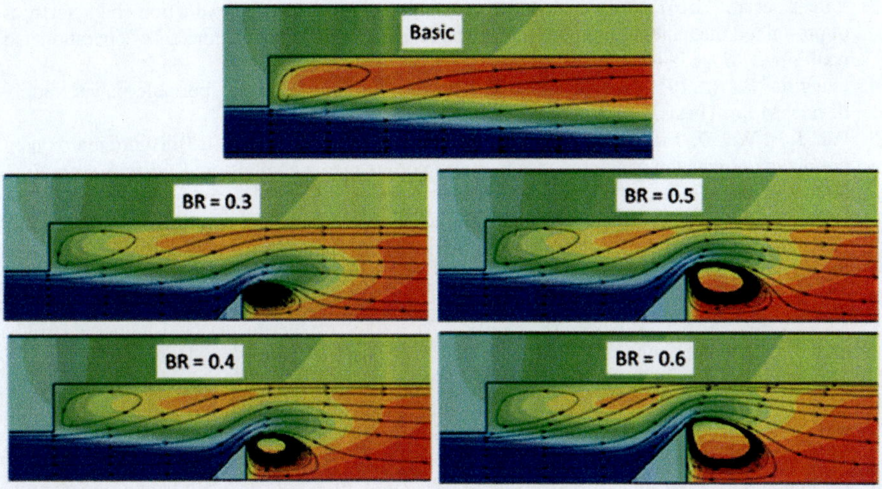

Fig. 6 Streamlines superimposed on temperature contour plots for inlet velocity u = 48 m/s

similar and the recirculation zones are only slightly varying in nature causing a lack of significant difference in temperature distribution with respect to BR.

4 Conclusions

The effect of blockage ratio on the outer wall temperature in an axisymmetric two-dimensional backward-facing stepped micro-combustor was numerically investigated. From the results, it can be concluded that

- The presence of block improves the performance of the combustor as compared to the case without any block.
- Size of the block to influences the outer wall temperature.
- Similar temperature fields are observed in the micro-combustors with different blockage ratios.
- The recirculation zones behind the blockage increase in size with the blockage ratio of the obstruction.

References

1. Ju, Y., & Maruta, K. (2011). Microscale combustion: Technology development and fundamental research. *Progress in Energy and Combustion Science, 37*(6), 669–715.
2. Jejurkar, S.Y., & Mishra, D. P. (2009). A review of recent patents on micro-combustion and applications a review of recent patents on micro-combustion and applications. *Recent Patents on Engineering*, 194–209.
3. Yan, Y., et al. (2016). Numerical study on catalytic combustion and extinction characteristics of pre-mixed methane-air in micro flatbed channel under different parameters of operation and wall. *Fuel, 180*, 659–667.
4. Guggilla, B. R. (2019). A study of catalytic microcombustion for a portable power supply device. M.Sc. Thesis, Rowan University.
5. Pan, J. F., Wu, D., Liu, Y. X., Zhang, H. F., Tang, A. K., & Xue, H. (2015). Hydrogen/oxygen premixed combustion characteristics in micro porous media combustor. *Applied Energy, 160*, 802–807.
6. Peng, Q., Jiaqiang, E., Chen, J., Zuo, W., & Zhao, X. (2018). Investigation on the effects of wall thickness and porous media on the thermal performance of a non-premixed hydrogen fueled cylindrical micro combustor. *Journal of Micromechanics and Microengineering, 155*, 276–286.
7. Li, J., Chou, S. K., Yang, W. M., & Li, Z. W. (2009). Experimental and numerical study of the wall temperature of cylindrical micro combustors. *Journal of Micromechanics and Microengineering, 19*(1), 015019.
8. Akhtar, S., Khan, M. N., Kurnia, J. C., & Shamim, T. (2017). Investigation of energy conversion and flame stability in a curved micro-combustor for thermo-photovoltaic (TPV) applications. *Applied Energy, 192*, 134–145.
9. Faramarzpour, H., Mazaheri, K., & Alipoor, A. (2018). Effect of backward facing step on radiation efficiency in a micro combustor. *International Journal of Thermal Sciences, 132*(June), 129–136.

10. Yahagi, Y., Sekiguti, M., & Suzuki, K. (2007). Flow structure and flame stability in a micro can combustor with a baffle plate. *Applied Thermal Engineering,, 27*, 788–794.
11. Amani, E., Alizadeh, P., & Moghadam, R. S. (2018). Science direct micro-combustor performance enhancement by hydrogen addition in a combined baffle-bluff configuration. *International Journal of Hydrogen Energy, 43*(16), 8127–8138.
12. Ansari, M., & Amani, E. (2018). Micro-combustor performance enhancement using a novel combined baffle-bluff configuration. *Chemical Engineering Science, 175*, 243–256.
13. Hosseini, S. E., & Wahid, M. A. (2014). Investigation of bluff-body micro-flameless combustion. *Energy conversion and management, 88*, 120–128.
14. Bagheri, G., Hosseini, S. E., & Wahid, M. A. (2014). Effects of bluff body shape on the fl ame stability in premixed micro-combustion of hydrogen e air mixture. *Applied Thermal Engineering, 67*(1–2), 266–272.
15. Pan, J. et al. (2020). Investigation on the effect of bluff body ball on the combustion characteristics for methane/oxygen in micro combustor. *Energy*, 190, 116465.
16. Peng, Q., Jiaqiang, E., Zhang, Z., Hu, W., & Zhao, X. (2018). Investigation on the effects of front-cavity on flame location and thermal performance of a cylindrical micro combustor. *Applied Thermal Engineering, 130*, 541–551.
17. Li, L., Yang, W., & Fan, A. (2018). Science direct effect of the cavity aft ramp angle on combustion efficiency of lean hydrogen/air flames in a micro. *International Journal of Hydrogen Energy, 44*(11), 5623–5632.
18. Yang, W., Xiang, Y., Fan, A., & Yao, H. (2017). Science direct effect of the cavity depth on the combustion efficiency of lean H_2/air flames in a micro combustor with dual cavities. *International Journal of Hydrogen Energy, 42*(20), 14312–14320.
19. Li, J., Chou, S.K., Yang, W.M., & Li, Z.W. (2009). A numerical study on premixed micro-combustion of CH4—air mixture : effects of combustor size, geometry and boundary conditions on flame temperature. *Chemical Engineering Journal, 150*, 213–222.
20. Ansys, I. (2009). ANSYS FLUENT theory guide. *Knowledge Creation Diffusion Utility, 15317*(November), 724–746.
21. Akhtar, S., Kurnia, J. C., & Shamim, T. (2015). A three-dimensional computational model of H_2—air premixed combustion in non-circular micro-channels for a thermo-photovoltaic (TPV) application. *Applied Energy, 152*, 47–57.

Thermodynamic Study for Performance Enhancement of Supercritical Carbon Dioxide Cycle

Amit Kanaujiya and Onkar Singh

Nomenclature

A_{he}	Heliostat Areab (m^2)
Cond.	Condenser
CPR	Cycle Pressure Ratio (dimensionless)
CSP	Concentrated solar power
C_p	Specific heat (KJ/Kg-K)
G	Generator
h	Enthalpy (KJ/Kg)
HPC	High-Pressure compressor (bar)
HPT	High-Pressure turbine (bar)
HRSG	Heat recovery steam generator
I	Solar radiation (W/m^2)
ISCCPP	Integrated solar combined cycle power plant
LPT	Low-Pressure turbine (bar)
LPC	Low-Pressure compressor (bar)
MSHE	Molten salt heat exchanger
m	Mass (kg)
P	Pressure (bar)
p	Pump
Q	Heat (KW)
rec	Receiver
SRC	Steam Rankine cycle
s	Entropy (KJ/Kg-K)
sCO_2	Supercritical carbon-di-oxide
ST	Steam turbine
T	Temperature (^oC)

A. Kanaujiya (✉) · O. Singh
Department of Mechanical Engineering, Harcourt Butler Technical University, Kanpur, India

P. Mahanta et al. (eds.), *Advances in Thermofluids and Renewable Energy*, Lecture Notes in Mechanical Engineering, https://doi.org/10.1007/978-981-16-3497-0_33

TIT	Turbine inlet temperature (°C)
$T_{sa,out}$	Molten salt outlet temperature from central receiver (°C)
$T_{sa,in}$	Molten salt inlet temperature to the central receiver (°C)
Turb	Turbine
w	Work (KJ/Kg)
1, 2, 3....12	Cycle states as shown in the diagram schematic

Greek Symbol

η	Efficiency (dimensionless

Subscripts

air	Air
abs	Heat absorbed by molten salt
amb.	Ambient
c	Compressor
cond.	Condenser
h_e	Heliostat
isen	Isentropic
loss	Heat loss
m_{sa}	Molten salt
m_s	A mass flow rate of supercritical carbon dioxide
poly_c	Polytropic for compressor
poly_turb	Polytropic for turbine
ST	Steam turbine
st.	Steam
sa	Molten salt
su.	Supercritical carbon dioxide
fw	Feed water
int.	Inter cooling

1 Introduction

In the present world, the regulation of the greenhouse effect is the biggest challenge that leads to different environmental disasters like global warming, a rise in sea level, climate change, etc. The efforts are made to reduce the greenhouse effect. Among various measures, the reduction of carbon dioxide (CO_2) emission is a potential approach to minimize environmental impact. The reduction of carbon dioxide emission can be realized by the carbon capture from existing emissions, and/or replacing fossil fuel-based plants with renewable energy-based plants to reduce the emissions. The use of supercritical carbon dioxide cycles run on carbon dioxide as power cycles have the capability to consume a certain amount of carbon dioxide. In the present study, solar-powered supercritical carbon dioxide cycle has been integrated with the steam Rankine cycle for effective utilization of the energy available in the cycle. The solar-powered supercritical carbon dioxide cycle working in a cycle has zero emissions from it. The combined cycle under study here has the topping cycle working on supercritical carbon dioxide (sCO_2) which is obtained from the carbon dioxide available in emissions of fossil fuel-based power plants. The availability of heat even after expansion in sCO_2 cycle up to 73.773 bar and 31.13 °C, i.e., limiting state for sCO_2 permits for combining steam Rankine cycle with it and thus augmenting the performance. The consumption of a certain amount of CO_2 in running the sCO_2 cycle also contributes to reducing the carbon footprints.

This section details some of the pioneering work done in this field. Shahin et al. [1] proposed a model that integrates a solar harvesting system of collectors that heat fluid into a conventional Rankine cycle for the utilization of thermal energy. The thermal efficiency of the heliostat field ranges from 74 to 92% and the maximum efficiency of 90%. Ho and Iverson [2] presented a receiver design with different substances like gas, liquid and solid which work under high temperatures and also demonstrate the thermal tower model that improved the efficiency of the system. Xu et al. [3] carried out a theoretical analysis of the solar tower power plant using molten salt as the heat transfer fluid and found the energy losses in the solar tower power plant. The smallest energy loss is reported as 9.7%, and the largest exergy loss is stated as 44.2%. Turchi et al. [4] studied the solar towers in the temperature range of 600–650 °C using molten salts of different combinations of $MgCl_2$/NaCl/KCl. Anagnostopoulos et al. [5] analyzed the variation in salt temperature with a different combination of solar salts and concluded that the (60% $NaNO_3$—40% KNO_3) gives the optimal temperature range for better heat transfer. The melting temperatures are considered to be 310 °C and 335 °C for $NaNO_3$ and KNO_3, 250 °C for the 60–40% and 40–60% and 220 °C for the 50–50% mixture. Binottia et.al. [6] stated that the combination of sCO_2 and CSP solar tower plant yields better results with KCl-$MgCl_2$ molten salt as the heat transfer fluid. Crespi et al. [7] presented a comparison between transcritical simple recuperated and supercritical partial cooling cycle in terms of the efficiency and economic analysis. Neises and Turchi [8] studied the closed-loop supercritical CO_2 cycle with an emphasis on CSP applications. This study compared the thermal efficiencies of the cycle with recompression and partial cooling. Shaaban [9] presented an analysis of the combined integrated solar combined cycles (ISCC) with steam and organic Rankine cycle. The cycle with R1234ze(z) as a working fluid showed an increase in the output power by 19.5% with solar and 23.1% without solar. The literature review shows that there is a need for further analysis of the supercritical carbon dioxide cycle for better energy utilization in it. Thermodynamic analysis of the solar-powered sCO_2 cycle operating with the steam Rankine cycle is pertinent for studying the effect of different parameters on the performance of this combination.

2 System Description of sCO_2—SRC Combined Power System

Figure 1 details the schematic diagram of heliostat field-based solar-powered intercooler, reheat type sCO_2 cycle coupled with the steam Rankine cycle (SRC) for power generation. The heliostat reflects solar radiation to the central reservoir carrying molten salt. In the present analysis, a mixture of $NaNO_3$ (59.66%) and KNO_3 (40.44%) is used as the salt. The boiling temperature of the salt is 1400 °C and the melting temperature is 310–335 °C, while the heliostat field offers temperature up to 1500 °C [5]. Carbon dioxide enters the LPC at state 1 for being compressed to state 2. For perfect intercooling, a heat exchanger is used between states 2 and 3 such that the temperature of state 3 is similar to that at state 1.

The compressed carbon dioxide then enters to HPC at state 3 for getting compressed up to the state 4. The compressed supercritical carbon dioxide then enters a molten salt heat exchanger, where it is heated using molten salt heat to state

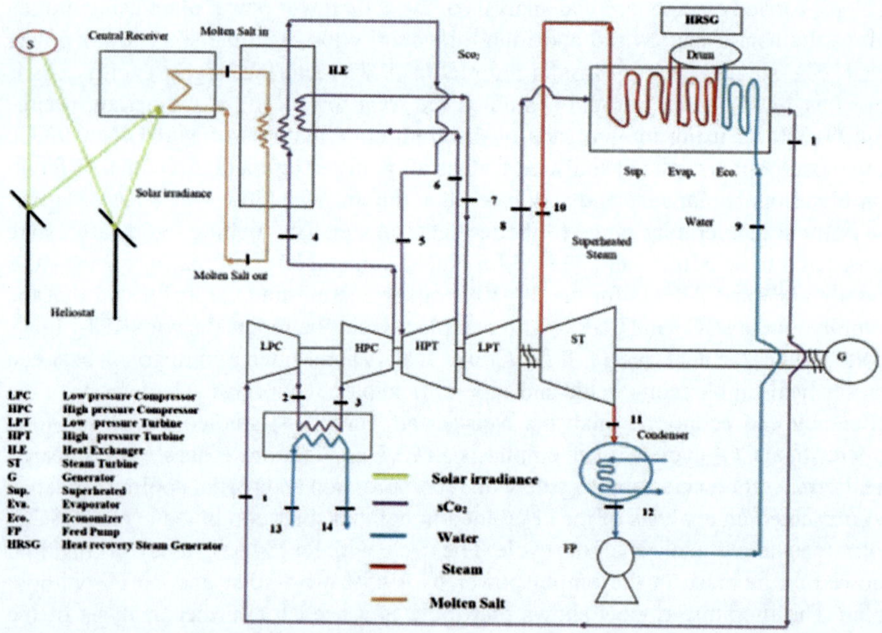

Fig. 1 Solarized combined sCO_2-SRC cycle with intercooling and reheating arrangement in the sCO_2 cycle

5. The high pressure and high-temperature supercritical carbon dioxide enters at state 5 and is then expanded in the HPT to state 6. Then, the sCO_2 is reheated to state 7 in the molten salt heat exchanger and expanded in LPT to state 8. The expanded supercritical carbon dioxide is sent to the heat recovery steam generator (HRSG) at state 8 for steam generation from the water coming from the feed pump of the steam Rankine cycle at state 9. The superheated steam expands in a steam turbine to state 11. After expansion steam enters the condenser of the steam Rankine cycle at state 11 where it loses heat to arrive at state 12. Water entering to feed pump at state 12 is pumped up to turbine inlet pressure at state 9. sCO_2 leaves HRSG at state 1 and enters the LPC.

3 Thermodynamic Modeling

Thermodynamic modeling of combined cycle components has been carried out based on the first law of thermodynamics. The cycle model and mathematical modeling for different other components of the cycle are detailed given below. The following assumptions are considered for thermodynamic modeling.

1. All processes considered here are at a steady-state.
2. Direct normal irradiation is considered constant with a value of 1000 W/m^2.
3. No variation is considered in kinetic energy and potential energy.
4. The chemical energy of the material is neglected.
5. The pressure loss at different locations is negligible for all system components.
6. No pressure loss is considered.
7. Inefficiency in compressor and turbine is considered through polytropic efficiency.

Generalized equations used for the mass and energy in components of the considered cycle are detailed below.

Mass balance across components-

$$\sum m_{\text{inlet}} = \sum m_{\text{outlet}}$$

Energy balance across components-

$$\sum E_{\text{inlet}} = \sum E_{\text{outlet}}$$

3.1 Carbon Dioxide Properties

The specific heat kJ/kg-K of supercritical CO_2 varies with a temperature significantly. Here the, specific heat of supercritical CO_2 is taken from REFPROP 9.0

Energy balance for heliostat field is given by:

$$Q_{\max} = Q_{\text{rec.}} + Q_{\text{lost}} \tag{1}$$

where

$$Q_{\max} = I.\, A_{\text{he}} \tag{2}$$

$Q_{\max}$ is the total heat available in a given area (A_{he}) on incident solar radiation (I).

Due to some environmental factors (Irreversibility, conduction, convection, and radiation), a part of the heat is lost and remaining goes to receiver (solar isolation). After considering energy losses, the energy efficiency of heliostat (η_{h}) is given by:

$$\eta_{\text{h}} = Q_{\text{rec}}/Q_{\max} \tag{3}$$

Energy analysis for the receiver is given by:

$$Q_{\text{rec}} = Q_{\text{rec, abs}} + Q_{\text{rec, loss}} \tag{4}$$

where $Q_{\text{rec, abs}}$ are the absorbed energy by receiver, and $Q_{\text{rec, loss}}$ is energy lost in emission, reflection convection and conduction.

The efficiency of receiver is given as,

$$\eta_{\text{rec}} = Q_{\text{rec, abs}}/Q_{\text{rec.}} \tag{5}$$

Specific heat of the molten salt in kJ/kg.K is given as [7],

$$C_{\text{p, sa}} = 0.172T + 1443 \tag{6}$$

The rate of molten salt absorbed heat can be expressed as:

$$Q_{\text{rec, abs}} = m_{\text{sa}}.C_{\text{p,sa.}}(T_{\text{sa,out}} - T_{\text{sa,in}}) \tag{7}$$

$$Q_{\text{rec,loss}} = Q_{\text{loss,cond}} + Q_{\text{loss,conv}} + Q_{\text{loss,ems}} + Q_{\text{loss,ref}} \tag{8}$$

The receiver efficiency is given as:

$$\eta_{\text{rec}} = Q_{\text{rec,abs}}/Q_{\text{rec.}} \tag{9}$$

$$Q_{\text{total}} = Q_{\text{max}} \cdot \eta_{\text{h}} \cdot \eta_{\text{r}} \tag{10}$$

3.2 *Topping Cycle*

The topping cycle is the Brayton cycle with supercritical carbon dioxide as its working fluid and is powered by solar energy, thus no requirement of burning the fuel. The enthalpy of supercritical carbon dioxide is taken from the software REFPROP9.0 based on pressure, temperature and entropy at specific states.

3.3 *Compressor*

Considering the polytropic efficiency of the compressor, the enthalpy at the exit of the compressor is obtained as,

$$\eta_{\text{poly-c}} = (h_2 - h_1)/(h_{2'} - h_1) \tag{11}$$

Work required for LPC is,

$$w_{\text{c1}} = h_{2'} - h_1 \tag{12}$$

For perfect intercooling,

$$T_3 = T_1 \tag{13}$$

Intercooling heat is,

$$Q_{\text{int}} = m_{\text{s}}.(h_{2'} - h_3) \tag{14}$$

Between states 3 and 4,

$$\eta_{\text{poly-c}} = (h_4 - h_3)/(h_{4'} - h_3) \tag{15}$$

Work required for HPC is,

$$w_{\text{c2}} = h_{4'} - h_3 \tag{16}$$

Total work required (KJ/kg) for the compressor is

$$w_{\text{cnet}} = w_{\text{c1}} + w_{\text{c2}} \tag{17}$$

3.4 Turbine

Considering polytropic efficiency, the enthalpy estimations are based on,

$$\eta_{\text{poly-Turb}} = (h_5 - h_{6'})/(h_5 - h_6) \tag{18}$$

Work obtained from HPT is

$$w_{T1} = h_5 - h_{6'} \tag{19}$$

For perfect reheating,

$$T_5 = T_7 \tag{20}$$

Between states 7 and 8,

$$\eta_{\text{poly-Turb}} = (h_7 - h_{8'})/(h_7 - h_8) \tag{21}$$

Work obtained from LPT is

$$w_{T2} = h_7 - h_{8'} \tag{22}$$

Total work available from turbines is

$$w_{\text{Tnet}} = w_{T1} + w_{T2} \tag{23}$$

Mass flow rate in sCO_2 cycle

$$Q_{\text{total}} = m_s(h_5 - h_{4'}) + m_s(h_7 - h_{4'}) \tag{24}$$

work output of the topping cycle is given by:

$$w_{\text{su.net}} = w_{\text{Tnet}} - w_{\text{cnet}} \tag{25}$$

supercritical carbon dioxide cycle's energy efficiency is expressed by:

$$\eta_{\text{s.u}} = (m_{s*} w_{\text{su.net}})/Q_{\max} \tag{26}$$

3.5 Heat Recovery Steam Generator

The energy balance across HRSG is given as,

$$Q_{st} = (m_s.h_8 - m_s.h_1) \tag{27}$$

$$\varepsilon_{HRSG}.(m_s.h_8 - m_s.h_1) = (m_{st}.h_{10} - m_{fw}.h_9) \tag{28}$$

Here, the mass flow rate of steam and feed water is the same.

$$m_{st} = m_{fw} \text{ and } m_8 = m_s \tag{29}$$

For 15 °C temperature difference between temperature inlet T_8 and steam turbine inlet temperature T_{10}.

$$T_{10} = T_8 - 15 \tag{30}$$

3.6 Bottoming Cycle

The steam Rankine cycle (SRC) is the bottoming cycle as shown in Fig. 1.

3.7 Steam Rankine Cycle

Here, SRC is run by the steam generated in HRSG which uses waste heat from sCO_2 cycle. Steam is added in the steam turbine from HRSG at state 50 bar and the temperature of steam generated depends upon the temperature of the exhaust from the sCO_2 turbine. Here, 15 °C approach temperature is considered in HRSG and limiting temperature is $T_{fw} < 32$ °C because of sCO_2 compressor inlet temperature is 32 °C.

3.8 Turbine Work

Work output from the steam turbine is given as

$$w_{st} = h_{10} - h_{11} \tag{31}$$

3.9 Condenser

Heat rejected in the condenser is given as,

$$Q_{con} = m_{fw}(h_{11} - h_{12}) \quad (32)$$

3.10 Pump Work

Work required by the pump is given as

$$w_p = h_9 - h_{12} \quad (33)$$

$$W_p = m_{fw}.(h_9 - h_{12}) = \int v.dp \quad (34)$$

The net work output of the bottoming cycle is given by

$$w_{st.net} = w_{st} - w_p \quad (35)$$

Steam Rankine cycle's energy efficiency is given by,

$$\eta_{st} = (m_{st}.w_{st.net})/Q_{st} \quad (36)$$

3.11 Overall Combined Cycle

The network output from considered combined cycle comes from the work of intercooled—reheat type supercritical carbon dioxide cycle and SRC.

Total work output is given as:

$$\begin{aligned} W_{total} = {} & m_s(h_5 - h_{6'}) + m_s(h_7 - h_{8'}) + m_{st}(h_{10} - h_{11'}) \\ & - m_s(h_{2'} - h_1) - m_s(h_{4'} - h_3) - m_{st}(h_9 - h_{12}) \end{aligned} \quad (37)$$

Overall cycle efficiency is given as:

$$\eta_{overall} = W_{total}/Q_{total} \quad (38)$$

Based on the mathematical model described above a computer program is written in C language for performance assessment with various input parameters. Thermodynamic properties required in the thermodynamic analysis are obtained by REFPROP 9.0 [10–12].

4 Results and Discussion

Results are obtained from the thermodynamic modeling and computer simulation of the solarized intercooled-reheat sCO_2 cycle combined with the steam Rankine cycle for carbon-free power based on the input parameters Table 1 and the property values from REFPROPand e-Thermo. The thermodynamic analysis is based on a particular set of parameters for solar radiations including the receiver efficiency of 80% in the current analysis and its utilization considering different operating parameters in the topping cycle and bottoming cycles, the variation of cycle pressure ratio, and TIT. Results obtained for different parametric variations have been plotted graphically and analyzed suitably.

Table 1 Input parameters [3, 4, 13–16]

Parameters	Symbol, Unit	Value
Solar irradiation	I, W/m^2	1000
Heliostat field area	A_{hf}, m^2	10,000
Heliostat efficiency	η_h, %	75
Receiver efficiency	η_{rec}, %	75
Polytropic efficiency of compressor	η_{poly_c}, %	89
Polytropic efficiency of turbine	η_{poly_turb}, %	93
Inlet temperature of the compressor	T_1, °C	32
Inlet pressure of compressor	P_1, bar	75
Cycle pressure ratio	–	3.33, 3.46, 3.6, 3.73, 3.86
Effectiveness of HRSG	ϵ_{HRSG}, %	95
Steam generation pressure	P_{ST}, bar	50
Condenser temperature	T_{con}, °C	31
Effectiveness of MSHE	ϵ_{MSHE}, %	85

Figure 2a-e shows the variations of specific work output given in kJ per kg of sCO_2 entering topping cycle, for the maximum pressure of topping cycle varying as 250 bar, 260 bar, 270 bar, 280 bar and 290 bar for steam generation states at 50 bar. It is seen that the work outputs from the sCO_2 cycle and steam Rankine cycle exhibit the increasing trend with an increase in cycle pressure ratio, so the combined specific work output is also increasing with respect to the increasing sCO_2 cycle pressure ratio. Because of the constant pressure at the exit of sCO_2 turbine, the increasing pressure and temperature at the inlet to sCO_2 turbine offers so the availability of heat in HRSG increases. The larger heat availability in HRSG yields more output from SRC. The

Fig. 2 **a–e** Variation of specific work output with TIT of the topping cycle for steam generation at 50 bar in steam Rankine cycle at different maximum pressures of topping cycle

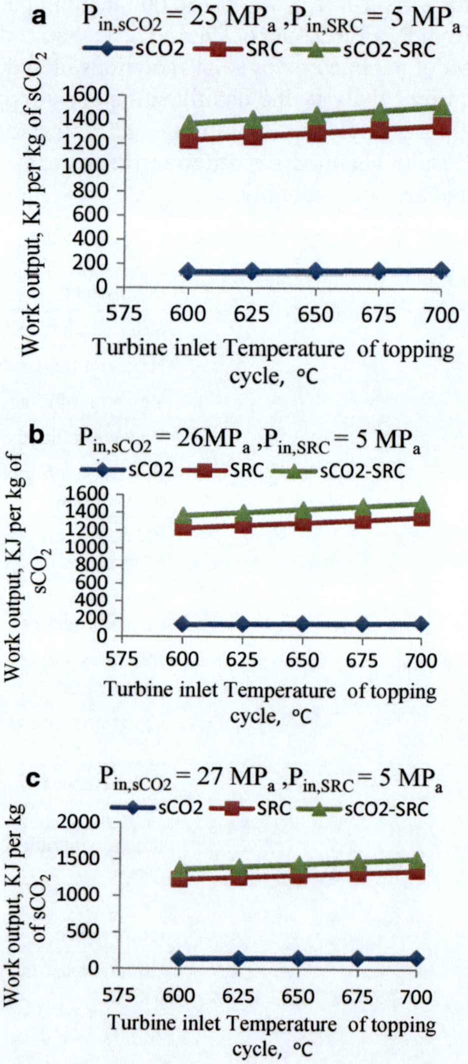

Fig. 2 (continued)

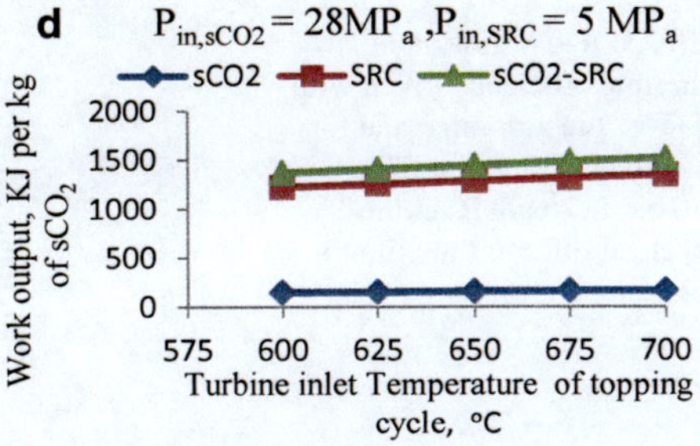

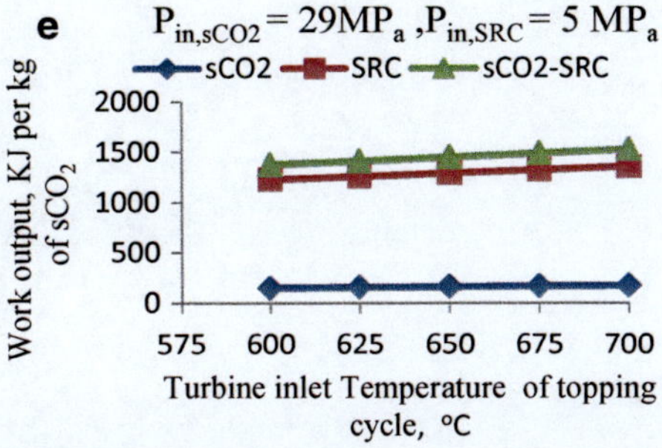

increase in turbine inlet temperature of topping cycle from 600 °C to 700 °C results in the overall work output increasing by10.99%, 10.91%, 11.04%, 11% and 11.05% for the sCO_2 cycle maximum pressure of 250 bar, 260 bar, 270 bar, 280 bar, and 290 bar, respectively at steam generation pressure of 50 bar. The maximum specific work output from the combined cycle is obtained as 1520.68 kJ per kg of sCO_2 at 290 bar pressure at the inlet to sCO_2 turbine and steam generation at 50 bar.

Figure 3a-e shows the variation of different efficiencies with respect to the cycle pressure ratio of topping cycle varying as 3.33,3.46,3.6,3.73,3.86 for steam generation at 50 bar. The efficiency of sCO_2 cycle, steam Rankine cycle, and their combination increases with an increase in cycle pressure ratio. The increase in turbine inlet temperature of topping cycle from 600 °C to 700 °C results in the overall efficiency increasing by 4.20%, 3.85%, 4.24%, 4.12% and 4.20% for the sCO_2 cycle pressure of 250 bar, 260 bar, 270 bar, 280 bar and 290 bar respectively at steam generation pressure of 50 bar. The maximum efficiency is obtained as 43.87% at 290 bar pressure at the inlet to sCO_2 turbine and steam generation at 50 bar while the sCO_2 cycle efficiency and steam Rankine cycle efficiency individually are 16.56% and 38.32%, respectively at these states. Here, the change is observed in the performance of the steam Rankine cycle due to different exit temperatures of the sCO_2 cycle with a change in pressure ratio of sCO_2 cycle.

Figure 4 shows the variation of the sCO_2 cycle's thermal efficiency and specific work output at different maximum pressure and maximum temperature of the sCO_2 cycle. Here, the sCO_2 cycle efficiency increases with the increase in cycle pressure

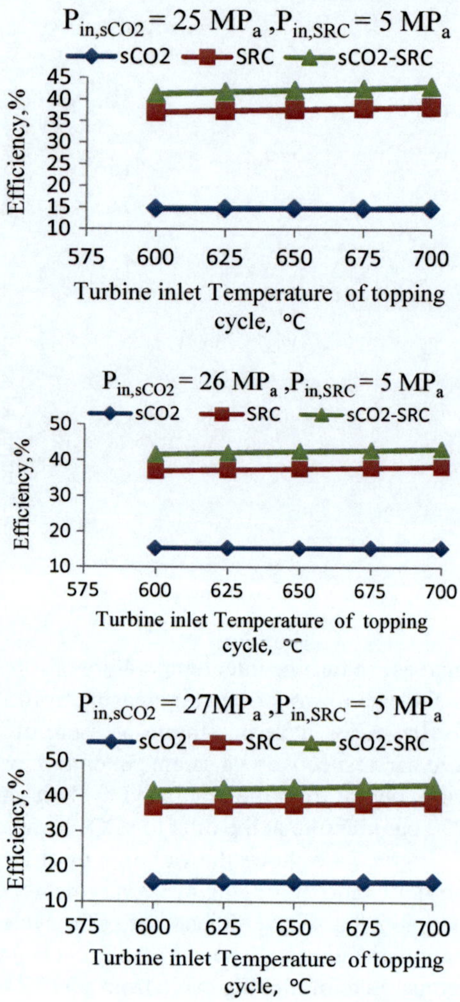

Fig. 3 a–e Variation of thermal efficiency given with TIT of the topping cycle for steam generation at ST 50 bar in steam Rankine cycle at different maximum pressures of topping cycle

ratio along with the increment in TIT of the sCO_2 cycle. The maximum sCO_2 efficiency is obtained as 16.56% at 3.86 cycle pressure ratio & 700 °C TIT. This is the selection diagram that helps in getting the efficiency and specific work output at any TIT and cycle pressure ratio in the considered range.

Figure 5 shows the sCO_2-SRC combined cycle's overall efficiency and specific work output in kJ/kg of sCO_2 for different pressure and temperature values at the inlet to sCO_2 turbine. The result depicts the continuous variation of efficiency and work output in the considered range of sCO_2 cycle's maximum temperature and the maximum pressure of the sCO_2 cycle.

Fig. 3 (continued)

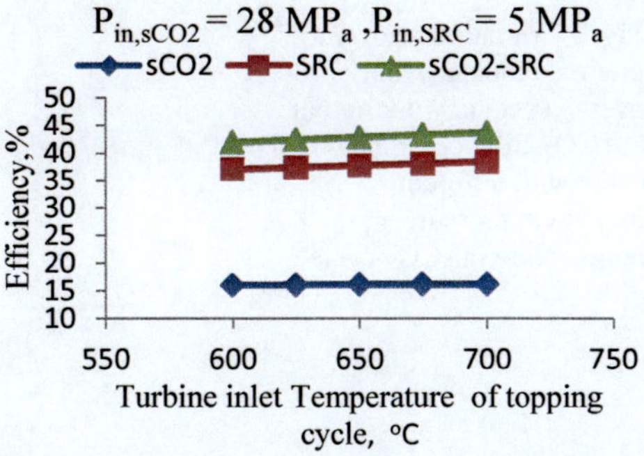

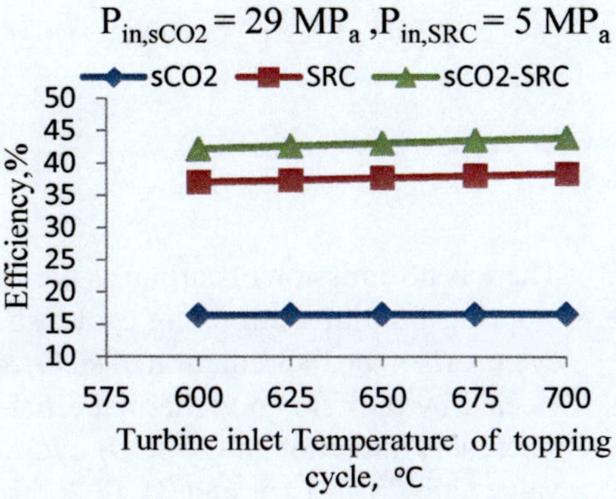

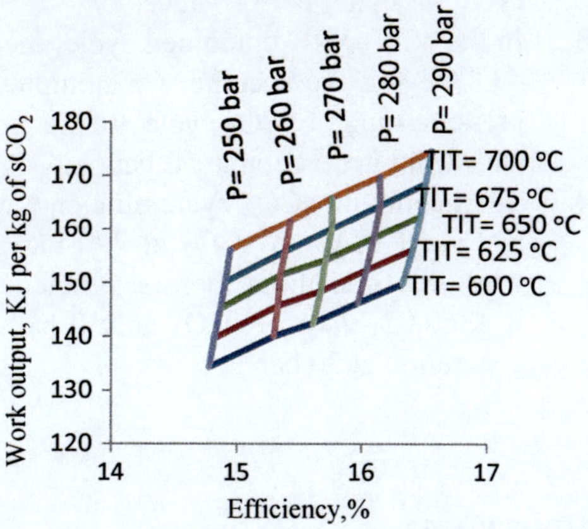

Fig. 4 Variation of thermal efficiency and work output of sCO_2 cycle for different maximum pressure and temperature of sCO_2 cycle

5 Conclusion

The following conclusions have been drawn from the considered solarized combined cycle having intercooling and reheating in the sCO_2 cycle and the waste heat at the exit of the sCO_2 cycle being used for running the steam Rankine cycle.

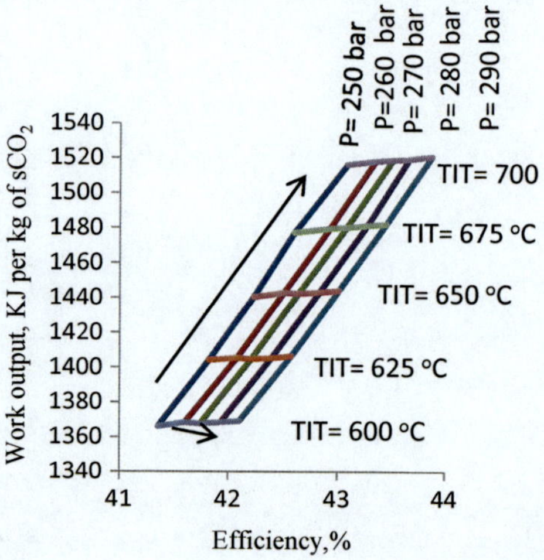

Fig. 5 Variation of overall thermal efficiency and overall specific work output of sCO_2-SRC combined cycle with different maximum pressure and temperature in sCO_2 cycle

1. There is no emission of carbon as the supercritical carbon dioxide is the primary working fluid in the topping cycle, so it is an environment-friendly combined cycle. Also, the recirculation of sCO_2 working fluid after harnessing the energy carried by it offers good thermal efficiency.
2. There is a limitation of the sCO_2 cycle as the limiting pressure and temperature values are 73.773 bar and 31.13 °C for maintaining the supercritical stage, so heat available with the exhaust is used to harness that heat in steam Rankine cycle for better performance.
3. In the sCO_2-SRC combined cycle, the maximum overall thermal efficiency is 43.87% and the specific work output is 1520.68 kJ per kg of sCO_2 at 3.86 pressure ratio of sCO_2 cycle, the maximum pressure in sCO_2 cycle of 290 bar, and steam generation at 50 bar.
4. The maximum sCO_2 cycle efficiency is 16.56% and the specific work output is 174.59 kJ/kg of sCO_2 at 290 bar maximum pressure in the sCO_2 cycle, while the maximum thermal efficiency is 38.47% and specific work output is 1360.02 kJ/kg of sCO_2 at 250 bar for steam Rankine cycle having steam generation at 50 bar.

References

1. Shahin, M. S., Orhan, M. F., & Uygul, F. (2016). Thermodynamic analysis of parabolic trough and heliostat field solar collectors integrated with a Rankine cycle for cogeneration of electricity and heat. *Solar Energy, 136*, 183–196.
2. Ho, C. K., & Iverson, B. D. (2014). Review of high-temperature central receiver designs for concentrating solar power. *Renewable and Sustainable Energy Reviews 29*, 835–846.
3. Xu, C., Wang, Z., Li, X., & Sun, F. (2011). Energy and exergy analysis of solar power tower plants. *Applied Thermal Engineering. 31*, 3904–3913.

4. Turchi, C.S., Vidal, J., & Bauer, M. (2018). Molten salt power towers operating at 600–650 °C salt selection and cost benefits. *Solar Energy, 164*, 38–46
5. Anagnostopoulos, A., Alexiadis, A., & Ding, Y. (2019). Molecular dynamics simulation of solar salt (NaNO3-KNO3) mixtures.*Solar Energy Materials and Solar Cells 200*, 109897
6. Binottia, M., Astolfia, M., Campanaria, S., Manzolinia, G., & Silvaa, P. (2017). Preliminary assessment of sCO_2 power cycles for application to CSP solar tower plants. *Energy Procedia, 105*, 1116–1122.
7. Crespi, F., Sanchez, D., Rodriguez, J.M., & Gavagnin, G. (2017). Fundamental thermo-economic approach to selecting sCO_2 power cycles for CSP applications. *Energy Procedia, 129*, 963–970.
8. Neises, T., & Turchi, C. (2014). A comparison of supercritical carbon dioxide power cycle configurations with an emphasis on CSP applications. *Energy Procedia, 49*, 1187–1196.
9. Shaaban, S. (2016). Analysis of an integrated solar combined cycle with steam and organic rankine cycles as bottoming cycles. *Energy Conversion Management, 126*, 1003–1012.
10. Bianchi, M., Branchini, L., & De Pascale, A. (2014). Combining waste-to-energy steam cycle with gas turbine units. *Applied energy.*
11. Li, X., Kong, W., Wang, Z., Chang C., & Bai, F. (2010). Thermal model and thermodynamic performance of molten salt cavity receiver. *Renewable Energy,35*, 981–988.
12. Angelino, G. (1968). Carbon dioxide condensation cycles for power production. ASME Paper No. 68-GT-23.
13. Sachdeva, J., & Singh, O. (2019). Thermodynamic analysis of solar-powered triple combined Brayton. Rankine and organic rankine cycle for carbon-free power. *Renewable Energy, 139*, 765–780.
14. Shahin, S. M., Orhan, M. F., & Uygul, F. (2016). Thermodynamic analysis of parabolic trough and heliostat field solar collectors integrated with a rankine cycle for cogeneration of electricity and heat. *Solar Field ,136*, 183–196.
15. Rabbani, M., Ratlamwala, T.A.H., & Dincer, I. (2015). Transient energy and exergy analyses of a solar based integrated system. *Journal of Solar Energy Engineering, 137*.
16. Deethayat, T., Kiatsiriroat, T., & Thawonngamyingsakul, C. (2015). Performance analysis of an organic Rankine cycle with an internal heat exchanger having zeotropic working fluid. *Case Studies in Thermal Engineering, 6* 155–161.

The Influence of Pressure and Bed Inventory in Hydrodynamics of Fluidized Bed Unit

A. Kumar, N. Mahanta, and A. Mahapatro

1 Introduction

The growth of any country depends on its utilization of resources and energy. For any nation, energy is the most important factor. The demand for energy is increasing day by day due to fast urbanization and industrialization which is ultimately raising the concern for depleting fossil fuel and environmental pollution. So there is a need for alternative fuel sources and technology that can be efficient, renewable, and clean by nature. Biomass is one of the major attractions as an alternate source of energy or chemicals. Every year, a huge quantity of biomass produces through photosynthesis by absorbing CO_2 from the environment. After burning biomass releases CO_2 that the plants had absorbed from the environment. Hence, the burning of biomass does not add the Earth's CO_2 inventory [1]. Fluidized bed technology is one of the promising technologies. Fluidized beds are used in many industrial applications like gasification and combustion, catalytic cracking of fluid, hydroforming of fluid as well as polyethylene production [2]. The hydrodynamic condition influences the parameters such as bed inventory, bed voidage, suspension density, temperature distribution, heat absorption, combustion condition, and erosion of the bed. The hydrodynamic behavior of fluidized bed is extremely complex, and from the last two decades, researchers are working extensively to enhance the performance of fluidized bed (FB).

The atmospheric fluidized bed is one of the suitable technologies for handling low calorific value fuels for gasification and combustion [3]. Louge et al. [4] compare the hydrodynamics of ACFB and PCFB riser. They have used glass and plastic powder along with helium, carbon dioxide, air, and a mixture of sulfur hexafluoride and

A. Kumar (✉) · N. Mahanta
Department of Mechanical Engineering, NIT Arunachal Pradesh, Yupia, Arunachal Pradesh 791112, India

A. Mahapatro
Department of Mechanical Engineering, IIT Guwahati, Guwahati, Assam 781039, India

P. Mahanta et al. (eds.), *Advances in Thermofluids and Renewable Energy*, Lecture Notes in Mechanical Engineering, https://doi.org/10.1007/978-981-16-3497-0_34

observed shorter acceleration region in PCFB when compared to ACFB for the same solid flux and superficial gas velocities. The excellent mixing of the solid–solid and gas–solid phase enhances the use of low-grade fuels in PCFB [5, 6]. Casleton et al. [7] used the image analysis technique while [8] used the only radioactive particle tracking technique for measuring the bed voidage within the fluidized bed systems. Cheng and Basu [9] determine the pressure effect on loop seal operation for Geldart A and B particles of sand in PCFB. They found an enhancement of solid recycle flow with a rise in aeration airflow and system pressure while it decreased with a rise within the diameter of particles. Gupta and Reddy [10] discuss the pressure effect within the riser column of PFB. They found that the concentration of particle within the upper region increases with an enhancement of pressure also as more concentration of particle were observed near the wall at elevated pressure. Zhiping et al. [11] investigated the glass bead and quartz sand minimum fluidization velocity in the PCFB. They observed that with an enhancement of pressure Umf decreases and the influence was greater of larger particle than smaller particle. With an increase in temperature, Umf of smaller particle decreases, larger particle increased while for medium particle there was no effect of temperature observed. Gupta and Nag [12] investigate the PCFB hydrodynamics with and without bed material. They have used sand of dp 264 μm as a bed material and noticed axial bed voidage increases along with the riser height and minimum at the bottom zone while when system pressure was increased the bed voidage was found maximum at the bottom zone and minimum at the top zone of the riser. With an increase in pressure, suspension density of particle and solid circulation rate increase whereas voidage of the bed decreases. However, with an enhancement of bed inventories bed shows the voidage decreases and solid circulation rate increased [13, 14].

Several works are reported on the behavior of flow between gas and solid, while only a few authors have studied with higher pressure conditions. As far as we all know, only a couple of researches are focused on the flow behavior of the local solids holdup under elevated pressure with sugarcane bagasse. Within the present work, investigation of cold bed hydrodynamic behavior in the PFB unit has studied and therefore the flow behavior of gas–solid profile for axial voidage and suspension density has calculated for various operating pressure and inventory.

2 Setup and Procedure

Figure 1 indicates the pictorial representation of the pressurized fluidized bed (PFB) unit. The PFB unit is formed from components like riser, distributer plate, cyclone separator, and a downcomer. The riser is manufactured with 100 mm ID (inner diameter), 110 mm OD (outer diameter), and 3000 mm height and connected to the cyclone separator through the connecting pipe. The cyclone separator is formed from chrome steel and having 160 mm barrel ID and 660 mm length. The highest of a cyclone separator is employed to gather the syngas while rock bottom of the cyclone separator is connected to a downcomer of ID 50 mm. further, the return leg

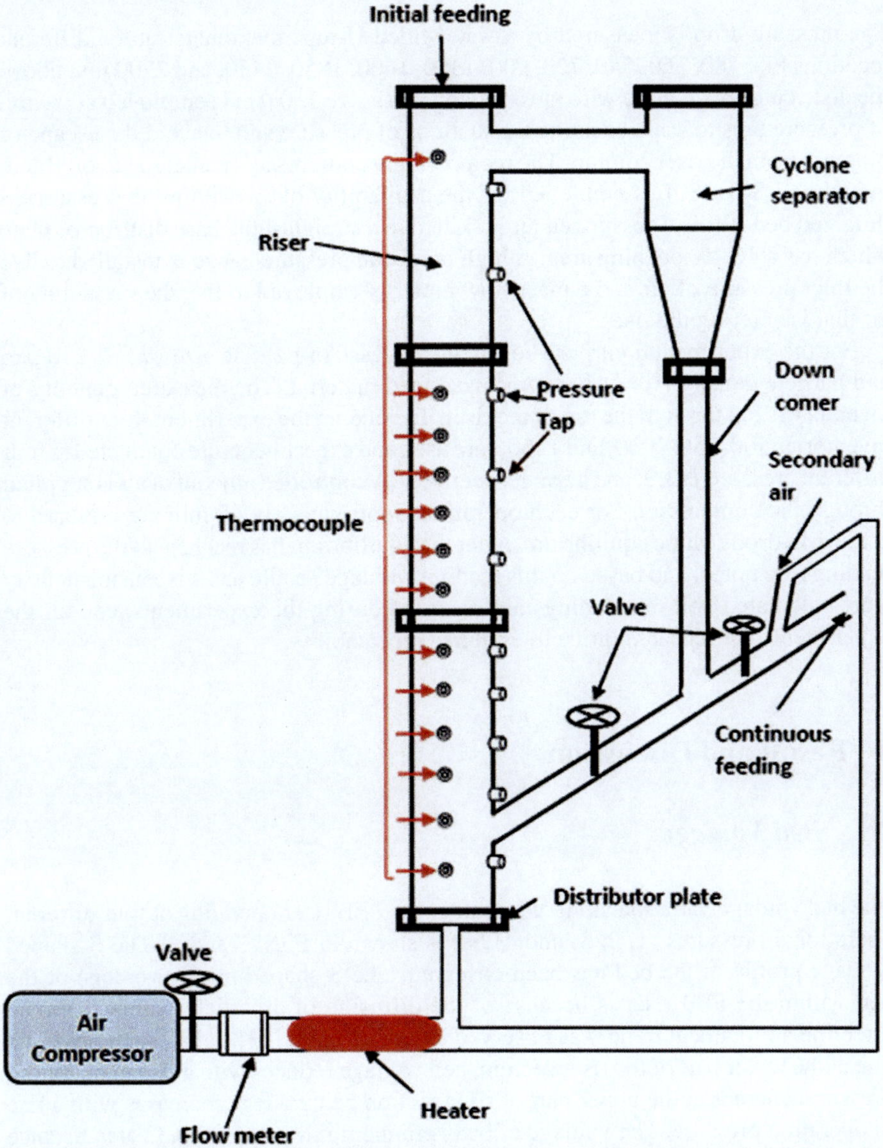

Fig. 1 Pictorial representation of a fluidized bed

is connected to the rock bottom of the riser column at 150 mm above the distributor plate. To avoid backflow from the riser, a gateway valve is established within the downcomer. To watch the bed temperature, thirteen thermocouples are arranged across the peak of the riser from the distributor plate. Ten pressure taps are installed on the riser above the distributor plate to live the pressure drop along the riser.

The pressure drop is measured by a water-filled U-tube manometer at ten different locations like 180, 360, 540, 720, 900, 1350, 1600, 1850, 2450, and 2700 mm above the distributor plate. Fine wire mess having mess size 150 μm is employed at the top of pressure taps to scale back the fluctuations of pressure and to avoid the escape of biomass from the riser column. The reciprocating compressor (make: Ingersoll Rand, model no.: S-01480), is employed for the availability of air within the pressurized fluidized bed (PFB). Pressurized air is skilled in a straight-hole base distributor plate which has a 16.4% opening area. A high precision pressure gauge is installed to live the inlet pressure of air and a mass flow meter is employed to live the mass flow of air that is entering the riser.

For the experimental purpose sugarcane bagasse of particle sizes, $d_p = 210\ \mu\text{m}$ and particle density $P_p = 410\ \text{kg/m}^3$ used feed material. The measured quantity of inventory is fed through the top of the riser. To conduct the experiment, four different inventories 500, 750, 1000, and 1250 g are used and experiments are conducted at four different pressure 1, 2, 3, and 4 bar, respectively. A controlled amount of air is supplied through the compressor. For each operation, approximately 45 min are required to achieve hydrodynamic equilibrium. After the equilibrium has reached all the pressure readings are noted, and based on this reading, voidage profile and suspension density were calculated. All the readings are recorded during the experiments, and all the experiments are repeated thrice to establish repeatability.

3 Result and Discussion

3.1 *Bed Voidage*

The bed voidage variation along the peak of the PFB riser operating at four different operational pressures (1, 2, 3, and 4) bar is shown in Figs. 2 and 3. The S-shaped voidage profile of the bed has been observed. The S-shaped profile voidage of the bed within the PFB riser is because of the diffusion of a particle within the axial direction an identical trend was observed by Kalita et al. [13]. It has been observed that at the lower part of the riser section, bed voidage reduces with a rise in operating pressure whereas at the upper part of riser section bed voidage increases with a rise in operating pressure. The voidage of bed is found minimum ($\varepsilon = 0.32$ at a distance 900 mm from distributor plate) at 1250 g of sugarcane bagasse and the operating pressure 4 bar. At elevated pressure, the drag force applied on the particle escalations which results in a rise in particle concentration at the upper splash region of the riser; therefore, the voidage of the bed increases at the exit of the riser with a rise in pressure.

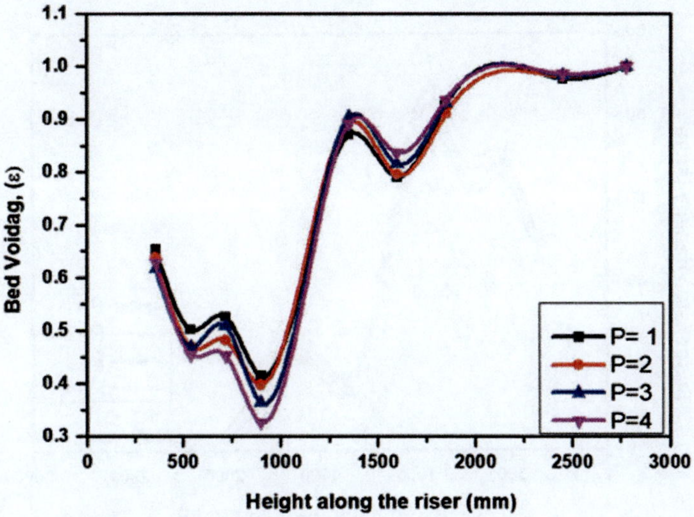

Fig. 2 Effect of working pressure on bed voidage at $I = 1250$ g

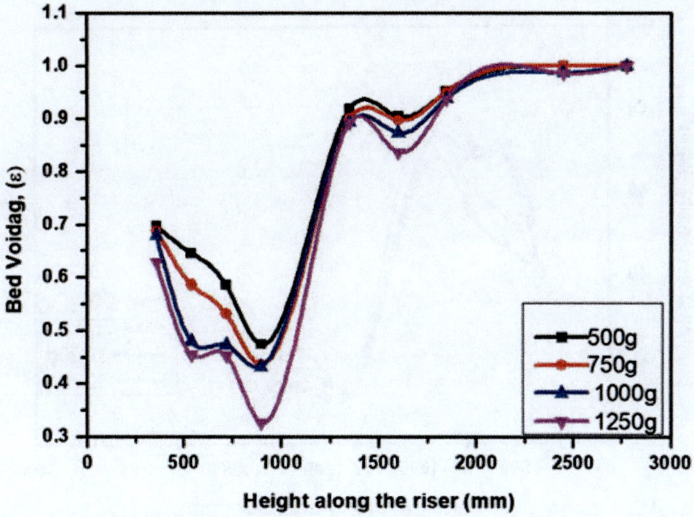

Fig. 3 Effect of inventory on bed voidage at $P = 4$ bar

3.2 Suspension Density

Figures 4 and 5 show the suspension density variation along with the height of the PFB riser operating at four different working pressures (1, 2, 3, and 4 bar) and the inventory of (500, 750,1000 and 1250 g, respectively. The suspension density is found to have

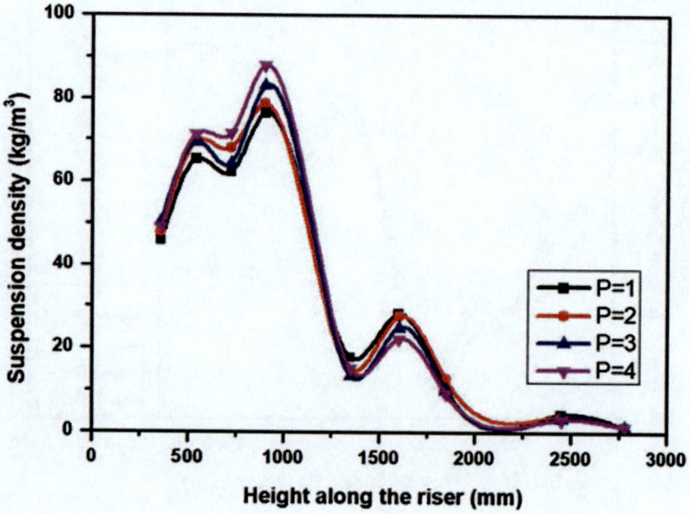

Fig. 4 Effect of working pressure on suspension density at $I = 1250$ g

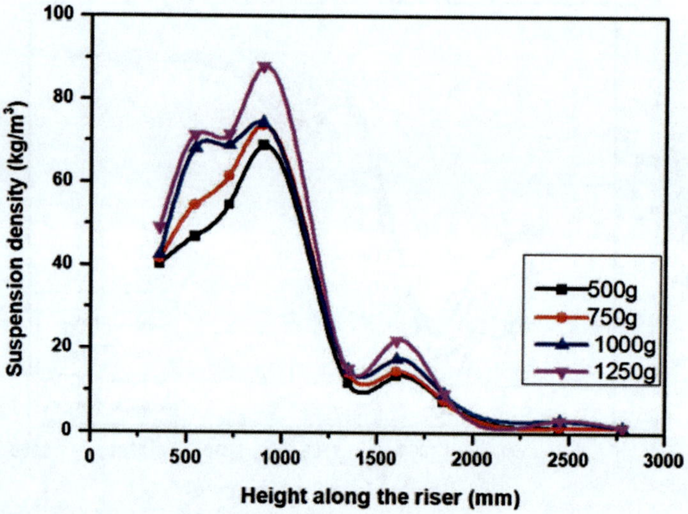

Fig. 5 Suspension density variation at $P = 4$ bar

an increasing trend concerning rising in pressure for all the inventory. The suspension density is found maximum to the location of 900 mm above the distributor plate at a pressure of 4 bar. The maximum suspension density is found 88 kg/m^3 for 1250 g of inventory. At the inlet of riser, minimum suspension density was found 24.86 kg/m^3 at a pressure of 1 bar having 500 g inventory, whereas maximum suspension density was found 50.45 kg/m^3 at a pressure of 3 bar having 1250 g inventory. This kind of trend is found due to minimum bed voidage at that location of the riser. Higher operating pressure and a higher concentration of sugarcane bagasse particles at higher bed inventory may be the reason due to which suspension density. A similar trend has been observed by Basu and Nag [15] and Mastellone and Arena [16].

4 Conclusions

In this present paper, the experimental investigation of the pressurized fluidized unit having sugarcane bagasse as an inventory material has been studied. The influence of bed inventory and operating pressure on the profile of bed voidage and suspension density is calculated. The following observations can be drawn based on this work:

- At the lower a part of the riser section, bed voidage reduces with an enhancement in operating pressure whereas at the upper a part of riser section bed voidage increase with a rise in operating pressure.
- Bed voidage decreases with an increase in inventory and found minimum ($\varepsilon = 0.32$) at 1250 g of inventory at 4 bar pressure.
- The suspension density increases with pressure and at the inlet of riser minimum suspension density was found 24.86 kg/m^3 and maximum suspension density was found 50.45 kg/m^3 at a pressure of 1 bar and 3 bar respectively.
- With an increase in inventory suspension density increases and found a maximum 88 kg/m^3 at an inventory of 1250 g of sugarcane bagasse.

References

1. Basu, P.: Biomass gasification, pyrolysis and torrefaction: practical design and theory (2013)
2. Wen, C. Y., Deole, N. R., & Chen, L. H. (1982). A study of jets in a three-dimensional gas fluidized bed. *Powder Technology, 31*, 175–184.
3. Basu, P. (1999). Combustion of coal in circulating fluidized-bed boilers: a review. *Chemical Engineering Science, 54*, 5547–5557.
4. Bricout, V., Louge, M.Y.: Measurements of cyclone performance under conditions analogous to pressurized circulating fluidization. Chem. Eng. Sci. **59** (2004)
5. Mahapatro, A., Mahanta, P., Jana, K.: Hydrodynamic study of low-grade Indian coal and sawdust as bed inventory in a pressurized circulating fluidized bed. Energy **189**, 116234 (2019)
6. Mahapatro, A., Kumar, A. Mahanta, P.: Parametric study and exergy analysis of the gasification of sugarcane bagasse in a pressurized circulating fluidized bed gasifier. J. Thermal Anal. Calorime. 1–11 (2020)
7. Casleton, D. K., Shadle, L. J., & Ross, A. A. (2010). Measuring the voidage of a CFB through image analysis. *Powder Technology, 203*, 12–22.
8. Mahmoudi, S., Baeyens, J., & Seville, J. (2011). The solids flow in the CFB-riser quantified by single radioactive particle tracking. *Powder Technology, 211*, 135–143.
9. Cheng, L., & Basu, P. (1999). Effect of pressure on loop seal operation for a pressurized circulating fluidized bed. *Powder Technology, 103*, 203–211.

10. Gupta, A.V.S.S.K.S., Reddy, B.V.: Effect of pressure on thermal aspects in the riser column of a pressurized circulating fluidized bed. Int. J. Energy Res. **30**, 149–162 (2006)
11. Zhiping, Z., Yongjie, N., & Qinggang, L. (2007). Effect of pressure on minimum fluidization velocity. *J. Thermal Sci., 16*, 264–269.
12. Gupta, A.V.S.S.K.S., Nag, P.K.: Bed-to-wall heat transfer behavior in a pressurized circulating fluidized bed. Int. J. Heat Mass Transf. **45**, 3429–3436 (2002)
13. Kalita, P., Saha, U. K., & Mahanta, P. (2013). Parametric study on the hydrodynamics and heat transfer along the riser of a pressurized circulating fluidized bed unit. *Experimental Thermal and Fluid Science, 44*, 620–630.
14. Mahapatro, A., Kalita, P., Mahanta, P., Saha, U.K., Mallick, S.S.: Numerical simulation of gas-solid flow in a pressurized circulating fluidized bed riser. In: CFB-11: Proceedings of the 11th International Conference on Fluidized Bed Technology, pp. 323–328 (2014)
15. Basu, P., & Nag, P. K. (1996). Heat transfer to walls of a circulating fluidized-bed furnace. *Chemical Engineering Science, 51*, 1–26.
16. Laura Mastellone, M., Arena, U.: The effect of particle size and density on solids distribution along the riser of a circulating fluidized bed. Chem. Eng. Sci. **54**, 5383–5391 (2002)

Study of Agricultural Product Drying in a Rotating Fluidized Bed in Static Geometry Dryer With and Without Slits

Pavitra Singh, Pankaj Kalita, and Pinakeswar Mahanta

Nomenclature

A	Cross-sectional area (cm^2)
C_D	Drag coefficient
D	Diameter of the vortex chamber (cm)
d_p	Particle mean diameter (m)
g	Acceleration due to gravity (m/s^2)
H	Height of the vortex chamber (cm)
I	Solid inventory (g)
k	Thermal conductivity (W/m °C)
n	Number of slots
T_a	Temperature of Fluidization air (°C)
T_o	Ambient Temperature (°C)
v_a	Velocity of the fluidizing air (m/s)
X_{initl}	Initial moisture content (kg water/kg dry paddy)
X_{req}	Required moisture content (kg water/kg dry paddy)

P. Singh
Center for Energy, Indian Institute of Technology, Guwahati, India

P. Kalita
Department of Mechanical Engineering, Indian Institute of Technology Guwahati, Guwahati, India

P. Mahanta (✉)
Department of Mechanical Engineering, National Institute of Technology Arunachal Pradesh, Yupia, India
e-mail: pinak@iitg.ac.in

P. Mahanta et al. (eds.), *Advances in Thermofluids and Renewable Energy*, Lecture Notes in Mechanical Engineering, https://doi.org/10.1007/978-981-16-3497-0_35

Subscripts

a	Air
i	Inlet
o	Ambient
p	Particle

Greek letters

ρ_g	Density of dry gas (kg/m^3)
ρ_s	Density of solid (kg/m^3)
λ	Force Ratio
θ	Injection angle (°)

1 Introduction

Rice is a staple food source derived from the post-paddy harvesting process, which serves more than half of the global population. To preserve and store the food grains, it is required to reduce the moisture content of the fresh grain to reach its safe moisture content level (13–14% for paddy), and this operation of moisture reduction is called the drying process. Processes such as drying the grains are highly needed to reduce as well as preserve the nutritional losses of food grains. Nygaard and Pellett [1] reported that post-harvest grain loss was found to be about 10%, while 1–5% of food grain was lost drying operations. By reducing the percentage of moisture content in the paddy, biological activities, as well as harmful elements such as germ pests and fungi, can be reduced [2]. In order to reduce grain-loss of the drying process, a promising technique of rotating fluidized bed with a static geometry dryer has been developed. RFB-SG can not only be used in food grain drying, but also this technique can be used in many operations, such as mixing, separation, agglomeration, particle coating, etc. The main advantage of this reactor is its static geometry, which has low maintenance, and the particle breakage is found to be very less. Many researchers have conducted various research on food grain drying by using various newer and traditional dryers. Fluidization is a technique of making solid particles behave like a fluid. The dryers developed, using this technique are called fluidized bed dryers. Fluidized bed dryers are of two kinds, namely direct and indirect dryers. In the case of indirect drying technique heat transfer to grains via the heat exchanging wall, while in the case of direct drying, the heat is directly transferred to the food grains surfaces. Drying the paddy at a temperature of 55–65°°C in a conventional dryer showed that both the nutritional value and texture of the paddy at temperatures above 65°°C were affected [3]. Making use of solar energy, an experimental investigation on paddy drying was successfully conducted in a forced-convection dryer [4]. In the drying operation, the hot air is the most commonly used medium, while the application of steam for commercial purposes is limited due to the complexity of technology and higher cost [5, 6]. To overcome the limitations and system complexity, the rotating fluidized beds (RFBs) were found better performing dryers as compared to the conventional dryers [7]. However, RFBs have excellent heat and mass transfer characteristics. Still, a few demerits are incorporated with these dryers such as oil leakage from bearings, vibration, wear-tear of moving parts, etc. In order to reduce the drawbacks of RFBs, a new concept of a rotating fluidized bed with static geometry (RFB-SG) was introduced by De Wilde and De Broqueville [8]. In this concept, the vortex was created by the injection of high-velocity gas in the reactor tangentially through multiple gas inlets. On confirming the vortex in the chamber, the solids were injected, which start moving under the effect of the centrifugal force in the tangential direction in the form of a rotating bed. Also, they had fixed guidelines for the reactor design [9, 10]. Pati et al. [11] had conducted paddy drying experimentation in the RFB-SG reactor and observed that the drying of paddy was faster as compared to the conventional fluidized bed dryer due to higher process intensity. Singh et al. [12, 13] had investigated a slit-less rotating fluidized bed (SL-RFBSG) dryer numerically, and they reported that overall drying cost, drying air requirement, and system complexity were reduced.

In the present investigation, the performance of the rotating fluidized bed with static geometry (RFB-SG) dryer with slit and without slits is evaluated experimentally, as shown in the Figs. 1, 2, and 3. Also, the comparative study of RFB-SG with slit and RFB-SG without slit has been proposed in the present study.

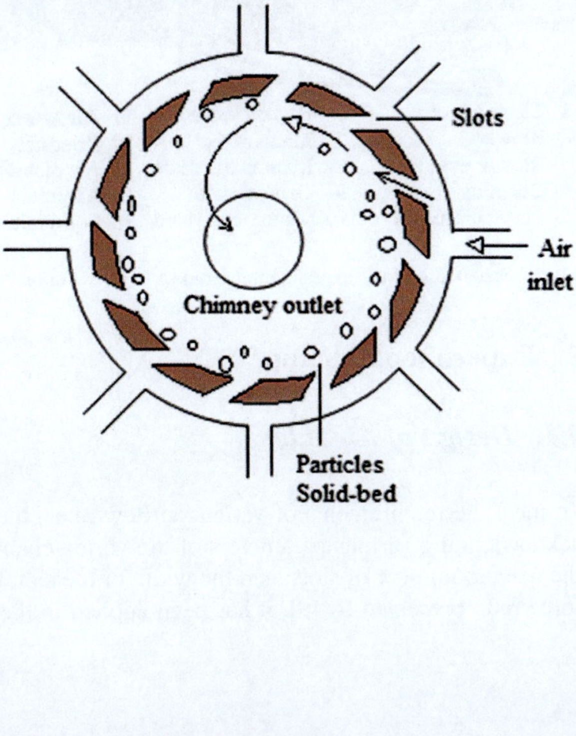

Fig. 1 Schematic of RFB-SG dryer with slit

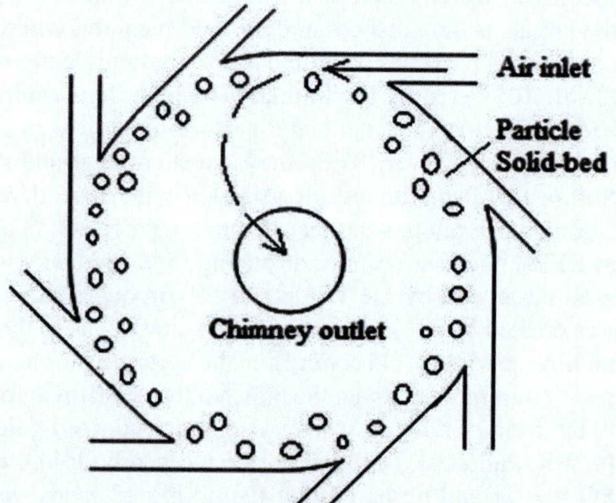

Fig. 2 Schematic of RFB-SG dryer without slit

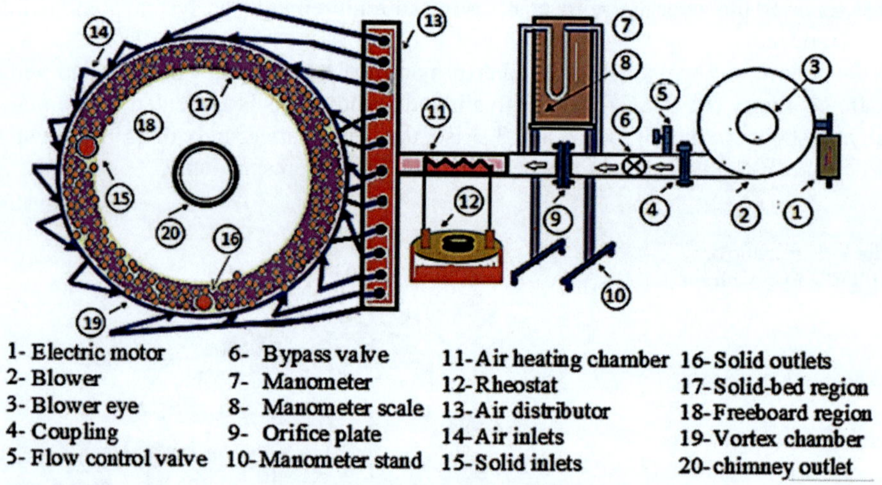

Fig. 3 Schematic of the experimental setup of RFB-SG dryer

2 Experimental Setup

2.1 *Design of Air Inlet*

To meet the requirements of various airflow rates, the air inlet slots are designed, acknowledging various parameters of the vortex chamber, such as the diameter of the reactor, number of slots, and the width of the slot. Based on a vital factor (λ) or force ratio presented by [9], it has been applied in the present design, as shown in

Fig. 1. This force ratio varies in the range of 0.025–0.038, and it can be written in Eq. (1) as;

$$\text{Force ratio, } \lambda = \frac{\text{Centrifugal force}}{\text{Drag force}} = \frac{n.s}{\pi.D} \quad (1)$$

where *n*, *s*, and *D*, are the number of gas slots, slot width, and vortex chamber diameter, respectively.

Rotating fluidized bed with static geometry (RFB-SG) dryer consists of various components, as listed in Fig. 3. Initially, the air comes from the blower is heated first in the air heating unit, and the hot air is distributed to multi-air inlets by the air distribution unit. Further, this hot air is injected into the vortex chamber via multi-air inlets in a tangential direction. On confirming vortex formation in the reactor, the solid particles (paddy) were injected via a solid inlet port at the higher pressure. High-velocity fluidization air causes the food grains to move the tangential direction, forming a solid rotating bed. Hence, the hot air first evaporates the moisture of the grain, and then, it is carried out into the surroundings via a chimney outlet. In the present experimentation, the samples were collected at the time interval of 5 min to observe the value of the moisture contents of the paddy. The entire drying operation has been carried out to reach its safe moisture content of paddy (13–14%) from its initial moisture content. As shown in Fig. 1 and Table 1, the RFB-SG drying chamber is made with slit, which is made up of 240-mm diameter, 50-mm length, 20-mm air inlet diameter, slit size of 3.14-mm, the diameter of 70-mm, and 8 numbers of air inlets.

Table 1 Input parameters for RFB-SG dryer and environmental conditions [9]

RFB-SG dryer with slit		RFB-SG dryer with slit
Dimensions	Vortex chamber diameter (D): 240 mm, chamber height (L):50 mm, air inlet diameter (s): 20 mm, and chimney diameter: 120 mm	Vortex chamber diameter (D): 240 mm, chamber height (L):50 mm, air inlet port width (s): 3.14 mm, and chimney diameter: 120 mm
Parameters	Particle diameter d_p =0.0025 mm, Initial value of MC (X_{intl}) = 0.33 kg water/kg dry paddy and required value of MC (X_{req}) = 0.13 kg water/kg dry paddy)	
Operating conditions		
Inlet air temperature	55 °C, 60 °C and 65 °C	
Inventory	300, 400, and 500 g	
Ambient temperature	27 ± 5 °C	
Airflow rate	400, 500, and 600 m^3/h	

3 Results and Discussion

The experimental investigation conducted on paddy drying in the RFB-SG dryer with slits and without slit has been presented in Figs. 4, 5, and 7, while comparative outcomes of RFB-SG with and without slits are shown in Fig. 8. Hence, the effects of drying temperature, airflow rate, and inventory variation are described in the following subsections.

3.1 *Effect of Air Temperature on Drying Time*

Figure 4 represents the effect of drying air temperature on the drying time at the airflow rate of 600 m^3/h and the inventory of 400 g paddy. It is seen that with the increase in drying air temperature from 55 to 65 °C, the drying time has been

Fig. 4 Effect of temperature on drying time, at airflow rate of 600 m^3/h and inventory of 400 g

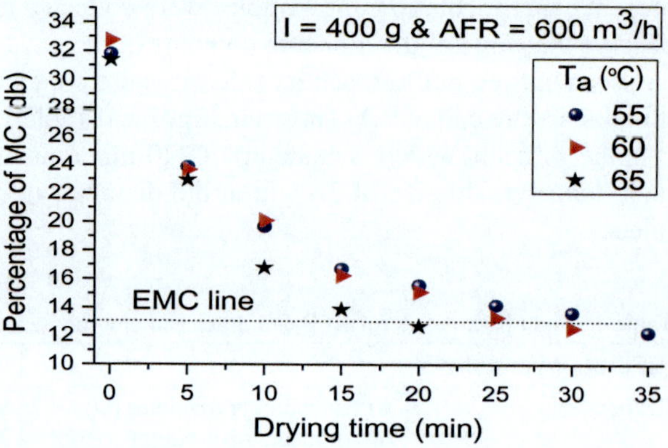

Fig. 5 Effect of airflow rate on drying time, at air temperature of 60 °C and inventory of 400 g

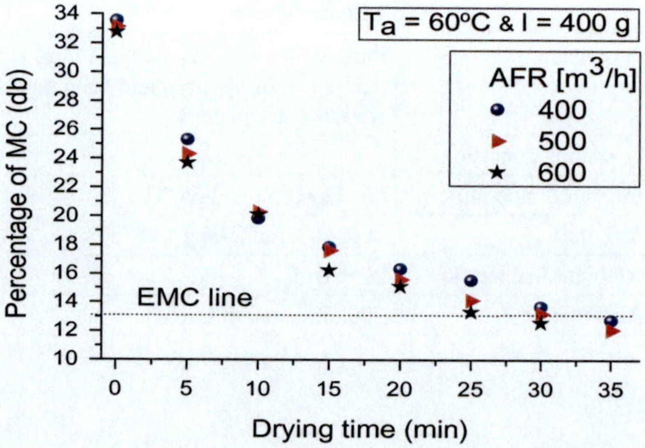

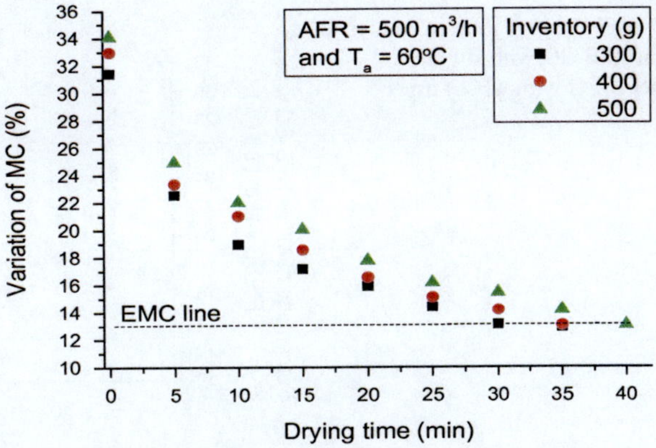

Fig. 6 Effect of inventory on drying time at air temperature 65 °C and inventory of 800 g

reduced by 46%. Also, when the paddy dried at a temperature of more than 65 °C, the nutritional value of paddy was found to be reduced [3].

3.2 *Effect of Airflow Rate on Drying Time*

Figure 5 represents the effect of airflow rate on the drying time at the air temperature of 60 °C, and the inventory of 400 g paddy. It has been observed that with the increase of airflow rate from 400 to 600 m^3/h, the drying time is found to be reduced by 20.6%. Hence, on increasing the drying air temperature and airflow rate, drying time decreases.

3.3 *Effect of Inventory on Drying*

Figure 6 represents the effect of inventory on the drying time at the air temperature of 60 °C and the airflow rate of 400 m^3/h. It has been observed that with the increase in inventory 300 g to 500 g, the drying time is found to be increased by 25%. However, the drying time of paddy increases with the increase in inventory, but the percentage increment of paddy is found to be higher than the increase in the drying time.

3.4 *Comparative Study on Drying Time of Paddy in RFB-SG with Slits vs. Without Slits Dryer*

Figure 7 shows the variation of moisture content in 400 g of paddy, at the airflow rate of 500 m^3/h, and the air inlet temperature of 55 °C and 65 °C, respectively. It

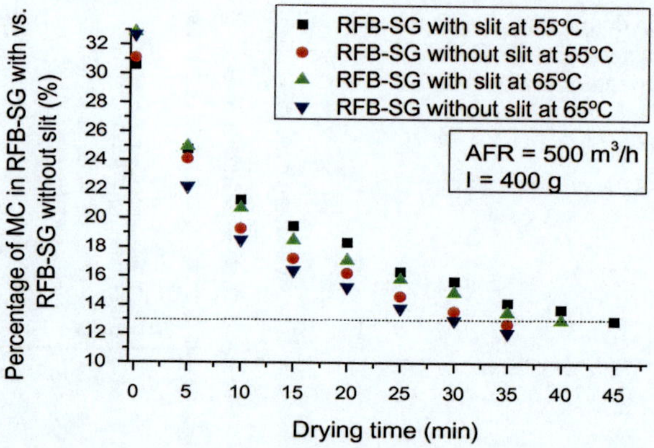

Fig. 7 Drying time of paddy in RFB-SG with slit and RFB-SG without slit dryer

has been observed that the drying time of paddy at the air temperature of 55 °C in RFB-SG dryer without slit is reduced by 24% as compared to the RFB-SG dryer with the slit. Similarly, the drying time of paddy at the air temperature of 65 °C in RFB-SG dryer without slit is reduced by 25% as compared to the RFB-SG dryer with the slit. Hence, RFB-SG dryer without slit is less time consuming as compared to RFB-SG with slits dryer. The reduction of drying time decreases the energy as well as the drying air requirement, which results in an improvement of drying efficiency.

4 Conclusions

A comparative study has been made with RFB-SG dryer with slits. By the effect of different parameters on the drying time, a few findings of the research are presented below:

- The drying time of paddy in RFB-SG without slits is found to be lower than the RFB-SG dryer with slits. Hence, the energy requirement is reduced due to time reduction.
- With the increase in the drying air temperature and airflow rate, the drying time is found to be decreased; hence the overall efficiency of the dryer is increased.
- Cyclic maintenance, system complexity, system cost, and drying air requirement have been reduced.
- A main cost-effective element slit is replaced by a tangential air inlet, which resulted in an improvement of the life of the dryer.

References

1. Nygaard, D., & Pellett, P. L. (1986). *Dry Area Agriculture, Food Science, and Human Nutrition.* Pergamon Press.
2. Mujumdar, A. S. (2007). *Handbook of Industrial Drying.* CRC Press, pp. 1133–1134.
3. Syahrul, S., Hamdullahpur, F., & Dincer, I. (2002). Energy analysis in fluidized-bed drying of large wet particles. *International Journal of Energy Research, 26*(6), 507–525.
4. Sreekumar, A., Manikantan, P. E., & Vijayakumar, K. P. (2008). Performance of indirect solar cabinet dryer. *Energy Conversion Management 49*(6), 1388–1395.
5. Smith, J .F. (1970). Microwave fluidized bed dryer. U.S. Patent 3,528,179, Issued 15 Sept 1970.
6. Kahyaoglu, L. Nesrin, S. S., & Sumnu, G. (2012). Spouted bed and microwave-assisted spouted bed drying of parboiled wheat. *Food Bioproducts Process, 90*(2), 301–308.
7. Quevedo, J., Pfeffer, R., Shen, Y., Dave, R., Nakamura, H., & Watano, S. (2006). Fluidization of nanoagglomerates in a rotating fluidized bed. *AIChE Journal, 52*(7), 2401–2412.
8. Wilde, J. D., & Broqueville, A. D. (2007). Rotating fluidized beds in a static geometry: experimental proof of concept. *AIChE Journal, 53*(4), 793–810.
9. Kochetov, L. M., Sazhin, B. S., & Karlik, E. A. (1969). Experimental determination of the optimal ratios of structural dimensions in the whirl chamber for drying granular materials. *Chemical Petroleum Engineering 5*(2), 106–108.
10. Dutta, A., Ekatpure, R. P., Geraldin, J. H., Broqueville, A. D., & Marin, G. B. (2010). Rotating fluidized bed with a static geometry: Guidelines for design and operating conditions. *Chemical Engineering Science, 65*(5), 1678–1693.
11. Pati, J. R., Dutta, S., Eliaers, P., Mahanta, P., Chatterjee, P. K., & Wilde, J. D. (2016). Experimental study of paddy drying in a vortex chamber. *Drying Technology, 34*(9), 1073–1084.
12. Singh, P., Kalita, P., & Mahanta, P. (2020). Numerical study on the gas-solid hydrodynamics and heat transfer in a rotating fluidized bed with static geometry dryer. Int. J. Heat Mass Transf. *153*, 119666
13. Singh, P., Kalita, P., & Mahanta, P. (2020). Numerical study of the hydrodynamics and heat transfer characteristics of gas–solid in a slit-less rotating fluidized bed in static geometry. Journal of Thermal Analysis and Calorimetry 1–10.

Recent Trends in Biomass Gasification in Germany

A. Gradel and Tobias Plessing

1 Introduction

Gasification of wood has a long history in Germany since it had been discovered for industrial applications in the nineteenth century. The first gasifiers were used for industrial heating applications and for the production of chemicals from the syngas [1, 2]. Since there was a shortage of fossil fuels, gasification of wood has also been used for powering freight traffic during the Second World War [3]. In the subsequent decades, biomass gasification was a technology of little interest as large amounts of fossil fuels were available at low cost. Only at the beginning of the twenty-first century, the technology gained importance again due to the upcoming climate debate. In 2000, the first German Renewable Energy Sources Act was adopted [4]. Within this law, a large governmental feed-in remuneration was granted for 20 years for power plants with renewable sources. Since then, more than ten considerable manufacturers of commercial gasifiers have raised on the domestic market. The following sections will give an overview of the historic growth of the market, promotions and restrictions for plant owners and the domestic technologies as wells as new approaches in research and development.

2 Legal Framework and Progress

The construction of new plants for the production of electrical energy from renewable sources in Germany was mainly triggered by the first Renewable Energy Sources Act

A. Gradel · T. Plessing (✉)
Institute for Water and Energy Management (Iwe), Hof University of Applied Sciences
Alfons-Goppel-Platz1, 95028 Hof, Germany
e-mail: tobias.plessing@hof-university.de
URL: http://iwe.hof-university.de

P. Mahanta et al. (eds.), *Advances in Thermofluids and Renewable Energy*, Lecture Notes in Mechanical Engineering, https://doi.org/10.1007/978-981-16-3497-0_36

in 2000. The act granted a feed-in remuneration of 10.23 ct kWh^{-1} (for plants <150 kW_{el}) for biomass plants (mainly digesters and gasifiers) for up to 20 years after commissioning [4]. In this time, commercial gasifiers with acceptable operating costs and an economical amount of annual operating hours were not available. However, the upcoming trend towards renewable energy sources and the increasing subsidies supported a faster development of the technology, and the first serious patents (e.g. Joos gasifier [5]) were applied. Since the production of energy from biomass was now an attractive field, the term "biomass" was defined by law at the same time. In the German Biomass Ordinance of 2001, biomass for the production of energy was limited to.

1. plants and parts of plants;
2. energy sources derived from plants or parts of plants, whose entire components and intermediate products were generated from biomass within the meaning of subsection (1);
3. waste and by-products of plant or animal origin from the agricultural, forestry or fishing industry;
4. biowaste within the meaning of § 2 no. 1 of the Ordinance on Biowastes;
5. gas produced from biomass within the meaning of subsection (1) through gasification or pyrolysis and secondary and by-products derived therefrom;
6. alcohols produced from biomass within the meaning of subsection (1), whose components, intermediate, secondary and by-products were generated from biomass [6].

This limitation excludes, for example, organic industrial waste like waste wood or bakery residuals from the remuneration. In 2004, the basic feed-in tariff was raised, and additional remuneration for different types of feedstocks (6 ct kWh^{-1} for residual wood, 7 ct kWh^{-1} for manure) was agreed in the first amendment of the Renewable Energy Sources Act [7]. By this change, the economical operation of gasifiers was realistic for the first time, and further development of the technology was encouraged. In 2008, the first commercial small-scale gasification plant was launched in Germany by the manufacturer Spanner Re^2 GmbH, which uses the principle of the Joos gasifier [8]. Further increments of the feed-in tariffs with the second and third amendment in 2009 and 2012 led to a large increase in the biomass gasification industry, and further manufacturers appeared on the market [9, 10]. The largest development of the market triggered by the legal framework was between 2012 and 2014, as there was the highest financial support from the government for plant operators. By 2014, the 4th amendment of the Renewable Energy Sources Act was passed, where the additional remunerations for the different feedstocks were removed again [11]. One of the aims of this amendment was to force the market strategies into the field of cogeneration applications instead of the pure generation of electricity to accelerate the development of renewable energy sources in both sectors. The last amendment of the act was made in 2017, where the maximum amount of subsidised plant capacity was limited by 150 MWel, and plant owners have to participate in a tendering procedure where the lowest price per unit energy wins (for plants up to 150 kW_{el}). Typical small-scale biomass gasifiers below 150 kW_{el} do not need to take part in the procedure and

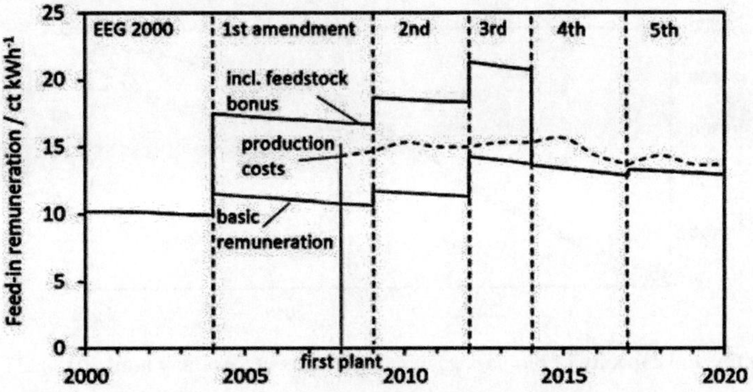

Fig. 1 Development of the feed-in remuneration and production costs per unit of electrical energy and production costs for electrical energy from biomass gasification in Germany

can still apply for the maximum amount of remuneration [12]. Figure 1 shows the development of the feed-in remuneration by law and the production cost per unit of electrical energy. Therefore, a typical amount of fixed costs (here 5 ct kWh^{-1}) and the domestic price development for wood chips was considered [13].

As biomass gasifiers have high-production costs in comparison with other renewable sources like solar power or wind, the pure production of energy was only economical until 2014. Since then, the trend shifted to heat-operated plants.

3 Market Size and Biomass Potentials

The market for biomass gasifiers is mainly limited by the feedstock quality and the technical requirements on the feedstock, given by the current technology. Recent reactor technologies for decentral applications mostly require feedstock moisture of below 10–12% and can only be fed with standardised wood chips or pellets. Technologies for challenging feedstock materials like cereal straw or sewage sludge are still not usable for continuous operation of plants as critical ingredients like chlorine cause corrosion of plant parts, or the melting point of ash is too low for stable operation. As applicable wood chips are comparatively costly, the market size is mainly affected on the feed-in remuneration given by law (Fig. 1). Figure 2 shows the development of installed plants and the correlating capacity in Germany. The 4th amendment of the Renewable Energy Sources Act 2014 is highlighted.

As clearly visible, the omission of the feedstock bonus caused a significant decrease in new plant commissions. Newly installed plants after 2014 are used for the cogeneration of heat and electricity in almost every case to provide an economic efficiency without the additional remuneration. The largest amount of biomass potential is used for the provision of heat in Germany. The combined generation of electrical

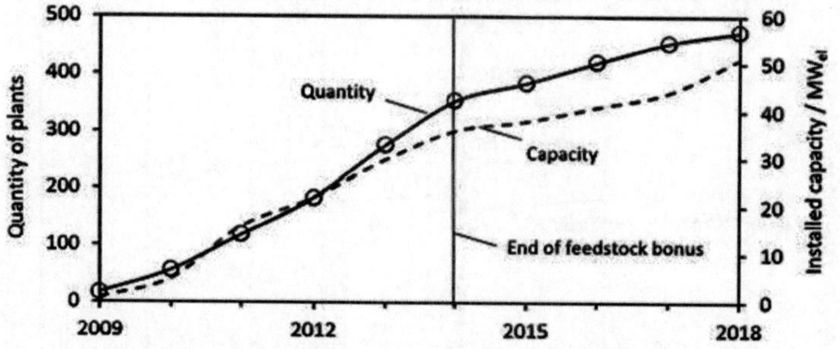

Fig.2 Quantity and capacity of biomass gasification plants in Germany until 2018 [15]

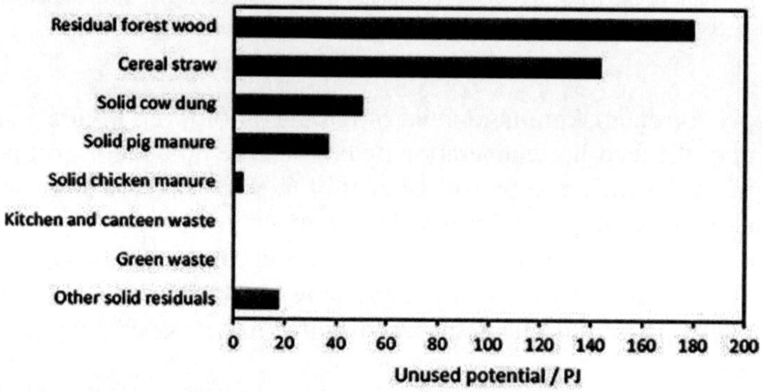

Fig.3 Potential of unused biomass residuals in Germany 2018 [14]

energy and heat from solid biomass was 446 PJ in Germany for 2018 [14]. Figure 3 shows the unused potential for different solid biomass residuals in Germany.

As the recent gasification technology is not able to handle straw or manure residues, the biomass potential that can be used as a feedstock for gasifiers is around 180 to 200 PJ (55 TWh). Assuming 8000 operating hours per year and a typical electrical efficiency of 30%, another 2 TW_{el} of maximum capacity could be installed to use this potential. This amount of capacity is around 40 times the current installed capacity.

4 Current Technologies

The current market is mainly dominated by six manufacturers, five based in Germany and one located in Austria. Figure 4 shows the installed capacity of the five market leaders and the cumulated capacity of other manufacturers (2018).

As the technological approaches and reactor types are different, the specifications, advantages and disadvantages will be discussed in the following sections.

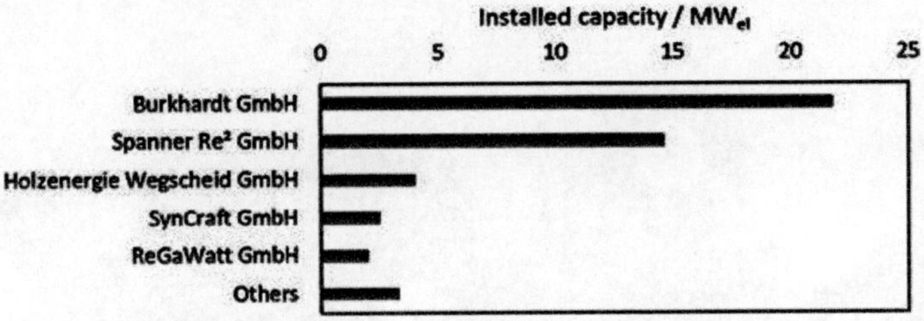

Fig.4 Installed electrical capacities of different manufacturers in Germany (2018) [15]

4.1 *Burkhardt GmbH*

The largest share of the installed capacity with around 45% is based on reactors of the Burkhardt GmbH, located in Bavaria, Germany. The main gasification unit is a fluidised bed reactor for standardised (6 mm) wood pellets. Figure 5 shows the main reactor unit of a Burkhardt V 3.90 plant with an electrical power output of 180 kW$_{el}$ [16].

As the temperature is very homogeneous in the fluidised bed, a comparatively clean gas is produced, and the maintenance intervals are long (>14 days) and more than 8000 operating hours per year are reached by nearly every plant. In comparison to wood chip reactors, the feedstock material is twice as costly. Therefore, the plants can only be operated economically if the provided electricity and heat is used completely. Plants of this type are mainly installed in industrial applications with appropriate demand for heat.

Fig. 5 Fluidised bed gasifier for wood pellets by Burkhardt GmbH (Bavaria, Germany) [16]

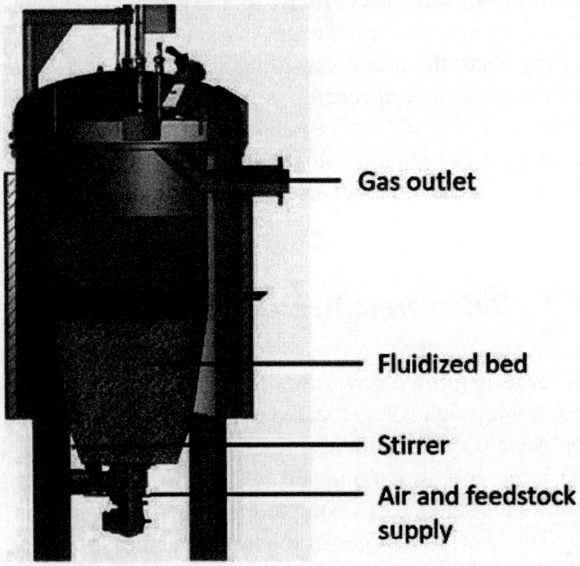

Fig. 6 Spanner Re^2 HKA 45 biomass gasifier for wood chips with 45 kW_{el} and 102 kW_{th} power output

4.2 Spanner Re^2 GmbH

A more common gasifier type for agricultural and forestry holdings is the stratified downstream reactor of Spanner Re^2. The plant can be fed by a large amount of variation of moisture in residual wood chips with a maximum moisture of 13 wt%. The current share of the total installed capacity of biomass gasifiers is around 30% for this plant type. Reactors are available with a power output of 9–68 kW_{el}. Figure 6 shows a Spanner Re^2 HKA 45 plant with 45 kW_{el} power output.

The downstream reactor (1) is based on the patented Joos gasifier [5] and equipped with a feedstock lock (2) and a textile filter (3) Residual charcoal after the gasification section is also converted to gas in a small post-reformer (4) to maximise the efficiency of the plant; the process residual is then only ash. The tar content in the product gas of the downstream reactor is comparatively low in comparison to countercurrent reactors (<1 g m^{-3} (STP) and no further filtering technology is needed for stable continuous operation. Most owners report more than 8000 annual operating hours for this reactor type but also challenging daily maintenance [17, 18].

4.3 Holzenergie Wegscheid GmbH

A larger stratified downstream gasifier for wood chips was developed by Holzenergie Wegscheid, also located in southern Bavaria. The power output of the plants is between 65 and 140 kW_{el}. In comparison with the small-scale gasifiers of Spanner Re^2, the system is equipped with a more complex system of filter units. Figure 7 shows the scheme of a complete gasification plant with filters and cogeneration unit.

The raw product gas of the main reactor (1), is first filtered by a metal-based hot gas filter (2), ash is removed there and stored in a separate bunker. Before the gas is

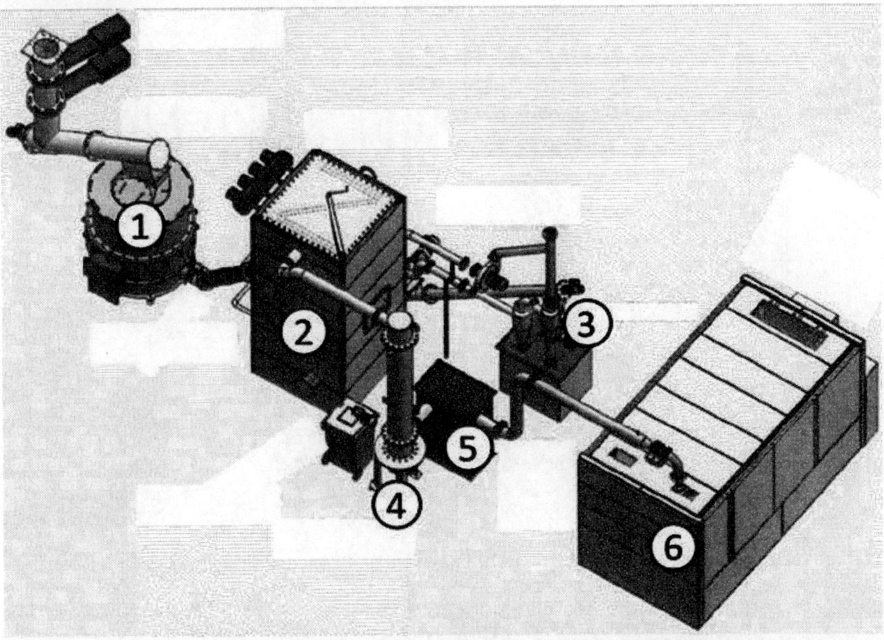

Fig. 7 Holzenergie Wegscheid biomass gasifier with filter system and cogeneration unit [19]

fed to the cogeneration unit (6), it is cooled down in a vertical heat exchanger (4), the main amount of tar species in the raw product gas condensates in this unit. The safety filter unit (5), protects the gas motor in any case of failure in the upstream filter system. Plant owners report an annual efficiency of more than 95% when the gasifier is well-adjusted, and the feedstock is dried homogeneously to less than 10 wt% moisture content [19, 20].

4.4 *SynCraft GmbH*

Around 5% of the recently installed capacity of biomass gasifiers is based on the technology of SynCraft (Austria). Syncraft plants are equipped with a large-scale (200–1000 kW_{el}) fluidised bed reactor for wood chips with up to 10 wt% moisture content after drying. Figure 8 shows the plant scheme of the SynCraft plant.

The SynCraft technology is based on two-stage gasification in an upstream pyrolysis stage ((1), $\lambda \approx 0.1$), and the subsequent main reactor ((2), $\lambda \approx 0.25$). Impurities in the feedstock are separated mechanically and collected in a storage (3). The hot gas is then purified in a cyclone filter unit (4) where ash and charcoal residuals are collected in a tray bunker (5) for subsequent applications (e.g. fertilisation). After cooling (6), the gas is scrubbed with circulating water (7) to meet the requirements on

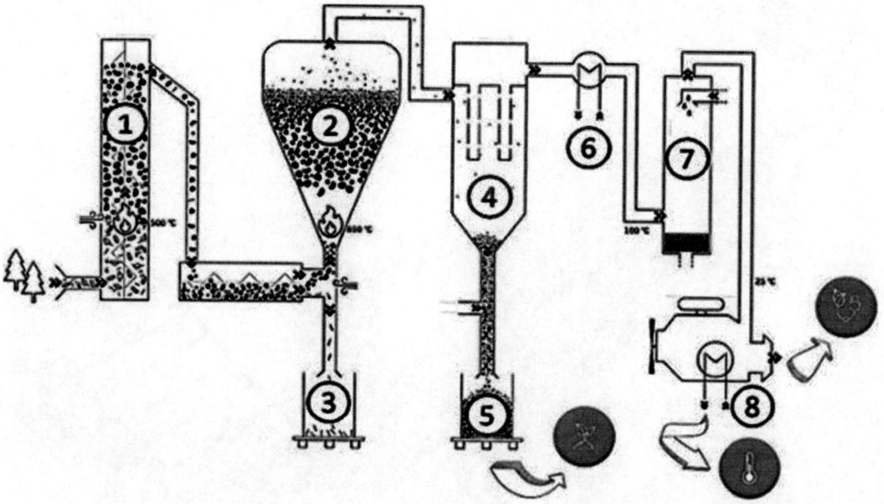

Fig. 8 SynCraft fluidised bed gasifier with filter system and cogeneration unit [21]

the tar content of the cogeneration unit (8). Plant owners report an annual efficiency of more than 90% (>8000 h a^{-1}). The relatively high-maintenance requirements are reasonable due to the high-power output, and therefore, high revenues per plant [22].

4.5 RegaWatt GmbH

A large-scale biomass gasifier with an electrical power output of 400 kW_{el} to 1 MW_{el} was developed by RegaWatt (Bavaria, Germany). The plants are based on an updraft gasifier and can be fed with nearly any quality grade of feedstock (up to 55 wt% moisture content), even the use of peat is currently tested. The plant configuration of RegaWatt gasifiers is very modular and depends on the demand of heat and electrical energy of the application. Figure 9 shows the scheme of a combined ORC and cogeneration application with an electrical power output of 890 kW_{el} and a thermal power of 3600 kW_{th} [23]. This plant is in operation for the supply of two district heating systems and the provision of up to 5.5 GWh per year for a large number of households and industrial consumers in Noville, Switzerland.

The feedstock (1) is directly fed into the 4.5 MW updraft gasifier (2) without further pretreatment. The product gas of the gasification unit is split into two strings, one for the supply of the combustion chamber (3) and for the supply of the cogeneration unit (9) with an upstream gas scrubber (8) Tar and charcoal residuals of the scrubber are also fed into the combustion chamber. A mixing chamber (4) regulates the 500 °C cycle for the ORC-Unit (6) which provides 12% of the overall electrical power. Two warm water boilers (5) for the 500 °C and the 900 °C cycle supply the

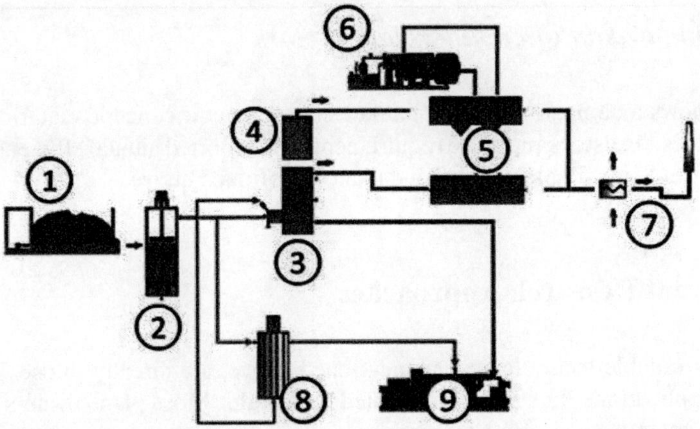

Fig. 9 RegaWatt biomass gasification plant with filter system, cogeneration unit, combustion chamber, ORC-turbine and warm water boilers [23]

district heating system with an overall thermal power output of 3.6 GW_{th}. Operators report a very high effort for the plant control and maintenance to reach a stable continuous operation. In return, very clean combustion is reported, and therefore, low-disposal costs for plant owners [24].

Table 1 Comparison of the current biomass gasification technologies in Germany

Manufacturer	Market share (inst. capacity) (%)	Electrical power output (kW_{el})	Max. moisture content (wt%)	Reported efficiency (%)
Burkhardt GmbH	45	50–180	<10	90–105[1]
Spanner Re[2] GmbH	30	9–68	<13	80–99
Holzenergie Wegscheid GmbH	8	65–133	<10	>95
SynCraft GmbH	5	200–1000	<10	>90
RegaWatt GmbH	4	400–1000	<55	>95

[1]efficiency of more than 100% is based on a higher electrical power output as listed for this plant

4.6 *Comparison of commercial reactors*

Table 1 shows a comparison of the market shares, reactor concept, electrical power output values, feedstock moisture requirements and reported annual efficiency values for the currently available commercial plants mentioned above.

5 Current Research Approaches

Recently available technologies, as mentioned above, are already in use in a large field of applications, but still complicated to handle. Most plant owners report a very high-maintenance effort to reach a high amount of annual operating hours. Therefore, different research approaches are recently in progress for the design of reliable gasification technologies with higher gas qualities and longer maintenance intervals. The most promising approaches and the latest results in research will be discussed in the subsequent sections.

5.1 *Biomass Gasification with Integrated Adsorption of Tar Species (UAS Hof)*

In a cooperative research project of the University of Applied Sciences Hof and WS Wärmeprozesstechnik GmbH (Renningen, Germany), a new technology for biomass gasification with integrated adsorption of tars by using the charcoal of the biomass is developed. The patented process approach is based on partial gasification of the raw biomass to get a residual amount of around 7 wt% charcoal for the use as an adsorption agent in a subsequent cooled section [25]. Figure 10 shows the process scheme.

Model results based on experimental data of the different mechanisms of gasification and adsorption showed a theoretical gas purity of below 100 mg m^{-3} (STP) tar content in the product gas. Therefore, the gas could be used directly for cogeneration without any additional filter units [27, 28].

First test series were performed on a 25 kW_{el} real-scale prototype of the gasifier for wood pellets in 2020. The results showed a tar content of only 25 mg m^{-3} (STP) for several hours. Figure 11 shows the latest experimental results for the gas quality, char conversion and load and the tar content in the clean gas after adsorption.

Further experiments with higher preheated air and long-term experiments are planned for the next months; recent model predictions calculate a maximum gas purity of below 1 mg m^{-3} (STP) for this process [29].

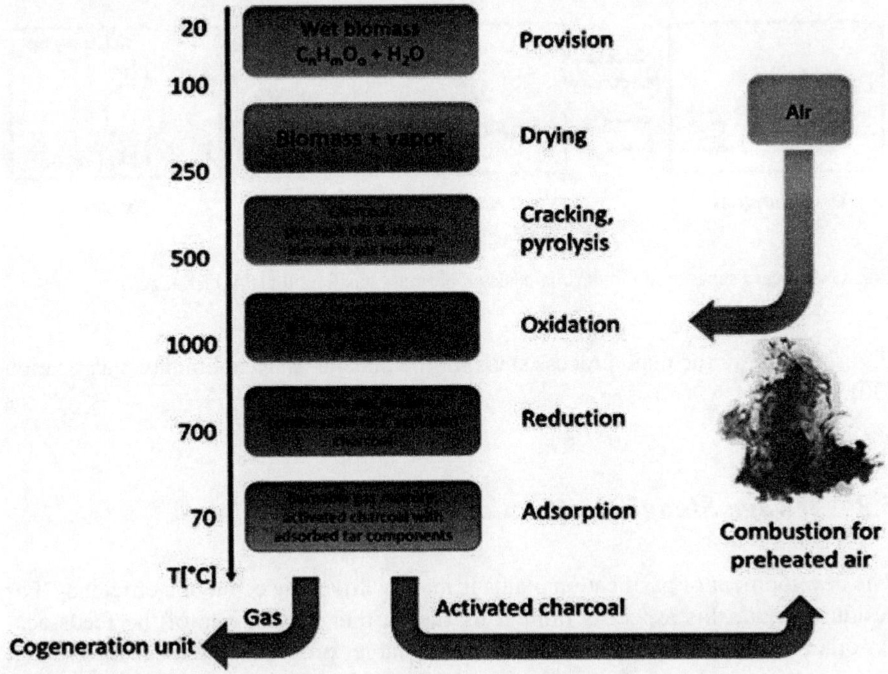

Fig. 10 Process scheme of the gasification with integrated adsorption of tar [26]

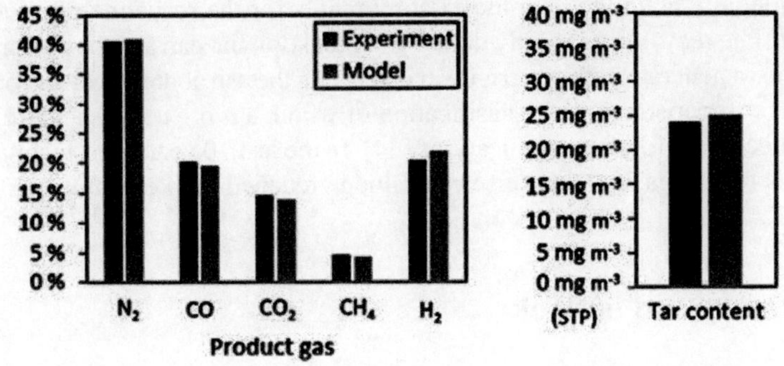

Fig. 11 Results of the first real-scale experiments on the gasification of biomass with integrated adsorption of tar (25 kg h^{-1} wood pellets, $\lambda = 0.2$, air preheating 200 °C) [29]

5.2 *Plasma-Assisted Biomass Gasification (FAU Erlangen)*

Another research approach for the gasification of biomass is the use of plasma generated from other renewable energy sources to improve the steam gasification process. The main aim of this research is to store renewable energy in syngas based on biomass with higher reaction rates (+33%) compared to an electric heated allothermal process.

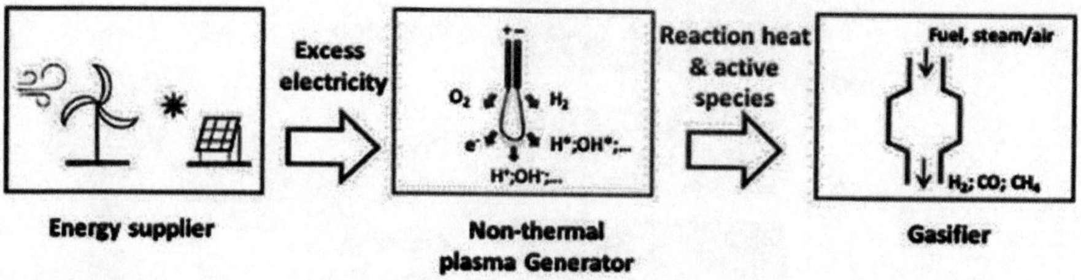

Fig. 12 Process scheme of the plasma-assisted biomass gasification (FAU Erlangen)

Figure 12 shows the main process steps for the plasma-assisted biomass gasification [30].

5.3 Sewage Sludge Gasification

The development of gasification plants is mainly driven by economic efficiency. For residual wood, this aspect is limited by the current market rate of the feedstock. As other residuals as waste or sludge provide larger proceeds, the focus of research and development has shifted to these topics [31, 32]. The main driver for the use of sewage sludge has been set with new legal regulations for the recycling of phosphorus in 2017 [33]. In the Sewage Sludge Ordinance of 2017, treatment plant operators have been legally bound to upgrade existing plants for the recycling of phosphorus until 2029. The recycling is mostly done by combustion but can also be performed by gasification to generate both electricity and heat. As the ash content and tar formation are high in comparison with the gasification of wood, a more complex gas clean-up system is necessary for these applications [34]. In the last 10 years, first technological approaches for the gasification of sewage sludge reached market maturity [35, 36].

6 Summary and outlook

The gasification of biomass has grown significantly in the last decade. However, it is still a small sector in comparison with other renewable energy sources like wind or solar power (<0.1% of the total installed capacity). As the production costs for electrical energy from biomass by thermal gasification are comparatively high, the main driver for the sector growth is the legal framework and the feed-in remuneration given by law. Newly installed plants have to be integrated into a well-designed concept, including the use of both thermal and electrical energy from the cogeneration unit to be profitable within the current legal framework. The upcoming tax for CO_2 emissions in Germany could give a new impulse to the market. Recent research approaches aim at the reduction of maintenance costs and sector coupling concepts. Therefore, new technologies that provide a more economical operation are also promising to be the next trigger for the growth of the market for biomass gasifiers.

References

1. Eckermann, E. (2016). *Fahren mit Holz – Geschichte und Technik der Holzgasgeneratoren und Ersatzantriebe*. Bielefeld: Delius Klasen Verlag.
2. Klass, D. L. (1998). *Biomass for renewable energy, fuels and chemicals*. San Diego: Academic Press.
3. Speer, A. (1942). Anordnung des Beauftragten für den VJP, Der Generalbevollmächtigte für Rüstungsaufgaben betr. Umstellung von Verbrennungsmotoren jeder Art auf den Betrieb mit Generator sowie Hoch- und Niederdruckgas, Reichs-Verkehrs-Blatt.
4. Federal Ministry of Justice and Consumer Protection. (2000). Gesetz für den Vorrang Erneuerbarer Energien, Berlin.
5. Joos, B. (2002). Vorrichtung zur Erzeugung eines brennbaren Gasgemisches, patent (AT000000472590E), Austria.
6. Federal Ministry of Justice and Consumer Protection. (2001). Verordnung über die Erzeugung von Strom aus Biomasse, Berlin.
7. Federal Ministry of Justice and Consumer Protection. (2004). Gesetz für den Vorrang Erneuerbarer Energien, Berlin.
8. Bleul, T. (2016). Die besten Ideen beginnen in der Garage, Holz-Kraft-Magazin 06.
9. Federal Ministry of Justice and Consumer Protection. (2009). Gesetz für den Vorrang Erneuerbarer Energien, Berlin.
10. Federal Ministry of Justice and Consumer Protection. (2012). Gesetz für den Vorrang Erneuerbarer Energien, Berlin.
11. Federal Ministry of Justice and Consumer Protection. (2014). Gesetz für den Vorrang Erneuerbarer Energien, Berlin.
12. Federal Ministry of Justice and Consumer Protection. (2017). Gesetz für den Vorrang Erneuerbarer Energien, Berlin.
13. C.A.R.M.E.N.e.V. Homepage. https://www.carmenev.de/infothek/preisindizes/hackschnitzel/graphiken. Last accessed 07 July 2020
14. Fachagentur Nachwachsende Rohstoffe e.V. (2019). Basisdaten Bioenergie Deutschland 2019, Gülzow.
15. Birth, T. (2019). Status Quo der Holzvergasung, 24. Fachtagung Nachwachsende Rohstoffe, Dresden.
16. Burkhardt, C. (2019). Burkhardt Energie und Gebäudetechnik, Fachkongress Holzenergie, Würzburg.
17. Bleul, T. (2019). Wie man solide Laufzeiten durch den richtigen Betrieb erreicht: Vortrag zu einer Holzgas-KWK-Anlagen von Spanner Re^2, Anwenderkonferenz Biomassevergasung, Innsbruck.
18. Fries, H. (2019). Strom und Wärmeerzeugung Fries GbR, Anwenderkonferenz Biomassevergasung, Innsbruck.
19. Schätzl, W. (2019). Holzenergie Wegscheid, Anwenderkonferenz Biomassevergasung, Inssbruck.
20. Thalmann, P. (2019). Immer über 8300 Betriebsstunden seit 4 Jahren, ist das wahr?, Anwenderkonferenz Biomassevergasung, Inssbruck.
21. SynCraft GmbH Homepage, https://www.syncraft.at/holzkraftwerke/uebersicht. Last accessed 13 July 2020
22. Trenkwalder, H. (2019). Laaser Eyrser Energiegenossenschaft, Anwenderkonferenz Biomassevergasung, Inssbruck

23. Röhrmoser, K. (2019). KWK mit Gegenstromvergasung—Status quo und Perspektiven, 19. Fachkongress Holzenergie, Würzburg.
24. Fuite, G. (2019). Wärme im großen Stil: Erfahrungen bei der Energieversorgung von Gewächshäusern mit einer Holzkraftanlage von ReGaWatt, Anwenderkonferenz Biomassevergasung, Inssbruck.
25. Wünning, J.A. (2015). Method and device for the gasification of biomass, European Patent EP000003088492B1.
26. Gradel, A. et al. (2019). Clean gasification of biomass with integrated tar adsorption. In *Circular Economy—A Fresh View on Petrochemistry (conference transcript)*.
27. Gradel, A., et al. (2019). Modeling of biomass gasification in a downdraft gasifier with integrated tar adsorption. *Chemical Engineering and Technology, 42*(9), 1895–1906.
28. Gradel, A. et al. (2019). Teerarme Vergasung von Biomasse durch Adsorption an prozesseigene Aktivkohle, Fachkongress Holzenergie, Würzburg
29. Gradel, A. et al. (2020). Vergasung von Holzpellets mit integrierter Adsorption von Teerkomponenten (doctoral thesis), University of Bayreuth.
30. Pang, Y., et al. (2019). Investigation on the influence of non-thermal plasma on reaction degree of wood gasification in a drop tube reactor. *Fuel, 253*, 95–105.
31. Vodegel, et al. (2015). Wissenschaftliche Untersuchungen thermochemisch schwieriger Biomassen unter Sauerstoffmangelatmosphären. *VDI-Berichte, 2267*, 155–164.
32. Meyer, B. (2017). Ein Konzept für die Kohlenstoffkreislaufwirtschaft Sektorkopplung zwischen Energie. *Chemie Abfall Chemie Ingenieur Technik, 90*(1), 241–248.
33. Federal Ministry of Justice and Consumer Protection: Verordnung zur Neuordnung der Klärschlammverwertung, Berlin (2017)
34. Hinnerk, B., et al. (2017). Thermochemische Verfahren zur Klärschlammverwertung—Übersicht der Entwicklungen in Deutschland. *Wasser und Abfall, 11*, 29–34.
35. Mey, S. (2016). Kopf—SynGas Klärschlammvergasung—Kraftwerk für die Kläranlage und saubere Klärschlammasche für die Phosphorrückgewinnung, Innovationsforum Thermolyphos, Halle
36. Appe, T. et al. (2017). Phosphordüngewirkung von industriell hergestellten Karbonisaten aus dem Klärschlamm einer kommunalen Kläranlage, Horizonte des Bodens, Göttingen.

Solar and Renewable Energy, Energy Storage

Mathematical Modeling and Simulation of the Drying Process of Tea Leaves in a Batch-Type Tray Dryer

M. Das and Deva Kanta Rabha

Abbreviations

A	Interficial area of contact (m^2)
c_a	Specific heat of the air (J/kg K)
c_{A1}	Specific heat of the liquid (J/kg K)
c_s	Specific heat of the solid (J/kg K)
c_v	Specific heat of vapor (J/kg K)
hg	Specific enthalpy of the humid air (J/kg)
hm	Specific enthalpy of wet solid (J/kg)
hv	Specific enthalpy of the moisture in the solid (J/kg)
Ls	The latent heat of sorption (J/kg)
Lv	The latent heat of evaporation of the liquid water (J/kg)
Ma	Mass of drying air hold up in the dryer (kg)
Ms	Mass of tea leaves hold up in the dryer (kg)
ma	Mass flow rate of drying air (kg/s)
m_s	Mass flow rate of solid (kg/s)
N_D	Drying rate (kg/m^2s)
q	Heat flux (kW/m^2)
t	Drying time (s)
T_g	Drying air temperature (°C)
Tm	Mean temperature of the tea leaves (°C)
X_1	Initial moisture content in dry basis (kg/kg)
X_2	Final moisture content in dry basis (kg/kg)
Y_1	Absolute humidity of air at the inlet (kg/kg)
Y_2	Absolute humidity of air at the inlet (kg/kg)
α	Volumetric heat transfer coefficient (W/m^3K)

M. Das · D. K. Rabha (✉)
Department of Mechanical Engineering, Jorhat Engineering College, Jorhat-7, India

P. Mahanta et al. (eds.), *Advances in Thermofluids and Renewable Energy*, Lecture Notes in Mechanical Engineering, https://doi.org/10.1007/978-981-16-3497-0_37

1 Introduction

Tea is one of the most consumed beverages and is produced from the leaves and buds of plant Camellia (Camellia sinesis L.) belonging to family Theaceae [1]. Chinese variety (C.s. Sinesis) and Assamese variety (C.s. Assamica) are the two major varieties of tea grown today. Tea has been a widely consumed beverage in China since ancient times. Cultivation of tea plant started in India around nineteenth century in the state of Assam [2]. There are numerous health benefits of green tea [3]. Numerous epidemiological studies and documented studies are available that highlights the health benefits of green tea consumption like reduction of high-cholesterol levels, diabetes, osteoporosis, cardiovascular diseases, and dental caries [4]. These benefits are attributed to the antioxidant property of polyphenolic compounds [5]. The tea manufacturing involves stages like withering, steaming, rolling, cutting, and drying depending upon the type of tea produced. The drying is the last stage of the tea manufacturing process. The tea leaves are dried with hot air at a temperature of between 90 °C and 110 °C. The leaves are dried from the moisture content of around 66% (w.b.) to the desired final moisture content of 3% (w.b.) [6]. Many sophisticated dryers are available for drying tea which are very costly and are not suitable for the small tea growers. Usually, the convective dryer is used for drying tea. In the convective drying, hot air or combustion gases is used to remove the moisture from the tea leaves. In the case of gaseous fuel like natural gases, the combustion gas is diluted with the excess air and then, it is directly used in the drying chamber for drying the tea leaves.

Nowadays, the small tea grower has also started manufacturing tea at a very small scale for which they need a small size and low-cost tea dryer. The small tea grower produces mainly green tea as the consumption of green tea is increasing day-by-day due to its enormous health benefits. The drying process of the green tea is similar to that of other types of tea. It is always necessary to predict the operating parameters of a dryer before developing its protype. In this study, a numerical model of a small batch tea dryer has been developed and solved in MATLAB to predict the operating parameters with the objective of designing and developing the physical model of the dryer.

Many numerical studies on different agricultural products have been reported in the literature. Basirat-Tabrizi et al. [7] in their work developed a model for the packed bed drying process and investigated the effects of particle size, inlet gas temperature on the moisture content and temperature of solid and gas at the outlet. Izadifar et al. [8] investigated the drying of paddy rice by taking an elementary volume of bed and applying conservation equations to it. Ratti and Fontaine [9] carried out investigation to know the drying curves for bananas, grapes, and mushrooms in a lab-scale dryer at different air velocities and temperature and compared the resuls with their model. Ratti and Crapiste [10] studied the drying kinetics of shrinkable biological materials. Tabarsa and Zarenezhad [11] investigated the drying behavior of tea leaves dried in a fluidized bed dryer. There is another reported study on drying behavior of green tea leaves by Shomalie and Souraki [12] in a multi-tray cabinet dryer.

2 Model Development

The model of a simple batch-type tray dryer for drying of tea leaves is shown in Fig. 1. Hot air is circulated throughout the dryer by a fan. A fraction of the exhaust air is recirculated and mixed with the fresh air. The tea leaves are placed on trays and loaded into the dryer. The moisture content and the temperature of the tea leaves vary only with time; therefore, the solid layer can be described by a lumped parametric model. The air is assumed to be perfectly well mixed within the dryer. While modeling the drying process, a few assumptions have been adopted which are summarized as follows.

a. The dryer operates in a batch mode.
b. Air enters the dryer sideways and parallel to the bed of leaves.
c. The shrinkage of the tea leaves is negligible.
d. The bed is described as a lumped parameter model.
e. The air is assumed to be well mixed.
f. The gas phase is also described by a lumped parameter model.

The mass conservation equation for solid can be written as follows [13]

$$m_S X_1 - m_S X_2 - N_D A = M_S \frac{dX}{d\tau} \tag{1}$$

Since the dryer is not operated in continuous mode, so $m_s = 0$ therefore, Eq. (1) reduces to

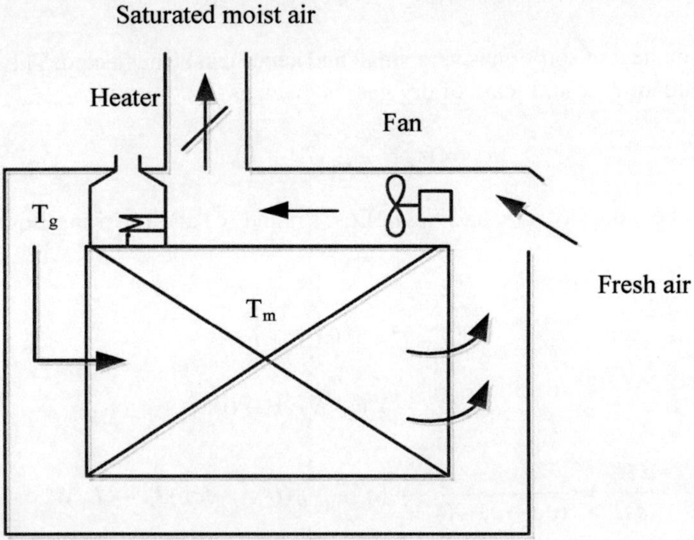

Fig. 1 Schematic of the batch tray dryer

$$N_D = -\left(\frac{M_S}{A}\right)\frac{dX}{d\tau} \tag{2}$$

The drying time required for reaching the final moisture content can be found out by integrating Eq. (2). From the conservation of mass principle in gas phase,

$$m_a Y_1 - m_a Y_2 + N_D A = M_a \frac{dY}{d\tau} \tag{3}$$

If the absolute humidity of the inlet air is taken as Y_o, then Eq. (3) can be written as

$$m_a Y_0 - m_a Y = M_S \frac{dX}{d\tau} + M_a \frac{dY}{d\tau} \tag{4}$$

Similarly, applying the conservation of energy principle in both the solid and gas phases [13].

$$M_S \frac{dh_m}{d\tau} = (q - N_D h_v)A \tag{5}$$

$$m_a h_{g0} - m_a h_g + \sum q = M_S \frac{dh_m}{d\tau} + M_a \frac{dh_g}{d\tau} \tag{6}$$

The enthalpy of the wet solid (per unit mass of dry solid) can be written as

$$h_m = (c_S + c_{A1} X)T_m - L_s X \tag{7}$$

The latent heat of sorption is very small and hence can be neglected. The enthalpy of the humid air (per unit mass of dry gas) is given as

$$h_g = (c_v Y + c_a)T_g + L_v Y \tag{8}$$

Putting the values of h_m and h_g in Eqs. (5) and (6) and rewriting Eqs. (2), (4), (5) and (6)

$$\frac{dX}{d\tau} = -N_D\left(\frac{A}{m_S}\right) \tag{9}$$

$$m_a(Y_0 - Y) + N_D A = 0 \tag{10}$$

$$\frac{dT_m}{d\tau} = \frac{1}{(c_S + c_{A1}X)}\frac{A}{M_S}[q + N_D((c_{A1} - c_A)T_m - L_v)] \tag{11}$$

$$m_a\left[(c_B + c_A H_0)T_{g0} - (c_B + c_A Y)T_{g0} + (Y - Y_o)c_A T_g\right] - A\left[\alpha\left(T_g - T_m\right) + N_D c_A\left(T_g - T_m\right)\right] + q = 0 \tag{12}$$

Table 1 Operating parameters and properties of tea leaves used in the simulation

Parameter	Value
Initial moisture content of tea leaves, $\boldsymbol{X}_0$ (kg/kg)	0.66
Equilibrium moisture content of tea leaves,$\boldsymbol{X}_{eq}$ (kg/kg)	0.03
Mass of tea leaves,$\boldsymbol{M}_S$ (kg)	20
Interfacial area of phase contact, $\boldsymbol{A}$ (m^2)	1.25
Mass flow rate of air, $\boldsymbol{m}_a$ (kg/s)	0.15
Absolute humidity of air at inlet, $\boldsymbol{Y}_0$ (kg/kg)	0.002
Specific heat capacity of the liquid phase, $\boldsymbol{c}_{A1}$ (kJ/kg K)	4.186
Specific heat of the vapor phase, $\boldsymbol{c}_A$ (kJ/kg	1.883
Specific heat capacity of tea leaves, $\boldsymbol{c}_S$ (kJ/kg) [11]	1.598
Specific heat capacity of dry air, $\boldsymbol{c}_B$ (kj/kg K)	1.02
Convective heat transfer coefficient, $\boldsymbol{\alpha}$ (W/m^2K) [14]	34.85
Temperature of air at the inlet of the dryer, $\boldsymbol{T}_{g0}$ (K)	353
Initial temperature of the tea leaves, $\boldsymbol{T}_{m0}$ (K)	293
Energy input, $\boldsymbol{q}$ (kW)	2.8

Equations (9) to (12) are the model equations of the batch drying process. Equations (9) and (11) are simulated with an ODE solver for a given set of data and initial conditions. The gas-phase parameters Y and T_g are found out by solving Eqs. (10) and (12) with each time step. The properties given in Table 1 are used in the simulation process.

3 Results and Discussions

Figure 2 depicts the variation of the moisture content of the tea leaves with time. The initial moisture content of leaves is assumed to be 66% (w. b) before loading it into the dryer. The ordinary differential Eq. (9) is simulated with an ODE solver. The tea leaves are reduced to a moisture content of 3% (w.b). As the air passes over the bed, the water in the leaves starts evaporating and its moisture content goes on decreasing in a rapid manner before finally coming down to its equilibrium moisture content. It is seen that the tea leaves take approximately 45 min to come down to its equilibrium moisture content.

The variation of the absolute humidity of the drying air at the outlet of the dryer with time at varying mass flow rates (0.15 kg/s, 0.25 kg/s and 0.35 kg/s) is shown in Fig. 3. The ambient air absolute humidity at the inlet is assumed to be 0.002 kg/kg. Initially, the humidity of the drying air at the outlet remains high because of the large amount of the moisture released from the tea leaves. The absolute humidity of the air at the outlet of the dryer goes on decreasing with time due to decreasing moisture content in the tea leaves. Finally, the absolute humidity of the drying air reaches the

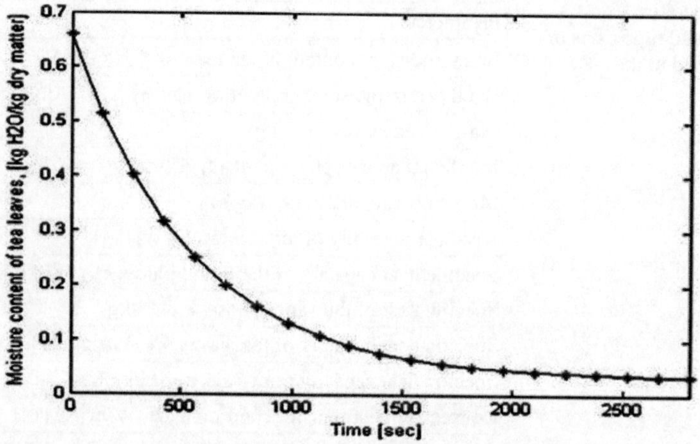

Fig. 2 Variation of the moisture content of the tea leaves with time

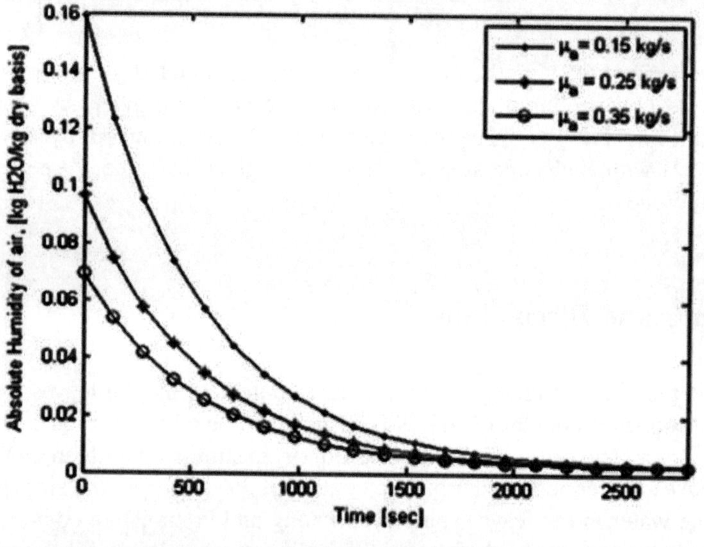

Fig. 3 Variation of the absolute humidity of the drying air at the outlet of the drying chamber with time at different mass flow rates

ambient humidity. When the mass flow rate of the air is increased, the drying process becomes faster, and the drying time is also shortened.

The change in the temperature of the tea leaves is shown in Fig. 4. The initial temperature of the tea leaves before loading is taken to be 20 °C. It is observed that with the evaporation of the moisture from the tea leaves, the temperature of the leaves increases gradually and finally, when the moisture content is reduced to the desired level, the temperature of the tea leaves becomes equal to around 74 °C. This occurs

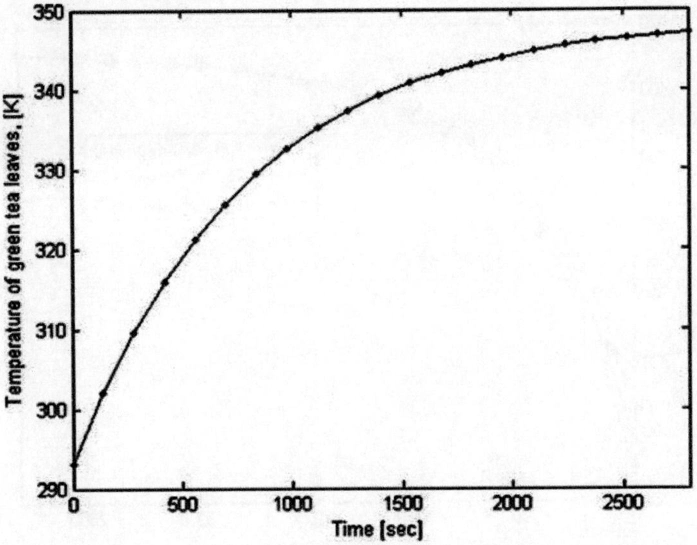

Fig. 4 Variation of the temperature of leaves with time

mainly due to recirculation of some percentage of the exhaust air. With elapse of the drying time, the drying rate decreases and the energy available with the drying air is underutilized. As the heat is supplied at a constant rate, and the energy available with the drying air is not used fully. The exhaust air temperature increases gradually and exceeds the desired temperature with the progress of the drying process. To reduce the wastage of energy in a dryer with exhaust air recirculation facility, the heat supplied for heating the air has to be reduced in the later period of drying.

Figures 5 shows the changes in the air temperature at the outlet of the drying chamber for different intlet air temperatures. Initially, the exhaust air temperature rises sharply, and in the later part of the drying period, the rise of temperature slows down. The exhaust air temperature exceeds the desired drying air temperature after a certain period of drying.

4 Conclusion

A model to predict the drying behavior of tea leaves dried in a batch-type tray dryer has been developed and solved in MATLAB. The developed model is a very simple model which helps in predicting the exit temperature of the dryer, drying rate at different mass flow rates, and drying time which are required for proper designing and operating a dryer.

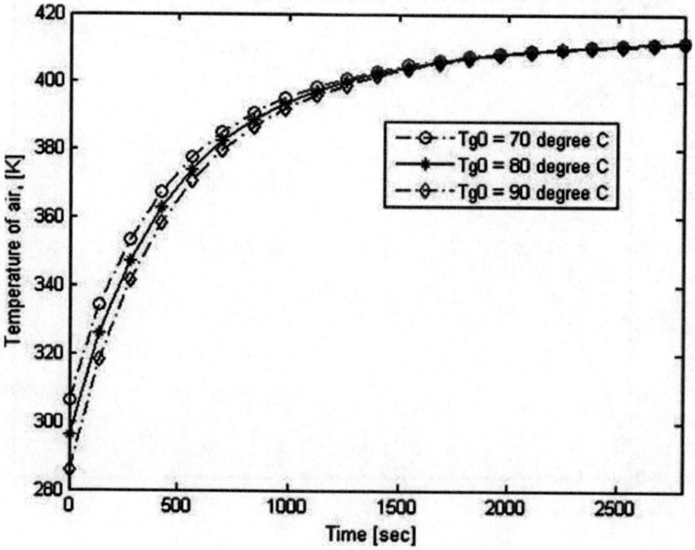

Fig. 5 Variation of the outlet air temperature with time at different inlet air temperatures

References

1. Naheed, Z., Barech, A. R., Sajid, M., Khan, N. A., & Hussain, R. (2007). Effect of rolling, fermentation and drying on the quality of black tea. Agric, *23*(3).
2. Chakravarty, A., Mazumdar, A. S., Raghavan, G. S. V., & Ramaswamy, H. S. (2003). *Handbook of Postharvest Technology*. New Work, Basel: Marcel Deker Inc.
3. He, X. Y., Liu, J. F., & Huang, Z. H. (2011). Preparation of cold brew tea by explosion puffing drying at variable temperature and pressure. *Drying Technology, 29*, 888–895.
4. Petti, S., & Scully, C. (2009). Polyphenols, oral health and disease: A review. *Journal of Dentistry 37*, 413–423.
5. Yang, C. S., Lambert, J. D., & Sang, S. M. (2009). Antioxidative and anticarcinogenic activities of tea polyphenols. *Archives of Toxicology, 83*, 11–21.
6. Willson, K. C., & Clifford, M. N. (1992). *Tea Cultivation to consumption, Springer*. London: Chapman & Hall.
7. Basirat-Tabrizi, H., Saffar-Avval, M., Assarie M,R., (2002). Two-dimensional mathematical model of a packed bed dryer and experimentation. *Proceedings of the Institution of Mechanical Engineers, Part A: Journal of Power and energy ,216*, 161–168.
8. Izadifar, M., Baik, O., & Simonson, C. J. (2006). Modeling of the packed bed drying of paddy rice using local volume averaging (LVA) approach. *Food Research International, 39*, 712–720.
9. Fontaine, J., & Ratti, C. (1999). Lumped–parameter Approach For prediction of drying Kinetics In Foods. *Journal of Food Process engineering, 22*, 287–305.
10. Ratti, C., & Crapsite, G. H. (2008). Modeling of Batch Dryers for Shrinkable Biological Materials. *Food and Bioprocess Technology, 2*, 248–256.
11. Tabarsa, M., & Zarenezhad, B. (2016). Modeling and investigating drying behavior of solid in a fluidize bed dryer. *Journal of chemical and Pharmaceutical Research, 8*, 682–689.
12. Shomali, A., & Souraki, B. A. (2019). Experimental investigation and mathematical modeling of drying of green tea leaves in a multi-tray cabinet dryer. *Heat and Mass Transfer, 55*, 3645–3659.

13. Pakowski, Z., & Mujumdar, A. S. (2006). Basic process calculations and simulations in drying. Taylors and Francis Group LLC.
14. Wami E. N., Ibrahim M. O., (2014). Model equation for heat transfer coefficient of air in a batch dryer. *International Journal of Scientific & Engineering Research* 5(11).

Design and Development of a Beach Sand Solar Collector

Animesh Borgohain and Kaustav Bharadwaz

Nomenclature

I	Rate of total radiation incident on the absorber surface
A_C	Collector area (m^2)
Q_U	Rate of useful energy collected by the air (W)
q_u	Energy per unit area (W/m^2)
Q_{COND}	Rate of conduction losses from the absorber (W)
Q_R	Rate of conduction losses from the absorber (W)
Q_{CONV}	Rate of convective losses from the absorber (W)
Q_ρ	Rate of reflection losses from the absorber (W)
Q_L	Overall heat transfer coefficient of the absorber (W)
T_C	Temperature of the collector's absorber (K)
T_A	Ambient air temperature (K)
τ	Transmittance of glass cover
Nu	Nusselt number
P	Reflection coefficient of absorber
Gr	Grashoff number
c_{pa}	Specific heat capacity of air
η	Thermal efficiency
m	Mass of air per unit time
U	Overall heat transfer coefficient
K	Thermal conductivity of air

A. Borgohain (✉)
Department of Mechanical Engineering, Assam Engineering College, Jalukbari, Guwahati 781013, India

K. Bharadwaz
Department of Material Science (Advanced Renewable and Sustainable Energy Engineering), Monash University, Melbourne, Australia

P. Mahanta et al. (eds.), *Advances in Thermofluids and Renewable Energy*, Lecture Notes in Mechanical Engineering, https://doi.org/10.1007/978-981-16-3497-0_38

1 Introduction

With the advent of the twentieth century, renewable energy sources have completely transformed the energy market. These energy sources are in-exhaustible and environment friendly compared to conventional energy sources. Renewable energy sources include solar, wind, hydel, biomass, and tidal energy. Almost 26.2% of entire electricity is generated from renewable energy sources, out of which 2.4% are from solar photovoltaic and 0.4% are from concentrated solar power [1]. Solar thermal energy systems work by capturing the heat generated from the sun, and further, this heat is used for heating and/or production of electricity. Solar collectors fall under the category of solar thermal energy systems. Solar collectors can be classified into two categories—concentrating and non-concentrating collectors. Concentrating solar collectors use reflective surfaces such as mirrors or lenses to concentrate sunlight into a particular area, where it is absorbed or further converted to heat. In a non-concentrating solar collector, the area which is collecting the energy from the sun is the same as the area which is absorbing the energy from the sun. A flat plate collector is a non-concentrating type of solar collector. It basically comprises three main components—a transparent cover, an absorber plate, and a heat transport system. The absorber plate is usually a sheet of copper or aluminum painted black to absorb the maximum amount of solar radiation which is used as a source to either heat air or water. The transparent cover is made out of either glass or plastic. Dust deposition density, angle of tilt as well as surface orientation in regard to the dominant wind direction influence glass normal transmittance [9]. Sand can be used as a medium for the transfer of heat from the absorber plate to the fluid inside the pipes. The heat transfer in the sand particles mostly takes place by conduction. The major thermal properties of sand are its specific heat capacity (c) and thermal conductivity (k). The specific heat determines the heat energy needed to change the temperature of the sand which further has an effect on the time required to reach the steady-state conditions. The thermal conductivity affects the temperature field and heat flow inside the sand at equilibrium. The major challenges that need to be taken into account while considering sand as a heat transfer media are the mass loss due to calcination and agglomeration phenomena [4].

1.1 Flat Plate Collector vs Evacuated Tube Collector

There are several advantages offered by flat plate collectors. They are comparatively easy to manufacture, and due to their simple design, the cost is also lower. They also require less maintenance due to no moving parts involved. Additionally, these collectors have the capability of capturing both beam and diffuse radiation. They offer more sunlight absorbing area. Moreover, the glass cover employed on flat plate solar collectors is tempered glass which is durable and weather resistant.

The evacuated tube collectors, on the other hand, have a complex design. The requirement for maintenance is high for these types of collectors. These collectors offer less amount of sunlight absorbing area. Overall, evacuated tube collectors are comparatively more expensive than flat plate collectors, but, however, these collectors have the capability of attaining higher temperatures and they can be installed easily on roofs.

2 Literature Review

Sun et al. 2016 [9] studied and reviewed the effects of nanostructured silicon on flexible photovoltaics, thermoelectric generators, and piezoelectric devices. They found that the optical and electrical properties of nanostructured silicon are heavily dependent on the surface effects. Surface modification in nanostructured silicon leads to enhanced supra-indirect gap absorption, reduced thermal conductivity, and induced piezoelectricity, which strengthens the ability of electricity generation for various generators. Dayan et al. 2017 mentioned that compared to a traditional parallel tube collector, a serpentine tube collector has the ability to function better in low-flow systems due to the prior start of turbulent flow. Additionally, it has also been mentioned that the difference in terms of performance between a conventional header-riser collector and a serpentine collector can be attributed to the internal heat transfer coefficient. Bhowmik et. al. 2017 [5] have used a solar reflector on the solar collector to improve its efficiency. Experiments showed that with the use of solar reflectors, the efficiency of the solar collector has increased to 10%. Diago et. al. 2015 [3] have thermally analyzed the suitability of desert sand samples as a thermal storage media. It was shown that there were relative mass losses in the sand samples during the calcination process due to the high calcite content in the sand. Further, it was seen that most of the sand particles tend to form agglomerates at higher temperatures rendering their ability to act as a heat transfer media. There were color changes from lighter tones to whiter tones which were also being observed. It suggested that it had an impact on the radiative properties of samples like solar absorptivity values. Saraf et al. 1988 [6] mentioned that the optimum tilt angle should be kept as low as 15 ° during the month of June and as high as 63 ° during the month of December to maintain the effectiveness of the solar thermal system. Gomariz et al. 2019 [8] mentioned that most of the flat plate solar collectors are used for applications such as drying of crops and space heating. Gomariz et al. 2019 [8] mentioned that serpentine type collectors have lower manufacturing costs, perform better at low flow rates and also, simplify the hydraulic connection of many collectors.

3 Theoretical Formulations

The energy balance on the absorber is obtained by equating the total heat gained to the total heat lost by the heat absorber of the solar collector. Therefore,

$$I\,A_c = Q_U + Q_{\mathrm{COND}} + Q_{\mathrm{CONV}} + Q_R + Q_\rho \tag{1}$$

where

I = rate of total radiation incident on the absorber's surface (W/m^2).
A_c = collector area (m^2).
Q_U = rate of useful energy collected by the air (W).
Q_{COND} = rate of heat losses from the absorber (W).
Q_{CONV} = rate of convective losses from the absorber (W).
Q_{R} = rate of longwave re-radiation from the absorber (W).
Q_{P} = rate of reflection losses from the absorber (W)

$$Q_L = Q_{\mathrm{COND}} + Q_{\mathrm{CONV}} + Q_R \tag{2}$$

If τ is the transmittance of the top glazing and I_T is the total solar radiation incident on the top surface, therefore,

$$I\,A_c = \tau\,I_T\,A_c \tag{3}$$

The reflected energy from the absorber is given by the expression

$$Q_\rho = \rho\tau\,I_T\,A_c \tag{4}$$

where

ρ = the reflection coefficient of the absorber.
Substitution of Eqs. (2), (3), and (4) in Eq. (1) yields:

$$\tau\,I_T A_c = Q_U + Q_L + \rho\tau\,I_T\,A_c$$

$$Q_R Q_U = \tau\,I_T\,A_c(1 - \rho) - Q_L$$

For an absorber $(1-\rho) = \alpha$, where α is solar absorptivity and hence,

$$Q_U = (\alpha\tau)\,I_T\,A_c - Q_L \tag{5}$$

Q_L is composed of different convection and radiation parts.
It is presented in the following form

$$Q_L = U_L A_C (T_C - T_A) \tag{6}$$

T_A = ambient air temperature (K).

From Eqs. (5) and (6), the useful energy gained by the collector is expressed as:

$$Q_U = (\alpha\tau)\, I_T A_C - U_L A_C(T_c - T_a) \tag{7}$$

Therefore, the energy per unit area

$$q_u = (\alpha\tau) I_T - U_L(T_C - T_A) \tag{8}$$

where

U_L = overall heat transfer coefficient of the absorber.

T_C = Temperature of the collector's absorber (K).

If Q_g is the heated air leaving the collector is at collector temperature, the heat gained by the air is:

$$Q_g = m_a C_{pa}(T_c - T_a) \tag{9}$$

where

m_a = mass of air per unit time (kg/sec).

C_{pa} = specific heat capacity of air (kJ/kg-K).

The collector heat removal factor F_R can be given by the equation:

$$F_R = m_a C_{pa}(T_c - T_a) / A_C[(\alpha\tau)\, I_T - U_L A_C(T_C - T_A)] \tag{10}$$

$$Q_R Q_g = A_C F_R[(\alpha\tau)\, I_T - U_{L.} A_C(T_C - T_A)] \tag{11}$$

The thermal efficiency of the collector is given by the collector

$$\eta = Q_g / A_c I_t \tag{12}$$

4 Design and Methodology

The main motive of the project is the design and construction of a beach sand solar collector. The collector as mentioned beforehand is of two main types: Flat plate collector and Evacuated tube collector. The collector that will be employed in the project will be a serpentine flat plate collector. Serpentine flat plate collector consists of single continuous tubes running from the inlet to the outlet of the collector. The collector will be made out of concrete because it is durable, maintenance-free, and also exhibits good thermal storage qualities because of its high specific heat capacity and density. The inner part of the collector will be painted matt black in order to absorb the maximum amount of sunlight. The collector will be inclined at an angle equivalent to the latitude of the location to receive the maximum amount of radiation from the sun. Being a serpentine flat plate collector, single continuous tubes would run from the inlet to the outlet of the collector. Copper tubes would be the best choice in this case because of its high temperature resistance, high mechanical strength, and also, it can be easily bent into tubes. In the proposed project, beach sand would be used for the generation of heat. A single-stage centrifugal air blower of power 200 Watts will be used for blowing the heated air from the sand into the inlet of the collector. The heated air from the inlet would move through the tubes into the outlet from where it can be stored or collected for different household or commercial applications. This nonlinear movement of air through the copper tubes has another advantage in the form that air will take more time to travel and thereby help in increasing and maintaining the temperature of the heated air. The temperature at the entry and exit of the solar collector will be measured using a digital LCD thermometer.

4.1 Components of a Beach Sand Solar Collector

A. **Solar Collector**: The heat absorber of the solar collector can be constructed using a 1 mm thick aluminum plate, painted black. The collector will be insulated with concrete of thickness 2.5 cm and having a thermal conductivity of 0.6 W/mK on all sides. Copper tubes would run in a single serpentine manner from the inlet to the outlet of the collector.

B. **Cover Plate**: The cover plate is made of transparent glass which transmits radiation to the absorber. It also prevents radiative and convective heat losses from the surface. Glass and plastic are the two most common materials for making cover plates. Plexiglass having a length of 170 cm, breadth of 110 cm, and a thickness of 0.3 cm can be employed. Plexiglass is more durable, fracture-resistant, and more resistant to erosion compared to normal glass. In order to attain maximum efficiency, absorption, and reflection should be less while transmission should be as high as possible. Plexiglass has an overall light transmission of 92%.

C. **Absorber Plate:** The absorber plates are flat, corrugated, or groove plates to which the tubes are attached. It is either made of copper, steel, or aluminum which have high thermal conductivity. The surface of the absorber plate is covered with black material in order to increase its absorptance. In order to maximize the solar absorptance, a selective coating can be applied on the absorber plate. Aluminum can prove to be much better compared to copper or steel in terms of heat dissipation capacity.

D. **Insulation:** Insulation is used to prevent heat loss from the system. Different types of insulating materials exist in the form of foam rubber, polystyrene, mineral wool, etc. Insulation can be applied over the tubes running inside the collector for any undesirable heat loss as well as over the outer portion.

E. **Air Blower:** The air blower that will be used will be a single-stage centrifugal air blower having a rated power of 200 W and a maximum speed of 2800 rpm.

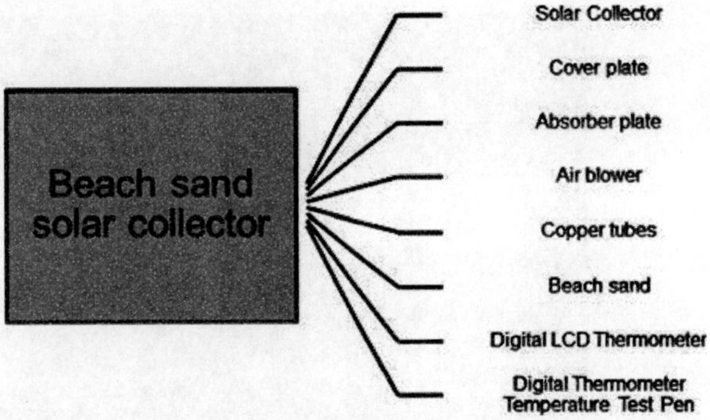

Fig. 1 Beach sand solar collector theoretical model

The main aim of an air blower will be to force the heated air from the sand up through the collector.

F. **Copper Tubes:** The copper tubes are mounted directly to the absorber plates by soldering, brazing, or bonding. The copper tubes are used to carry the heat transfer fluid such as water or air.

G. **Beach sand:** Sand will be the source of heat generation. The concept of generation of heat from beach sand is quite a new concept. The general properties of beach sand are: However, the heat capacity of the sand will vary depending upon the relative proportions of the constituent materials making up the sand grain. Dry quartz sand which makes up the desert has a specific heat of about 0.19 cal/gram °C. Beach sand for instance is not entirely composed of quartz. It also contains considerable amounts of calcium carbonate, feldspar, gypsum, etc. But, however, wet beach sand has a high heat capacity since the specific heat of water is higher compared to any other substances which are around 4,186 Joules/kg-°C.

H. **Digital LCD thermometer**: These will be used for measuring the temperature at the inlet and outlet of the collector. The specifications are listed below: (Figs. 1 and 2; Tables 1 and 2).

5 Results and Discussion

A flat plate solar collector consists of a glazing surface, and an absorber plate painted black to maximize the heat input and an insulated cover (Figs. 3 and 4 ; Tables 3 and 4).

The glazing surface is plexiglass which has a high transmissivity value of about 92%.

If the angle of incidence is 60 °, then the absorptance of the black plate will be 0.85.

The solar irradiance I_{in} incident on the cover glass is given by

$$I_{\text{in}} = I_b \cos(\theta) + I_d \quad (13)$$

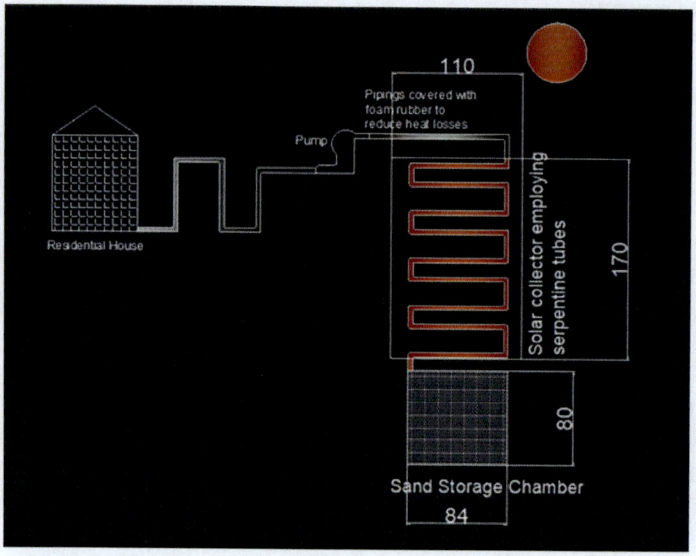

Fig. 2 A 2-D model of the proposed beach sand solar collector

Table 1 Properties of beach sand

Thermal conductivity: 0.6 W/m/C°
Electrical conductivity: 41.37 ± 1.25μS/cm
Specific gravity: 2.6–2.8
Fineness modulus: 2.64
Dry bulk density: 1468 kg/cm
pH: 8.6
Moisture content: (0.5–2) %
Dry density: (100–110) pcf
Optimum moisture content: (8–10)%
Liquid limit: 37%
Plastic limit: 16%

where

I_b= beam solar irradiance

θ = angle of incidence.

I_d = diffuse irradiance.

If there is only one glazing surface, then the solar irradiance on the black plate will be given by

$$(\theta) = \cos\theta \cdot \mathrm{I_{b+}}\tau \cdot \mathrm{I_d}$$

where τ_m is the mean value of $\tau(\theta)$.

Table 2 Specification of the digital LCD thermometer

Operating temperature: 40–70 °C
Measuring humidity range: 10–99% RH
Humidity accuracy: 5%
Humidity display resolution: 1% RH
Temperature accuracy: 1°C
Operating voltage: 1.5 V, or LR44 batteries (included)
Dimensions: 48 × 28.5 × 15.2 mm
LCD dimension: 40 × 17 mm
Installation dimensions: 46 × 27 × 13.5 mm
Color: White

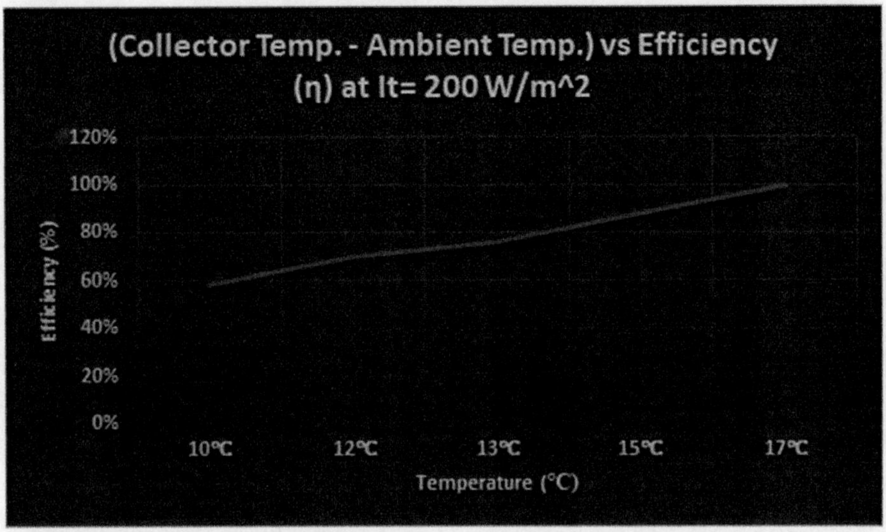

Fig. 3 Efficiency at different values of collector temperature (in °C) at It = 200 W/m^2

The glazing cover, i.e., the plexiglass acts as a black body for longwave radiation. We have the emittance of glass cover (plexiglass) = 0.85 and the emittance of the absorber plate for longwave radiation depend upon whether the surface is selective or non-selective. So, we have an emittance of absorber plate = 0.92 for a non-selective surface and 0.10 for the selective surface.

Let

$$T_a = \text{Ambient temperature}$$
$$T_b = \text{Black plate (absorber) temperature}$$
$$T_c = \text{Glass cover temperature}$$

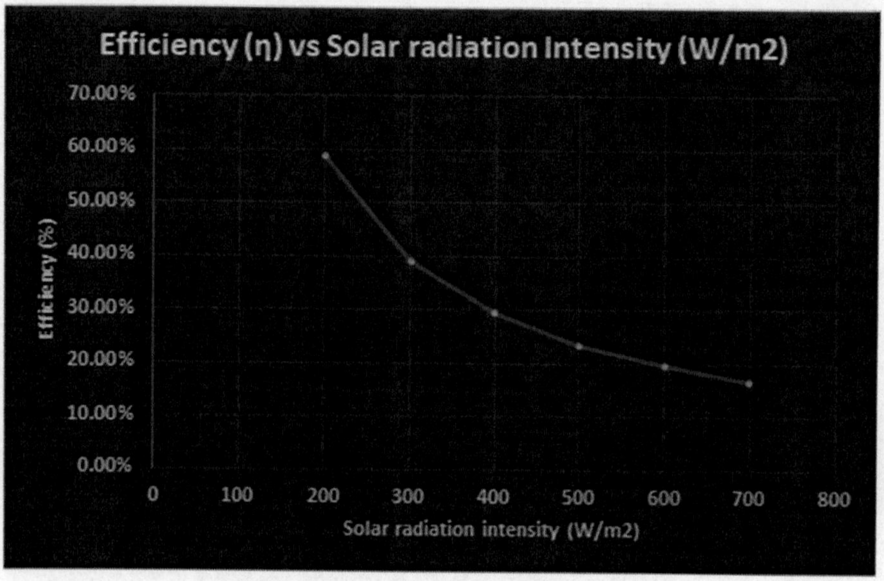

Fig. 4 Efficiency vs Solar Radiation Intensity at Normal Temperature

Table 3 The efficiency of solar radiation intensity

Solar radiation Intensity (W/m^2)	200	300	400	500	600	700
Efficiency (η)	58.7%	39.17%	29.38%	23.5%	19.6%	16.79%

Table 4 Efficiency at different values of collector temperature (in °C) at $I_t = 200$ W/m^2

Collector temperature (T_C)—ambient temperature (T_a)	10°C	12°C	13°C	15°C	17°C
Efficiency at $I_t = 200$ W/m^2	58%	70%	76%	88%	99%

Heat is lost from the absorber plate to the glass cover by convection and radiation.
For free convection, the Nusselt number (Nu) is given as.
$\mathrm{Nu} = 0.152Gr^{0.281}$ for horizontal plate.
$\mathrm{Nu} = 0.093Gr^{0.310}$ for plates tilted at an angle of 45°.
Grashoff number is given by $= g \cdot \beta \cdot (Tb - Tc) \cdot L^3/U^2$.

where g = 9.81 m/s^2 (acceleration due to gravity).

Coefficient of thermal expansion, $\beta = 1/T = 0.0034/K$ (at normal standard conditions)

$$T_b - T_c = (80 - 35)°\mathrm{C} = 45°\mathrm{C} = 318\,K$$
$$L = 0.075m \text{ (spacing between plates)}$$

Kinematic Viscosity, $\nu = 1.562 \times 10^{(-5)}$ m^2/sec (at 25 °C which is the standard temperature)

$$Gr = (9.81 \times 0.0034 \times 318 \times 0.075^3)/(1.562 \times 10^{-5})^2$$
$$= 18.3 \times 10^6$$

Assuming a tilt angle of 45°, then

$$Nu = 0.093 Gr^{0.310}$$
$$= 0.093 \times (18.3 \times 10^6)^{0.310}$$
$$= 16.58$$

$$\text{Nu} = h \times L/k$$
$$h = \text{Nu} \times k/L$$
$k = 0.02624\, W/mK$ is the thermal conductivity of air at $25\,^\circ C$
$$h = (16.58 \times 0.02624)/0.075 = 5.8 W/m^2K$$

If $I_t = 200 W/m^2$ (transmitted solar irradiance)

$$1/U_l = 1/h \times A_c + k \times A_c/L + 1/h \times A_c$$
$$1/U_l = 0.828$$
$$U_l(\text{overall heat transfer coefficient}) = 1.21 W/m^2K$$
$$Q_l = U_l \times A_l \times (T_c - T_a) = 1.21 \times 1.83 \times (35 - 25) = 22.14\, W$$

If $I_t = 200$ W/m^2.
Rate of useful energy collected by air

$$Q_u = (\alpha \times \tau) \times I_t \times A_c - Q_l$$
$$= (0.92 \times 0.85) \times 200 \times 1.83 - 22.14$$
$$= 264.072\, W$$

The collector heat removal factor, F_r is given by

$$F_r = \left(m_a \times C_{pa} \times (T_c - T_a)\right)/A_c \times (\alpha \times \tau \times I_t - U_l \times (T_c - T_a))$$
$$F_r = (0.0214 \times 1005 \times (35 - 25)/1.85 \times (0.85 \times 0.92 \times 200 - 1.21 \times (35 - 25))$$
$$F_r = 0.815$$

Now the heated air leaving the collector is denoted by Q_g, so heat gained by air

$$Q_g = A_c \times F_r \times ((\alpha \times \tau) \times I_t - U_l \times A_c \times (T_c - T_a))$$
$$Q_g = 1.83 \times 0.815((0.85 \times 0.92) \times 200 - 1.21 \times 1.83 \times (35 - 25))$$
$$Q_g = 215\ W$$

The efficiency of the collector

$$\eta = Q_g / A_c \times I_t$$
$$= 215/1.83 \times 200$$
$$= 58.7\%$$

When performing thermal analysis, the factors which affect the solar collector performance include the operating temperature, the solar radiation intensity, and the severity of the ambient outdoor temperature. From Fig. 3, it can be observed that when the difference between the collector and ambient temperature is high, the efficiency of the solar collector increases. The effectiveness of a solar collector depends upon whether they can produce a temperature which is sufficient to meet the needs of any attached heating job at any moment. It can be further observed that with the increase in solar radiation intensity, the efficiency of the solar collector falls. When designing solar heating systems, it is always paramount to maintain a balance between temperature and energy output. If a proper balance is not maintained, then it can ultimately have negative influence over the overall performance level.

6 Conclusion

After completion of the experiment, it can be concluded that with the increase in temperature, the collector efficiency also increases. Moreover, the angle at which the collector is inclined also plays a critical role. In order to attain maximum amount of solar radiation, the collector should be inclined at an angle equal to the latitude of the place. With a solar radiation intensity of 200 W/m2, an overall solar collector efficiency of 58.7% can be achieved. Wet beach sand has a high heat retention capacity, so, it has a potential to generate a good amount of heat which can be used for residential/commercial purposes in a sustainable manner.

References

1. Roberts, D. (2019). The global transition to clean energy, explained in 12 charts, Vox, June 26, 2019. Retrieved from: https://www.vox.com/energy-and-environment/2019/6/18/18681591/renewable-energy-china-solar-pv-jobs
2. Sun, B., Shao, M., & Lee, S. (2016). Nanostructured silicon used for flexible and mobile electricity generation. *Advanced Materials*.

3. Miguel Diago, Alberto Crespo Iniesta, Thomas Delclos, Tariq Shamim, Nicolas Calvet, "Characterization of desert sand for its feasible use as thermal energy storage medium" The 7th International Conference on Applied Energy – ICAE2015
4. Dayan, M., Klein, S., & Beckman, W. (1998). Analysis of serpentine collectors in low flow systems. Solar Energy Laboratory University of Wisconsin-Madison, 1500 Engineering Drive Madison, WI 53706.
5. Bhowmik, H., & Amin, R. (2017). Efficiency improvement of flat plate collector using reflector. Energy Reports, vol. 3, pp. 119–123. Retrieved from: https://doi.org/10.1016/j.egyr.2017.08.002
6. Saraf, G.R., & Hamad, F-A-W. (1988). Optimum tilt angle for a flat-plate.
7. Aldabbagh, L. B. Y., Egelioglu, F., & Ilkan, M. (2010). Single and double-pass solar air heaters with wire mesh as packing bed. *Energy, 35*(9), 3783–3787.
8. Francisco Plaza Gomariz, Jose Manuel Cezudo Lopez, Fernando Dominguez Munoz. (Feb, 2019) An analysis of low flow for solar thermal system for water heating. Solar Energy *179*, 67–73. Retrieved from: https://doi.org/10.1016/j.solener.2018.12.060
9. Sun, B., Shao, M., & Lee, S. (2016). Nanostructured silicon used for flexible and mobile electricity generation. advanced materials. Retrieved from: https://doi.org/10.1002/adma.201601012

Spectrum-Based 3D Thermal Modeling of the PV Module: Simulation and Experimental Study

H. Brahma and N. Sarmah

Nomenclature

AM	Air mass
c_p	Specific heat capacity (J/kg.K)
d	Thickness (cm)
EQE	External quantum efficiency
EVA	Ethyl–vinyl acetate
GTI	Global tilted irradiance (W/m^2)
$G_{pyranometer}$	Global irradiance from pyranometer (W/m^2)
G_{SMARTS}	Global irradiance from SMARTS (W/m^2)
$G_{spectrometer}$	Global irradiance from spectrometer (W/m^2)
h	Convective heat transfer coefficient (W/m^2 K)
k	Thermal conductivity (W/m K)
n	Normal vector on the boundary
q	Conduction heat flux (W/m^2)
q_o	Inward heat flux (W/m^2)
q_r	Radiation heat flux (W/m^2)
Q_{ted}	Thermoelastic damping (W/m^3)
Q	Heat source (W/m^3)
RH	Relative Humidity (%)
S	Second Piola–Kirchhoff stress tensor (Pa)
T	Temperature (K)
T_b	Module back temperature (K)
T_c	Module cell Temperature (K)
U_{trans}	Velocity vector of the translational motion (m/s)
Ws	Wind speed (m/s)

H. Brahma (✉) · N. Sarmah
Department of Energy, Tezpur University, Sonitpur, India

P. Mahanta et al. (eds.), *Advances in Thermofluids and Renewable Energy*, Lecture Notes in Mechanical Engineering, https://doi.org/10.1007/978-981-16-3497-0_39

Greek Symbols

α	Coefficient of the thermal expansion (l/K)
ρ	Density (kg/m^3)

Subscripts

amb	Ambient
b_sim	Back simulation
c_exp	Cell experimental
c_sim	Cell simulation
conv	Convection
ext	External
g_sim	Glass simulation

1 Introduction

The growth rate in energy production from solar photovoltaic (PV) in India has started in the recent decade, encouraged by the Target 2022. On average yearly, 64% increase in solar power generation was reported during the year 2013–2017. Total installed solar capacity of 32.5 GW was announced at the end of 2019 as reported by IEA [1]. Ramachandra et al. [2] noted that based on the geographical location, rotation of the earth, earth–sun movement, and atmospheric attenuations, the incident solar radiation on the earth's surface defines the solar hotspot of the location. So, to support the installation of a PV system in a specific location, the pre-installation investigation of PV performance for a location based on its atmospheric conditions and geographical location is important as suggested by Kant et al., and Weiss et al. [3, 4]. Many a time, while calculating the energy output from the PV system, the effect of the spectrum of radiation is neglected which may lead to inaccurate estimation of energy output as reported in the work of Okullo et al., Alonso-Abella et al., and Gottschalg et al. [5–7]. The spectrum of radiation changes with the time of the day as the path traveled by the sun shifts from sunrise to sunset whereas the standard calculation only considers the absolute broadband irradiation at AM1.5 spectrum. Nagae et al. [8] noted that the spectral irradiance and temperature are the main parameters which affect the performance evaluation of the PV module. King et al. [9] showed that the temperature has a negative impact on the performance of the PV module. Jonas and Underwood and Lobera and Valkealahti [10, 11] studied the cell temperature varying dynamically with the time of the day depending on the integrated power of the spectrum of radiation, atmospheric conditions, material properties, configurations, and type of PV module. Therefore, it is important to study the impact of transient temperature on the performance of the PV module. Amstrong and Hurley and Skoplaki et al. [12–14] developed empirical formulas, and kant et al., Du et al., Aly et al., and Barroso et al. developed computational modeling [3, 15–17] for thermal model by considering assumptions to investigate the effect of temperature on the PV performance.

In this work, a spectrum-based 3D thermal model has been proposed to predict the transient cell temperature of the PV module as a function of varying atmospheric parameters (ambient temperature, spectral irradiance, relative humidity, and wind speed). The thermal model analysis of the PV module has been done using COMSOL Multiphysics, heat transfer module. The model is validated with the experimental results. In addition, to examine the effect of integrated power as heat on solar cell, a comparative study has been carried out considering the spectrum of radiation-based heat power and solar radiation-based heat power.

2 Materials and Methods

Monocrystalline PV module was mounted in a fixed open rack outdoor at an angle 26 ° (coordinates of the location under study: 26.7 °N and 92.8 °E) facing toward south. The ambient parameters such as temperature, relative humidity, and wind speed were recorded using a *k*-type thermocouple, hygrometer, and anemometer, respectively. The front and back temperatures of the module were measured using *k*-type thermocouples connected in the Agilent 34972A data logger. The variation in the spectrum of radiation and solar radiation was measured using the BLUE-Wave spectrometer and Kipp and Zonen CMP3 pyranometer, respectively. The experiment was conducted during the winter season on clear sky day from 7:30 till 16:30 with 5 min interval in order to obtain the range of variation in spectrum throughout the day.

The simulation tools employed are SMARTS2.9.5 (Simple Model for the Atmospheric Radiative Transfer of Sunshine version 2.9.5) developed by Gueymard [18, 19] for the generation of the synthetic spectrum and COMSOL Multiphysics version 5.4 for design and development of the 3D thermal model. The real-time diurnal variations in the atmospheric parameters such as relative humidity, ambient temperature, and wind speed are fed as an input to both SMARTS and COMSOL Multiphysics. The generated spectrum from SMARTS for the time of the day is multiplied with the EQE of the monocrystalline silicon solar cell to obtain the spectrum used in electricity generation, and the remaining spectrum is assumed to generate heat in the PV module. Thereafter the part of spectrum which contributes to heat generation in PV module is set to act as the heat source in the thermal model.

The cell temperature of the module can be equated as in Eq. (1) as proposed by King et al. [14]:

$$T_c = T_b + \frac{G}{1000}\Delta T \tag{1}$$

Table 1 Material properties of the materials used in the thermal model

Material	k(W/m K)	ρ(kg/m^3)	c_p(J/kg K)	d(cm)
Glass	2	2450	500	0.32
EVA	0.29	935	2400	0.1
Silicon	130	2329	700	0.03
Tedlar	0.14	1200	1250	0.05

where T_c is the cell temperature of the module (°C), T_b is the temperature of the module back surface (°C), G is the measured solar irradiance (W/m^2), and ΔT is the temperature difference between the cell and the back surface at reference solar irradiance.

In 3D thermal modeling, the geometry of the PV module is designed having the same dimension (64.8 x 53 cm) as the module used in the experiment. The layers of the module arranged in the top to bottom order are glass, EVA sheet, solar cells, EVA sheet, and backsheet (Tedlar).The geometry also includes the aluminum frame; however, the electrical contacts such as fingers and busbars are not included in the model to simplify the designed model. The thermophysical properties of the materials used in simulation of the thermal model are presented in Table 1.

The module is modeled using heat transfer in solids. The heat transfer in solids, Eq. (2), is expressed by:

$$\rho c_p\left(\frac{\partial T}{\partial t} + u_{\text{trans}} \cdot \nabla T\right) + \nabla.(q + q_r) = Q_{\text{ted}} + Q \tag{2}$$

here, T is the absolute temperature (K), u_{trans} is the velocity vector of the translational motion (m/s), q is the heat flux by conduction (W/m^2), q_r is the heat flux by radiation (W/m^2), and Q is the additional heat source (W/m^3). The term $Q_{ted} = -\alpha T : \frac{dS}{dt}$ is the thermoelastic damping (W/m^3) which accounts thermoelastic effect in solids, α is the coefficient of the thermal expansion (l/K), and S is the second Piola–Kirchhoff stress tensor (Pa).

The boundary conditions are set for thermal analysis by defining heat flux at the front surface and the back surface of the module. The governing equation for the heat flux boundary condition is equated in Eq. (3):

$$-n \,.\, q = q_o \tag{3}$$

where n is the normal vector on the boundary.

$$q = -k \, \nabla T \tag{4}$$

$$q_o = h\,(T_{ext} - T) \tag{5}$$

where q_o is the inward heat flux (W/m^2) normal to the boundary, h is the convective heat transfer coefficient (W/m^2 K), T_{ext} and T are the external temperature and boundary temperature, respectively. The convective heat transfer considered is as follows:

$$h_{\text{conv}} = 6W_s + 4 \tag{6}$$

where$h = h_{conv}$ and W_s is the wind speed (m/s).

The gaping between each layer of the module has been neglected because the size of the gapping is comparatively very small compared to the overall size of the module as noted in the work of Du et al. [15]. In order to investigate the effect of spectrum variation on the thermal performance of the PV module, the part of the spectrum which remains unused during electricity generation is considered as the power to heat source in the thermal analysis model. Solar cells are considered as the heat source. Moreover, the effect of solar radiation when considered as a heat source is also modeled for thermal performance evaluation. However, unlike the case of spectrum-based model where the spectrum of solar radiation which accounts to heat is considered, in the case of the solar radiation-based model, the whole broadband solar radiation is considered as the heat source.

Uniform physics-controlled finer mesh has been used to create a mesh in the designed model as shown in Fig. 1. Time-dependent solver with time-stepping backward differentiation formula (BDF method) is used to evaluate the model. This method solves implicit equation and known for their stability. The maximum and minimum BDF orders are 2 and 1, respectively, with event tolerance of 0.01. The time period for the computation is from 7:45 till 15:55 in steps of 5 min. Therefore, the computation was run using time-dependent solver with time range of (0, 0.08, 8.2) in hours.

Fig. 1 3D mesh design of the PV module

3 Results and Discussion

The simulated and experimental GTI values are compared as shown in Fig. 2. The percentage change in GTI obtained through integration of the spectral irradiation generated using SMARTS compared to the measured spectrum using spectrometer, and the percentage change between the spectral irradiation GTI generated using SMARTS and the broadband radiation obtained using pyranometer was found to be 1.5%. The percentage deviation in the GTI from the pyranometer and spectrometer was obtained to be 0.06%. The modeled GHI is validated with the experimental GTI, and Fig. 3 represents the linear regression analysis that was performed to show a good agreement between the modeled GTI (generated spectrum from SMARTS) and the measured GTI with R^2 adjusted value of 0.98.

Temperature distribution of the PV module using COMSOL Multiphysics for different time of the day from morning till evening has been shown in Fig. 4. It could be seen that with variation in time and atmospheric parameters throughout the day the temperature profile in each layer of the PV module changes. As the sun moves through its path from sunrise to sunset, the temperature of the PV module surfaces increases gradually and reaches a maximum during solar noon and then decreases being subjected to the atmospheric parameters including wind speed, relative humidity, ambient temperature, and spectral irradiance.

The average cell temperature (T_{c_sim}), average front glass surface temperature (T_{g_sim}), and average back surface temperature (T_{b_sim}) were obtained to be 48.2 °C, 47.5 °C, and 47.4 °C, respectively, for the recorded average ambient temperature (T_{amb}) was 25 °C. The front surface temperature of the module was higher compared

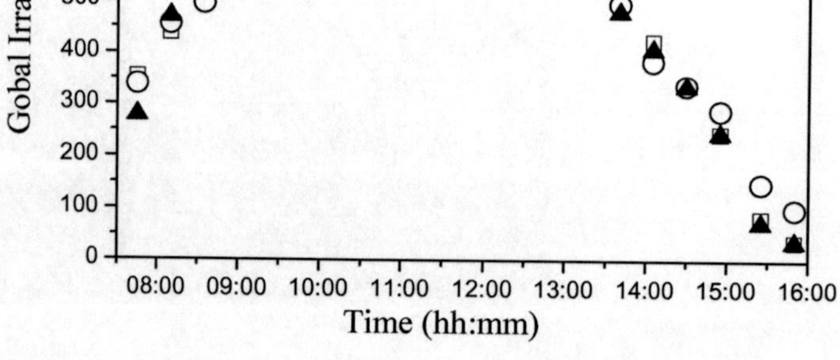

Fig. 2 Comparison of the global radiation measured using pyranometer, spectrometer, and simulated spectrum from SMARTS

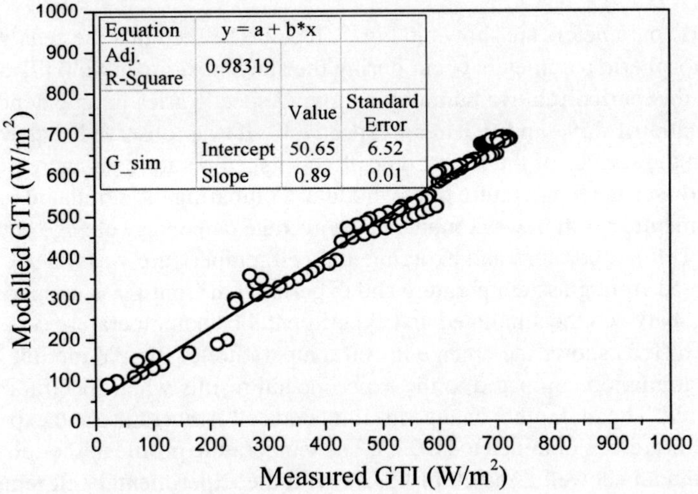

Fig. 3 Linear fit of modeled GTI versus measured GTI using linear regression analysis

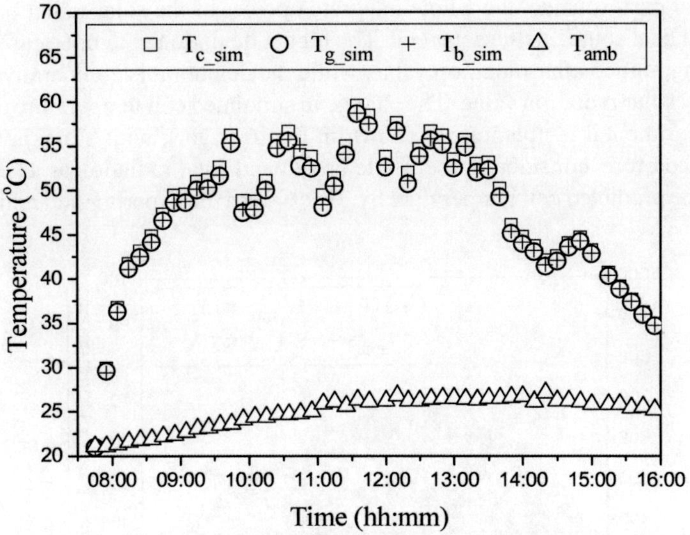

Fig. 4 Temperature distribution of the PV module during different time of the day

to the back surface temperature this might be due to the tilt angle of the module and free stand mount of the PV module with wind flowing in both front and back surface of the module. The average percentage change between the simulated cell temperature and front glass temperature was found to be 1.47%, between the cell temperature and back temperature was 1.68%. The variation in cell temperature of the module with the change in time of the day and the respective variations in atmospheric parameters are shown in Fig. 5. It could be seen that the peak variations in the atmospheric parameters occur during the sun's position at 8:30 till 14:30 h,

during this period relative humidity and wind speed varies to an extent with the average value of 55% and 0.3 m/s, respectively. It was observed that with wind flowing on the surface of the PV module affects its temperature, an increase in wind speed decreases the temperature of the module. Comparing the simulated result with the experimental result, it was found that the average percentage change between the simulated cell temperature and experimental cell temperature was 4.54%, between the simulated front glass temperature and experimental front glass temperature was 2.54% and between the simulated and experimental back temperature was 2.63%.

Figure 6 (left) shows the change in cell temperature of the PV module obtained from the simulation compared to the experimental results when spectral irradiance is considered. The maximum change in simulated cell temperature and experimental cell temperature was obtained to be 0.23. The temperature profile of the cell obtained from the model showed a match of 95.6% with the experimental cell temperature. Thus, it is evident that for better prediction of the thermal performance of the PV module spectrum-based thermal model is a better option.

Moreover, for a better understanding of the effect of broadband solar radiation on the thermal performance, the whole integrated power of the solar radiation was fed as an input heat source to the solar cell. The higher deviation in temperature could be seen during higher solar radiation value, while the deviation is comparatively lower during low solar radiation value. The change in simulated cell temperature compared to the experimental temperature is shown in Fig. 6 (right), with a maximum value of 0.75. Therefore, considering the whole broadband solar radiation as a heat source deviates the predicted cell temperature by 31.9% from the experimental temperature.

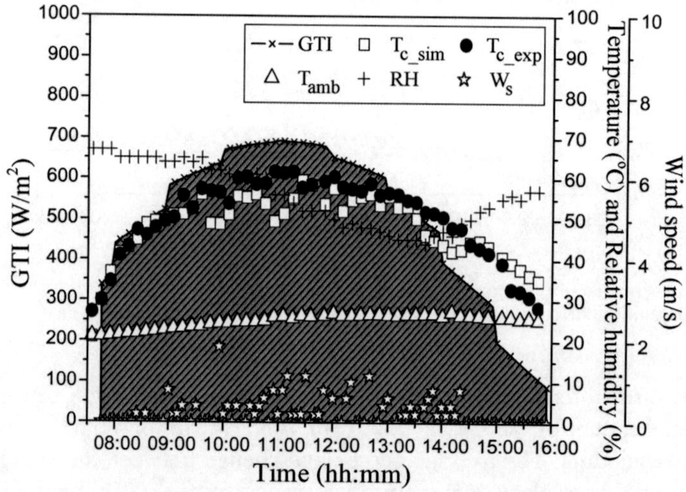

Fig. 5 Variation in cell temperature and the respective atmospheric parameters changing with time

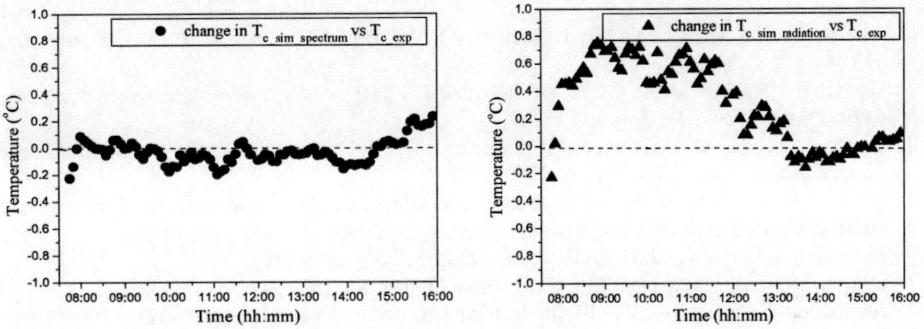

Fig. 6 Comparison between the experimental cell temperature, T_{c_exp} and the simulated cell temperature considering: the spectrum, $T_{c\_sim\_spectrum}$ (left) and the solar radiation, $T_{c\_sim\_radiation}$ (right)

4 Conclusion

In the present work, a 3D thermal model has been successfully developed considering the real-time variations in parameters such as the spectrum of radiation, ambient temperature, relative humidity, and wind speed. The developed model outputs the transient cell temperature which has a great contribution in estimating the thermal loss associated with the PV system. It was observed that the temperature in each layer of the PV module varies with the parameters under consideration, and comparatively PV cells have the maximum temperature. The temperature of the PV module is comparatively higher in the center compared to the corner parts of the PV module. The average cell temperature was recorded to be 48.2 °C, average front surface temperature 47.5 °C and average back surface temperature 47.4 °C. The results obtained from the simulated were validated with the experimental results. A match of 95.7% has been obtained between the simulated cell temperature and the experimental cell temperature while considering the spectrum of radiation as the source of heat power in the PV cell. Moreover, the whole broadband of solar radiation as heat power showed a deviation of 31.9% in the simulated cell temperature from the experimental cell temperature.

References

1. IEA, India, (2020). Energy Policy Review.
2. Ramachandra, T., Jain, R., & Krishnadas, G. (2011). *Renewable and Sustainable Energy Reviews, 15*(6), 178–3186.
3. Kant, K., Shukla, A., Sharma, A., & Biwole, P. H. (2016). *Solar Energy, 134*, 147–155.
4. Weiss, L., Amara, M., & Ménézo, C. (2016). *Prog Photovolt, 24*(1), 12–27.
5. Okullo, W., Munji, M., Vorster, F., & Van Dyk, E. (2011). *Solar Energy Materials and Solar Cells, 95*(2), 759–764.

6. Alonso-Abella, M., Chenlo, F., Nofuentes, G., & Torres-Ramírez, M. (2014). *Energy, 67*, 435–443.
7. Gottschalg, R., Infield, D. G., & Kearney, M. J. (2003). *Solar Energy Materials and Solar Cells, 79*(4), 527–537.
8. Nagae, S., Toda, M., Minemoto, T., Takakura, H., & Hamakawa, Y. (2006). *Solar Energy Materials and Solar Cells, 90*(20), 3568–3575.
9. King, D. L., Boyson, W. E., & Kratochvil, J. A. (2002). In Conference Record of the Twenty-Ninth IEEE Photovoltaic Specialists Conference, 2002, (pp. 1356–1361). IEEE.
10. Jones, A., & Underwood, C. (2001). *Solar Energy, 70*(4), 349–359.
11. Lobera, D. T., & Valkealahti, S. (2013). *Solar Energy, 93*, 183–194.
12. Armstrong, S., & Hurley, W. (2010). *Applied Thermal Engineering, 30*(11–12), 1488–1495.
13. Skoplaki, E., Boudouvis, A., & Palyvos, J. (2008). *Solar Energy Materials and Solar Cells, 92*(11), 1393–1402.
14. King, D. L., Kratochvil, J. A., & Boyson, W. E. (2004). Photovoltaic array performance model. Department of Energy, United States.
15. Du, Y., Tao, W., Liu, Y., Jiang, J., & Huang, H. (2017). *Solar Energy, 146*, 257–263.
16. Aly, S. P., Ahzi, S., Barth, N., & Figgis, B. W. (2018). *Solar Energy Materials and Solar Cells, 180*, 289–302.
17. Barroso, J. S., Barth, N., Correia, J., Ahzi, S., & Khaleel, M. (2016). *Solar Energy Materials and Solar Cells, 148*, 73–86.
18. Gueymard, C. (1995). SMARTS2: A simple model of the atmospheric radiative transfer of sunshine: Algorithms and performance assessment. Florida Solar Energy Center Cocoa, FL.
19. Gueymard, C. (1995). Simple model of the atmospheric radiative transfer of sunshine, version 2 (SMARTS2): Algorithms description and performance assessment, Florida Solar Energy Center.

Duckweed Assisted Sediment Microbial Fuel Cell for Powering Small Scale Devices

Arup Dutta, Lepakshi Barbora, Aparupa Thakuria, Pranab Goswami, and Davard Stom

Nomenclature

DO	Dissolve oxygen (mg/L)
R_{int}	Internal Resistance (Ohm)
R_{Load}	Load Resistance (Ohm
SMFC	Sediment Microbial Fuel Cell
V	Voltage (Volts)
V_{ocv}	Open Circuit Voltage (mV)
V_{Load}	Voltage across Load (mV)

Symbols

Ω	Resistance (Ohm)
°C	Temperature (Degree Celsius)

A. Dutta · L. Barbora (✉) · A. Thakuria · P. Goswami
Centre for Energy, IIT Guwahati, Assam 781039, India
e-mail: lopab@iitg.ac.in

P. Goswami (✉)
Department of Biosciences and Bioenegineering, IIT Guwahati, Assam 781039, India
e-mail: pgoswami@iitg.ac.in

D. Stom
Laboratory of the Aquatic Toxicology, Irkutsk State University, Irkutsk, Russia

P. Mahanta et al. (eds.), *Advances in Thermofluids and Renewable Energy*, Lecture Notes in Mechanical Engineering, https://doi.org/10.1007/978-981-16-3497-0_40

1 Introduction

Energy generation in the form of electricity has been a major concern for the society since several decades. Meanwhile there have been many techniques or procedures for the production of electricity. However, there is the need of finding alternatives for energy generation since the traditional resources like fossil fuels are depleting due to the ever increasing need [1]. The renewable energy sources possess the potential to replace the conventional sources for production of energy. In this regard, the fuel cells are the best example for the production of electricity of which, the microbial fuel cells (MFCs) are one of the types.

A fuel cell is an electrochemical cell that converts the chemical energy of fuel into electricity by virtue of an electrochemical reaction in the presence of an oxidation and a reduction catalyst [2]. In an MFC, the microorganisms convert the chemical energy associated with organic substances to electricity and if the organic source is a water body then during the electrochemical reactions the source is simultaneously bio-remediated by the microorganisms [3].

Sediment microbial fuel cells (SMFCs) are such type of MFCs where along with electricity generation, bio-remediation also takes place [4] (Fig. 1). The SMFCs differ from the basic MFC technology in that they are completely anoxic and lack a membrane to separate the two electrode chambers [1]. They possess an extensive range of operational parameters and have the ability to utilize a wide variety of organic substances as fuel, such as glucose, glutamic acid, river water, cysteine, acetate and starch to name a few [1].

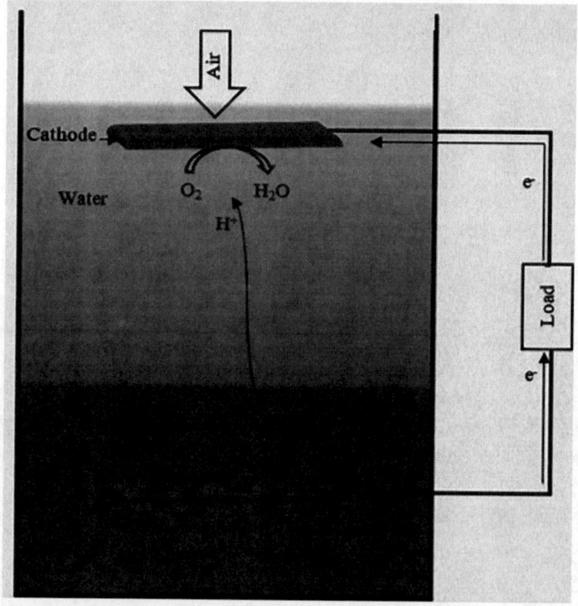

Fig. 1 Working of a Sediment Microbial Fuel Cell

The SMFCs over the years are gaining a substantial scientific interest due to the long-term advantages associated with energy production. They can produce electrical energy resiliently without high-maintenance need and without production of any environment impairing components. Additionally, their ability to remediate the sediments of heavy metals has rendered them prominence. However, the low potential and current production of the SMFCs are limiting their prevalent use for practical applications. In most scenarios, the SMFCs are installed in mainstream rivers, ponds and seas [4–7].

Donovan et. al. (2008, 2011) installed SMFC setups in mainstream river [4, 5]. Meehan et. al. (2009) deployed a SMFC setup inside sea to power hydrophones with the help of power management systems [6]. Installing setups in such locations is also a challenging task and there are limitations in power production even with large sized anodes [4–8]. To increase the power production, connections such as electrical series or parallel was also not possible since the SMFCs were deployed in same water body [6, 7]. There are reports of SMFCs being constructed in laboratory scale also, however, they reported about limitations in power production and the systems were not self-sustainable [9].

This work proposes for a SMFC system with introduction of duckweed (*Spirodela sp.*) into the system. Duckweed generally grows in lakes and ponds. Scientifically, these group of aquatic plants belong to the family of monocotyledon floating plants called *Lemnaceae* [10]. Duckweed possesses some additional exceptional properties like phenomenal growth rate, high-protein content, waste water treatment capability and their potential to be used for bio-fuel production. Moreover, duckweed directly effects the DO, pH and conductivity of the water body [11]. It is anticipated that introduction of duckweed into the system would help create an anaerobic environment in the water level below and thereby facilitate the conditions for the anodic chamber. The dissolved oxygen (DO) is an important parameter in SMFCs as it effects the cathode performance [12]. The duckweed assisted SMFC system is expected to help enhance the cathode potential and impart self-sustainability. This SMFC configuration would be very useful to power low-scale devices in rural areas due to its low cost and simple mode of construction.

2 Methodology

2.1 *Construction of the SMFC*

The duckweed assisted SMFC was constructed in a 250 ml glass beaker. The beaker was filled with 175 g of sediment to cover a height of 2.54 cm inside the glass beaker. This was, followed by addition of water into the system which, covered a height of 5 cm. The height of the water layer was maintained by replenishing water into the SMFC from time to time. Two isomolded graphite plate (purchased from GraphiteStore.com, USA) having dimensions 6 cm × 2 cm × 0.3 cm and 3 cm ×

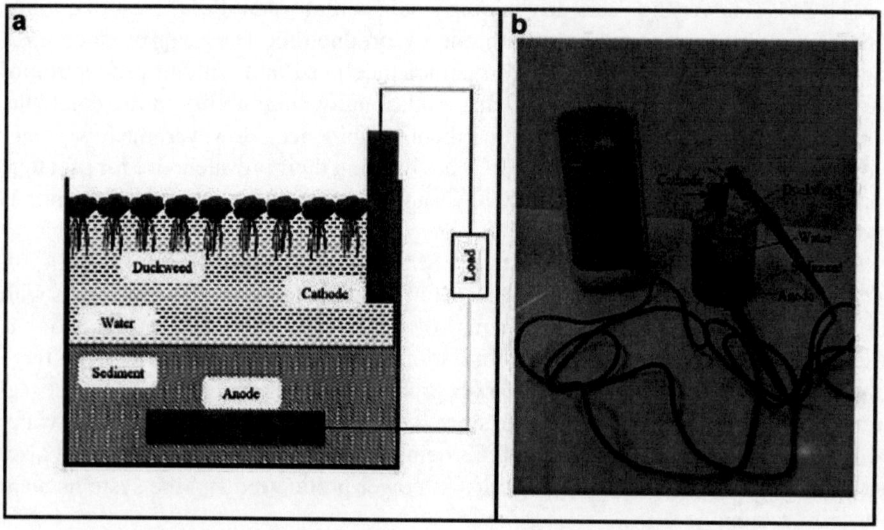

Fig. 2 **a** Schematic representation of the laboratory scale duckweed assisted SMFC. **b** Snapshot of the constructed SMFC

4 cm × 0.3 cm were used as cathode and anode, respectively. The anode was placed horizontally inside the sediment layer and the cathode was placed vertically at the top of the beaker with a support to enable 50% of the cathode to be immersed in water and the rest of the portion to be in direct contact with air. Eventually, duckweed was added to complete the arrangement of the system. A natural pond nearby the Indian Institute of Technology, Guwahati campus (26.1903° N, 91.6920° E) was the source for the sediment, water and duckweed. Figure 2 shows a schematic of the laboratory scale SMFC.

2.2 *Operational Conditions*

A 10–14 h day-night condition was maintained inside the laboratory for the entire period of SMFC operation, which facilitated the duckweed in the SMFC system to grow naturally. The duckweed assisted SMFC was operated at ambient temperature, which was recorded to vary between 25 and 31 °C during the 31 days of operation.

2.3 *Characterization of the SMFC*

The open circuit potential of the duckweed assisted SMFC was monitored under the aforementioned laboratory conditions for a duration of 31 days. When the open circuit potential of the SMFC system achieved a stationary potential level, a 10 KΩ resistive load was connected and the stable potential across the load was recorded. The internal resistance of the SMFC system was calculated by applying Eq. 1.

$$Internal\ resistance = \frac{OCV - Voltage\ across\ load}{Voltage\ across\ load} \times Load\ resistance$$

$$\text{or,}\quad R_{int} = \frac{V_{OCV} - V_{Load}}{V_{Load}} \times R_{Load} \tag{1}$$

The introduction of duckweed was anticipated to influence the content of dissolved oxygen (DO) in the SMFC system and was monitored using a DO700 (Eutech Instruments).

3 Results and Discussions

3.1 Effect of Duckweed on DO Content of the SMFC System

Figure 3 shows the dissolved oxygen content of the SMFC system at various levels from the surface to a depth of 5 cm before and after addition of duckweed. The DO content at the depth of 5 cm was reduced by 11.19% after introduction of duckweed to the SMFC system. Thus, an anaerobic condition with DO content of 3.7 mg/L could be achieved just at the interface of sediment and water layer. This is very

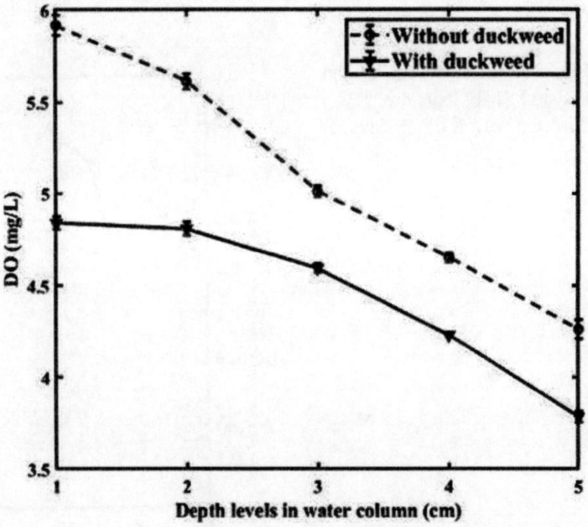

Fig. 3 Dissolved oxygen content at various depths of the SMFC system before and after addition of duckweed

encouraging as the dissolved oxygen content is reported to be one of the parameters hindering the SMFC operation and such arrangement enables the replication of natural water systems for construction of SMFC systems [12]. Another advantage of having duckweed in the SMFC system is that being a green plant it performs photosynthesis and a part of the produced food is released by the roots, which gets deposited at the bottom and helps maintain the nutrient level as well as the microbial diversity in the sediment layer [13]. So, the introduction of duckweed empowers the SMFC system self-sustainability.

3.2 *Characterization of the SMFC*

Figure 4 shows the open circuit potentials of the duckweed assisted SMFC system monitored for a period of 31 days. The potential achieved maximum value of 796 mV on the12th day of operation and from day 22 onwards, the output potential was relatively stable. On addition of duckweed the cathode potential increased to 264 mV where it was 259 mV prior to addition of duckweed. The reduction in dissolve oxygen helped the anodic reactions to occur and a maximum of −519 mV anode potential was recorded during the experimental period (Fig. 4). Furthermore, it has been observed that the OCP could be enhanced by connecting additional SMFC systems in series or parallel depending upon our requirement. Since internal resistance is the factor which governs the output characteristics of the fuel cells [16, 17]; after 31 days of open circuit operation the SMFC was connected to a 10 KΩ load. The potential across the load was documented over a period of 30 min (until the potential was steady). Finally, against the steady voltage across the load the internal resistance of the cell was calculated as 9.8 KΩ (average of triplicate data).

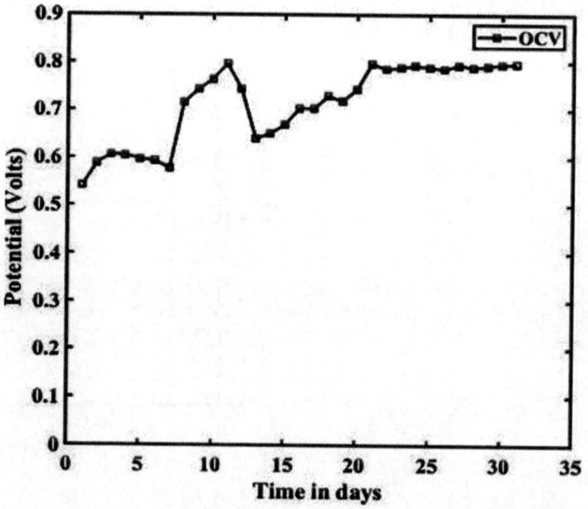

Fig. 4 OCP of the duckweed assisted SMFC system during a period of 31 days

4 Conclusion

A self-sustainable duckweed assisted sediment microbial fuel cell system was constructed in this work. Introduction of duckweed into the system helped to create an anaerobic condition at a depth of 5 cm water level and helped to maintain a stable OCP of maximum 796 mV. The proposed SMFC configuration can be connected in series or parallel as per need to power low-power consuming electronic devices.

Acknowledgements The authors gratefully acknowledge the financial support of the Department of Science and Technology, Government of India, for the above project (DST/INT/RMES/P-10/2016).

References

1. Syed, Z. A., Rafatullah, M., Norli, I., & Muhammad, I. S. (2017). A review on sediment microbial fuel cells as a new source of sustainable energy and heavy metal remediation: Mechanisms and future prospective. *International Journal of Energy Research, 41*, 1242–1264.
2. Khurmi, R. S (2014). *Material Science.* edn.
3. Bond, D. R., Holmes, D. E., & Tender, L. M., Lovley, D. R., (2002). Electrode-reducing microorganisms that harvest energy from marine sediments. *Science 295*. https://doi.org/10.1126/science.1066771
4. Donovan, C., Dewan, A., Heo, D., & Beyenal, A. (2008). Batteryless, wireless sensor powered by a sediment microbial fuel cell. *Environmental Science and Technology, 42*, 8591–8596.
5. Donovan, C., Dewan, A., Peng, H., Heo, D., & Beyenal, H. (2011). Power management system for a 2.5W remote sensor powered by a sediment microbial fuel cell. *Journal of Power Sources, 196*, 1171–1177.
6. Meehan, A., Gao, H., & Lewandowski, Z. (2009). Energy harvest with microbial fuel cell and power management system. IEEE 978-1-4244-2893-9/09/$25.00. https://doi.org/10.1109/ECCE.2009.5316034
7. Gao, H., Meehan, A., & Lewandowski, Z. (2011). New microbial fuel cell power system for efficiency improvement. International Conference on Electrical Machines and Systems, Beijing, pp. 1–5.
8. Wanga, G., Yua, M., Xieb, K., Zhaoa, R., Fua, Y., & Chen, T. (2019). Graphene modified polyacrylonitrile fiber as high-performance cathode for marine sediment microbial fuel cells. *Journal of Power Sources 438*, 227002.
9. Song, T.-S., Yan, Z.-S., Zhao, Z.-W., & Jiang, H.-L. (2011). Construction and operation of freshwater sediment microbial fuel cell for electricity generation. *Bioprocess and Biosystems Engineering, 34*, 621–627.
10. Renslow, R., Donovan, C., Shim, M., Babauta, J., Nannapaneni, S., Schenk, J., & Beyenal, H. (2011). Oxygen reduction kinetics on graphite cathodes in sediment microbial fuel cells. *Physical Chemistry Chemical Physics: PCCP, 13*, 21573–21584.
11. Hubenova, Y., & Mitov, M. (2012). Conversion of solar energy into electricity by using duckweed in Direct Photosynthetic Plant Fuel Cell. *Bioelectrochemistry, 87*, 185–191.
12. Landolt, E. (1986). The family Lemnaceae, a monographic study. Vol 1 Morphology, karyology, ecology, geographic distribution, systematic position, nomenclature, descriptions. Veroff. Geobot. Inst. ETH, Stiftung Rubel, 71: 566.
13. Sajana, T. K., Ghangrekar, M. M., Mitra, A. (2013). Effect of pH and distance between electrodes on the performance of a sediment microbial fuel cell. *Water Science & Technology 68*(3). https://doi.org/10.2166/wst.2013.271

14. Arif, M., Cheung, S. C. P., & Andrews, J. (2020). A systematic approach for matching simulated and experimental polarization curves for a PEM fuel cell. *International Journal of Hydrogen Energy, 45*, 3-2206-2223.
15. Huang, X., Zhang, Z., Jiang, J. (2006) Fuel cell technology for distributed generation: An overview. Industrial Electronics, IEEE International Symposium on, vol. 2. https://doi.org/10.1109/ISIE.2006.295713
16. Miller, A., Singh, L., Wang, L., & Liu, H. (2019). Linking internal resistance with design and operation decisions in microbial electrolysis cells. *Environment International, 126*, 611–618.
17. Vazquez-Larios, A. L., Solorza-Feria, O., Vazquez-Huerta, G., & Ríos-Leal, E. (2011). Internal resistance and performance of microbial fuel cells: influence of cell configuration and temperature. *Journal of New Materials for Electrochemical Systems, 14*(2), 67–139.

Investigation and Modeling for Energy Consumption During Conventional Machining: A Case Study

Shubhajit Das, Sangeeta Das, and Kakoli Roy

Nomenclature

ANOVA	Analysis of variance
d	Depth of cut (mm)
EC	Energy consumed (kWh)
f	Feed (mm/rev)
PCD	Polycrystalline diamond
RSM	Response surface methodology
S	Spindle speed (rpm)
S/N	Signal to noise

1 Introduction

Manufacturing is an essential part of the industrial segment and plays a dynamic role in the global economy. It basically converts raw materials into useful products by exhausting electrical energy while concurrently producing wastes and emissions. Since resources of electrical energy are mostly produced through the scorching of the fossil fuels, the depletion of electrical energy in manufacturing produces carbon footprints [1, 2]. Understanding and approximating the energy expended by machining are important as the energy consumption on machining is accountable for a considerable part of environmental problems in manufacturing industry. There is an urgent

S. Das (✉) · K. Roy
Department of Mechanical Engineering, NIT Arunachal Pradesh, Arunachal Pradesh, India
e-mail: shubhajit@nitap.ac.in

S. Das
Department of Mechanical Engineering, GIMT Guwahati, Assam, India

P. Mahanta et al. (eds.), *Advances in Thermofluids and Renewable Energy*, Lecture Notes in Mechanical Engineering, https://doi.org/10.1007/978-981-16-3497-0_41

need to decrease the energy consumption in manufacturing in order to cut down the CO_2 emissions.

The energy consumption for different manufacturing processes has been studied by different researchers over the years. An increasing interest in an effort to discover new ways for investigating and modeling energy consumption of machining is encouraged by various needs associated to energy efficiency improvement. Khanna et al. [3] studied the evaluation of energy consumption along with tool wear and surface roughness during turning of Inconel 718 using sustainable machining technique. They found that, energy consumption in cryogenic turning reduced by up to 8–17% when compared with dry and wet turning. Shang et al. [4] investigated toward less energy-intensive heavy-duty machine tools for power consumption characteristics and energy saving strategies. They developed a generic power consumption model for heavy duty machine tools and proposed a new method to simplify the study of cutting power characteristics. Xu et al. [5] developed a novel intelligent reasoning system to estimate energy consumption toward sustainable machining. They found that cutting speed has the greatest influence on the cutting power and vibration. The proposed intelligent reasoning system was able to reduce energy consumption, maintaining machine tool stability and improve machining efficiency. Zhao et al. [6] studied on reduction strategy and prediction of energy consumption during machining. They classified energy consumption at the process, machine, and system levels. There are many constituents that consume a lot of energy in machine tools, such as hydraulic pumps and motors. In a machining process, energy consumption can be estimated at different levels such as process levels, machine tool, and spindle. At the machine level, the energy consumed by the whole machine tool (e.g., drive systems, cooling and lubrication units, spindle motor, manufacturing process, control systems, etc.) is considered. Understanding the relationship between the cutting conditions and the energy consumption is important for improving the overall efficiency of the machine tools.

In the present work, an attempt has been made to study the effect of the turning process parameters of spindle speed (S), feed rate (f), and depth of cut (d) on energy consumed during conventional machining of Inconel 625 using a polycrystalline diamond (PCD) tool. The variation of the input parameters was studied with respect to energy consumption. Further, response surface methodology (RSM) was used to develop a relation between the energy consumption and the machining parameters.

2 Experimentation

The machining of Inconel 625 using a polycrystalline diamond (PCD) tip tool under dry condition was performed on a semi-automatic lathe (Panther 1350/2) present in the mechanical workshop, NIT Arunachal Pradesh. The semi-automatic lathe had a power rating of 1.5 kW housed with a three jaw self-centering chuck for holding the work material. CNMG 16 0402 PCD insert was used with CNMG 16 0402 tool holder for machining. Inconel 625 used during experimentation was of

200 mm length and 30 mm diameter with major alloying constituents of Ni 58%, Cr 23%, Fe 5%, and Mo 8% and weighted about 1.3 kg. Inconel 625 is a nickel-based non-magnetic super-alloy and shows prominent toughness and strength in the temperature range of 2000 °C. Due to its resistance to high temperature, superior mechanical properties, and good weld ability, it is extensively used in gas turbine blades, aerospace components, and nuclear plants. The workpiece material, PCD tool inserts, and the machine tool used during experimentation are shown in Fig. 1. Three input process parameters namely spindle speed (*S*), feed rate (*f*), and depth of cut (*d*) were considered to measure the energy consumed (*EC*) during machining. The input parameters considered were of four levels each. The levels of each of the process parameters were considered after performing trial experiments and depending on machine limitations. The input parameters and the corresponding value of the levels are shown in Table 1.

A clamp on power meter was used to measure the power consumed by the semi-automatic lathe for different combination of input parameters during machining. Clamp meters uses the principle of magnetic induction on coils from alternating current. In a clamp meter, a ferrous coil is contained inside the loop which is opened to allow the wire placement inside, then closed to complete the coil, exactly like a transformer. A secondary coil inside the body of the meter can determine the input current by reading the generated voltage over a small fixed load resistor. Since none of the circuit current actually interfaces with the meter at all, the circuit need not be broken open and hence the reading is much faster. Based on three parameters of four

Fig. 1 **a** Inconel 625, **b** PCD insert, **c** semi-automatic lathe

Table 1 Input parameters and their levels

Factors	Parameter	Units	Levels			
			1	2	3	4
A	Spindle speed (*S*)	rpm	288	384	598	938
B	Feed rate (*f*)	mm/rev	0.06	0.12	0.18	0.24
C	Depth of cut (*d*)	mm	0.2	0.4	0.6	0.8

Table 2 Measured output response based on L_{16}

Exp. no.	Spindle speed (rpm)	Feed rate (mm/rev)	Depth of cut (mm)	Energy consumed (EC) (kWh)	*S/N* ratio
1	288	0.06	0.2	151.4	– 43.60
2	288	0.12	0.4	143.8	– 43.15
3	288	0.18	0.6	152.0	– 43.64
4	288	0.24	0.8	158.8	– 44.02
5	384	0.06	0.4	160.6	– 44.11
6	384	0.12	0.2	155.4	– 43.83
7	384	0.18	0.8	156.0	– 43.86
8	384	0.24	0.6	159.2	– 44.04
9	598	0.06	0.6	152.4	– 44.66
10	598	0.12	0.8	150.2	– 43.53
11	598	0.18	0.2	160.0	– 44.08
12	598	0.24	0.4	166.2	– 44.41
13	938	0.06	0.8	159.6	– 44.06
14	938	0.12	0.6	162.4	– 44.21
15	938	0.18	0.4	171.6	– 44.69
16	938	0.24	0.2	178.5	– 45.03

levels each, Taguchi's L_{16} orthogonal array was used for performing the machining experiments. L_{16} orthogonal array indicates that a total of 16 number of experimental runs need to be performed. Moreover, orthogonal array has been considered so as to give equal weight to all the factor levels across the entire design. The different combinations obtained based on L_{16} during machining along with the measured response of energy consumption during dry machining of Inconel 625 using PCD insert tool are shown in Table 2.

3 Results and Discussion

3.1 *Effect of Input Parameters on Energy Consumed*

The main effect plot for means of energy consumed during machining is shown in Fig. 2. It shows the deviation of spindle speed (*S*), feed rate (*f*), and depth of cut (*d*) on energy consumed. It can be seen that spindle speed (*S*) is the most dominating factor affecting the energy consumed during dry machining followed by feed rate (*f*) and depth of cut (*d*). At the spindle level, the energy expended by the spindle motor is measured. The electricity expended by the spindle motor rotates the work material in the chuck. The spindle energy is dependent on the motor which widely

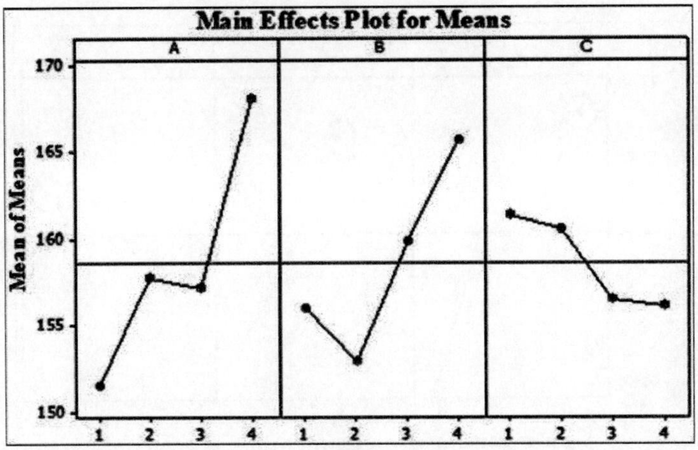

Fig. 2 Deviation of spindle speed, feed rate, and depth of cut on energy consumed

Table 3 Response table for means of energy consumption

Level	*A*	*B*	*C*
1	151.5	156.0	161.3
2	157.8	153.0	160.5
3	157.2	159.9	156.5
4	168.0	165.7	156.2
Delta	16.5	12.7	5.2
Rank	1	2	3

differs across machine tools. The effect of the input parameters on the response can also be confirmed from the response table for means of energy consumption as shown in Table 3. The response table was obtained by considering the average value of energy consumed for each factor at a particular level. Accordingly, rank was assigned based on the delta value which is the difference between the maximum and minimum average values at a particular level. More is the delta value, higher is the rank which indicates greater slope. The parameter having greater slope designates significant contribution affecting the measured response value.

3.2 *Effect of Input Parameters Based on* S/N *Ratio*

The effect of different input parameters (S, f, d) on the output response of energy consumed is determined by Taguchi's signal-to-noise (*S*/*N*) ratio which checks the variation of responses at different noise conditions. Certain quality characteristics

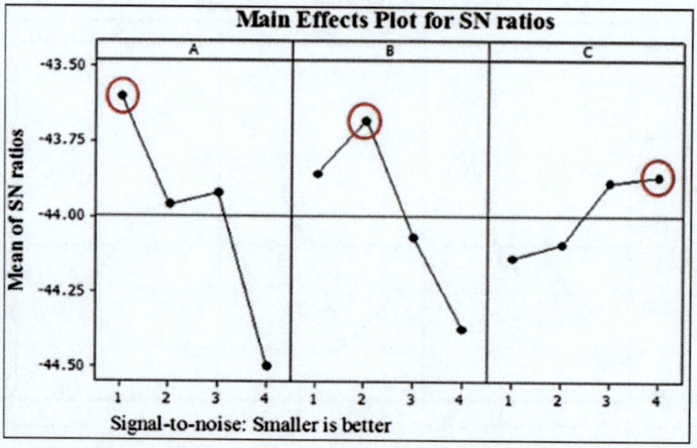

Fig. 3 *S*/*N* ratio plot for energy consumption during dry machining

are assigned to the measured responses for analyzing *S*/*N* ratios. The quality characteristics actually indicates the desirability of the output response. It is quite desirable to minimize the energy consumption value during machining and providing quality products at a higher production rate. Thus, quality characteristics "smaller is better" has been used for analyzing the energy consumed values during dry machining. The *S*/*N* ratio values for "smaller is better" criterion is measured using eq. 1.

$$\frac{S}{N} = -10 \times \log \frac{\sum Y^2}{n} \tag{1}$$

where "*Y*" is the responses for the given factor level combination and "*n*" is the number of responses in the factor-level combination. The main effect plot for *S*/*N* ratio of energy consumption during dry machining is shown in Fig. 3.

The *S*/*N* ratio for energy consumption is shown in Fig. 3 where the *X*-axis indicates the levels of the input parameters and the *Y*-axis represents the mean of *S*/*N* ratio. The levels of input parameters marked with circle will offer the least variability to the response of energy consumed. The combination of input parameters of A1-B2-C4, i.e., ($S = 288$ rpm; $f = 0.12$ mm/rev; $d = 0.8$ mm) will result in minimum energy consumption during dry machining of Inconel 625 using a PCD insert tool. Experimentally, the energy consumed at the optimum combination of A1-B2-C4 was found to be 146.3 kWh.

Table 4 ANOVA for energy consumption during dry machining

Source	DF	Seq. SS	Adj. SS	Adj. MS	*F*	*p*
Regression	9	990.33	990.34	110.04	26.97	0.014
Linear	3	815.19	351.41	117.14	27.42	0.019
A	1	479.71	316.08	316.08	40.03	0.004
B	1	258.84	20.79	20.78	21.32	0.035
C	1	76.64	14.54	14.54	20.92	0.044
Square	3	98.54	98.54	32.85	22.08	0.024
$A \times A$	1	20.48	20.48	20.48	21.30	0.029
$B \times B$	1	77.88	77.88	77.88	24.94	0.068
$C \times C$	1	0.18	0.18	0.18	20.01	0.091
Interaction	3	76.61	76.61	25.54	21.62	0.028
$A \times B$	1	2.94	2.94	2.94	20.19	0.068
$A \times C$	1	38.35	38.35	38.35	22.43	0.017
$B \times C$	1	35.32	35.32	35.32	22.24	0.018
Residual Error	6	94.66	94.66	15.78		
Total	15	1084.99				

3.3 *Analysis of Variance (ANOVA)*

ANOVA is a collection of statistical models that is used to analyze the difference among the group mean in a sample. It was developed by Robert Fisher, a statistician in 1918. ANOVA helps in identifying the significant parameters contributing effectively to the measured response depending on the "*F*" and "*p*" values in the analysis. The analysis was done at a confidence level of 95%. The ANOVA for energy consumption is shown in Table 4. The ANOVA table includes the linear, square, and interaction terms which were considered during quadratic model development of energy consumption during machining. From Table 4, it can be seen that the factor spindle speed (*S*), feed rate (*f*), and depth of cut (*d*) are significant factors as their corresponding "*p*" value is less than 0.05, as the analysis was performed at 95% confidence level. Also, it can be seen that spindle speed (*S*) contributes more toward energy consumed followed by feed rate (*f*) and depth of cut (*d*) due to higher "*F*" value of the corresponding parameter.

3.4 *Response Surface Methodology (RSM) Predictive Model*

RSM can be termed as a collection of statistical and mathematical technique used for developing co-relation between the input parameters and the output responses [7]. RSM is more promising due to its giving very low average error toward modeling and experimental validation. Also, significance of interactions and square terms of parameters are more clearly predicted in RSM. Taguchi technique is normally used in linear interactions only. This is due to the fact that in Taguchi design, interactions between control factors are aliased with their main effects. In the present study, a quadratic model consisting of linear, square, and interaction terms was considered for developing a predictive model for energy consumption. The developed RSM predicted model for energy consumption is shown in Eq. 2.

$$\begin{aligned} EC = {} & 161.484 + 4.81705 \times A - 15.7670 \times B \\ & - 1.75795 \times C + 1.13125 \times A^2 + 2.20625 \times B^2 \\ & + 0.106250 \times C^2 + 0.578409 \times A \times B \\ & - 2.08750 \times A \times C + 2.00341 \times B \times C \end{aligned} \quad (2)$$

The coefficient of determination (R^2) value for the developed model was found to be 98.67% which shows good fit for the developed model while the predicted R^2 value was found to be 95.84%. The developed RSM model has been validated to find out the model accuracy. Five numbers of experimental datasets were used to validate the model. These datasets were however not used during the RSM model development. The comparison of the experimental values and the predicted values used for validating the model is shown in Table 5. It was found that the maximum percentage error was 1.71%, the minimum percentage error was 0.98%, while the average percentage error was found to be 1.32%. The RSM developed model accuracy was found to be 98.68% which is quite acceptable. Thus, the developed RSM predicted model for energy consumption can be sensibly used in industries for predicting energy consumption during dry machining without performing actual experimentation thus saving time and money.

Table 5 Validation of the developed RSM equation

Exp. no.	Spindle speed (rpm)	Feed rate (mm/rev)	Depth of cut (mm)	Energy consumed (EC) (kWh)		% error
				Experimental value	Predicted value	
1	288	0.18	0.6	153.7	151.78	1.25
2	384	0.06	0.8	159.4	157.83	0.98
3	384	0.18	0.2	157.2	155.01	1.39
4	598	0.24	0.6	162.0	159.97	1.25
5	938	0.06	0.8	156.6	153.92	1.71

4 Conclusion

In the current study, a method was presented to estimate the energy consumption during dry machining of Inconel 625 using a PCD insert tool. The contribution of this work is to investigate and develop a predictive model to evaluate the energy consumed by considering the correlation between the input parameters and the responses. The following outcomes can be highlighted from the present work:

- Spindle speed (*S*) is the most dominating factor affecting the energy consumed during dry machining followed by feed rate (*f*) and depth of cut (*d*).
- The combination of input parameters of A1-B2-C4, i.e., ($S = 288$ rpm; $f = 0.12$ mm/rev; $d = 0.8$ mm) will result in minimum energy consumption during dry machining of Inconel 625 using a PCD insert tool. Experimentally, the energy consumed at the optimum combination of A1-B2-C4 was found to be 146.3 kWh.
- The RSM developed predicted model accuracy was found to be 98.68% for energy consumption and can be sensibly used in industries for predicting energy consumption during dry machining thus saving time and money.

References

1. Yuan, C., Zhai, Q., & Dornfeld, D. (2012). *CIRP Annals- Manufacturing Technol., 61*(1), 39–42.
2. Balogun, V. A., & Mativenga, P. T. (2013). *J. Cleaner Prod., 41*, 179–186.
3. Khanna, N., Agrawal, C., Dogra, M., & Pruncu, C. I. (2020). *J. Mater. Res. Technol., 9*(3), 5794–5804.
4. Shang, Z., Gao, D., Jiang, Z., & Lu, Y. (2019). *Energy, 178*, 263–276.
5. Xu, L., Huang, C., Li, C., Wang, J., Liu, H., & Wang, X. (2020). *J. Cleaner Prod., 261*, 121160
6. Zhao, G. Y., Liu, Z. Y., He, Y., Cao, H. J., & Guo, Y. B. (2017). *Energy, 133*, 142–157.
7. Das, S., Chandrasekaran, M., & Samanta, S. (2019). AIP Conference Proceedings 2128, 040008 (2019). https://doi.org/10.1063/1.5117970

Investigation of Thermo-Hydraulic Performance for Different Arrangements of Ribs in Rectangular Solar Air Channel

J. Singh, A. Sharma, and R. Chauhan

Nomenclature

D	Diameter (m)
d	Discrete distance from outer edge of the rib (m)
d/x	Relative gap position (dimensionless)
e	Thickness of rib (m)
e/D_h	Relative roughness height (dimensionless)
f	Friction factor (dimensionless)
g	Discrete width (m)
g/e	Relative gap width (dimensionless)
H	Height of the channel (m)
Nu	Nusselt number (dimensionless)
P	Pitch of the ribs (m)
P/e	Relative roughness pitch (dimensionless)
Re	Reynolds number (dimensionless)
W	Width of the channel (m)
W/w	Relative roughness width (dimensionless)
W/H	Duct aspect ration (dimensionless)
w	Width of the ribs (m)
x	Half length of the rib

J. Singh · A. Sharma (✉) · R. Chauhan
Department of Mechanical Engineering, Dr B R Ambedkar National Institute of Technology, Jalandhar, Punjab 144011, India
e-mail: ashutoshs.me.19@nitj.ac.in

P. Mahanta et al. (eds.), *Advances in Thermofluids and Renewable Energy*, Lecture Notes in Mechanical Engineering, https://doi.org/10.1007/978-981-16-3497-0_42

Greek Symbols

α	Angle of attack (dimensionless)
$\alpha/90$	Relative angle of attack (dimensionless)

Subscripts

ar	Arc ribs
dar	Discrete arc ribs
dmar	Discrete
dVr	Discrete V-shape ribs
dmVr	Discrete multiple V-shape ribs
H	Hydraulic
Mar	Multiple arc ribs
mVr	Multiple V-shape ribs
S	Smooth surface
Vr	V-shape ribs

1 Introduction

The effective utilization is very much necessary for energy converting machine in order to achieve maximization in desired form of energy yield. Solar air channels are generally used for heating the air by utilizing solar energy. In solar air channel, the air passes beneath an absorber plate which receives energy from the solar radiations [1]. Then heat energy is transferred to air which is further used for different applications. The effectiveness of solar air channel is quite low due to very poor convective heat transfer between the air and heated plate, i.e., absorber plate of solar air channel [2]. Its thermal performance can be increased by inducing different turbulent geometries within the air flow channel which separates the laminar sublayer on the absorber plate [3]. However, modifications in flow channel with intention to increase thermal performance also offer significant amount of hurdles to air flow. So while designing the turbulent geometries both of these parameters, i.e., thermal and hydraulic performance should be taken into consideration. For increasing the performance of solar air channel, various shapes and types of geometries have been studied in the literature. Most of these turbulent geometries are in the form of ribs [4–11], baffles, fins, protrusions, dimples, etc. [12]. Generally, these different studies had carried out to identify the thermal and hydraulic performance individually at different flow conditions. For identifying the overall performance of any roughness geometry, effect of both thermal and hydraulic performance should be considered concurrently and at same flow conditions. Thermo-hydraulic performance parameter

(THPP) accounts together the hydraulic and thermal performance of the roughness geometry simultaneously under same flow conditions. In the present study, different arrangements of ribs as roughness geometry have considered for estimating thermal, hydraulic, and thermo-hydraulic performance. These parameters have been evaluated for different flow conditions, i.e., Reynolds number. Correlations provided in literature for Nusselt number (*Nu*) and friction factor (f) for different geometries have been used to estimate performance parameters.

2 Rib Roughness Geometries

Various geometries of ribs as roughness over the absorber plate of solar air channel have been studied in literature. In this study, rib geometries in arc and V-shape have been selected because of their effective performance. The selected geometries have been presented in this section.

2.1 Arc Ribs

The influence of arc ribs made from circular wire, in air flow channel of solar air channel was studied by Saini and Saini [4]. The schematic of the geometry with potential parameters is shown in Table 1. These parameters were varied in a definite range and optimum values were found out on the ground of performance. These parameters along with optimum values have been presented in Table 2. An enhancement of 3.8 and 1.75 was found for *Nu* and f, respectively from smooth channel. Based on the variation in performance with parameters, correlations that were developed for Nu_{ar} and f_{ar}, are presented in Table 1.

2.2 Discrete Arc Ribs

Hans et al. [5] investigated the effect of discrete arc ribs in solar air channel. Schematic of the discrete arc ribs and potential parameters has been presented in Tables 1 and 2, respectively. Developed correlations for Nu_{dar} and f_{dar} are also presented in Table 1. Authors observed that maximum increment in Nu_{dar} and f_{dar} was 2.63 and 2.44, respectively.

Table 1 Schematics and correlations for Nu and f of different geometries in solar air channel

Sr. no.	Geometry	Schematic	Correlations
1	Arc ribs	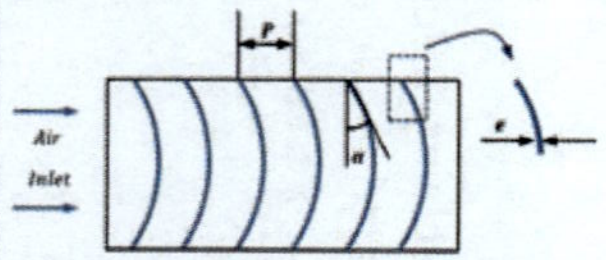	$Nu_{ar} = 0.001047\mathrm{Re}^{1.3186}(e/D_h)^{0.3772}(\alpha/90)^{-0.1198}$ $f_{ar} = 0.14408\mathrm{Re}^{-0.17103}(e/D_h)^{0.1765}(\alpha/90)^{0.1185}$
2	Discrete arc ribs	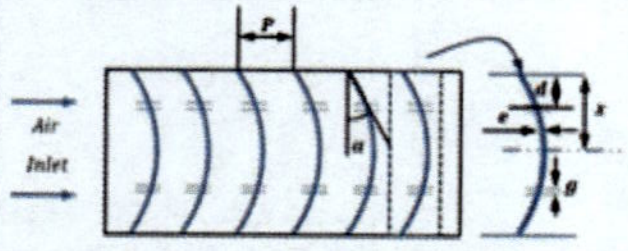	$Nu_{dar} = 1.014 \times 10^{-3}\mathrm{Re}^{1.036}(P/e)^{2.522}(\alpha/90)^{-0.293}(d/x)^{-0.078} \times (g/e)^{-0.016} \times (e/D_h)^{0.412}$ $\times \exp(-0.567(Ln(P/e))^2) \times \exp(-0.114(Ln(\alpha/90))^2) \times \exp(-0.077(Ln(d/x))^2)\exp\left(-0.133(Ln(g/e))^{220527}\right)$ $f_{dar} = 8.1921 \times 10^{-2}(\mathrm{Re})^{-0.147}(P/e)^{1.191}(\alpha/90)^{-0.292}(d/x)^{-0.067} \times (g/e)^{-0.006}(e/D_h)^{0.528}\exp(-0.255(\ln(P/e))^2)\exp(-0.110(\ln(\alpha/90))^2)$ $\times \exp(-0.063(\ln(d/x))^2)\exp(-0.158(g/e)^2)$
3	Multiple arc ribs	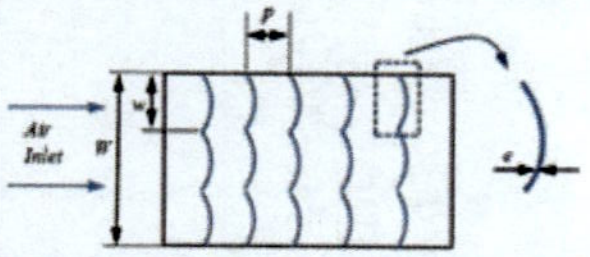	$Nu_{mar} = 1.564 \times 10^{-4} Re^{1.343}(e/D)^{0.048}(W/w)^{0.407} \times exp(-0.099(Ln(W/w))^2)(P/e)^{0.572}$ $\times exp(-0.148(Ln(P/e))^2) \times (\alpha/90)^{-0.355} \times exp(-0.272(Ln((\alpha/90)))^2)$ $f_{mar} = 0.063 \times (e/D)^{0.102} \times \mathrm{Re}^{-0.16} \times (W/w)^{0.277} \times (p/e)^{0.562} \times (\alpha/90)^{0.023}$ $\times \exp(-0.140(Ln(p/e))^2) \times \exp(-0.013(Ln(\alpha/90))^2)$
4	Discrete multiple arc ribs	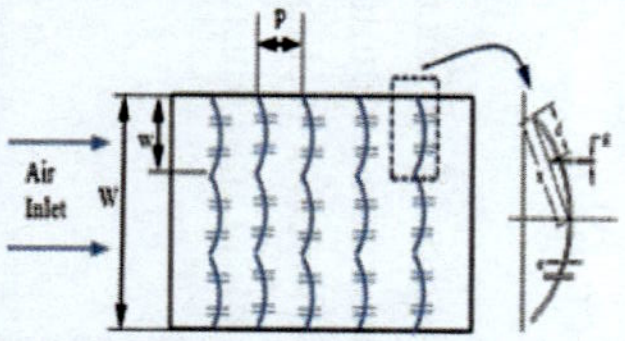	$Nu_{dmar} = 1.39 \times 10^{-4} \times \mathrm{Re}^{1.3701} \times (d/x)^{-0.4997} \times (g/e)^{-0.0292} \times (W/w)^{0.4017} \times (p/e)^{0.5854} \times (\alpha/60)^{-2.235} \times (e/D)^{0.0931}$ $\times \exp(-0.3989(Ln(d/x))^2) \times \exp(-0.2013(Ln(g/e))^2) \times \exp(-0.129(Ln(W/w))^2) \times \exp(-0.142(Ln(P/e))^2) \times \exp(-0.5614(Ln(\alpha/60))^2)$ $f_{dmar} = 2.11 \times 10^{-1} \times \mathrm{Re}^{-0.25} \times (d/x)^{-0.888} \times (g/e)^{-0.079} \times (W/w)^{0.032} \times (p/e)^{0.643} \times (\alpha/60)^{-2.546} \times (e/D)^{0.145} \times \exp(-0.662(Ln(d/x))^2)$ $\times \exp(-0.496(Ln(g/e))^2) \times \exp(-0.160(Ln(p/e))^2) \times \exp(-3.96(Ln(\alpha/90))^2)$

(continued)

Table 1 (continued)

Sr. no.	Geometry	Schematic	Correlations
5	V-shape ribs	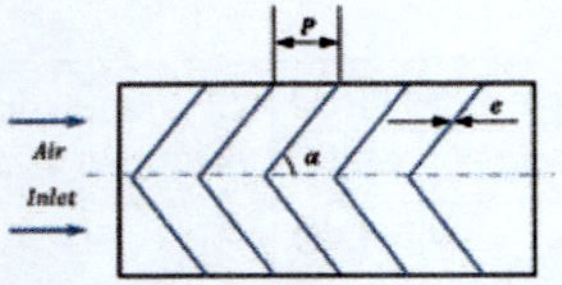	$Nu_{Vr} = 0.067 \times \mathrm{Re}^{0.888}(e/D_h)^{0.424}(\alpha/60)^{-0.077} \times \exp[-0.782 \times (\ln\alpha/60)^2]$ $f_{Vr} = 6.266 \times \mathrm{Re}^{-0.425} \times (e/D_h)^{0.565} \times (\alpha/60)^{-0.093} \times \exp(-0.719(Ln(\alpha/60))^2)$
6	Discrete V-shape ribs	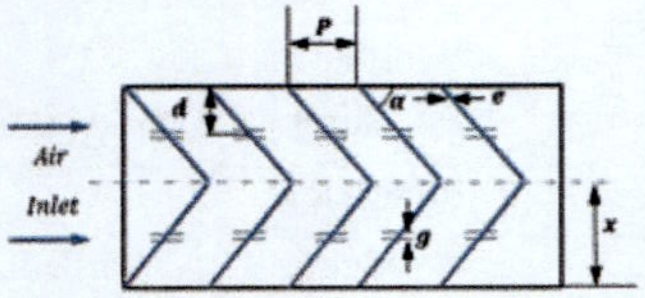	$Nu_{dVr} = 2.36 \times 10^{-3}\mathrm{Re}^{0.9}(P/e)^{3.5}(\alpha/60)^{-0.23}(d/x)^{-0.043} \times (g/e)^{-0.014} \times (e/D_h)^{0.47}$ $\times \exp(-0.84(Ln(P/e))^2) \times \exp(-0.72(Ln(\alpha/60))^2) \times \exp(-0.05(Ln(d/x))^2)\exp\left(-0.15(Ln(g/e))^{220527}\right)$ $f_{dVr} = 4.13 \times 10^{-2}\mathrm{Re}^{-0.126} \times (P/e)^{2.74} \times (\alpha/60)^{-0.034} \times (d/x)^{-0.058} \times (g/e)^{0.031} \times (e/D_h)^{0.70}$ $\times \exp(-0.685(Ln(P/e))^2) \times \exp(-0.93(Ln(\alpha/60))^2) \times \exp(-0.058(Ln(d/x))^2)\exp\left(-0.21(Ln(g/e))^{220527}\right)$
7	Multiple V-shape ribs	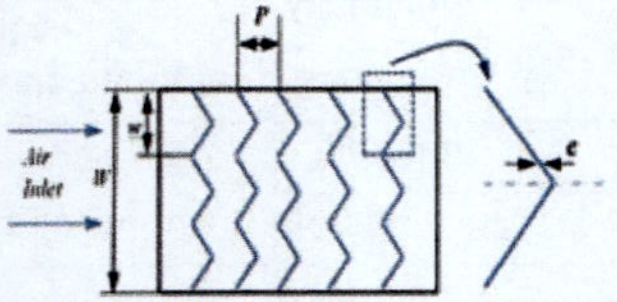	$Nu_{mVr} = 3.35 \times 10^{-5}\mathrm{Re}^{0.92}(e/D)^{0.77}(W/w)^{0.43}(\alpha/90)^{-0.49}$ $\times \exp^{(-0.117(Ln(W/w))^2)} \times \exp^{(-0.61(Ln(\alpha/90))^2)}(P/e)^{8.54} \times \exp(-2.0407(Ln(P/e))^2)$ $f_{mVr} = 4.47 \times 10^{-4}\mathrm{Re}^{-0.3188} \times (e/D)^{0.73} \times (W/w)^{0.22} \times (\alpha/90)^{-0.39}$ $\times \exp^{(-0.52(Ln(\alpha/90))^2)} \times (P/e)^{8.9}\exp^{(-2.133(Ln(P/e))^2)}$
8	Discrete multiple V-shape ribs	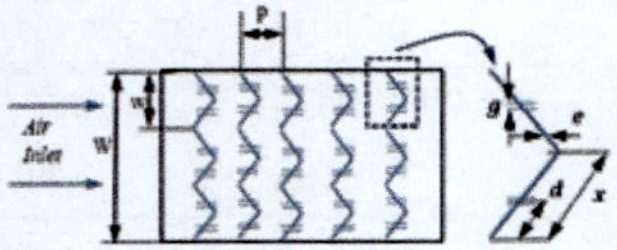	$Nu_{dmVr} = 8.532 \times 10^{-3} \times \mathrm{Re}^{0.932} \times (e/D)^{0.175} \times (W/w)^{0.506} \times \exp(-0.0753(Ln(W/w))^2)(d/x)^{-0.0348}$ $\times \exp(-0.0653(Ln(d/x))^2)(g/e)^{-0.0708} \times \exp(-0.223(Ln(g/e))^2)(\alpha/60)^{-0.0239}$ $\times \exp(0.1153(Ln(\alpha/60))^2)(P/e)^{1.196} \times \exp(-0.2805(Ln(p/e))^2)$ $f_{dmVr} = 3.19 \times \mathrm{Re}^{-0.3151} \times (e/D)^{0.268} \times (W/w)^{0.1132} \times \exp(-0.0974(Ln(W/w))^2)(d/x)^{0.0610}$ $\times \exp(-0.1065(Ln(d/x))^2)(g/e)^{-0.1769} \times \exp(-0.6349(Ln(g/e))^2)(\alpha/60)^{0.1553}$ $\times \exp(-0.1527(Ln(\alpha/60))^2)(P/e)^{-0.7941} \times \exp(0.1486(Ln(p/e))^2)$

Table 2 Potential parameters of different arrangements of ribs as turbulent geometry

Sr. no.	Arrangements of ribs	Potential parameter	Re	*e/d*	*α/90*	*P/e*	*d/x*	*W/w*	*g/e*	*W/H*
1	Arc ribs	Range	2000–17,000	0.0213–0.0422	0.333–0.666	10	–	–	–	12
		Optimum	–	0.0422	0.333	–	–	–	–	–
2	Discrete arc ribs	Range	2000–16,000	0.022–0.043	0.1666–0.8333	4–12	0.2–0.8	–	0.5–2.5	–
		Optimum	–	0.043	0.333	10	0.65	–	1	–
3	Multiple arc ribs	Range	2200–22,300	0.018–0.045	0.1666–0.8333	4–16	–	1–7	–	11
		Optimum	22,300	0.045	0.6667	8	–	5	–	–
4	Discrete multiple arc ribs	Range	2100–21,000	0.016–0.044	0.1666–0.8333	4–16	0.25–0.85	1–7	0.5–2.0	10
		Optimum	21,000	0.044	0.6667	8	0.65	5	1	–
5	V-shape ribs	Range	2500–18,000	0.02–0.034	0.333–1	10	–	–	–	10.15
		Optimum	–	0.034	0.6667	–	–	–	–	–
6	Discrete V-shape ribs	Range	3000–15,000	0.015–0.043	0.1666–0.8333	4–12	0.2–0.8	–	0.5–2.0	–
		Optimum	–	0.043	0.6667	8	0.65	–	1	–
7	Multiple V-shape ribs	Range	2000–20,000	0.019–0.043	0.1666–0.8333	6–12	–	1–10	–	–
		Optimum	–	0.043	0.6667	8	–	6	–	–
8	Discrete multiple V-shape ribs	Range	2000–20,000	0.022–0.043	0.1666–0.8333	6–12	0.24–0.80	1–10	0.5–1.5	10
		Optimum	–	0.043	0.6667	8	0.69	6	1	–

2.3 Multiple Arc Ribs

Effect of multiple arc ribs inside the solar air channel was observed by Singh et al. [6]. Schematic of the roughened absorber plate with potential parameters and developed correlations of Nu_{mar} and f_{mar} are shown in Table 1. Range of the potential parameters studied and optimum value found is depicted in Table 2. Maximum enhancement in Nu_{mar} and f_{mar} was founded as 5.07 and 3.71, respectively.

2.4 Discrete Multiple Arc Ribs

In continuation to the study of arc shapes in solar air flow channel, discrete multiple arc ribs were studied by Pandey et al. [7]. Schematics of the geometry and range of potential parameters is presented in Tables 1 and 2, respectively. The enhancement in Nu_{dmar} and f_{dmar} was founded 5.85 and 4.96, respectively.

2.5 V-shape Ribs

The study of ribs made of circular wire in V-shape was carried out by Momin et al. [8]. V-shape upstream ribs were studied for performance enhancement from smooth duct and it was found that Nu_{Vr} and f_{Vr} increases by 2.3 and 2.83 respectively. Schematic of the geometry used and correlations of Nu_{Vr} and f_{Vr} are presented in Table 1. The ranges of geometrical parameters with founded optimum values are shown in Table 2.

2.6 Discrete V-shape Ribs

Singh et al. [9] carried out research to investigate the effect of discrete V-shape ribs onto the performance of solar air channel. Top view of the plate used for experimentations is shown in Table 1. Potential parameters of the geometry with their range and founded optimum values have been presented in Table 2. From this study, it was concluded that Nu_{dVr} and f_{dVr} upsurges by 3.04 and 3.11, respectively.

2.7 Multiple V-shape Ribs

Variation in heat transfer rate and flow friction by introducing multiple V-shape ribs in rectangular air channel was studied by Hans et al. [10]. Table 1 depicts the schematic of roughness geometry onto the absorber plate and ranges of potential parameters are presented in Table 2. By using the multiple V-shape ribs of optimum value, authors found maximum enhancement in Nu_{mVr} and f_{mVr} of 6 and 5, respectively, from smooth plate air channel. Correlations for Nu_{mVr} and f_{mVr} were also developed with an error of ± 10% and are presented in Table 1.

2.8 *Discrete Multiple V-shape Ribs*

Extending the research on V-shape ribs, discrete multiple V-shape ribs were studied by Kumar et al. [11]. Schematics of the discrete multiple V-shape ribs in air flow channel and range of potential parameters are shown in Tables 1 and 2, respectively. Developed correlations for Nu_{dmVr} and f_{dmVr} are also presented in Table 1. Authors found that by using this geometry the Nu_{dmVr} and f_{dmVr} can be increased by 6.74 and 6.37 times from smooth plate solar air channel, respectively.

3 Effect of Potential Parameters

The turbulent geometries that have been studied in the solar air channels can be standardized by defining the various geometrical parameters. In this section, these otential parameters associated with arc, discrete arc, multiple arc, discrete multiple arc, V-shape, discrete V-shape, multiple V-shape, and discrete multiple V-shape ribs along with their effect on the performance of solar air channel have been discussed.

3.1 *Relative Roughness Pitch* (P/e)

The relative roughness pitch is the ratio of pitch of ribs over the absorber plate to that of the height of the ribs. For ribs made up of circular wire, its thickness is equal to its height. From the previous studies, this has been found that the heat transfer rate upsurges with increase in *P/e* up to certain limit (generally 8–10) and afterward it decreases. The flow gets deviated toward the absorber plate as the fluid flow past a rib and resultingly flow reattachment points are created. Over these reattachment points, the heat transfer rate is maximum. As the fluid approaches toward the downstream rib, the boundary layer gets developed which results in lower convective heat transfer. So, the heat transfer rate decreases with the further increase in the *P/e*.

3.2 *Relative Roughness Height (e/D_h)*

The relative roughness height is expressed as the ratio of the rib height to that of the height of the rectangular channel. The relative roughness height affects the heat transfer enhancement in a duct flow of solar air heater. It is found that the heat transfer increases with increasing the relative roughness height. As the e/D_h increases, the turbulence due to more protrusions in flow increases. Because of this turbulence, the shear layer got reattached from the heated surface and the contact of main fluid stream occurred with the hot surface which result in heat transfer enhancement.

3.3 *Relative Angle of Attack* (A/90)

Relative angle of attack is the ratio of angle of attack to the right angle. Heat transfer rate is varied as the value of $\alpha/90$ is varied. The heat transfer rate firstly increases with increase in $\alpha/90$ and afterward decreases. Corresponding to 0.666 value of $\alpha/90$ from the centerline, the heat transfer rate has found maximum (0.333 from the edge). The arc and V-shape rib act as transverse ribs at larger values of $\alpha/90$ so significant heat transfer increment observed [13]. As the value of $\alpha/90$ decreases, the secondary fluid stream formation diminishes and results in comparatively less heat transfer augmentation.

3.4 *Relative Gap Position* (D/x)

Relative gap position (d/x) is the ratio of discrete distance from the leading edge of rib to the length of single limb. Heat transfer enhancement is greatly affected by variation in d/x. When the gap is placed near the trailing edge, the secondary flow contributes much in energizing the main flow and heat transfer rate increases resultingly. This effect of heat transfer augmentation diminishes as the discrete slice moves too near the apex. The maximum heat transfer is observed for $d/ \times 0.65$–0.69 but beyond this point the fluid mixing becomes again insignificant due to the increased length of the rib which results in truncation of secondary flow contribution in heat transfer.

3.5 *Relative Gap Width* (g/e)

The ratio of gap width to the height or thickness of the rib is called relative gap width (g/e). A significant effect of variation in g/e has been seen over the heat transfer. The heat transfer rate firstly increases with the increase in g/e until it attains value 1.0 and afterward starts decreasing. The secondary fluid stream gets developed along the length of the rib and escape from the gap in the rib which results in energizing the main fluid stream and thereby heat transfer rate increases. The two contrasting effects of each other which govern the energy of secondary fluid stream passing through the discrete slice are the area of discrete slice and the tangential velocity of secondary stream. As the discrete area increases, the tangential velocity decreases and vice versa. So, because of this the average value of *g/e* attains maximum heat transfer and decreases at higher and lower values.

3.6 Relative Roughness Width (W/w)

Relative roughness width is the ratio of the roughness width to the single arc or V-shape rib width. The *W/w* affects the heat transfer and friction factor enhancement in duct flow of solar air channel. It is found that the performance of the solar air channel increases with increase in *W/w* up to certain limit (generally 5–6) and afterward starts decreasing. The reason for this trend could be as the roughness in form of arcs and Vs increases in the channel, and the secondary flow formation on the upstream side of the rib increases and thereby heat transfer rate also increases. As the number of arcs and Vs increase beyond a certain limit, the secondary flow streams increases but due to flow over a very small rib region their energy level decreases. So heat transfer rate decreases after this limit of *W/w* and friction factor increases.

4 Thermal and Friction Characteristics

Nusselt number and friction factor for different arrangements of ribs have been reproduced at varying values of Re by using the correlations provided in the literature. The prime values of the potential parameters suggested by the authors for different parameters of the geometries have been selected as standard of comparison. The obtained results are much close to the real experimental results of different geometries. Nusselt number for selected arrangement of ribs is presented in Fig. 1.

It can be clearly seen from Fig. 1 that the Nusselt number which signifies the heat transfer, increase with increase in flow rate, i.e., Re for all the selected rib geometries. The discrete V-shape ribs yields maximum Nusselt number from rest of the selected geometries at almost every value of Reynolds number in the chosen range. Multiple V-shape ribs and discrete multiple V-shape ribs shows quite high increment from other geometries. Similarly, values of friction factor for selected geometries have also been computed and presented in Fig. 2 w.r.t. Re.

Figure 2 shows the variation of friction factor with respect to the *Re*. The graph depicts that the friction factor decreases at higher flow rates. It can be clearly seen that the maximum hurdles to the flow to happen, i.e., maximum friction factor are offered by discrete multiple V-shape ribs while arc shape ribs offers least within the selected flow range, i.e., *Re*. The friction factor for all the geometries follows almost

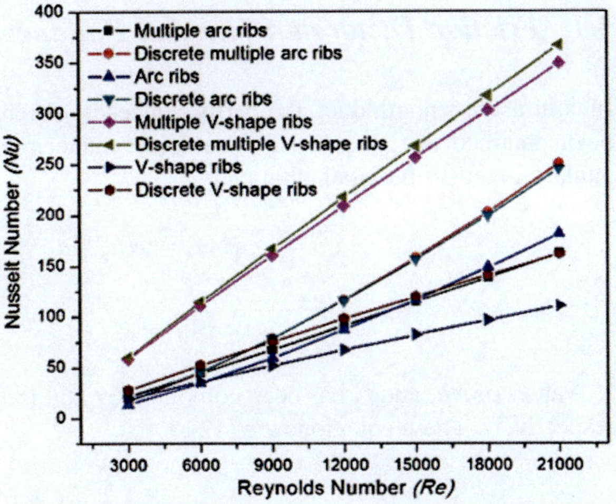

Fig. 1 Variation of *Nu* with *Re* for different turbulent geometries

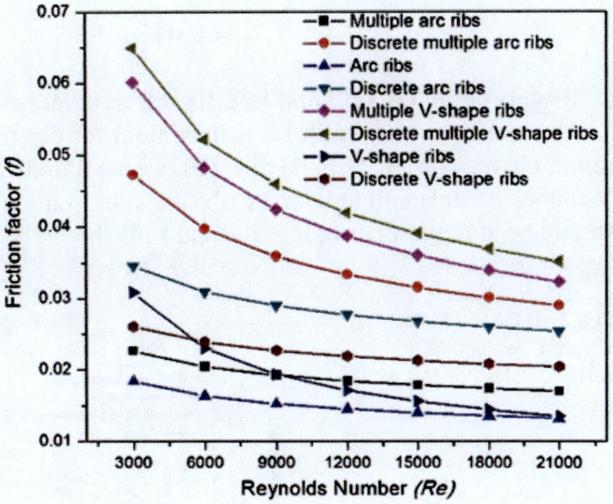

Fig. 2 Variation of friction factor with Reynolds number for different geometries

same trend of decreasing with Re due to destruction of laminar boundary layer at high flow velocities.

5 Performance Parameter

The performance parameters are standard of comparison for performance evaluation of different geometries of turbulators. In this section these parameters have been discussed for selected geometries.

5.1 Friction Factor and Thermal Enhancement

In contrast to smooth duct, the performances of different turbulent geometries have been obtained in terms of Heat transfer enhancement (HTE) and Friction factor enhancement (FFE). Mathematically,

$$\text{HTE} = Nu/Nu_s \tag{1}$$

$$\text{FFE} = f/f_s \tag{2}$$

Values of Nu_s and f_s has been computed by Dittus-Boelter and Blasius equations respectively. These equations are [12],

$$Nu_s = 0.023 \times \text{Re}^{0.8} \times \text{Pr}^{0.4} \tag{3}$$

$$f_s = 0.085 \times \text{Re}^{-0.25} \tag{4}$$

Results of the HTTHE and FFE for different *Re* have been presented in Fig. 3.

Thc graphs shows that HTE is maximum for discrete multiple V-shape ribs and minimum for upward V-shape ribs. FFE is also found maximum for discrete multiple V-shape ribs and minimum for arc ribs. In order to attain maximum performance HTE should be at highest possible value and FFE should be at least.

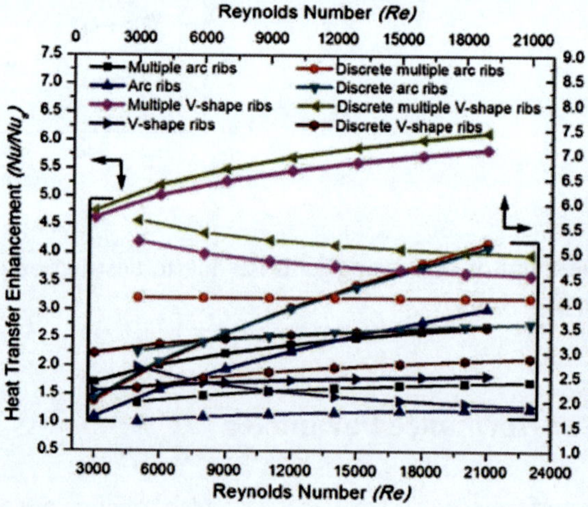

Fig. 3 HTE and FFE for different geometries at variable values of Reynolds number

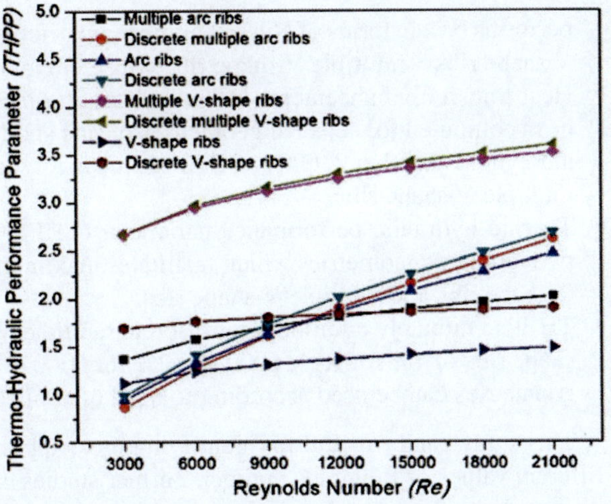

Fig. 4 THPP for different arrangements of ribs

5.2 *Thermo-Hydraulic Performance Parameter (THPP)*

The combined effect of HTE and FFE can be studied by computing the thermo-hydraulic performance parameter (THPP). Mathematically, THPP is given as [11],

$$THPP = HTE/(FFE)^{1/3} \tag{5}$$

The results of THPP for different arrangements of ribs have been presented in Fig. 4.

Figure 4 shows that the THPP is maximum for discrete multiple V-shape ribs and is followed by multiple V-shape ribs. The minute difference between THPP of these leading two roughness geometries can be seen only at high values of Re but for low flow rates both these performs almost equally. For *Re* below 6000, THPP is 1% more and for *Re* between 6000 and 21,000 THPP is 2% more for discrete multiple V-shape ribs than multiple V-shape ribs.

6 Conclusion

This study has been conceded to find the most appropriate arrangement of ribs in shape of arc and V-shape, placed in single-pass solar air channel which can offer maximum heat transfer rate at minimum expense of friction factor. The broad conclusions of this study are,

- Nusselt number and friction factor have been estimated by using correlations provided in literature for *Re* range of 3000–21,000. The top three geometries that performs best in terms of Nusselt number and friction factor are, discrete multiple V-shape ribs > multiple V-shape ribs > discrete multiple arc ribs.

- Heat transfer enhancement (HTE) and friction factor enhancement (FFE) have been computed for selected geometries of ribs and found maximum HTE of 6.13 at Reynolds number 21000 and FFE of 5.64 at Reynolds number 3000 for discrete multiple V-shape ribs.
- Thermo-hydraulic performance parameter (THPP) has been calculated and top two turbulent geometries which facilitates maximum THPP are discrete multiple V-shape ribs and multiple V-shape ribs.
- THPP is minutely additional for discrete multiple V-shape ribs from multiple V-shape ribs of 1% for Re < 6000 and 2% for 6000 < Re < 21,000. So any of these geometries can be used according to need of application.

This study can be useful for identifying most appropriate arrangement of ribs at different values of Reynolds number. Further studies can be carried out to compare other turbulent geometries to found optimum one for different flow conditions according to demand of applications.

References

1. Sharma, A., Chauhan, R., Singh, T., Kumar, A., Kumar, R., Kumar, A., & Sethi, M. (2017). *Renewable Energy, 106*, 310–320.
2. Prasad, B. N., & Saini, J. S. (1988). *Solar Energy, 41*, 555–560.
3. Kumar, A., & Layek, A. (2019). *Renewable Energy, 130*, 687–699.
4. Saini, S. K., & Saini, R. P. (2008). *Solar Energy, 82*, 1118–1130.
5. Hans, V. S., Gill, R. S., & Singh, S. (2017). *Experimental Thermal and Fluid Science, 80*, 77–89.
6. Singh, A. P., Varun & Siddhartha (2014). *Experimental Thermal and Fluid Science, 54*, 117–126.
7. Pandey, N. K., & Bajpai, V. K. (2016). *Solar Energy, 134*, 314–326.
8. Momin, A. M. E., Saini, J. S., & Solanki, S. C. (2002). *International Journal of Heat and Mass Transfer, 45*, 3383–3396.
9. Singh, S., Chander, S., & Saini, J. S. (2011). *Energy, 36*, 5053–5064.
10. Hans, V. S., Saini, R. P., & Saini, J. S. (2010). *Solar Energy, 84*, 898–911.
11. Kumar, A., Saini, R. P., & Saini, J. S. (2013). *Renewable Energy, 58*, 151–163.
12. Singh, I., & Singh, S. (2018). *Renewable and Sustainable Energy Reviews, 92*, 405–425.
13. Kumar, R., Chauhan, R., Sethi, M., Sharma, A., & Kumar, A. (2016). *Advances in Mechanical Engineering, 8*, 1–12.

Repowering of Wind Farm and Its Economic Feasibility: A Case Study

Suriti Kumari, K. Boopathi, and Nabin Sarmah

Nomenclature

Greek Symbols

% Percentage (dimensionless)
σ Standard deviation (dimensionless)

Superscripts

(*n*) *n*Th iteration

1 Introduction

Renewable energy is the solution to provide cleaner energy and meet the growing demand for energy. However, this sector needs to be developed in order to use the resources to its maximum potential. Repowering (replacing old turbines with new ones) is a way to boost resource utilisation. The repowered sites maintain the transmission asset while, reducing the cost in the repowering of farm, also minimising the payback time and levelized cost of electricity (LCOE).

S. Kumari (✉) · N. Sarmah
Department of Energy, Tezpur University, Tezpur, India

K. Boopathi
Research and Development, National Institute of Wind Energy, Chennai, India

P. Mahanta et al. (eds.), *Advances in Thermofluids and Renewable Energy*, Lecture Notes in Mechanical Engineering, https://doi.org/10.1007/978-981-16-3497-0_43

Repowered sites are more efficient with fewer machines [1]. The government target of 175 GW by 2022, of which 60 GW are to be achieved by wind energy projects (39.2 GW as on 31.03.2021 GW has been achieved to date).

Old wind turbines below 500 kW range are still in operation in states such as Tamil Nadu, Maharashtra, Gujarat, Madhya Pradesh and, according to studies, about 10% of the total wind installation in India is less than 500 kW rating totalling around 3,500 MW [2]. Therefore, considering the repowering of older wind turbines, it is necessary to enhance the additional capacity of onshore wind energy.

Financial modelling [3] indicates that the profitability of repowering is highly dependent on the specific performance of the project and the cost of operation and maintenance. The repowering can be partially or fully repowered. Completely repowered implies the decommissioning of turbines and the replacement of new turbines by newer assets at the same project site, though part of this includes leaving some turbines and replacing some existing parts, either the turbine height or the rotor diameter is expanded to generate further output [4]. Repowering tends to increase the power generation and performance of turbines, as has been done in some of the studies [4, 2], partly because the sites with the best potential have been used in the early years and the current infrastructure can be used to harness more energy [5].

According to General Electric (GE), the leading turbine installer, repowering can increase fleet output by 25% and adds 20–25 years to the life of the turbine [6]. The repowering market in India is high, as 0.3 GW of turbines are 20 years old and older [5].

2 Problem Formulation

The investigation was carried to clarify the economic viability of the site-specific risks in the repowering of the plant. The risk-associated research was carried out for the analysis of uncertainty, and the economic viability was evaluated independently using the effects of uncertainty.

2.1 Uncertainty Analysis

The economic feasibility of the wind farm needs to have a strong evaluation of the risks associated with energy production uncertainties. The increasing share of wind power generation enhances the stochastic nature of the power system and thus has a significant impact on energy markets [7]. Owing to the intermittent nature of the RES, there is some ambiguity associated with it. Resource estimation is an unpredictable process, and a wide range of variables, ranging from measurement errors to resource variations, lead to uncertainty and occur on a several time scale. The probability of a wind energy system may not be feasible without taking into account the uncertainties involved with the systems [8]. The uncertainty in the event of a site assessment is the uncertainty associated with the input (i.e. wind), the methods employed, the models used, the results observed (i.e. energy) and the sensitivity factor implemented to the transfer of uncertainty from wind to energy [9]. As per the IEC standards [10], the standard uncertainty is expressed as given by Eq. (1).

$$U_P = \frac{\sigma_P}{\sqrt{N}} \tag{1}$$

σ_P is the standard deviation, N is the number of samples.

As per the IEC standards [10], the overall uncertainty, however, is the RMS of all the individual uncertainty as expressed in Eq. (2) [11].

$$U_{\text{overall}} = \sqrt{\left(U_1^2 + U_2^2 + \cdots + U_n^2\right)} \tag{2}$$

2.2 *Economic Analysis*

The capital cost in establishing a wind turbine is high [12, 1]. Thus, carrying out the economic feasibility of the project is essential in determining the profitability of the project. Calculations indicate the costs, NPV (net present value), IRR (internal rate of return) and payback period for the various alternatives.

NPV is defined as the sum of present values of the individual cash flows of the same entity. The **IRR** measures the rate of return of a project while disregarding the absolute amount of money to be earned. Finally, **Payback Period** calculates the time required for the cash inflows to equal the original outlay, measuring risk, not return [13].

The equation for the determination of NPV, IRR and payback period are given in Eq. (3), (4) and (5), respectively, with the rate of interest being (r), the time period (t) and the total life-span of the project being (n), for the series of annual cash flows (CF), the IRR is assumed and follows an iterative method for the determination of IRR to get 0 NPV [14].

$$\textbf{NPV} = \sum_{t=1}^{n} \frac{\text{CF}t}{(1+r)^t} \tag{3}$$

$$\textbf{IRR},\ 0 = \sum_{t=0}^{n} \frac{\text{CF}t}{(1+\text{IRR})^t} \tag{4}$$

$$\textbf{Payback Period} = \frac{\text{Initial Investment}}{\text{Annual Cash Inflows}} \tag{5}$$

3 Analysis

3.1 The Wind Farm Specifications

The site considered is a flat terrain and has latitude and longitude 8.962236°N and 77.720203°E, zone 43P. With an elevation of 90m above mean sea level and 7m above ground level. The UTM projection is taken for analysis. The region is a flat terrain with minimal hindrances. The RIX (Ruggedness Index) is 0, indicating the nature of the terrain (flat terrain). The site is having the predominant wind direction west. The vector map for the site is given in the Fig. 1.

3.2 Classification of Site Uncertainties Associated with the Wind Farm Development

See Fig. 2.

3.3 Quantification of Uncertainties

The uncertainties connected with the site, instrument, modelling, etc. are quantified in Table 1; even the reference used for the quantification is cited, and the overall uncertainty is the root-mean square of the individual uncertainties.

To obtain the uncertainty in the energy production, the uncertainty in wind-speed is scaled by the sensitivity factor which was obtained to be 2.5 for the existing turbines giving uncertainty in energy production to be 19.28%, and taking that as a reference, the uncertainty for the repowered site is scaled by a factor of 2, giving the uncertainty to be 15.43% for the repowered case.

The annual energy production (AEP) can be written as:

$$\text{AEP} = \text{AEP(simulated)} - \text{Losses} - \text{Uncertainty}$$

Considering the losses to be 18.08% for flat terrain according to the standards by National Institute of Wind Energy.

Existing Turbines

The existing turbine generators are at 30 m hub height with 200 kW capacity. The distance between the turbines is 0.12 km. The wind power density of the site at 30 m hub height is 224 W/m^2, the capacity utility factor (CUF) for the region is 19.45%, and total energy production was obtained to be 4091.38 MWh.

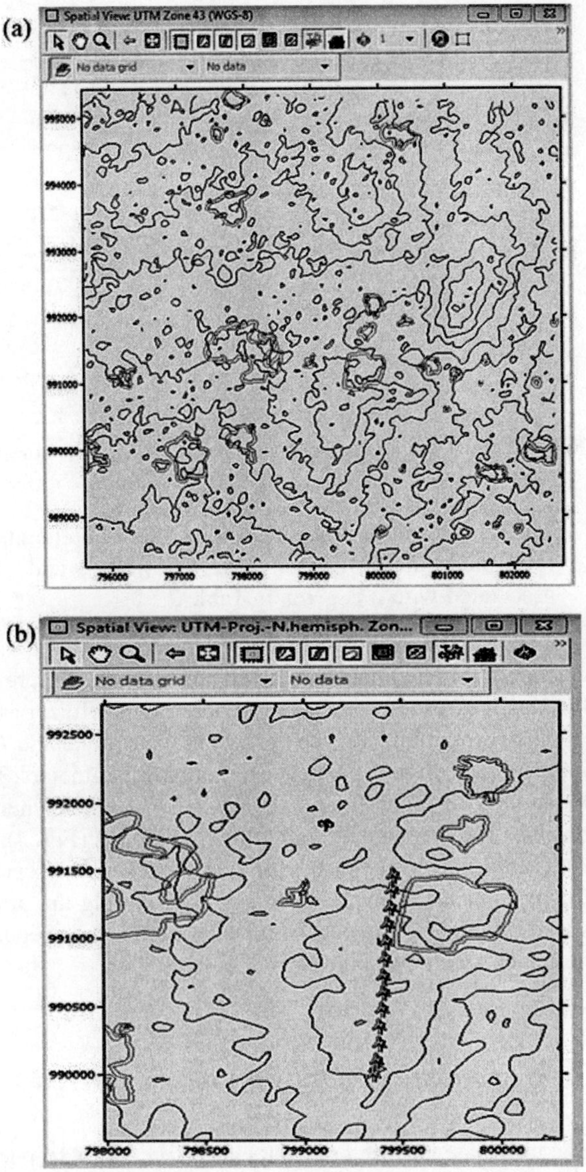

Fig. 1 **a** Vector map for the considered site. **b** The turbine configuration for the site

Repowering of the given turbines for Kayathar region, the current configuration, however, is having turbines of ratings-1 turbine = 225 kW and 12 turbines = 200 kW.

AEP at 30 m hub height considering the losses and uncertainty = **2562.85 MWh.**

Repowered Turbines

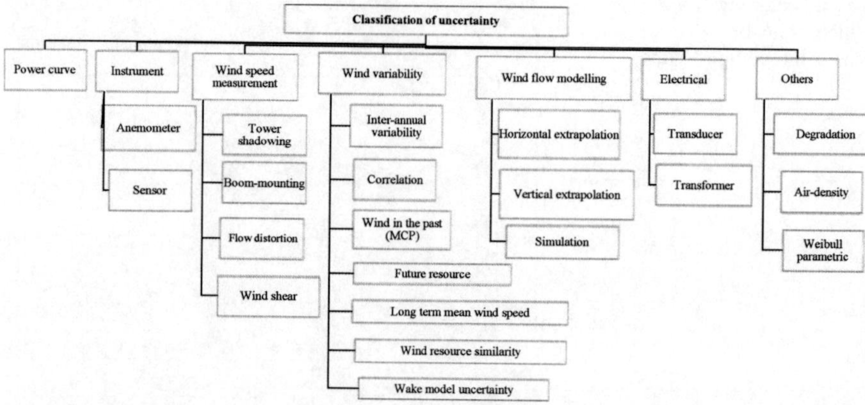

Fig. 2 The uncertainties are classified based upon the literature survey of many standard sources

The repowering for the sites for different hub heights to find the best hub height (HH) and turbine capacity for repowering and to understand the wake loss associated with it is given in Table 2.

The repowering is considered for the area with the same turbine location taking in consideration the finance involved and to make it more financially viable; therefore, the wake losses are considered to observe the effect of wake of the turbines (Fig. 3).

The correlation between the two graphs (wake and AEP) is observed to be 0.979 indicating the high dependence of both the parameters. The number of turbines in the location is 13 represented as $T1$, $T2\ldots$, $T13$. The energy production by individual turbines is represented in the following graphs (Fig. 4).

The turbine power curves for the turbines with 1, 2 and 3 MW capacity considered for repowering is given in the Fig. 5 providing the power produced at given wind speed, specifying the cut-in, cut-out, rated wind speed, the rotor diameter and hub height of specific turbines.

3.4 Economic Viability of the Repowered Wind Farm

The tariff rate has been set to Rs 4.16/kwh (Rs 4.16 per unit) as per the Tamil Nadu electricity regulation. The financial assumptions considered based on the recommendations made by Tamil Nadu electricity regulatory commission 2018 are described in Table 3.

The NPV, IRR and payback are determined considering the above economic constants. It was observed at 2 MW (120 m) hub height, the payback period obtained is lower than the other configuration at the given investment; also, the IRR is high at the assumed discount rate as provided in Table 3. Since the turbine configuration of 2 MW (120 m) is having comparatively high IRR, NPV and lower payback period, it

Table 1 The quantification of uncertainties derived from standard literature

Sl. no	Uncertainty	Reference No	Wind speed (%)
1	*Instrument*		*1.14*
1.1	*Anemometer*		*1.0004*
1.1.a	Calibration	[15]	1.00
1.1.b	Dynamic overspeeding	[16]	0.03
1.2	*Sensors*		*0.54*
1.2.a	Sensor assembly	[17, 18]	0.50
1.2.b	Pressure sensor	[10]	0.0007
1.2.c	Data logger	[17]	0.20
2	*Wind speed measurement*		*3.08*
2.1	Tower shadowing	[19]	0.05
2.2	Boom-mounting	[10]	0.50
2.3	Flow distortion	[19]	3.00
2.4	Wind shear	[17, 20]	0.48
3	*Wind Variability*		*6.08*
3.1	Inter annual variability	[17, 21]	5.79
3.2	Correlation	[6]	1.00
3.3	Wind in the past (MCP)	[22, 23]	0.36
3.4	Wind resource similarity	[14]	1.50
4	*Wind flow modelling*		*3.08*
4.1	Vertical extrapolation	[10, 11, 17]	–
4.2	Horizontal extrapolation	[11, 17]	0.68
4.3	Simulation	[11, 21]	3.00
5	*Electrical*		*0.0140*
5.1	Transducer	[16]	0.0037
5.2	Transformer	[16]	0.0135
6	*Others*		*1.50*
6.1	Degradation		1.00
6.2	Air density	[24, 25]	1.00
6.3	Weibull parametric		0.50
	Total		*7.71*

Table 2 Turbines configuration for repowering and the energy production, CUF and wake loss associated

Configuration	HH (m)	CUF (%)	Wake Loss (%)	AEP simulated (GWh)	AEP with losses and uncertainty (GWh)
Bonus-1 MW	90	28.04	1.15	31.93	21.23
Bonus-1 MW	120	30.95	1.73	35.24	23.43
Kenersys-2 MW	90	36.83	3.33	83.88	55.77
Kenersys-2 MW	120	40.16	4.15	91.47	60.82
Vestas-3 MW	90	35.18	4.07	120.19	79.92
Vestas-3 MW	120	38.34	5.13	130.99	87.11

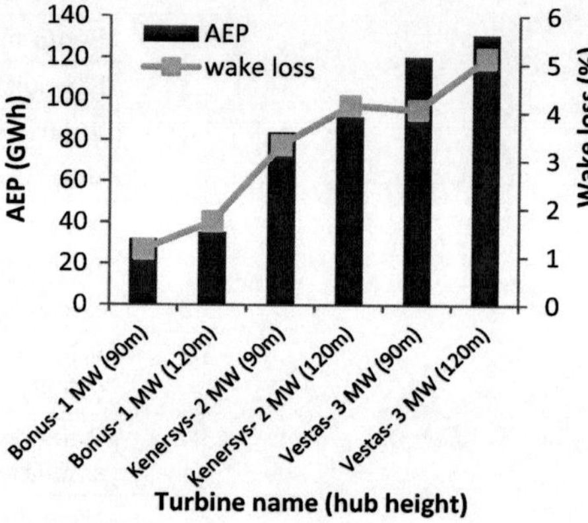

Fig. 3 Annual energy production vs. the wake loss

can be considered as the best option for the repowering of the wind farm as provided in Table 4.

By replacing the current turbines with a 2 MW turbine with a hub height of 120 m, it can lead to greater economic viability and lower tariff. The NPV vs IRR graph for the viable turbine groups is represented in Fig. 6, indicating the NPV at various discount rates and the point at which the NPV is 0 is the IRR for the project. As shown in the figure, the NPV decreases exponentially with the increase in the discount rate, as observed the projects having discount rate lower than IRR are economically acceptable while the projects with discount rate more than the IRR are economically not acceptable. As assumed, the discount rate is 9.95% in this case, all the turbines configuration with IRR lower than 9.95% are economically not feasible as provided in Table 4.

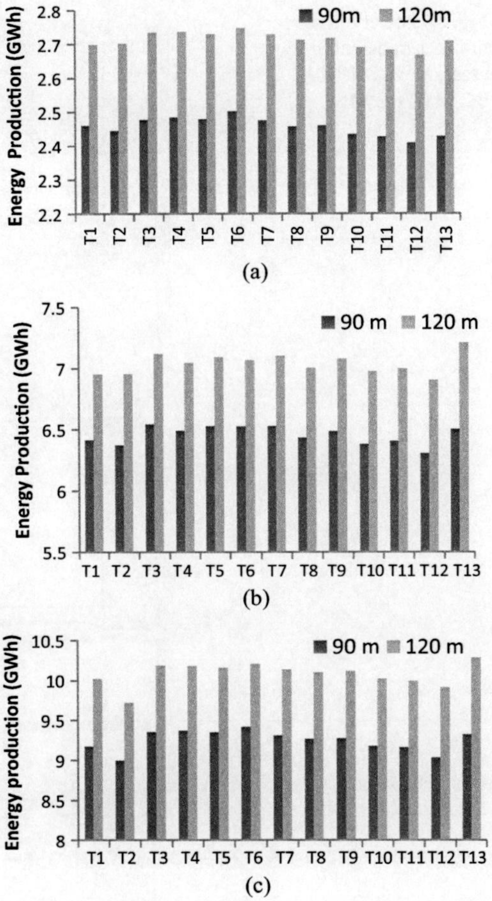

Fig. 4 Annual energy production for individual turbines of different capacity at hub heights 90 and 120 m. **a** 1 MW. **b** 2 MW. **c** 3 MW

4 Conclusions

The study highlights the need for repowering and suggests that the effectiveness of the estimation can be improved by carrying out the uncertainty estimate. The uncertainty in wind speed was 7.71%. The study takes into account the repowering of the turbines installed in the Kayathar area and indicates that the turbine with a power of 2 MW at a height of 120 m is economically viable. The results show that the new repowered projects are 4 times the height, 3.42 times the diameter of the rotor and 11.67 times that of the swept area. Repowering could boost annual energy generation by 23.73 times, bringing the CUF from 19.28–40.16. %.

Fig. 5 Power curves for turbines considered for assessment. **a** 1 MW. **b** 2 MW. **c** 3 MW

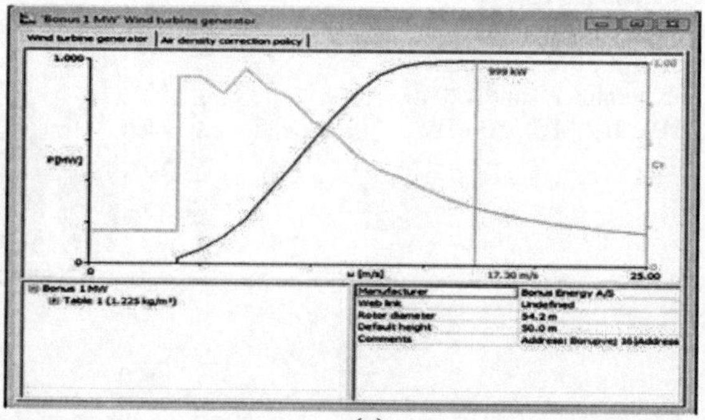

(a)

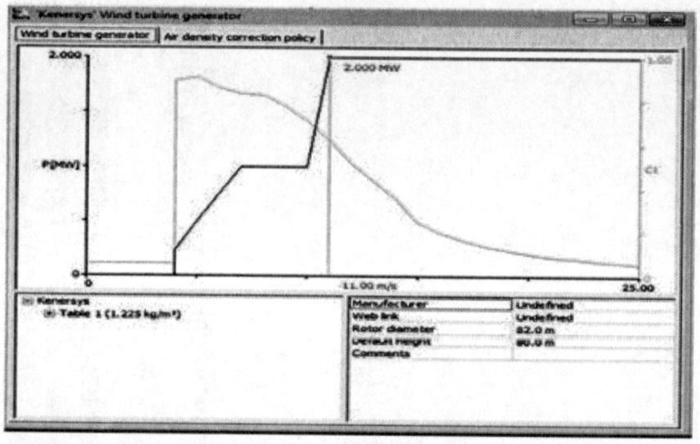

(b)

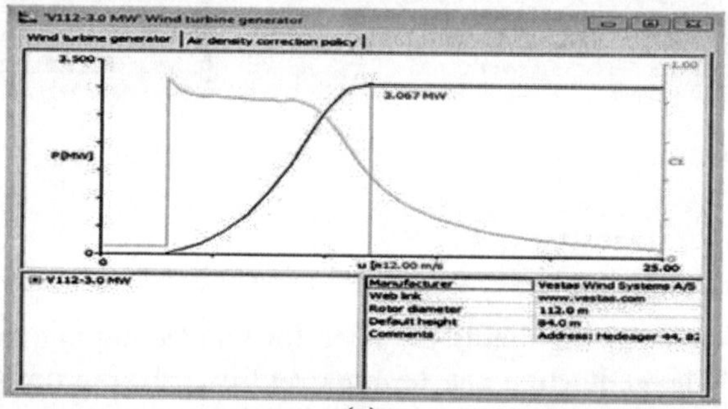

(c)

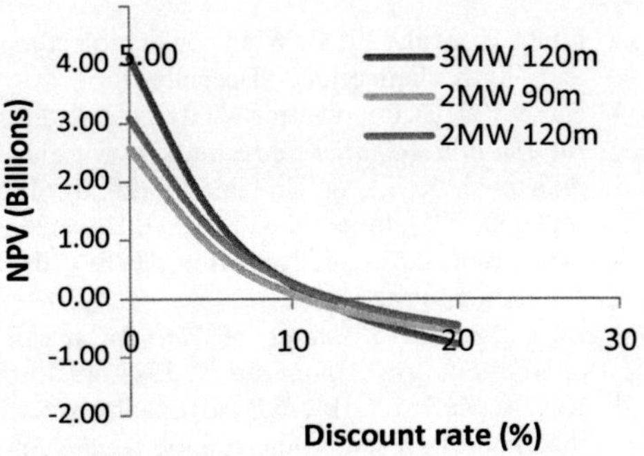

Fig. 6 NPV vs IRR graph for the viable turbine group

Table 3 Financial constraints considered for the economic feasibility

Investment	Amount
Capital cost	Rs. 5.25 crores/MW
Operation and maintenance expenses	1.1% on 85% of capital investment and 0.22% on 15% of the capital investment with an escalation of 5%
Insurance	0.75% on 85% of the capital cost for the first year and to be reduced by 0.5% every year
Life of the plant machinery	25 years
Depreciation	3.60% p.a
Discount rate	9.95%
For salvage value of 13 old WTGs and scrap	Rs 15,000,000 (source: Indo–German energy forum)

Table 4 Economic feasibility for the project with specific turbine configuration

Turbine capacity (MW)	HH (m)	IRR (%)	NPV (Billion)	Project feasibility	Payback period (Yrs)
1	90	5.49	−0.12	Not feasible	–
1	120	7.38	−0.21	Not feasible	–
2	90	10.73	0.08	Feasible	8.5
2	120	12.51	0.27	Feasible	7.5
3	90	9.69	−0.04	Not feasible	–
3	120	11.43	0.23	Feasible	8.0

References

1. Blanco, I. (2009). *The Economics of Wind Energy*. https://doi.org/10.1016/j.rser.2008.09.004
2. Idam Infrastructure Advisory Private Limited. (2018). Repowering of old wind turbines in India. http://idaminfra.com/wp-content/uploads/2018/10/Repowering-of-Old-Wind-Turbines-in-India.pdf.

3. Lantz, E., et al. (2013). Wind power project repowering : Financial feasibility, decision drivers, and supply chain effects (December).
4. Ray S. (2017). Repowering wind turbines adds generating capacity at existing sites. *U.S. Energy Information Administration.* https://www.eia.gov/todayinenergy/detail.php?id=33632.
5. Arántegui, R. L., & Uihlein, A. (2020). Technology effects in repowering wind turbines (pp. 660–675). https://doi.org/10.1002/we.2450.
6. Mortensen, N. G., & Jørgensen, H. E. (2011). Comparison of Resource and Energy Yield Assessment Procedures.
7. Zucker, A. (2015). *Impact of Wind Power Uncertainty Forecasting on the Market Integration of Wind Energy in Spain q, 159*, 334–349. https://doi.org/10.1016/j.apenergy.2015.08.104
8. Roy, A., Kedare, S. B., & Bandyopadhyay, S. (2010). Optimum sizing of wind-battery systems incorporating resource uncertainty. *Applied Energy, 87*(8), 2712–2727. https://doi.org/10.1016/j.apenergy.2010.03.027
9. MEASNET Procedure. (2016). Evaluation of site specific wind conditions. Version 2.
10. IEC 61400-12-1:2005. Wind Turbines-Part 12-1 : Power performance measurements of electricity producing wind turbines.
11. Zhang, M. H. (2015) Wind park production estimate. In *Wind Resource Assessment and Micro-Siting* (pp. 143–168). https://doi.org/10.1002/9781118900116.ch7.
12. Yang, J., et al. (2012). Economic feasibility analysis of a renewable energy project in the rural China. *Procedia Environmental Sciences, 13*, 2280–2283. https://doi.org/10.1016/j.proenv.2012.01.217
13. Castro-Santos, L., et al. (2016). Costs and feasibility of repowering wind farms. *Energy Sources, Part B: Economics, Planning, and Policy, 11*(10), 974–981. https://doi.org/10.1080/15567249.2014.907845
14. Bierman, H. (2009). Comparing net present value and internal rate of return. *Economics & Sociology, 4*(January), 46–53.
15. Coquilla, R., Obermeier, J., & White, B. (2007). Calibration procedures and uncertainty in wind power anemometers. *Wind Engineering, 31*, 303–316. https://doi.org/10.1260/030952407783418720
16. Busch, N., & Kristensen, L. (1976). Cup anemometer overspeeding. *Journal of Applied Meteorology, 15*, 1328–1332. https://doi.org/10.1175/15200450(1976)015%3c1328:CAO%3e2.0.CO;2
17. Brower, M. C. (2012). Uncertainty in wind resource assessment. In *Wind Resource Assessment* (pp. 220–232). Wiley Online Books. https://doi.org/10.1002/9781118249864.ch15.
18. Indasi, V. S., Lynch, M., Mcgann, B., Sutton, J., Yu, F., & Jeanneret, F. (2017). Use of mast and remote sensing data for wind resource assessment in Kenya *3*(03), 106–115.
19. Zhang, M. H. (2015), Measuring the Wind. In *Wind Resource Assessment and Micro-Siting* (pp. 169–200). Wiley Online Books. https://doi.org/10.1002/9781118900116.ch8.
20. Durišić, Ž, & Mikulović, J. (2012). A model for vertical wind speed data extrapolation for improving wind resource assessment using WAsP. *Renewable Energy, 41*, 407–411. https://doi.org/10.1016/j.renene.2011.11.016
21. Neves, D., Brito, M. C., & Silva, C. A. (2016). Impact of solar and wind forecast uncertainties on demand response of isolated microgrids. *Renewable Energy, 87*, 1003–1015. https://doi.org/10.1016/j.renene.2015.08.075
22. Zhang, M. H. (2015). Measure–Correlate–Predict. In *Wind Resource Assessment and Micro Siting* (pp. 121–142). Wiley Online Books. https://doi.org/10.1002/9781118900116.ch6.
23. Lackner, M. A., Rogers, A. L., & Manwell, J. F. (2008). Uncertainty analysis in MCP-based wind resource assessment and energy production estimation. *Journal of Solar Energy Engineering, Transactions of the ASME, 130*(3), 0310061–03100610. https://doi.org/10.1115/1.2931499

24. Becerra, L. O., & Hern, I. (2006). Evaluation of the air density uncertainty : The effect of the correlation of input quantities and higher order terms in the Taylor series expansion *17*, 2545–2550. https://doi.org/10.1088/0957-0233/17/10/003.
25. Afanasyeva, S., et al. (2016). Technical, economic and uncertainty modelling of a wind farm project. *Energy Conversion and Management, 107*, 22–33. https://doi.org/10.1016/j.enconman.2015.09.048

Computational Performance Analysis of the Perforated and Flat Plates Double Pass Solar Air Heaters

P. P. Dutta, P. Goswami, A. Sharma, Polash P. Dutta, and M. G. Baruah

Nomenclature

A	Area (m^2)
a	Length of specimen (m)
D_h	Hydraulic diameter of SAH (m)
FVM	Finite volume method
f	Coefficient of friction
h	Heat transfer coefficient (W/m^2K)
k	Thermal conductivity (W/mK)
Nu	Nusselt number
ΔP	Pressure drop (Pa)
Pr	Prandtl number
R_e	Reynolds number
SAH	Solar air heater
u, v, w	Velocity components in X, Y, Z axes

Greek Symbols

α	Volume fraction (dimensionless)
η	Efficiency (dimensionless)
ρ	Density of air (kg/m^3)
μ	Coefficient of viscosity (kg/ms)
δ	Delta

P. P. Dutta (✉) · A. Sharma · P. P. Dutta · M. G. Baruah
Department of Mechanical Engineering, Tezpur University, Napaam, Sonitpur, India

P. Goswami
Indian Institute of Technology, Kharagpur, India

P. Mahanta et al. (eds.), *Advances in Thermofluids and Renewable Energy*, Lecture Notes in Mechanical Engineering, https://doi.org/10.1007/978-981-16-3497-0_44

σ Diffusion Prandtl number
ε Dissipation kinetic energy, m^2/s^3

Superscripts

(n) nth iteration

Subscripts

a Air

1 Introduction

The natural non-renewable energy resources are diminishing each day due to the ever increasing population and industrialization. During 1970, solar energy was considered as one of the finest alternatives for fulfilling the snowballing demand for energy in the world. As a result of intensification in oil prices in the 1970s, a committed and sustainable research work was supported for extraction of maximum benefit from the solar energy [1]. Of all the renewable energy sources available, solar energy has the paramount reliability and sustainability. It has the potential to cope with the current energy demand to a considerable extent. As the solar energy radiation is elevated to a higher temperature than the ambient air, it may be utilized in a more beneficial ways [2]. A good part of Indian terrain is positioned in the tropical region of the globe and it has a decent hand access to the solar energy. Therefore, this territory of India receives a fair amount solar insolation. As a result, it is an ideal place for generating useful heat and electricity from solar potential. India's location on globe is fortunate enough to receive solar energy equivalent of more than 5000 trillion kWh per year. This is much more than the total annual consumption by the population. The daily global radiation is around 5 kWh per square meter per day with sunshine ranging between 2300 and 3200 h per year in the most parts of India [3].

Different solar thermal conversion devices are available. Dutta et al. [4–6] have reported energy efficient solar dryer, compact corrugated heat exchanger, different types of air heaters like flat plate, corrugated plate etc. *Garcinia pedunculata*, a locally available medicinal fruit, was dried in an innovative free convection corrugated solar dryer [7]. Sharma et al. [8] studied performance of aluminum can solar air heater. They observed that thermal performance of the aluminum solar air heater was better in terms of thermal efficiency and outlet air temperature. Begum et al. [9] studied parabolic trough collector for purification of drained water. Appropriate design of solar air heaters (SAHs) is a very crucial aspect for its optimal performance particularly the selection of absorber plate geometry. Ekechukwa and Nortan [10] reported research work on principles of operations, construction, and design of the various SAHs. SAHs may be classified based on the operational modes such as passive, hybrid, and active. Tyagi et al. [11] classified SAHs according to the tracking, non-tracking, with rock bed and PCM materials as the heat storages and the number of glass covers used. Air was heated in various space of duct, and it was directed into the one end of heating applications. It is termed as active SAH. Heat storage may be utilized in an active SAH, and the stored heat is used in the night time. Passive SAHs are rugged and very simple in design, construction and cost are very attractive in contrast to the construction cost of the active SAHs. Air heater may also be classified according to number of air passes; i.e., uni-pass, double-pass and multi-pass SAHs. The air flows in one approach either above or below the absorber plate from air inlet to outlet of the channel in a single pass SAH. While for a double-pass SAH, air flows in two ways that may be either in counter or parallel flow. Air passes above the collector plate support to shrink the top loss to the environment appreciably. At the other hand, side, edge and the bottom losses are reduced by using insulation material of low-thermal conductivity; e.g., wood, glass wool, saw dust etc. Top cover glass should be of low absorptivity and high transmissivity are preferred. Various researchers modified the main components, e.g., absorber plate and configurations of SAHs for analysis [12–15].

Mahmood [17] studied energy, exergy efficiency of a perforated plate solar air heater and a wire mesh cover at Baghdad. The maximum thermal efficiency with 90° bed slope was 63% for glazed and 47% for unglazed SAH. Corresponding exergy efficiencies were 4.8 and 2.3%, respectively. In another study, Mahmood [18] obtained thermal efficiency 86% at an air mass flow rate 0.018 kg/s. Wang et al. [19] performed thermal performance study with a black-coated perforated aluminum collector. At the air mass flow rate of 0.057 kg/s, maximum thermal efficiency was 75%. Shetty et al. [20] experimentally studied performance of a cross flow solar air heater with perforation. Maximum thermal efficiency was 75.55% at Reynolds number 9,000–15,000 at a perforation diameter 8 mm. Promvonge and Skullong [21] studied enhanced heat transfer in a solar receiver heat exchanger with the punched holes. Nusselt number over the flat plate duct increased around 3.91–5.52 times with a substantial friction loss. Skullong et al. [22] studied thermal efficiency in a solar air heater with the perforated-winglet-type vortex. Perforation diameter 1 mm gave the maximum heat transfer and friction factor up to 6.78 and 84.32 times. Khanlari et al. [25] studied experimentally and numerically a perforated baffle solar air heater for drying applications. They observed best instantaneous efficiency 84.30% at an enhanced mass flow rate. Dutta and Goswami [28] studied thermal performance of two different configurations of solar air heater absorber plates. Therefore, the present literature review addresses various researches and developments work on the efficient solar air heaters. Dutta et al. [29, 30] studied thermal performance of corrugated solar air heater. Sharma and Dutta [31] studied corrugated solar air heater for green tea partial drying.

Based on above literature, the objective of present work considers performance analysis of a perforated and flat plate solar air heaters. The finite volume method (FVM) is used to discretize the 3D equations in the turbulent regime, and the effect of radiation is considered. The behavior of the air flow strictly depends on the flow regime. The problem is simulated for the Tezpur University with latitude 26°65′28" N and longitude of 92°79′26" E with an average altitude of 73 m from the sea level. Moreover, solar radiation data for simulation study was considered for the month of November 2018.

2 Methodology

Both the absorber plates had dimensions of 1800 × 900 × 1 mm^3. The distances between the glazing and absorber plate and the bottom and absorber plate were maintained at 40 mm for both the cases. Perforated plate circular holes pitch was 10 mm, and perforation diameter was 1.5 mm accordingly. The geometry of the perforated and solid mild steel solar air heater absorber plates are presented in Fig. 1a, b.

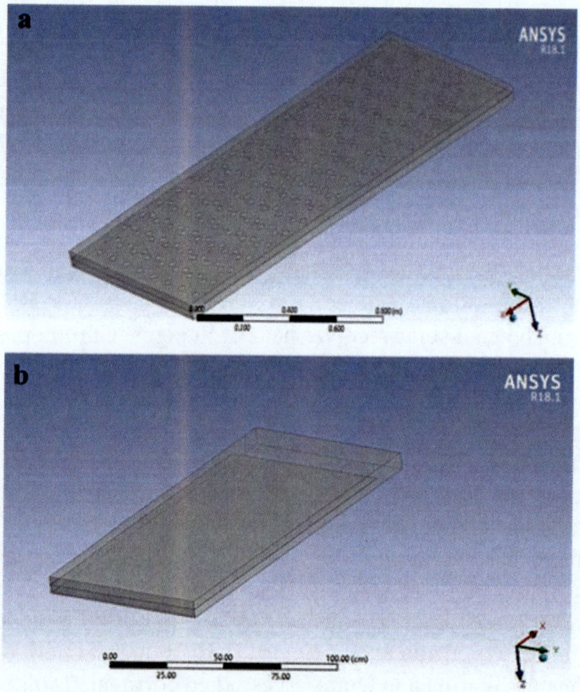

Fig. 1 **a** Perforated plate double pass. **b** Flat plate double pass

2.1 Governing Mathematical Relationships

The assumptions made for simulation in ANSYS: FLUENT 18.1 are given below. The detailed boundary conditions are presented in Table 1.

i. The working fluid air flow is assumed as incompressible.
ii. Constant thermal as well as physical properties of the working fluid.
iii. Air flow for the heat transfer study is steady and no slip is at the wall.

The general mathematical equations of continuity, momentum, and energy are given as below [24]:

$$\frac{\partial}{\partial x_i}(\rho v_i) = 0 \quad (1)$$

$$\frac{\partial}{\partial x_j}(\rho v_i v_j) = \left[-\frac{\partial p}{\partial x_i} + \frac{\partial}{\partial x_j}\left[\mu\left(\frac{\partial v_i}{\partial x_j} + \frac{\partial v_j}{\partial x_i} \right) \right] - \frac{2}{3}\mu \frac{\partial v_k}{\partial x_k} \partial_{ij} \right] \quad (2)$$

$$\frac{\partial}{\partial x_j}\left(\rho v_j C_p T - k\frac{\partial T}{\partial x_j} \right) = u_j \frac{\partial p}{\partial x_j} + \mu\left(\left(\frac{\partial v_i}{\partial x_j} + \frac{\partial v_j}{\partial v_i} \right) - \frac{2}{3}\frac{\partial v_k}{\partial x_k}\partial_{ij} \right) \quad (3)$$

The k-ε model was used for study solar air heater performance. The boundary conditions enforced on the perforated and flat plates duct. The local flow was considered turbulent and incompressible. The velocity and pressure boundary conditions were put at the inlet and at the outlet.

The governing mathematical relationship for the standard k-ε model [24] is:

$$\frac{\partial}{\partial x_i}(\rho k v_i) = \frac{\partial}{\partial x_j}\left[\left(\frac{\mu + \mu_t}{\sigma_k} \right) \frac{\partial k}{\partial x_j} \right] + \mu_t \left[\frac{\partial v_i}{\partial x_j} + \frac{\partial v_j}{\partial v_i} \right] \frac{\partial v_i}{\partial x_j} - \rho\varepsilon \quad (4)$$

$$\frac{\partial}{\partial x_i}(\rho \varepsilon v_i) = \frac{\partial}{\partial x_j}\left[\left(\frac{\mu + \mu_t}{\sigma_\varepsilon} \right) \frac{\partial \varepsilon}{\partial x_j} \right] + C_{1\varepsilon}\frac{\varepsilon}{k}\mu_t \left[\frac{\partial v_i}{\partial v_j} + \frac{\partial v_j}{\partial v_i} \right] \frac{\partial v_i}{\partial v_j} - C_{2\varepsilon}\rho\frac{\varepsilon^2}{k} - \alpha\rho\frac{\varepsilon^2}{k} \quad (5)$$

Modeling of turbulent Viscosity:

Table 1 Boundary condition for computation of perforated duct solar air heater

Boundary region	Type of region	flow rate (kg/s)	Pressure (Pa)	Temp. (K)
Inlet	Velocity	0.08	101,315	298
Outlet	Pressure	Out flow	101,305	340
Duct wall	Wall	No slip	–	–
Absorber	Wall	No slip	–	370

Table 2 Grid independency test

Refining levels	Number of elements	Percentage variation of air temperature	Percentage variation pressure drop
1	120,000	3.9	4.9
2	180,000	2.5	3.0
3	320,000	0.54	0.60
4	360,000	0.40	0.55

$$\mu_t = \rho C_\mu \frac{k^2}{\varepsilon} \tag{6}$$

The constants have been reported for the experimental work on fundamental turbulent and shear flows similar to boundary, mixing layers and jets or diminishing isotropic grid turbulence [24]. It applies fine in case of good range of wall-bounded and free shear flows.

2.2 *Grid Convergence Test*

Grid convergence test is used to define the improvement of the computational results by using uninterruptedly smaller cell sizes. It should converge to the precise results as the mesh become finer. The usual CFD practice is to start with a coarse mesh size and gradually improve it until the deviations perceived in the results are lesser than a predefined acceptable error (Table 2).

For the analysis of the perforated duct, and in order to evaluate the precision of the numerical procedure, the domain for a 2D channel of length 1800 mm, width 900 mm, channel height 40 mm and was tested. To describe the number of elements required, four different meshes were confirmed. The grid independence test was reinforced out in the analysis by accepting dissimilar grid elements as shown in Table 2. The grid individuality tests show that the grids of 320,000 elements assure a satisfactory solution. It is clear that after 320,000 there is not noteworthy percentage difference in appraised parameters.

2.3 *Data Reduction*

Reynolds, Nusselt numbers, average heat transfer coefficient, hydraulic diameter and pressure drop in the perforated and flat plate ducts are given by Eqs. (7)–(11) [23].

$$R_e = \frac{\rho \mathrm{V} \mathrm{D}_h}{\mu} \tag{7}$$

$$\overline{Nu} = \frac{\bar{h} D_h}{k} \quad (8)$$

$$\bar{h} = \frac{Q_u}{A_{\text{duct}}\left(\overline{T}_{\text{wall}} - \overline{T}_{\text{fluid}}\right)} \quad (9)$$

$$D_h = \frac{4A}{P} \quad (10)$$

$$\Delta P_{\text{duct}} = \frac{\rho V^2 4 f L}{2 D_h} \quad (11)$$

3 Results and Discussion

The perforated and a solid mild steel plates were considered for a double pass solar air heater. The absorbers were designed and investigated for computation of its performance. Keeping all the operating parameters identical for simulation in both the cases, observations were made. Mass flow rate and thermal performance are the most important variables to investigate the thermal efficiency of a solar air heater. Theoretical solar thermal energy potential for the region considered (Tezpur University Campus) throughout the year is depicted in figures below [16]. From Fig. 2, average solar radiation 750 W/m^2 is considered.

After performing simulation, using ANSYS Fluent version 18.1 the results that are obtained as follows:

For the perforated plate, simulation was done using the S2S radiation model with longitude and latitude as 26°65′28" N and 92°79′26" E. and time zone of +5.5 h. The constants of mathematical modeling are presented in Table 3.

Figure 3 shows temperature distribution in a perforated plate solar air heater. It is clear that for an average solar radiation of 750 W/m^2, the minimum air temperature 325 K and maximum 335 K near the plate surface. The maximum temperature achievable near perforation (330 K) is less than 335 K because of enhanced heat transfer into the flowing air caused by local turbulence produced.

Since the air velocity through the perforated duct was small, as a result pressure drop from inlet to the outlet is small. Average pressure drop from the inlet to outlet was 0.20 Pa as shown in Fig. 4.

Figure 5 represents overall increasing air temperature from inlet to out of the solar air heater. Maximum air temperature very near to the absorber plate is in the range of 335–338 K. Middle air layer temperatures are around 328 K, and top air layer temperature is around 318 K. Under similar operating conditions, flat plate solar air heater was simulated. The performance results are presented below.

Figure 6 represents temperature distribution in a flat plate solar air heater. It is clear that for an average solar radiation of 750 W/m^2, the minimum air temperature

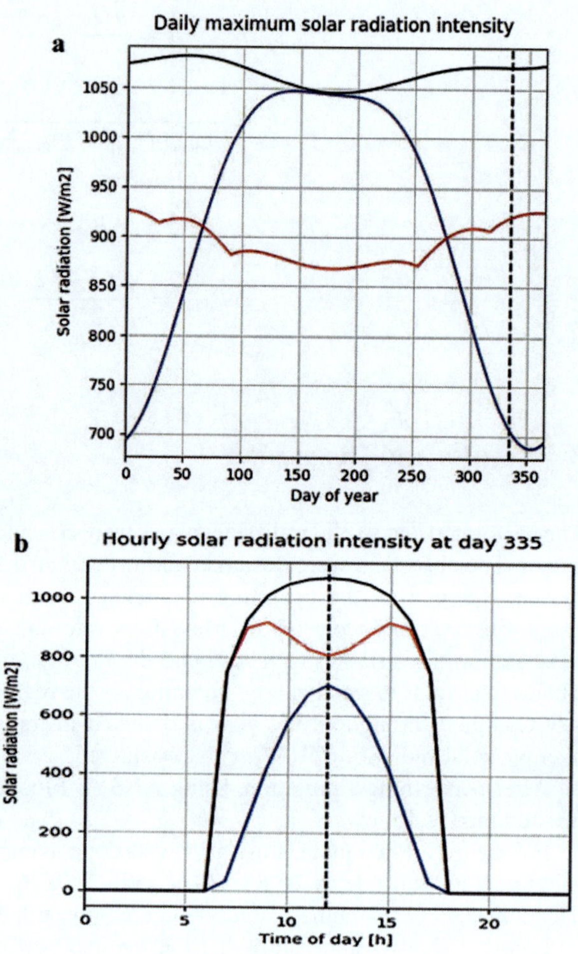

Fig. 2 **a** Maximum solar radiation intensity on a horizontal surface (blue), on a vertical surface (red), and on a surface perpendicular to the sunlight (black). **b** Hourly solar radiation intensity on a horizontal surface (blue), on a vertical surface (red), and on a surface perpendicular to the sunlight (black) on a particular day

Table 3 K-ε model constants

$C_{1\varepsilon}$	C_2	$C_{3\varepsilon}$	σ_ε	σ_k	C_μ
1.42	1.8	1	1.2	1	0.1

305 K and maximum 320 K near the plate surface. Average pressure drop at outlet is 0.11 Pa as shown in Fig. 7.

Figure 8 represents overall increasing air temperature from inlet to out of the flat plate solar air heater. The average air temperature just above the absorber plate is 325 K, middle layer is 320 K, and the top average temperature 315 K. There is a average overall rise in air temperature (318 K) at the outlet.

Figure 9 represents point to point rise in air temperature from inlet to outlet for both the solar air heaters. For both the air heaters, inlet temperature is almost constant.

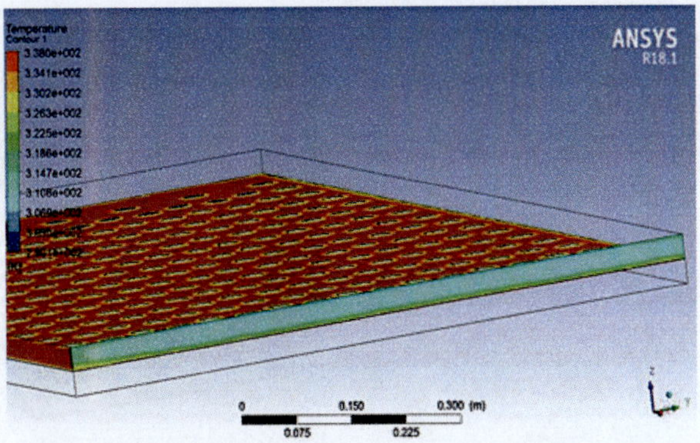

Fig. 3 Temperature distribution from inlet to outlet

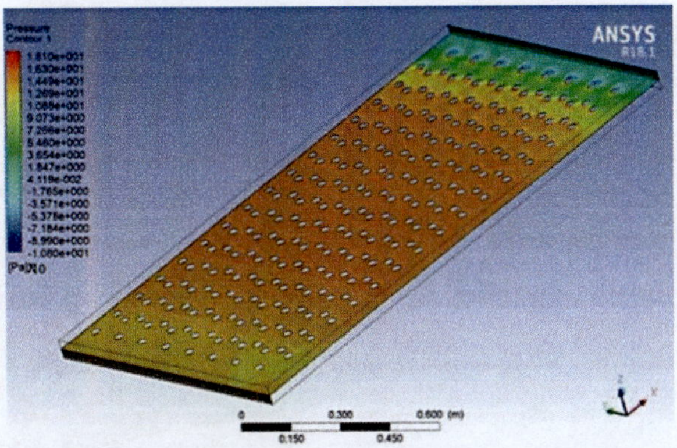

Fig. 4 Pressure distribution from inlet to outlet

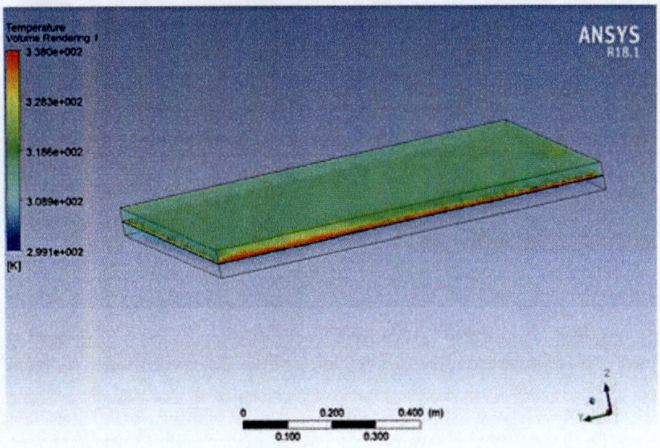

Fig. 5 Temperature distribution across the volume of a perforated plate solar air heater

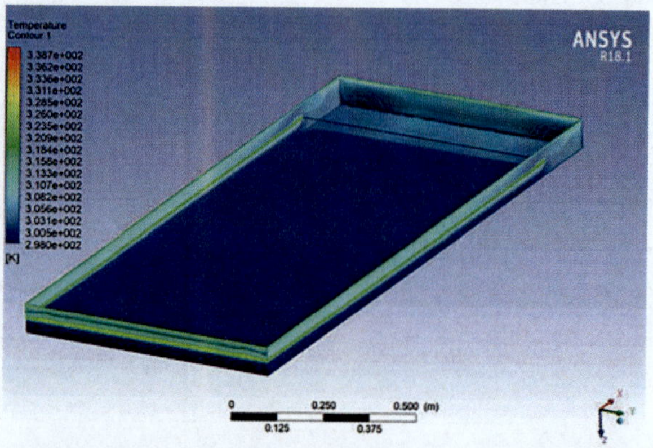

Fig. 6 Temperature distribution across inlet and outlet (K)

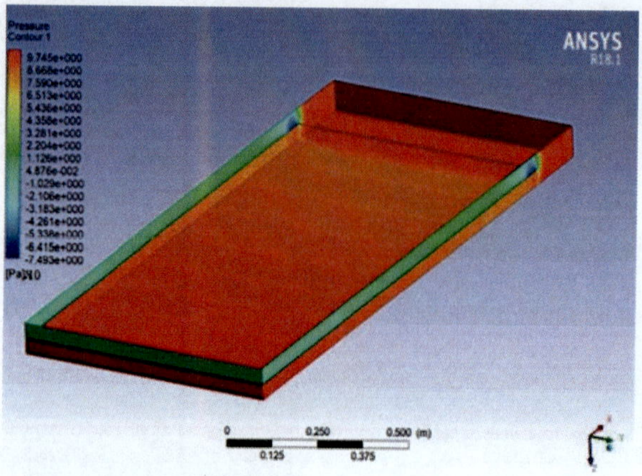

Fig.7 Pressure distribution at the inlet and outlet

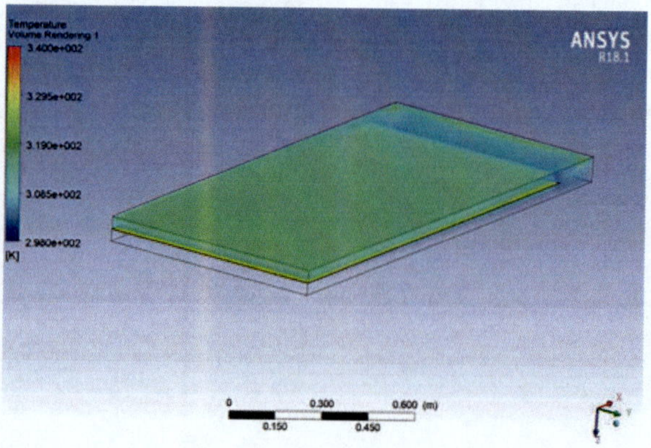

Fig. 8 Temperature distribution across the volume of flat plate solar air heater

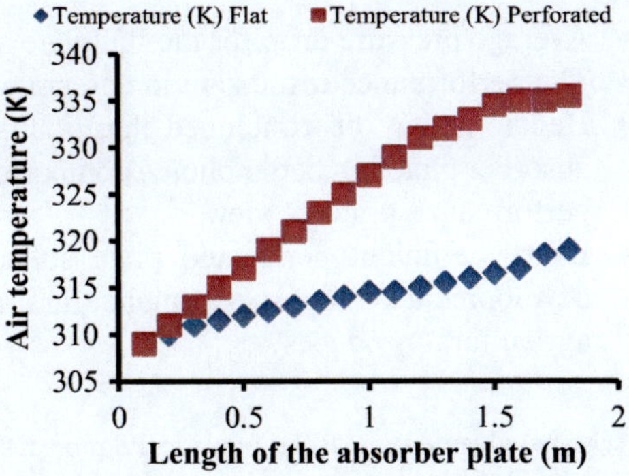

Fig. 9 Average variation of air temperature from inlet to outlet of flat solar air heater

The average air temperatures increase linearly for both the air heater. The average outlet temperature for the flat air heater was 315 and 335 K for the perforated solar air heater. Hence, there is in an average 20 K more temperature rise for the perforated solar air heater under similar operating conditions. Hence, the perforated double pass solar air heater works satisfactorily than flat double pass solar air heater.

3.1 Validation

The results obtained by, Abuska [9] for energy and exergy analysis of solar air heater have in agreement with present studies. Zheng et al. [26] reported similar temperature rise (average 23 °C) for outlet temperature comparing to inlet temperature. Promvonge et al. [27] also observed enhancement in efficiency for punched delta winglet absorber plate SAH.

4 Conclusion

CFD analysis of a double pass perforated plate solar air heater (SAH) and a flat plate SAH have been performed using Ansys FLUENT 18.1. Both the absorber plates had dimension of 1800 × 900 × 1 mm^3. Finite volume method (FVM) is used to discretize the governing equations. The computational results show that there is a significant improvement in thermal efficiency of a perforated plate double pass comparing to that of a flat plate one for the given mass flow rate of air of 0.08 kg/s.

- For an average solar radiation of 750 W/m^2, the minimum air temperature 325 K and maximum 335 K near the perforated plate surface.
- Average pressure drop at outlet for the perforated plate SAH was 0.20 Pa.
- For the flat plate solar air heater, minimum air temperature 305 K and maximum 325 K near the plate surface.

- Average pressure drop for the flat plate solar air heater at outlet is 0.11 Pa.
- The performance results are in agreement with previously published works.
- Hence, it may be concluded that double pass perforated plate solar air heater absorber plate is a better choice comparing to that of a flat plate from heat transfer performance point of view.
- Energy efficient perforated plate solar air heater may be recommended for development of high-performance indirect type solar dryer for drying of local agricultural produces.

Acknowledgements AICTE funded RPS project "Design Development and Performance Testing of Solar Dryer for Drying *Garcinia pedunculata* with Thermal energy Storage" 2019–2022.

References

1. Yadav, A. S., & Bhagoria, J. L. (2014). A CFD based thermo-hydraulic performance analysis of an artificially roughened solar air heater having equilateral triangular sectioned rib roughness on the absorber plate. *International Journal of Heat. Mass Transfer, 70*, 1016–1039.
2. Bhushan, B., & Singh, R. (2010). A review on methodology of artificial roughness used in duct of solar air heaters. *Energy, 35*, 202–212.
3. Khare, V., Nema, S., & Baredar, P. (2013). Status of solar & wind energy in India. *Renewable and Sustainable Energy Reviews, 27*, 1–10.
4. Dutta, P., Dutta, P. P., & Kalita, P. (2019). Thermohydraulic investigation of different channel height on a corrugated heat exchanger. In *AIP Conference Proceedings*. https://doi.org/10.1063/1.5096502.
5. Dutta, P.P., & Kumar, A. (2017). Development and performance study of solar air heater for solar drying applications. In O. Prakash, A. Kumar (Eds.), *Solar drying technology: Concept, design, testing, modeling, economics and environment* (pp. 579–601). Springer. https://doi.org/10.1007/978-981-10-3833-4_21.
6. Dutta, P. P. (2014). *Prospect of renewable thermal energy in black tea processing in Assam: An investigation for energy resources and technology* (Ph.D. diss). Tezpur University. https://shodhganga.inflibnet.ac.in/handle/10603/37571.
7. Dutta, P., Dutta, P. P., & Kalita, P. (2021). Thermal performance studies for drying of *Garcinia pedunculata* in a free convection corrugated type of solar dryer. *Renewable Energy, 163*, 599–612.
8. Sharma, A., Dutta, P., Goswami, P., Dutta P. P., & Das, H. (2019). Possibility of waste aluminium can for solar air heater through thermal performance studies. In: *Proceeding of International Conference on Waste management towards circular economy from 27/11/2019 to 30/11/2019.* KIIT University.
9. Begum, S. S., Goswami, A. P., Jangid, H., & Dutta, P. P. (2021). Design and modelling of a parabolic trough solar collector. In: B. Kakati, B. R. Phukan, T. Rajbongshi, & D. Bora (Eds.), *Advances in science and technology* (Vol. II, pp. 197–203). Tata Mc Graw Hill.
10. Ekechukwu, O., & Norton, B. (1999). Review on solar energy drying technology. *Energy Conversion and Management, 40*, 615–655.
11. Tyagi, V. V., Panwar, N. L., Rahim, N. A., & Kothari, R. (2012). Review on solar air heating system with and without thermal energy storage system. *Renewable Sustainable Energy Reviews, 16*, 2289–2303.
12. Chamoli, S., Chauhan, R., Thakur, N. S., & Saini, J. S. (2012). A review of the performance of double pass solar air heater. *Renewable Sustainable Energy Review, 16*, 481–492.

13. Singh, S., & Dhiman, P. (2016). Thermal performance of double pass packed bed solar air heaters- a comprehensive reviews. *Renewable able Energy Review, 53*, 1010–1031.
14. Kabeel, A. E., Hamed, M. H., Omara, Z. M., & Kandeal, A. W. (2017). Solar air heaters: Design, configurations, improvement methods, and applications-a detailed review. *Renewable Sustainable Energy Review, 70*, 1189–1206.
15. Abuska, M. (2018). Energy and exergy analysis of SAH having new design absorber plate with conical surface. *Applied Thermal Engineering, 131*, 115–124.
16. https://www.cableizer.com/tools/solar_radiation/ (Dated: 01/12/2018 13:16).
17. Mahmood, A. J. (2019). An experimental study of energy and exergy for glazed and unglazed solar system with perforated absorber plate and wire mesh layers. *International Journal of Renewable Energy Research, 9*(4), 1901–1911.
18. Mahmood, A. J. (2020). Thermal evaluation of a double pass unglazed solar air heater with perforated plate and wire mesh layers. *Sustainability, 12*, 3619. https://doi.org/10.3390/su1209 3619
19. Wang, Y., Boulic, M., Robyn Phipps, R., Plagmann, M., & Cunningham, C. (2020). Experimental performance of solar air collector with a perforated black plate in New Zeland. *Energies, 13*, 1415. https://doi.org/10.3390/en13061415
20. Shetty, S. P., Paineni, A., Kande, M., Madhwesh, N., Sharma, N. Y., & Vasudeva, K. K. (2020). Experimental investigations on a cross flow solar air heater having perforated circular absorber plate for thermal performance augmentation. *Solar Energy, 197*, 254–266.
21. Promvonge, P., & Skullong, S. (2019). Heat transfer augmentation in solar receiver heat exchanger with hole-punched wings. *Applied Thermal Engineering, 155*(5), 59–69.
22. Skullong, S., Promthaisong, P., Promvonge, P., Thianpong, C., & Pimsam, M. (2018). Thermal performance in solar air heater with perforated-winglet-type vortex generator. *Solar Energy, 170*, 1101–1117.
23. Nakhchi, M. E. (2019). Experimental optimization of geometrical parameters on heat transfer and pressure drop inside sinusoidal wavy channels. *Thermal Science and Engineering Progress, 9*, 121–131.
24. Akdag, U., Akcay, S., & Demiral, D. (2017). Heat transfer enhancement with nanofluids under laminar pulsating flow in a trapezoidal-corrugated channel. *Progress in Computational Fluid Dynamics, 17*, 302–312.
25. Khanlari, A., Guler, H. O., Tuncer, A. D., Sirin, C., Bilge, Y. C., Yalmiz, Y., & Gungor, A. (2020). Experimental and numerical study of the effect of integrating plus-shaped perforated baffles to solar air collector in drying application. *Renewable Energy, 145*, 1677–1692.
26. Zheng, W., Li, B., Zhang, H., You, S., Li, Y., & Ye, T. (2016). Thermal characteristics of a glazed transpired solar collector with perforating corrugated plate in cold regions. *Energy, 216*, 781–790.
27. Promvonge, P., & Skullong, S. (2020). Enhanced heat transfer in rectangular duct with punched winglets. *Chinese Journal of Chemical Engineering, 28*, 660–671.
28. Dutta, P. P., & Goswami, P. (2021). A Comparative thermal performance studies between two different configurations of absorber plates in a double pass solar air heater using CFD and experimental analysis. Tata Mc Graw HillIn B. Kakati, B. R. Phukan, T. Rajbongshi, & D. Bora (Eds.), *Advances in science and technology* (Vol. II, pp. 178–183)
29. Dutta, P. P., Kakati, H., Bardalai, M., & Dutta, P. P. (2021). Performance studies with trapezoidal, sinusoidal, and square corrugated aluminium alloy (AlMn1Cu) plate ducts. B. Das, R. Patgiri, S. Bandyopadhaya, V. M. Balas (Eds.), *Modeling, simulation and optimization: Proceedings of CoMSO 2020* (pp. 751–774, Springer, Singapore). https://doi.org/10.1007/978-981-15-9829-6_59.
30. Dutta, P. P., Sharma, D., Das, H., Sharma, A., Dutta, P., & Dutta, P. P. (2021). Computational fluid dynamics and experimental analysis of a corrugated plate duct. *Mathematical Forum, 28*(1), 1–18.
31. Sharma, A., & Dutta, P. P. (2021). Exergy analysis of a solar thermal energy powered tea withering trough. *Materials Today: Proceeding*. https://doi.org/10.1016/j.matpr.2021.06.

Numerical Analysis and Performance Evaluation of Cast Steel Based Sensible Heat Energy Storage System

Sanjeev K. Vishwakarma, Sukumar Pati, and Siddhartha Kosti

1 Introduction

The idea behind the thermal energy storage system changing the way users generate the numerous amounts of heating and cooling capacity that consume lot of conventional energy from the grid. The problem is that much of the grid power used for heating and cooling buildings is generated by energy from fossil fuels such as coal, oil and natural gas. This can be addressed using TES, which can provide heating and cooling solutions simply by carry out the distributed heat in a natural landscape or cycle, for example, by applying heat stored in solar collectors or by distributing cold water or air from underground to cool a building space.

2 Governing Equation

In order to study the thermal storage behavior of SHS system, three physical processes have been simulated, i.e., fluid flow, heat transfer by conduction and convection. In this section, a summary of the mathematical model used for simulation of transient behavior of SHS units is presented. Assumptions taken are

- The inlet velocity profile of HTF is fully developed, but heat transfer occurs in the thermal entry region.
- The pressure drop of HTF through the tube is neglected.

S. K. Vishwakarma (✉)
Department of Mechanical Engineering, Jorhat Engineering College, Jorhat 785007, India

S. Pati
Department of Mechanical Engineering, NIT Silchar, Assam 788010, India

S. Kosti
Department of Mechanical Engineering, Rajkiya Engineering College, Banda, U.P. 210001, India

P. Mahanta et al. (eds.), *Advances in Thermofluids and Renewable Energy*, Lecture Notes in Mechanical Engineering, https://doi.org/10.1007/978-981-16-3497-0_45

- SHSM is isotropic.
- No axial conduction in HTF.
- Neglected radiation effects.

2.1 Continuity Equation

$$\frac{\delta \rho_f}{\delta t} + \rho_f \nabla \vec{v} = 0 \tag{1}$$

2.2 Navier Stokes Equation in Cylindrical Co-Ordinate

$$\rho_{f}.\frac{\mathrm{D}v}{\mathrm{D}t} = -\nabla P + \mu \nabla^{2} \tag{2}$$

2.3 Heat Transfer in Solid Fluid Interface

$$\rho_f C_{\mathrm{Pf}} \frac{\mathrm{D}T}{\mathrm{D}t} = k_s \nabla^2 T_f \nabla^2 \tag{3}$$

2.4 Heat Transfer in Solid Region

$$\rho_f C_{\mathrm{Pf}} \frac{\delta T}{\delta t} = k_s \nabla^2 T_f \nabla^2 \tag{4}$$

3 Material and Geometry

Cast steel is considered as TES material due to its favorable properties. Figures 1 and 2 show the geometry of the cylindrical model examined in the present work.

Properties: Thermal conductivity—40 W/m, Density—7800 kg/m3, Specific heat—600 J/kgk.

Dimension of the model: Length—740 mm, Diameter—267 mm, Total no. of tube-19, Tube hole diameter—2.7 mm.

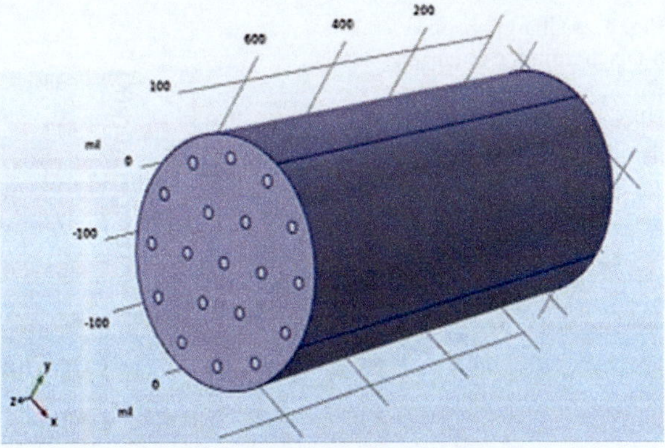

Fig. 1 Model geometry in 3D

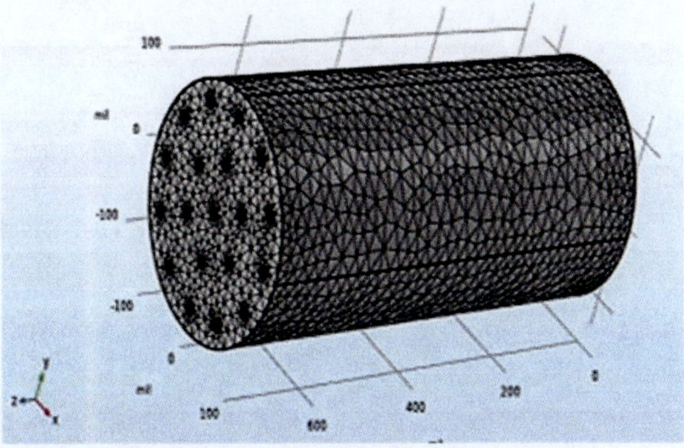

Fig. 2 Mesh in XYZ plane

The TES system has been designed to consist maximum flexibility basis on the criterion such as topology, orientation, quantity of storage material and placement of temperature sensors for testing various TES prototypes.

4 Validation

Numerical simulation of cast steel as sensible heat storage material on COMSOL software is done and charging and discharging results of the storage bed is obtained after simulation. The charging and discharging time at different HTF velocities and at different temperature ranges are obtained and studied. The results obtained after study is compared with a previously conducted research using cast steel as sensible heat storage material by K. Vigneshwaran, Gurpreet Singh Sodhi, P. Muthukumar, Anurag Guha, S. Senthilmuruga. Figure 3a, b represents the volume average bed temperature variation of cast steel during charging and discharging at $v = 2.5$ m/s.

Fig. 3 **a** Charging. **b** Discharging. **c** Charging efficiency. **d** Discharging efficiency

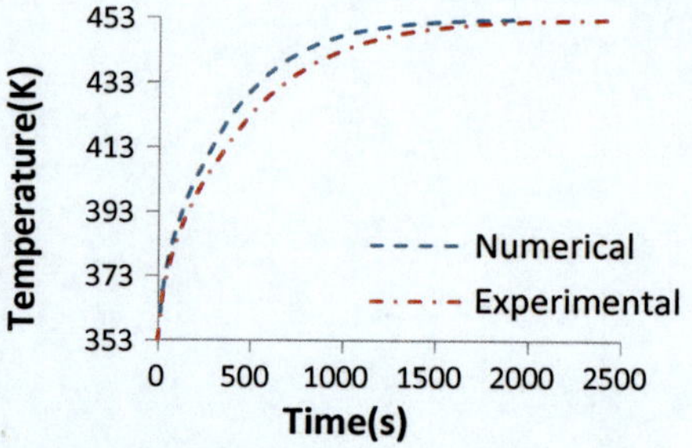

(a): Charging

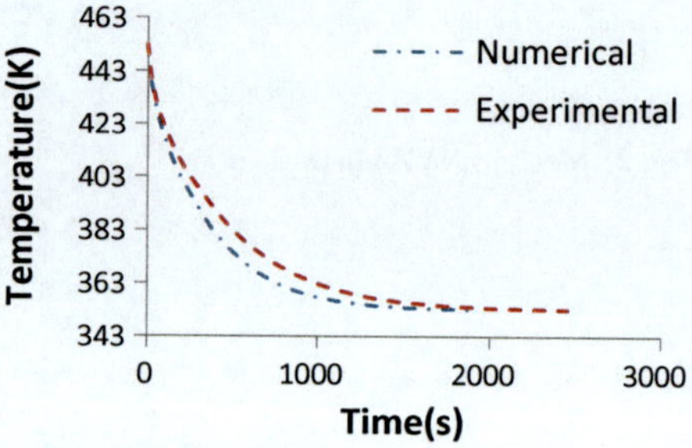

(b): Discharging

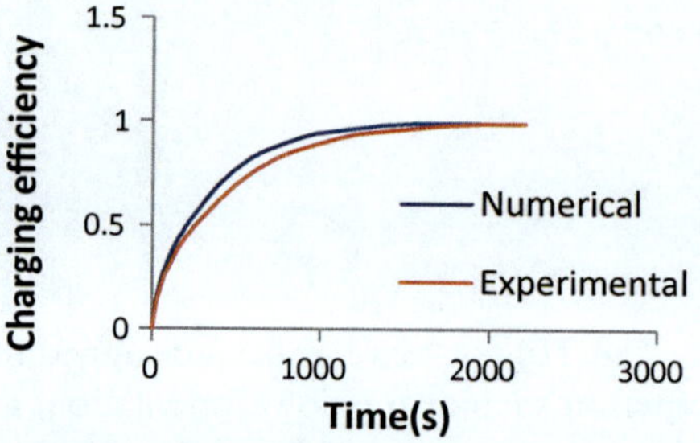

(c) Charging efficiency

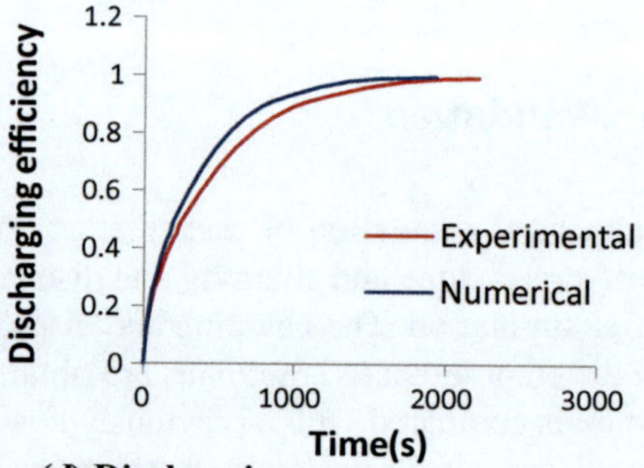

(d) Discharging

5 Result and Discussion

5.1 Convergence Validation of the Problem

The dependent variables are temperature, velocity and pressure for which equations are compiled and results are computed. It is observed that initially the result diverges for charging cycle after that it converges smoothly up to end of the solution. Convergence plot for discharging cycle is smooth from starting to end of the solution. The convergence plots for 2.5 m/s velocity and temperature range 493–573 K are shown in Fig. 4a, b.

Fig. 4 **a** Charging cycle **b** Discharging cycle

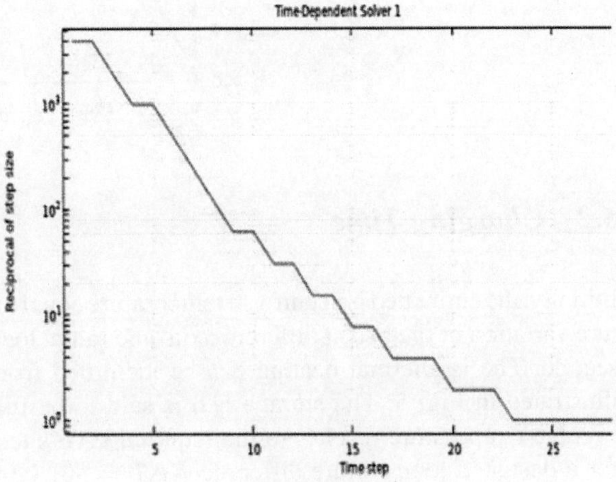

(a) Charging cycle

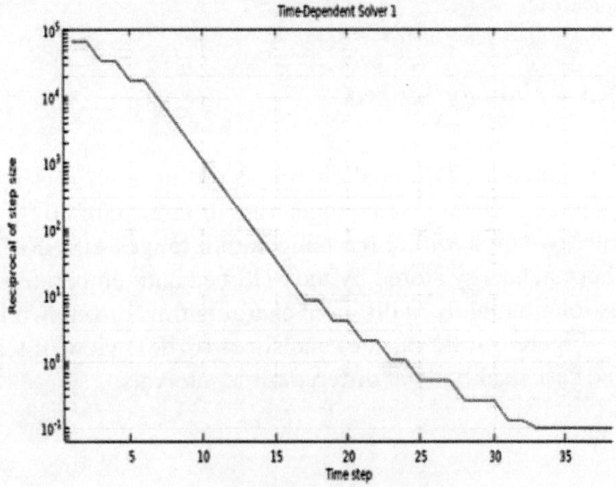

(b) Discharging cycle

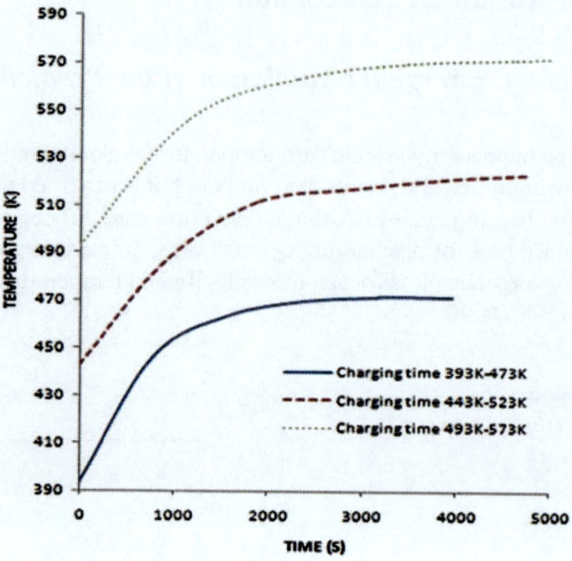

Fig. 5 Volume average bed temperature variation of CAST STEEL during charging

5.2 *Charging Time*

Initially, the entire bed is at uniform temperature equal to room temperature. Temperature variation of the bed at different axial and radial locations was recorded for every second. The isothermal heating can be identified from the HTF inlet temperature illustrated in Fig. 5. The storage bed is said to be fully charged when its volume average temperature reaches to the required HTF's temperature. It is observed that for a designed temperature difference ($\Delta T = 80$ °C) with the range of 493–573 k with different velocities.

5.3 *Energy Stored*

The amount of thermal energy stored in the bed at its respective charging time is estimated. For a given temperature difference of 80 °C (ΔT), the amount of thermal energy stored within the temperature ranges of 493–573 K. Numerical analysis of thermal energy stored by the SHS bed and temperature distribution radially as well as longitudinally at different charging time are shown in Fig. 6.

Figure 6 solid (a, c, e) and slices (b, d, f) view of temperature distribution inside the cast steel body at different time intervals.

Fig. 6 **a** $t = 0$ s. **b** $t = 0$ s .**c** $t = 3005.6$ s. **d** $t = 3005.6$ s. **e** $t = 5780$ s. **f** $t = 5780$ s

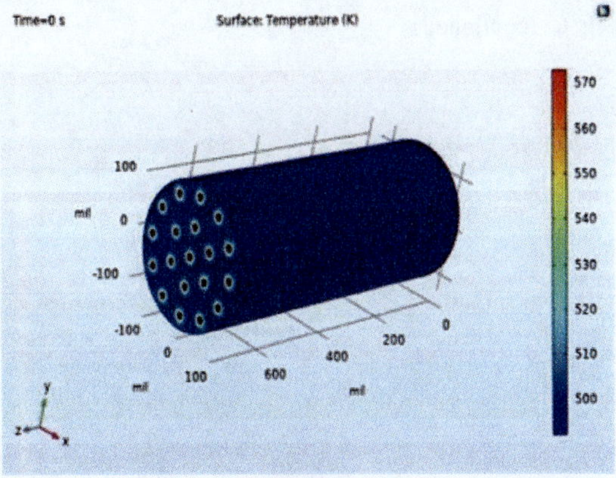

(a) t=0s

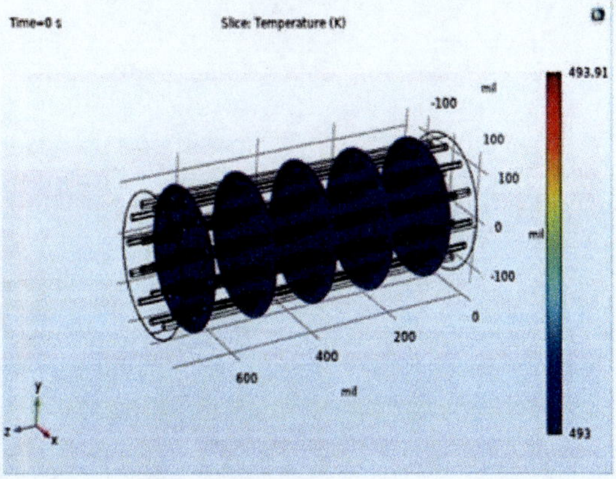

(b) t=0s

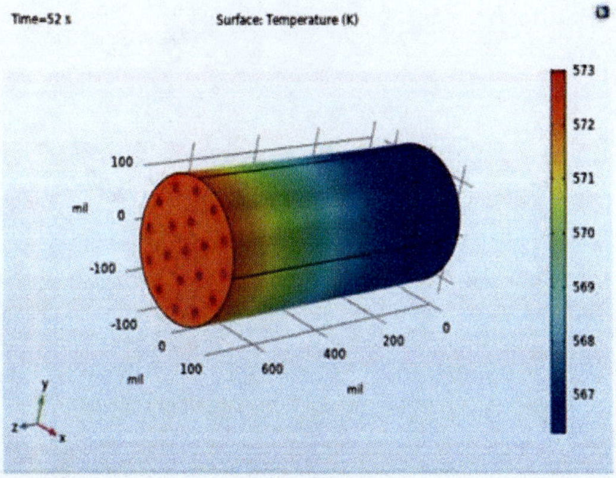

(c) t=3005.6s

Fig. 6 (continued)

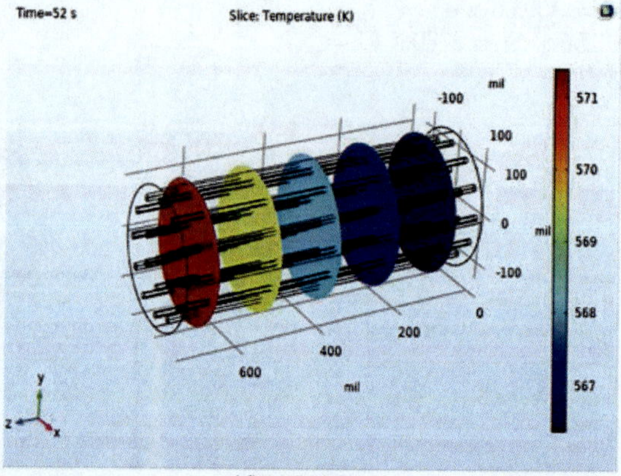

(d) t=3005.6s

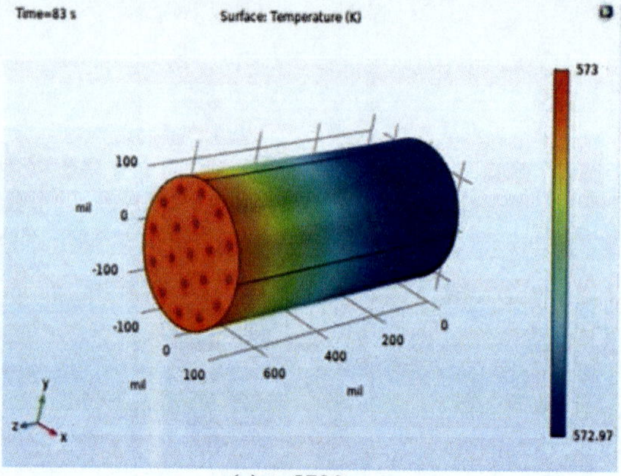

(e) t=5780s

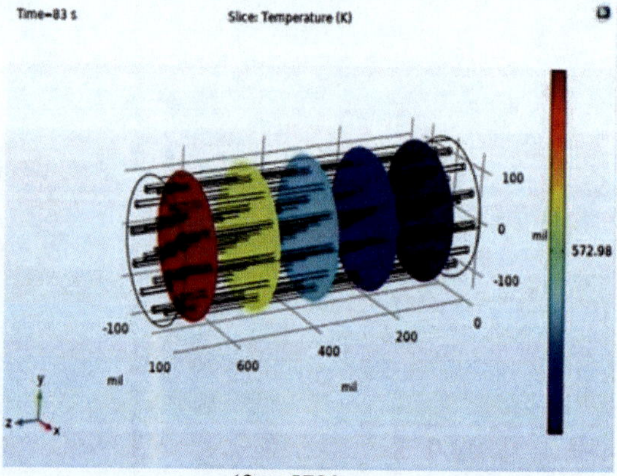

(f) t=5780s

5.4 Charging Energy Efficiency

Charging energy efficiency for the range of 493–573 K is shown in Fig. 7. η_{ch} of bed is initially zero when the heat transfer fluid temperature is 493 K. Once the storage bed is fully charged, for the temperature range of 493–573 K, was almost about 92.78% as because the volume average temperature of the storage bed attains the value of the Tinlet inlet within their respective charging times.

Fig. 7 **a** Charging energy efficiency (η_{ch}). **b** Energy storage rate in cast steel

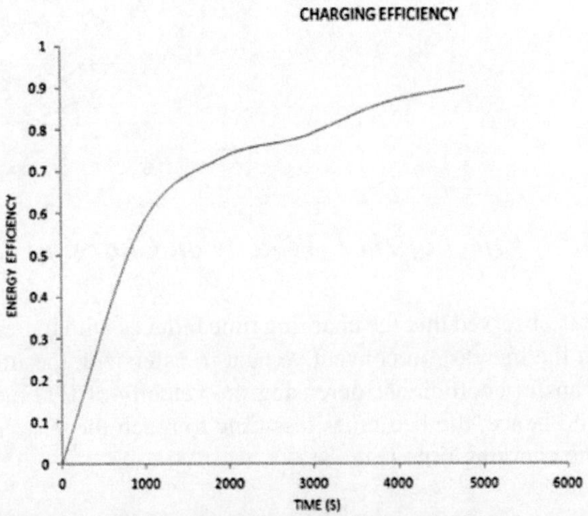

(a) Charging energy efficiency (η_{ch})

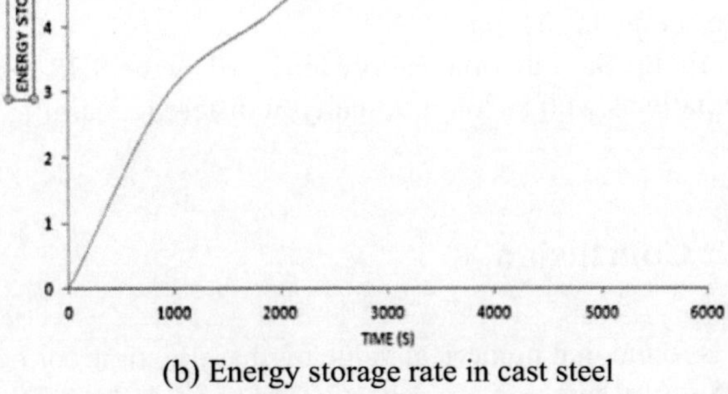

(b) Energy storage rate in cast steel

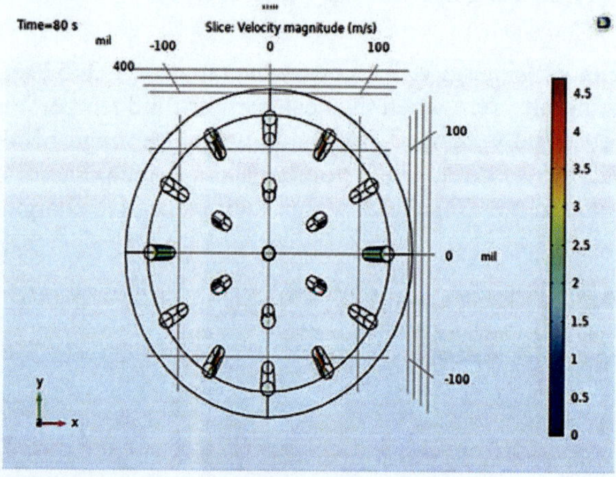

Fig. 8 Velocity profile

5.5 *Effect of HTF Velocity on Charging Time*

It is observed that the charging time reduces with increase in velocity. It is mainly due to the increase in convective heat transfer with the improvement in the overall heat transfer coefficient. Increasing the velocity of HTF accelerates the rate of charging and hence, the bed takes less time to reach the T_{inlet} resulting into the reduction of the charging time (Fig. 8).

5.6 *Energy Retrieved*

The amount of thermal energy released from the bed at its respective discharging time is estimated by using Eq. (vi). For a given temperature difference of 80 °C (ΔT), the amount of thermal energy released with the temperature drift ranges of (573–493 K). Numerical analysis of thermal energy retrieved by the SHS bed and temperature distribution radially as well as longitudinally at different charging time are shown in Fig. 9a–f.

Figure 9a–f thermal energy retrieved by the SHS bed and temperature distribution radially as well as longitudinally at different charging time.

6 Conclusion

It is found that numerical value of charging time for our cast steel model at velocity 2.5 m/s at temperature range 393–473 K, 443–523 K, 493–573 K are 3951, 4775, 5780 s, respectively, similarly discharging times at corresponding temperature are

Fig. 9 **a** $t = 0$ s. **b** $t = 0$ s. **c** $t = 816.14$ s. **d** $t = 816.14$ s. **e** $t = 6278$ s. **f** $t = 6278$ s

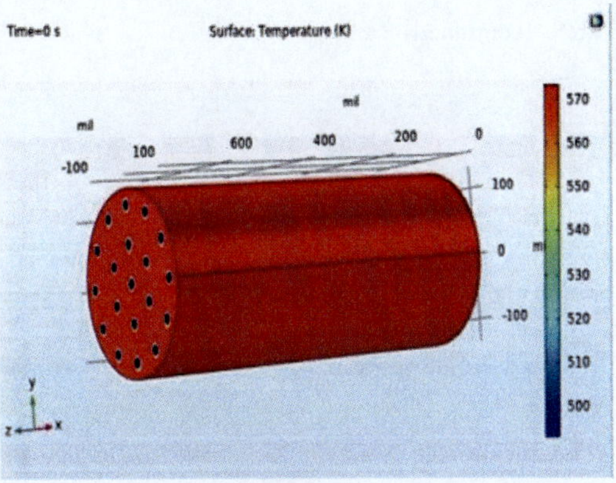

(a) t = 0 s

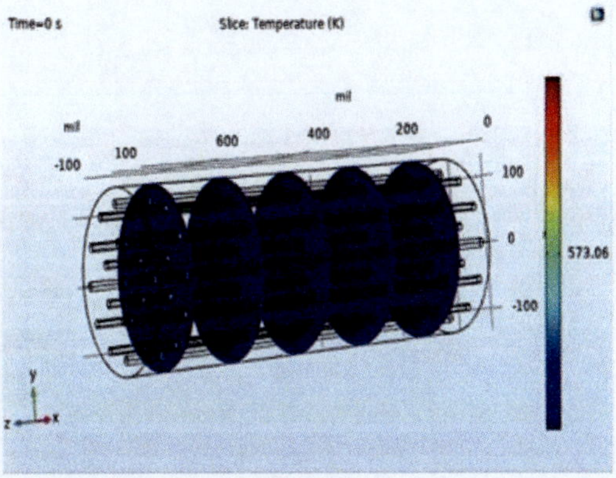

(b) t = 0 s

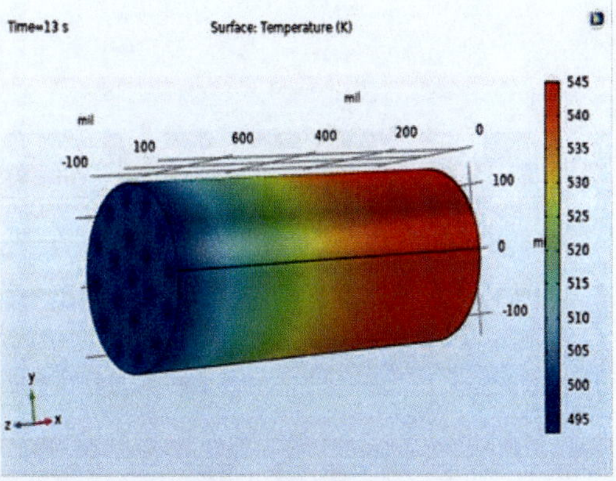

(c) t = 816.14s

Fig. 9 (continued)

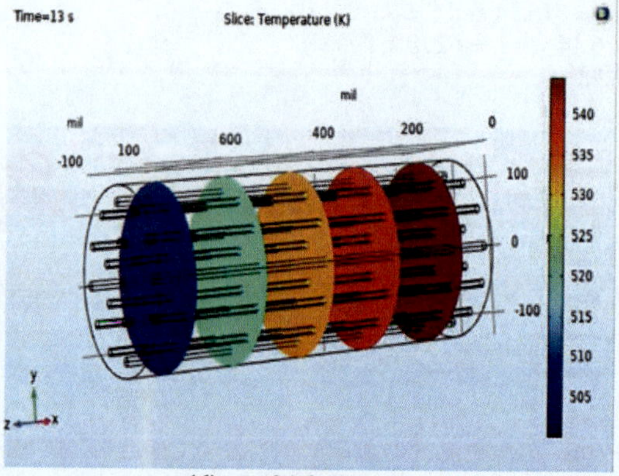

(d) t = 816.14s

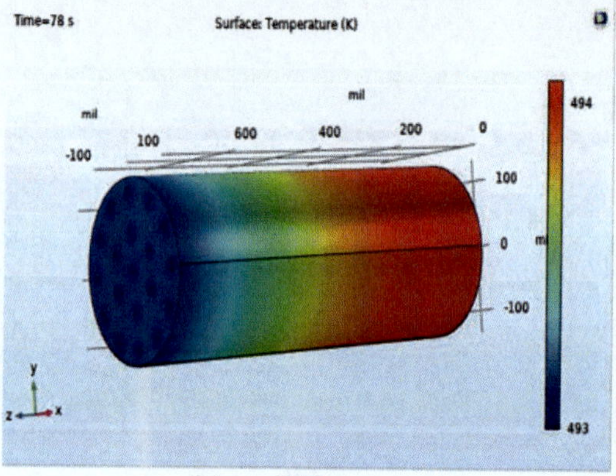

(e) t = 6278 s

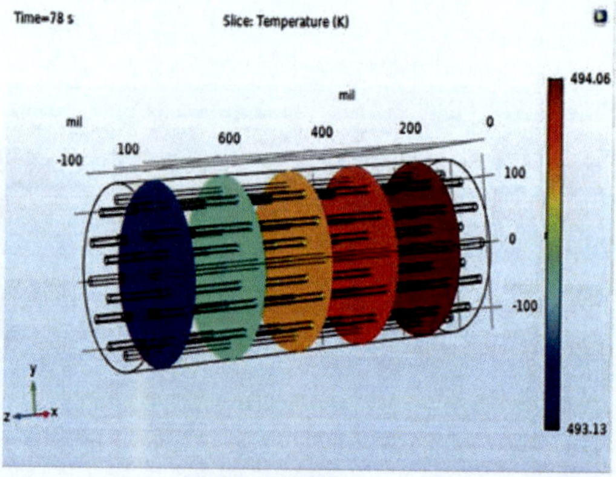

(g) t = 6278 s

4150, 4980, 6278 s, respectively. Further, the charging efficiency of this model at temperature range 493–573 K is 92.78%. Moreover, energy stored and retrieved from this cast steel model is found to be 7.3–7.1 MJ, respectively. To the contrary, the cast steel model which was tested before for velocity 2.5 m/s at temperature rang 473–393 K, 523–443 K, 573–493 K are 4993, 5907, 7012 s, respectively, for charging. Similarly, it took 4080, 2300 and 2240 s, respectively, for discharging at corresponding velocities.

Thus, we can conclude that the experimental and numerical value are almost similar. In all the above experiments, the time taken by the charging process to reach steady state condition is quite higher than that of discharging. The modified module shows a great efficiency along with all the essential conditions. The evaluated result helps us to continue our experiments in near future using our own parameters and boundary conditions.

Development in the field of SHS system can potentially act as an important utility component in the real time application. It has a numerous advantage of using the SHS system are been highlighted by integrating with solar heating, energy production and waste heat recovery application.

References

1. TIAN, Y. (n.d.). (School of Engineering, University of Warwick, CV4 7ALCoventry, United Kingdom). C Y ZHAO.
2. A. R. (2018). Ultra-high temperature thermal energy storage.
3. A. M. (2002). A review on energy conservation in building applications with thermal storage by latent heat using phase change materials.
4. Fernandez, A. I., & M. (2010). Selection of materials with potential in sensible thermal energy storage.
5. Al-Abidi, A. A., & S. (2012). A review on thermal energy storage system for airconditioner.
6. Adam Robinson, A. (2017). Concepts renewable energy sourced from the sun, wind, waves or tides is clean and secure.
7. Rainer Tamme. (n.d.). Advanced thermal energy storage technology for parabolic trough.
8. De Gracia, A., & L. F. (2014). 4th generation district heating.
9. De Gracia, A., & L. F. (2015). Thermal energy storage system for buildings.
10. De Gracia, A., & L. F. (2015). Phase change materials and thermal energy storage for buildings.
11. Khudhair, A. M, & M. M. (2002). 2002. A review on energy conservation in buildings.
12. Evans, A., & V. S. (2012). Assesment of utility energy storage.
13. Gil, A., & M. M. (2009). State of the art on high temperature thermal energy storage for power generation.
14. Sharma, A., & V. T. (2007). Review on thermal energy storage with phase change materials and application.
15. Zalba, B., & J. M. (2002). A review on thermal energy storage in phase change materials.
16. Burkhard Sannera, C. K. (2003). Geothermal heat pumps.
17. Zhao, C. Y., & Z. W. (2010). Heat transfer enhancement of high temperature thermal energy storage using metal foams and expanded graphite.
18. Das, D. C. (2012). GA based frequency controller for solar thermal–diesel–wind hybrid energy generation/energy storage system.
19. Rao1, C. R. C., & H. N. (2017). Performance tests on lab–scale sensible heat storage prototypes.

20. CLARK. (1983). Transient response of a packed bed for thermal energy storage An analysis, with experimental results, for the transient response of a packed bed thermal storage unit is presented.
21. Clark, D. E. (1983). Transient response of a packed bed for thermal energy storage.
22. Fernandes, D., & F. P. (2012). Thermal energy storage.
23. Zhou, D., & C. Z. (2011). Review on thermal energy storage with phase change materials (PCMs) in building applications.
24. Malan, D. J., & R. D. (2015). Solar thermal energy storage in power generation using phase change material with heat pipes and fins to enhance heat transfer.
25. Das, D. C., & A. R. (2012). GA based frequency controller for solar thermal diesel wind hybrid energy storage system.
26. Mettawee, E. B. S. & G. M. (2006). Thermal conductivity enhancement in a latent heat storage system.
27. Faghrlt, Y. Z. (1994). Semi analytical solution of thermal energy storage system with conjugate laminar forced convection.
28. Regin, F., & S. S. (2008). Analysis of a packed bed latent heat thermal energy storage system using pcm capsules.
29. Garimella, ♣. Z. (2010). Thermal analysis of solar thermal energy storage in a molten salt thermocline.
30. Paksoya, H. O., & O. A. (n.d.). Heat transfer enhancement for thermal energy storage using metal foams embedded with pcm.
31. Paksoya, H. O., & O. A. (2000). Heating and cooling of a hospital using solar energy.
32. Chen, H., & T. (2009). Progress in electric energy storage system.
33. Niyas, H., & L. P. (2014). Performance investigation of high-temperature sensible heat thermal energy storage system during charging and discharging cycles.
34. Lafdi, K., & O. M. (2007). Phase changing materials.
35. Vigneshwarana, K., & G. S. (2019). Cast steel based high temperature thermal energy storage system.
36. Vigneshwarana, K., & G. S. (2019). Experimental and numerical investigations on high temperature cast steel based sensible heat storage system.
37. Vigneshwarana, K., & G. S. (2018). Experimental investigation of a cast-steel based thermal energy storage system.
38. Kenisarin, M. M. (2009). High temperature phase change material.
39. Ki, J. J.-H.-G. (2011). Application of pcm thermal energy storage system.
40. Kody, M., & Powell, T. (2011). Medelling and control of solar power plants.
41. Lafdi, K. (2007). Graphite foams infiltrated with phase change materials as alternative materials for space and terrestrial Thermal energy storage applications.
42. Lalit, M., & Bal, S. S. (2010). Solar dryer with thermal energy storage for drying agricultural food products.

Comparison of System Performance of Different Types of Ice-Battery for Cooling Applications in Industries

Tushar Sharma, Robert Honke, and Tobias Plessing

Nomenclature

ρ	Density (kg/m^3)
c_p	Specific heat capacity (J/kgK)
T	Temperature (°C)
t	Time (s)
M	Mass (kg)
V	Volume of the storage tank (m^3)
y	Coordinate along the height of tank (m)
λ	Thermal conductivity (W/mK)
h_f	Enthalpy of fusion of water (J/kg)
$\dot{q}_{\mathrm{surr}}$	Heat flux per unit volume between storage fluid and surroundings (W/m^3)
$\dot{q}_{\mathrm{hx}}$	Heat flux per unit volume between storage fluid and heat exchanger (W/m^3)
$\dot{Q}_{\mathrm{lat}}$	Rate of latent heat energy (W)
ΔT_{LM}	Logarithmic mean temperature difference (K)
Δr_{ice}	Change in ice thickness on a tube (m)
r'_{ice}	Radius of ice layer of previous time step (m)
Δs_{ice}	Change of ice thickness on a flat plate (m)
A_{plate}	Total heat transfer surface area of a flat plate (m^2)
L	Length of tube (m)
U	Overall heat transfer coefficient (W/m^{2}K)
$\mathrm{UA_{HX}}$	Overall heat transfer from heat exchanger per unit temperature difference (W/K)
SC	Solar control (0–1)
SP	Solar PV power production (W)

T. Sharma (✉) · R. Honke · T. Plessing
Institute for Water and Energy Management (Iwe), Hof University of Applied Sciences
Alfons-Goppel-Platz1, 95028 Hof, Germany
e-mail: tushar.sharma@hof-university.de

P. Mahanta et al. (eds.), *Advances in Thermofluids and Renewable Energy*, Lecture Notes in Mechanical Engineering, https://doi.org/10.1007/978-981-16-3497-0_46

f	Instantaneous ice mass fraction (%)
SPF	System performance factor
$\dot{Q}_c$	Yearly cooling demand of system (W)
P	Electric power consumption (W)
R_{aut}	Degree of self-sufficiency (%)
FD	Flow diverter
PV	Photovoltaic
FP	Flat plate heat exchanger
CM	Capillary mat heat exchanger

1 Introduction

In 2015, around 35 TWh/yr of electrical energy was needed for process cooling in German industries and around 3 TWh/yr was required for space cooling. Although this amounts to just 5% of the total share of electrical energy demand, it is predicted to increase rapidly in the coming years [1]. A considerable amount of energy consumption in industrial bakeries is accounted for cooling-related activities. Some activities are quick cooling of the products coming out of the baking oven into deep-freezers, cooling of the products to be shipped daily or storing specific products for longer durations. A survey done by our institute for one of the leading bakeries in Bavaria showed that around 45% of their electrical energy consumption accounted for cooling-related activities on an annual basis. In addition to that, a vast amount of solar PV, about 1 MWp of installed capacity remained locally unutilized as the generated energy was fed directly to the grid.

Several cooling processes take place inside this bakery, for different kinds of products coming in and going out at different temperatures. Therefore, walk-in coolers for deep freezing (for longer durations) and normal cooling are used (for daily operations). They are being maintained at temperatures around −20 °C and +5 °C to + 10 °C, respectively. Due to the unavailability of the cooling load profiles for the walk-in cooler, appropriate assumptions have been made for the cooling load calculations [2]. The hypothetical walk-in cooler considered for the present study is 3 m high with 26 m^2 of floor area and contains 78 m^3 of air volume. It has been assumed to be maintained at 6 °C throughout its period of operation. The ambient conditions have been assumed to be constant at 20 °C and 60% relative humidity (RH), representing a typical location in an industrial hall.

1.1 Definition of the Ice-Battery

The significance of using the term 'battery' in 'ice-battery' lies in the conception that it can replace electrochemical batteries as an energy backup by transforming electrical power to cold-thermal power. Two different heat exchanger geometries have been analyzed for this project: flat plates and capillary mats. Tank and heat exchanger geometry details with all essential parameters are described in Fig. 1. The ice storage tank selected for the present study is made from glassfiber reinforced plastic (GRP) with thermal conductivity of 0.25 W/mK and thickness 0.01 m. The insulation around the tank has a thermal conductivity of 0.034 W/mK and a thickness of 2 cm. The thermal conductivity of the flat plate heat exchanger is 0.23 W/mK with a thickness of 5 mm. The thermal conductivity of the capillary mat heat exchanger is 0.33 W/mK with 2.75 mm inside diameter and 4.25 mm outside diameter placed 2 mm apart (x_{tubes}) along the width of the tank. The total heat exchanger surface area for the simulation experiments is varied by changing the distance between the heat

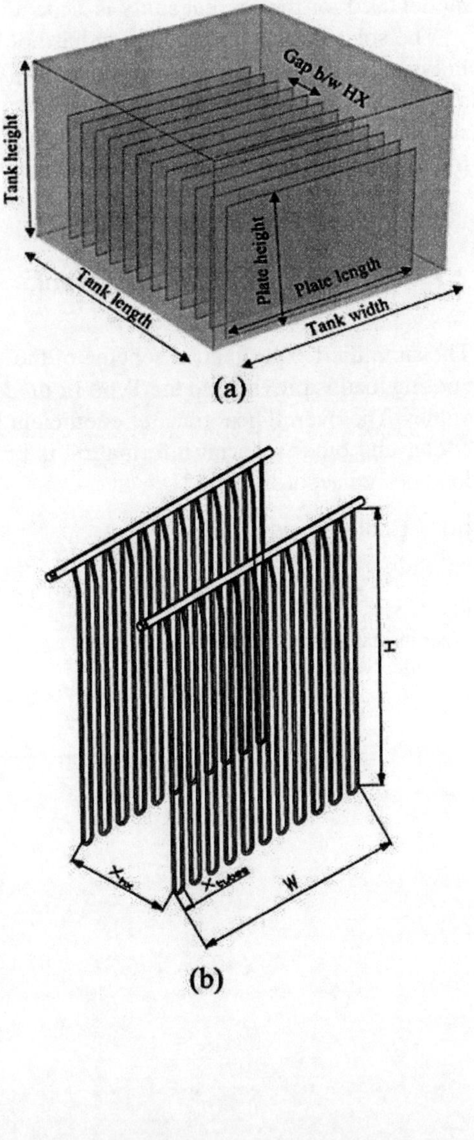

Fig. 1 Ice-battery tank and heat exchanger geometries. **a** Flat plate. **b** Capillary mat [3]

exchangers (x_{hx}) along the length of the tank. Details about the parametric variations will be discussed in the subsequent sections.

2 Methodology

Dynamic system simulations using TRNSYS [5] simulation software have been performed to assess the feasibility of the control strategy at the system level. In this section, the model definitions of all the components and the experimental scheme will be explained briefly. The essential component models used for the system simulation are chiller, walk-in cooler, cooling tower, ice-battery, and solar PV. The ice-battery model used for the present study is Type 861 [4].

The solar PV power production has not been simulated within TRNSYS, but in Polysun [6] simulation software since only the latter could provide weather data for a location near to the campus of Hof University in Hof, Germany. The time step for simulation is 120 s, and the convergence criteria have been set to 1e-2, which was found suitable for the present level of analysis.

2.1 *TRNSYS Simulation Model*

The simplified experimental scheme of the system analyzed is shown in Fig. 2. The cooling load is provided to the Type 12 model (walk-in cooler) in the form of hourly values. The overall heat transfer coefficient (U-value), the initial temperature of the cooler and other relevant information is provided to the model as per the cooling load design according to [2].

(a) Cooling load

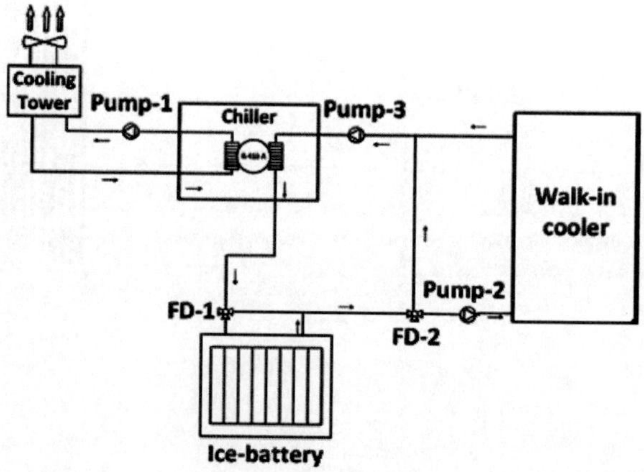

Fig. 2 Simplified experimental scheme for the hydraulic system analyzed in TRNSYS

The cooling load for the present study has been assumed to repeat in a period of 4 hours. In the initial 10 mins of the time period, there is a distinct peak of thermal load due to hot products coming in the walk-in cooler. For the rest of the time period, the walk-in cooler is subjected to the thermal load due to the transmission losses through the envelope of the cooler, infiltration losses through the door and internal heat gains due to lighting, occupancy, etc. A yearly cooling energy demand of around 16 MWh was calculated for the given conditions. An initial room temperature of 20 °C was assumed inside the walk-in cooler, and the desired room temperature is 6 °C. Pump-2 (as shown in Fig. 2) is controlled based on the room temperature of the walk-in cooler and tries to maintain it at 6 °C with a dead band of 2 °C (assumed to be reasonable for the present study).

(b) Solar PV

A solar PV module of 7.98 kWp electrical power has been considered for the present study. Several parametric studies were performed to determine the optimum capacity for the given load profile. PV data is provided to TRNSYS as an external file in the form of hourly values in the simulation model.

(c) Chiller

A chiller with a rated thermal capacity of 10 kW and a rated coefficient of performance (COP) of 4.8 was selected for the present study. Several parametric studies were performed to determine the optimum chiller capacity for the given load profiles. The performance data for the chiller is provided in the form of external files in the simulation model.

(d) Ice-battery (Type 861)

The ice-battery model used for the present study is Type 861 [4]. The main concept of the model is based on the following governing equation:

$$\rho c_p \frac{\partial T}{\partial t} + \left(\rho c_p\right)_{\text{ice}} \frac{\partial T_{\text{ice}}}{\partial t} = \frac{\partial}{\partial y}\left(\lambda \frac{\partial T}{\partial y}\right) + \frac{h_f}{V} \frac{\partial M_{\text{ice}}}{\partial t} + \dot{q}_{\text{surr}} + \dot{q}_{\text{hx}} \quad (1)$$

where c_p is the specific heat capacity of water, h_f the enthalpy of fusion, M_{ice} the mass of ice, $\dot{q}_{\text{surr}}$ and $\dot{q}_{\text{hx}}$ are the heat fluxes per unit volume between the storage fluid, the surroundings and heat exchanger, respectively. T is the water temperature, t is the time, V is the water volume of the storage, y is the coordinate along with the height of the tank, and λ and ρ are the thermal conductivity and density of water. The first and second term of the equation represents the accumulated sensible heat in the storage fluid. The first term on the right side represents heat flow due to conduction in the storage fluid. The second term on the right side represents the latent heat of solidification and melting, and the last two terms are heat flux due to the surroundings and due to the heat exchange from the heat exchangers, respectively.

For the present study, a correlation of the amount of latent heat extracted from the water and the amount of heat exchanged with the heat exchanger was determined. As can be seen from Eq. 1, we assume that the accumulated heat in the storage, heat due to conduction and heat transfer from the surroundings is independent of the types of heat exchanger geometry. Thus, the rate of latent heat extraction from the storage is directly proportional to the rate of heat exchanged between the heat exchanger and water:

$$\dot{Q}_{\text{lat}} \propto (\text{UA})_{\text{HX}} \Delta T_{\text{LM}} \tag{2}$$

where $\dot{Q}_{\text{lat}}$ is the rate of latent heat energy extracted from water, UA_{HX} is the overall heat transfer coefficient-area product, and the term HX is used for the particular heat exchanger in use. ΔT_{LM} is the logarithmic mean temperature difference across the heat exchanger fluid (brine) and storage fluid (water).

From Eq. 1, we have

$$\dot{Q}_{\text{lat}} = h_f \rho_{\text{ice}} \frac{\partial V_{\text{ice}}}{\partial t} \tag{3}$$

Based on Eq. 2, formulas for ice growth on a flat plate and a capillary mat can be simplified to the following equations:

$$\frac{\Delta r_{\text{ice}}}{\Delta t} = \frac{\dot{Q}_{\text{lat}}}{h_f \rho_{\text{ice}} 2\pi L r'_{ice}} \tag{4}$$

$$\frac{\Delta s_{\text{ice}}}{\Delta t} = \frac{\dot{Q}_{\text{lat}}}{h_f \rho_{\text{ice}} A_{\text{plate}}} \tag{5}$$

Δr_{ice} is the change in the ice thickness for a tube in the radial direction, L is the length of the tube, r'_{ice} is the radius of the ice layer of the previous time step, Δs_{ice} is the change in the ice thickness on a flat plate, and A_{plate} is the net heat transfer surface area of a flat plate.

UA_{HX} values have been calculated based on the formulae mentioned in [4]. The values, as shown in Fig. 3, are for the total surface area of 8 m^2 for both types of heat exchangers. As seen in the figure, UA_{HX} values for a capillary mat (CM) are way higher than that observed for the flat plate (FP). The values for a capillary mat start at the same point as a flat plate but decrease at a slower rate as seen in the

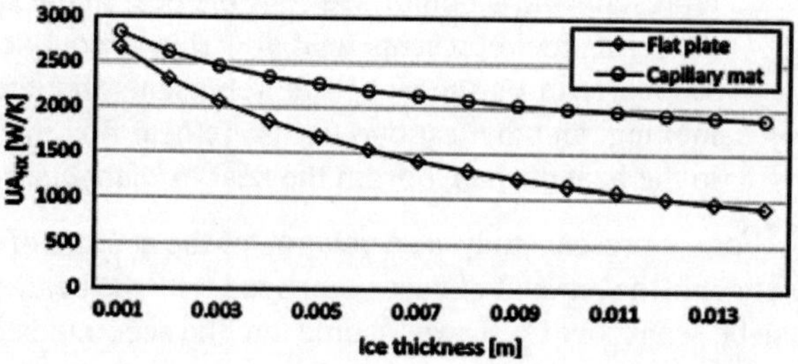

Fig. 3 Variation of UA_{HX} value with ice thickness for an FP and CM heat exchanger

Table 1 Parameter variation for the ice-battery (Type 861)

Parameter	Flat plate	Capillary mat
Tank volume	2–5 m^3	2–5 m^3
A/V* ratio	4–25	4–25
Distance b/w heat exchangers	6–67 cm	4–25 cm

*Total surface area of all the heat exchangers per unit tank volume

figure. The values presented in this figure are calculated under the assumption that natural convection effects are negligible in comparison to the conduction effects. Actual variation of UA_{HX} values with ice thickness will depend on the ΔT_{LM} values and the mass flow rate of brine. The variations of $\dot{Q}_{lat}$ and ice mass fractions for flat plate and capillary mat geometries will be discussed in the results section. Following parameters for the ice-battery model are varied as listed in Table 1:

2.2 System Control

As the solar PV power is fluctuating every day throughout the year. In our control strategy, the solar PV power generated is compared with a threshold value to control the chiller and flow diverter (FD-1) as explained in Table 2:

In Table 2, **SP** is the solar PV power production (in watts), $\boldsymbol{f}$ is the instantaneous ice mass fraction (in percentage), $\boldsymbol{f_{min}}$ is the minimum permissible ice mass fraction (in percentage), and $\boldsymbol{f_{max}}$ is the maximum permissible ice mass fraction (in percentage).

Table 2 Control strategy for system simulation

Control signal	Value	Condition	Comments
Solar control (**SC**)	0	SP < 1950 W	Comparison of solar power with the maximum chiller power consumption
	1	SP > = 2050 W	
Chiller control	OFF (0)	{((SC = 0) **AND** ($f \in [f_{min}, f_{max})$)} **OR** ($f >= f_{max}$)	Conditions to turn the chiller OFF
	ON (1)	($f < f_{min}$) **OR** {($f \in [f_{min}, f_{max})$ **AND** (SC = 1)}	Conditions to turn the chiller ON
Flow diverter valve (controlling the flow of coolant through ice-battery)	0.0001	(SC = 0) **AND** ($f < f_{min}$)	Ice-battery bypass
	1.0	Default condition	Fully opened

The ice mass fraction is evaluated in the simulations as per the following equation:

$$f[\%] = \frac{\text{Mass of ice}[\text{kg}]}{\text{Tank volume}[\text{m}^3] \times \text{Density of ice}\left[\frac{\text{kg}}{\text{m}^3}\right]} \times 100 \quad (6)$$

The threshold value used in the control strategy is the maximum electrical power consumed by the chiller during its operation, around 2 kW_{el} for a 10 kW chiller operating at 4.8 COP. Values of f_{min} and f_{max} are chosen as 10 and 90%, respectively, for the present analysis.

2.3 Performance Indicators

The following set of performance indicators are used to compare the system efficiency under different circumstances:

1. **System performance factor (SPF)**
   As per Malenkovic.et.al [7], system performance factor is defined as follows:

$$\text{SPF} = \frac{\int(\dot{Q}_c)\text{d}t}{\int(\sum P_{\text{el},T})\text{d}t} = \frac{\text{Yearly cooling demand [kWh]}}{\text{Annual lectrical power consumption. [kWh]}} \quad (7)$$

   where Q_c is the yearly cooling demand of the system and $P_{\text{el},T}$ is the total annual electric power consumption of the whole system defined as follows:

$$P_{\text{el},T} = P_p + P_{\text{ch}} + P_{\text{pen}} + P_{\text{backup}} \quad (8)$$

   where *p*, ch, pen, and backup refer to circulation pumps, chiller, penalties, and backup, respectively. For the sake of simplicity, penalties and backup have not been considered for the present study.
2. **Degree of self-sufficiency (R_{aut})**
   The degree of self-sufficiency is a measure of the amount of energy generated by the solar PV, which is utilized to meet the electrical energy demand of the components.

$$R_{\text{aut}} = \frac{\text{Total self consumption [kWh]}}{\text{Total electrical consumption [kWh]}} \quad (9)$$

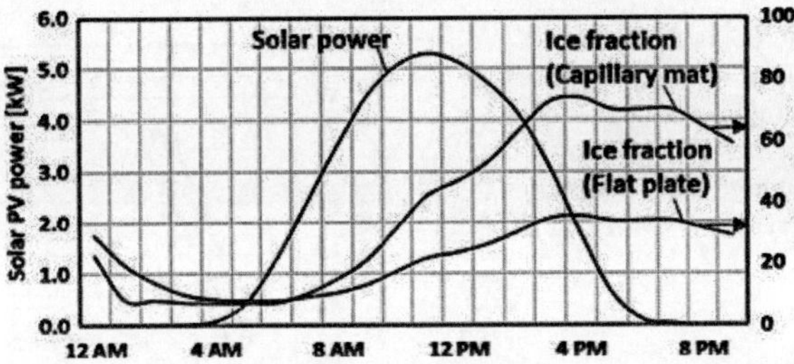

Fig. 4 Variation of ice mass fraction (%) for a flat plate and a capillary mat ice-battery for a typical summer day

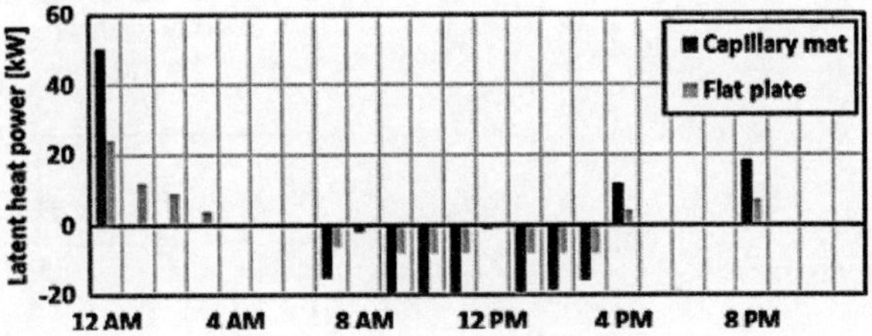

Fig. 5 Variation of latent heat power (kW) for flat plate and a capillary mat heat exchanger-based ice-battery for an ideal summer day

3 Results and Discussion

3.1 Typical Summer day

For the present study, the suitability of the controls for an ideal summer day for the current location was considered. The criteria for selection of an ideal summer day are based on the maximum solar PV power generation in a day. For the chosen PV capacity, 28th May was found to be an appropriate typical summer day.

The results presented in Figs. 4 and 5 are for the 2 m^3 tank with a total surface area of 50 m^2 for both the type of heat exchangers. As can be seen in these two figures, a CM can produce much higher ice fractions and is able to charge and discharge at a much faster rate than an FP of the same area per tank volume which agrees with Eqs. (2), (4), and (5).

3.2 Degree of Self-Sufficiency (R_{aut})

As shown in Fig. 6, a capillary-mat-based ice-battery gives better self-sufficiency ratios than a flat plate-based ice-battery up to a specific A/V ratio, found to be around 10 for the present case. Beyond that, the flat plate is found to be slightly better. The reason for this behavior lies in the fact that for the same A/V ratio, the distance between adjacent capillary mats needs to be quite less to achieve the same surface area as a flat plate. This restricts the ice growth beyond a certain point, as a result of which chiller cannot extract more energy out of the ice-battery. On the other hand,

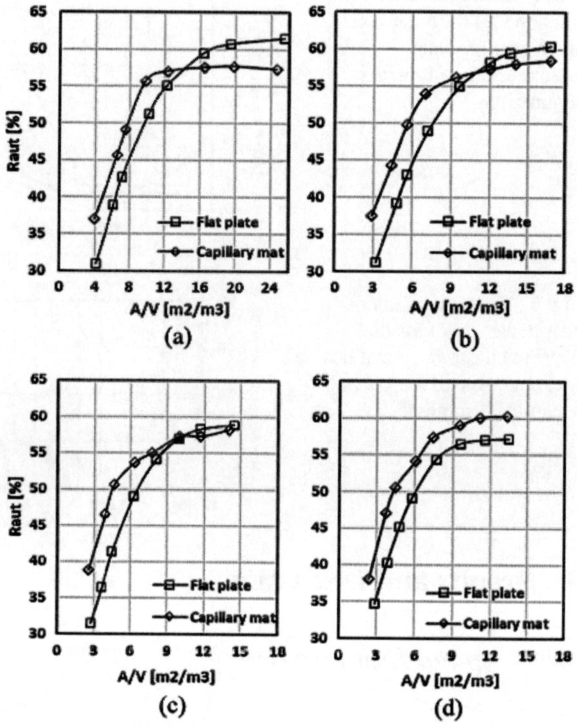

Fig. 6 Yearly values of degree of self-sufficiency as a function of heat exchanger area per tank volume (A/V) for three tank volumes. **a** 2 m^{3}. **b** 3 m^{3}. **c** 4 m^{3}. **d** 5 m^{3}

a flat plate provides more space between adjacent plates and, thus, can go further. With the increasing tank volume, as a flat plate needs more time to make the same amount of ice than a capillary mat, thus capillary mat gives better self-consumption values than a flat plate for higher tank volumes. The R_{aut} values reach a maximum value of about 60% for most of the cases as it is limited by the chiller capacity for the present study. Thus, by having a 60% self-consumption from solar PV, we can save around 3000 kWh grid power annually, which amounts to savings of about 1872 kg of CO_2 production annually [8].

3.3 SPF (System Performance Factor)

The results for the yearly system performance in Fig. 7 show better values for the capillary-mat-based ice-battery than for the flat plate-based ice-battery beyond a specific heat exchanger area per tank volume ratio. For smaller tank sizes, this ratio is found to be around 16, and it keeps reducing for bigger tanks. The reason for that can be seen in Fig. 6, where the self-consumption increases for the capillary mat up to a particular ratio, and then it becomes lesser than that of the flat plate. As SPF is the ratio of yearly cooling energy provided per annual electrical consumption, the more

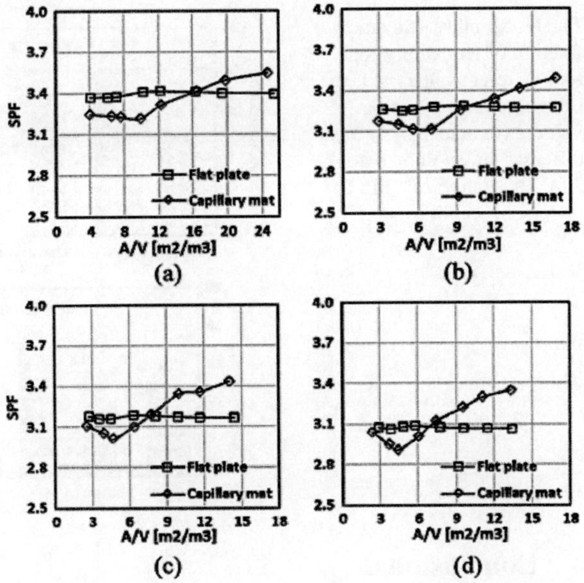

Fig. 7 Yearly system performance as a function of heat exchanger area per tank volume (A/V) for three tank volumes. **a** 2 m^3. **b** 3m^3. **c** 4 m^3. **d** 5 m^3

the electrical consumption, the less the SPF value. Also, as the tank size increases, the net electrical consumption (self-consumption from solar PV and energy taken from the grid) increases as the chiller runtime is increased to produce the same amount of ice fraction as needed in smaller tanks.

3.4 Utilization of Ice-Battery

This section explains the amount of time the ice-battery was in use to provide the cooling energy for the walk-in cooler. Ice-battery with a flat plate and a capillary mat heat exchangers of different A/V ratios and different tank volumes were simulated, and the results are shown in the Fig. 8.

The results for the monthly utilization fraction given in Fig. 8 show that the ice-battery for any given configuration is less utilized for the most part of the month. The reason for this is linked to the amount of time the pump connected to the walk-in cooler runs to meet the demand which is just 5 h (20% of the full day) on any given day due to the present loading conditions. Thus, the maximum amount of time an ice-battery can be utilized to meet the cooling demand solely is 20%. We can observe that a capillary-mat-based ice-battery is used more than its flat plate counterpart for any given configuration, the reason for that being the faster charging and discharging rates for capillary mat than a flat plate. The utilization fraction can be improved by increasing the cooling demand, which will increase the run time of the pump and thus, in turn, increase the time for energy extraction from ice-battery.

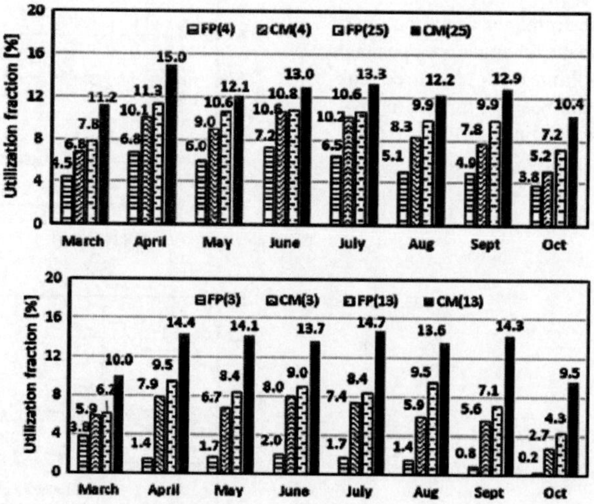

Fig. 8 Monthly utilization fraction of ice-battery for four different values of heat exchanger area per tank volume (in brackets) for two different tank volumes (top) 2 m^3 (bottom) 5 m^3. FP: flat plate, CM: capillary mat

4 Conclusions.

In the present study, the system performance of two types of ice-batteries based on two different heat exchanger geometries, a flat plate, and a capillary mat was compared. The present analysis has been done for two objectives: To maximize the self-consumption from the solar PV power production and to maximize the utilization of an ice-battery.

From the present study, the following points can be concluded:

1. Capillary mat-based ice-battery shows better charging and discharging rates than a flat plate-based ice-battery for any given configuration.
2. Capillary mat-based ice-battery shows a better degree of self-sufficiency (R_{aut}) until a specific A/V ratio, beyond which it is limited by the distance between the mats to achieve the same surface area as needed for a flat plate.
3. Better SPF values are seen in flat plates than in capillary-mat-based ice-battery up to a specific A/V ratio beyond which shows better values for capillary mats due to reverse relation of SPF with net electrical consumption.
4. In terms of the amount of the time, the ice-battery was utilized to meet the cooling demand of the walk-in cooler, capillary-mat-based ice-battery exceeds the flat plate-based ice-battery for any given configuration.
5. For the present study, it can be said that around 60% of the yearly energy demand can be met from solar PV power production, which saves about 3000 kWh of grid electricity annually.

Future work will be focused on optimizing the size of the ice-battery (tank volume and the net surface area of heat exchanger) for any given loading conditions based on the same optimization objectives. The key variables that have been identified from the present study are chiller capacity, solar PV capacity, and the cooling demand.

Acknowledgements The authors would like to thank Europäische Fonds für Regionale Entwicklung (EFRE) for the financial support received for the project GTW (Grüne Technologiewerkstatt Hof) within which this work has been performed.

References

1. Heat Roadmap Europe 4(HRE4). Heating and Cooling-facts and figures—The transformation towards a low-carbon Heating and Cooling sector. https://heatroadmap.eu/.
2. Breidert, H. J. (2013). Projektierung von Kälteanlagen Berechnung, Auslegung, Beispiele (4th ed.).
3. Carbonell, D., Battaglia, M., Philippen, D., & Haller, M. Y. (2018). Numerical and experimental evaluation of ice storages with ice on capillary mat heat exchangers for solar-ice systems. *International Journal of Refrigeration, 88*, 383–401.
4. Carbonell, D., Battaglia, M., Philippen, D., Haller, M. Y., (2017a). Ice-Ex—heat exchanger analyses for ice storages in solar and heat pump applications. Institut für Solartechnik SPF for Swiss Federal Office of Energy (SFOE). Research Programme Solar Heat and Heat Storage, CH-3003 Bern.
5. Klein, et al. (2010). Trnsys 18: A transient system simulation program, a solar energy laboratory. In *Technical report*. University of Wisconsin. http://sel.me.wisc.edu/trnsys.
6. https://www.velasolaris.com/polysun/
7. Malenković, I. (2013), Definition of main system boundaries and performance figures for reporting on SHP systems deliverable B19, IEA SHC Task 44 / HPP Annex 38. http://www.iea-shc.org/task44.
8. https://www.eumayors.eu/IMG/pdf/technical_annex_en.pdf

An Overview of the Methods of Enhancement of the Reaction Kinetics of Al-Water Reaction in an Aqueous Medium and the Prospect of the Economic Viability

Biswajyoti Das, Pinakeswar Mahanta, and P. S. Robi

1 Introduction

The environmental pollution caused by the burning of fossil fuel is a major challenge in the present scenario. The fossil fuel is getting reduced due to its wide applications mainly for power generation. Hydrogen gas is assumed to be a source for future power production. Among the various method of production of hydrogen gas, the chemical method has dominated the market. The production of hydrogen gas from Al-water reaction in the presence of aqueous NaOH solution is considered as a low-cost production method. The raw material used in the reaction is Al scrap. Moreover, Al has several advantages, e.g. high-energy density, lightweight, etc. However, ball milling of aluminum and mechanical alloying with other elements can enhance the yield of hydrogen. The production of hydrogen gas also depends on the size of the Al particles, initial reaction temperature, and concentration of NaOH solution.

2 Activated Al Metal

The Al-water reaction is a surface reaction process. Al alloyed with other metals like In, Ga, Sn, Bi, etc., have a fair production rate as reported [1]. Ball milled Al with salt particles can accelerate the reaction and increase the yield of hydrogen gas. Maximum hydrogen production of 75 ml/min per gram of Al has been reported for Al particles milled with salt particles [2]. Fan *et. al.*[3] have reported the hydrolysis effectiveness of Al metal alloyed with other metals. It is stated that the ball milling

B. Das · P. Mahanta (✉) · P. S. Robi
Department of Mechanical Engineering, Indian Institute of Technology Guwahati, Assam Guwahati, India
e-mail: pinak@iitg.ac.in

P. Mahanta et al. (eds.), *Advances in Thermofluids and Renewable Energy*, Lecture Notes in Mechanical Engineering, https://doi.org/10.1007/978-981-16-3497-0_47

of Al with Zn, Ca, Ga, Bi, Mg, In, and Sn is much more productive than making alloys of these metals by the process of melting.

3 Size of Al Particles

One of the main parameters that affect the production of hydrogen is the size of the Al particles. Al particles with smaller size are found to be profound in the liberation of hydrogen gas [4]. It was reported that 60–80% of hydrogen gas can be obtained using 6 μm size of Al particles. The research carried out by Elitzur S. *et. al.*[5] shows that the hydrogen production rate depends on the specific surface area (m^2/g) of the Al particles. It also implies that the rate of kinetics is dependent on the shape of the raw Al.

4 The Effect of the Characteristics of Water

The type of water used for the reaction has a prominent role in the production rate. The amount of salt content in the water affects the reaction kinetics. The reaction rate is faster for the water that contains more salt particles [6]. Wang *et.al.*[7] have carried out experiments to investigate the effect of deionized water and tap water in the production of hydrogen gas. The same set of the experiment conducted with tap water has very less yield of hydrogen. Phung *et.al.*[8] have carried out experiments with distilled water, deionized water, tap water, and ultrapure water to observe the effect on the hydrogen production rate.

5 Transition to the Hydrogen Economy

The optimum combination of the aforesaid parameters can contribute to the production of the maximum amount of hydrogen gas at a faster rate. However, the analysis of economic aspects of hydrogen energy is crucial in entering into the market. The critical evaluation of the usage of hydrogen gas for commercial purposes is discussed in the following subsections.

5.1 Hydrogen Transportation and Distribution

Demir *et.al.* [9] have assessed the cost-effectiveness and the performance of the hydrogen delivery system. They have adopted three different methods of hydrogen transportation and distribution. Hydrogen transmission by a pressurized tank, gas pipeline and cryogenic liquid transport are three modes as mentioned. The hydrogen gas transmission through the existing natural gas pipeline is envisaged by another researcher [10]. However, to realize the hydrogen economy and marketing breakthrough some challenges have to be overcome. The problem related to the embrittlement of hydrogen gas does not only depend on the material property but also the pipeline history. A thorough examination is required to stream hydrogen through the natural gas pipe. It is suggested that the transition of traditional energy to hydrogen energy is possible by combining hydrogen with natural gas and delivery through the natural gas pipelines. The delivery infrastructure of hydrogen gas concedes a substantial cash outflow. The storage and transmission cost assessment in the European perspective have been carried out by Tzimas *et.al.* [11]. It is proposed that pipelines of a 1–4 million km length are necessary to enter the hydrogen economy in the European energy system. Out of which 35,000 km of pipelines are required for high-pressure hydrogen transport and 400,000 km for a medium range of transmission by 2050. The production cost for this infrastructure is estimated to be 700–2200 thousand million euros. The successful entry of the commercial use of hydrogen gas is mostly depending upon infrastructure development. This is also one of the issues that restrict the technical modification of automobile engines fuelled by hydrogen gas. The cost of the hydrogen delivery system is directly related to the span of the H_2 transportation and distribution network. However, the market demand for hydrogen is a key point for the wide-spread of the delivery system. The efficient design of a hydrogen delivery network depends on many factors like the geographical location of the place, market demand, and number of refueling stations within the city radius [12]. Optimum design of the hydrogen delivery network considers cost-effectiveness, ease of distribution, environmentally benign character, and easy operation process by the end-users. A substantial amount of exploration has been conducted in this aspect [13, 14]. A various optimization model for the economic benefits of the hydrogen delivery system has been proposed [15, 16]. The commercial feasibility of hydrogen gas has been studied by considering the cost associated with the production of hydrogen gas, transportation, and distribution through various channels to the refueling stations. Weisberg and co-workers [17] have constructed a glass tube fiber channel that operates at 200 K and 700 bar can transport hydrogen gas with a minimum cost 1/kg USD. The optimization of the location of refueling stations for the convenience of the automobiles has been done by applying the statistical tool called, mixed-integer linear model by another researcher [18]. The infrastructure expenditure for hydrogen fuel stations has been surveyed by Weinert and Lipman [16].

5.2 *Hydrogen Refueling Station Setting Model*

The work carried out by Weinert and Lipman implicates that a small power station of hydrogen contain a capacity 30 kg/day has a setup cost is only 0.5 million USD. However, for a refueling station of capacity 1000 kg/day, the cost rises to five million USD.

The traditional power system can be replaced by a brand new technology that incorporates the mixing of hydrogen gas with natural gas. The power drawn by the integrated gas-energy system is cost-effective and environmentally friendly. However, the operational cost and the CO_2 emission level are under study. The investigation regarding the optimal cost of the hybrid gas-energy system is carried out by many researchers [19]. The economic and environmental issues relating to the integration of hydrogen and natural power system are not addressed yet. The operating cost and the CO_2 emission level are greatly reduced in the case of gas-fuelled power systems than coal-fuelled system [20]. The market demand and the production capacity of hydrogen are very less compared to gasoline. The fostering of the hydrogen economy into the energy market is hindered owing to the high-capital cost of the hydrogen station. Lee *et.al.*[21] have shown a cost comparative study between different fuels. The analysis includes all types of cost starting from the capital to the labor cost and from the distribution cost to maintenance cost. The study also illustrates the social and environmental costs owing to the emission of NO_x, CO_2, etc. However, the result so obtained is varied from geographical locations. For the life cycle assessment of the different fuel, the effect of social cost has been pivotal. The cost of fuel for future reference is also demonstrated. The way to cut down the price of fuel cell vehicles by reducing the production cost is described in this literature.

5.3 *Fuel Cell Vehicle and Hydrogen Energy*

It is anticipated that by 2030 most of the hydrogen vehicles will be powered by fuel cells [22]. The evolution of fuel cell vehicle makes a pavement to the development of hydrogen supply chain infrastructure. The growing demand in the market encourages the hydrogen economy. It is predicted that within the next few years the hydrogen vehicle will be running on the road due to the sustainable development in the hydrogen delivery system [23]. The fuel efficiency of the fuel cell vehicle is estimated to be 30 km/l by Lee *et.al* [21]. The hydrogen-fuelled vehicles have good fuel efficiency and maintenance cost is low. There are less numbers of moving parts in the vehicle. In spite, the fuel cell vehicle has demerits of being very expensive to penetrate the hydrogen economy. Moreover, due to the limited hydrogen energy infrastructure available the technology has not grown as expected. The price of an average model of a hydrogen car is almost 50% higher than a gasoline car. The various drawbacks of the fuel cell vehicles impede the hydrogen economy and lacking it to qualify with the fossil fuel energy system. A cost comparative assessment has been presented in Table 1 [24].

Table 1 Comparison of the average price of hydrogen vehicle with petrol vehicle [24]

Sr No	Brand	Category of vehicle	Average price ($)	The average price of the counterpart ($)
1	Hyundai ix 35 (fuel cell)	SUV	68,309–70,000	33,000
2	Hyundai Tuscon (FCEV)	SUV	68,309–70,000	33,000
3	Toyota Mirai	D class	58,365–65,00	25,000
4	Honda Clarity	D class	59,385–59,385	25,000

6 Conclusions

The production of hydrogen gas from an Al-water reaction is a promising method. The evolution of hydrogen gas can be enriched by optimizing the factors that affect the reaction kinetics. It is environmentally friendly fuel for power production. However, the hydrogen delivery system is very costly and as a result, is not able to acquire the momentum in the market. Nevertheless, one key point to be addressed is the technological expansion to distribute hydrogen energy to the end-users. The various technical aspects and delivery infrastructure have to be explored for the transition. The injection of hydrogen gas with natural gas and transmit through the existing pipeline is an emerging field of study. Thus, sustainable hydrogen economy can be fostered by adopting technical modification of the delivery channel to penetrate the market and liberalize the dependency on traditional fuel.

References

1. Wang, H. Z., Leung, D. Y. C., Leung, M. K. H., & Ni, M. (2009). A review on hydrogen production using aluminum and aluminum alloys. *Renewable and Sustainable Energy Reviews, 13*(4), 845–853. https://doi.org/10.1016/j.rser.2008.02.009
2. Alinejad, B., & Mahmoodi, K. (2009). A novel method for generating hydrogen by hydrolysis of highly activated aluminum nanoparticles in pure water. *International Journal of Hydrogen Energy*. https://doi.org/10.1016/j.ijhydene.2009.07.028
3. Fan, M. Q., Xu, F., & Sun, L. X. (2007). Studies on hydrogen generation characteristics of hydrolysis of the ball milling Al-based materials in pure water. *International Journal of Hydrogen Energy*. https://doi.org/10.1016/j.ijhydene.2006.12.020
4. Ivanov, V. G., Safronov, M. N., & Gavrilyuk, O. V. (2001). Macrokinetics of oxidation of ultradisperse aluminum by water in the liquid phase. *Combustions Explosives Shock Waves*. https://doi.org/10.1023/A:1017505709456
5. Elitzur, S., Rosenband, V., & Gany, A. (2014). Study of hydrogen production and storage based on aluminum-water reaction. *International Journal of Hydrogen Energy, 39*(12), 6328–6334. https://doi.org/10.1016/j.ijhydene.2014.02.037
6. Chai, Y. J., et al. (2014). Hydrogen generation by aluminum corrosion in cobalt (II) chloride and nickel (II) chloride aqueous solution. *Energy*. https://doi.org/10.1016/j.energy.2014.03.011
7. Wang, H. W., Chung, H. W., Te Teng, H., & Cao, G. (2011). Generation of hydrogen from aluminum and water—Effect of metal oxide nanocrystals and water quality. *International*

Journal of Hydrogen Energy, 36(23), 15136–15144. https://doi.org/10.1016/j.ijhydene.2011.08.077
8. Khai Phung, K., Sethupathi, S., & Siang Piao, C. (2018). Production of H_2 from aluminium/water reaction and its potential for Co_2 methanation. In *IOP Conference Series Earth Environment Science*, (Vol. 140). https://doi.org/10.1088/1755-1315/140/1/012020.
9. Demir, M. E., & Dincer, I. (2018). Cost assessment and evaluation of various hydrogen delivery scenarios. *International Journal of Hydrogen Energy*. https://doi.org/10.1016/j.ijhydene.2017.08.002
10. Haeseldonckx, D., & D'haeseleer, W. (2007). The use of the natural-gas pipeline infrastructure for hydrogen transport in a changing market structure. *International Journal Hydrogen Energy, 32*, 1381–1386. https://doi.org/10.1016/j.ijhydene.2006.10.018
11. Tzimas, E., Castello, P., & Peteves, S. (2007). The evolution of size and cost of a hydrogen delivery infrastructure in Europe in the medium and long term. *International Journal of Hydrogen Energy*. https://doi.org/10.1016/j.ijhydene.2006.10.017
12. Yang, C., & Ogden, J. (2007). Determining the lowest-cost hydrogen delivery mode. *International Journal of Hydrogen Energy*. https://doi.org/10.1016/j.ijhydene.2006.05.009
13. Seymour, E. H., Murray, L., & Fernandes, R. (2008). Key Challenges to the introduction of hydrogen-European stakeholder views. *International Journal of Hydrogen Energy*. https://doi.org/10.1016/j.ijhydene.2008.01.042
14. Gerboni, R. (2016). Introduction to hydrogen transportation. In *Compendium of Hydrogen Energy*.
15. Yang, C., & Ogden, J. M. (2013). Renewable and low carbon hydrogen for California-Modeling the long term evolution of fuel infrastructure using a quasi-spatial TIMES model. *International Journal of Hydrogen Energy*. https://doi.org/10.1016/j.ijhydene.2013.01.195
16. Sperling, D., & Yeh, S. (2009). An assessment of the near-term costs of hydrogen refueling stations and station components. *Institute Transport and Studies*. https://doi.org/10.1007/s11116-007-9132-x
17. Weisberg, A. H., Aceves, S. M., Espinosa-Loza, F., Ledesma-Orozco, E., & Myers, B. (2009). Delivery of cold hydrogen in glass fiber composite pressure vessels. *International Journal of Hydrogen Energy*. https://doi.org/10.1016/j.ijhydene.2009.09.051
18. Kim, J. G., & Kuby, M. (2012). The deviation-flow refueling location model for optimizing a network of refueling stations. *International Journal of Hydrogen Energy*. https://doi.org/10.1016/j.ijhydene.2011.08.108
19. Mukherjee, U., Maroufmashat, A., Narayan, A., Elkamel, A., & Fowler, M. (2017). A stochastic programming approach for the planning and operation of a power to gas energy hub with multiple energy recovery pathways. *Energies*. https://doi.org/10.3390/en10070868
20. G. P. H., Liu, J., Sun, W. (2019). The economic and environmental impact of power to hydrogen/power to methane facilities on hybrid power-natural gas energy systems. *International Association Hydrographic Energy*.
21. Lee, J. Y., Yoo, M., Cha, K., Lim, T. W., & Hur, T. (2009). Life cycle cost analysis to examine the economical feasibility of hydrogen as an alternative fuel. *International Journal of Hydrogen Energy*. https://doi.org/10.1016/j.ijhydene.2009.03.012
22. Hetland, J., & Mulder, G. (2007). In search of a sustainable hydrogen economy: How a large-scale transition to hydrogen may affect the primary energy demand and greenhouse gas emissions. *International Journal of Hydrogen Energy*. https://doi.org/10.1016/j.ijhydene.2006.08.011
23. Mulder, G., Hetland, J., & Lenaers, G. (2007). Towards a sustainable hydrogen economy: Hydrogen pathways and infrastructure. *International Journal of Hydrogen Energy*. https://doi.org/10.1016/j.ijhydene.2006.10.012
24. Turoń, K. (2020). Hydrogen-powered vehicles in urban transport systems—current state and development. *Transport Research Procedia, 45*, 835–841.

Control Mechanism for Constant Power Generation, Distributed Generation, Electric Vehicle, Hybrid Power System, Hydro Power System

A Novel Architecture for Cluster Based False Data Injection Attack Detection and Location Identification in Smart Grid

S. Mallikarjunaswamy, N. Sharmila, G. K. Siddesh, K. R. Nataraj, and M. Komala

1 Introduction

In today's world, the smart grid contains various physical devices and the technology, which works together to serve various purposes in cyber security and other day-to-day challenges such has traffic management system, banking security management system and many more. The failure of these results in huge divesting effects such as traffic jam reported on china national highway 110 on August 2010 that last for two weeks and many more like these. With the increased challenges today the integration of the new emerged technologies that includes sensors, power efficient resources and control systems in the smart grid makes it a powerful, efficient and more reliable.

Cyber physical system (CPS). As these cyber physical systems (CPS) are becoming more powerful and are used to handle the data securely inside and outside of the organizations all over the world, these systems are becoming an attacking targets by the attackers to snip the secured data by causing a potential threat for the organization or the persons who are handling this confidential data, For example An Ex-employee in Queensland, Australia launches a dangerous attack on the sewage system causing a threat. Thus, it is very important to monitor the components of the smart grid in a timely manner to identify any attacks and quarantine to the compromised nodes. The threats to a CPS are mainly classified into the three main categories: Availability of the network, integrity of the data and privacy of the Information/

S. Mallikarjunaswamy · G. K. Siddesh (✉)
Department of ECE, JSS Academy of Technical Education, Bangalore, Karnataka, India

N. Sharmila
Department of EEE, RNS Institute of Technology, Bangalore, Karnataka, India

K. R. Nataraj
Department of ECE, Don Bosco Institute of Technology, Bangalore, Karnataka, India

M. Komala
Department of ECE, SJB Institute of Technology, Bangalore, Karnataka, India

P. Mahanta et al. (eds.), *Advances in Thermofluids and Renewable Energy*, Lecture Notes in Mechanical Engineering, https://doi.org/10.1007/978-981-16-3497-0_48

confidentiality [1]. Recent literature studies have discussed about the attacks such as denial of service (DoS) [2], load distribution attack and reply attack on the network to make the network resources unavailable, injection of false data into the smart grid [3, 4], and many such attacks to compromise the data integrity and other attacks such as eavesdropping, identity spoofing attack which compromises the confidentiality of the attack etc. are discussed in [5]. The attacks which compromises the above-mentioned security requirements on control systems and power grid system using DC are discussed in [6]. Load redistribution attack is a kind of false data injection attack where the load bus power injection and measurements of line power flow are compromised [7]. The minimum sized unobservable attacks can be found using the graph theoretic approach is studied in [8]. Exactly, two power injection meters and arbitrary number of power meter can be compromised using a sparse data injection attack is discussed in [9]. Various methods and techniques to detect and defend against the attacks on smart grid are discussed in [10]. A selected set of sensor measurements are strategically observed and secured to investigate about false data injection attack is discussed in [11]. Algorithms, which are used, to embed the encrypted devices to enhance the security and other utilities of the system are discussed in [12]. Injection of an incorrect data into the smart grid can be identified using together the reconfigurable Kalman Filter and detectors such as χ^2-detector and Euclidean detector, according to the work shown in [13]. The attacks on the Cyber physical systems like smart grids are very significant and need to address by detecting the attacks and isolating the failed nodes is the smart grid. However, identifying the specific attack and the failed nodes in CPS is still quite open and challenging [14]. The faults and the faulty locations in control systems are identified using the techniques called as FDI (Fault Detection and isolation) technique. There are two types of FDI technique, fault detection and isolation based on signal processing and model based fault detection and isolation. In [15], the decision-making process in model-based fault detection and isolation are facilitated based on the consideration of measurement residual obtained from the schemes or observer schemes considers the measurement residual generated from Kalman filter. In traditional control systems, the faults are identified and are quarantined using the techniques called DOS (Dedicated observer Scheme) and generalized observe scheme (GOS). The fault identification in doubly fed induction generators using generalized observe scheme is discussed in [16]. Similarly, the GOS scheme to identify faults and quarantine them in fractional order systems are proposed in [13]. In case of DOS, DOS can identify and quarantine faults at multiple places at the same time. However, according to the study in [17], it is not feasible to implement DOS due to its complication and its cost of implementation in large system since each observations are compelled by a single instrument and needs the system to be noticeable with any single output.

In this paper, a scheme called IOS (Iterative Observer Scheme) is proposed which identifies the arbitrarily attacked sensors and faults in the smart grid. The iterative observer scheme works by clustering the sensors in the smart grid into subset of sensors and when the attack or fault is sensed in the system, a subset of sensor consisting of misbehaving sensors are selected and the input from all the sensor are iteratively processed by kalman filter and $\chi 2$-detector to identify attacked or faulty nodes. Based on the proposed model, The GOS scheme (Generalized Observer Scheme) is implemented, and then, it is compared with iterative observer method (IOS) to show that proposed IOS scheme effectively identifies the attack or faults on sensor at multiple location, in contrast to the GOS scheme which is capable of identifying the attacks or faults on single sensor. Our study has also shown that using 'S' observers. The IOS detects all the attacked or misbehaving nodes in $O(\log\lambda S)$, whereas with the same amount of observers the GOS is capable of detecting a single misbehaving sensor in constant time. The value of the λ can be set to any value depending on the specific systems, for our implementation the value of λ is set to 2. It is also observed that a larger λ value combined with parallel processing can speeds up the faults isolation process significantly. Further throughout our implementation, the value of λ is considered to be 2. The contents in this paper are organized as follows. In Sect. 2, the system model is derived. The Kalman filter is described in Sect. 3. The attacks/ failure detection using $\chi 2$- detector and the other fault detecting schemes are discussed in Sect. 4. Section 5 shows the simulation results and at the last we conclude this in Sect. 6.

2 System Model

The smart grid consists of sensors and other physical devices to collect the sensitive data and communicate the data to the central controller, there are various types of sensors that sense different types of data, such as PMUs that are responsible of uninterrupted operation of the power system by verifying the system state at multiple location and different time by measuring the current phase and amplitude [18]. The sensors communicate to the central controller, the measurements such as bus voltage, angles and magnitudes through the wired/wireless communication infrastructure. If we encounter any change in the actual value of either the current, phase or voltage [19], then we can say the power system is compromised. The DC power flow model is used to model the power system as discussed in [20].

$$P_{ij}=\frac{Y_i Y_j}{Z_{ij}}\sin(\theta_i-\theta_j) \tag{1}$$

where *Pij* is the active power transmitted from bus *i* to *j*, the magnitude of the voltage is represented using *Yi* and The bus *i*, phase angle is represented by θi, the reactance of the transmission line from bus *i* to bus *j* is *Zij*. The assumption is that the voltage amplitude of the buses is close to unity, keeping in mind that the DC power flow model the difference in the phase between voltages in two buses is also assumed to be small, so further we can simplify Eq. (1) as represented by Eq. (2).

$$P_{ij} = \frac{\theta_i - \theta_j}{Z_{ij}} \tag{2}$$

The '*r*' real time measurements are used to estimate '*q*' phase angles (θi). According to the work in [21] the reference bus voltage phase angle (θi) in the power flow studies is fixed and known hence we need to estimate *q*-1 phase angles only. The θ, called as state vector is defined as $\theta = [\theta 1 \ldots \theta \mathrm{q}]^w$.

Similarly, we can describe the observation vector as shown in Eq. (3).

$$k = p(\theta) + e \tag{3}$$

where the active power in the transmission lines is measured using the vector, $k = [k1, \ldots., kr]^w$, the non-linear relation between measurement vector *k* and state vector θ is defined as $P(\theta)$ and the Gaussian measurement error vector is given as $e = [e1, \ldots., er]^w$. m active power measured data is used by the control station to calculate the vector *Z*.

Jacobian matrix $G \in L^r$ is defined as

$$G = \frac{\mu P(\theta)}{\mu\theta}|\theta = 0 \tag{4}$$

The phase difference $(\theta_i - \theta_j)$ is assumed to be small in Eq. (2) to rewrite, the linear approximation model described in Eq. (3) as Eq. (5).

$$k = G\theta + e \tag{5}$$

The states in time series for the system can be described using the following equation.

$$\theta_{t+1} = \theta_t + h_t \tag{6}$$

where the state vector θ at time $t + 1$ is given by θ_{t+1}.

3 Reconfigurable Kalman Filter

The state space vector *zt*, estimates is derived using the high-switching speed technique known as reconfigurable kalman filter (RKF) [22]. The RKF uses *zt* to represent θt, the state equation is given as follows,

$$z_{t+1} = \boldsymbol{A} z_t + \boldsymbol{d}_t \tag{7}$$

where, unity matrix is denoted as *A* and $zt + 1$ is equals to $\theta t + 1$. Using Eq. (5), the observation equation for RKF can be written as:

$$\boldsymbol{f}_t = \boldsymbol{U}_t\boldsymbol{x}_t + \boldsymbol{v}_t \tag{8}$$

Here, the measurement vector k is represented as ft, $Ut = Gt$ and vt, a noise measurement which does not depend on initial conditions and a process noise. We also assume that dt and vt are the white Gaussian noise with standard deviation η and zero mean.

The state estimations $\hat{z}_t$ can be computed using RKF. For this, we define covariance of the estimates and mean as follows:

$$\begin{aligned}
\hat{z}_{t|t} &= F[z_t, f_o, \ldots, f_t] \\
\hat{z}_{t|t-1} &= F\left[z_t, f_o, \ldots, f_{t-1}\right] \\
\gamma_{t|t-1} &= \sum_{t|t-1} \cdot \\
\gamma_{t|t} &= \sum_{t|t} \cdot
\end{aligned} \tag{9}$$

The measurements for estimates at time t and $t-1$ is defined as follows. Using measurements up to time t and $t-1$ the estimates at time t is given by $\hat{z}_{t|t}$ and $\hat{z}_{t|t-1}$. Similarly, using data up to time t and $t-1$ the covariance of the estimates at time t is given by $\gamma_{t|t}$ and $\gamma_{t|t-1}$. Now, the RKF iterations can be written as:

Time Update:

$$\begin{aligned}
\hat{z}_{t-1|t} &= A\hat{z}_t \\
\gamma_{t|t-1} &= A\gamma_{t|t-1}A^W + F
\end{aligned} \tag{10}$$

Here, the processing noise covariance matrix is denoted using F and the state space model in Eq. (5) gives A. The state and covariance estimates at $t+1$ time step from t time step is shown in Eq. (10).

Measurement Update:

$$\begin{aligned}
L_t &= \gamma_{t|t-1}U_t^W(U_t\,\gamma_{t|t-1}U_t^W + H)^{-1} \\
\gamma_{t|t} &= \gamma_{t|t-1} - L_tU_t\gamma_{t|t-1} \\
\hat{z}_t &= \hat{z}_{t-1|t} + L_t\left(f_t - U_t\,\gamma_{t|t-1}\right)
\end{aligned} \tag{11}$$

The reconfigurable Kalman filter's measurement updates are given in Eq. (11) as shown in Fig. 1. The Lt and H give the kalman gain and measurement noise covariance matrix. The more accurate estimate can be obtained by incorporating the measurements ft in the formulas given in the above equations.

The condition $\hat{z}_{0|-1} = 0$ and $\gamma_{0|-1} = \Sigma$ is considered initially. The Kalman gain is assumed to converge into a steady state in few steps. Finally, a training period is conducted assuming that Kalman gain is already known to filter before the estimation,

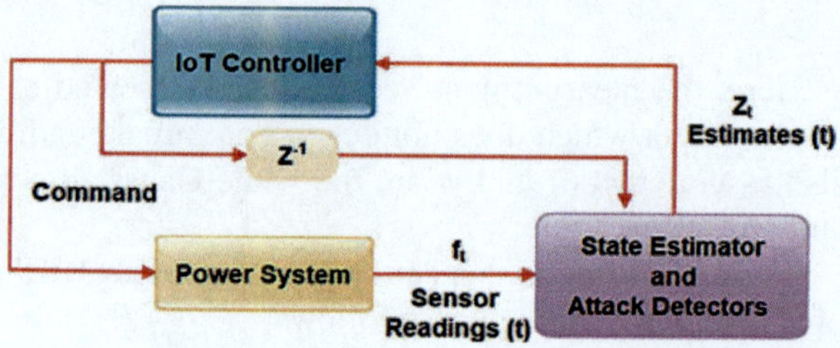

Fig. 1 Power smart grid device

then,

$$\gamma \triangleq \lim_{k\to\infty} \gamma_{t|t-1}$$
$$L = \gamma U^{W}(U\gamma UW + H)^{-1} \tag{12}$$

Further we can rewrite Eq. (10) as:

$$\hat{z}_{t+1} = A\hat{z}_t + L\left[f_{t+1} - U\left(A\hat{z}_t\right)\right] \tag{13}$$

The estimation error e_t is defined as:

$$e_t = \hat{z}_t - z_t \tag{14}$$

4 Failures/Attacks Detector and Identifier

In this section, we discuss about three important concepts of the study, χ2–detector, GOS and IOS. After the next state of the system is calculated by the RKF and the readings from the sensor are available, the detector compares the readings from the sensor and the estimates to identify any disagreement.

4.1 χ^2-detector

The χ^2–*detector* which is used with kalman filters generates a χ^2 test statistics from them as mentioned in [23] and are checked against the threshold which are already computed and available using χ^2-detector-table to identify the faulty area or the attack. At time $t + 1$, the residue $\sigma t + 1$ is defined as:

$$\boldsymbol{\sigma}_{t+1} \triangleq \boldsymbol{f}_{t+1} - \hat{\boldsymbol{f}}_{t-1|t} \tag{15}$$

Using Eqs. (8), (10) and (15) we write,

$$\boldsymbol{\sigma}_{t+1} \triangleq \boldsymbol{f}_{t+1} - \boldsymbol{U}(\boldsymbol{A}\hat{\boldsymbol{z}}_t) \tag{16}$$

Then, the scalar test statistics are compared in the test conducted by the χ2-detector and is given by:

$$\boldsymbol{v}(\boldsymbol{t}) = \boldsymbol{\sigma}_t^W \boldsymbol{B}_t \boldsymbol{\tau}_t \tag{17}$$

The covariance matrix of $\boldsymbol{\sigma}_t$ is represented by $\boldsymbol{B}_t$, the failures or the attack on the smart grid is detected by comparing $v(t)$ with an already calculated threshold which are availed using the χ2-detector-table as discussed in [24]. The unique property χ2–detector is that it performs a long-term test, by each time considering system starting time and all integrated effects at every detection steps. This property of χ2-detector is very useful for fault identification in the smart grid that consists of large number of sensors, which are susceptible for instrument bias shift like soft failures.

Fault detection process using χ2-detector is fast and straight forward which makes it another advantage, because the RKF generates all the statistics that are required to run the test and test results are directly compared with the pre-computed threshold which are available in χ2- detectors table.

4.2 Filter Bank for Attacks/Faults Isolation

Identifying and quarantining the attacks or faults on the smart grid is very essential. To do so we use filter banks, these filter banks evaluates the dynamic behavior of the system [25] to detect and isolate the attacks. In this paper, two of such filter banks are studied and implemented.

4.3 Generalized Observer Scheme (GOS)

GOS is one of the filter banks, which is used to detect and isolate the attack [27]. GOS uses many observers/kalman filters to do the job, kalman filters. Each observer accepts all of the inputs to the grid and a subset of $\boldsymbol{r}_i$ outputs from the grid to compute residual vector, which are then fed into χ2-detector (decision unit) to detect and isolate faults or attacks. The mechanism of GOS is depicted in Fig. 2.

Here, the smart grid system produces $\boldsymbol{f}_i$ outputs and consists of 'r' observers. Each ith observer accepts the input 'u' and all the 'r' output from the grid except $\boldsymbol{f}_i$ to produce the residual vector $\boldsymbol{\sigma}_i$, the value of $\boldsymbol{\sigma}_i$ is dependent on all, except the ith sensor reading. Each residual vector consists of m elements $\boldsymbol{\sigma}_i = [\boldsymbol{\sigma}_{i1}, \boldsymbol{\sigma}_{i2}, \ldots\ldots\ldots\ldots, \boldsymbol{\sigma}_{ir}]$ where $\boldsymbol{\sigma}_{ii} = 0$. Then, the decision unit that consists of χ2-detector accepts these residual vectors as input to detects and isolates the identified faults or attacks.

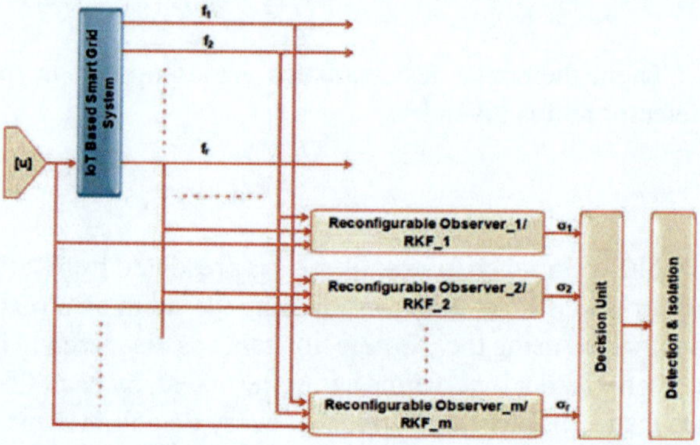

Fig. 2 Generalized observer scheme using IoT

4.4 *Iterative Observer Scheme (IOS)*

The GOS scheme is able to identify attacks or faults on a single sensor at a time, which is a time-consuming process, in contrast to GOS we propose another scheme called as IOS, which can detect and isolate attack on multiple nodes. The mechanism of GOS is illustrated in Fig. 3. Initially, the combination of RKF and χ^2-detector is used to check for the faults by considering all the 'r' outputs and n inputs. If the attack is identified then in the next level the 'r' outputs are divided into two subgroups. First subgroup containing 1 to $\frac{r}{2}$ outputs and second subgroup contains $\frac{r}{2}+1$ to r. Each subgroup along with n input is then used to drive the observer separately to check for the attacks or faults. If the attack is detected, then the above procedure is repeated iteratively until the subset size is less than or equal to β sand the value of $\beta \geq 1$.

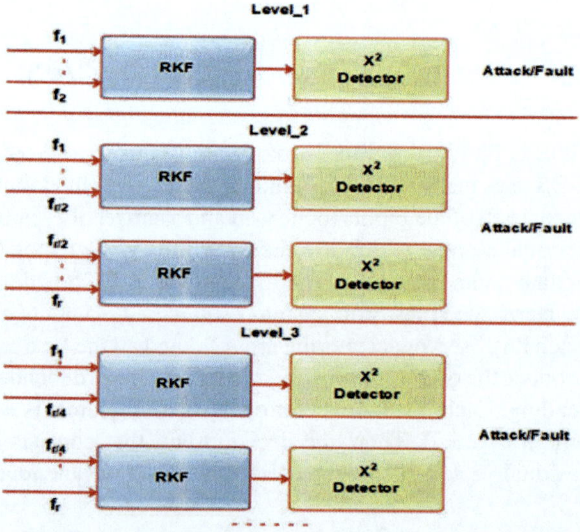

Fig. 3 Iterative observer scheme using IoT

Where β is a user defined cut off parameter. If the attack is not detected then we discontinue the process of dividing the output.

When $\beta = 2$ the algorithm ends in the implementation. It is also observed that for r measurements the maximum computation required is only $2r - 1$, and there are at most $\log_2(r)$ levels. The maximum number of steps required in each iteration if we divide the set of inputs into λ subsets will be $O \log_\lambda r$. To speed up the isolation process, the parallel processing technique can also be used.

Figure 3 shows IOS Scheme using RKF and χ^2-detector for identifying attack/faults. The three level of iterations are shown when attack is detected. In level 1, all the 'r' outputs and n inputs are considered. After the attack is identified, in level 2 the 'r' outputs are divided into two subgroups. First subgroup containing 1 to $\frac{r}{2}$ outputs and second subgroup contains $\frac{r}{2} + 1$ to r each subgroup along with n input are inputted into two separate observer to check for the attack. If the attack is identified the above procedure is repeated as shown in level 3 iteration.

5 Implementation and Performance Evaluation

We have used Matlab software for implementation process of the RKF estimator, χ^2-detector and observers. The bus system of IEEE 9 and IEEE 14 [28] was solved using the package called as MATPOWER. Figure 4 shows the bus system of IEEE 9. The state estimation uses matrix H was obtained using MATPOWER. The function of Matlab, randn() is used to generate a distributed noise with mean value zero. The χ^2 table gives the threshold.

5.1 *Fault Detection and Isolation Using GOS*

GOS scheme that is used on bus system of IEEE 9 to identify an attack/ fault on single bus (Bus 5) is depicted in Fig. 5a. This figure clearly indicates that the measurement

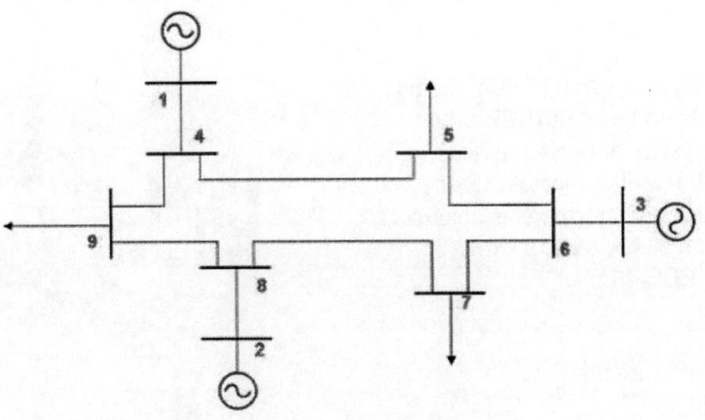

Fig. 4 The vulnerability level of IEEE 9-bus system

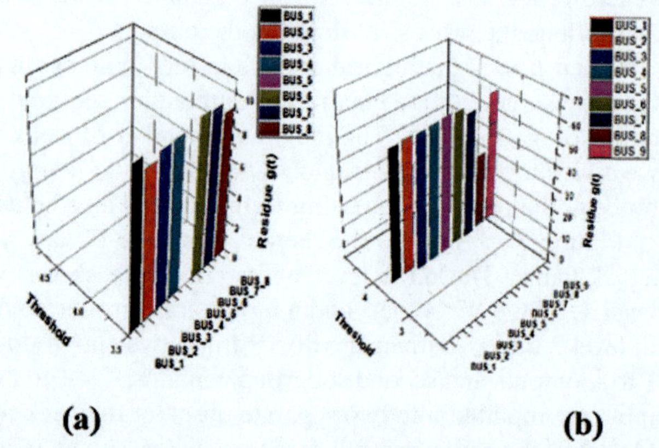

Fig. 5 **a** The simulation result shows the detection process of attack on bus 5 of bus system IEEE-9 using GOS. **b** The simulation results show the GOS scheme has failed to identify the attack on bus system of IEEE-9

y_5 does not affect the observer 5 and the straight line indicates that the $v(t)$ which is a residue function does not exceed the threshold. However, the rest of the observers are affected due to the attack. Hence, GOS identifies the attack on the Bus 5. Figure 5b indicates two attacks on IEEE 9-bus system, the calculate residue function $v(t)$ in all the GOS observers exceeds the threshold. Hence the multiple attacks or the faults in the systems are not recognized by the GOS.

5.2 *Fault Detection and Isolation Using IOS*

Figure 6a shows an application of IOS scheme on bus system of IEEE 9, the top most graph in figure represents the entire system after implementing IOS scheme. Since there is attack the $v(t)$, a residue function exceeds the threshold and hence the IOS splits the 9 output to two subsets having two halves output 1 to 5 in first half and output 6 to 9 in the second half.

Through graph, it is observed that both the subsets are compromise by the attack. Hence the above splitting procedure is continued, the IOS splits the subsets again into halves resulting in four subsets containing the output 1 to 3, 4 to 5, 6 to 7 and 8 to

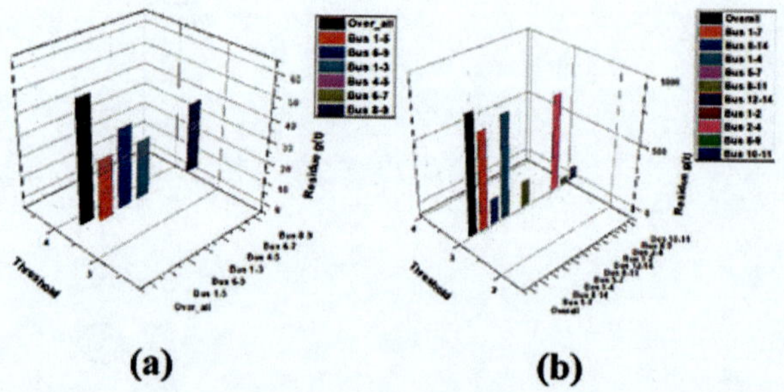

Fig. 6 **a** In this case, the bus 2 and bus 8 of IEEE-9 bus system are under attack. **b** Result indicates that the IOS scheme detects an attack on bus 3, 8 and 11 for IEEE-14

Table 1 Comparison of GOS and IOS

	GOS	IOS
Number of worst-case detection rate	r	2 r-1
Number of best-case detection rate	r	2 log m-1
Simulation of multiple attacks detection	No	Yes

9. Again these subsets are fed into the observers to generate the residue function $v(t)$, which is plotted in the last row of the figure. Now graph clearly shows that the residue function of the first (buses 1–3) and last figure (buses 8–9) of this row exceeds the threshold indicating the attack on this set of buses. Similarly, the simulation graphs that are obtained by implementing IOS scheme on bus IEEE 14 is depicted in Fig. 6b According to figure, three buses namely 3, 8 and 11 are under attack.

5.3 *Comparison of GOS and IOS*

The comparison between GOS and IOS schemes are done based on the number of computations required in the best-case scenario and the worst-case scenarios in Table 1. In GOS both the scenario, the best case and the worst case are same as it can detect a single attack at a time. There are m observers in GOS, based on the assumption each does unit computation and hence needs m computation for all the cases. The worst case is defined when all the buses in the system are compromised. In case of IOS, all the subsets are verified for the attack in every iteration indicating $2\boldsymbol{r} - 1$ computation, in the best case it is assumed that only one bus is compromised, only two subsets are to be checked for the attack resulting in $2 \log \boldsymbol{r} - 1$ computations.

6 Conclusion

In this paper, we have discussed two schemes GOS and IOS to detect the attacks on smart grid. The GOS uses m computations in the best and the worst case to detect the attacks and it is also noted that the GOS scheme is capable of detecting attack on only single sensor and fails if the attack is on multiple sensors in contrast to GOS we propose other scheme called IOS which detects the attack on multiple servers by clustering the sensors into the subset whenever residue function exceeds the threshold and iteratively processing them to find out the faulty area or attacked node. The IOS scheme needs at most $r - 1$ computations in the worst case and only $2 \log r - 1$ computations in the best case, where m is the number of measurements in the system. In future work, the unobservable attacks like false data injection which compromises the data integrity can be identified and isolated with new scheme called DOS which uses the same context as of IOS.

Acknowledgements The authors would like to thank JSS Academy of Technical Education, Technical Education Quality Improvement Programme (TEQIP) Cell and Visvesvaraya Technological University (VTU), Belagavi for all the support and encouragement provided by them to take up this research work and publish this paper.

References

1. Wang, X., & Luo, X. (2019). Distributed detection and isolation of false data injection attacks in smart grids via nonlinear unknown input observers. *Electrical Power and Energy Systems, 110*, 208–222.
2. Li, B., & Rongxing, Lu. (2016). Distributed host-based collaborative detection for false data injection attacks in smart grid cyber-physical system. *Journal of Parallel and Distributed Computing, 12*, 1–42.
3. Anwar, A., & Mahmood, A. N. (2017). Modelingand performance evaluation of stealthy false data injection attacks on smart grid in the presence of corrupted measurements. *Journal of Computer and System Sciences, 83*(1), 58–72.
4. Luo, X., Wang, X. Distributed detection and isolation of bias injection attack in smart energy grid via interval observer. Applied Energy *256*(3), 1–19.
5. Mohammadpourfard, M., & Weng, Y. (2020). Ensuring cybersecurity of smart grid against data integrity attacks under concept drift. *Electrical Power and Energy Systems, 119*(3), 1–9.
6. Yang, L., & Li, Y. (2017). Improved-ELM method for detecting false data attack in smart grid. *Electrical Power and Energy Systems, 91*(4), 183–191.
7. Lee, L., & Hu, P. (2019). Vulnerability analysis of cascading dynamics in smart grids under load redistribution attacks. *Electrical Power and Energy Systems, 111*(4), 182–190.
8. Chen, R., Li, X., & Zhong, H. (2019). A novel online detection method of data injection attack against dynamic state estimation in smart grid. *Neurocomputing, 344*(3), 73–81.
9. Yanguang, J., & Yi, H. (2017). On designing consistent extended Kalman filter. *Journal of Systems Science and Complexity Springer Publication, 30*(4), 751–764.
10. Yang, Y., & Gao, W. (2006). An optimal adaptive Kalman filter. *Journal of Geod esy, 80*(5), 177–183.
11. Oussalah, M., & Zaatri, A. (2002). Kalman filter approach for lane extraction and following. *Journal of Intelligent and Robotic Systems, 34*, 195–218.
12. Rajan, M. P., & Mathew, J. (2012). Kalman filter and financial time series analysis. *International Conference on Eco-friendly Computing and Communication Systems, 305*, 339–351.
13. Zhang, J., & Welch, G. (2015). Kalman filters for dynamic and secure smart grid state estimation. *Intelligent Industrial Systems, 1*, 29–36.
14. Nemanja, Ž, & Andrija, T. S. (2018). Detection of false data injection attacks using unscented Kalman filter. *Journal of Modern Power Systems and Clean Energy, 6*, 847–859.
15. Gel, Q., & Jiang, H. (2020). Power load forecast based on fuzzy BP neural networks with dynamical estimation of weights. *International Journal of Fuzzy Systems, 22*, 956–969.
16. Ansari, M. H., & Vakili, V. T. (2019). Evaluation of big data frameworks for analysis of smart grids. *Journal of Big Data, 6*, 1–14.
17. Galád, M., & Špánik, P. (2017). Analysis of state of charge estimation methods for smart grid with VRLA batteries. *Electrical Engineering, 99*, 1233–1244.
18. Jiongcong, C., & Gaoqi, L. (2016). Impact analysis of false data injection attacks on power system static security assessment. *Journal of Modern Power Systems and Clean Energy, 4*, 496–505.
19. Pilz, M., & Baghaei Naeini, F. (2019). Security attacks on smart grid scheduling and their de fences: A game-theoretic approach. *International Journal of Information Security, 23*, 1–17.

20. Ansari, M. H., & Tabataba, V. (2018). Graph theoretical defense mechanisms against false data injection attacks in smart grids. *Journal of Modern Power Systems and Clean Energy, 6*, 860–871.
21. Yufei, S., & Xuan, L. (2018). Graph theoretical defense mechanisms against false data injection attacks in smart grids. *Journal of Modern Power Systems and Clean Energy, 6*, 630–641.
22. Ansari, M. H., Vakili, V. T. (2019). Evaluation of big data frameworks for analysis of smart grids. Journal of Big Data *6*(109).
23. ŽIvkovic, N., & Saric, A. T. (2018). Detection of false data injection attacks using unscented Kalman filter. *Journal of Modern Power Systems and Clean Energy, 6*, 847–859.
24. Nirmal-Kumar, C., & Zhang, L. (2009). SmartGrid: Future networks for New Zealand power systems incorporating distributed generation. *Energy Policy, 37*, 3418–3427.
25. Hardan, F., & Norman, R. (2019). Balancing loads of rotating generators utilizing VSC direct power controllers in a ship AC/DC smart grid. *Electric Power Systems Research, 182*, 1–17.

Speed Control of PMBLDC Motor Using Rotor Position Rotor Speed PWM Technique

Rajen Pudur and Brajagopal Datta

Nomenclature

A Area (m^2)
a Length of specimen (m)
D Diameter of rod (m)

Greek Symbols

α Volume fraction (dimensionless)
η Efficiency (dimensionless)

Superscripts

(*n*) nth iteration

Subscripts

a Air

R. Pudur · B. Datta (✉)
Department of Electrical Engineering, NIT Arunachal Pradesh, Yupia, India
e-mail: brajagopal@nitap.ac.in

P. Mahanta et al. (eds.), *Advances in Thermofluids and Renewable Energy*, Lecture Notes in Mechanical Engineering, https://doi.org/10.1007/978-981-16-3497-0_49

1 Introduction

Recent technology advancement in DC motors for electric bike resulted in new energy efficient derives using PMBLDC motors. A BLDC motor has windings on stator and alternate permanent magnets on rotor. As the name suggests, BLDC motor does not have brushes for commutation, but electronic commutation of stator windings is based on rotor position with respect to the stator windings. The Brushless Direct Current (BLDC) motor is widely used in many industrial applications, automotive, instrumentation, etc., due to its several advantages like better speed vs. torque characteristics, high dynamic response in transient states, high efficiency than conventional motor drives, long operation life irrespective of working conditions, noiseless operation over high-speed operations, higher speed ranges and easy controls [1]. Also, the torque delivered with respect to the size of the motor is higher, making it useful in applications where space and weight are critical factors like electric vehicles [2]. Torque of the BLDC motor is controlled by the waveform of back-EMF (the voltage induced into the stator winding due to rotor movement). Generally, the BLDC motors have trapezoidal back-EMF waveforms and are fed with rectangular stator currents, which give a theoretically constant torque. But, in practice, torque ripple exists, mainly due to back-EMF waveform imperfections, current ripple and phase current commutation. The current ripple result is from PWM generation technique or hysteresis control. The back-EMF waveform imperfections result from variations in the shapes of slot, skew and magnet of BLDC motor and are dependent on designing methods. An error can occur between actual value and the simulation results. Several simulation models have been proposed for the analysis of BLDC motor [3–6]. But simultaneously, two controls can be carried out to have smooth control over speed of motor. Inverter triggering is controlled by PWM generated by comparing the actual speed of rotor and the rotor position, and same rotor speed can be used to control speed of motor using controlled voltage source. In this paper, using this method PMBLDC motor speed is controlled and compared with the system uses inverter triggering using SPWM.

2 Mathematical Model

The BLDC motor is working as same as the traditional DC motor, but it is construction wise different from it. It is a self-rotating synchronous machine whose stator is similar to that of an induction motor, and the rotor has permanent magnet which is rotating part.

The circuit model of BLDC is as shown in Fig. 1, and the voltage equations of BLDC motor is given as:

$$V_a = R_a i_a + \frac{\mathrm{d}}{\mathrm{d}t}(L_{aa} i_a + L_{ab} i_b + L_{ac} i_c) + \frac{\mathrm{d}\lambda_{ar}(\theta)}{\mathrm{d}t}$$

Fig. 1 Motor circuit model

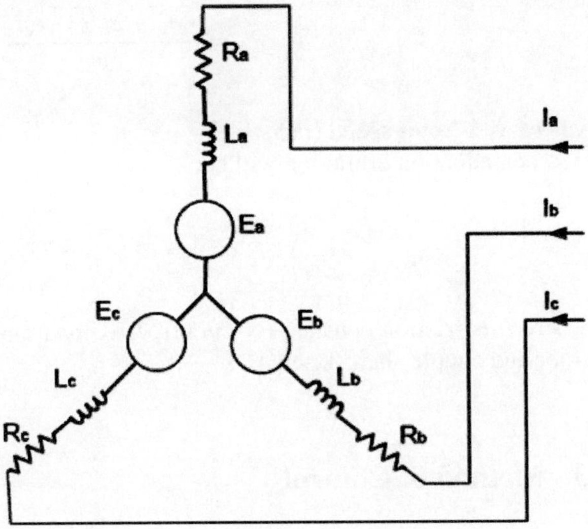

$$V_b = R_b i_b + \frac{\mathrm{d}}{\mathrm{d}t}(L_{ba} i_a + L_{bb} i_b + L_{bc} i_c) + \frac{\mathrm{d}\lambda_{br}(\theta)}{\mathrm{d}t}$$
$$V_c = R_c i_c + \frac{\mathrm{d}}{\mathrm{d}t}(L_{ca} i_a + L_{cb} i_b + L_{cc} i_c) + \frac{\mathrm{d}\lambda_{cr}(\theta)}{\mathrm{d}t}$$

Let us assume magnets have high sensitivity, all stator resistances are equal and rotor induced currents are neglected, then in matrix form for balanced system we have:

$$\begin{pmatrix} V_a \\ V_b \\ V_c \end{pmatrix} = \begin{pmatrix} R & 0 & 0 \\ 0 & R & 0 \\ 0 & 0 & R \end{pmatrix}\begin{pmatrix} i_a \\ i_b \\ i_c \end{pmatrix} + \frac{d}{dt}\begin{pmatrix} L_a & L_{ba} & L_{ca} \\ L_{ba} & L_b & L_{cb} \\ L_{ca} & L_{cb} & L_c \end{pmatrix}\begin{pmatrix} i_a \\ i_b \\ i_c \end{pmatrix} + \begin{pmatrix} e_a \\ e_b \\ e_c \end{pmatrix}$$

The above equations described the mathematical model of motor; hence, the rotor reluctance does not change with angle.

If,

$$L_a = L_b = L_{c|} = L$$

That is self and mutual inductance is constant, then the voltage equations become

$$\begin{pmatrix} V_a \\ V_b \\ V_c \end{pmatrix} = \begin{pmatrix} R & 0 & 0 \\ 0 & R & 0 \\ 0 & 0 & R \end{pmatrix}\begin{pmatrix} i_a \\ i_b \\ i_c \end{pmatrix} + \begin{pmatrix} L-M & 0 & 0 \\ 0 & L-M & 0 \\ 0 & 0 & L-M \end{pmatrix}\frac{\mathrm{d}}{\mathrm{d}t}\begin{pmatrix} i_a \\ i_b \\ i_c \end{pmatrix} + \begin{pmatrix} e_a \\ e_b \\ e_c \end{pmatrix}$$

And the electrostatic torque is given as:

$$T_e = \frac{e_a i_a + e_b i_b + e_c i_c}{w_r}$$

where *wr* is rotor speed (rad/sec).

Then equation of motion will be:

$$\frac{dw_r}{dt} = (T_e - T_l - Bw_r)/J$$

where *B* is friction constant (N-m/rad), *Tl* is load torque (N-m) and *J* is inertia of rotor and couple shaft (kg-m^2).

3 Method of Control

3.1 Rotor Speed and Rotor Position PWM

The schematic diagram of system is shown in Fig. 2, and MATLAB model is shown in Fig. 3. Two control loops are used; the inner loop synchronizes the inverter gate signals with the actual speed of rotor. The outer loop controls the motor's speed by varying the DC bus voltage using PI controller.

The gates signals for MOSFET-based inverter are derived from the error signal generated by comparing rotor actual speed and position of rotor sensed by Hall sensor which is decoded. Three-phase windings use one Hall sensors each, which are provided three overlapping signals giving a 60o wide position range. Whenever

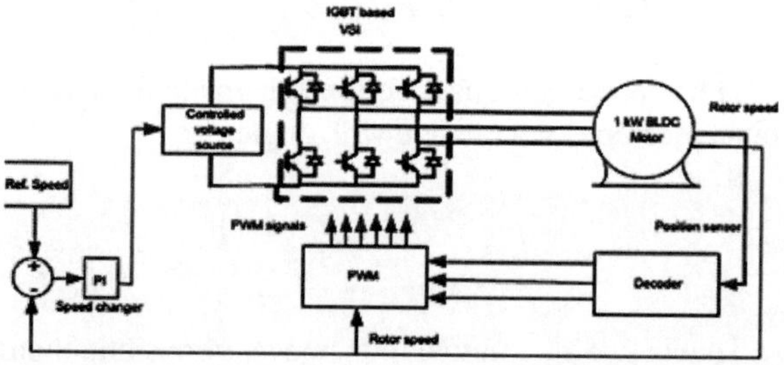

Fig. 2 Schematic diagram of system

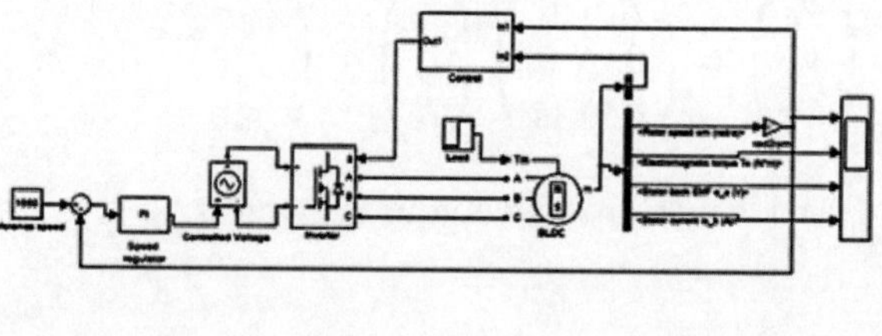

Fig. 3 MATLAB model of system

Fig. 4 Position of rotor decoded

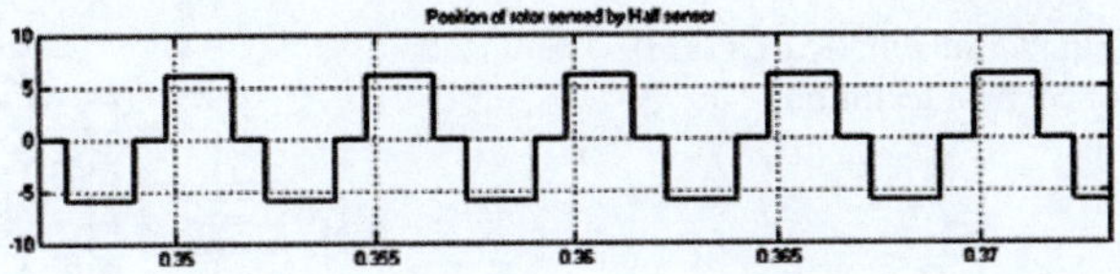

Fig. 5 Gate pulse control

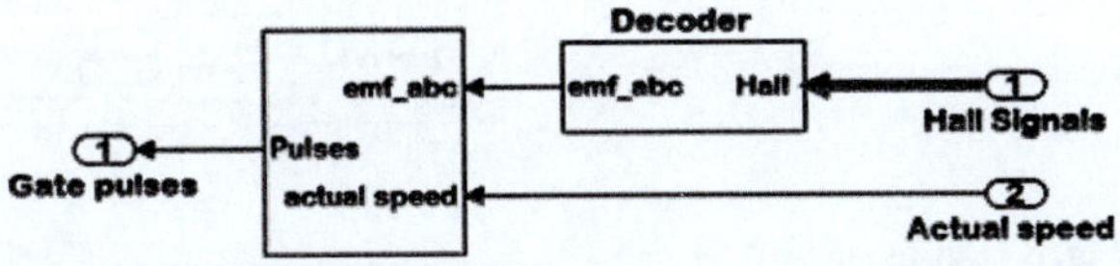

Fig. 6 Gate pulse to inverter

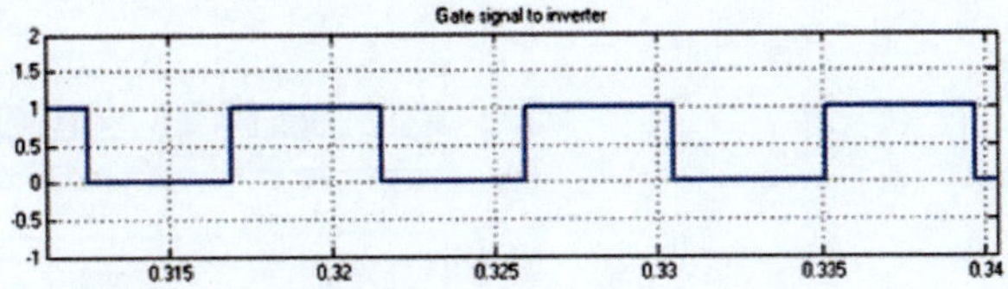

the magnetic poles pass near the sensors, they give high or low signal showing South and North Pole as shown in Fig. 4.

By varying the voltage across the motor terminals, we can control the speed of the BLDC, and when pulse is fed to six switches of the three-phase bridge, variation of motor voltage can be obtained by varying the duty cycle of the PWM signals. Inner closed-loop control is seemingly automatic as the position of rotor and speed of rotor are resulting due to rotation of rotor; hence, the gate signals to inverter will try to switch inverter in such a way to control the voltage input to motor, which will be corresponding to speed and position of rotor. The final speed control is fully done by outer closed loop, which is with speed changer or reference speed.

This method is more suitable for speed control for electric bike where the speed of wheel is in full control of throttle (Fig. 5).

The required speed is controlled by a speed controller. This is achieved by implementing conventional proportional-integral (PI) controller. PI was input by the difference in actual rotor speed and required speed. Based on this data, PI controller controls the duty cycle of PWM pulses which correspond to the voltage amplitude required to maintain the desired speed. Figure 6 shows the PWM pulses of 50% duty cycle.

3.2 *Sinusoidal Pulse Width Modulation*

Second method used is SPWM technique, where carrier wave is taken as triangular with variable frequency, and modulating wave is taken as decoded signal of rotor position sense by Hall sensor. The PWM generated by comparing these two waves

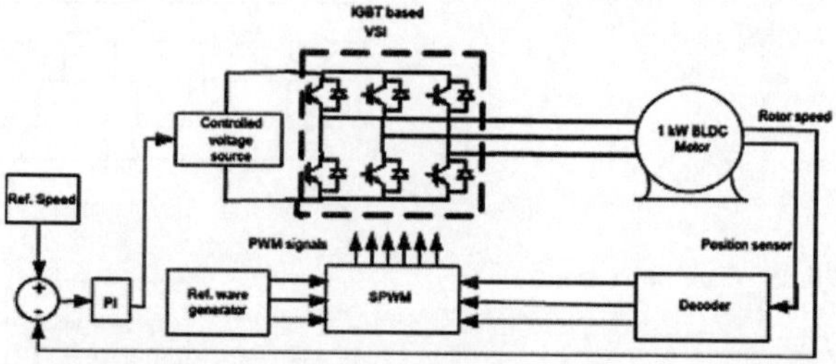

Fig. 7 Schematic diagram of SPWM technique

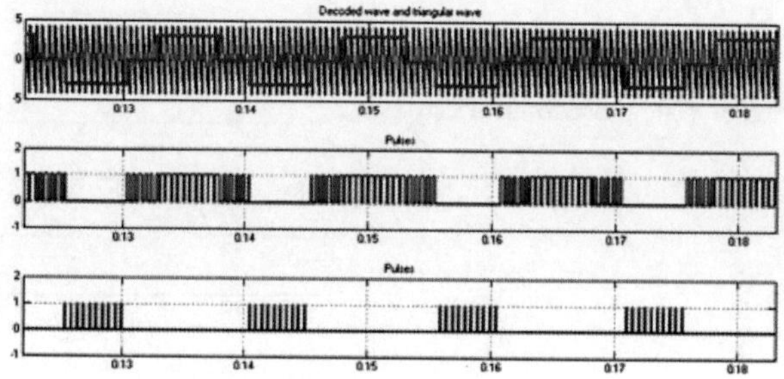

Fig. 8 Pulses of SPWM technique

is fed to inverter, as shown in Fig. 8. Figure 7 shows the schematic diagram of this technique.

4 Results and Discussions

A three-phase motor rated 1 kW, 500 Vdc, 3000 rpm is fed by a six step voltage inverter. The inverter is realized by MOSFET bridge. The three-phase output of the inverter is applied to the PMBLDC motor's stator windings. Load torque of 6 N-m is applied to the machine's shaft at $t = 0.5$ s, and differences are observed.

Figures 9 and 10 show the motor speed when required speed is set to 3000 rpm, and load torque of 9 N m is applied at $t = 0.5$ s, the differences in smoothness are clearly seen

The motor is tested further on different speed requirement such as 2500 rpm and 1650 rpm, and results are shown in Figs. 11, 12, 13and 14.

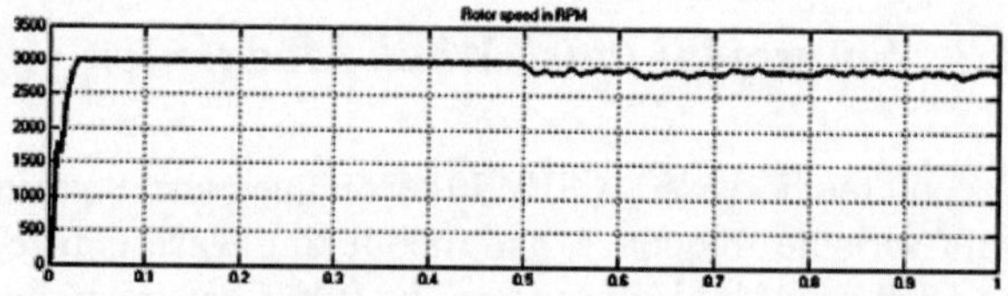

Fig. 9 Motor speed using RSRPPWM technique

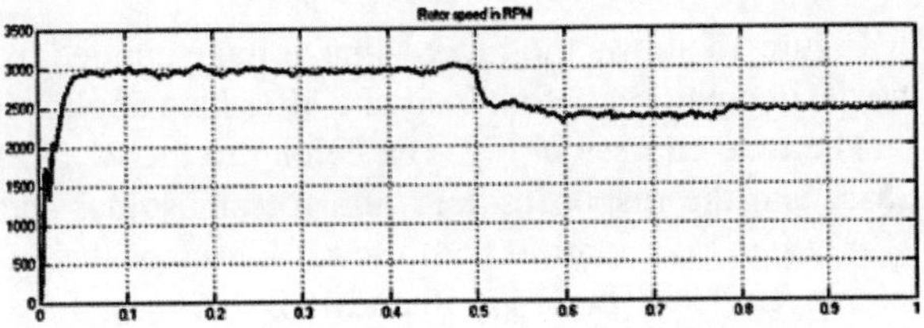

Fig. 10 Motor speed using SPWM technique

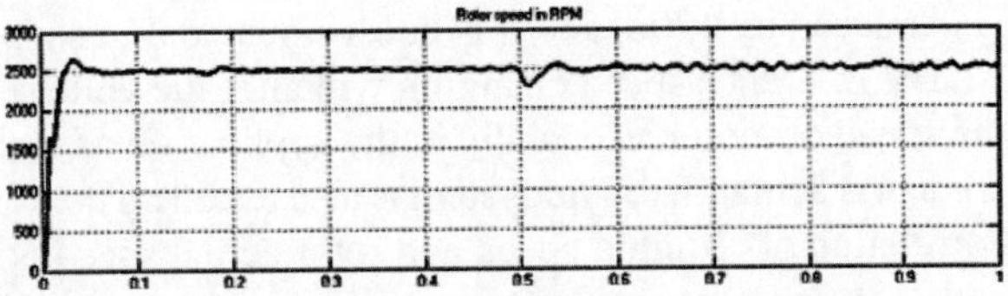

Fig. 11 Rotor speed with 2500 rpm using RSRPPWM technique

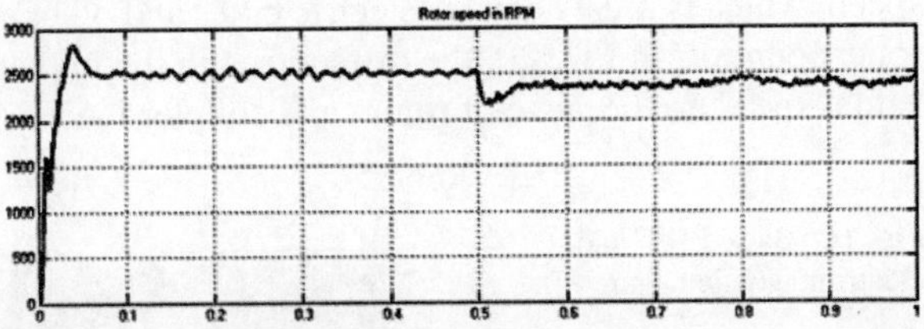

Fig. 12 Rotor speed with 2500 rpm using SPWM technique

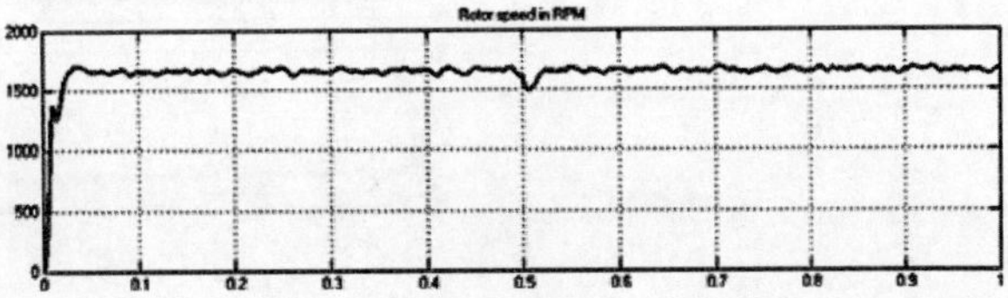

Fig. 13 Rotor speed with 1650 rpm using RSRPPWM technique

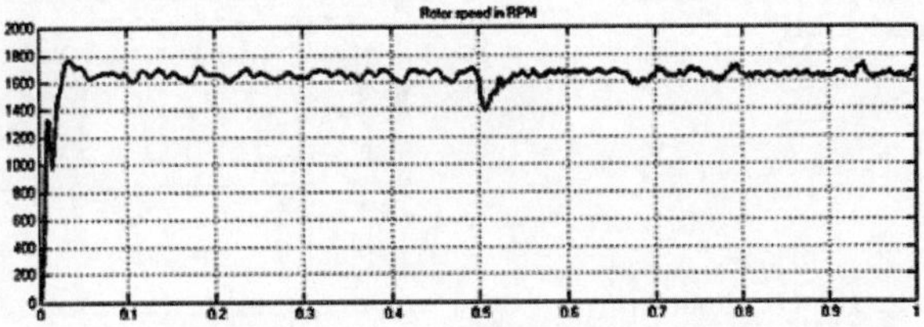

Fig. 14 Rotor speed with 1650 rpm using SPWM technique

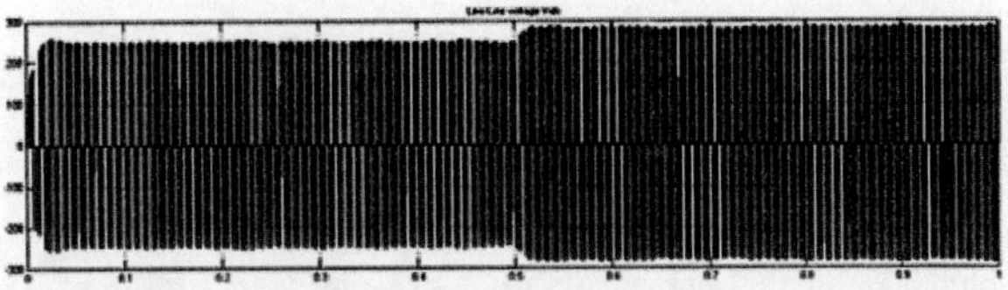

Fig. 15 Line-Line voltage to motor terminal

Figure 15 shows the line-to-line voltage applied to BLDC motor, which follows the desired pattern before load and after load application.

The time courses in Fig. 16a belong to the BLDC motor with trapezoidal back-EMF, and the responses are valid for sinusoidal back-EMF BLDC motor. There is very good correlation with the responses of the BLDC motor models known in references [7, 8] (Figs. 17, 18 and 19).

At closer look at the line-to-line voltage applied to motor reveals that the waveform is displaced by 120o, and it is quasi sinusoidal in shape.

The reference speed changing will alter the voltage applied to motor and thereby the speed of motor, especially in the application of motor for electric bike the robust control is demanded. The system is also tested on the variable required speeds and the correlation of required speed and rotor actual speed is studied, as shown in Figs. 20 and 21. It can be seen that using RSRPPWM technique has superior control over rotor speed, which is must need for electric bikes and other automobiles requiring fast and robust control. In Fig. 20, the differences in the higher speed between required and actual speed are because of rotors and coupler inertia (Figs. 22 and 23).

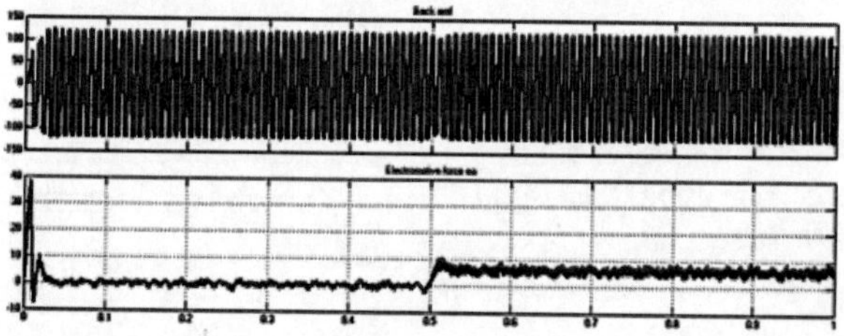

Fig. 16 Back-EMF and electromagnetic torque

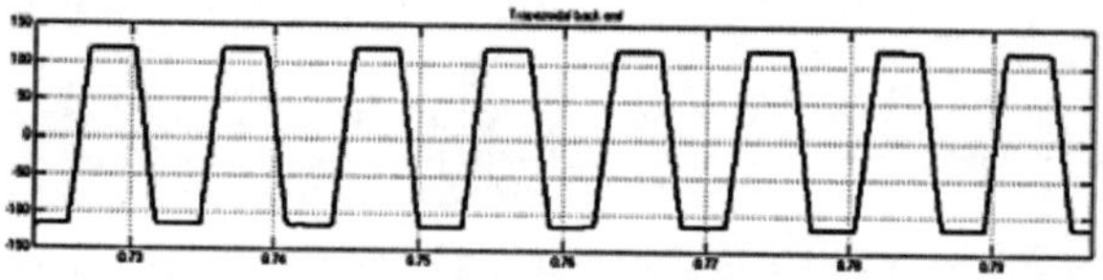

Fig. 17 Trapezoidal back-EMF

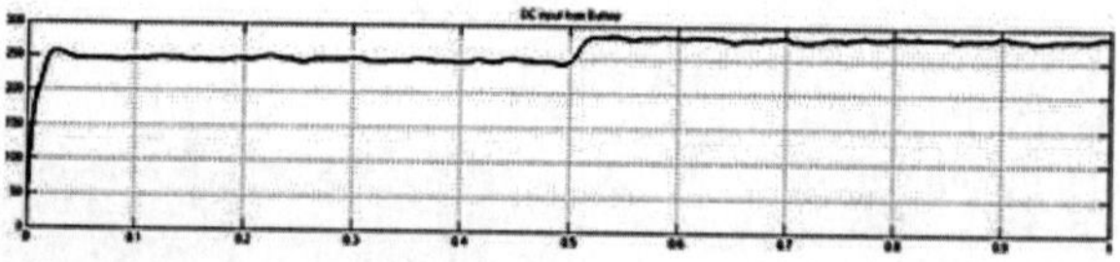

Fig. 18 DC input from battery

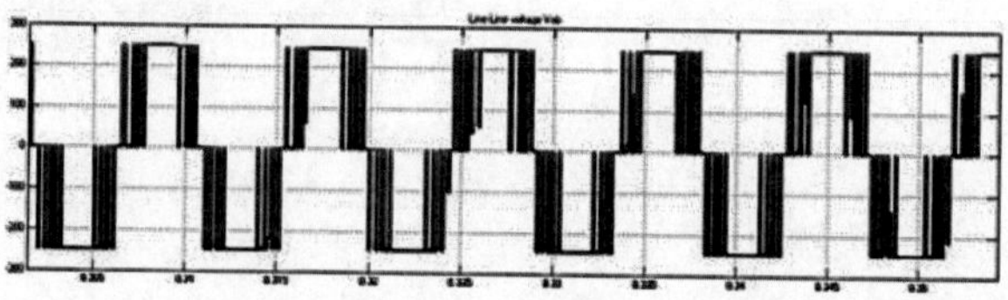

Fig. 19 Close look to voltage applied to motor

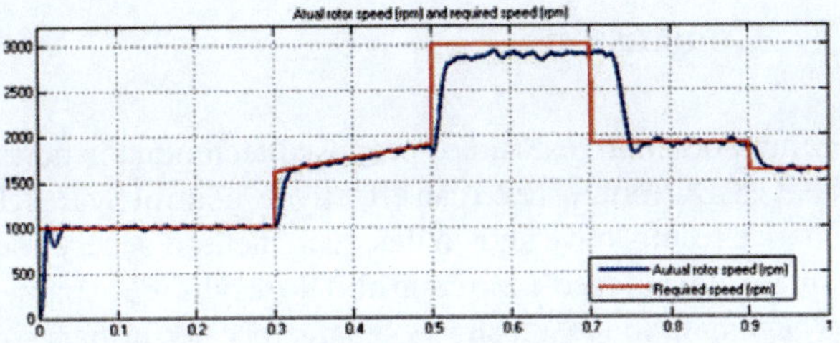

Fig. 20 Actual rotor speed and required speed using RSRPPWM technique

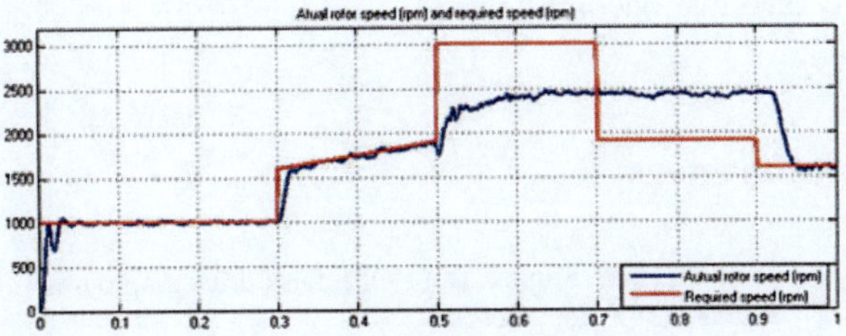

Fig. 21 Actual rotor speed and required speed using SPWM technique

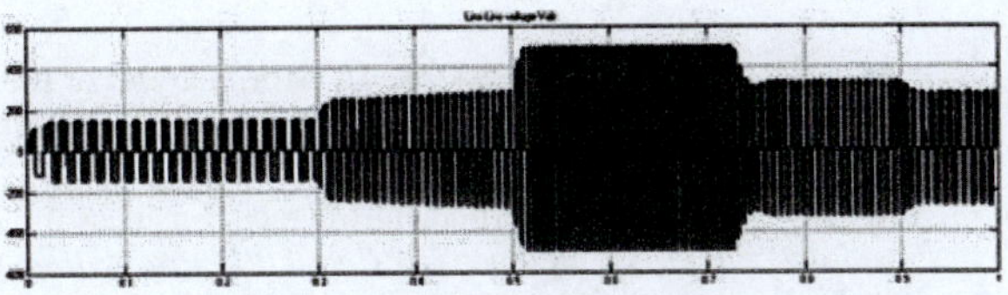

Fig. 22 Variation of motor input terminal voltages using RSRPPWM technique

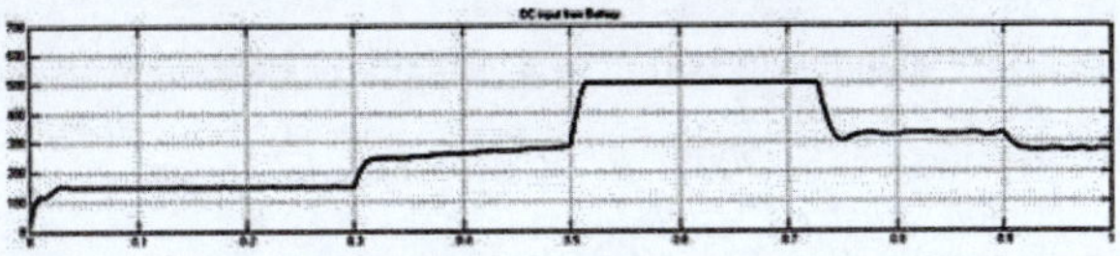

Fig. 23 Variation of DC input from battery using RSRPPWM technique

It is observed that using RSRPPWM technique is the best suited for controlling the speed of BLDC motor in automobile applications. The pulses to inverter will decide the motor speed and in turn, the speed of rotor will decide the pulse; hence, the technique of PWM generation performs vital role here. Figure 21 illustrates the out of relation of actual speed of rotor with required one, in between $t = 0.5$ s and $t = 0.9$ s the rotor speed literally constant without responding to required speed sent through speed changer. And at $t = 0.9$ s the rotor responded lately to the requirement of $t = 0.7$ s by speed changer. Also as shown in Fig. 21, at higher speed there exist large gap between two, which could be because of improper pulses.

5 Conclusion

Simulation and test run of proposed technique is done and presented, and it is seen that BLDC motor speed can effectively control by RSRPPWM technique. The performance results have shown that such method is very useful for applications requiring single outer closed-loop control. The results are at par with the theoretical predictions. The simulation result shows satisfactory performance and hardware implementation can be done. Further, the required speed input can be scaled down to 1 to 10 so that controlling becomes easier.

References

1. Civilian, R., & Stupak, D. (1995). Disk drive employing multi-mode spindle drive system. *U. S. patent 5471353*, Oct 3.
2. Jang, G. H., & Kim, M. G. (2005). A bipolar-starting and unipolar-running method to drive an HDD spindle motor at high speed with large starting. *IEEE Transactions on Magnetics, 41*(2), 750–755.
3. Kim, T. H., & Ehsani, M. (2003). An error analysis of the sensorless position estimation for BLDC motors 3. In *IEEE Conference of Industry Applications Conference, 38th IAS Annual Meeting* (Vol.1, pp. 611–617).
4. Hui, T. S., et al. (2003). Permanent Magnet Brushless motor control techniques. In *National Power and Energy Conference (PECon) 2003 Proceeding*. Bangi Malaysia.
5. Bianchi, N., et al. (2006). Comparison of PM motor structure and sensor less control techniques for zero-speed rotor position detection. *IEEE Transaction on Power Electronics, 22*(6), 2466–2475.
6. Thirusakthimurungan, P., et al. (2006). *A new control scheme for the speed control of MBLDC motor drive*. IEEE
7. Baldursson, S. (2005). *BLDC Motor Modelling and control-A MATLAB/Simulink Implementation (Master Thesis).*
8. Hong, W., Lee, W., & Lee, B. K. (2007). Dynamic simulation of brushless DC motor drives considering phase commutation for automotive application. In *IEEE International Electric machines and drives conferences* 2007 (pp 1377–1383). IEMDC 3–5 May 2007.
9. Kotrba, A., Yetkin, A., Gough, B., Gundogan, A., Mastbergen, D, & Paterson, C. (2011). *Performance characterization of a thermal regeneration unit for exhaust*. Presented at the Commercial Vehicle Engineering Congress, Rosemont, IL, USA, Paper SAE 2011-01-2208.
10. Song, J.-H., & Choy, I. (2004). Commutation torque ripple reduction in brushless DC motor drives using a single DC current sensor. *IEEE Transactions on Power Electronics, 19*(2), 312–319.
11. Tsai, M.-F., PhuQuy, T., Wu, B.-F., & Tseng, C.-S. (2011). Model construction and verification of a BLDC motor using MATLAB/SIMULINK and FPGA control. In *Proceedings 6th IEEE Conference Industrial Electronic Application* (pp. 1797–1802). Beijing.

Analysis of Savonius Rotor with Multiple Blades for Hydrokinetic Application

Rajen Pudur, Mrinal Kanti Rajak, and Shadab Zafar

Nomenclature

A	Frontal area (m^2)
As	The swept area of Savonius rotor (m^2)
C_D	Drag coefficient
C_p	Power coefficient of a Savonius turbine
D_f	Drag force
f	Frequency of the rotation (Hz)
U	Water speed (m/sec)
V	Velocity of the rotor tip (m/sec)
ρ	Water density (Kg/m^3)
ω_s	Angular velocity of Savonius rotor
λ	Tip speed ratio

1 Introduction

Use of Savonius rotor for hydrokinetic application is not new, but its uses for power generation with the application of water is new and recent research, only the hurdle is its lowest power coefficient. However, the drag mechanism of Savonius rotor is taken as advantageous in terms of water as a medium to rotate the same. Many research activities are going on to increase the power coefficient of this rotor, and Gorlov et al. [1] suggested that the maximum power coefficient with proper experiments can be thought of around 0.35. Khan et al. [2] tested the same experiments on single-stage, two-stage, and three-stage conventional Savonius turbine in a water channel.

R. Pudur · M. K. Rajak (✉) · S. Zafar
Department of Electrical Engineering, National Institute of Technology Arunachal Pradesh, Yupia 791112, India

P. Mahanta et al. (eds.), *Advances in Thermofluids and Renewable Energy*, Lecture Notes in Mechanical Engineering, https://doi.org/10.1007/978-981-16-3497-0_50

Nakajima et al. [3, 4] investigated the conventional Savonius turbine with a horizontal axis in a water channel. They concluded that the maximum coefficient of power is 0.25 at a Reynolds number of 1.1 × 105. Golecha et al. [7] carried out an experimental investigation on the performance improvement of the modified Savonius rotor by providing a deflector plate on the returning blade side. Hence, there are many types of research worked and still working to make Savonius rotor better for hydrokinetic applications, such as Kamoji et al. [8], and can be seen in [5, 6, 9–22]. But they have not tested or had findings on the position of the rotor in the free-flowing river with different blades arrangements, which is prime important to extract electricity from free-flowing river or stream using this rotor or turbine.

In this paper, an experimental finding is an attempt for speed variation with the various positions at different blades. A rotor prototype is made with PVC pipe, and model is tested in the free-flowing river, and speed differences of various arrangements are observing and also validates with the theoretical results.

2 Savonius Rotor

Water is the best medium for power generations, which is a little predictable as compared with the wind. Extractions of electricity from free flow water or river demand turbine which works on drag mechanism not lift mechanisms, and among all wind turbines, Savonius rotor is the best suited for this work. Invented by Finnish engineer Sigurd Johannes Savonius in the year 1922, which was also called an ‘S’ type rotor because of its shape. This rotor has the lowest efficiency in the aerodynamic application the reason being characteristics, it works on drag force, and with sufficient wind velocity only it rotates, but with simply 2.5–5 m/s velocity of free-flowing water can rotate this turbine. The research of this paper is only to compare the different positions for Savonius for electricity production in rural areas for their socio-economics developments. Research and developments works are in full swing, and literature suggests that this rotor is gaining popularities [22, 23].

This rotor uses stagnation pressure on one side to facilitate rotation around the shaft, the blade turning redirects water around itself with its rounded shape. Because of the concave shape of the rotor, any tangential flow of water will produce a positive force on the rotor. But one should understand that this rotor cannot exceed the velocity of water, and it will operate at lower RPM range with huge torque which can give enough speed wit gearbox arrangements.

If C_p is the power coefficient of a Savonius, then the power (P) can be given from water as:

$$p = \frac{1}{2} C_p \rho \mathrm{As} V^3 \tag{1}$$

where

P is the output power (W), ρ is the water density in kg/m^3, As is the swept area of the rotor in m^2, and V is the velocity of water in m/s.

The tip peripheral velocity of the rotor,

$$U = \omega_s \times R_s \tag{2}$$

where

ω_s is the angular velocity of the rotor and R is the radius of the rotor.

2.1 *Modifications of Savonius Rotor*

Since the Savonius rotor is drag operated device, it has the lowest efficiency among all other wind turbines available, and the power coefficient (C_p) is marginally less than 0.5. All possible modifications are under test, as the use of multiple buckets, use of staged rotors in one axis and use of guided plates or vanes in the incoming lines, etc. Savonius rotor gives high drag force to overcome the rotor inertia, and the drag force can be seen as:

$$D_f = 0.5C_D \times \rho \times A \times V^2 \tag{3}$$

where

D_f = Drag force, C_D = Drag coefficient,

The drag force depends on the Reynolds number (Re). The power coefficient of Savonius rotor is very low as it operates on drag mechanism, rather than aerodynamics, researchers have shown that the power coefficient of Savonius rotor can be improved to 0.35, with various arrangements of guiding vanes, and with different rotor structures as shown in Fig. 1.

2.2 *Aspect Ratio*

Power generation using Savonius rotor also depends on aspect ratio, which is sometimes denoted by As or Aα and defined as the ratio of active length or height (m) of the rotor to active diameter (m) of the rotor. The aspect ratio of the Savonius rotor should always be larger than unity,

$$A\alpha = \frac{\text{Hs}}{\text{Ds}} > 1$$

That means Ds should be less than Hs, shown in figure two different types of rotors with different aspect ratios (Fig. 2).

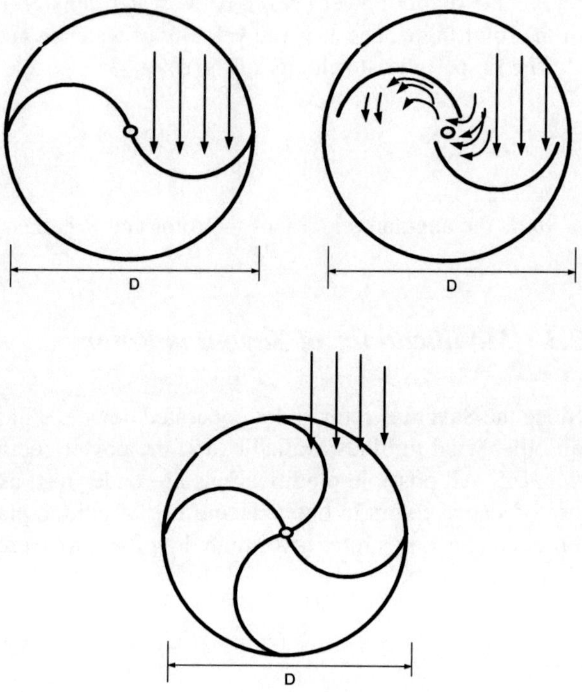

Fig. 1 Three different types of modified Savonius rotor

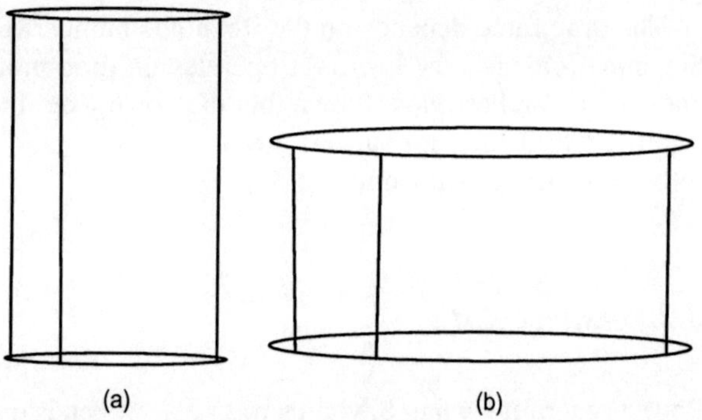

Fig. 2 Savonius rotor with **a** Hs > Ds **b** Hs < Ds

2.3 *Tip Speed Ratio (TSR)*

The relation between the water speed and the rate of rotation of the rotor is characterized by a non-dimensional factor, known as tip speed ratio or lambda (λ).

$$\text{TSR: } \lambda = \frac{\text{speed of the rotor tip}}{\text{water speed}} = \frac{U}{V} = \frac{\omega r}{V} \tag{4}$$

where
U is the water speed (m/s)
V is the velocity of the rotor tip (m/s)
R is the rotor radius (m)
ω is the angular velocity (rad/s)
f is the frequency of the rotation (Hz).

2.4 *Optimal Rotor Tip Speed Ratio*

The optimal tip speed ratio for maximum power extraction is incidental by relating the time for the disturbed water to resettable, itself t_w to the time taken for the blade of rotational frequency ω to move into the position occupies by its predecessor t_s.

The time for the blade to move to its predecessor's position for 'n' number of blades will be given as:

$$T_s = 2\prod/n\omega(\text{s})$$

If the length of the strongly disturbed water of the rotors' then the time for the water to return to normal is given by:

$$t_w = s/U(\text{s})$$

If $t_s > t_w$, then some water is unaffected, and if $t_s < t_w$, then some water is not allowed to flow through the rotor. The maximum power extraction occurs when two time periods are about to equal $t_s = t_w$

$$2\prod/n\omega = s/U,\ n\omega/U = 2\prod/s$$

From which the optimal rotational frequency is;

$$\omega_{\text{opt}} = \frac{2\pi U}{ns} \tag{5}$$

3 Construction of Savonius Rotor

This study aims to design a device that generates energy using the Savonius rotor. To accomplish this, the following steps are made: First, the researchers will survey a location (river) to serve as the basis of the design and testing site of the hydroelectric generator using Savonius turbine as the rotor. The essential parts are the flow rate and width of the river, and these would determine whether or not electrical generation would be possible to the location. After a location is chosen and gathered its data, the design and list of materials to be used for the system will be prepared. The next step will be the fabrication of a prototype; then after, the researchers will try to implement the prototype based on the data gathered from the qualified location. The last step will be the analysis and interpretation of data obtained for evaluation and recommendation.

4 Location of the Site

Pachin is a river in Arunachal Pradesh where the experiment is done. Because of the rainy season, the water of the river is dirty but we have an advantage of little higher density because of the dirtiness of water. Pachin River is at 27° 5′0″ N latitude and 93° 42′0″ E, longitude near Pachin Colony. The temperature of the river is recorded as 24 °C during the observation. The depth of the river is calculated as 0.7 m, and the average speed of the river is calculated as 1.06874 m/s. The material used for the rotor is PVC whose details are given in Table 1.

The blade is made up of PVC pipe, which is strong and rigid. It is resistance to a variety of acids and bases, but it may damage by some solvent and chlorinated hydrocarbons. PVC can be used in the maximum temperature of 140 °F (60 °C) can be used for water, gas, and drainage system. For study purpose PVC material is used but in the actual turbine can be constructed with water-resistant materials. Use of PVC material for study can also advocate that it have little elasticity, which helps in gaining more efficiency, as water striking the concave side of rotor blade will elongate the blade to increase its torque. All rotor parts can be produced cheaply by a simple manufacturing process. The shaft is made up using wooden for a prototype for easy assembling into frame also it makes the whole process fast, and there will be no any effect on the actual performance of rotor, also to the fact that metal shaft may rust and corrosion may take place. On the shaft, required slots are made and blades are fixed with the nail. The main reason to choose wood for the shaft is it is rigid in mature and light in weight. The manufacturing process can be done very easily on the wood so according to the requirement of dimension, we can easily get that dimension. The shaft is adjusted in the bearing which is fixed in the two legs of the support using the iron key.

We need not require heavy support as it is difficult to handle; therefore, support is also made up of wood which is rigid in nature and light in weight, and according to the requirement, we can easily do manufacturing process on the wood. As the system

Table 1 Parameter of rotor used

Nominal pipe size (inches)	Outside diameter (inches)	Minimum wall thickness (inches)	Inside diameter (inches)	Weight (kg/m)
6	6.625	0.280	6.065	5.25

is used in the water, there is the possibility of corrosion in the metallic structure which will decrease the life of the structure, so the support is made by using wooden material which can easily be manufacture according to the requirement. Within the support to the hole is made for the ball bearing which will be fixed in the two legs of the support.

Dimension of the support.

Height of the support = 30 inch = 1.062 m.
Width of the support = 13 inch = 0.4602 m.

The complete assembly is shown in Fig. 3.

For smooth rotational outputs, a ball bearing is used, and it works excellent, the purpose of using the ball bearing is to reduce the rotational friction and support radial and axial loads, by using at least two races to contain the balls and transmit the loads through the balls. In most applications, one race is stationary and the other is attached to the rotating assembly because balls are rolling, and they have a much lower coefficient of friction than if two surfaces were sliding against each other. The wooden frame is fixed with the stationary race of bearing, and shaft is fixed on the movable race of bearing for smooth and frictionless rotation when the force applied.

The ball bearing is fixed into the two legs of the support as shown in Fig. 4d, and the rotating shaft is adjusted into the bearing during the experiment with the help of key.

Fig. 3 Rotor holding frame

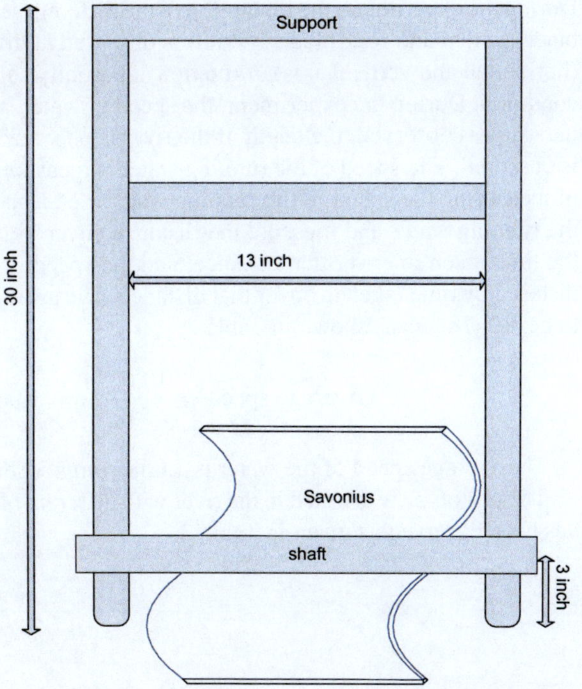

Fig. 4 **a** Two blades, **b** three blades, **c** four blades and **d** complete set

The internal diameter of the bearing = 2.5 cm.

The external diameter of the bearing = 3.5 cm.

The design is made up of both wood and plastic. The blades are made with PVC pipe, shaft, bottom support, and the upper support are made up of wood. This design was synthesized from numerous past Savonius wind turbine application. The plastic material is classified as both super lightweight and has great elasticity. This is useful for it follows or bends to the direction of water flow rather than resist it. The wooden part is both harder than plastic and lighter than galvanized steel which is great for the supports.

5 Result and Discussion

During the experiment, the revolution of the different set up (two-blade system, three-blade system and four-blade system) is observed at different positioning of the axis (horizontal and vertical axis) of rotation at partially submerged and fully submerged condition. During the experiment, the speed of water is observed, the temperature of the water is observed, the depth of the river is observed, and the density of the water is observed. The speed of the running water is measured by using the simple method of measuring the speed of the running water; i.e., a sponge ball is used to flow over the running water and measure the distance cover by the sponge ball and calculate the time taken to cover that distance, and the speed is so calculated by dividing the distance by time taken to cover that distance. The average speed of the water is found to be 1.0678 m/s as shown in Table 2.

$$\text{Average speed} = \frac{10.6874}{10} = 1.06874\ \text{m/s}$$

The average speed of the water is 1.06874 m/s. (Though surface velocity).

The prototype was tested in the river with different blades, and the rotational speed has been observed as given in Table 3.

Table 2 Observation table for calculating the speed of running water of Pachin River

S. No	Distance (m)	Time (s)	Speed (m/s)
1	10	9.2	1.0869
2	10	9.3	1.0752
3	10	8.9	1.1235
4	10	9.3	1.0752
5	10	10.5	0.9523
6	10	9.9	1.0101
7	10	9	1.1111
8	10	9	1.1111
9	10	9.7	1.0309
10	10	9	1.1111

Table 3 Observation table for rotational speed with different blades combinations

Inclination of blades	Position of a blade in the water	Rotational speed (rpm) (two blades)	Rotational speed (rpm) (three blades)	Rotational speed (rpm) (four blades)
Horizontal axis	Partially submerged	63	58	38
Horizontal axis	Fully submerged	68	53	49
Vertical axis	Partially submerged	52	50	39
Vertical axis	Fully submerged	56	52	46

6 Calculation

The theoretical calculation for optimal rotor tip speed ratio for two blades is shown below:

Angular velocity of the two-blade system

$$\omega = 2\pi f = \frac{2\pi N}{60} = \frac{2 \times 3.14 \times 68}{60} = 7.117 \text{ rad/s}$$

Velocity of rotor, V = $\omega r = 7.117 \times 0.3998/2 = 1.4227$ m/sec

$$\lambda_1 = \frac{\text{speed of the rotor tip}}{\text{water speed}} = \frac{V}{U} = \frac{\omega r}{U} = \frac{1.4227}{1.06874} = 1.331$$

ω_{opt} for two-blade,
Speed of water $U = 1.068$ m/s,
Number of blade $n = 2$,
Length of blade $s = 12$ inch $= 12 \times 0.0254 = 0.3048$ m.

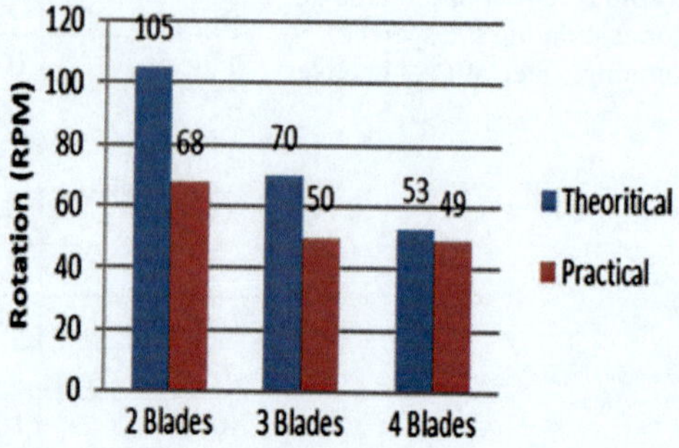

Fig. 5 Comparison between the theoretical and practical rotational speed

Now,

$$\omega_{\text{opt}} = \frac{2\pi U}{ns}$$

$$\omega_{\text{opt}} = \frac{2 \times 3.14 \times 1.068}{2 \times 0.3048} = 11.0023.89 \text{ rad/s}$$

$$\omega_{\text{opt}} = \frac{2\pi N}{60} = 11.0023.89 \text{ rad/s}$$

$$N = 105.118 \text{ rpm}$$

And the speed of the other two arrangements is also been calculated in the same fashion and presented in Fig. 5.

It is observed from the above result that two-blade systems are best as compared to the three and four-blade system. But during the observation, it is found that two-blade systems are more unbalanced than the three-blade system and four-blade system, because of the turbulence of the water during the striking of water on the blade. Theoretically, if the number of the blade increases, the rotational speed of the rotor decreases. It is also observed during the investigation that in the fully submerged condition, the rotational speed of the rotor is more than the partially submerged condition.

7 Conclusion

This paper details the possibility of extraction of hydropower from free-running rivers, by run-off-river. A Savonius rotor, which is commonly used as a wind turbine, is a unique fluid mechanical device, which works on drag force rather than lift force. Among all known wind turbines, it has a low power coefficient. Experiments have shown that the Savonius rotor can be effectively used to generate power using hydrodynamics rather than aerodynamic. Generation of electricity using flowing river water is a good option as it does not require the construction of dams or water storage system. The generating scheme uses the kinetic energy of the natural flow of river water requiring zero head rotors. It is also being shown that for remote areas, this scheme of the generation of electricity may give added helps to the inhabitants. The possibility of distribution of power to remote location produced by generation plant established with simple locally available materials is more vital than any other technologies. Local manufacturer, local entrepreneurs, and employments generation will be its main advantages; it will, not only boost socioeconomic development, but will also transformation of their lifestyles.

The appropriate tip speed ratio for power generation is advocated, keeping the factual conditions of the velocity of water in hilly areas, hilly areas like state of Arunachal Pradesh, which comes under seismic zone Vth, construction of the large dam is always been a big question mark. Hydro-potential of this beautiful state is widely known, especially the surface flowing rivers. Mini- and micro-hydropower plants using the Savonius rotor will be encouraging one.

From the above studies, two blades Savonius rotor was found to be unique and more efficient in harvesting hydrokinetic energy as compared to the three blades and four blades Savonius rotor, with proper gear system power can be extracted from the free-flowing river. It was also found during the observation that the balancing of the two blades Savonius rotor is least because of high turbulence of water; but for the four blades Savonius, the turbine is rotated smoother than the three blades and two blades Savonius rotor. Theoretically, also it was found that as the number of blades increases, the revolution of the turbine decreases. The major differences in theoretical and practical reading for two blades are because in two blades system, there is huge jerking effect as the subsequent blades are at a far distance. So the experiment is validating the theoretical concept.

Acknowledgements Authors would like to thank the researchers or academicians whose works have been cited in this paper. Authors are also grateful to National Institute of Technology Arunachal Pradesh for funding the experiments and also the local public of Pachin colony for their support.

References

1. Gorlov, A. M. (1998). Helical turbines for the Gulf Stream: Conceptual approach to design of a large-scale floating power farm. *Marine Technology and Sname News, 35*(3), 175.
2. Khan, M., Islam, N., Iqbal, T., Hinchey, M., & Masek, V. (2009). Performance of Savonius rotor as a water current turbine. *The Journal of Ocean Technology, 4*(2), 71–83.
3. Nakajima, M., Iio, S., & Ikeda, T. (2008). Performance of Savonius rotor for environmentally friendly hydraulic turbine. *Journal of Fluid Science and Technology, 3*(3), 420–429.
4. Nakajima, M., Iio, S., & Ikeda, T. (2008). Performance of double step Savonius rotor for environmentally friendly hydraulic turbine. *Journal of Fluid Science and Technology, 3*(3), 410–419.
5. Hayashi, T., Li, Y., & Hara, Y. (2005). Wind tunnel tests on a different phase three-stage Savonius rotor. *JSME International Journal Series B Fluids and Thermal Engineering, 48*(1), 9–16.

6. Huda, M. D., Selim, M. A., Islam, A. K. M. S., & Islam, M. Q. (1992). The performance of an S-shaped Savonius rotor with a deflecting plate. *RERIC International Energy Journal, 14*(1), 25–32.
7. Golecha, K., Eldho, T. I., & Prabhu, S. V. (2011). Influence of the deflector plate on the performance of modified Savonius water turbine. *Applied Energy, 88*(9), 3207–3217.
8. Kamoji, M. A., Kedare, S. B., & Prabhu, S. V. (2009). Experimental investigations on single stage modified Savonius rotor. *Applied Energy, 86*(7–8), 1064–1073.
9. Shaughnessy, B. M., & Probert, S. D. (1992). Partially-blocked Savonius rotor. *Applied Energy, 43*(4), 239–249.
10. Ogawa, T., Yoshida, H., & Yokota, Y. (1989). Development of rotational speed control systems for a Savonius-type wind turbine. *Journal of Fluids Engineering, Transactions of the ASME, 111*(1), 53–58.
11. Alexander, A. J., & Holownia, B. P. (1978). Wind tunnel tests on a Savonius rotor. *Journal of Wind Engineering and Industrial Aerodynamics, 3*(4), 343–351.
12. Gorelov, D. N., & Krivospitsky, V. P. (2008). Prospects for development of wind turbines with orthogonal rotor. *Thermophysics and Aeromechanics, 15*(1), 153–157.
13. Islam, M., Ting, D. S. K., & Fartaj, A. (2008). Aerodynamic models for Darrieus-type straight-bladed vertical axis wind turbines. *Renewable and Sustainable Energy Reviews, 12*(4), 1087–1109.
14. Sahim, K., Santoso, D., & Sipahutar, R. (2015). Performance of combined water turbine Darrieus-Savonius with two stage Savonius buckets and single deflector. *International Journal of Renewable Energy Research (IJRER), 5*(1), 217–221.
15. Mabrouki, I., Driss, Z., & Abid, M. S. (2014). Experimental investigation of the height effect of water Savonius rotors. *International Journal of Mechanics and Applications, 4*(1), 8–12.
16. Chauvin, A., & Benghrib, D. (1989). Drag and lift coefficients evolution of a Savonius rotor. *Experiments in Fluids, 8*(1–2), 118–120.
17. Fujisawa, N. (1992). On the torque mechanism of Savonius rotors. *Journal of Wind Engineering and Industrial Aerodynamics, 40*(3), 277–292.
18. Fujisawa, N., Taguchi, Y., Satoh, T., Uemura, T., Nagaya, K., & Ikai, S. (1993). Investigation of flow field inside a Savonius rotor by image processing technique with conditional sampling. *Nippon Kikai Gakkai Ronbunshu, B Hen/Transactions of the Japan Society of Mechanical Engineers, Part B, 59*(567), 3519–3523.
19. Islam, A. S., Islam, M. Q., Mandal, A. C., & Razzaque, M. M. (2017). Aerodynamic characteristics of a stationary Savonius rotor. *International Energy Journal, 15*(2), 125–136.
20. Rajkumar, M. J., & Saha, U. K. (2006). Valve-aided twisted Savonius rotor. *Wind Engineering, 30*(3), 243–254.
21. Saha, U. K., & Rajkumar, M. J. (2006). On the performance analysis of Savonius rotor with twisted blades. *Renewable Energy, 31*(11), 1776–1788.
22. Khan, M. J., Bhuyan, G., Iqbal, M. T., & Quaicoe, J. E. (2009). Hydrokinetic energy conversion systems and assessment of horizontal and vertical axis turbines for river and tidal applications: A technology status review. *Applied Energy, 86*(10), 1823–1835.
23. Anyi, M., & Kirke, B. (2010). Evaluation of small axial flow hydrokinetic turbines for remote communities. *Energy for Sustainable Development, 14*(2), 110–116.

A Sensor-Based Improved MPPT Technique Under Partial Shading Condition

Boni Satya Varun Sai, Sarang A. Khadtare, and Debashis Chatterjee

1 Introduction

As day by day, the population is expanding quickly, it prompts huge exploitation of fossil fuels to satisfy the regular vitality need. This emerges issues like consumption of non-renewable energy sources, unsteady costs alongside extreme ecological contamination issues. These all deficiencies made researchers to follow the paths of renewable energy generation schemes. Sun energy is greatly engaged and settled among the available renewable energy sources, because of its heavy accessibility and relative simpler control strategies [1]. Indeed, in modern days, the establishment charges for solar panel system is very high. Therefore, increase in MPPT efficiency is the good method of reducing photovoltaic (PV) implementation cost. So, better MPPT methods required for the ideal activity in maximum power point (MPP) detection, prompting decreased expenses.

The adaptability in development and accuracy of developing performance characteristics of practical PV panel established the single diode model is extremely employed for designing the solar cell [2]. The power produced by single solar cell is minute in quantity so the PV panels are arranged in the cascaded outlines to meet the load characteristics. Due to various hindrances for instance trees and birds, etc., outcomes in the uneven irradiation over the surface of solar panel resulting in PS condition. During the uniform irradiation, output characteristics of PV panel display a single maximum peak resulting in the easier tracking [3, 4]. Under the influence of PS condition, solar panel characteristics display multiple peaks in output characteristics making it difficult in tracking global maxima. The conventional MPPT methods are inefficient in tracking maximum power under PS condition [5].

For satisfying load parameters, boost converter is employed because of its simple structure and lower cost [6, 7]. Different metaheuristic-based MPPT techniques available with efficient operation with little complication for tracking maximum power

B. S. V. Sai (✉) · S. A. Khadtare · D. Chatterjee
Department of Electrical Engineering, Jadavpur University, Kolkata, India

P. Mahanta et al. (eds.), *Advances in Thermofluids and Renewable Energy*, Lecture Notes in Mechanical Engineering, https://doi.org/10.1007/978-981-16-3497-0_51

under PS condition [8, 9]. Artificial vision-based MPPT scheme is introduced for PS MPPT tracking, where partial shading detection is done using web cam. Besides being shown good operating characteristics compared to other schemes, it requires high initial cost [10]. MPP scanning (MPPS)-based MPPT scheme is discussed where a group selection optimizer along with P&O is employed for improved operation purpose; but for improving the dynamic characteristics, a lot of development should be done in controlled voltage source [11]. Fusion firefly algorithm (FFA) shows enhanced characteristics compared to other MPPT schemes, but it is moderately difficult coming to hardware implementation [12]. MPPT based on optical isolation is able to track the global maxima economically with good tracking efficiently [13]. PSO and CS are frequently used algorithms for tracking maximum power under PS condition [14, 15]. CS shows better dynamic properties compared to PSO, resulting in smooth tracking. Even though optimization schemes show better tracking efficiency and dynamic properties, they require higher mathematical burden for operation. Also, they are very complex while coming to real-time implementation [16].

In present work, sensor-based (SB) MPPT scheme is projected for overpowering the major disadvantage of the optimization-based schemes. It can be workable for any configuration of the PV farm with proper initial work. The SB MPPT structure is similar to that of conventional constant voltage (CV) MPPT where PV panel voltage is compared with reference voltage in tracking maximum power. The projected SB MPPT is considered to 3S configuration, in which output characteristics displays multiple peaks. The paper framed as proposed SB MPPT algorithm in Sect. 2, simulation results and experimental setup in Sect. 3, and conclusion in Sect. 4.

2 Proposed SB MPPT Algorithm

2.1 Flow Chart for SB MPPT Algorithm

The projected SB MPPT scheme is demonstrated as Fig. 1. Even though optimization-based MPPT techniques can track the maximum power more efficiently, they lack the good dynamic properties. Coming to real time, optimization-based algorithms are difficult to implement as they are too complex and takes more time to settle. Thus, the requirement is high for an alternate MPPT scheme which will possesses good dynamic properties and easy real-time implementation. The proposed SB MPPT will be able to overcome the disadvantages of all available optimization—based techniques by using good initial work. In proposed SB MPPT, the tracking error between PV panel voltage and reference voltage is driven into PI controller for generating switching pulses to gate of DC-DC converter.

PI controller is employed for producing the good dynamic properties. Increasing the proportional gain (Kp) will reduce the rise time, and it has the ability to reduce the steady state error (ess), but will not make it completely zero, and also there is

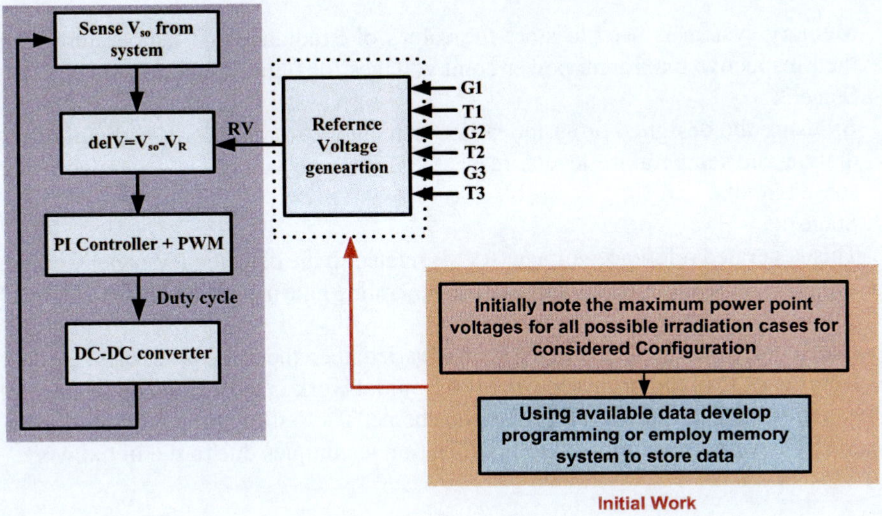

Fig. 1 Proposed SB MPPT (Vso is solar panel voltage and V_R is reference voltage)

an increase in overshoot. Increasing integrator gain (Ki) has the advantage to make almost equal to zero, but same time it will increase the settling time and overshoot of the system. The respective control gain values designed employing trial and error method for getting desired output characteristics at STC, and they are 1 and 6, and same values are employed for all patterns. SB MPPT undergo four stages to track the MPP, they are:

Stage 1:
For the configuration available, initially note down the values of the voltage where the maximum power is seen.
Stage 2:
Using the irradiation, temperature levels and their respective maximum power point voltages write the sample programming as follows:

```
function RV = fcn (G1, T1, G2, T2.................)
RV = 0;
If G1 == 100&&G2 == 200&&...&&T1 = 25&&T2 = 2...
RV = RV + 60.6891;
elseif G1 == 100&&G2 == 200&&...&&T1 = 25&&T2 = 2...
RV = RV + 61.7343;
elseif
...................................................................................................
..........................................................
end
end
(or)
```

Memory system is used to store the values of irradiance and temperature with their respective maximum power point voltages for future usage.
Stage 3:
By using the designed program earlier or memory storage, for a particular irradiation and temperature levels, respective reference voltage is generated using sensed inputs.
Stage 4:
This generated reference voltage (RV) is related to the detected PV panel voltage, and error is driven into PI controller for generating gate triggering pulse to DC-DC converter.
If the configuration of existing system is varied, then the same process is repeated from stage 1. In the proposed SB MPPT, initial work is to be done up to stage 2, which is the heart of overall proposed scheme. The settling time required for SB MPPT is very less compared to optimization techniques due to the initial work.

2.2 Overall Block Diagram for Proposed System

Figure 2 denotes the overall schematic diagram for the proposed system where PV panel is designed using single diode model. Table 1 represents the parameters of

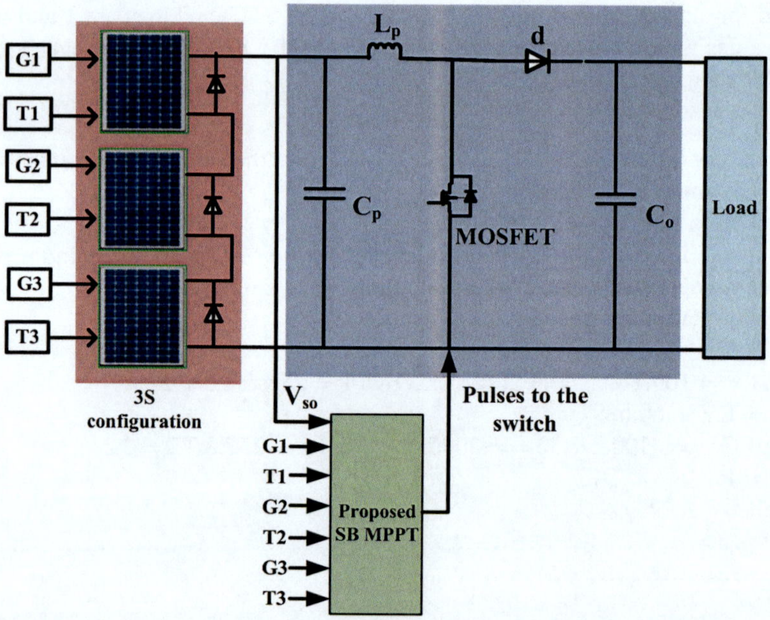

Fig. 2 Overall proposed system (L_P is inductance, (G1, G2, and G3) are irradiances, (T1, T2 and T3) is temperatures, C_P is input capacitor, d is diode, and C_o is output capacitor)

Table 1 Photo voltaic cell parameters

Variable	Quantity
Maximum power output (P_m)	249 W
MPP voltage (V_m)	30 V
MPP current (I_m)	8.3 A
Short circuit current (I_s)	8.83 A
Open circuit voltage (V_{op})	36.8 V
Temperature coefficient of short circuit current (I_s)	0.04%/°C
Temperature coefficient of open circuit voltage (V_{op})	−0.325%/°C

solar panel provided by manufacturer. Boost converter is designed using data given by manufacturer, load voltage, current, and switching frequency (50 kHz). Anti-parallel diodes are connected across PV panels to protect them from overheating and hot-spots (Table 2).

3 Simulation Results and Experimental Setup

3.1 PV Panel Characteristics

The proposed system is tested for different shading outlines are displayed in Fig. 3. For the considered shading patterns, power-voltage (*P–V*) characteristics are obtained as Fig. 4. Using this curve, it is possible to note down the voltages at which maximum power is seen like Table 1. Since in case of two peaks (pattern 1), the global peak is at first position, conventional MPPT techniques like P&O and I&C, etc., can track the power. The main problem arises when the peak is anywhere except the first position.

3.2 MPPT Using Proposed SB algorithm for Sudden Change in irradiation

For the proposed SB MPPT, MPPT operation under step change in irradiation is studied for the case of Pa2-Pa1-Pa3 step variation.

For these step variations, the proposed SB MPPT is employed for tracking maximum power. From Fig. 5, it can be clearly observed that the SB MPPT is able in detecting the power smoothly without any delay in tracking compared to PSO and CS algorithms. Converging period is very low when there is change in irradiation, and this quick convergence leads to the improved characteristics of overall system. It is

Table 2 Maximum power point voltages for different irradiation stages at STC temperature for all panels

G1	G2	G3	V_m	G1	G2	G3	V_m	G1	G2	G3	V_m
100	**200**	**300**	**60.6891**	100	500	600	60.8559	200	300	900	28.5734
100	200	300	60.6891	100	500	700	61.5512	200	300	1000	28.4441
100	200	400	61.7343	100	500	800	62.1442	200	400	500	60.9058
100	200	500	28.1614	100	500	900	62.5432	200	400	600	61.9067
100	200	600	28.2777	100	500	1000	62.8501	200	400	700	62.4167
100	200	700	28.8362	100	600	600	58.7663	200	400	800	62.5768
100	200	800	29.0341	100	600	700	60.6068	200	400	900	62.9817
100	200	900	28.4992	100	600	800	61.4803	200	400	1000	28.5432
100	200	1000	28.5297	100	600	900	61.974	200	500	600	60.8581
100	300	400	60.9627	100	600	1000	62.411	200	500	700	61.6487
100	300	500	61.7126	100	700	700	60.1392	200	500	800	62.2141
100	300	600	62.3026	100	700	800	60.4604	200	500	900	62.5377
100	300	700	28.6175	100	700	900	61.2608	200	500	1000	62.8693
100	300	800	28.6205	100	700	1000	62.2402	200	600	700	60.7051
100	300	900	28.5047	100	800	800	59.4511	200	600	800	61.4599
100	300	1000	28.4792	100	800	900	60.3134	200	600	900	62.0394
100	**400**	**400**	**59.8118**	100	800	1000	61.0485	200	600	1000	62.361
100	400	500	60.9049	100	900	900	59.4495	200	700	800	60.5088
100	400	600	61.9057	100	900	1000	60.1624	200	700	900	61.3015
100	400	700	62.4175	200	300	400	94.7184	200	700	1000	61.742
100	**400**	**800**	**62.6125**	200	300	500	95.2338	200	800	900	60.2759
100	400	900	62.9233	200	300	600	95.7571	200	800	1000	61.07

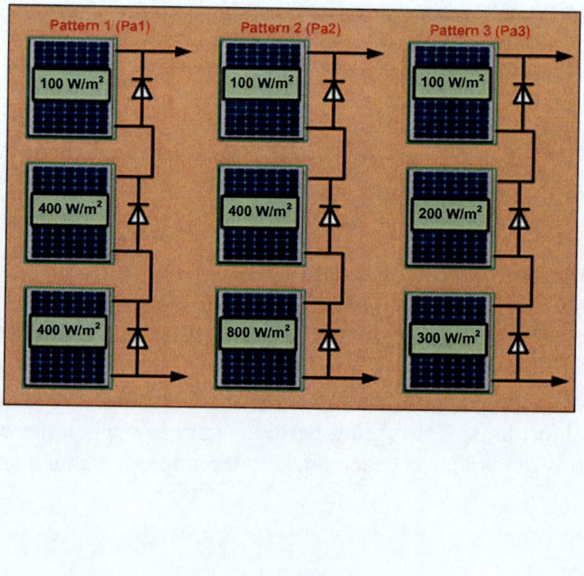

Fig. 3 Different shading patterns employed in the system

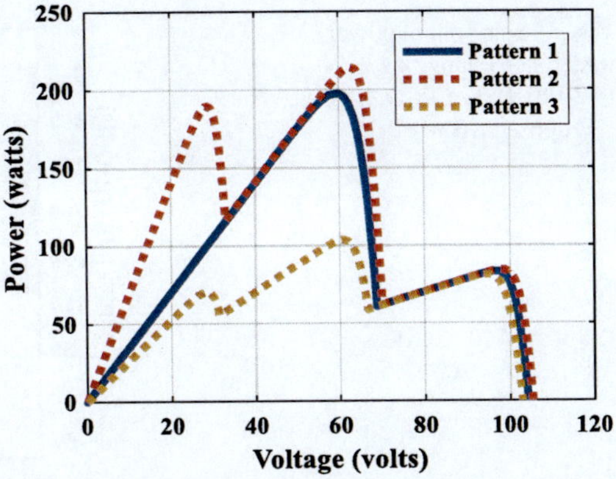

Fig. 4 *P–V* curves for considered shading patterns

found recently that MPPT schemes used in PV systems are also a main reason behind introduction of interharmonics into system. Reduction of interharmonics can be done by employing a MPPT, which possess low settling and searching time during sudden change in irradiation [23]. The proposed SB MPPT have all characteristics requires to reduce the interharmonics. Overall, comparing to PSO and CS, SB MPPT has the major advantages of easy real-time implementation, simple structure, and good dynamic characteristics.

3.3 Overall Experimental Setup for Proposed System

The overall hardware setup for proposed system is represented as in Fig. 6, in which three PV panels are coupled in series outline. PV panel voltage is sensed and given as an input to PIC18F452 controller. LM35 is used for sensing temperature, and the equivalent voltage is given as input to the PIC controller. Irradiance is sensed using a pyranometer, in which the irradiance is calculated as an equivalent voltage. Since the voltage levels of the temperature sensor and irradiation sensor are of lower levels, they passed through an amplifier and driven into PIC18F452 controller. Since the output voltage of PIC controller in the range 5 V, it is passed through a driver circuit to enhance it to 12 V peak to peak voltage. This boosted pulse is driven to triggering the gated of K2611.

4 Conclusion

Even though optimization-based algorithms are good at tracking global maximum under PS condition, they fail for the case coming to real-time operation due to their higher mathematical burden. In this paper, the proposed SB MPPT technique, which

Fig. 5 Solar panel power under step change of Pa2-Pa1-Pa3, **a** PSO, **b** CS, **c** Proposed SB MPPT

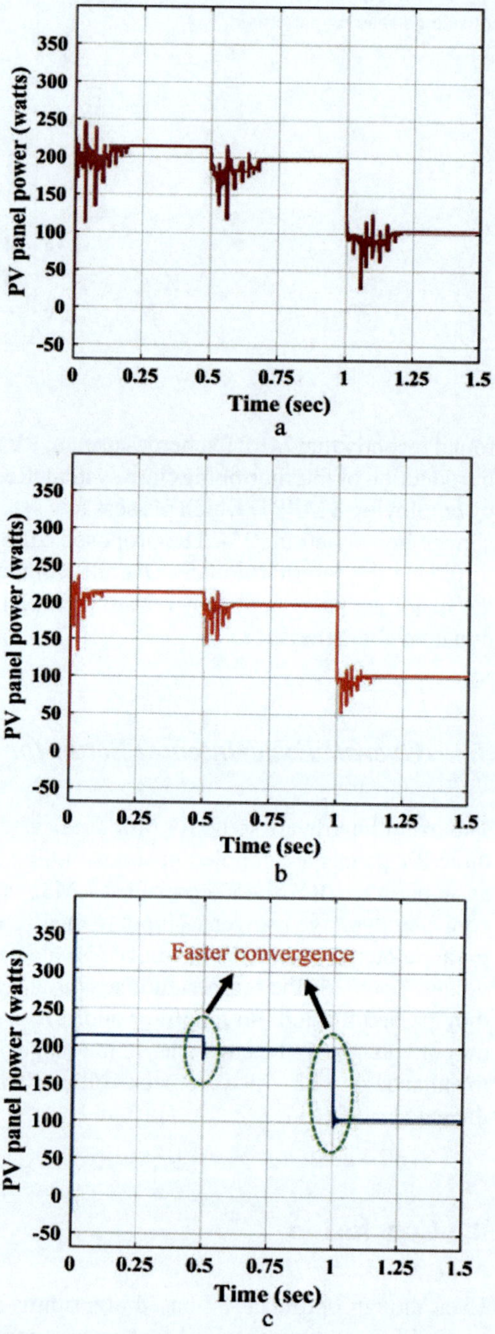

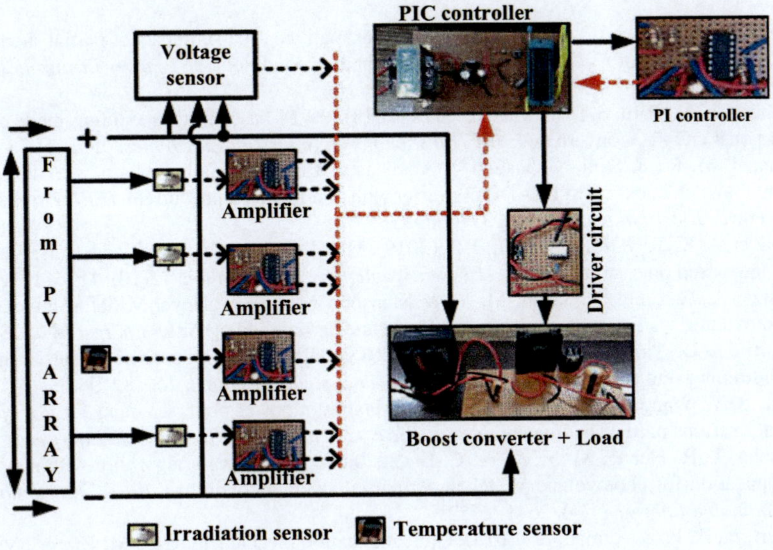

Fig. 6 Experimental setup for overall system

is simple and able to produce good dynamic properties. With a good initial work for considered configuration, data is collected regarding maximum power point voltages. This data can be programmed or stored in memory for future usage as a reference voltage generator. Due to the developed dynamic characteristics, it can be applied in interharmonic reduction.

Acknowledgements The authors are appreciative for the economic aid in research work by All India Council for Technical Education-National Doctoral Fellowship (AICTE-NDF).

References

1. Aminul Islam, Md., Merabet, A., Beguenane, R., & Ibrahim, H. (2013). *Modeling solar photovoltaic cell and simulated performance analysis of a 250W PV module*. IEEE Electrical Power & Energy Conference.
2. Ding, K., Bian, X. G., Liu, H. H., & Peng, T. (2012). A MATLAB-simulink-based PV module model and its application under conditions of non uniform irradiance. *IEEE Transactions on Energy Conservation, 27*(4), 864–872.
3. Patel, H., & Agarwal, V. (2008). MATLAB-based modelling to study the effects of partial shading on PV array characteristics. *IEEE Transactions on Energy Conservation, 23*(1), 302–310.
4. Ma, J., Pan, X., Man, K. L., Li, X., Wen, H., & Ting, T. O. (2018). Detection and assessment of partial shading scenarios on photovoltaic strings. *IEEE Transactions on Industry Applications, 54*(6), 6279–6289.

5. Balasubramanian, I. R., Ganesan, S. I., & Chilakapati, N. (2014). Impact of partial shading on the output power of PV systems under partial shading conditions. *IET Power Electronics, 7*(3), 657–666.
6. Pandey, S. K., Patil, S. L., & Phadke, S. B. (2018). PWM-based adaptive sliding-mode control for boost DC–DC converters. *IEEE Transactions on Industrial Electronics, 65*(6), 5078–5080.
7. Banaei, M. R., & Sani, S. G. (2018). Analysis and implementation of a new SEPIC-based single-switch buck–boost DC–DC converter with continuous input current. *IEEE Transactions on Power Electronics, 33*(12), 10317–10325.
8. Podder, A. K., Roy, N. K., & Pota, H. R. (2019). MPPT methods for solar PV systems: A critical review based on tracking nature. *IET Renewable Power Generation, 13*(10), 1615–1632.
9. Mirza, A. F., Ling, Q., Javed, M. Y., & Mansoor, M. (2019). Novel MPPT techniques for photovoltaic systems under uniform irradiance and Partial shading. *Solar Energy, 184*, 628–648.
10. Martin, A. D., Vazquez, J. R., & Cano, J. M. (2018). MPPT in PV systems under partial shading conditions using artificial vision. *Electric Power Systems Research, 162*, 89–98.
11. Lin, B. Y., Wang, L., & Wu, Q. H. (2020). Maximum power point scanning for PV systems under various partial shading conditions. *IEEE Transactions on Sustainable Energy*.
12. Huang, Y.-P., Huang, M.-Y., & Ye, C.-E. (2020). A fusion firefly algorithm with simplified propagation for photovoltaic MPPT under partial shading conditions. *IEEE Transactions on Sustainable Energy, 11*(4), 2641–2652.
13. Murtaza, A. F., & Ahmad, R. (2019). Optical isolation mechanism based MPPT for PV array under partial shading condition. *Solar Energy, 185*, 516–524.
14. Li, G., Jin, Y., Akram, M. W., Chen, X., & Ji, J. (2018). Application of bioinspired algorithms in maximum power point tracking for PV systems under partial shading conditions—A review. *Renewable and Sustainable Energy Reviews, 81*, 840–873.
15. Li, H., Yang, D., Su, W., Lu, J., & Yu, X. (2019). An overall distribution particle swarm optimization mppt algorithm for photovoltaic system under partial shading. *IEEE Transactions on Industrial Electronics, 66*(1).
16. Ding, J., Wang, Q., Zhang, Q., Ye, Q., & Ma, Y. (2019). A Hybrid Particle Swarm Optimization-Cuckoo search algorithm and its engineering applications. *Mathematical Problems in Engineering*.

A Modified ERA Algorithm for Mode Estimation in Power System

Angelica Wartina and Shekha Rai

1 Introduction

In power system, the main agenda is to convert mechanical form of energy to electrical form and route it to the consumer through a proper arrangement. The instability issues arise when there is an imbalance in power generation and power requirement. The rotor angle has to be maintained with adequate damping for maintaining stability. In energy management system, blackout might be a result of poorly damped oscillations [1]. So by identifying the critical modes, a proper controller can be designed which in turn improves the damping of the system. There are traditional methods for analysing modes which are based on Eigenvalue analysis [2]. But the conventional methods have limitations in determining the non-linear characteristics of the system.

With the deployment of wide area monitoring, a signal is transmitted from phasor data concentrator (PDC) to phasor measurement unit (PMU) through a transmission link in a phasor network. PMU has the ability to accomplish synchronized data with high precision on a global positioning system (GPS) [3] which facilitates real-time monitoring. The online mode estimation techniques used for low-frequency mode identification are fast Fourier transform (FFT) [4], Kalman filter [5, 6], ERA [7–9], Prony analysis [10, 11] and TLS-ESPRIT [12]. The FFT has the advantage of less computational complexity and also less sensitive to noise. But this technique has the drawback of resolution of frequency. Kalman filter has been applied in positioning of GPS; it is fast and efficient but has the limitation that it is only worthiness to linear system. The signal's frequency and attenuation factor can be directly calculated by Prony algorithm, but this method is sensitive to noise. TLS ESPRIT can withstand high variance Gaussian noise, and it can identify modes at different SNR.

A. Wartina (✉)
Department of Electrical Engineering, Jorhat Engineering College, Jorhat 785007, India

S. Rai
Department of Electrical Engineering, NIT Rourkela, Rourkela, Odisha 769008, India

P. Mahanta et al. (eds.), *Advances in Thermofluids and Renewable Energy*, Lecture Notes in Mechanical Engineering, https://doi.org/10.1007/978-981-16-3497-0_52

Because of network crowding and particular hardware issues, the data from PDC may not always transmit to PMU [13] within the specific time frame, and this is classified as missing values. The presence of missing values in PMU might be a reason for less precise results and may lead to biasness in mode estimations. To overcome these problems, an online mode estimation technique, i.e. a modified Eigen-system Realisation (ERA) algorithm, has been introduced in this paper to deal with the insufficient data. The effectiveness of the proposed ERA method has been compared with the other two algorithms like ERA and Prony. The rest of the paper is standardized as follows: Sect. 2 illustrates the proposed method, Sect. 3 describes online identification of proposed method; the simulation results have been discussed in Sect. 4, and finally, the conclusion is presented in Sect. 5.

2 Description of Proposed Modified ERA Algorithm

2.1 Eigensystem Realization Algorithm [7–9]

For a discrete system, the state-space representation can be considered as,

$$\boldsymbol{p}(r+1) = \boldsymbol{A}\boldsymbol{p}(r) + \boldsymbol{B}\boldsymbol{u}(r) \tag{1}$$

$$\boldsymbol{q}(r) = \boldsymbol{C}\boldsymbol{p}(r) + \boldsymbol{D}\boldsymbol{u}(r) \tag{2}$$

where $\boldsymbol{p}(r)$ is the state vector, $\boldsymbol{u}$ is the system input, r is the step size; $\boldsymbol{A}, \boldsymbol{B}, \boldsymbol{C}, \boldsymbol{D}$ are the matrices for discrete-time-state-space. To achieve a state-space representation the Eigensystem, realisation algorithm approaches the minimum realisation principle and only the smallest number of states are taken. The matrix $\boldsymbol{A}$'s Eigenvalues are complex conjugate, and it is used to determine the damping and natural frequency. Matrix $\boldsymbol{A}$ gives the modal parameter of the system. The steps to apply ERA are,

Step1: The first and foremost step is to form the Hankel matrix.

$$\boldsymbol{H}(r) = \begin{bmatrix} \boldsymbol{q}(r+1) & \boldsymbol{q}(r+2) & \cdots & \boldsymbol{q}(r+m) \\ \boldsymbol{q}(r+2) & \boldsymbol{q}(r+3) & & \boldsymbol{q}(r+m+1) \\ \vdots & & \ddots & \vdots \\ \boldsymbol{q}(r+n) & \boldsymbol{q}(r+n+1) & \cdots & \boldsymbol{q}(r+m+n) \end{bmatrix} \tag{3}$$

Here, $n \times n$ and m are the number of rows and columns of Hankel matrix.

Step2: The singular value decomposition of the Hankel matrix $\boldsymbol{H}(r)$ at $r = 0$ has to be performed.

$$\boldsymbol{H}(0) = \boldsymbol{R}\sum \boldsymbol{S}^T \tag{4}$$

Here, **R** and **S** are the orthonormal matrices of order $m \times m$ and $n \times n$, respectively. The $\sum$ is the $m \times n$ matrix whose diagonal has non-negative values.

In ideal case,

$$\sum = \begin{bmatrix} \sum_s & 0 \\ 0 & 0 \end{bmatrix} \tag{5}$$

$\sum_s$ is the matrix of order $s \times s$, s is the order of the system.

The matrix **A** can be calculated as

$$\boldsymbol{A} = \sum^{-\frac{1}{2}} \boldsymbol{R}^T \boldsymbol{H}(1) \boldsymbol{S} \sum^{-\frac{1}{2}} \tag{6}$$

2.2 Improved ERA with Non-linear Filter

ERA algorithm involves the formation of Hankel matrix which may contain missing values and lead to bias in the estimation. Thus, to preserve the non-linear characteristics, data restoration using a probable value in place of missing measurement is highly essential. This paper imputes the missing data by using a non-linear filtering process. At first, the median of the non-missing entries in a column are determined and then the missing values are replaced within each column separately as described below [14].

If $\boldsymbol{x}(n)$ is a vector of values

$$\boldsymbol{x}(n) = \{x(n), x(n-1), \ldots x(n-m)\}^T \tag{7}$$

The median filter output $y(n)$ can be obtained by middle value from the particular collection of elements of vector $\boldsymbol{x}(n)$.

$$y(n) = \text{med}\{x(i)\}, \text{ where } i = n, n-1, \ldots, n-M \tag{8}$$

where M + 1 is taken as an odd number and is termed as the length of the median filter. The Hankel matrix, thus, obtained via the filtering process is free from missing measurements.

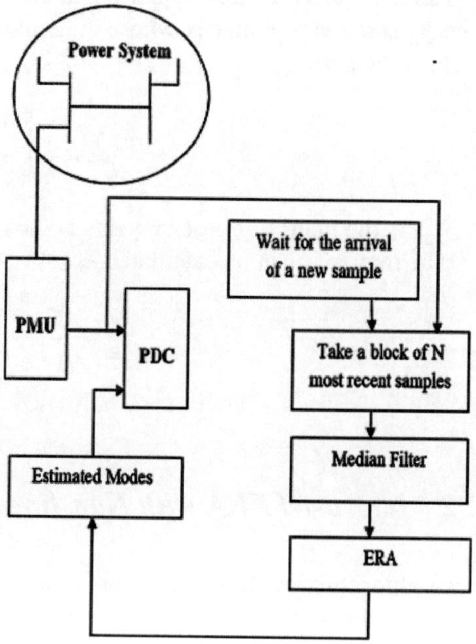

Fig. 1 Block diagram of online detection of modes for proposed modified ERA

3 Online Identifications of Modes Using Proposed Modified ERA

Figure 1 represents the general block diagram of the proposed method which is used for online detection of oscillatory modes. There is a wired link between phasor measurement unit (PMU) and phasor data concentrator (PDC), and through this link, the power signal has been transferred from PMU to PDC. The power signals that have been received from PDC can estimate the critical modes. These critical modes occur because of the imbalance between power generation and power requirement. As there are some incomplete data present in PMU, this may lead to some unwanted results. So to minimise this incomplete measurement issues, modified ERA algorithm has been anticipated. In Fig. 1, a block of N most recent samples has been retained. Then Hankel matrix **A** has been formed. In the formation of Hankel matrix, there are some particular missing data. So, to deal with these incomplete data, a non-linear median filter has been introduced. The output of the median filter is fed into the ERA algorithm for obtaining modal parameters.

4 Simulation Result

Two test signals with known damping and frequency are taken for local mode as well as for inter-area mode for the comparison of performance of the proposed modified

ERA algorithm with different methods such as ERA and Prony. The different methods are estimated by simulating 20,000 Monte Carlo cycles at some particular SNR values. For the accomplishment of simulation of the two test signals, a sample window of 250 samples and a sampling frequency equal to 12.5 Hz are used. Later, to establish the effectiveness of the proposed scheme, actual PMU signal attained from WECC system is also tested.

4.1 Test Signal Interrelated to Local Mode

Comparison of the proposed method, i.e. Modified ERA with ERA and Prony algorithm

The simulation results that have been provided in Table1 show the mean and variance of the attenuation factor and frequency for local modes. For local modes, the attenuation factor has been considered as −0.20. At 50 dB SNR, the attenuation factor of ERA, Prony and modified ERA are −0.1839, −0.1740 and 0.1901, respectively. So, in percentage, the experimented value of ERA deviates 8.05% from the actual value, and for Prony, it is deviated 13% from the true value, whereas for modified ERA, it is 4.95% deviation from the considered value. And also, at SNR 40, 30 and 20,

Table 1 Mean and variance for the ERA, Prony and proposed modified ERA at particular SNR values

SNR	Modes		ERA	Prony	Modified ERA
50 dB	d	μ	−0.1839	−0.1740	−0.1901
		σ^2	1.1787×10^{-08}	1.0277×10^{-07}	1.2273×10^{-08}
	f	μ	1.2000	1.2000	1.2000
		σ^2	1.7391×10^{-10}	6.0017×10^{-10}	1.7842×10^{-10}
40 dB	d	μ	−0.1839	−0.1742	−0.1901
		σ^2	1.1833×10^{-07}	1.0120×10^{-06}	1.2286×10^{-07}
	f	μ	1.2000	1.2000	1.2000
		σ^2	1.7251×10^{-09}	5.8569×10^{-09}	1.7650×10^{-09}
30 dB	d	μ	−0.1839	−0.1758	−0.1901
		σ^2	1.1816×10^{-06}	9.2751×10^{-06}	1.2360×10^{-06}
	f	μ	1.2000	1.2000	1.2000
		σ^2	1.7271×10^{-08}	5.7858×10^{-08}	1.7571×10^{-08}
20 dB	d	μ	−0.1840	−0.1817	−0.1903
		σ^2	1.1789×10^{-05}	4.6895×10^{-05}	1.7571×10^{-05}
	f	μ	1.2000	1.2000	1.2000
		σ^2	1.7466×10^{-07}	5.4812×10^{-07}	1.8196×10^{-07}

True Values: damping, $d = -0.2$ frequency, $f = 1.2$ Hz

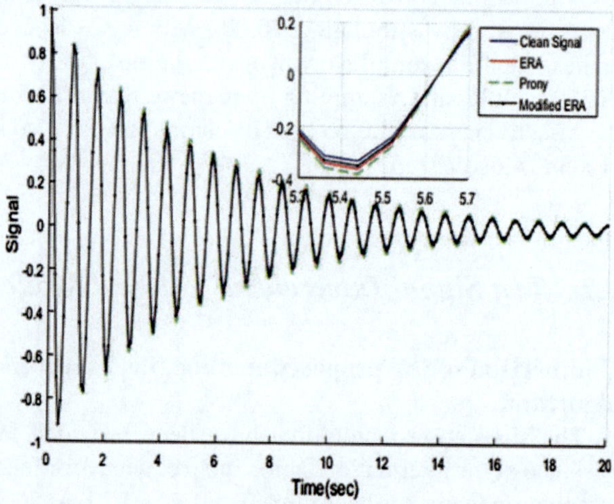

Fig. 2 Signal reconstructed using different methods

the proposed method gives more nearby value than that of the other two algorithms. For higher SNR, the variance is less and for lower SNR, it is more. The frequency has been considered as 1.2 Hz for local mode. The ERA, Prony and modified ERA give comparative results at all taken SNR values.

The signals reconstructed using these methods are shown in Fig. 2. It is seen that the signal obtained from proposed scheme almost matches with the clean signal, while the other schemes deviate (Fig. 3).

Graphs

4.2 *Test Signal Interrelated to Inter-Area Mode*

For the inter-area mode, the signal with attenuation factor has been considered as −0.06 and the frequency 0.6 Hz and corrupted with missing values are taken for the calculation.

Comparison of the Proposed modified ERA algorithm with ERA and Prony algorithm:

For three of the methods, the mean and variance of the attenuation factor and frequency for inter-area modes at particular SNR are provided in Table 2.

In inter-area mode, the true value of attenuation factor is −0.06. At SNR 50 dB, the attenuation factor of ERA, Prony and modified ERA are −0.0617, −0.0580 and −0.0597, respectively. For attenuation factor estimation, the proposed algorithm gives more accurate results than that of ERA and Prony. In percentage, the attenuation factor of ERA is deviated −2.83% from the true value, and for Prony, it is deviated 3.33% from the actual value. And in case of proposed modified ERA, it is just 0.5% deviated from the actual value.

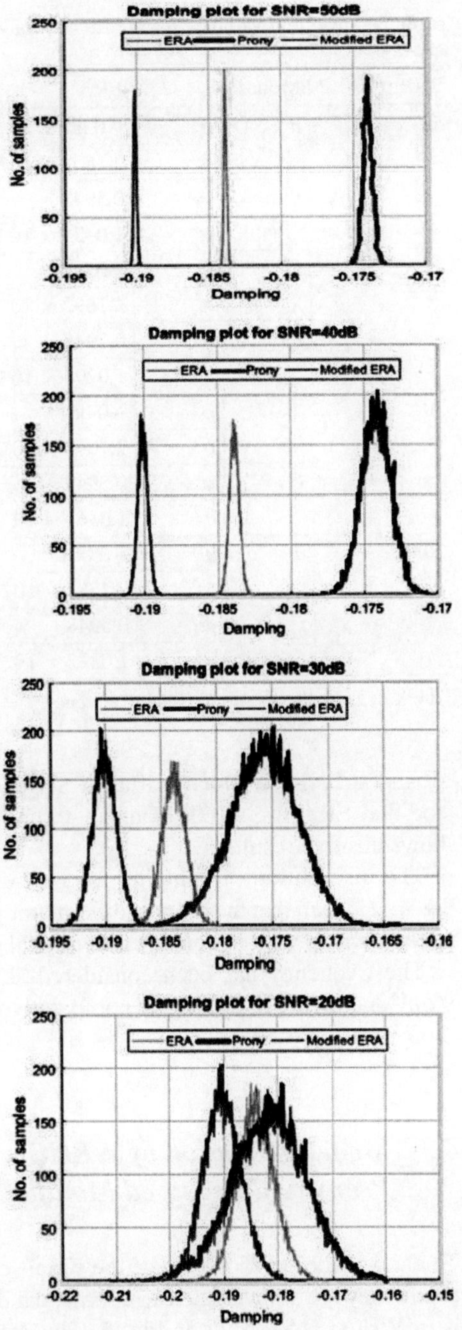

Fig. 3 Damping plots for SNR different SNR for ERA, prony and modified ERA

Table 2 Mean and variance for the ERA, Prony and proposed modified ERA at particular SNR values

SNR (dB)	Modes		ERA	Prony	Modified ERA
50	*d*	μ	−0.0617	−0.0580	−0.0597
		σ^2	2.1898×10^{-8}	9.7310×10^{-8}	2.6508×10^{-8}
	f	μ	0.5948	0.6001	0.5999
		σ^2	1.0424×10^{-9}	2.0193×10^{-8}	7.5456×10^{-10}
40	*d*	μ	−0.0617	−0.0399	−0.0597
		σ^2	2.1692×10^{-7}	6.7420×10^{-6}	2.6889×10^{-7}
	f	μ	0.5948	0.6006	0.5999
		σ^2	1.0298×10^{-8}	2.0235×10^{-7}	7.6769×10^{-9}
30	*d*	μ	−0.0617	0.1395	−0.0597
		σ^2	2.1697×10^{-6}	6.3724×10^{-4}	2.6631×10^{-6}
	f	μ	0.5948	0.6051	0.5999
		σ^2	1.0361×10^{-7}	2.1693×10^{-6}	7.5489×10^{-8}
20	*d*	μ	−0.0617	1.8457	−0.0597
		σ^2	2.1783×10^{-5}	0.0570	2.6578×10^{-5}
	f	μ	0.5948	0.5843	0.5951
		σ^2	1.0322×10^{-6}	1.9390×10^{-4}	7.5923×10^{-7}

True Values: damping, $d = -0.06$, frequency, $f = 0.6$ Hz.

And at SNR = 30 as well as at SNR = 20, the attenuation factor of ERA and modified ERA are −0.0617 and −0.0597, respectively, and the Prony algorithm shows positive results.

The distribution of damping for various methods at different SNR is shown in Fig. 4. It is seen that the proposed estimator gives better mean value with less variance in comparison with the other conventional schemes.

The frequency has been considered as 0.6 Hz for inter-area mode. The ERA, Prony and modified ERA give comparative results at all taken SNR values.

Graphs

4.3 Mode Estimation of WECC with Practical Data for ERA, Prony and Proposed Modified ERA

To analyse the performance of the proposed modified ERA, ERA and Prony algorithm, they are all evaluated with practical data from phase measurement unit which is in WECC [15, 16] as in Fig. 5. The data which is used for analysis by various methods is listed in Table 3 [3]. For window 1, the ERA gives 9.17%, the Prony

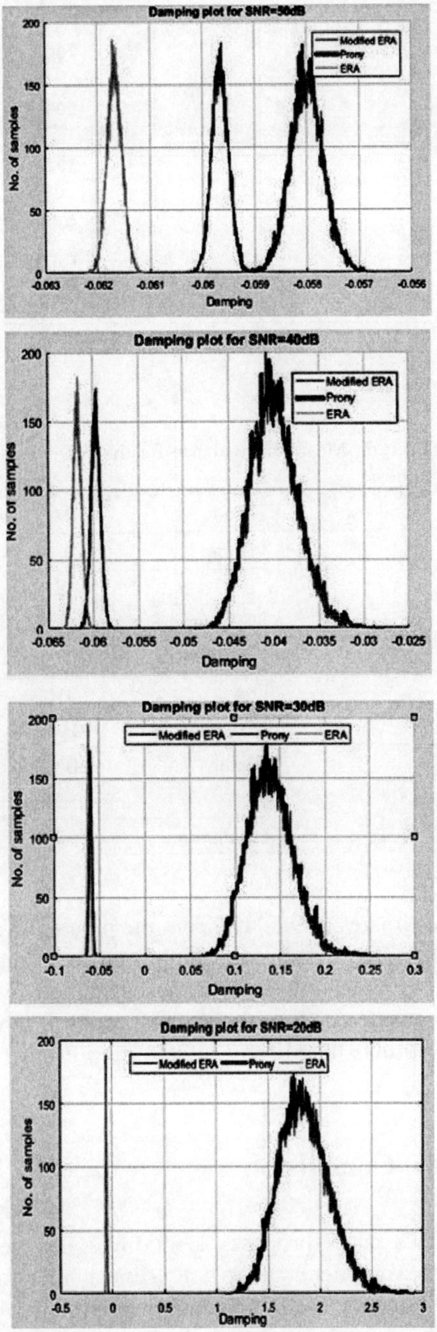

Fig. 4 Damping plots for SNR different SNR for ERA, Prony and modified ERA

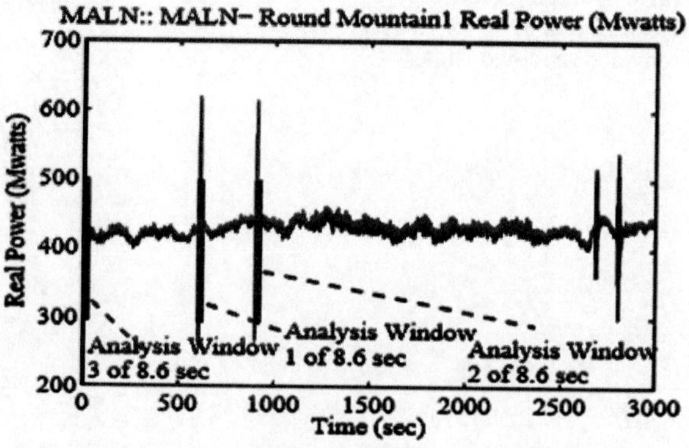

Fig. 5 Probing data of WECC

Table 3 Mean and variance for the ERA, Prony and proposed modified ERA at particular SNR values

Wind-ow	Mode	ERA	Prony	Modified ERA
1	Damp	−0.1917	−0.1938	−0.1650
	fr	0.3312	0.3301	0.3287
	% damp	9.17	9.304	7.96
2	damp	−0.2140	−0.1190	−0.1951
	fr	0.3102	0.3162	0.3107
	% damp	10.912	5.982	9.945
3	damp	−0.0568	0.1062	−0.0622
	fr	0.3038	0.3074	0.3019
	% damp	2.976	−5.48	3.275

estimates as 9.304% from the practical value, whereas the proposed modified ERA has 7.96% damping which is closer to real value (8.3% damping).

For window 2, the proposed method gives better modal estimates than the other two considered algorithms. The proposed scheme provides superior damping (3.275%) estimate than ERA (2.976%) and Prony (−5.48%).

5 Conclusion

This paper proposes a modified ERA estimator to deals with incomplete PMU measurements using a non-linear filter for the detection of critical modes for low-frequency oscillations in power system. A comparative study has been done among the algorithms, such as Prony and ERA. From the results, it can be concluded that the modified ERA method gives better performance as the other estimators do not account for the missing PMU measurements. Thereby, this algorithm can be implemented for real-time applications in power systems.

References

Journals

1. Venkatasubramanian, V. M., & Li, Y. (2004). Analysis of 1996 Western American Electric Blackouts. In *Bulk Power System Dynamics and Control* (pp. 685–721).
2. Byerly, R. T., Bennon, R. J., & Sherman, D. E. (1982). Eigenvalue analysis of synchronizing power flow oscillations in large electric power systems. *IEEE Transactions on Power Apparatus Systems, 101*(1), 235–243.
3. Liao, Y., & Kezunovic, M. (2009). Online optimal transmission line parameter estimation for relaying applications. *IEEE Transactions on Power Delivery, 24*(1), 96–102.
4. Girgis, A. A., & Ham, F. M. (1980). A Quantitative Study of Pitfalls in the FFT. *IEEE Transactions Aerospace Electronic Systems, 16*(4), 434–439.
5. Korba, P., Larsson, M., & Rehtanz, C. (2003). Detection of oscillations in power systems using Kalman filtering techniques. In *Proceedings of the IEEE Conference Control Applications* (Vol. 1, pp. 183–188).
6. Li, Q., Li, R., Ji, K., & Dai, W. (2015). Kalman filter and its application. In *8th International Conference on Intelligent Networks and Intelligent Systems* (pp. 74–77).
7. Caicedo, J. M. (2014). Practical guidelines for the Natural Excitation Techniques (NExT) and the Eigensystem Realisation Algorithm (ERA) for modal identification using ambient vibrant. *Experimental Techniques*, 52–58.
8. Juang, J.-N., & Pappa, R. S. (1985). An Eigensystem realization algorithm for modal parameter identification and model reduction. *Journal of Guidance, 8*, 620–627.
9. Juang, J.-N., & Pappa, R. S. (1986). Effects of noise on modal parameters identified by the Eigensystem realization algorithm. *Journal of Guidance, 9*, 294–303.
10. Hauer, J. F., Demeure, C. J., & Scharf, L. L. (1990). Initial results in Prony analysis of power system response signals. *IEEE Transactions on Power Systems, 5*(1), 80–89.
11. Hauer, J. F. (1991). Application of Prony analysis to the determination of modal content and equivalent models for measured power system response. *IEEE Transactions on Power Systems, 6*(3), 1062–1068.
12. Rai, S., Tripathy, P., & Nayak, S. K. (2014). A Robust TLS-ESPIRIT method using covariance approach for identification of low-frequency oscillatory mode in power systems. In *Eighteenth NPSC*, IEEE.
13. Fernandes, E. R., et al. (2017). Application of a phasor—only state estimator to a large power system using real PMU data. *IEEE Transactions on Power Systems, 32*(1), 411–420.
14. Koschman, A., & Abidi, M. (2001). A comparison of median filter techniques for noise removal in color images. *Report University of Erlangen-Nurnberg, 34*(15).

Online documents

15. PDCI Probe Testing Plan. (2005). (Online). Available: http://www.transmission.bpa.gov/business/operations/SystemNews/
16. Report and data of WECC (Online). Available: ftp://ftp.bpa.gov/pub/WAMS Information/

A Wind-PV Hybrid Generation Scheme for Grid-Isolated Remote Application

Sarang A. Khadtare, Boni Satya Varun Sai, and Debashis Chatterjee

1 Introduction

Nowadays, the world is facing a severe energy crisis due to rapid growth in industrialization as well as in population. The scarcity of fuel, its polluting effects on environment, and continuously increasing prices are some of the actualities related to conventional fuel. Thus, free and large availability of renewable energy sources with non-polluting nature increases the interest of the researchers around the world. There are many generation schemes based on renewable energy sources [1–5]. These schemes are designed either by considering single source such as wind, PV, hydro or with the combination of these for grid-connected or grid-secluded mode [1–4]. The single renewable source-based schemes are easy to design but the reliability of system under the variable environmental conditions is the matter of concern for grid-isolated operation. Hybrid renewable energy generating system has two or more distributed energy sources with or without battery. For the electrification of remote and difficult locality, the hybrid generation system with battery is considered as most potent [3–6]. In last few decades, the wind and PV become very popular among all the other renewable energy sources due to their suitability for ease of conversion [1–3]. For the grid-isolated and grid-connected wind-PV hybrid scheme, the SCIG is very common [2–4]. The SCIG is robust in construction and has a negligible maintenance requirement with excellent controllability; therefore, it is widely preferred for grid-secluded operation at inaccessible locations [4]. The single-phase induction generator supplies the domestic loads [5–8]. In case of SCIG, the value of excitation capacitor affects the output generated voltage. The problem of voltage unbalance arises if the SCIG is provided with fixed reactive power under change in wind velocity and load. Under such circumstances, for maintaining the generated voltage constant, the technique of capacitor switching [9] or auxiliary series-connected capacitor can be used [10]. However, this technique introduces harmonics in output voltage as switching is not

S. A. Khadtare (✉) · B. S. V. Sai · D. Chatterjee
Department of Electrical Engineering, Jadavpur University, Kolkata, India

P. Mahanta et al. (eds.), *Advances in Thermofluids and Renewable Energy*, Lecture Notes in Mechanical Engineering, https://doi.org/10.1007/978-981-16-3497-0_53

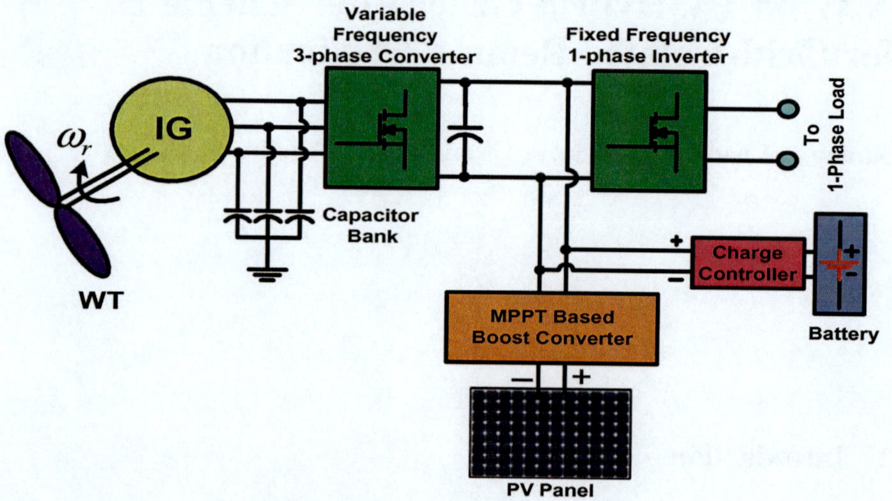

Fig. 1 Block diagram of proposed generation scheme

smooth. Doubly fed induction generator (DFIG) offers better control and generation capabilities for grid-connected system [11–13]. But this system is affected by several issues like high maintenance, poor reliability (due to the use of mechanical contacts), higher cost, and poor power factor. The various control techniques for asymmetrical operation of three-phase SCIG are presented in [14–19]. Few articles demonstrate the hybrid system with wide wind speed operation of SCIG [14, 20]. The hybrid generation schemes with complex control methodologies, e.g., vector control with loss minimization, dynamic voltage restorer (DVR)-based control, have been demonstrated in [21–24].

Among different hybrid control schemes, the wind-PV-based systems are most popular as these sources are complementary in nature. The major drawback of such system for grid-isolated operation is the optimal battery capacity selection considering low-speed operation. The battery is a component which increases system installation cost as well as maintenance cost which reflects in effective energy pricing.

In this paper, a method has been described to design optimum battery capacity considering grid-secluded operation. The starting condition for the IG with low wind speed and low solar insolation has been considered for optimal battery capacity selection. The proposed method has been tested in a 1 kW hybrid generation system which showed promising results.

2 Proposed Generating Scheme

Figure 1 represents the schematic diagram of proposed hybrid generation scheme. It consists of two renewable energy source of wind and PV along with the power converters and battery. In order to generate the energy from wind, aero turbine coupled with three-phase SCIG is employed. Initially, essential reactive power to the SCIG under no load condition is provided by the stator-connected capacitor bank. These excitation capacitors act as the fixed source of reactive power. Under variable load and wind speed conditions, the variable reactive power is required by SCIG. This necessity is fulfilled by a variable frequency bidirectional converter connected at stator terminals of SCIG. To obtain the maximum energy from the wind, the three-phase bidirectional converter is to be strategically triggered. The triggering pulses of converter are generated corresponding to maximum power points of the turbine characteristics as shown in Fig. 4. The solar panel is connected across the common DC bus. The PV module is interfaced with DC bus through a boost converter, which can also supply power to load during no or low wind condition. In order to obtain the maximum power from PV, the triggering of boost converter is controlled by MPPT controller. The DC power available at common bus is converted into single-phase AC by means of fixed frequency single-phase inverter. This generated AC voltage is applied across the terminals of single-phase domestic loads. The battery is incorporated as a standby storage for reactive power source for the generator during starting or low wind condition. In case of no wind and no PV condition, the battery can also supply critical loads for a limited period of time. The bidirectional DC-DC converter is connected as a charge controller for a battery.

3 Proposed Control Strategy

The inverter connected with the SCIG provides variable reactive power at variable frequencies depending on the wind speed. Thus, during starting, as the turbine has low speed, the impressed frequency by inverter to the stator of SCIG is low. The line frequency equivalent circuit for the SCIG has been provided in Fig. 2a. On the other hand, low frequency equivalent circuit for the SCIG is given below in Fig. 2b. In this paper, the battery capacity has been calculated based on the requirement that the battery is only required for maintaining the wind generation during the starting of turbine at low speed when the PV generation is absent of feeble. The battery is not for continuous operation to supply load. This can reduce installation and replacement cost which ultimately results in considerably reduced energy pricing.

The selection of stator side capacitor C is usually based on no load and rated conditions where it can supply the reactive power required for establishing the machine flux. Thus, at no load the reactive power requirement is mainly for the magnetizing flux. So the values of C in per phase basis can be approximately given by,

$$C = \frac{1}{\omega_{\text{rated}}^2 L_m} \tag{1}$$

For the proposed method of battery capacity selection, an extreme case is considered where PV power availability is negligible and the wind speed is low. The machine

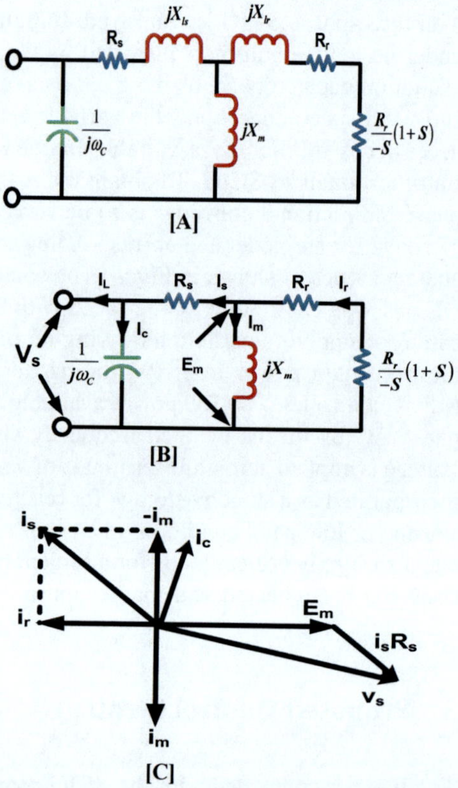

Fig. 2 **a** Equivalent circuit of SCIG **b** Equivalent circuit of SCIG at low frequency **c** Phasor diagram under load

is started at no load with low wind speed where the operating slip is also very low. Therefore, at low slip near $S = 0$ the rotor current i_r will be negligible. Thus, from equivalent circuit in Fig. 2b, the fixed capacitor current i_c will be lesser than the magnetizing current $i_{m,}$ i.e., $i_c < i_m$. This is because at low speed the generated voltage and frequency will be low as evident from the V–f plot shown in Fig. 3. The fixed capacitor current i_c is given by,

$$i_c = v\omega C \tag{2}$$

Or,

$$i_c = \frac{V_\omega}{\omega_{\text{rated}}^{2L_m}} \tag{3}$$

Also,

$$V = \frac{V_{\text{rated}} \times \omega}{\omega_{\text{rated}}} \tag{4}$$

Fig. 3 Voltage V_s frequency (v–f) plot

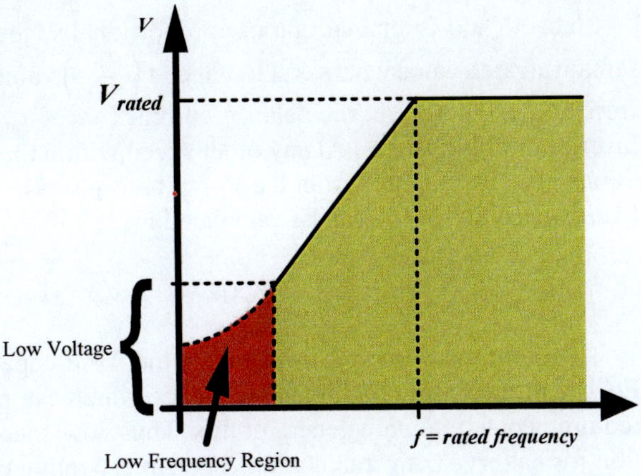

Fig. 4 Wind turbine characteristics at different wind velocities

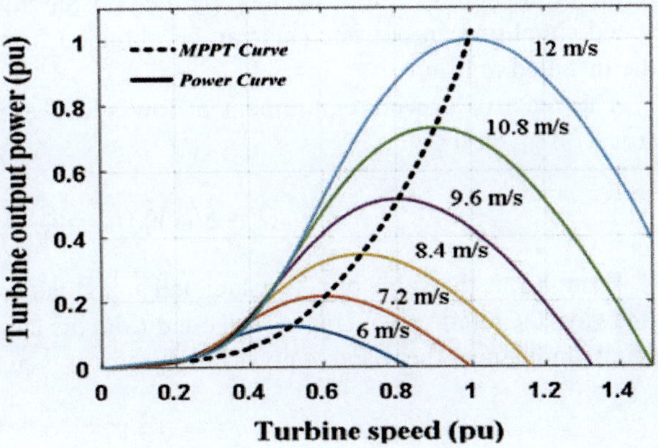

Hence, the above Eq. 3 is given by,

$$i_c = \left(\frac{\omega}{\omega_{\text{rated}}}\right)^2 \times \left(\frac{V_{\text{rated}}}{\omega_{\text{rated}} L_m}\right) \tag{5}$$

$$i_c = \left(\frac{\omega}{\omega_{\text{rated}}}\right)^2 i_m \tag{6}$$

On the other hand, the magnetizing current i_m can be expressed as,

$$i_m = \frac{V}{\omega L_m} \tag{7}$$

where V is the stator side voltage of SCIG and ω is the generated angular frequency.

Since V and ω change almost proportionally, the magnetizing current i_m can remain at rated value whereas at low speed $\left(\frac{\omega}{\omega_{\text{rated}}}\right)$ value is less giving i_c to a low value from Eq. 2. Therefore, the balance current $(-i_c + i_m)$ has to be supplied from the inverter through battery. At any other speed within the operating range, the terminal voltage can be obtained from the V–f plot as given in Fig. 3. Thus, the approximate battery capacity in VA can be calculated as,

$$VA_{\text{Battery}} = \sqrt{3}\, V_L i_m \tag{8}$$

Therefore, as observed from Fig. 3 the fixed capacitor current calculated from Eq. 2 will increase with higher frequency which can result in lesser battery current requirement to maintain generator flux. Thus, low-speed situation is the worst condition for battery sizing calculation. The battery rating can be decided by the voltage at IG terminal obtained from Fig. 3 multiplied by current requirement. The amp-hrs. rating of the battery can be decided by the consideration of time for which the low-speed situation can persist. This can be obtained from the meteorological data for the installed region.

The reactive power requirement at low speed during starting when current is $(i_m - i_c)$ given by,

$$Q = \sqrt{3}\, V_L (i_m - i_c) \tag{9}$$

From Eq. 1, the value of C is calculated at no load condition. Thus, the value of i_m is always the rated value. Under such condition, the capacitor current i_c is negligibliy small, and hence, the reactive power requirement is given by,

$$Q = \sqrt{3}\, V_L i_m \tag{10}$$

So, battery volt–ampere rating (VA) can be calculated by considering the standard battery ratting of 48 V. The battery current rating can also be determined. This low-speed situation will not exist for a long duration as wind flow increases suddenly. Therefore, for analysis purpose we are considering a short duration of 2 h. Thus, the current ratting of a battery can be calculated by using Eq. 10 as,

$$\text{Battery current rating} = \frac{\sqrt{3}\, V_L i_m}{\text{Battery Voltage}} \tag{11}$$

Similarly, the amp-hr capacity of a battery with safety factor of 1.2 can be calculated as for 2 h,

$$\text{Amp-hr Capacity} = \text{Current Capacity of Battery} \times 2\ \text{Amp-hr}$$

On the other hand, the power that can be supplied through stator under this condition is low and can be available from the turbine data. Thus, I_r can be calculated

Table 1 SCIG parameters

KW	500 W
Voltage	415 V
Rated speed	1500 rpm
Power factor	0.8 lag
Stator resistance	13.2 Ω
Rotor resistance	19.1 Ω
Magnetising inductance	950 mH
Stator and rotor leakage inductances	85 mH

with corresponding slip data. The battery capacity can be effectively calculated and compared with higher speed requirement. For the proposed 1 kW system with SCIG of 500 W and PV system of 500 W, the calculated minimum battery capacity is 50Ah, 12 V. This battery can supply the required magnetizing current for 16 h at no load at wind speed of one-third of the rated value with no PV power generated. Since the wind power generated at lower speed can be also low, the effect of load on this battery size is almost negligible with only reduction of time duration by marginal amount (Fig. 4).

4 Simulation Result

The MATLAB–Simulink package is used to simulate the proposed scheme. For simulation purpose a three-phase 500 W, 415 V, 1500 rpm squirrel cage induction machine is used as SCIG. The various parameters of simulated SCIG and wind turbine are given in Table 1. Figure 5a shows the plot for generator speed of value 500 rpm specifying low frequency operation of SCIG. Figure 5b shows the plot for inverter phase voltage having peak value of 108 V. Figure 5c shows the plot for capacitor current having value of 0.25 amp. Figure 5d shows the plot of generator speed of value 800 rpm. Figure 5e shows the plot for inverter generated voltage having peak value of 175 V. Figure 5f shows the plot for capacitor current having value of 0.5 amp at 800 rpm. MATLAB–Simulink model of the proposed scheme is shown in Fig. 6 (Table 2).

5 Conclusion

In case of hybrid generation, the battery sizing is one of the important factors along with the various controlling strategies, and it has to be addressed. In this research work, the methodology is proposed for calculating the optimum sizing of the battery considering the worst scenario of low-speed operation with no PV insolation. The

Fig. 5 Simulation results **a** Generator speed 500 rpm. **b** Inverter output phase voltage (108 V). **c** Capacitor current (0.25 amp). **d** Generator speed 800 rpm. **e** Inverter output phase voltage (174 V). **f** Capacitor current (0.5 amp)

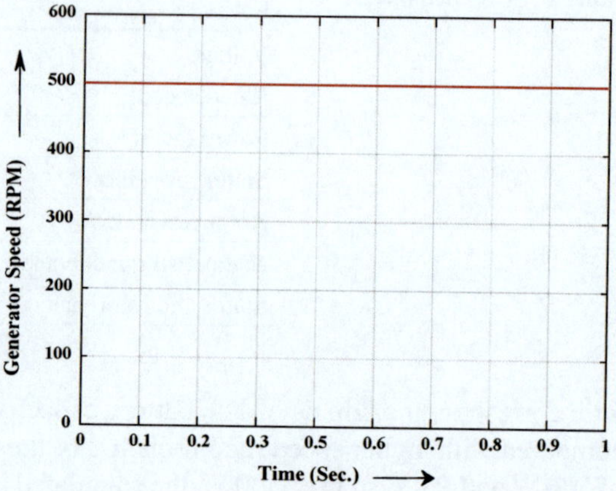

[A]

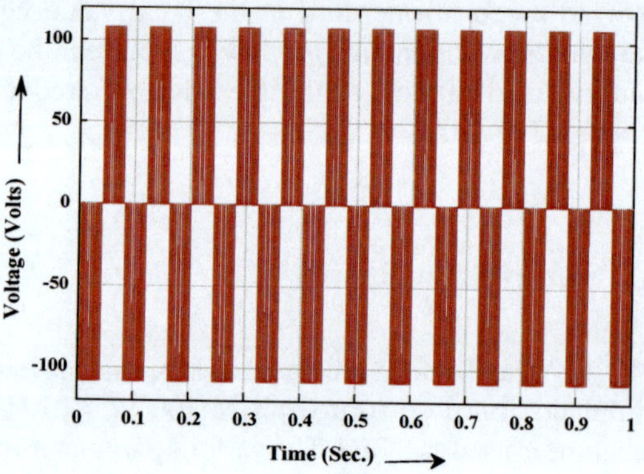

[B]

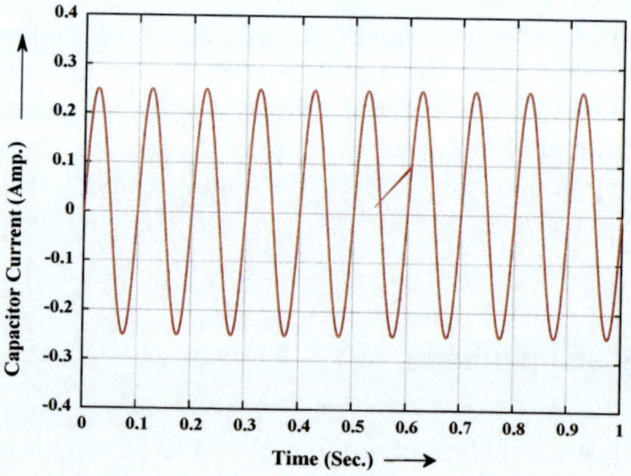

[C]

Fig. 5 (continued)

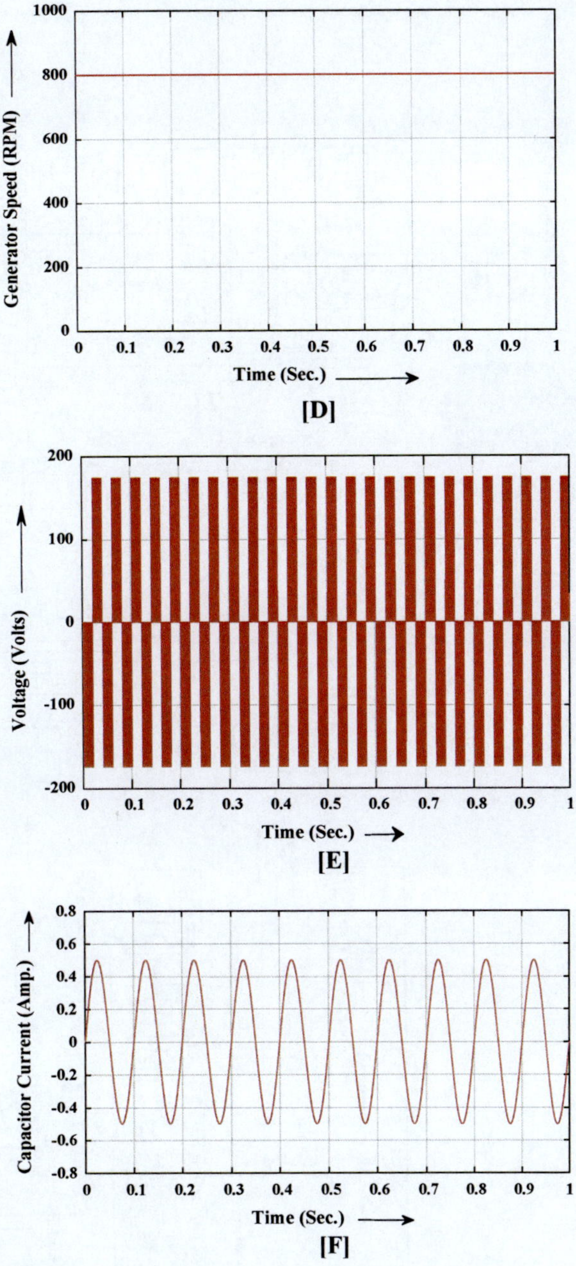

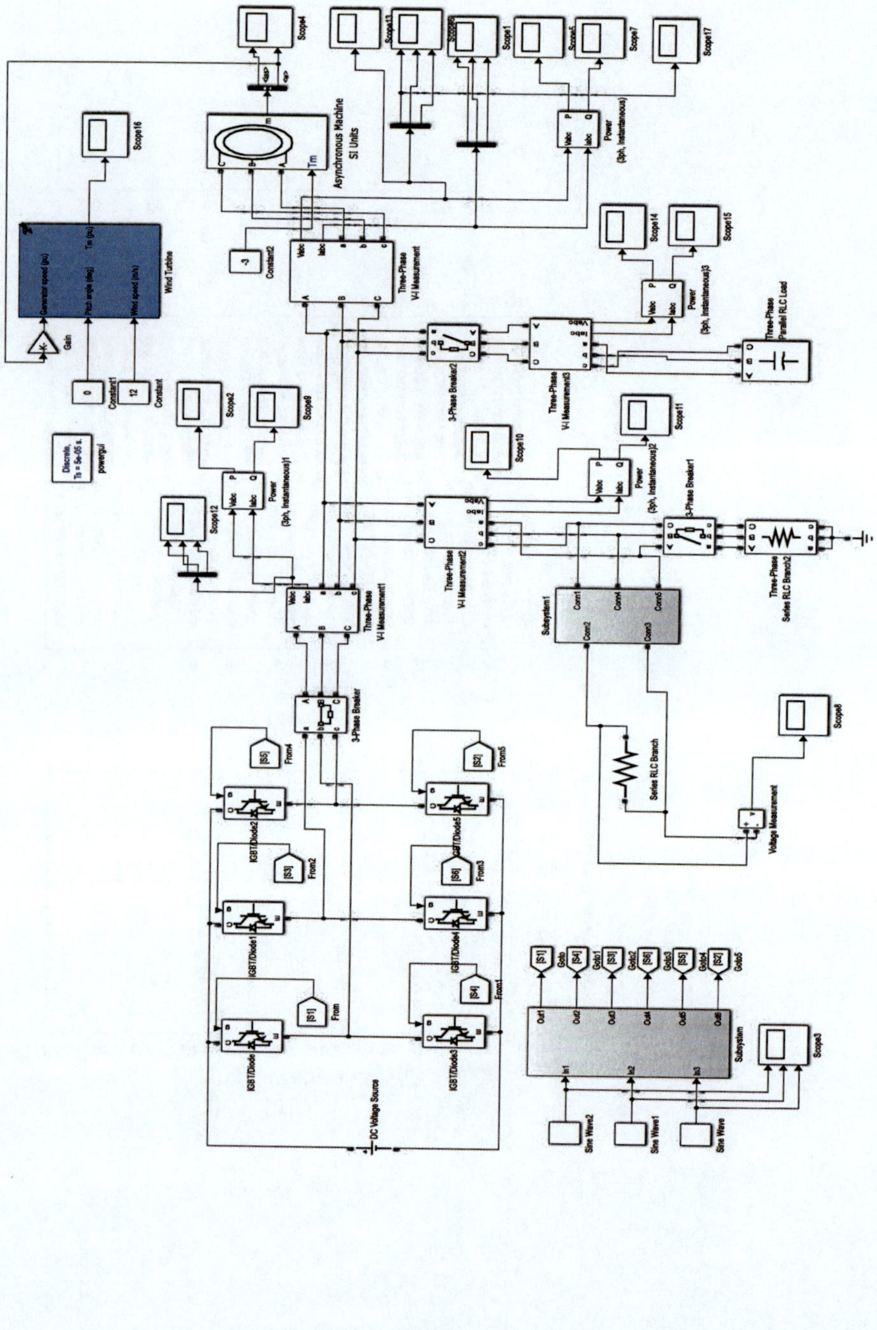

Fig. 6 MATLAB simulation model of proposed scheme

Table 2 Wind turbine parameters

Diameter	2.75 m
Optimal tip speed ratio	8.5
Optimal power coefficient	0.41
Power output at 12 m/s	2000 W
Cut-in wind speed	4.5 m/s

proposed method for battery sizing has been practically tested to verify simulation results.

Acknowledgements The authors acknowledge the Ministry of New and Renewable Energy, India for funding this research work with file no.- 24/22/2016-SWES(R&D).

References

1. Blaabjerg, F., & Ma, K. (2013). Future on power electronics for wind turbine systems. *IEEE Journal of Emerging and Selected Topics in Power Electronics, 1*(3), 139–152.
2. Patel, M. R. (2006). *Wind and solar power systems: Design, Analysis and Operation*. Boca Raton, FL, USA: CRC Press.
3. Bansal, R. C., Bhatti, T. S., & Kothari, D. P. (2005). A bibliographical survey on induction generators for application of nonconventional energy system. *IEEE Transactions on Energy Conversion, 18*(3), 433–439.
4. Bansal, R. C. (2003). Three-phase self-excited induction generators: An overview. *IEEE Transactions on Energy Conversion, 20*(2), 292–299.
5. Murthy, S. S. (1993). Anovel self-induced self-regulated single phase induction generator. *IEEE Transactions on Energy Conversion, 8*(3), 377–382.
6. Rahim, Y. H. A., Alolah, A. I., & Al Mudaiheem, R. I. (1993). Performance of single phase induction generators. *IEEE Transactions on Energy Conversion, 8*(3), 389–395.
7. Ojo, O., & Bhat, I. (1995). An analysis of single phase self-excited induction generators. *IEEE Transactions on Energy Conversion, 10*(2), 254–260.
8. Ojo, O. (1995). The transient and qualitative performance of a self-excited single-phase induction generator. *IEEE Transactions on Energy Conversion, 10*(3), 493–501.
9. Ahmed, T., Noro, O., Hiraki, E., & Nakaoka, M. (2004). Terminal voltage regulation characteristics by static Var compensator for a three-phase self-excited induction generator. *IEEE Transactions on Industry Application, 40*(4), 978–988.
10. Chauhan, P. J., & Chatterjee, J. K. (2019). A novel speed adaptive stator current compensator for voltage and frequency control of standalone SEIG feeding three-phase four-wire system. *IEEE Transactions on Sustainable Energy, 10*(1), 248–256.
11. Vicatos, M. S., & Teqopoulos, J. A. (1989). Steady state analysis of a doubly-fed induction generator under synchronous operation. *IEEE Transactions on Energy Conversion, 4*(3), 495–501.
12. Xing, X., Meng, H., Xie, L., Yue, L., & Lin, Z. (2019). Switching performance improvement based on model—predicive control for wind turbine covering the whole wind speed range. *IEEE Transactions on Sustainable Energy, 10*(1), 290–300.
13. Xu, H., Hu, J., & He, Y. (2012). Integrated modeling and enhanced control of DFIG under unbalanced and distorted grid voltage cnditions. *IEEE Transactions on Energy Conversion, 27*(3), 725–736.

14. Chatterjee, A., & Chatterjee, D. (2019). An improved current balancing technique of two-winding IG suitable for Wind-PV-based grid-isolated hybrid generation system. *IEEE Systems Journal, 4*(4), 4874–4882.
15. Smith, O. J. M. (1987). Three phase induction generator for single phase line. *IEEE Transactions on Energy Conversation, 2*(3), 382–387.
16. Chan, T. F., & Lai, L. L. (2001). Steady-state analysis and performance of stand-alone three-phase induction generator with asymmetrically connected load impedances and exciyation capacitances. *IEEE Transactions on Energy Conversion, 16*(4), 327–333.
17. Gao, S., Bhuvaneswari, G., Murthy, S. S., et al. (2014). Efficient voltage regulation scheme for three-phase self-excited induction generator feeding single-phase load in remote locations. *IET Renewable Power Generation, 8*(2), 100–108.
18. Arthishri, K., Kumaresan, N., & Gounden, N. A. (2019). Analysis and application of three phase SEIG with power converters for supplying single phase grid from wind energy. *IEEE System Journal, 13*(2), 1813–1822.
19. Chatterjee, A., & Chatterjee, D. (2016). Analysis and control of photovoltaic assisted three-phase induction machine operating as single-phase micro-wind generator. *IET Generation Transmission and Distribution, 10*(9), 2156–2176.
20. Wandhare, R. G., & Agarwal, V. (2015). Novel integration of a PV-wind energy system with enhanced efficiency. *IEEE Transactions on Power Electronics, 30*(7), 3638–3649.
21. Castillo, J. P., Mafiolis, C. D., Escobar, E. C., Barrientos, A. G., & Segura, R. V. (2015). Design, construction and implemention of a low cost solar-wind hybrid energy system. *IEEE Latin America Transactions, 13*(10), 3304–3309.
22. Leidhold, R., Garcia, G., & Valla, M. I. (2002). Field oriented controlled induction generator with loss minimization. *IEEE Transactions on Industry Electronics, 49*(1), 147–156.
23. Mesemanolis, A., Mademlis, C., & Kioskeridis, I. (2013). Optimal efficiency control strategy in wind energy conversion system with induction generator. *IEEE Journal of Emerging and Selected Topics in Power Electronics, 1*(4), 238–246.
24. Leon, A. E., Farias, M. F., Battaiotto, P. E., et al. (2011). Control strategy of a DVR to improve stability in wind farms using squirrel-cage induction generators. *IEEE Transactions on Power Systems, 26*(3), 1609–1617.

Comparison of Multi-carrier MLI, CHB-MLI, SHEPWM and SPWM Inverters for PV-Grid Integration

Swati Suman, Rupali Mohanty, and Debashis Chatterjee

Nomenclature

A,*C*	Position of grey wolves
m,*N*	level of output voltage of inverter (dimensionless)
M	Modulation index (dimensionless)
V	Voltage (Volts)
α	Switching angles (degree)
η	Efficiency (%)
b_n	Amplitude of harmonics nth iteration
G	Search agents in GWO algorithm

1 Introduction

Among the various sources of alternative energy, solar energy is the most efficient and reliable that can be applied to smart grids. The factors that made them more popular over conventional energy are abundance, less cost, pollution-free and easy availability. To put these renewable energies to use, converters are required to supply the demand side. Therefore, in order to fulfil the need efficiently, a comparative analysis of different types of inverters has been carried by Mohan et al. in [1].

There are diversified topologies related to MLI such as symmetrical and asymmetrical dc sources for each bridge, reduced switch MLI, Bridges connection with transformers addressed by Mohanty et al. and Mohammad in [2–5]. There are different papers which include several other optimization techniques which were used for CHB-MLI inverters stated by Kangarlu and Babaei in [6]. Ye et al. and Laali et al.

S. Suman (✉) · R. Mohanty · D. Chatterjee
Department of Electrical Engineering, Jadavpur University, Kolkata, India

P. Mahanta et al. (eds.), *Advances in Thermofluids and Renewable Energy*, Lecture Notes in Mechanical Engineering, https://doi.org/10.1007/978-981-16-3497-0_54

in [7, 8] introduces the hybrid multicarrier PWM scheme to satisfy the deficiencies provided by carrier phase shift and in phase disposition.

A set of transcendental equations has been used to find the optimal solution using metaheuristic techniques. Fundamental component is assigned the desired values and harmonics are set to zero. It generates multiple solutions and makes the process problematic which was described by Ray et al. and Wells et al. in [9–15].

As in MLI topologies level of voltage decides the number of components, e.g. dc sources, diodes and MOSFET/ IGBTs, to be used. As the level of inverter increases complexity, cost, size and installation area also increases. Losses also increase as the number of switches used increases; therefore, the efficiency of the system decreases. In the present paper, the inverter used from the MLI family is multicarrier MLI and conventional CHB-MLI topology. The proposed multicarrier MLI technique has addressed these issues and managed to give better results which marks the novelty of the paper. Metaheuristic technique has been used to produce angles that can drive the SHEPWM and CHB-MLI inverters. This optimization technique uses an objective function that reduces or eliminates the undesired harmonics to a large extent which plummets the THD. Apart from that, the other motive was to maintain the power quality with least voltage and current harmonics to be fed in the demand side which has been included.

The paper has been introduced in different sections. Section 2 describes the proposed schemes. Different types of inverters with detailed optimization techniques have been elaborated in Sect. 3. MATLAB and simulation results along with different comparative analyses have been carried out in Sect. 4. Efficiency calculation was carried out in Sect. 5 and lastly, conclusion has been given in Sect. 6.

2 Proposed Scheme

Solar panel has been used in the present scheme as a source of generation. It was connected to the boost converter to increase the voltage which was controlled by perturb and observe (P&O) method.

The dc voltage obtained was used as an input to the inverter. Inverters that are chosen for comparison are CHB-MLI, multicarrier MLI, SHEPWM and SPWM. Constant dc voltage was obtained by the DC–DC boost converter which was given to the inverters. Different topologies can be seen in Fig. 1a–d. GWO techniques were used to calculate switching angles in SHEPWM and CHB-MLI. The PWM pulses are obtained by sinusoidal and triangular comparison, whereas in multicarrier MLI switching, different carrier signal was compared with sinusoidal and inverse sinusoidal reference wave. This multicarrier topology has been inspired by the survey of carrier level shifted like phase disposition (PD), and carrier phase-shifted techniques such as Modified Reference Phase Disposition PWM (MRPD-PWM) etc.

The system has been obtained for three-phase. In this case, triple-n harmonics are absent in line to line voltage. As the level of output voltage increases in MLI, THD decreases as it almost reflexes equivalent to sinusoidal wave. Hence, no filter

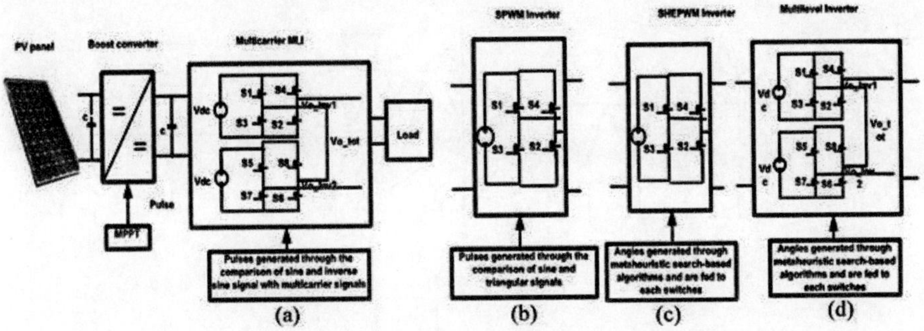

Fig. 1 Proposed scheme of PV integrated **a** multicarrier MLI **b** SPWM inverter **c** SHEPWM inverter **d** multilevel inverter

requirement is there. RL load signifies the grid in the present scheme. This system is apt for a wide range of applications thus reliable for serving industrial and domestic demands.

3 Types of Inverter

3.1 Sinusoidal Pulse Width Modulation Inverter

In the present paper, SPWM inverter is designed for the two-level. This gating pulses are used to control the triggering of the switches as shown in Fig. 2a.

If the value of M is selected wisely, then it could also minimize the THD by reducing or eliminating harmonics to large extent. The THD of unipolar is found out to be better than bipolar because the levels are more obtained for the same number of carriers and reference signals given by Ray et al. and Wells et al. in [10, 15].

3.2 Multicarrier MLI

In the present paper, dc sources used were symmetrical in nature. In this topology, for generating 'N' level inverters, number of carrier signals used would be '$N - 1$', number of dc sources can be calculated by $((N - 1)/2)$ and the number of switches is given by $2(N - 1)$. In the conventional techniques, single sine wave (reference) was compared with triangular (carrier) wave to generate pulses. But, for multilevel topologies, it got extended to multicarrier waves with a single reference wave, therefore, called multicarrier MLI as shown in Fig. 2b. This type of topology was considered to generate multilevel in the output voltage waveform which would automatically reduce the THD which was addressed by Laali et al. in [8]. As per the conventional

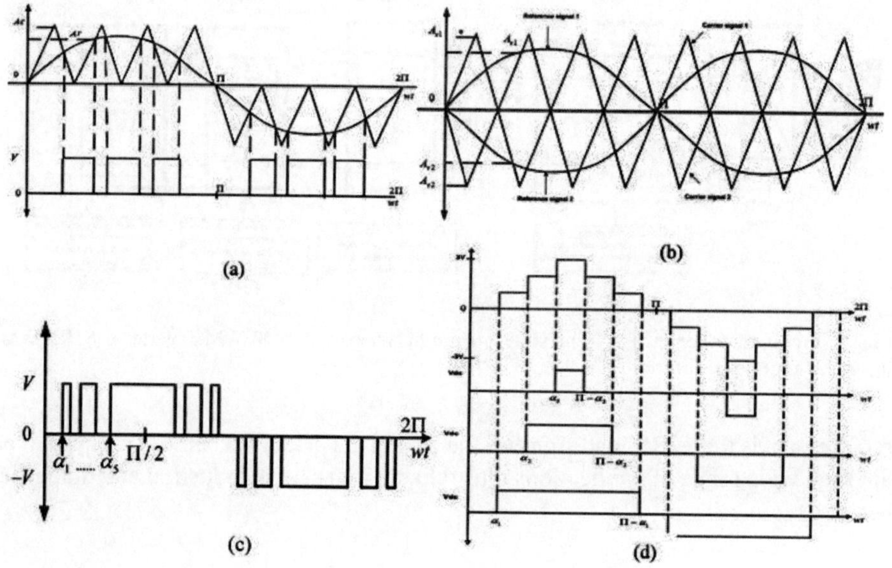

Fig. 2 **a** Pulse generation for SPWM **b** pulse generation for multicarrier MLI **c** unipolar voltage output for SHEPWM inverter **d** output voltage waveform for CH-MLI

methods, the harmonics are obtained in the sidebands around carrier frequency 'Fc' which are resolved by using filters, but in this case, it is solved automatically.

3.3 Selective Harmonic Elimination PWM Inverter

SHEWPM works at low switching frequency which is having the benefit of lower switching loss as compared to the conventional converters. Angles are calculated by solving the non-linear equations given by Ray et al. in [9, 10]. If the switching angles are considered to be of '*d*' number, then it is efficient enough to remove '$(n-1)$' number of harmonics.

Generally, odd harmonics are present and even harmonics are mostly zero. In the present case, harmonics such as 5th, 7th, 11th and 13th would be taken care of by the five switching angles, and rest of higher order harmonics would be controlled by the objective function in Eq. (3). Three-phase inverter has been taken into consideration for the unipolar case of output voltage as shown in Fig. 2c.

As per the Fourier series, the mathematical equation of output voltage waveform of a SHEPWM inverter can be given as

$$v(wt) = \sum_{n=1}^{\infty} (a_n \cos nwt + b_n \sin nwt) \tag{1}$$

According to the quarter wave symmetry, only odd harmonics are present i.e. ($a_n = 0$).

where

$$b_n = -\frac{4}{n\pi}\sum_{d=1}^{m}(-1)^d \cos n\alpha_d \tag{2}$$

The objective function can be given as

$$y(\alpha) = (b_1 - M)^2 + K_5 * (b_5 - e_5)^2 + \cdots + K_n * (b_n - e_n)^2 \tag{3}$$

Subjected to

$$0 < \alpha_1 < \alpha_2 .... < \alpha_m < \Pi/2 \tag{4}$$

where n = harmonic order, $\alpha_1 \ldots \alpha_d$ are the switching angles, M = modulation index, y = objective function, e = value which is generally chosen very close to zero for the harmonics to follow that value, k = priority setter value which is usually set high for those harmonics who should be bound to follow the restricted 'e' value, b_1, b_5, b_7 ...b_n = amplitude of harmonics voltage. The model has been tested for three-phase system, therefore triple-n and even harmonics are absent in line to line voltage. By using Eqs. (2) and (3) in search-based algorithm, minimum THD possible was generated in Eq. (5) and corresponding angles were further considered

$$\%\text{THD} = \sqrt{\frac{b_5^2 + b_7^2 + \cdots + b_n^2}{b_1^2}} \tag{5}$$

Metaheuristic technique was used to find angles for SHEPWM and CH-MLI inverter.

Grey wolf Algorithm

Mirjalili et al. in [11] have invented this algorithm which involves searching and hunting by grey wolves for the survival of their pack. There are different levels that define the position and power of wolves. 'Alpha' is the strongest and superior amongst the whole group, basically the decision maker. 'Beta' holds the next position which helps alpha in decision making and then comes the delta who listens to the order placed by the dominant members and rest candidate comes under the grade of 'Omega' given by Dzung in [12]. The equations which describe the algorithm are as follows

$$\vec{A} = 2\vec{a}\,r_1 - \vec{a} \tag{6}$$

$$\vec{C} = 2r_2 \tag{7}$$

$$\vec{D} = \left| \vec{C} \cdot \vec{G_P}(t) - \vec{G}(t) \right| \tag{8}$$

$$\vec{G}(t+1) = \vec{G_P}(t) - \vec{A} \cdot \vec{D} \tag{9}$$

Steps which was followed for executing GWO are:

Step1: Search agents were initialized randomly.
Step2: Fitness of search agents was analysed in the search space.
Step3: Fitness values of the wolves were updated using iteration loop using Eq. (9).
Step4: The parameters a, A, C were updated along with the position of the grey wolves using Eqs. (6 and 7).
Step5: If the criteria was met, best values were displayed; otherwise, the process was repeated from step 2.

When $|\vec{A}| > 1$, the wolves leave the prey in search of better prey which is called global search in optimization, but when $|\vec{A}| < 1$, grey wolf converges to the prey they found. Equations (6–8). The values $\vec{a}$ decreases linearly from 2 to 0.

3.4 *Cascade H Bridge MLI*

In the present paper, MLI is powered by a set of PV panel and boost converter. The switches are fed using the angles found by metaheuristic technique. According to CHB-MLI topology, if the number of levels are '*m*', number of bridges can be found by '$(m - 1)/2$' for each phase. The number of angles can be calculated by, $A_i = i*180/\ m$, where $i = 1,2,3\ldots.\ (m - 1)/2$ by Kala and Arora in [3]. The switching pulses can be seen in Fig. 2d. Table 1 shows the switching pattern of Phase-A. In case of three-phase, it would be shifted by 120°. As the input voltage (V_{dc}) is symmetric in nature for each cell/ bridge, the voltage can be expressed in terms of Fourier series.

Table 1 Switching pattern for phase-A

Switching states	Output voltage levels								
	0 V	V	2 V	V	0	−V	−2 V	−V	0
S1	0	1	1	1	0	0	0	0	0
S2	0	1	1	1	0	0	0	0	0
S3	0	0	0	0	0	0	1	0	0
S4	0	0	0	0	0	1	1	1	0
S5	0	0	1	0	0	0	0	0	0
S6	0	1	1	1	0	0	0	0	0
S7	0	0	0	0	0	1	1	1	0
S8	0	0	0	0	0	1	1	1	0

$$V(wt) = \sum_{n=1,5,7...}^{\infty} \frac{4V_{dc}}{n\Pi}(\cos(n\alpha_1) + \cos(n\alpha_2) + \cdots + \cos(n\alpha_n))\sin(wt) \quad (10)$$

With the condition applied as mentioned in Eq. (4), the objective function or the fitness function in the search-based algorithm is to minimize the THD as far as possible as given in Eq. (5).

$$V_{ph} = V_{o1} + V_{o2} + \cdots + V_{on} \quad (11)$$

As the bridges are series connected, phase voltage (V_{ph}) is the sum total of individual output voltage of each bridges.

4 MATLAB and Simulation Results

There are two blocks: one is the supply block, i.e. PV panel connected to the boost converter, and other block, i.e. the inverter connected load, as shown in Fig. 1. The simulation was carried out for a load resistance and inductance of 50 Ω and 0.01 mH, respectively. The output waveform of the DC–DC Boost converter can be seen in Fig. 3.

The output voltage waveform and the FFT spectrum for the full-bridge SPWM inverter can be seen in Fig. 4a, b, respectively. THD was found out to be 9.32%. It was found that in the overmodulation case, THD gets better. It can be observed that in the unipolar output voltage waveform, lower-order harmonics are mostly dominant.

The seven-level output voltage waveform for multicarrier MLI can be observed in Fig. 5a, and its corresponding THD was found to be 1.56% as shown in Fig. 5b.

As the level of the output voltage increases, THD gets decreased. FFT spectrum was considered up to 29th order for all cases. Lower and higher order harmonics are mostly reduced or eliminated by the three-phase multicarrier MLI.

In case of SHEPWM inverter, search-based techniques were used for angle calculation which was determined using MATLAB program. Angles generated using GWO are $\alpha_1 = 44.40°, \alpha_2 = 54°, \alpha_3 = 67°, \alpha_4 = 82°, \alpha_5 = 105°$ at a modulation index ($M = V_1/V_{max}$) of 0.85. The output voltage using SHEPWM inverter can be

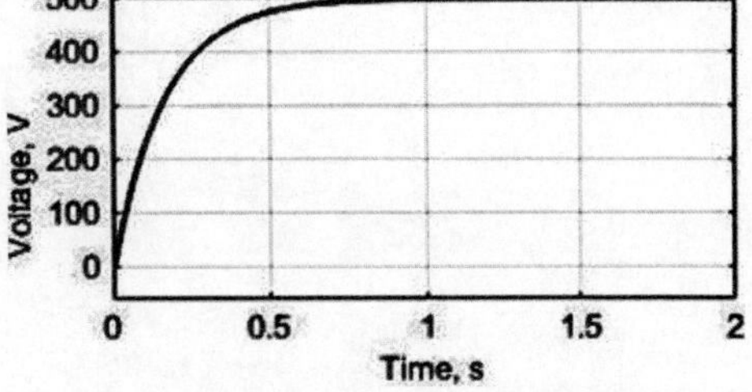

Fig. 3 Output voltage waveform of Boost converter

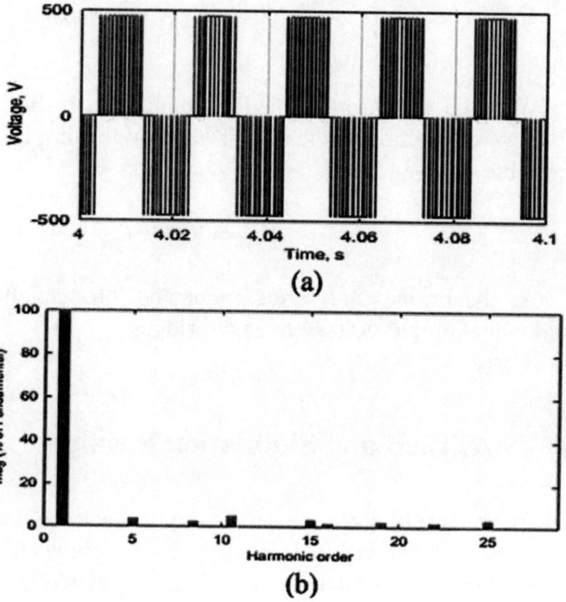

Fig. 4 **a** Three-phase output voltage of SPWM inverter **b** FFT spectrum of the output voltage waveform

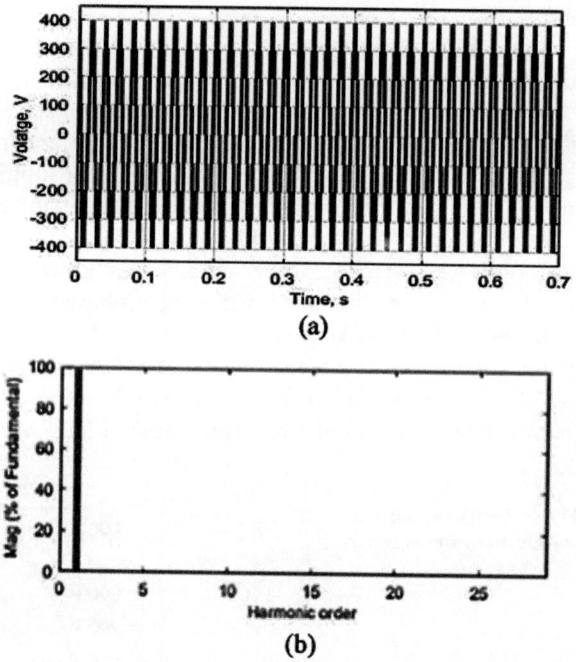

Fig. 5 **a** Output voltage of multicarrier MLI inverter **b** FFT spectrum of the output voltage waveform

seen in Fig. 6a. The iteration curve for GWO is shown in Fig. 6b. It attains the convergence in merely 0.28 s. THD found out was 3.05% as seen in Fig. 6c. Figure 6c, d shows the FFT analysis of the output voltage waveform taken by MATLAB/Simulink

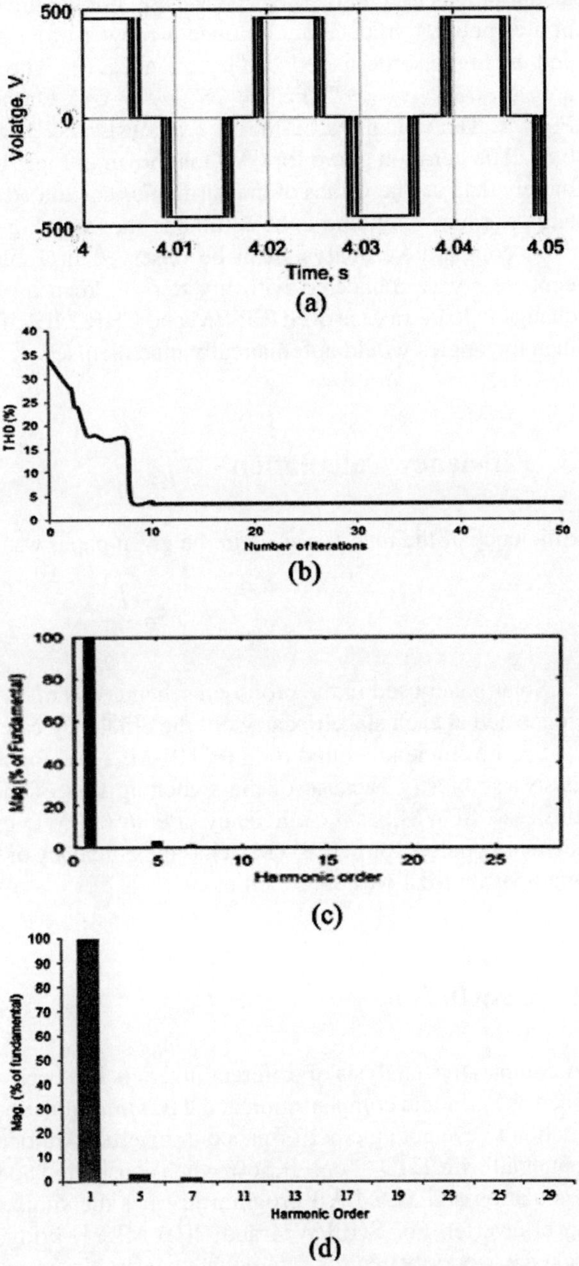

Fig. 6 **a** Output voltage of SHEPWM inverter **b** convergence behaviour of GWO search-based techniques **c** FFT spectrum of the output voltage waveform for GWO obtained by MATLAB Simulink **d** FFT spectrum obtained by MATLAB program

and MATLAB program. It was found that the FFT analysis by calculated values of MATLAB program and simulated values of Simulink are almost matching.

In case of H bridge CHB-MLI inverter, for generating five-level output, the number of angles considered were two. This was calculated by using the same optimization techniques as used in the SHEPWM case. For the three-phase inverter, it was appropriate enough to reduce or eliminate the lower-order harmonics such as 5th and 7th, and the higher order is reduced by using Eq. (3). The angles calculated using GWO are $\alpha_1 = 36°, \alpha_2 = 72°$; $\alpha_1 = 28°, \alpha_2 = 64°$. Output line voltage can be seen in Fig. 7a. THD obtained by the FFT analysis was 3.1% at the modulation index of 0.85. The iteration curve for GWO is shown in Fig. 7b. It attains the convergence in merely 0.45 s. The values of the harmonics calculated using the MATLAB Simulink and program was found to be similar as shown in Fig. 7c, d.

A comparative analysis can be observed in Table 2. A closed-loop feedback controller was connected with inverters to keep a check on the response for the change in load. In case of SHEPWM and CHB-MLI, if there was any change in load, then the angles would automatically tune themselves to maintain low THD.

5 Efficiency Calculation

Efficiency of the inverters used in the given paper was calculated by Eq. (12)

$$\eta = \frac{P_{\text{out}}}{P_{\text{s}}} \tag{12}$$

Solar panel used in the proposed scheme was of 260 W rating. Power block was connected at each stage to carry out the efficiency calculation.

So, the efficiency found for the CHB-MLI was found to be 89.8%. The efficiency is so less mostly because of the switching loss of the switches (IGBT's) used in this case. SPWM has the efficiency of 90.6%. GWO technique was used to find the switching pulses for SHEPWM. It has the efficiency of 93.4%, whereas the proposed multicarrier MLI was 88.4% efficient.

6 Conclusion

A comparative analysis of different inverters has been carried out in various aspects such as harmonic compensation and THD minimization has been taken into consideration to evaluate the performance of inverter. Multicarrier MLI was found to give out of all with THD of only 1.56% when considered up to 29th order harmonic. Both simulation and MATLAB program justifies the similarity of the THD for the five-level inverter. For SHEPWM and CHB-MLI H-bridge inverters, GWO switching scheme works better for removing undesirable harmonics. The controller would

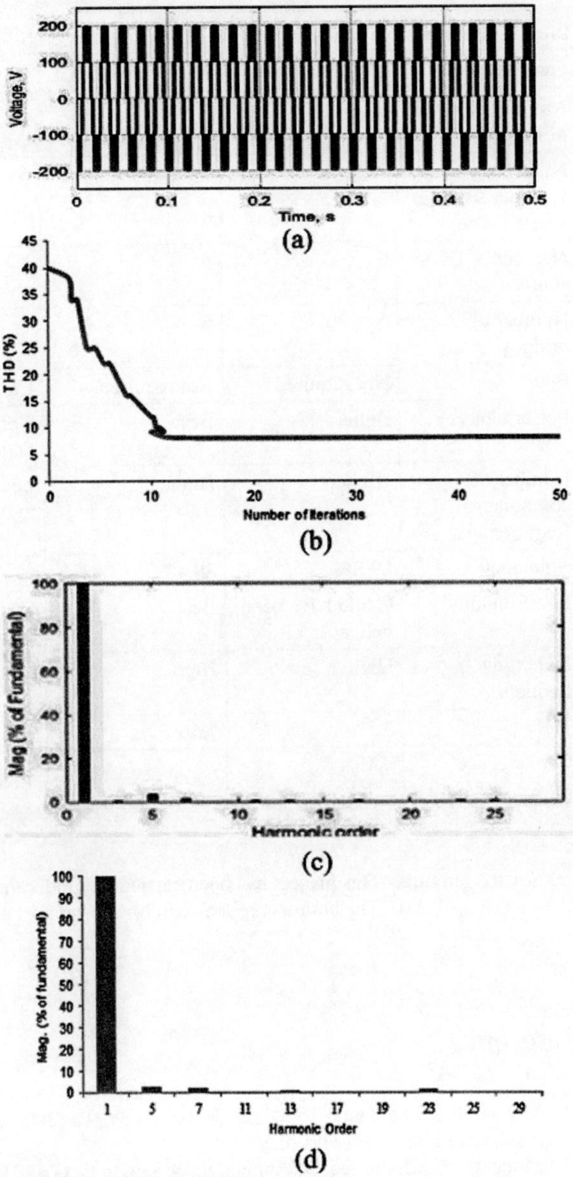

Fig. 7 **a** Output voltage of CHB-MLI inverter **b** Convergence behaviour of GWO search-based techniques **c** FFT spectrum of the output voltage waveform for GWO obtained by MATLAB Simulink **d** FFT spectrum obtained by MATLAB program

automatically adjust itself with the variation in load. The proposed method shows benefits in terms of power quality and switching frequency and the complex circuits designs, thus, reducing the overall costing. The purpose to find the best inverter for the proposed topology was successfully attained.

Table 2 Comparative analysis of CHB-MLI, Multicarrier, SPWM, SHEPWM inverter

Parameters	CHB-MLI	Multicarrier MLI	SPWM	SHEPWM
Number of levels of inverter	5	7	3	3
Number of switches used (in single phase)	24	24	6	6
Number of DC sources	6	6	1	1
Number of bridges	6	6	–	–
Filter	Not required	Not required	Can be used	Not required
Performance	Better	Best	Could have been better	Better
Number of components Requirements	Higher	Higher	Lesser compared to others	Lesser compared to others
Efficiency	89.8%	88.4%	90.6%	93.4%
Power quality	Could have been better	Best	Could have been better	Best
Switching frequency	High	High	Low	Low
THD	3.1%	1.56%	9.32%	3.05%
Convergence time	0.45 s	–	–	0.28 s

Acknowledgements This project has been carried out under the grant Rashtriya Ucchatar Siksha Abhiyan (RUSA 2.0). The authors are thankful for the support provided for the work.

References

1. Mohan, N., Undeland, T. M., & Robbins, W. P. (2003). *Power electronics: Converters, applications, and design*. Wiley.
2. Mohanty, R., Chatterjee, D., Suman, S., & Sengupta, G. (2019). PSO based improved topology for MLI considering low THD and low switching loss. *In 2019 10th International conference on computing, communication and networking technologies (ICCCNT-19)-IEEE* (pp. 1–5). India: Kanpur.
3. Kala, P., & Arora, S. (2017). A comprehensive study of classical and hybrid multilevel inverter topologies for renewable energy applications. *Renewable and Sustainable Energy Reviews, 76*, 905–931.
4. Hasan, N. S., Rosmin, N., Osman, D. A. A., A. H., & Musta'amal. (2017). Reviews on multilevel converter and modulation techniques. *Renewable and Sustainable Energy Reviews, 80*, 163–174.

5. Mohammed, J. A. K. (2016). Economical design of H-bridge multilevel inverter drive controlled by modified fast algorithm. *Microelectronics Reliability, 65*, 89–97.
6. Kangarlu, M. F., & Babaei, E. (2013). Cross-MLI: An innovative topology. *IET Power Electronics, 6*, 642–651.
7. Ye, M., Chen, L., Kang, L., Li, S., Zhang, J., & Wu, H. (2019). Hybrid multi-carrier PWM technique based on carrier reconstruction for cascaded H-bridge inverter. *IEEE Access, 7*, 53152–53162.
8. Laali, S., Abbaszadeh, K., & Lesani, H. (2010). Development of multi-carrier PWM technique for multilevel inverters. In *2010 International Conference on Electrical Machines and Systems-IEEE* (pp. 77–81).
9. Ray, R. N., Chatterjee, D., & Goswami, S. K. (2010). Reduction of voltage harmonics using optimisation-based combined approach. *IET Power Electronics, 3*(3), 334–344.
10. Ray, R. N., Chatterjee, D., & Goswami, S. K. (2009). An application of PSO technique for harmonic elimination in a PWM inverter. *Applied Soft Computing, 9*(4), 1315–1320.
11. Mirjalili, S., Mirjalili, S. M., & Lewis, A. (2014). Grey wolf optimizer. *Advances in Engineering Software, 69*, 46–61.
12. Dzung, P. Q., Tien, N. T., Tuyen, N. D., & Lee, H. H. (2015). Selective harmonic elimination for cascaded multilevel inverters using Grey Wolf Optimizer algorithm. In *2015 9th International Conference on Power Electronics and ECCE Asia -ICPE-ECCE Asia*, IEEE (pp. 2776–2781).
13. Steczek, M., Chudzik, P., & Szeląg, A. (2017). Combination of SHE-and SHM-PWM techniques for VSI DC-link current harmonics control in railway applications. *IEEE Transactions on Industrial Electronics, 64*(10), 7666–7678.
14. Nalcaci, G., Yildirim, D., & Ermis, M. (2020). Selective harmonic elimination for light-rail transportation motor drives using Harris Hawks Algorithm. In *2020 IEEE International Conference on Environment and Electrical Engineering and 2020 IEEE Industrial and Commercial Power Systems Europe (EEEIC/I&CPS Europe)* (pp. 1–6).
15. Wells, J. R., Geng, X., Chapman, P. L., Krein, P. T., & Nee, B. M. (2007). Modulation-based harmonic elimination. *IEEE Transactions on Power Electronics, 22*(1), 336–340.

Comparison Analysis of THD Optimization Using PSO and Jaya Algorithm in Reduced H-bridge Multilevel Inverter

Rupali Mohanty, Swati Suman, and Debashis Chatterjee

Nomenclature

M	Modulation Index
MLI	Multilevel Inverter
S	Number of voltage source
t	Number of iteration
δ	Switching angle
X	Accelarion facor
i	Population number
n	nth order harmonic

1 Introduction

Rapid increase of energy demand and current existence of conventional energy sources have the main motivation of research for extracting energy from renewable energy sources. Considering the solar power that fulfilling the power demand today, the major issue of power conversion and power quality improvement has drawn the attention in different research areas. The conventional H-bridge inverters were previously used with transformers and filters for PV-grid integration. To avoid the use of transformers and filters, multilevel inverters have been proposed in different research areas. In Letha [1], cascade multilevel inverter has been used for power quality improvement in photovoltaic, and the harmonics has been eliminated using SHE techniques with different algorithms. As in multilevel inverter, the harmonics across output voltage can be minimized by increasing the output voltage level so it

R. Mohanty (✉) · S. Suman · D. Chatterjee
Electrical Engineering Department, Jadavpur University, Kolkata, West Bengal, India

P. Mahanta et al. (eds.), *Advances in Thermofluids and Renewable Energy*, Lecture Notes in Mechanical Engineering, https://doi.org/10.1007/978-981-16-3497-0_55

leads to increase the number of switching devices, switching power loss and cost. To overcome this problem, different topology of reduced multilevel inverter has been proposed for grid integration. Review of different topology of reduced multilevel inverter and its operation has been explained in Gupta et al. [2]. In Agrawal and Jain [3], a reduced topology of multilevel inverter has been proposed with different control scheme for low-, medium-, and high-voltage grid that can be interfaced with renewable sources. In Babaei [4], a cascade multilevel inverter with reduced switch has been proposed. Comparitive analysis of symmetric and assymetric reduced multilevel inverter using unipolar PWM method has been discussed in Prabaharan [5]. Some of the reduced and modular topology of MLI and its control techniques have been mentioned in Hagiwara and Akagi [6] and Mohanty et al. [7]. A novel single-phase cascaded multilevel inverter has been discussed in Kaliamoorthy et al. [8] which has been used for both solar and wind (hybrid) energy sources. In this paper, a reduced H-bridge inverter has been implemented that referred from Gupta et al. [2] for power conversion with minimum output voltage THD. This reduced multilevel inverter has been operated with very small number of switches as compared to cascade multilevel inverter. This inverter is using (S + 4) number of semiconductor switches with 'S' number of sources to provide 2S + 1 level output voltage. Different control scheme and different algorithm have been implemented in different topology of multilevel inverter to reduce THD across output voltage. Selective harmonic elimination in multilevel inverter using Newton–Raphson method has been discussed in Kumar [9]. But Newton–Raphson method required a close initial guess to get the better convergence. For harmonics elimination, SPWM techniques have also been implemented, and this technique also reduces the use of filters at the load end as in Mihalic et al. [10]. But this technique is not able to reduce higher order harmonics so far. Heuristic and metaheuristic algorithms have drawn the interest of researchers to solve the optimization problem with better convergence criterion such as genetic algorithm, particle swarm optimization algorithm, ant colony algorithm, and bacterial foraging optimization algorithm. Among these particle swarm optimization algorithm is the basic metaheuristic technique which has proved its efficiency to solve different optimization problem in various fields. In Mohanty et al. [7] Panda and Panda [11], Ray et al. [12] PSO algorithm has been implemented successfully to eliminate the harmonics in different topology of multilevel inverter. In Suman et al.[13], comparison of PSO and GWO techniques has been done for SHEPWM inverter to observe the efficiency of both the algorithms. This paper focuses on minimization of total harmonics distortion from output voltage of a reduced H-bridge MLI with two metaheuristic algorithms. Here, PSO has been implemented in a single-phase 11-level reduced MLI to minimize the THD up to a desire level. Another algorithm known as Jaya optimization algorithm referred from Rao [14], which has been used in different field as Das and Roy [15], Wadood [16], to solve the problem with very minimum controlling or tuning parameter. This is a newly proposed algorithm which has been used to minimize THD from H-bridge reduced MLI in this paper. The circuit design and its control using the optimization algorithms has been carried out in MATLAB simulation to verify the result. The comparison analysis has been described here elaborately, and the result has been discussed with advantages and disadvantages for both the algorithms.

2 Multilevel Inverter

In normal conventional inverter, the harmonic contents are more, and the THD is around 40%. This inverter requires filtering devices to eliminate the harmonics from the output. This makes the circuit complicated and needs huge maintenance. These disadvantages overcome by multilevel inverter, which increases the level of output voltage by using number of DC sources in input side and provides a smooth output voltage waveform close to the fundamental waveform. This waveform contents less harmonics voltage in the output side. The harmonic contents can be reduced efficiently by increasing the level of output voltage or application of different control techniques to operate the semiconductor switches of the MLI. The level of voltage can be determined by $2S + 1$, where $S$is the number of voltage sources(DC) across input side. The output voltage waveform of a 11-level MLI has been shown below.

In a multilevel inverter, the voltage is varied from –V5 to V5, which are considered as the minimum and maximum voltage, respectively. If here each voltage value will be taken as 10 V, then the minimum and maximum voltage value will be −10 and 10, respectively. In some multilevel inverter, unequal voltage sources have been used, and the levels of each voltage are varied accordingly. In MLI, the output voltage is symmetric in nature (Fig. 1).

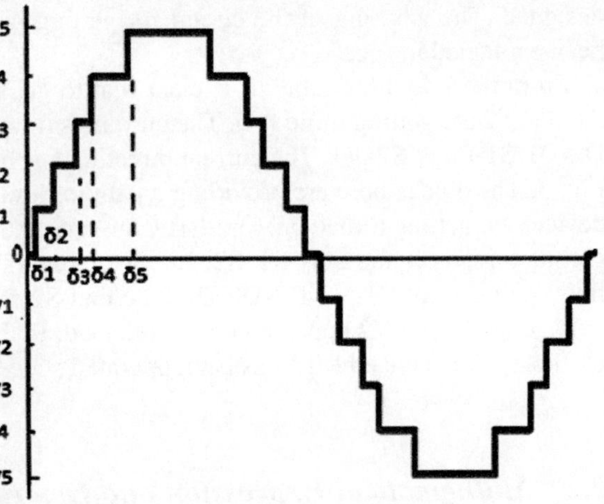

Fig. 1 Output voltage waveform of a multilevel inverter (11 level)

2.1 Reduced H-Bridge Multilevel Inverter

This reduced MLI topology has been employed here to minimize THD from the output voltage waveform across load. This topology has been referred from Gupta et al. [2]. The detail circuit diagram and its operation has been displayed in Figs. 2 and 3. This reduced inverter is designed with a very less number of switching devices in comparison with conventional cascade H-bridge multilevel inverter. For getting 2S + 1 level of voltage across load, a cascade H-bridge inverter MLI is using 4S switching devices where reduced H-bridge MLI is using only S + 4 number of switching devices with few number of diodes, which makes the circuit compact and provides minimum switching loss. In this paper, an 11-level reduced MLI has been used with 5 similar DC voltages, and for getting this voltage level, the reduced MLI has used 9 switching devices with 4 diodes. This reduced MLI can be implemented in medium- and high-voltage grid with suitable arrangements.

Here, similar value of DC sources has been used to energize the reduced MLI. The voltage sources are V1=V2=V3=V4=V5=100 Volts. The switching table of the inverter has been shown in the table.

Here, *Sw* indicates the number of switches in the inverter.

This switching table describes the working of above 9 semiconductor devices to get the 11-level voltage waveform across the load. Here '1' indicates turning on condition and '0' indicates turning off of each switch. These devices are operated in different time interval for both positive and negative half cycle. As shown in Fig. 1 for achieving the 11-level output voltage, 5 switching angles are playing the major role for this operation, and the equivalent time has been calculated and the tabulation has been done. According to this table, the switching pulses for all the switches were designed. The working of the circuit for getting its first level voltage (100 V) has been explained in Fig. 3.

From the switching table, it is clear that to get 100 voltage across the load S5, S1, and S2 are getting turned on. The current is flowing from supply V1-S5-D4-D3-D2-D1-S1-Load-S2-V1. The current direction has been shown through the arrow in Fig. 3. The diodes here are providing a path to flow the current when the switching devices are getting turned on. Similarly, for getting the second level, i.e., 200 V S5 is getting turned off and only S1, S2, and S6 are operated.The current in this case will flow from + ve of V1-V2-D3-D2-D1-S1-Load-S2 to −ve of V1. Here, the additional voltage of V1 and V2 appeared across the load. In this manner, taking the reference of Table 1 the reduced MLI has been operated successfully.

2.2 Mathematical Expression and Determination of Optimization Function

The output staircase voltage waveform based on fundamental frequency has been calculated using Fourier series analysis. Here, an 11-level reduced MLI has been

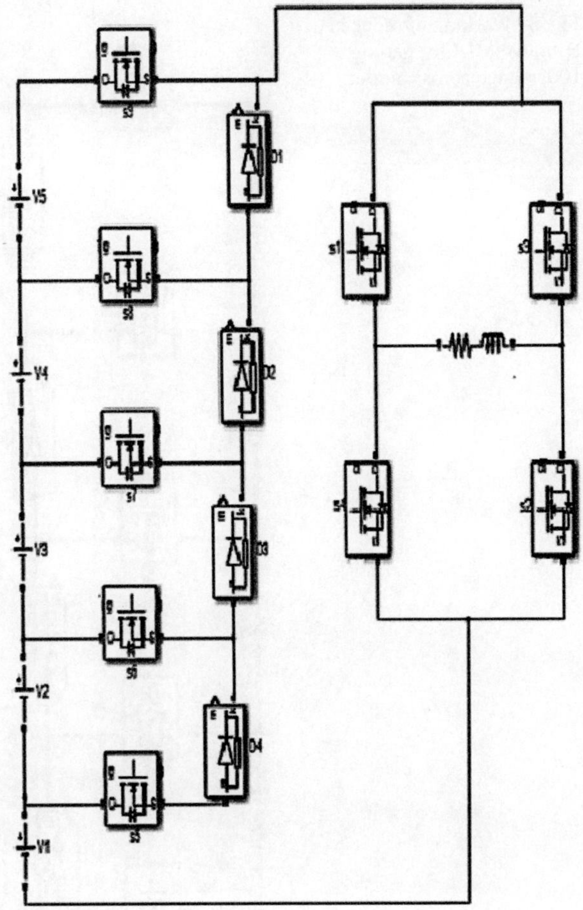

Fig. 2 Circuit diagram of an 11-level reduced H-bridge multilevel inverter

implemented with similar DC voltage sources. The expression of output voltage has been given below.

$$V(\omega t) = \sum_{n=1,3,5}^{\infty} \frac{4V_{DC}}{n\pi}(\cos(n\delta_1) + \cos(n\delta_2) + \cos(n\delta_3) + \cdots + \cos(n\delta_n))\sin(\omega t) \tag{1}$$

Here V_{DC} is the common DC voltage applied across the input side. The voltage value of 5 DC sources is taken as 100 V to get medium range voltage at load side. δ_1, δ_2, δ_3, δ_4, δ_5 are the switching angles at which the semiconductor switches ($S1$, $S2$... $S9$) will get turned on. These angles must have to satisfy the below condition.

$$0 < \delta_1 < \delta_2 < \delta_3 < \delta_4 < \delta_5 < \pi/2 \tag{2}$$

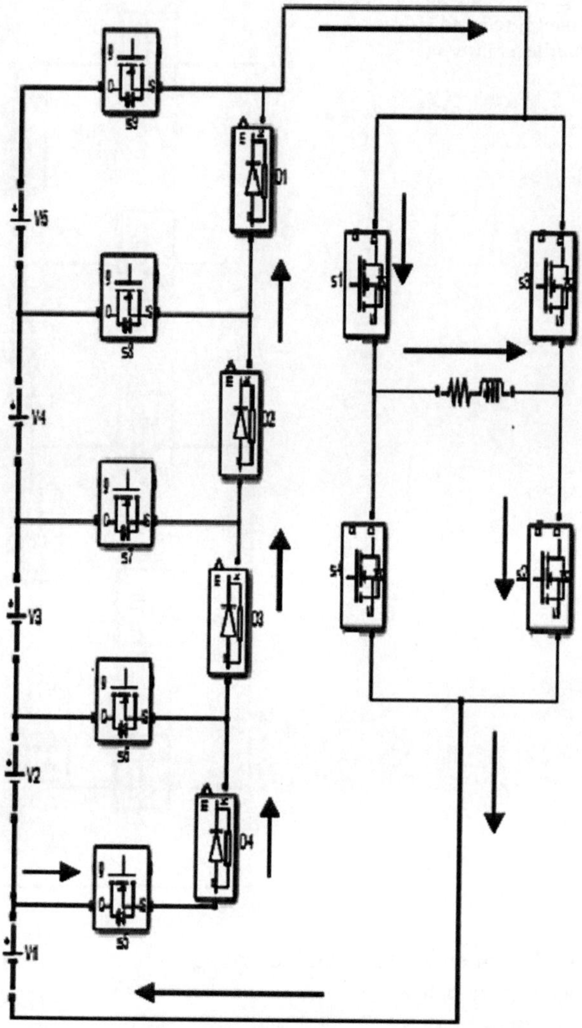

Fig. 3 Working of reduced H-bridge MLI for getting 100 voltage across output

Variation of these switching angles varies the value of output voltage, the fundamental voltage, and the harmonics voltage value accordingly. The total harmonics distortion (THD), which is treated as the optimization function, has been affected significantly by the variation of the switching angles. The optimization function can be expressed as:

$$\%\,\text{THD} = \frac{\sqrt{\sum_{n=3,5,7\ldots}^{49} V_n^2}}{V_1} \times 100 \quad (3)$$

Table 1 Switching table of reduced H-bridge multilevel inverter

Sw	V	2 V	3 V	4 V	5 V	−V	−2 V	−3 V	−4 V	−5 V
S1	1	1	1	1	1	0	0	0	0	0
S2	1	1	1	1	1	0	0	0	0	0
S3	0	0	0	0	0	1	1	1	1	1
S4	0	0	0	0	0	1	1	1	1	1
S5	1	0	0	0	0	1	0	0	0	0
S6	0	1	0	0	0	0	1	0	0	0
S7	0	0	1	0	0	0	0	1	0	0
S8	0	0	0	1	0	0	0	0	1	0
S9	0	0	0	0	1	0	0	0	0	1

Here V_n is the nth order harmonic voltage, and V_1 is the fundamental voltage of the waveform that defines THD. In this paper, THD has been determined for various modulation index, which can be determined by the following expression.

$$M = \frac{V_1}{V_{\max}} \tag{4}$$

where $V_{\max}$ is the maximum output voltage. For this voltage waveform shown in Fig. 1 $V_{\max} = 5 \times V_{DC}$.

3 Particle Swarm Optimization Algorithm (PSO)

This metaheuristic algorithm was invented in the year 1995 by Eberhart and Kennedy. This algorithm has optimized the problem in different area due to its random search technique and modification of search agent in each step from their personal experience and global experience as well. The main motivation of the algorithm is sociological practice of the swarm like bird flocks and fish schools. The individual from the group is called particle and each having individual velocity in the search space. The particles share their personal experience and experience of swarm among each other and finding out the best search path to converge the target. The particles modify their velocity from their experience, and this modified velocity accelerates the particle position to be modified. Based on the personal and global experience, equations have been developed to modify the velocity and position of the particle.

$$V_i(t+1) = X[V_i(t) + C_1 R_1 (P_i\text{best} - P_i(t)) + C_2 R_2 (P_G\,\text{best} - P_i(t))] \tag{5}$$

$$P_i(t+1) = P_i(t) + V_i(t+1) \tag{6}$$

Here 'i' represents individual particle and 't' determines the iteration count. $P_i$best and $P_G$best are the personal and global best position of the particles. R_1 and R_1 are the random numbers varied between 0 and 1.

C_1 and C_2 are the cognitive parameters which satisfy: $C_1 + C_2 = C\ C > 4$. X is accelerating the convergence of PSO algorithm by tuning the following equation.

$$\chi = \frac{2k}{\left|2 - C - \sqrt{C^2 - 4C}\right|} \tag{7}$$

PSO algorithm has been programmed in MATLAB to optimize the THD. Let δ_1, δ_2..., δ_i are the ith particles which are generated randomly to achieve the best value of fitness function. Here, the switching angles are treated as particles, which are needed to be modified through PSO algorithm to get the minimum THD (fitness function) from the output voltage.

The following steps are structured from the PSO algorithm to find out the optimized result.

Steps:

1 Initially, all the tuning parameters need to be initialized. The maximum iteration, population size, and the cognitive parameters have to be set.
2 In this step, the particles are randomly initialized between 0 and $\pi/2$, and the velocity vector that the voltage which accelerates the position of the particles is also initialized between $V_{\max}$ and $-V_{\max}$.
3 Now the fitness function is calculated and each particle needs to be evaluated from the minimum value of fitness function. The fitness function is expressed as:

$$\%\text{THD} = \frac{\sqrt{\sum_{n=3,5,7\ldots}^{49} V_n^2}}{V_1} \times 100 \tag{8}$$

4 As we have discussed in the literature, the personal best position and global best position of the ith particle are determined and the condition which needs to be satisfied is: if the personal best position is better that the current position of particle, then P_{best} replace to PP_i and if global best is better than the personal best then $P_{G\text{best}}$ replace to P_{best}.
5 The velocity vector will be updated by the help of personal best and global best of the particle using Eq. 5, and then, the position vector will be updated by the updated velocity vector using Eq. 6.
6 In this step, the maximum iteration count is checked: if iteration is reached to its maximum value, the steps need to terminate or else the iteration count is increased by one and return back to step 3.

4 Jaya Algorithm

R.Venkata Rao has proposed a novel optimization algorithm in 2015,which has been able to solve many constraint and unconstraint problem successfully. The best part of this algorithm is it consumes less time to solve the problem with less number of parameters. In comparison with other heuristic techniques, this has very less tuning parameter before conducting the experiment. The propensity of this algorithm is to move toward the best and avoid the worst solution in the iteration. The victorious nature of this algorithm makes it powerful globally. Here, the population is initiated randomly to get the best solution, and the position of the population is modified by the following equation.

$$X_i(t+1) = X_i(t) + r_1(X_{i\,\text{best}} - X_i(t)) - r_2(X_{i\text{worst}} - X_i(t)) \quad (9)$$

Here $X_{i\text{best}}$ and $X_{i\text{worst}}$ are the best and worst position of the population, respectively, determined from the best and worst value of fitness function. r_1 and r_2 are the random numbers between 0 and 1. The steps which are followed to optimize the THD using Jaya algorithm are expressed in detail.

Here, the 5 switching angles $\delta_{1,}\ \delta_2 \ldots \delta_5$ are the population that has to be initialized and evaluated to get the minimum THD value.

Steps:

1 The population size, maximum iteration number, and number of design variables are to be initialized at first.
2 The population is randomly generated between 0 and $\pi/2$.
3 In this step, the objective function (THD) is calculated from the generated population using Formula 3.
4 The best value and the worst value of solution have been determined from the objective function, and the population is modified using Eq. 9.
5 Now the modified population has been compared with the previous population: if $X_i(t+1)$ is giving the better solution than $X_i(t)$, then the solution has been updated otherwise the solution is not updated
6 If the termination condition satisfied, the algorithm has been stopped or moved to step 3.
7 The optimal solution is displayed at the end.

5 MATLAB Simulation Result

Here, a single-phase 11-level reduced H-bridge multilevel inverter has been designed with similar DC voltage sources using MATLAB simulation. All the DC sources are of 100 V that has been produced maximum 500 V and RMS of 360.6 V across the RL load. In simulation, a load of 100Ω and 0.3mH has been used to observe the result as shown in Fig. 2. The desired switching angles to operate the semiconductor devices have been obtained from PSO and Jaya algorithm which has been successfully optimized by the THD across output voltage. The THD up to 39th order harmonics which has been achieved through PSO algorithm is 5.4% with modulation index 0.8 and the switching angles are $\delta_1 = 1°$, $\delta2 = 15.89°$, $\delta_3 = 30.11°$, $\delta_4 = 45.30°$, $\delta_5 = 62.44°$. The output voltage THD through Jaya algorithm is 5.1% with modulation index 0.79 and the required switching angles are: $\delta_1 = 4.07°$, $\delta= 17.91°$, $\delta_3 = 29.66°$, $\delta_4 = 44.46°$, $\delta_5 = 60.40°$. The output voltage and current waveforms from the simulation are displayed below (Fig. 4).

The FFT analysis has been carried out to determine the voltage THD for both the algorithms, and the bar graph is shown in Fig. 5.

From the above figure, the reduction of each harmonic voltage has been shown clearly, and the overall THD has also been determined. The lower-order harmonics in both the cases has been minimized successfully.

Fig. 4 **a** Output voltage waveform with respect to time, **b** load current waveform with respect to time

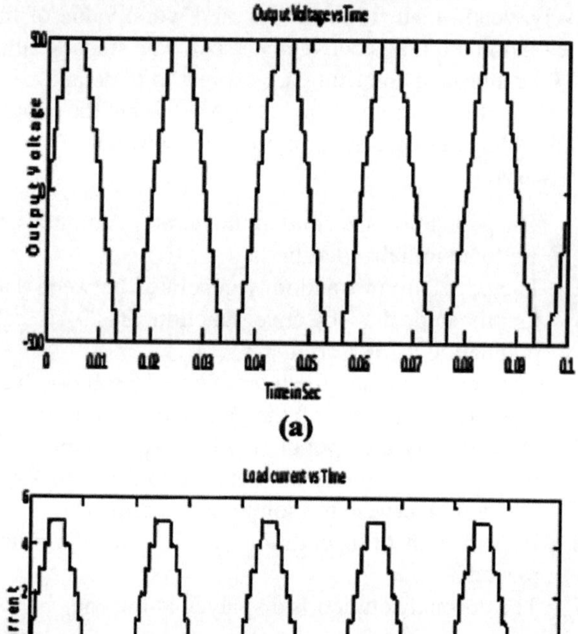

(a)

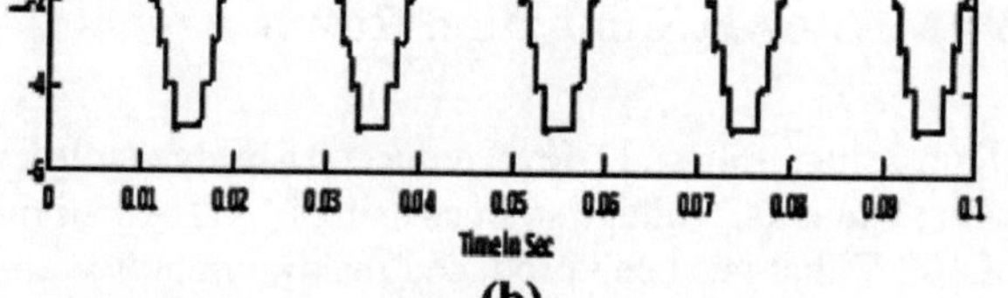

(b)

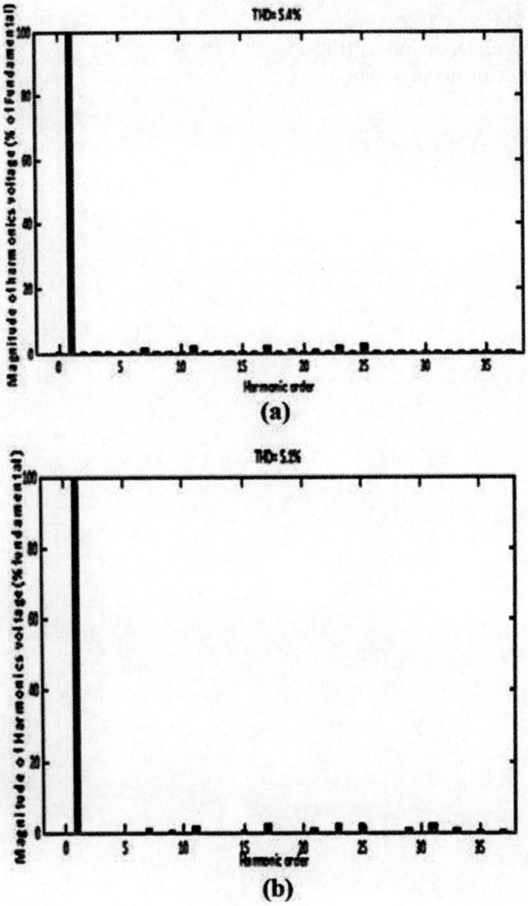

Fig. 5 **a** Magnitude of harmonic voltages in percentage of fundamental versus harmonic order from PSO algorithm, **b** magnitude of harmonic voltages in percentage of fundamental versus harmonic order from Jaya algorithm

5.1 Comparative Analysis

In this paper, the performance and the efficiency of both the algorithms have been observed, and some comparative analysis has been done, which verified the suitable algorithm for this reduced topology of multilevel inverter to minimize the THD. The THD of output voltage has been measured for different value of modulation index and different iteration number for both the algorithm and the comparative graph has been shown below.

From Fig. 6b, it is clear that, to get the optimized result Jaya algorithm has used less iteration in comparison with PSO algorithm. Some comparative analysis has been described elaborately in the following Table .

In this paper, the experiment was carried out by taking same population size and same maximum iteration count for both PSO and Jaya algorithm, and the result

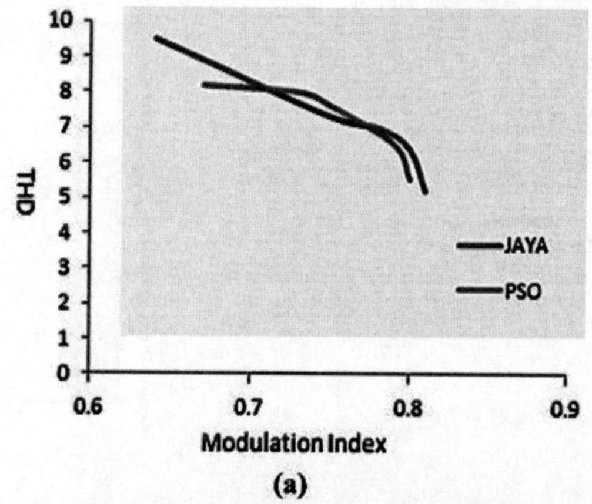

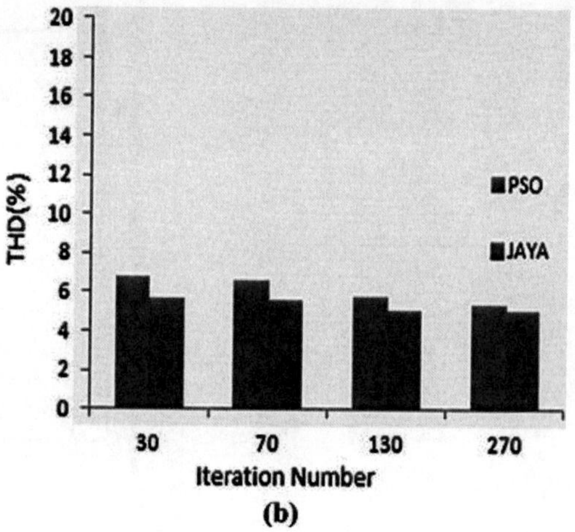

Fig. 6 **a** THD versus modulation index, **b** THD versus iteration number

Table 2 Comparison analysis of PSO and Jaya algorithm

Algorithm	Overall THD (%)	Third-order harmonic (%)	Fifth-order harmonic (%)	Elapse time	Maximum iteration to get the best result
PSO	5.4	0.08	0.16	0.16 sec	270
Jaya	5.1	0.06	0.12	0.26 sec	90

that we observed is: Jaya algorithm has taken less time to compute the problem in comparison with PSO and minimized the harmonics successfully.

6 Conclusion

In this paper, an 11-level single-phase reduced H-bridge MLI has been used and simulated to observe the percentage of THD content across the load. This MLI has been designed with less than half number of semiconductor switches as compared to cascade H-bridge inverter. Here, few diodes are connected instead of semiconductor switches those provide the path to current. This design minimizes the switching loss and volume of the inverter in a greater value. The triggering pulses are designed for only 9 switches as shown in Table 1 instead of 20. Here, two types of metaheuristic algorithms are implemented on the reduced MLI to get the desired switching angles for achieving the minimum THD across load voltage. The verification of the algorithm has been successfully done using MATLAB12 tool. Jaya algorithm has used a smaller number of controlling parameter, less iteration count, and fewer equations than PSO to achieve the better convergence. In single-phase case Jaya algorithm resulted 5.1% and PSO algorithm resulted 5.4% THD across the load voltage as discussed in Table 2. The reduced MLI is a suitable topology that can be implemented in high-voltage application with high rating of switching device as MOSFET or IGBT with proper controlling techniques. From the result analysis, this can be stated that both the algorithms can be able to minimize the THD below 3% in three-phase reduced multilevel inverter, and this can be suitable for grid interface of different voltage level.

References

1. Letha, S. S., Thakur, T., & Kumar, J. (2018). Hardware implementation of power quality improvement in photo-voltaic fed cascaded h-bridge multilevel inverter. In IEEE Conference on Industrial Electronics, (pp. 401–406).
2. Gupta, K. K., Ranjan, A., et al. (2016). multilevel inverter topologies with reduced device count: A review. *IEEE Transactions on Power Electronics, 31*, 135–149.
3. Agrawal, R., & Jain, S. (2017). Multilevel inverter for interfacing renewable energy sources with low/medium-and high voltage grid. *IET Renewable Power Generation, 11*, 1822–1831.
4. Babaei, E. (2008). A Cascade multilevel converter topology with reduced number of switches. *IEEE Transactions on Power Electronics, 23*, 2657–2664.
5. Prabaharan, N., & Palanisamy, K. (2016). comparative analysis of symmetric and asymmetric reduced switch MLI topologies using unipolar pulse width modulation strategies. *IET Power Electronics, 9*, 2808–2823.
6. Hagiwara, M., & Akagi, H. (2009). Control and experiment of pulse width modulated modular multilevel converters. *IEEE Transactions on Power Electronics, 24*, 1737–1746.
7. Mohanty, R., & Chatterjee, D. et al. (2019). PSO based improved topology for MLI considering low THD and low switching loss. In 10th International Conference on Computing, Communication and Networking Technologies- ICCCNT'19, Kanpur (pp. 1–5).
8. Kaliamoorthy, M., Rajasekaran,V., & PraveenRaj, G. (2015). A novel single phase cascaded multilevel inverter for hybrid renewable energy sources. In International Conference On Advanced Computing and Communication Systems- Icaccs'15, Coimbatore (pp. 01–10).
9. Kumar, J., Das, B., & Agarwal, P. (2008). Selective harmonic elimination technique for a multilevel inverter. In Fifteenth National Power Systems Conference-NPSC'08, IIT Bombay, (pp. 608-613).

10. Mihalic, F., Milanovic, M., & Hren, A. (2017). Third harmonic elimination by SPWM for filter reduction in a over modulated single-phase inverter. In International Conference on Electrical Drives And Power Electronics- EDPE'17, Dubrovnik (pp. 37–45).
11. Panda, K. P., & Panda, G. (2018). Application of swarm optimization based modified algorithm for selective harmonic elimination in reduced switch count multilevel inverter. *IET Power Electronics, 11*, 1472–1482.
12. Ray, R. N., Chatterjee, D., & Goswami, S. K. (2009). Harmonics elimination in a multilevel inverter using the particle swarm optimization technique. *IET Power Electronics, 2*, 646–652.
13. Suman, S., Chatterjee, D., & Mohanty, R. (2020). Comparison of PSO and GWO techniques for SHEPWM inverter. In International Conference on Computer, Electrical and Communication Engineering-ICCECE'20, Kolkata, in press.
14. Rao, R.V., & Jaya Algorithm, Rao. (2016). Jaya: A simple and new optimization algorithm for solving constrained and unconstrained optimization problems. *International Journal of Industrial Engineering Computations, 7.1*, 19–34.
15. Das, T., & Roy, R. (2018). Optimal reactive power dispatch using JAYA algorithm-Emerging trends in electronics devices and computational techniques-EDCT'18. Kolkata, (pp. 1–6).
16. Wadood, A., Farkoush, S. G., et al. (2019). Application of the JAYA algorithm in solving the problem of the optimal coordination of overcurrent relays in single- and multi-loop distribution systems (pp. 1–13). Wiley.

Stability Improvement of Renewable Source Integrated Hybrid System Using SVC Controller and ALO Technique

Gomar Bam, Pabitra Kumar Guchhait, and Abhik Banerjee

1 Introduction

Access of renewable energy sources (RESs) is increased steadily which in term reduced the use of percentage fossil fuels [1]. The RESs in geographically backward locations of India (*i.e.,* North East) is also significantly increased. The growth of RESs rapidly increases in this location because of centralized grid are far away from this area. Distributed generation (DG) consists of RESs is the most effective in this region due to the low cost, low-transmission loss and reliable power generation [2]. The DG system consists of more than one energy sources often referred as hybrid power system (HPS). Perhaps for its huge natural resources in this NE region, it will be probable next industrial hubs and therefore, more RESs consisted HPS is essential for fulfilling the power demands [3]. Recently, the RESs in this region are growing enormously. Recent data about development of RESs in this region is given in [4]. Hydro energy is one of the kinds of promising RES in this region. There is also a massive possibility for wind RESs. As the hydro and wind have the capability to complement each other [1], in this work, these two RESs have consider as the sources of energy in this region with a synchronous generator (SG) drive diesel engine. The considered isolated HPS consisted of wind, micro-hydro and diesel might have capable to fulfill the peak load demands of this NE location. In the considered isolated HPS, induction generators (IGs) are attached with micro-hydro and wind turbine for the energy conversion [5]. IG coupled RESs has the advantages (like, without brush and DC excitation source, ruggedness, cheap etc.), but it desires reactive power for its uninterruptible operation [6]. This is also required to drive most of the commercial and industrial loads, whereas SG is the single supplier of reactive power. Therefore, the difference in reactive power between the demands and generations in the studied system causes unmanageable voltage fluctuations in the

G. Bam · P. K. Guchhait · A. Banerjee (✉)
Department of Electrical Engineering, National Institute of Technology, Yupia, Arunachal Pradesh, India

P. Mahanta et al. (eds.), *Advances in Thermofluids and Renewable Energy*, Lecture Notes in Mechanical Engineering, https://doi.org/10.1007/978-981-16-3497-0_56

terminal which also reduces the stability of the said system. This problem may take severe faults in the considered system. So, this should be avoided with the proper management of reactive power in the considered system. SVC is such kind of device which may properly manage and control of the reactive power [7].

Recently, the meta-heuristic algorithms have been used diversely in the purpose of technology applications for their fabulous optimization capability. It also makes a suggestion about real circumstances of the system. Different nature inspired algorithms have been applied to optimize the parameters of differently configured HPSM. Seeker optimization algorithm, symbiosis organisms search, BGA and also many other soft computing techniques (refer [8–10]) have been successfully applied for this purposes. In this considered, antlion optimization (ALO) (see [11]) has successfully applied for the better optimization of studied system parameters which conversely helped to improve stability of the studied system.

The most important purposes of this work are that.

(a) Properly maintained the reactive power of the system under any perturbed case such that the stability should be well maintained.
(b) Intelligently optimize the parameters of the studied system using ALO algorithm and compare the outcomes (like, convergence and terminal voltage profiles) with BGA.

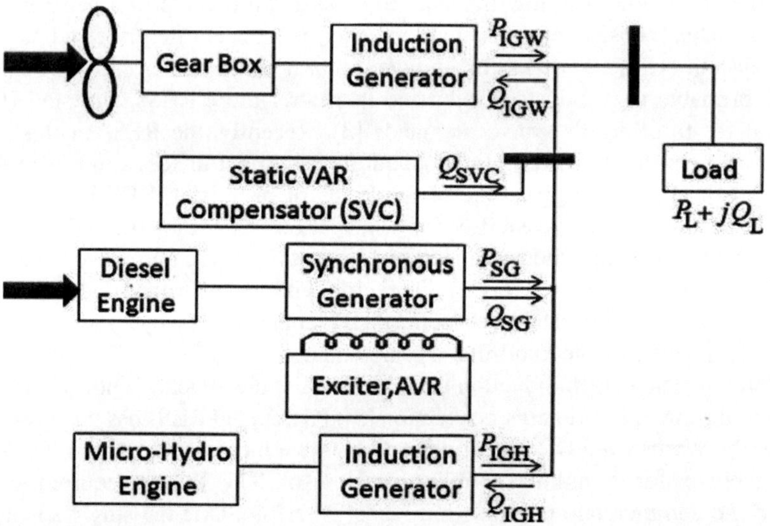

Fig. 1 Structure of studied wind, micro-hydro and diesel HPSM

2 Mathematical Modeling of the System

Figure 1 presents the single line structure of studied system, whereas the modeling has been portrayed in Fig. 2. When any small perturbation has been eventuated in the studied system (as in Fig. 2), the load reactive power demand (ΔQ_L) enhanced which on the contrary causes the voltage fluctuations in the bus. On contrary, it has also raised the reactive power demands by other system components (*i.e.*, ΔQ_{IGW} rise for wind system and ΔQ_{IGH} rise for micro-hydro system). The difference between the generations and demands in reactive power for this unwilling incident is calculated as ($\Delta Q_{SG} + \Delta Q_{SVC} - \Delta Q_{IGH} - \Delta Q_{IGW} - \Delta Q_L$) (see Fig. 2). The enhancement of the bus voltage ($\Delta U(s)$) is derived from the modeling of the studied system (see Fig. 2) as

$$\Delta U(s) = \frac{K_u}{1+sT_u}(\Delta Q_{SG}(s) + \Delta Q_{SVC}(s) - \Delta Q_{IGH}(s) - \Delta Q_{IGW}(s) - \Delta Q_L(s)) \quad (1)$$

Here, in Eq. (1), K_u and T_u are energy balance loop constant and time constant.

The reactive power demands and generations by the IGs and SG are given in Eqs. (2–4) (See [5]),

$$\Delta Q_{IGH}(s) = K_{5H}\Delta U(s) \quad (2)$$

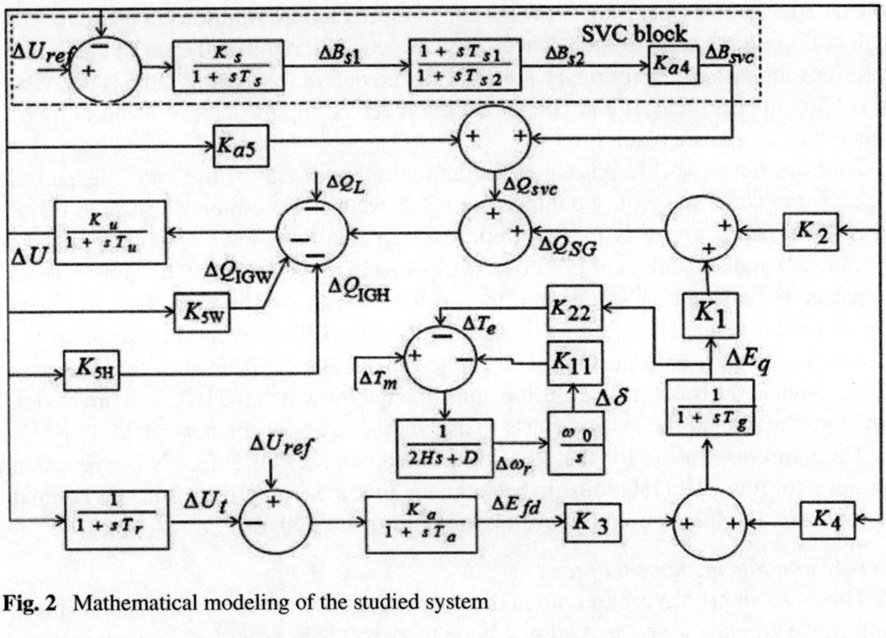

Fig. 2 Mathematical modeling of the studied system

$$\Delta Q_{IGW}(s) = K_{5W}\Delta U(s) \tag{3}$$

$$\Delta Q_{SG}(s) = K_1\Delta E_q(s) + K_2\Delta U(s) \tag{4}$$

whereas the amount of reactive power is delivered by SVC device (refer [5]) under perturbed situation is

$$\Delta Q_{SVC}(s) = K_{a5}\Delta U(s) + K_{a4}\Delta B_{SVC}(s) \tag{5}$$

In the studied system, the other system components connected by IGs, SG, SVC and load are decrypted and considered constant as well as ranges are given in Appendix A.1.

3 Details of ALO Technique

The Antlion Optimizer is an algorithm inspired by nature (population-based algorithm) which imitates the hunting method of antlions. It is a meta-heuristic algorithm and operates with stochastic operators, which means that it utilizes non-deterministic operators on the lookout for global optima in search spaces. Meta-heuristic algorithms are better than deterministic algorithms since stagnation of local optima can be evaded. Random operators allow for one or more random solutions in the search of a global optimum in the search space. They are characterized as the set of candidate solutions. In case of evolutionary algorithms, the derivation of problems is not vital in solving the problem. It all depends on the received inputs and produced outputs; hence they work as a black box.

Antlions have two life phases larvae and adulthood. Out of the total lifespan of 3 years, the adult stage only comprises of 3–5 weeks. They hunt during the larval phase. The adult phase is just for reproduction. Antlions have a unique hunting mechanism and ants are their preferred prey. A sand pit is dug up by an antlion larva which is in the shape of a cone and then it lies in wait for the prey. The prey gets trapped in the sand. The insects fall into the sand trap. To prevent the escape of the prey from the trap, antlions hurl sand to the trap's edge. This ensures that the prey slides down to the bottom. The antlion consume it after which the leftovers are hurled out from the pit and the antlion starts to dig another trap for the next meal.

The main inspiration for the algorithm is the behavior of antlion's larvae when foraging for food. The relationship between the predator and prey is first represented mathematically; this is used to develop the optimization algorithm.

***Step 1**: Random movement of ants*

The ALO algorithm is based upon the interaction between a predator and its prey. Ants move around the space. Antlions catch them by the use of traps. The movement, of ants, while searching for food, is stochastic in nature. Thereby, the modeling of the ants' movement is taken to be random and described as shown below:

$$X(p) = [0,\ C_{\text{sum}}(2r(p_1 - 1), C_{\text{sum}}(2r(p_2 - 1), \cdots, C_{\text{sum}}(2r(p_n - 1)] \tag{6}$$

Here, in Eq. (6), C_{sum} = cumulative sum, n = No of iterations, p = Random walk step.

Here, $r(p)$ is stochastic and can be given as:

$$r(p) = \begin{cases} 1 \ if \ \text{rand} > 0.5 \\ 0 \ if \ \text{rand} < 0.5 \end{cases} \tag{7}$$

where, rand is a number generated arbitrarily, and it has uniform distribution encompassing [0,1].

The location of ants can be described by the M_{ant} matrix and the elements of this matrix are defined as follows:

ant_{ij} = jth variable for ith number of ant, n = Ants number, d = Number of variables.

Random movements in a search space are depicted by Eq. (6). The position of ants is updated at each level of optimization. Equation (6) should be modified for updating the position of each ant. Hence, we use min–max normalization to limit the range

$$X_i^p = \frac{(X_i^p - a_i) \times (d_i^p - c_i^p)}{(b_i - a_i)} + c_i^p \tag{8}$$

where, a_i and b_i represent min and max values of random walks for p^{th} variables.

***Step 2**: Trap in antlion's pits.*

The traps made by antlions determine the random walk of the prey. The assumption is depicted mathematically by the following Eqs. (9–10)

$$c_i^p = \text{Antlion}_j^p + c^p \tag{9}$$

$$d_i^p = \text{Antlion}_j^p + d^p \tag{10}$$

Here, c^p, d^p = Min and max value of all variables at the pth iteration, c_i^p, d_i^p = Min and max of the variables at the ith number of ant and Antlion_j^p = location of jth antlion at pth iteration.

***Step 3**: Building of antlion's trap*

Antlions move toward a particular ant and the pair is determined by use of a roulette wheel. During this process, the algorithm selects the predator which is the antlion on the basis of their fitness. This allows for the fitter antlions better chances in catching ants.

***Step 4**: Sliding the prey (ant) toward the antlion*

When an ant gets trapped in the sand pit, then antlions throw sand on them so that they cannot escape. This behavior is modeled mathematically, wherein the radius of the arbitrary walks of ants in the hyper-sphere is shrunk accordingly. Thereby, Eqs. (11–12) proposed are given as follows:

$$C^p = \frac{c^p}{I} \tag{11}$$

$$D^p = \frac{d^p}{I} \tag{12}$$

Here, I = Ratio between current and max iterations, C^p, D^p = Radius of ants walk.

***Step 5**: Consuming the ants and re-building the trap*

In the final level, the ants get caught and preyed upon by the antlions. The ant gets fitter than the corresponding antlion, and this occurs when the antlion has caught an ant. The antlion then assumes the position of the trapped ant, waiting for the next prey. The following Eq. (13) gives the updated position of the antlions:

$$\text{Antlion}_j^p = \text{Ant}_i^p \text{ if } f\left(\text{Ant}_i^p\right) > f\left(\text{Antlion}_j^p\right) \tag{13}$$

***Step 6**: Elitism*

The best solution(s) in any optimization process is maintained by the concept of Elitism. The fittest antlion has the capability to make bigger traps and hence influence the movement of the prey, thereby making it elite. Hence, we operate under the assumption that every ant walks arbitrarily around a specified antlion and therefore, the elite described as in Eq. (14).

$$\text{Ant}_i^p = \frac{R_A^p + R_E^p}{2} \tag{14}$$

Here, R_A^p, R_E^p = Random walks of the nearer antlion elite at the pth iteration by the roulette wheel.

4 Mathematical Problem Statement

The intention of this present work is to control the reactive power of the studied system in an intelligent method such that the voltage deviations at the bus are minimized and also maintained whole system as well stabilized. To maintain this, an objective function (J) which is totally interrelated with the system eigenvalues has been defined (as in Eq. (15)) so that the all eigenvalues are in negative axis.

$$J = 10J_A + 10J_B + 0.01J_C + J_D \qquad (15)$$

J_A, J_B, J_C and J_D are the weighting factors and connected with the system eigenvalues (see [12]). Therefore, these factors should be appropriately optimized.

5 Results and Discussions

The results of interest are being carried out in MATLAB by simulated the above described system. The observations are followed with small per unit varying in load and equivalent reactance (U and X_{eq}). Under the small incremental input step perturbation (*i.e.*, 0.01p.u.) in reference voltage or in mechanical torque or in both (*i.e.*, U_{ref} and T_m), the transient voltage response (ΔU_t) has been observed. The optimization of the parameters (*i.e.*, K_s, T_s, T_{s1}, T_{s2}) for this studied system has been done within the range ($10 \le K_s \le 100$, $0.005 \le T_s \le 0.05$, $0.005 \le T_{s1} \le 1.5$ and $0.005 \le T_{s2} \le 1.05$) by the help of above described ALO algorithm and compared the results with BGA based technique, considering the same running time (10 s) and number of iterations (500). The results and its discussions are given to the next subsections.

5.1 *Analysis on the Convergence Profile of Objective Function* (J)

The optimizing parameters of the used SVC controller (K_s, T_s, T_{s1}, T_{s2}) and the corresponding defined J values for the studied isolated micro-hydro wind and diesel model are given in Table 1. The table also forecasted the results with above said applied optimizing techniques. It is noticed that in every kind of input conditions, and the J value is lesser for ALO than BGA technique. It is easily inferred that in the studied system, and the parameters are being well optimized for the ALO technique. Figure 3 also shows the comparative convergence profile curve. The convergence profile is also better in order for the ALO technique. For the ALO technique, it is converged with less number of iterations and also in less value. So, it may be drawn a conclusion from the results that the reactive power has well optimized and it may improve the voltage profile of the system under any disturbed conditions.

5.2 *Analysis on Transient Voltage Profile*

Figure 4 shows the voltage profiles of the studied system under mentioned 0.01 p.u. perturbation. The response is in better order for the ALO technique. This is only

Table 1 Optimized parameters value and corresponding J value for applied ALO and BGA techniques

Inputs	Algorithms	Optimized SVC parameters	6-value
$U = 0.97$ and $X_{eq} = 1.0$	BGA	99.6484 0.4336 0.4492 1.7506	180.3283
	ALO	50.0146 0.0478 0.0050 1.0358	176.8960
$U = 1.0$ and $X_{eq} = 1.05$	BGA	55.3516 0.9713 0.1998 0.1920	181.3014
	ALO	52.9476 0.0378 0.0050 1.0364	178.5623
$U = 1.01$ and $X_{eq} = 1.0$	BGA	63.4375 0.0673 0.0440 1.1895	179.6981
	ALO	51.9022 0.0459 0.0050 1.0500	176.4625

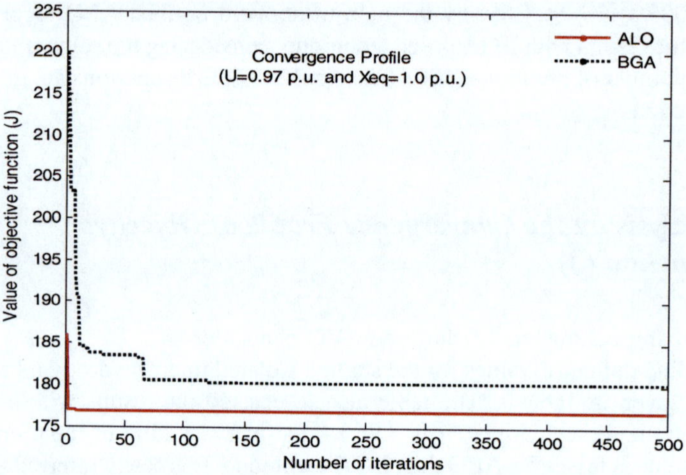

Fig. 3 Comparative convergence profile for the inputs $U = 0.97$ and $Xeq = 1.0$ are in per unit

possible when all the voltage response specifications are minimized properly. This specifications value is also given in Table 2 for some chosen sample input conditions. The value of every specification, i.e., rise time, overshoot and steady state error are also low for the applied ALO technique. From Fig. 4 and Table 2, it is also stated that the stability improvement will be better for the applied SVC controller with the intelligent ALO computing technique.

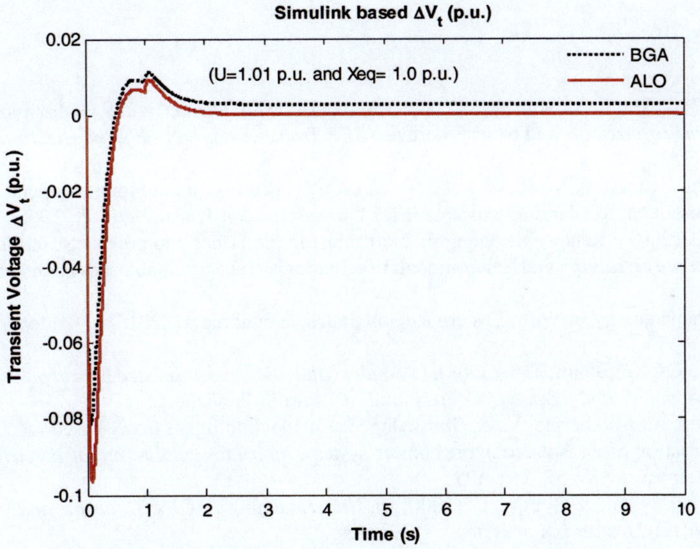

Fig. 4 ALO and BGA based transient voltage profile

Table 2 Response parameters value

Inputs are in p.u. system	Algorithms	%Overshoot	Steady state error	Rise time	Settling time
$U = 0.97; X_{eq} = 1.0$	BGA	0.0146	0.0097	9.9465	10.4560
	ALO	0.0092	0.0004	9.8404	10.6714
$U = 1.01; X_{eq} = 1.0$	BGA	0.0114	0.0052	9.8945	10.5396
	ALO	0.0093	0.0019	9.8553	10.6532

6 Conclusion

In this paper, the reactive power difference under input disturbed situation has successfully checked and overcome. It has been compensated intelligently with the ALO and BGA techniques and also seen that ALO is fruitful for this purpose. The stability of the studied system has been also enhanced and better in order for ALO as compared to BGA. Therefore, ALO computing technique may be the promising for the further studied of the system before designing and implementing of the said hybrid system in the NE region.

References

1. Goel, P. K., Singh, B., Murthy, S. S., & Kishore, N. (2009). Isolated wind–Hydro hybrid system using cage generators and battery storage. *IEEE Transactions on Industrial Electronics, 58*(4), 1141–1153.
2. Katiraei, F., Iravani, M. R., & Lehn, P. W. (2005). Micro-grid autonomous operation during and subsequent to islanding process. *IEEE Transactions on Power Delivery, 20*(1), 248–257.
3. Palit, D. (2003). Renewable energy in North East India: Issues and prospects. In International Conference on Energy and Environmental Technologies for Sustainable Development, (pp. 85–93).
4. Renewable energy in North Eastern Region States, annual report (2016–17). https://mnre.gov.in
5. Bansal, R. C., & Bhatti, T. S. (2008). In *Small signal analysis of isolated hybrid power systems: Reactive power and frequency control analysis*. Narosa Publishers.
6. Banerjee, A., Mukherjee, V., & Ghoshal, S. P. (2014). Intelligent fuzzy-based reactive power compensation of an isolated hybrid power system. *International Journal of Electrical Power and Energy Systems, 57*, 164–177.
7. Hingorani, N. G., & Gyugyi, L. (2000). In *Understanding FACTS: Concepts and technology of flexible AC transmission systems*. IEEE press.
8. Banerjee, A., Mukherjee, V., & Ghoshal, S. P. (2014). An opposition-based harmony search algorithm for engineering optimization problems. *Ain Shams Engineering Journal, 5*(1), 85–101.
9. Dai, C., Chen, W., Zhu, Y., & Zhang, X. (2009). Seeker optimization algorithm for optimal reactive power dispatch. *IEEE Transactions on Power Systems, 24*(3), 1218–1231.
10. Banerjee, A., Mukherjee, V., & Ghoshal, S. P. (2013). Modeling and seeker optimization based simulation for intelligent reactive power control of an isolated hybrid power system. *Swarm and Evolutionary Computation, 13*, 85–100.
11. Veera Reddy, V.C. (2018). Ant Lion optimization algorithm for optimal sizing of renewable energy resources for loss reduction in distribution systems. *Journal of Electrical Systems and Information Technology, 5*(3), 663–680.
12. Ghoshal, S. P., Chatterjee, A., & Mukherjee, V. (2009). Bio-inspired fuzzy logic based tuning of power system stabilizer. *Expert Systems with Applications, 36*(5), 9281–9292.